Medical Virology

Medical Virology

Fourth Edition

David O. White
Department of Microbiology
University of Melbourne
Parkville, Victoria, Australia

Frank J. Fenner
The John Curtin School of Medical Research
The Australian National University
Canberra, ACT, Australia

Academic Press
San Diego New York Boston London Sydney Tokyo Toronto

Cover photo: Computer graphic representation of the virion of human rhinovirus 14 along the icosahedral 3-fold axis of symmetry, highlighting the topographic details of the surface. Lighter colored structures are situated further away from the virion center, showing that the icosahedral 5-fold axis region is the most prominent feature. The "canyon" is clearly seen as a dark blue depression around the 5-fold axis and is the binding site for the cellular receptor ICAM-1. A depression is visible at the icosahedral 2-fold axis of symmetry (equidistant between two 5-fold vertices), but has no known role. Antibody binding sites determined by escape mutations are shown in magenta and clearly appear in more exposed (white or light blue) areas on "domes" and "ridges," suggesting that the dark blue areas are not within the reach of antibodies. (Courtesy of Dr. Jean-Yves Sgro.)

This book is printed on acid-free paper. ♾

Academic Press, Inc.
A Division of Harcourt Brace & Company
525 B Street, Suite 1900, San Diego, California 92101-4495

United Kingdom Edition published by
Academic Press Limited
24–28 Oval Road, London NW1 7DX

Library of Congress Cataloging-in-Publication Data

White, David O.
Medical virology / by David O. White, Frank J. Fenner.
p. cm.
Order of authors reversed on previous eds.
Includes index.
ISBN 0-12-746642-8
1. Medical virology. I. Fenner, Frank, DATE. II. Title.
QR360.F43 1994
616'.0194--dc20 93-48068
CIP

PRINTED IN THE UNITED STATES OF AMERICA
94 95 96 97 98 99 MM 9 8 7 6 5 4 3 2 1

Contents

Part II Viruses of Humans

Preface

The aim of this text is to present the fundamental principles of medical virology to students of science and medicine. It is also hoped that it will serve as a useful resource for teachers of virology, specialists in infectious diseases, and postgraduate students. The pace of change in the eight years since the previous edition has been so great that the book has necessarily been substantially rewritten and expanded. The essential plan of previous editions has been retained, but our account of the molecular biology of viral infections is more detailed than in the third edition. Part I of the book presents an overview of the principles of animal virology, while Part II, entitled *Viruses of Humans* and arranged by virus family, is oriented toward the needs of medical students and clinicians. In order to focus on concepts, mechanisms, and basic facts, minutiae have been omitted except where necessary to understand an important phenomenon. Much of the factual material is consolidated into tables and figures. Statements have not been individually supported by references to research papers, but selective lists of recent authoritative books and reviews are provided at the end of each chapter to simplify the reader's entry into the scientific literature.

David White
Frank Fenner

Acknowledgments

The indirect but invaluable input of our coauthors of *Veterinary Virology*, Drs. E. P. J. Gibbs, F. A. Murphy, R. Rott, and M. J. Studdert, to this fourth edition of *Medical Virology* is gratefully acknowledged. One of us (D.W.) is particularly grateful to Drs. F. A. Murphy and B. W. J. Mahy for their hospitality during a three-month visit to the Centers for Disease Control and Prevention, Atlanta, Georgia, during the early stages of the preparation of this book. Helpful discussions with the following scientists in America or Australia are acknowledged: Drs. G. L. Ada, M. J. Alter, D. A. Anderson, L. J. Anderson, G. M. Baer, M. J. Beach, W. J. Bellini, R. V. Blanden, D. W. Bradley, C. H. Calisher, N. J. Cox, L. Dalgarno, R. Doherty, D. and M. J. Dyall-Smith, S. P. Fisher-Hoch, H. A. Fields, T. M. Folks, J. R. L. Forsyth, K. Hayes, K. L. Herrmann, J. C. Hierholzer, I. H. Holmes, A. P. Kendal, O. M. Kew, S. A. Locarnini, B. W. J. Mahy, H. S. Margolis, I. D. Marshall, T. P. Monath, S. S. Monroe, F. A. Murphy, M. A. Pallansch, P. E. Pellett, H. G. Pereira, C. J. Peters, M. A. Purdy, W. C. Reeves, A. Simmons, and E. G. and J. H. Strauss. We are also indebted to those scientists and publishers, too numerous to mention here, who responded generously to our requests for illustrative material; appropriate acknowledgments accompany the legends to figures. We are particularly grateful to Dr. Jean-Yves Sgro, who provided the beautiful color image of a rhinovirus particle that decorates the cover. We are grateful to Mr. Kevin Cowan for once again helping with the preparation of line drawings and to the Photographic Section of the John Curtin School of Medical Research for preparing the illustrations for publication. Last but not least, Dr. White would like to record his appreciation of the forbearance of his research colleagues Drs. E. M. Anders, L. E. Brown, and D. C. Jackson, and of the graduate students in the laboratory during the three-year gestation period of this book and to thank Nadia Puglielli and Belinda Lightfoot for their perseverance and devotion in the preparation of the manuscript. Shirley Light and Michael Early of Academic Press have been most cooperative throughout the production of this edition.

David White
Frank Fenner

Part I

Principles of Animal Virology

Chapter 1

Structure and Composition of Viruses

The unicellular microorganisms can be arranged in order of decreasing size and complexity: protozoa, fungi, and bacteria—the latter including mycoplasmas, rickettsiae, and chlamydiae, which, like viruses, replicate within eukaryotic cells. These microorganisms, however small and simple, are cells. They always contain DNA as the repository of genetic information, they also contain RNA, and they have their own machinery for producing energy and macromolecules. Unicellular microorganisms grow by synthesizing their own macromolecular constituents (nucleic acid, protein, carbohydrate, and lipid), and most multiply by binary fission.

Viruses, on the other hand, are not cells. They possess no functional organelles and are completely dependent on their cellular hosts for the machinery of energy production and synthesis of macromolecules. They contain only one type of nucleic acid, either DNA or RNA, but never both, and they differ from nonviral organisms in having two clearly defined phases in their life cycle. Outside a susceptible cell, the virus particle is metabolically inert; it is the transmission phase of the virus. This extracellular transmission phase alternates with an intracellular reproductive phase, in which the viral genome exploits the metabolic pathways of the host to produce progeny genomes and viral proteins that assemble to form new virions. Further, unlike any unicellular microorganism, many viruses can reproduce themselves even if nothing but the viral genome is introduced into the cell.

The key differences between viruses and unicellular microorganisms are listed in Table 1-1. Several important practical consequences flow from these differences. For example, some viruses can persist in cells by the integration of their DNA (or a DNA copy of their RNA) into the genome of the host cell.

Table 1-1
Contrasting Properties of Unicellular Microorganisms and Viruses

Property	Bacteria	Mycoplasmas	Rickettsiae	Chlamydiae	Viruses
> 300 nm diameter[a]	+	±	+	±	−
Growth on nonliving media[b]	+	+	−	−	−
Binary fission	+	+	+	+	−
DNA and RNA	+	+	+	+	−
Nucleic acid infectious	−	−	−	−	+[c]
Ribosomes	+	+	+	+	−
Metabolism	+	+	+	+	−
Sensitivity to antibiotics	+	+	+	+	−[d]

[a] Some mycoplasmas and chlamydiae measure around 300 nm or less.
[b] Chlamydiae and most rickettsiae are obligate intracellular parasites.
[c] Some among both DNA and RNA viruses.
[d] With very few exceptions.

Moreover, viruses are not susceptible to antibiotics that act against specific steps in the metabolic pathways of bacteria.

The simplest conventional viruses consist of a nucleic acid genome and a protein coat. However, there exists a class of even simpler infectious entities known as *viroids,* which are infectious RNA molecules that lack a protein coat; as viroids have so far been found only in plants they are not discussed further in this book. In contrast, *prions,* such as the agents that cause the spongiform encephalopathies in humans, appear to be a filamentous protein with no associated nucleic acid.

Viral Morphology

Physical Methods for Studying Viral Structure

It has been known for many years that viruses are smaller than microorganisms. The first unequivocal demonstration of this for an animal virus occurred in 1898, when Loeffler and Frosch demonstrated that foot-and-mouth disease, an important infectious disease of cattle, could be transferred by material which could pass through a filter of average pore diameter too small to allow passage of bacteria. The new group of "organisms" became known as the "filterable viruses." For a time they were also called "ultramicroscopic," since most viruses are beyond the limit of resolution of light microscopes [200 nanometers (nm) = 2000 angstroms (Å)]. Only with the advent of the electron microscope was it possible to study the morphology of viruses properly. It then became apparent that they range in size from about the size of the smallest microorganisms down to little bigger than the largest protein molecules.

Early electron microscopic studies of viruses by Ruska in 1939–1941 were expanded during the 1950s to include thin sectioning of infected cells and metal shadowing of purified virus particles. Then in 1959 our knowledge of viral ultrastructure was transformed when *negative staining* was applied to the electron microscopy of viruses. A solution of potassium phosphotungstate, which is an electron-dense salt, when used to stain virus particles, fills the interstices of the viral surface, giving the resulting electron micrograph a degree of detail not previously possible (Fig. 1-1). Electron micrographs of negatively stained preparations of virions representing the families of viruses that cause human infections are shown in the chapters of Part II of this book.

Viral Structure

The *virion* (infectious virus particle) of the simplest viruses consists of a single molecule of nucleic acid surrounded by a protein coat, the *capsid;* the capsid and associated nucleic acid constitute the *nucleocapsid.* The nucleocapsid of some viruses is surrounded by a lipoprotein *envelope* (Figs. 1-1B and 1-2C,D). In some of the more complex viruses the capsid surrounds a protein core, which encloses the viral nucleic acid.

The capsid is composed of a defined number of morphological units

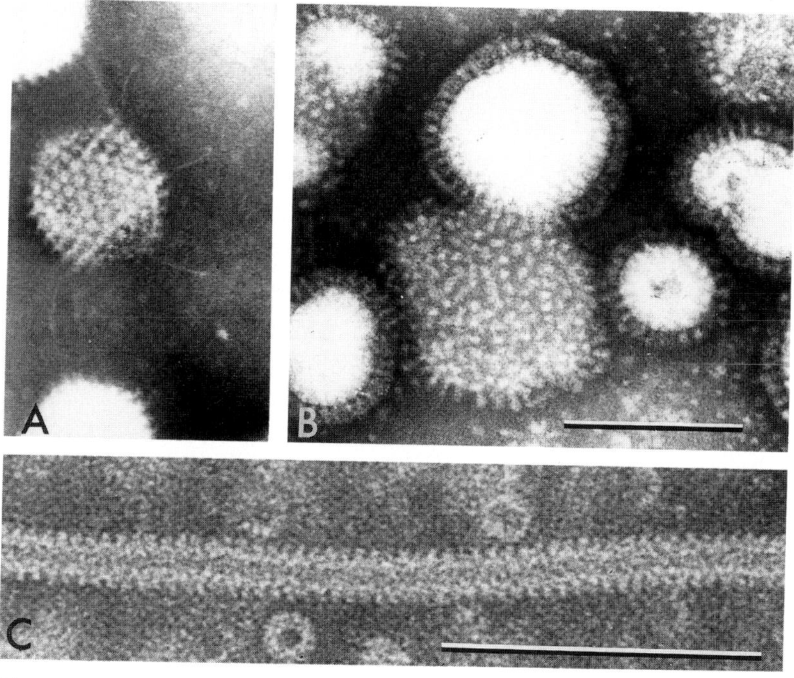

Fig. 1-1 Morphological features of viral structure revealed by negative staining and electron microscopy (bars, 100nm). (A) Virion of an adenovirus, showing icosahedral capsid composed of hexons, pentons, and fibers projecting from vertices (compare with Fig. 1-2A). (B) Enveloped virion of influenza virus. The two types of peplomers, hemagglutinin and neuraminidase, are visible but not distinguishable in this electron micrograph (compare with Fig. 1-2D), nor are the helical nucleocapsids usually visible (but see Fig. 31-1). (C) Nucleocapsid of parainfluenza virus. The RNA is wound within and protected by a helical capsid composed of thousands of identical capsomers (compare with Fig. 1-2B). The complete nucleocapsid is 1000 nm long, but in the intact particle it is folded within a roughly spherical envelope about 180 nm in diameter. (A, B, Courtesy Dr. N. G. Wrigley; C, courtesy Dr. A. J. Gibbs.)

called *capsomers* (Figs. 1-1A and 1-2A,B), which are held together by noncovalent bonds. Within an infected cell, the capsomers undergo self-assembly to form the capsid. The manner of assembly is strictly defined by the nature of the bonds formed between individual capsomers, which imparts symmetry to the capsid. Only two kinds of symmetry have been recognized, cubical (icosahedral) and helical (Fig. 1-2A,B).

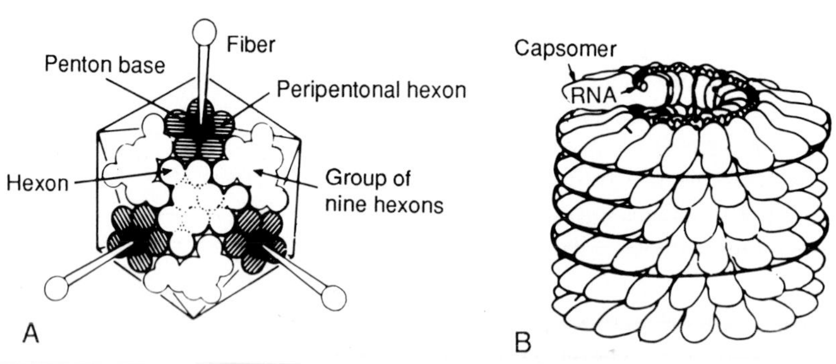

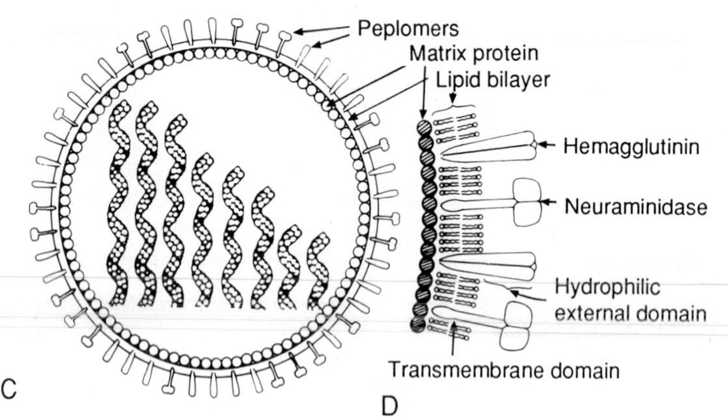

Fig. 1-2 Features of virion structure, exemplified by adenovirus (A), tobacco mosaic virus (B), and influenza A virus (C, D). (A) Icosahedral structure of an adenovirus virion. All hexon capsomers are trimers of the same polypeptide, distinguished as "peripentonal" or "group of nine" by their location in the capsid. The penton base is a pentamer of another polypeptide; the fiber is a trimer of a third polypeptide. (B) The structure of helical nucleocapsids was first elucidated by studies of a nonenveloped plant virus, tobacco mosaic virus, but the principles apply to animal viruses with helical nucleocapsids, all of which are enveloped. In tobacco mosaic virus, a single polypeptide forms a capsomer, and 2130 capsomers assemble in a helix. The 6-kb RNA genome fits in a groove on the inner part of each capsomer, and is wound to form a helix which extends the length of the virion. (C) Structure of virion of influenza A virus. All animal viruses with a helical nucleocapsid and some of those with an icosahedral capsid are enveloped. The nucleocapsids with helical symmetry are long and thin (see Fig. 1-1C and 28-1) and in influenza A virus occur as eight segments, which may be loosely connected (not shown). The viral RNA is wound helically within the helically arranged capsomers of each segment, as shown for tobacco mosaic virus. (D) The envelope of influenza virus consists of a lipid bilayer in which are inserted several hundred glycoprotein peplomers or spikes; beneath the lipid bilayer there is a virus-specified matrix protein. The glycoprotein peplomers of influenza virus comprise two different proteins, hemagglutinin (a rod-shaped trimer) and neuraminidase (a mushroom-shaped

Icosahedral Symmetry

The cubic symmetry found in viruses is based on that of an icosahedron, one of the five classical "Platonic solids" of geometry. An icosahedron has 12 vertices (corners), 30 edges, and 20 faces, each an equilateral triangle. It has axes of two-, three-, and fivefold rotational symmetry, passing through its edges, faces, and vertices, respectively (Fig. 1-3A–C). The icosahedron is the optimum solution to the problem of constructing, from repeating subunits, a strong structure to enclose a maximum volume. Before icosahedrons were discovered in viruses the same principles were applied by the architect Buckminster Fuller to the construction of icosahedral buildings ("geodesic domes"). An object with icosahedral symmetry need not appear angular in outline; the virions of many animal viruses with icosahedral symmetry appear spherical with a bumpy surface.

Only certain arrangements of the capsomers can fit into the faces, edges, and vertices of the viral icosahedron. The capsomers on the faces and edges of adenovirus particles, for example, bond to six neighboring capsomers and are called *hexamers;* those at the vertices bond to five neighbors and are called *pentamers* (Figs. 1-1A and 1-2A). In virions of some viruses both hexamers and pentamers consist of the same polypeptide(s); in those of other viruses they are formed from different polypeptides. The arrangements of capsomers on the capsids of virions of three small icosahedral viruses are shown in Fig.1-3D–F.

High-Resolution Structure

The recent demonstration by X-ray crystallography of the structure of the capsids of several picornaviruses (small RNA viruses), as well as a parvovirus and a papovavirus (small DNA viruses), at near atomic resolution has provided a remarkable insight into their organization and assembly, the location of antigenic sites, and aspects of their attachment and penetration into cells. In several picornaviruses examined, the amino acids of each of the three larger structural proteins are packaged so as to have a wedge-shaped eight-stranded antiparallel β-barrel domain (Fig. 1-4). The outer contour of the virion depends on the packing of these domains and on the way the loops project from the framework. The capsomers of the parvovirus consist of an unusually large wedge-shaped protein with a β-barrel core, hence the ability to form a 250 Å shell from only 60 subunits of a single protein. High-resolution studies with picornaviruses and polyomaviruses have revealed that the capsid proteins have flexible "arms" which interlock with arms of an adjacent structural unit to mediate assembly and stability of the virion. Cations may also stabilize the interface between subunits, and arms extending from internal proteins may interact with proteins of the outer capsid. In virions of

tetramer), each of which consists of an internal domain, a hydrophobic transmembrane domain, and a hydrophilic external domain. Some 50 molecules of a small membrane-associated protein, M2 (not shown), form a small number of "pores" in the lipid bilayer. [A, By John Mack, from R. M. Burnett, *in* "Biological Macromolecules and Assemblies: Virus Structures" (F. Jurnak and A. McPherson, eds.), Vol. 1, p. 337. Wiley, New York, 1984, B, From C. F. T. Mattern, *in* "Molecular Biology of Animal Viruses" (D. P. Nayak, ed.), p. 5. Dekker, New York, 1977.]

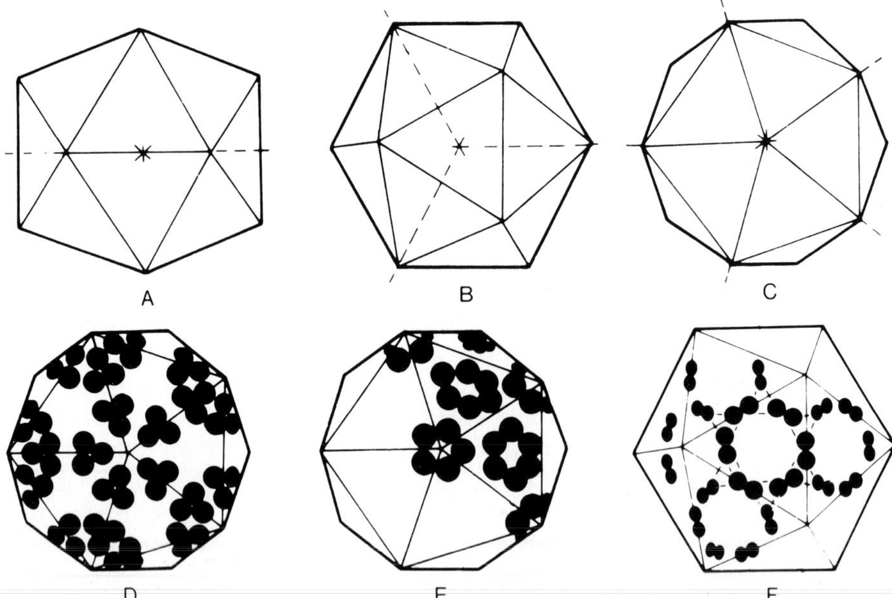

Fig. 1-3 (Top row) Icosahedron viewed along twofold (A), threefold (B), and fivefold (C) axes of symmetry. (Bottom row) Various clusterings of capsid polypeptides are responsible for the characteristic appearances of particular viruses as seen by negative contrast electron microscopy. For example, when capsid polypeptides are arranged as 60 trimers, capsomers themselves are difficult to define; this is the case with foot-and-mouth disease virus (D). When capsid polypeptides are grouped as 12 pentamers and 20 hexamers they form bulky capsomers as is the case with parvoviruses (E). When capsid polypeptides form dimers on the faces they produce ringlike features on the virion surface, as is the case with caliciviruses (F).

rhinovirus (see cover) the antigenic sites are located on the outermost part of the particle.

Ultrastructural studies have also provided some insight into how virions bind to cell receptors. In this connection the words "receptor" and "ligand" are often used in imprecise ways. In this book we use the term *receptor* to designate the specific molecule or structure on the surface of the cell membrane that is recognized by a specific extracellular molecule or structure that binds to it. The *ligand* is the receptor-binding molecule of the virus. For example, the hemagglutinin of influenza virus is the ligand which binds to the cellular receptor, a glycoconjugate terminating in *N*-acetylneuraminic acid. In some picornaviruses there is a deep canyon or pit encircling each fivefold axis (Fig. 1-5). The amino acids within the canyon are far more conserved throughout evolution of different strains of viruses than residues elsewhere on the surface; the conserved residues form the ligand that binds to the host cell receptor and are inaccessible to antibody binding.

Helical Symmetry

The capsomers and nucleic acid genomes of several RNA viruses undergo self-assembly as a cylindrical nucleocapsid in which the capsomers are ar-

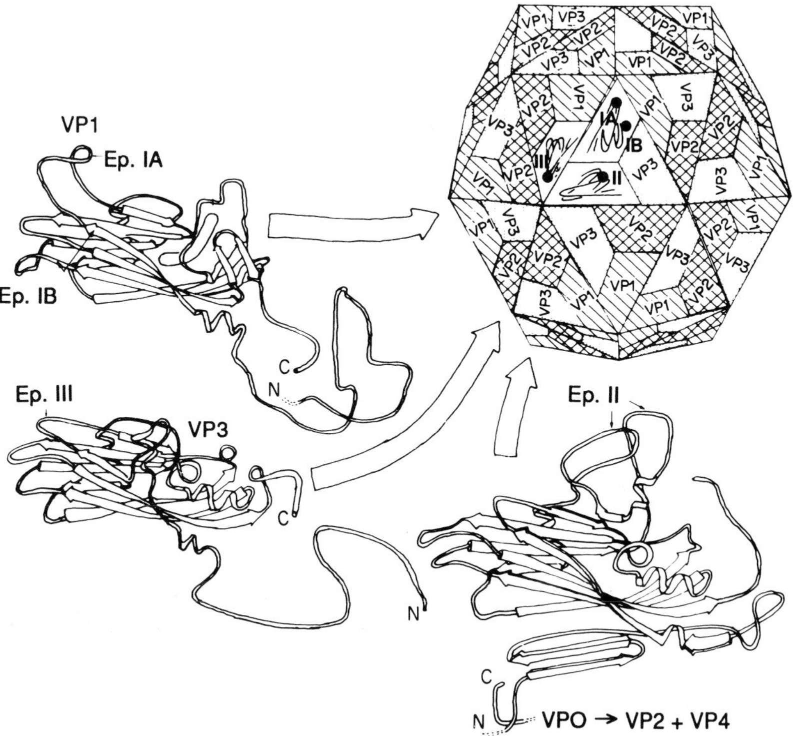

Fig. 1-4 Structure of a rhinovirus as determined by X-ray crystallography. Each of the viral structural proteins VP1, VP3, and VPO (VPO is cleaved during maturation to VP2 and VP4) has an eight-stranded antiparallel β-barrel structure. In the mature virion parts of VP1, VP2, and VP3 are intertwined to produce four neutralizing antibody-binding sites (neutralization epitopes Ep. IA, Ep. IB, Ep. II, and Ep. III), the locations of which in the structural proteins and on the virion are indicated. [Modified from M. G. Rossmann and R. R. Rueckert, *Microbiol. Sci.* **4,** 206 (1987).]

ranged like a spiral staircase, or helix; hence such viruses are said to display helical symmetry. The viral RNA forms a spiral within the nucleocapsid (Figs. 1-1C and 1-2B,C). Each capsomer of such viruses consists of a single polypeptide molecule. The plant viruses whose nucleocapsids have helical symmetry are rod-shaped and nonenveloped. However, in all animal viruses with helical nucleocapsids these are wound into a secondary coil and enclosed within a lipoprotein envelope.

Viral Envelopes

Virions acquire an envelope during maturation by the process known as "budding" from cellular membranes. The lipids of the viral envelope are derived directly from the cell, but the proteins in the envelope are virus-coded. One kind of viral envelope protein structure is the glycoprotein *peplomer* (*peplos,* envelope) or *spike* (Fig. 1-2C,D), which can often be seen in electron micrographs as projections from the surface of the envelope (Fig. 1-1B). Another kind of envelope protein, *matrix protein,* is nonglycosylated and is found as a layer on the inside of the envelope of virions of several

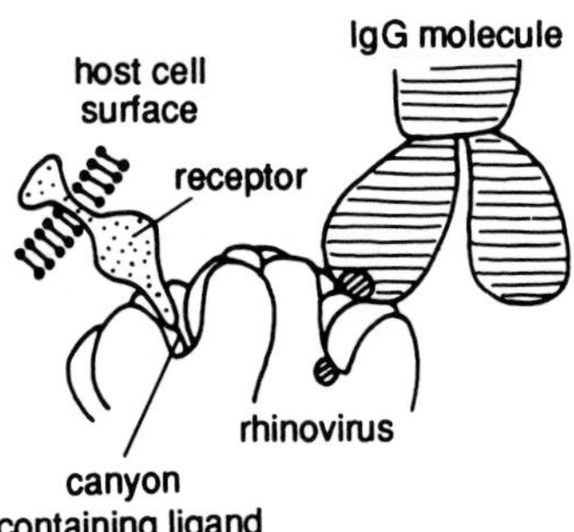

Fig. 1-5 Model of the interaction between receptor on host cells and ligand on a rhinovirus. The ligands are situated within surface depressions ("canyons") near axes of fivefold symmetry, a location which serves to prevent access of antibody to those crucial sites. However, antibodies specific for antigenic sites on the rim of the canyon can block virion–cell interaction by steric hindrance. [Modified from M. G. Rossmann and R. R. Rueckert, *Microbiol. Sci.* **4,** 206 (1987).]

families (Fig. 1-2C,D); matrix protein provides added rigidity to the virion. For example, the envelope of rhabdoviruses with its projecting peplomers is closely applied to a layer of matrix protein which in turn interfaces with a helical nucleocapsid within. Some enveloped viruses, including arenaviruses, bunyaviruses, and coronaviruses, have no matrix protein.

Envelopes are not restricted to viruses of helical symmetry; icosahedral viruses belonging to several families (herpesviruses, togaviruses, flaviviruses, and retroviruses) have envelopes. The infectivity of most enveloped viruses depends on the integrity of the envelope, but some poxviruses have an envelope which is not necessary for infectivity.

Chemical Composition of Virions

Viruses are distinguished from all other forms of life by their simple chemical composition, which includes a genome comprising one or a few molecules of either DNA or RNA, a small number of proteins which form the capsid or are present within the virion as enzymes, and in the case of enveloped viruses, a lipoprotein bilayer with associated glycoprotein peplomers and sometimes a matrix protein.

Viral Nucleic Acid

All viral genomes are *haploid,* that is, they contain only one copy of each gene, except for retrovirus genomes, which are *diploid.* Viral DNA or RNA can be *double-stranded* (*ds*) or *single-stranded* (*ss*). The genome of a representative member of most viral families has now been completely sequenced.

When carefully extracted from the virion, the nucleic acid of viruses of certain families of both DNA and RNA viruses is itself infectious, that is, when experimentally introduced into a cell it can initiate a complete cycle of viral replication, with the production of a normal yield of progeny virions. The essential features of the genomes of viruses of vertebrates are summarized in Table 1-2. Their remarkable variety is reflected in the diverse ways in which the information encoded in the viral genome is transcribed to RNA, then translated into proteins, and the ways in which the viral nucleic acid is replicated (see Chapter 3).

DNA

The genome of all DNA viruses consists of a single molecule, which is double-stranded except in the case of the parvoviruses, and may be linear or circular. The DNA of papovaviruses and hepadnaviruses is circular; the circular DNA of hepadnaviruses is only partially double-stranded. Within the virion, the circular DNA of the papovaviruses is supercoiled.

Most of the linear viral DNAs have characteristics that enable them to adopt a circular configuration, which is a requirement for replication by what is called a rolling circle mechanism. The two strands of poxvirus DNA are covalently cross-linked at their termini, so that on denaturation the molecule becomes a large, single-stranded circle. The linear dsDNA of several DNA viruses (and the linear ssRNA of retroviruses) contains *repeat sequences* at the ends of the molecule that permit circularization. In adenovirus DNA there are inverted terminal repeats; these are also a feature of the ssDNA parvoviruses.

Another type of terminal structure occurs in adenoviruses, hepadnaviruses, parvoviruses, and some ssRNA viruses such as the picornaviruses and caliciviruses. In all of these a protein, which has an essential function in replication of the genome, is covalently linked to the 5′ terminus.

Table 1-2
Structures of Viral Genomes

Family	Type and structure of virion nucleic acid
Parvoviridae	Linear ssDNA, minus sense; with palindromic sequences at ends
Papovaviridae	Circular supercoiled dsDNA
Adenoviridae	Linear dsDNA with inverted terminal repeats and a covalently bound protein
Herpesviridae	Linear dsDNA; unique sequences flanked by repeat sequences; different isomers occur
Poxviridae	Linear dsDNA; both ends covalently closed, with inverted terminal repeats
Hepadnaviridae	Circular dsDNA with region of ssDNA
Picornaviridae *Caliciviridae* *Togaviridae* *Flaviviridae* *Coronaviridae*	Linear ssRNA, plus sense; serves as mRNA; 3′ end polyadenylated (except *Flaviviridae*); 5′ end capped, or protein covalently bound (in *Picornaviridae, Caliciviridae*)
Paramyxoviridae *Rhabdoviridae* *Filoviridae*	Linear ssRNA, minus sense
Orthomyxoviridae	Segmented genome; 7 or 8 molecules of linear ssRNA, minus sense
Bunyaviridae	Segmented genome; 3 molecules of linear ssRNA, minus sense or ambisense; "sticky ends" allow circularization
Arenaviridae	Segmented genome; 2 molecules of linear ssRNA, minus sense or ambisense; "sticky ends" allow circularization
Reoviridae	Segmented genome; 10, 11, or 12 molecules of linear dsRNA
Retroviridae	Diploid genome, dimer of linear ssRNA, plus sense; hydrogen-bonded at 5′ ends; terminal redundancy; both 3′ termini polyadenylated, both 5′ ends capped
Deltavirus	Circular ssRNA, minus sense

The size of viral DNA genomes ranges from 3.2 *kilobase pairs* (*kbp*), for the hepadnaviruses, to over 200 kbp for the largest of the dsDNA herpesviruses and poxviruses. As 1 kilobase (kb) or 1 kbp contains enough genetic information to code for about one average-sized protein, it can be surmised that viral DNAs contain roughly between 4 and 200 genes, coding for some 4 to 200 proteins. However, the relationship between any particular nucleotide sequence and its protein product is not so straightforward. First, the DNA of most of the larger viruses, like that of mammalian cells, contains what appears to be redundant information, in the form of (1) repeat sequences and (2) *introns,* that is, regions which are noncoding and are spliced out from the primary RNA transcript to form the mRNA. On the other hand, a single such primary RNA transcript may be spliced or cleaved in several different ways to yield several distinct mRNAs, each of which may be translated into a different protein. Furthermore, a given DNA or mRNA sequence may be read in up to three different *reading frames,* giving rise to two or three proteins with different amino acid sequences. In addition, either or both strands of double-stranded viral DNA may be transcribed, in a leftward or a rightward direction.

Viral DNAs contain several other kinds of noncoding sequences, some of which are *consensus sequences* which tend to be conserved through evolution because they serve vital functions, including RNA polymerase recognition sites and promoters, enhancers, initiation codons for translation, termination codons, and RNA splice sites.

RNA

The genome of RNA viruses may be single-stranded or double-stranded. While some genomes occur as a single molecule, others are *segmented.* Arenavirus genomes consist of 2 segments, bunyavirus genomes of 3, orthomyxovirus of 7 or 8 (in different genera), and reovirus of 10, 11, or 12 (in different genera). Each of the molecules is unique (often a single "gene"). Except for the very small circular ssRNA of hepatitis D virus (the structure of which resembles that of viroids of plants), no animal virus RNA genome is a covalently linked circle. However, the ssRNAs of arenaviruses and bunyaviruses are "circular," by virtue of having hydrogen-bonded ends. The genomes of ssRNA viruses have considerable secondary structure, regions of base pairing causing the formation of loops, hairpins, etc., which probably serve as signals controlling nucleic acid replication, transcription, translation, and/or packaging into the capsid.

Single-stranded genomic RNA can be defined according to its *sense* (also known as *polarity*). If it is of the same sense as mRNA, it is said to be of *positive* (or *plus*) *sense.* This is the case with the picornaviruses, caliciviruses, togaviruses, flaviviruses, coronaviruses, and retroviruses. If, on the other hand, its nucleotide sequence is complementary to that of mRNA, it is said to be *negative* (or *minus*) *sense.* Such is the case with the paramyxoviruses, rhabdoviruses, filoviruses, orthomyxoviruses, arenaviruses, and bunyaviruses, all of which have an RNA-dependent RNA polymerase (*transcriptase*) in the virion, which in the infected cell transcribes plus sense RNA using the viral genome as template. With the arenaviruses and at least one genus of bunyaviruses one of the RNA segments is *ambisense,* that is, part plus sense, part minus sense. Where the viral RNA is plus sense, it is usually polyadenylated at its 3′ end (in picornaviruses, caliciviruses, togaviruses, and coronaviruses, but not in flaviviruses) and capped at its 5′ end (togaviruses, flaviviruses, coronaviruses).

The size of ssRNA viral genomes varies from 1.7 to 21 kb (M_r approximately 1–10 million) and that of the dsRNA viruses from 18 to 27 kbp, a much smaller range than found among the dsDNA viruses. Accordingly, the RNA viruses encode fewer proteins than many DNA viruses, generally less than a dozen. Most of the segments of the genomes of orthomyxoviruses and reoviruses are individual genes, each coding for one unique protein.

Anomalous Features of Viral Genomes

Viral preparations often contain some particles with an atypical content of nucleic acid. Several copies of the complete viral genome may be enclosed within a single particle, or viral particles may be formed that contain no nucleic acid (empty particles) or that have an incomplete genome (*defective interfering particles*). Moreover, host cell DNA may sometimes be incorporated into virions (e.g., papovavirus), while ribosomal RNA or part thereof has been found in orthomyxovirus virions.

Viral Proteins

Some virus-coded proteins are *structural,* that is, they are part of the virion; some are *nonstructural* and are concerned with various aspects of the replication cycle. An essential role for one class of structural proteins is to provide the viral nucleic acid with a protective coat. One of the surface proteins bears the ligand for binding to the host cell receptor molecule. The virions of all viruses of vertebrates contain several different proteins, the number ranging from 2 in the simplest viruses to over 100 in the most complex. For viruses with cubic symmetry, the structural proteins form an icosahedral capsid which sometimes encloses an additional layer, or *core,* composed of different, often basic histonelike polypeptides that are intimately associated with the nucleic acid.

Also associated with most virions are one or more enzymes, most of which are involved in nucleic acid transcription. These include various types of transcriptases which transcribe mRNA from dsDNA or dsRNA viral genomes or from genomes of viruses with minus sense ssRNA. Reverse transcriptase, which transcribes DNA from RNA, is found in retroviruses and hepadnaviruses, while other enzymes found in retrovirus particles are involved in the integration of the transcribed DNA into the cellular DNA. Poxviruses, which replicate in the cytoplasm, carry a number of enzymes involved in processing RNA transcripts and in replicating DNA. One of the glycoprotein spikes of the envelope of orthomyxoviruses and paramyxoviruses, the neuraminidase, has enzymatic activity.

Viral Glycoproteins

Most viral glycoproteins occur as membrane-anchored spikes or projections from the envelope of enveloped viruses, but the virions of some of the more complex viruses also contain glycosylated internal or outer capsid proteins. Oligosaccharide side chains (glycans) are attached by N-glycosidic, or more rarely O-glycosidic, linkage. Because the glycans are synthesized by cellular glycosyltransferases, the sugar composition of the glycans corresponds to that of host cell membrane glycoproteins.

Viral Envelope Lipids

Lipids constitute about 20–35% of the dry weight of enveloped viruses, the viral envelope being composed of cellular lipids and viral proteins. As a consequence, the composition of the lipids of particular viruses differs according to the composition of the membrane lipids of the host cells from which they came. Approximately 50–60% of the envelope lipid is phospholipid, and most of the remainder is cholesterol. Most lipid found in enveloped viruses is present as a typical lipid bilayer in which the virus-coded glycoprotein spikes (and occasionally trace amounts of residual cellular membrane proteins) are embedded.

Preservation of Viral Infectivity

In general, viruses are more sensitive than bacteria or fungi to inactivation by physical and chemical agents. A knowledge of their sensitivity to environmental conditions is therefore important for ensuring the preservation of the infectivity of viruses as reference reagents, and in clinical specimens collected for diagnosis, as well as for their deliberate inactivation for such practical ends as sterilization, disinfection, and the production of inactivated vaccines.

Temperature

The principal environmental condition that may adversely affect the infectivity of viruses is temperature. Surface proteins are denatured within a few minutes at temperatures of 55°–60°C, with the result that the virion is no longer capable of normal cellular attachment and/or uncoating. At ambient temperature the rate of decay of infectivity is slower but significant, especially in the summer or in the tropics. To preserve infectivity, viral preparations must therefore be stored at low temperature; 4°C (wet ice or a refrigerator) is usually satisfactory for a day or so, but longer term preservation requires much lower temperatures. Two convenient temperatures are −70°C, the temperature of frozen CO_2 ("dry ice") and of some mechanical freezers, or −196°C, the temperature of liquid nitrogen. As a rule of thumb, the half-life of most viruses can be measured in seconds at 60°C, minutes at 37°C, hours at 20°C, days at 4°C, and years at −70°C or lower. The enveloped viruses are more heat-labile than nonenveloped viruses. Enveloped virions, notably those of respiratory syncytial virus, are also susceptible to repeated freezing and thawing, probably as a result of disruption of the virion by ice crystals. This poses problems in the collection and transportation of clinical specimens. The most practical way of avoiding such problems is to deliver specimens to the laboratory as rapidly as practicable, packed without freezing, on ice-cold gel packs.

In the laboratory, it is often necessary to preserve virus stocks for years.

This is achieved in one of two ways: (1) rapid freezing of small aliquots of virus suspended in medium containing protective protein and/or dimethyl sulfoxide, followed by storage at −70°C or −196°C; (2) freeze-drying (lyophilization), that is, dehydration of a frozen viral suspension under vacuum, followed by storage of the resultant powder at 4°C or −20°C. Freeze-drying prolongs viability significantly even at ambient temperatures, and it is universally used in the manufacture of attenuated live-virus vaccines.

Ionic Environment and pH

On the whole, viruses are best preserved in an isotonic environment at physiologic pH, but some tolerate a wide ionic and pH range. For example, whereas most enveloped viruses are inactivated at pH 5–6, rotaviruses and many picornaviruses survive the acidic pH of the stomach.

Lipid Solvents and Detergents

The infectivity of enveloped viruses is readily destroyed by lipid solvents such as ether or chloroform and by detergents like sodium deoxycholate, so that these agents must be avoided in laboratory procedures concerned with maintaining the viability of viruses. On the other hand, detergents are commonly used by virologists to solubilize viral envelopes and liberate proteins for use as vaccines or for chemical analysis.

Further Reading

Bishop, D. H. L. (1986). Ambisense RNA viruses: Positive and negative polarities combined in RNA virus genomes. *Microbiol. Sci.* **3,** 183.

Dubois-Dalco, M., Holmes, K. V., and Rentier B. (1984). "Assembly of Enveloped RNA Viruses." Springer-Verlag, New York.

Harrison, S. C. (1990). Principles of virus structure. *In* "Fields Virology" (B. N. Fields, D. M. Knipe, R. M. Chanock, M. S. Hirsch, J. L. Melnick, T. P. Monath, and B. Roizman, eds.), 2nd Ed., p. 37. Raven, New York.

McGeoch, D. J. (1981). Structural analysis of animal virus genomes. *J. Gen. Virol.* **55,** 1.

Rossmann, M. G., and Johnson, J. E. (1989). Icosahedral RNA virus structure. *Annu. Rev. Biochem.* **58,** 533.

Rossman, M. G., and Rueckert, R. R. (1987). What does the molecular structure of viruses tell us about viral functions? *Microbiol. Sci.* **4,** 206.

Tsao, J., Chapman, M. S., Agbandje, M., Keller, W., Smith, K., Luo, M., Smith, T. J., Rossman, M. G., Compans, R. W., and Parrish, C. R. (1991). The three dimensional structure of canine parvovirus and its functional implications. *Science* **251,** 1456.

Wiley, D. C., and Skehel, J. J. (1990). Viral membranes. *In* "Fields Virology" (B. N. Fields, D. M. Knipe, R. M. Chanock, M. S. Hirsch, J. L. Melnick, T. P. Monath, and B. Roizman, eds.), 2nd Ed., p. 63. Raven, New York.

Chapter 2

Classification and Nomenclature of Viruses

There is good evidence to indicate that all organisms are infectable by viruses: vertebrate and invertebrate animals, plants, algae, fungi, protozoa, and bacteria; indeed, every species of animal, bacterium, and plant that has been intensively searched has yielded numerous different viruses belonging to several viral families. Because all viruses, whatever their hosts, share the features described in the previous chapter, viral taxonomists have developed a scheme of classification and nomenclature that is universal. In this book, however, we are concerned solely with the viruses that cause disease in humans. Some of these (called arboviruses) also replicate in insects, ticks, or other arthropods.

Several hundred distinguishable viruses have been recovered from humans, the best studied vertebrate host, and new ones are being discovered each year. Somewhat fewer have been recovered from each of the common species of farm and companion animals and from the commonly used laboratory animals. To simplify the study of this vast number of viruses we need to sort them into groups that share certain common properties.

Criteria for Viral Classification

Classification into major groups called families, and the subdivision of families into genera, has now reached a position of substantial international agreement. Recently three families (*Paramyxoviridae, Rhabdoviridae, and Filoviridae*) were grouped into a higher taxon, the order *Mononegavirales,* on the basis of the similarity in genome structure and strategy of replication of the member viruses. It is likely that in the future other families will be grouped to form orders.

The primary criteria for delineation of families are (1) the kind of nucleic acid that constitutes the genome (see Table 1-2), (2) the strategy of viral replication, and (3) the morphology of the virion. Subdivision of families into genera is based on criteria that vary for different families. Genera, usually defined by substantial differences in their genomes, contain from one to over a hundred species. The definition of species is more arbitrary, and virologists continue to argue about the criteria for the designation of species and about their nomenclature.

Serology, more recently strengthened by the use of monoclonal antibodies, is of great value in the differentiation of viruses below the species level, namely, types, subtypes, strains, and variants—terms that have no generally agreed taxonomic status. Characterization of the nucleic acid of the viral genome, as revealed by such techniques as nucleotide sequence analysis (partial or complete genome sequencing), restriction endonuclease mapping, electrophoresis in gels (especially useful for RNA viruses with segmented genomes), oligonucleotide fingerprinting, and molecular hybridization, is being used more and more to identify viruses and to distinguish differences between strains and variants.

Nomenclature

Since 1966 the classification and nomenclature of viruses, at the higher taxonomic levels (families and genera), have been systematically organized by the International Committee on Taxonomy of Viruses. The highest taxon used is the order, named with the suffix *-virales.* Families are named with the suffix *-viridae,* subfamilies with the suffix *-virinae,* and genera with the suffix *-virus.* The prefix may be another Latin word or it may be a *sigla,* that is, an abbreviation derived from initial letters. Order, family, subfamily, and generic names are capitalized and written in italics, for example, *Paramyxoviridae;* vernacular terms derived from them are written in roman letters, without an initial capital letter, for example, paramyxoviruses. Currently, viral species are designated by vernacular terms, for example, measles virus.

A brief description of each family of viruses of importance in human medicine (see Fig. 2-1) is given below, and their properties are summarized in Tables 2-1 and 2-2. Viral families or genera that infect vertebrates but do not include human pathogens are omitted from this book; they are described in the companion volume, *Veterinary Virology,* 2nd edition (Academic Press, 1993).

Families of DNA Viruses

Family: *Parvoviridae* (Parvoviruses)
Subfamily: *Parvovirinae*
Genus: *Parvovirus* (parvoviruses of mammals and birds)

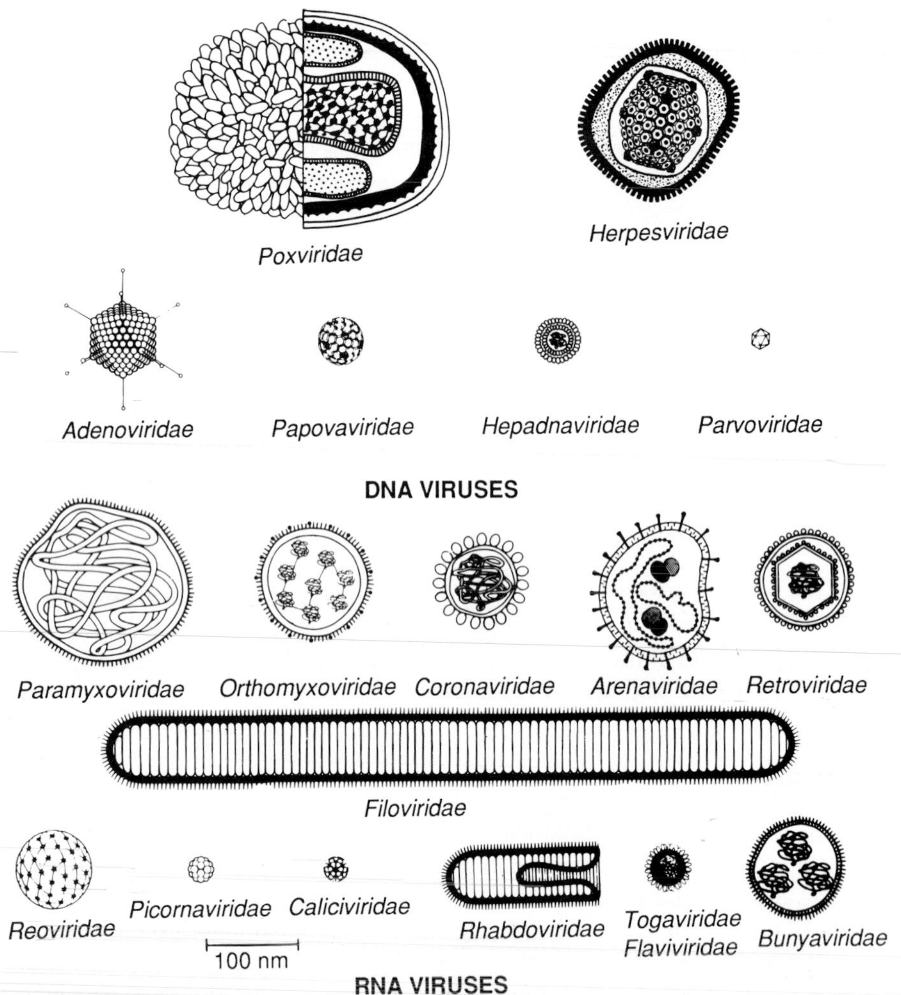

Fig. 2-1 Shapes and sizes of viruses of families that include human pathogens. The virions are drawn to scale, but artistic license has been used in representing their structure. In some, the cross-sectional structure of capsid and envelope is shown, with a representation of the genome; with the very small virions, only size and symmetry are depicted.

Genus: *Erythrovirus* (human parvovirus B19)
Genus: *Dependovirus* (adeno-associated viruses)

Parvoviruses (*parvus,* small) are about 20 nm in diameter, have icosahedral symmetry, and possess a genome of ssDNA, 5 kb (Table 2-1). The virions are relatively heat stable. Most species have a narrow host range and replicate preferentially in dividing cells. Members of the genus *Parvovirus* infect a number of species of animals, and one parvovirus (B19, the only member of the genus *Erythrovirus*) has been identified in humans. Members of the genus *Dependovirus* are defective viruses, which depend an adenovirus (or, experimentally, a herpesvirus) for replication. Five serotypes of these "adeno-associated viruses" occur in humans but are not known to cause disease.

Table 2-1
Properties of Virions of Families of DNA Viruses Infecting Humans

	Virion					Genome	
			Nucleocapsid				
Family	Diameter (nm)	Envelope	Symmetry	Capsomers	Transcriptase	Nature[a,b]	Size[c] (kb, kbp)
Parvoviridae	20	−	Icosahedral	32	−	ss, −, linear	5
Papovaviridae	45, 55[d]	−	Icosahedral	72	−	ds, circular	5, 8[d]
Adenoviridae	70	−	Icosahedral	252	−	ds, linear	36–38
Herpesviridae	150	+	Icosahedral	162	−	ds, linear	125–229
Poxviridae	250 × 200 × 200[e]	+[f]	Complex	—	+	ds, linear	130–250
Hepadnaviridae	42	−	Icosahedral	?	+[g]	ds, circular[h]	3.2

[a] All DNA virus genomes comprise a single molecule.
[b] ds, Double-stranded; ss, single-stranded; sense of single-stranded nucleic acid (+) or (−).
[c] For species that cause human infections.
[d] Lower figure, *Polyomavirus;* higher figure, *Papillomavirus.*
[e] *Orthopoxvirus,* brick-shaped; *Parapoxvirus,* ovoid, 260 × 160 nm.
[f] Not essential for infectivity.
[g] Reverse transcriptase.
[h] Circular molecule is double-stranded for most of its length but contains a single-stranded region.

Family: *Papovaviridae* (Papovaviruses)
Genus: *Papillomavirus* (papillomaviruses)
Genus: *Polyomavirus* (polyomaviruses)

The papovaviruses [sigla, from papilloma, polyoma, vacuolating agent (early name for SV40)] are small nonenveloped icosahedral viruses which replicate in the nucleus and may transform infected cells. In the virion the nucleic acid occurs as a cyclic double-stranded DNA molecule, which is infectious. There are two genera. Virions of *Papillomavirus* (wart viruses) are 55 nm in diameter and have a larger genome (8 kbp) which may persist in transformed cells in an episomal form. Virions of members of the genus *Polyomavirus* are 45 nm in diameter, have a smaller genome (5kbp), and may persist in cells via the integration of their genome into the host cell DNA.

Human papillomaviruses cause warts, and some of the several dozen types are associated with cancer of the cervix or the skin; the human polyomaviruses usually cause inapparent infections but may be reactivated by immunosuppression. Murine polyoma virus and simian virus 40 (SV40, from rhesus monkeys) have been useful models for the laboratory study of viral oncogenesis.

Family: *Adenoviridae* (Adenoviruses)
Genus: *Mastadenovirus* (mammalian adenoviruses)

The adenoviruses (*adenos*, gland) are nonenveloped icosahedral viruses 70 nm in diameter, with a single linear dsDNA genome of 36–38 kbp. They replicate in the nucleus. Nearly 50 serologically distinct types of human adenoviruses are currently recognized; all share a group antigen with adenovirus serotypes infecting other mammals (the genus *Mastadenovirus*).

Human adenoviruses are associated with infections of the respiratory tract, the eye, and the intestinal tract. Many infections are characterized by prolonged persistence and may be reactivated by immunosuppression.

Family: *Herpesviridae* (Herpesviruses)
Subfamily: *Alphaherpesvirinae* (herpes simplex-like viruses)
Genus: *Simplexvirus* (herpes simplex-like viruses)
Genus: *Varicellovirus* (varicella-zoster virus)
Subfamily: *Betaherpesvirinae* (cytomegaloviruses)
Genus: *Cytomegalovirus* (human cytomegalovirus)
Genus: *Roseolovirus* (human herpesvirus 6)
Subfamily: *Gammaherpesvirinae* (lymphoproliferative herpesviruses)
Genus: *Lymphocryptovirus* (Epstein-Barr virus)

The herpesviruses (*herpes*, creeping) have enveloped virions about 150 nm in diameter, with icosahedral nucleocapsids about 100 nm in diameter. The genome is a single linear molecule of dsDNA, 125–229 kbp. The herpesviruses replicate in the nucleus and mature by budding through the nuclear membrane, thus acquiring an envelope. This large family includes several important human pathogens and has been divided into three subfamilies. *Alphaherpesvirinae* includes herpes simplex types 1 and 2, varicella-zoster virus, and B virus of monkeys, which is pathogenic for humans. *Betaherpesvirinae* comprises the cytomegaloviruses, which are highly host-specific viruses of humans and other animals and produce low-grade, chronic infections. Human cytomegalovirus is an important cause of congenital abnormalities. Human herpesvirus 6 has been allocated to the subfamily *Betaherpesvirinae,* genus *Roseolovirus. Gammaherpesvirinae* includes the genus *Lymphocryptovirus,* which contains one species, Epstein-Barr (EB) virus, the cause of infectious mononucleosis.

A feature of all herpesvirus infections is life-long persistence of the virus in the body, usually in latent form. Excretion, especially in saliva or genital secretions, may occur continuously or intermittently without disease, or episodes of recurrent clinical disease and recurrent excretion may occur years after the initial infection, especially following immunosuppression. There is evidence that some herpesviruses may have a role in human cancers, notably EB virus in nasopharyngeal cancer and Burkitt's lymphoma.

Family: *Poxviridae* (Poxviruses)
Subfamily: *Chordopoxvirinae* (poxviruses of vertebrates)
Genus: *Orthopoxvirus* (vaccinia virus subgroup)
Genus: *Parapoxvirus* (orf virus subgroup)
Genus: *Molluscipoxvirus* (molluscum contagiosum virus)
Genus: *Yatapoxvirus* (yaba/tanapox virus subgroup)
Genus: *Avipoxvirus* (bird poxviruses)

The poxviruses (*poc, pocc,* pustule) are the largest and most complex viruses of vertebrates. The virions are brick-shaped, measuring about 250 by 200 by 200 nm in all genera that cause human infections except *Parapoxvirus,* the virions of which are ovoid and measure 260 by 160 nm. All poxviruses have an inner core which contains a single linear molecule of dsDNA, 130–250 kbp. Unlike most other DNA viruses of vertebrates, poxviruses replicate in the cytoplasm, mRNA being transcribed by a virion-associated transcriptase. A large number of other virion-associated enzymes are involved in DNA synthesis.

The family is divided into two subfamilies, one of which, *Chordopoxvirinae,* comprises the poxviruses of vertebrates. This subfamily contains four genera that include human pathogens. The genus *Orthopoxvirus* includes cowpox, ectromelia (mousepox), rabbitpox (a variant of vaccinia virus), and monkeypox viruses. Variola virus, which caused human smallpox, and vaccinia virus, used to control that disease, also belong to this genus. *Parapoxvirus* includes contagious pustular dermatitis virus of sheep and pseudocowpox (milker's node) virus, both of which produce skin lesions in humans. Molluscum contagiosum (genus *Molluscipoxvirus*) is a specifically human virus, and the genus *Yatapoxvirus* contains two viruses of African wildlife, yabapoxvirus and tanapoxvirus, both of which may infect humans. Fowlpox virus and canarypox viruses (genus *Avipoxvirus*) are being investigated as possible vectors for human vaccines.

Family: *Hepadnaviridae* (Hepatitis B-like Viruses)
Genus: *Orthohepadnavirus* (mammalian hepatitis B-like viruses)

Human hepatitis B virus and related viruses of other animals, all highly host-specific, comprise the family *Hepadnaviridae* (*hepar,* liver; *dna,* sigla for deoxyribonucleic acid). The virions are spherical particles 42 nm in diameter, consisting of a 27-nm icosahedral core within a closely adherent outer capsid that contains cellular lipids, glycoproteins, and a virus-specific surface antigen (HBsAg). The genome is a small, circular, partially double-stranded DNA molecule, which consists of a long (3.2 kb) and a short (1.7–2.8 kb) strand. Replication involves an RNA intermediate and requires a virus-coded reverse transcriptase.

The hepadnaviruses replicate in hepatocytes and cause hepatitis, which may progress to a chronic carrier state, cirrhosis, and primary hepatocellular carcinoma. The most important species is human hepatitis B virus, but hepadnaviruses also occur in woodchucks, ground squirrels, Pekin ducks, and herons.

Families of RNA Viruses

Family: *Picornaviridae* (Picornaviruses)
Genus: *Enterovirus* (enteroviruses)
Genus: *Hepatovirus* (hepatitis A-like viruses)
Genus: *Rhinovirus* (rhinoviruses)

The *Picornaviridae* (*pico,* "micro–micro"; *rna,* sigla for ribonucleic acid) comprise small nonenveloped icosahedral viruses 25–30 nm in diameter, which contain a single molecule of plus sense ssRNA (7.5–8.5 kb), and replicate in the cytoplasm (see Table 2-2). The genus *Enterovirus* includes 3 polioviruses, 32 human echoviruses, 29 coxsackieviruses, and a few other human enteroviruses. Most of these viruses usually produce inapparent enteric infections, but the polioviruses may also cause paralysis; other enteroviruses are sometimes associated with meningoencephalitis, rashes, carditis, myositis, conjunctivitis, and mild upper respiratory tract disease. The only human pathogen in the genus *Hepatovirus* is human hepatitis A virus. The genus *Rhinovirus* includes well over 100 serotypes that affect humans; they are the most frequent viruses causing the common cold.

Family: *Caliciviridae* (Caliciviruses)
Genus: *Calicivirus* (caliciviruses)

The caliciviruses (*calix,* cup) are icosahedral viruses whose virions are 35–40 nm in diameter and have 32 cup-shaped depressions on the surface. The genome consists of one molecule of plus sense ssRNA, size 8 kb. The Norwalk agent and related caliciviruses are important causes of gastroenteritis, and one cause of human hepatitis transmitted by the fecal–oral route, hepatitis E virus, is a calicivirus.

Family: *Astroviridae* (Astroviruses)
Genus: *Astrovirus* (astroviruses)

Astrovirus (*astron,* star) is a name accorded to small spherical virions with a characteristic star-shaped outline by negative staining. These viruses have been found in the feces of humans, calves, and lambs suffering from enteritis. The genome consists of one molecule of ssRNA about the same size as that of the picornaviruses, but unlike picornaviruses and caliciviruses they possess only two capsid proteins.

Table 2-2
Properties of Virions of Families of RNA Viruses Infecting Humans

	Virion					Genome	
			Nucleocapsid				
Family	Diameter (nm)[a]	Envelope	Symmetry	Capsomers	Transcriptase	Nature[b]	Size (kb, kbp)
Picornaviridae	25–30	−	Icosahedral	60	−	ss, +	7.5–8.5
Caliciviridae	35–40	−	Icosahedral	32	−	ss, +	8
Astroviridae	28–30	−	Icosahedral	?	−	ss, +	7.5
Togaviridae	60–70	+[c]	Icosahedral	60	−	ss, +	12
Flaviviridae	40–50	+[c]	Icosahedral	?	−	ss, +	10
Coronaviridae	75–160	+[c]	Helical	?	−	ss, +	27–33
Paramyxoviridae	150–300	+	Helical	?	+	ss, −	15–16
Rhabdoviridae	180 × 75	+	Helical	?	+	ss, −	13–16
Filoviridae	790–970 × 80	+	Helical	?	+	ss, −	12.7
Orthomyxoviridae	80–120	+	Helical	?	+	ss, 7–8, −	13.6
Arenaviridae	110–130	+[c]	Helical	?	+	ss, 2, −	10–14
Bunyaviridae	90–120	+[c]	Helical	?	+	ss, 3, −	13.5–21
Reoviridae	60–80	−	Icosahedral	32, 92[d]	+	ds, 10–12[e]	18–27
Retroviridae	80–100	+[c]	Icosahedral	?	+[f]	ss, +	7–10[g]
Deltavirus	32	−	Icosahedral	?	−	ss, −, circular	1.7

[a] Some enveloped viruses are very pleomorphic and sometimes filamentous.
[b] All genomes except that of *Deltavirus* are linear; ss, single-stranded; ds, double-stranded; 2 to 12, number of segments in segmented genomes; + or −, sense of single-stranded nucleic acid.
[c] No matrix protein.
[d] 32, inner capsid of *Orbivirus*, *Rotavirus*, and *Orthoreovirus*; 92, outer capsid of *Orthoreovirus*.
[e] *Orthoreovirus* and *Orbivirus*, 10; *Rotavirus*, 11; *Coltivirus*, 12.
[f] Reverse transcriptase.
[g] Genome is diploid, two identical molecules being held together by hydrogen bonds at their 5′ ends.

Family: *Togaviridae* (Togaviruses)
Genus: *Alphavirus* (formerly "group A" arboviruses)
Genus: *Rubivirus* (rubella virus)

The togaviruses (*toga*, cloak) are small spherical enveloped viruses 60–70 nm in diameter, containing plus sense ssRNA (12 kb) enclosed within an icosahedral core. They replicate in the cytoplasm and mature by budding from cell membranes. The genus *Alphavirus* contains many species, all of which are mosquito-transmitted. Important human pathogens include eastern, western and Venezuelan equine encephalitis viruses, Ross River virus, and chikungunya virus. In nature the alphaviruses usually produce inapparent viremic infections of birds, mammals, or reptiles. When humans are bitten by an infected mosquito the usual consequence is an inapparent infection, but generalized disease, often associated with arthritis or encephalitis, can result.

The only non-arthropod-borne togavirus is rubella virus (genus *Rubivirus*), a human pathogen important for its ability to cause congenital defects in the fetus when pregnant women are infected.

Family: *Flaviviridae* (Flaviviruses)
Genus: *Flavivirus* (formerly "group B" arboviruses)
Genus: *Hepatitis C* (hepatitis C virus)

Flaviviruses (*flavus*, yellow) have an enveloped icosahedral virion 40–50 nm in diameter and a genome of 10 kb. Viruses of the largest genus, *Flavivirus*, are arthropod-borne (mosquitoes and ticks), but hepatitis C virus is transmitted sexually and via human blood. The genus *Flavivirus* contains several important human pathogens including the viruses of yellow fever, dengue, and St. Louis, Japanese, Murray Valley, and Russian tick-borne encephalitides.

Family: *Coronaviridae* (Coronaviruses)
Genus: *Coronavirus* (coronaviruses of mammals and birds)

The coronaviruses (*corona*, crown) are somewhat pleomorphic viruses 75–160 nm in diameter, with widely spaced, pear-shaped peplomers embedded in a lipoprotein envelope. The envelope lacks a matrix protein, and it encloses a core of helical symmetry with a single linear molecule of plus sense ssRNA, 27–33 kb. Some coronaviruses cause common colds in humans, while others have been visualized in human feces.

Family: *Paramyxoviridae* (Paramyxoviruses)
Subfamily: *Paramyxovirinae*
Genus: *Paramyxovirus* (parainfluenzaviruses)
Genus: *Morbillivirus* (measleslike viruses)
Genus: *Rubulavirus* (mumps virus)
Subfamily: *Pneumovirinae*
Genus: *Pneumovirus* (respiratory syncytial viruses)

The paramyxoviruses (*para*, by the side of; *myxa*, mucus) have a large, pleomorphic, enveloped virion 150–300 nm in diameter, with a helical nucleocapsid. The genome consists of a single linear molecule of minus sense ssRNA (15–16 kb). The envelope contains two glycoproteins, a hemagglutinin (in most species with neuraminidase activity also) and a fusion protein.

Human pathogens in the genus *Paramyxovirus* include four types of parainfluenzaviruses, which cause respiratory disease. Measles (genus *Morbillivirus*) is an important generalized infection associated with a rash, and mumps virus is the only human pathogen in the genus *Rubulavirus*. Respiratory syncytial virus (subfamily *Pneumovirinae*, genus *Pneumovirus*) is a major cause of respiratory disease in infants.

Family: *Rhabdoviridae* (Rhabdoviruses)
Genus: *Vesiculovirus* (vesicular stomatitis-like viruses)
Genus: *Lyssavirus* (rabies-like viruses)

The rhabdoviruses (*rhabdos*, rod) are bullet-shaped viruses, about 180 by 75 nm, containing a single molecule of minus sense ssRNA (13–16 kb). The helical capsid is enclosed within a shell to which is closely applied an envelope with embedded peplomers. The virion matures at the plasma membrane. Animal pathogens in the genus *Vesiculovirus* include vesicular stomatitis, Chandipura, Piry, and Isfahan viruses, each of which is an occasional human pathogen. The genus *Lyssavirus* includes rabies virus and several serologically related viruses from Africa, which may cause severe disease in humans following animal bites.

Family: *Filoviridae* (Filoviruses)
Genus: *Filovirus* (Marburg, Ebola, and Reston viruses)

The virions of filoviruses resemble those of rhabdoviruses but are pleomorphic and sometimes very long (*filo*, threadlike), maximum infectivity being associated with a particle 790–970 nm long and 80 nm wide. The genome is a single molecule of minus sense ssRNA, 12.7 kb. Marburg and Ebola viruses cause sporadic infections and occasional nosocomial epidemics of severe hemorrhagic fever in humans in Africa. In 1989 another filovirus, called Reston virus, was isolated from monkeys imported from the Philippines, but it caused only subclinical infections in animal handlers.

Family: *Orthomyxoviridae* (Influenza viruses)
Genus: *Influenzavirus A, B* (influenza A and B viruses)
Genus: *Influenzavirus C* (influenza C virus)
Genus: unnamed Thogoto-like viruses (tick-borne orthomyxoviruses)

The orthomyxoviruses (*orthos*, straight; *myxa*, mucus) are spherical RNA viruses 80–120 nm in diameter, with a helical nucleocapsid enclosed within an envelope acquired by budding from the plasma membrane. The genome consists of seven (influenza C virus) or eight (influenza A and B viruses) segments of minus sense ssRNA (total size, 13.6 kb). The envelope is studded with spikes, which are of two kinds, a hemagglutinin and a neuraminidase in influenza A and B viruses, and of one kind, hemagglutinin–esterase, in influenza C virus.

Influenza A virus infects birds, horses, swine, mink, seals, and whales, as well as humans; influenza B virus is a human pathogen only. Influenza C virus infects humans and swine, but rarely causes serious disease. Influenza A viruses of birds and humans undergo genetic reassortment to generate novel subtypes ("antigenic shift") which cause major pandemics of human influenza. The tick-borne orthomyxoviruses, Dhori and Thogoto, which occasionally infect humans, have been allocated to a separate genus, so far unnamed.

Family: *Arenaviridae* (Arenaviruses)
Genus: *Arenavirus* (arenaviruses)

Arenaviruses (*arena,* sand) are so named because of the presence of particles resembling ribosomes (hence grains of sand) within the pleomorphic 110–130 nm enveloped virions. The genome consists of two segments of minus sense or ambisense ssRNA (total size 10–14 kb), each held in a circular configuration by hydrogen bonds.

Arenaviruses cause natural inapparent infections of rodents, and humans occasionally develop a serious generalized disease following accidental exposure to arenavirus-infected rodent urine. Lymphocytic choriomeningitis virus is an important laboratory model for the study of persistent infections and may cause human disease; Lassa, Machupo, Junin, and Guanarito viruses may cause severe hemorrhagic fever and (like the filoviruses) are Biosafety Level 4 pathogens.

Family: *Bunyaviridae* (Bunyaviruses)
Genus: *Bunyavirus* (Bunyamwera supergroup)
Genus: *Phlebovirus* (sandfly fever viruses)
Genus: *Nairovirus* (Nairobi sheep disease-like viruses)
Genus: *Hantavirus* (hemorrhagic fever with renal syndrome viruses)

The bunyaviruses (Bunyamwera, locality in Uganda), numbering over 100, comprise the largest single group of arboviruses. The enveloped virions are 90–120 nm in diameter, within which there are three tubular nucleocapsids, each in the form of a circle. The genome consists of three molecules of minus sense (or in *Phlebovirus,* ambisense) ssRNA (total size 13.5–21 kb), each held in a circular configuration by hydrogen bonds. The bunyaviruses replicate in the cytoplasm and bud from Golgi membranes. Because of their segmented genome, closely related bunyaviruses readily undergo genetic reassortment.

All members of the family *Bunyaviridae* except the hantaviruses are arboviruses which have wild animal reservoir hosts; some are transovarially transmitted in mosquitoes with a high frequency. The hantaviruses, which are enzootic in rodents, cause hemorrhagic fever with renal syndrome or pulmonary disease in humans. The genus *Phlebovirus* includes sandfly fever virus (transmitted by *Phlebotomus*) and Rift Valley fever virus, a mosquito-transmitted virus that is an important pathogen of sheep and humans. The genus *Bunyavirus,* most members of which are mosquito-transmitted, includes the California group arboviruses, some of which occasionally cause encephalitis in humans. Members of the genus *Nairovirus* are tick-borne and include the virus of Crimean–Congo hemorrhagic fever.

Family: *Reoviridae* (Reoviruses)
Genus: *Orthoreovirus* (reoviruses of animals)
Genus: *Orbivirus* (orbiviruses)
Genus: *Rotavirus* (rotaviruses)
Genus: *Coltivirus* (Colorado tick fever virus)

The family name *Reoviridae* is a sigla, respiratory enteric orphan virus, reflecting the fact that members of the first discovered genus, *Orthoreovirus*, were found in both the respiratory and enteric tract of humans and most animals, but were not associated with any disease (orphan viruses were "viruses in search of a disease"). The distinctive feature of the family is that the virions contain dsRNA, in 10–12 segments: 10 [*Orthoreovirus* (22 kbp total) and *Orbivirus* (18 kbp)], 11 [*Rotavirus* (16–21kbp)], or 12 [*Coltivirus* (27 kbp)]. The virion is a nonenveloped icosahedron 60–80 nm in diameter. Orbiviruses (*orbi*, ring) are arboviruses, some of which cause disease in humans, as does Colorado tick fever virus (genus *Coltivirus*). The rotaviruses (*rota*, wheel) include viruses that are important causes of diarrhea in humans and some domestic animals.

Family: *Retroviridae* (Retroviruses)
Genus: *Lentivirus* (HIV-like viruses; maedi/visna-like viruses)
Genus: *Spumavirus* (foamy viruses)
Genus: Unnamed; HTLV-BLV viruses (includes human T-cell leukemia viruses)

The name *Retroviridae* (*retro*, backwards) is used for a large family of enveloped viruses 80–100 nm in diameter, with a complex structure and an unusual enzyme, reverse transcriptase. Uniquely among viruses, the genome is diploid, consisting of an inverted dimer of plus sense ssRNA, 7–10 kb in size. In the life cycle of the exogenous retroviruses the dsDNA copy of the viral genome transcribed by the viral reverse transcriptase is circularized and integrates into the cellular DNA as an essential part of the replication cycle. Proviral DNA of endogenous retroviruses is found in the DNA of all normal cells of many species of animals and may under certain circumstances be induced to produce virus.

This important family is subdivided into seven genera, only two of which have been given official generic names, and only three of which contain viruses known to infect humans. The viruses of medical importance are the human T-cell lymphotropic viruses (HTLV-BLV viruses), which cause leukemia, and the human immunodeficiency viruses (genus *Lentivirus; lenti*, slow), which produce the acquired immunodeficiency syndrome (AIDS). Spumaviruses (foamy agents) have been isolated from humans but appear to be completely benign. Retroviruses of other genera have been important research tools for unraveling the mechanisms of oncogenesis and the nature of oncogenes.

Other Viruses

There are a number of known human viruses that are as yet not allocated to families, including hepatitis D virus and the unusual agents that cause the subacute spongiform encephalopathies.

Hepatitis D Virus

Hepatitis D virus, recently assigned to the genus *Deltavirus*, is a satellite virus, the replication of which is dependent on simultaneous infection of cells with a hepadnavirus. The virion is about 32 nm in diameter and consists of the 24 kDa delta (δ) antigen encapsidated by the surface antigen of the helper hepatitis B virus. The genome of only 1.7 kb of covalently closed circular minus sense ssRNA contains regions of extensive base pairing and probably exists as a partially double-stranded rodlike structure with self-cleaving ribonuclease (ribozyme) activity, similar to the plant virus "satellites." Known only as a virus of humans who are simultaneously infected with hepatitis B virus, hepatitis D virus causes severe disease which often progresses to chronic hepatitis and/or cirrhosis.

Prions (Causing Subacute Spongiform Encephalopathies)

The causative agents of the subacute spongiform encephalopathies, prions (proteinaceous infectious particles), are composed largely of a protein designated as the scrapie isoform of the prion protein, PrP^{Sc}. They are highly resistant to inactivation by physical and chemical agents, they are nonimmunogenic, and they are devoid of nucleic acid. All produce slow infections with incubation periods measured in years, followed by progressive disease, which leads inexorably to death from a degenerative condition of the brain characterized by a spongiform appearance. The prototype is scrapie, a disease of sheep, but there are several prion diseases in humans (e.g., kuru, Creutzfeldt-Jakob syndrome). Recently, cattle in several European countries have been infected by feeding on material containing scrapie-infected sheep offal, and there has been a very large outbreak of bovine spongiform encephalopathy (popularly known as "mad cow disease") in the United Kingdom.

Groupings Based on Epidemiologic/Pathogenic Criteria

In discussing the epidemiology and pathogenesis of viral infections it is often convenient to use groupings of viruses based on the routes of transmission. These are not taxonomic groupings, but it is important to understand the context in which the terms are traditionally employed.

Enteric viruses, which replicate primarily in the intestinal tract, are acquired by ingestion of material contaminated with feces. These viruses usually remain localized in the intestinal tract, rather than becoming generalized. Enteric viruses that cause gastroenteritis in humans include the rotaviruses, caliciviruses, astroviruses, and some adenoviruses and coronaviruses. Many enteroviruses that are acquired by ingestion and replicate first in the gut, such as the polioviruses, do not cause gastroenteritis but may cause generalized diseases.

Respiratory viruses are usually acquired by inhalation of droplets and replicate in the respiratory tract. The term is usually restricted to those viruses that remain localized in the respiratory tract, including the orthomyxoviruses, rhinoviruses, and some of the paramyxoviruses, coronaviruses, and adenoviruses. Viruses that infect via the respiratory tract but cause generalized infections, such as measles and mumps viruses, are not normally called respiratory viruses.

Arboviruses (arthropod-borne viruses) infect arthropods that ingest vertebrate blood; they replicate in the tissues of the arthropod and can then be transmitted by bite to susceptible vertebrates. Viruses that belong to five families are included: all orbiviruses (a genus of the reoviruses), most bunyaviruses, flaviviruses, and togaviruses, and some rhabdoviruses.

Sexually transmitted viruses include some herpesviruses and papillomaviruses that cause lesions in the genital tract, as well as certain retroviruses and hepatitis viruses that are often transmitted during sexual activity but cause generalized disease.

Hepatitis viruses are often considered together because the liver constitutes the principal target. Formerly including the viruses of yellow fever and Rift Valley fever, the term is now generally restricted to hepatitis A, B, C, D, and E viruses. Epidemiologically they are diverse, for hepatitis A and E are spread by the enteric route, whereas hepatitis B, C, and D are transmitted parenterally (by blood) or sexually. Because each hepatitis virus belongs to a taxonomically different family, each is considered in a different chapter in Part II of this book.

Further Reading

Fenner, F., and Gibbs, A. J., eds. (1988). "Portraits of Viruses." Karger, Basel.

Fenner, F., Gibbs, E. P. J., Murphy, F. A., Rott, R., Studdert, M. J., and White, D. O. (1993). "Veterinary Virology," 2nd Ed. Academic Press, San Diego.

Matthews, R. E. F., ed. (1983). "A Critical Appraisal of Viral Taxonomy." CRC Press, Boca Raton, Florida.

Murphy, F. A., Fauquet, C. M., Bishop, D.H.L., Ghabrial, S. A., Jarvis, A. W., Martelli, G. P., Mayo, M. A., and Summers, M. D. (1994). "Virus Taxonomy. The Classification and Nomenclature of Viruses. Sixth Report of the International Committee on Taxonomy of Viruses." Springer-Verlag, Wien, New York.

Chapter 3

Viral Replication

Unraveling the complexities of viral replication is the central focus of much of experimental virology. Studies with bacteriophages in the 1940s and 1950s provided the first insights. With the development of mammalian cell culture procedures, the techniques used for the study of bacteriophages were adapted to animal viruses. Progress has been such that the basic mechanisms of transcription, translation, and nucleic acid replication have now been characterized for all the major families of animal viruses, and attention has turned to the strategy of gene expression and regulation. Many important phenomena that have general application in biology, such as splicing, editing, and other types of posttranscriptional processing of RNA, posttranslational processing of proteins, reverse transcription, integration, and transposition of viral genes and cellular oncogenes, as well as replication of RNA, were first elucidated in studies of animal viruses.

Our knowledge of viral replication is now so detailed and progress so rapid that it is impossible to cover the unique replication strategy of every family of viruses in a single chapter. Here we present an overview of the subject. Further information on each family is provided in the viral replication sections of the Part II chapters.

The Viral Replication Cycle

One-Step Growth Curve

Most studies of the replication of animal viruses have been conducted using cultured mammalian cell lines growing either in suspension or as a monolayer adhering to a flat surface. Classic studies of this kind defined the "one-step growth curve," in which all cells in a culture are infected simultaneously, by using a high *multiplicity of infection,* and the increase in infectious virus over time is followed by sequential sampling and titration (Fig. 3-1). Virus that is free in the medium can be titrated separately from virus that remains cell-associated. Shortly after infection, the inoculated virus "disappears"; infectious particles cannot be demonstrated, even intracellularly. This *eclipse period* continues until the first progeny virions become detectable some hours later. Nonenveloped viruses mature within the cell and may be detectable for some time as infectious intracellular virions before they are released by cell lysis. Many enveloped viruses, on the other hand, mature by budding from the plasma membrane of the host cell (see Fig. 3-10) and are immediately released into the medium. The eclipse period generally ranges from 3 to 12 hours for viruses of different families (Table 3-1).

Early studies, relying on quantitative electron microscopy and assay of

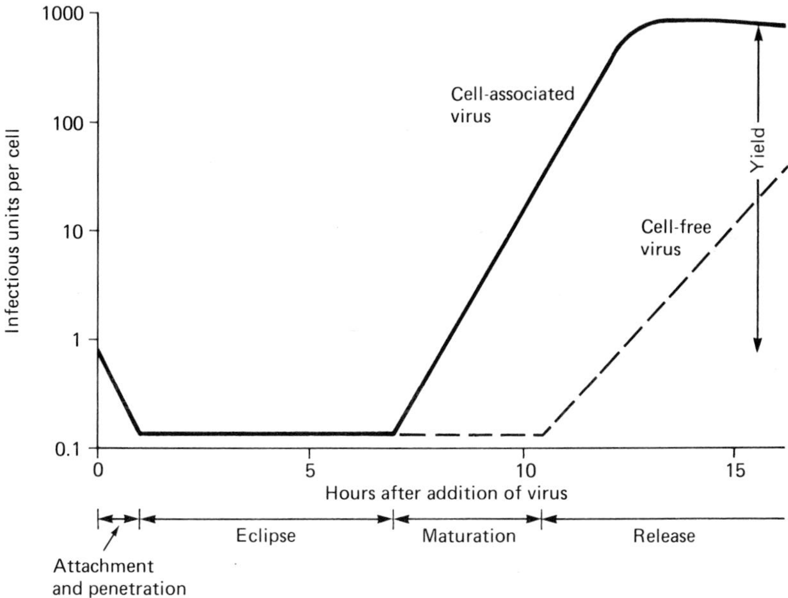

Fig. 3-1 One-step growth curve of a nonenveloped virus. Attachment and penetration are followed by an eclipse period of 2–12 hours (see Table 3-1) during which cell-associated infectivity cannot be detected. This is followed by a period during which maturation occurs. Virions of nonenveloped viruses are often released late and incompletely, when the cell lyses. Release of enveloped virions occurs concurrently with maturation by budding, generally from the plasma membrane.

Table 3-1
Characteristics of Replication of Viruses of Different Families

Family	Site of nucleic acid replication	Eclipse period (hours[a])	Budding (membrane)
Parvoviridae	Nucleus	6	None
Papovaviridae	Nucleus	12	None
Adenoviridae	Nucleus	10	None
Herpesviridae	Nucleus	4	Nuclear
Poxviridae	Cytoplasm	4	Golgi
Hepadnaviridae	Cytoplasm	?	Endoplasmic
Picornaviridae	Cytoplasm	2	None
Caliciviridae	Cytoplasm	3	None
Togaviridae	Cytoplasm	2	Plasma
Flaviviridae	Cytoplasm	3	Endoplasmic
Coronaviridae	Cytoplasm	5	Golgi
Paramyxoviridae	Cytoplasm	4	Plasma
Rhabdoviridae	Cytoplasm	3	Plasma
Orthomyxoviridae	Nucleus	4	Plasma
Arenaviridae	Cytoplasm	5	Plasma
Bunyaviridae	Cytoplasm	4	Golgi
Reoviridae	Cytoplasm	5	None
Retroviridae	Nucleus	10	Plasma

[a] Differs with multiplicity of infection, strain of virus, cell type and physiologic condition.

infectious virions, provided information about the early and the late events in the replication cycle (attachment, penetration, maturation, and release) but not about what happened during the eclipse period. Investigation of the expression and replication of the viral genome became possible only with the introduction of biochemical methods for the analysis of viral nucleic acids and proteins, and now all the sophisticated techniques of molecular biology are being applied to this problem.

Key Steps in the Viral Replication Cycle

Figure 3-2 illustrates in a greatly simplified diagram the major steps that occur during the viral replication cycle, using a DNA virus as an example. Following attachment, the virion is taken up by the host cell and is partially uncoated to expose the viral genome. Certain *early viral genes* are transcribed into RNA which may then be processed in a number of ways, including splicing. The early gene products translated from this *messenger RNA* (*mRNA*) are of three main types: proteins that shut down cellular nucleic acid and protein synthesis (see Chapter 5), proteins that regulate the expression of the viral genome, and enzymes required for the replication of viral nucleic acid. Following viral nucleic acid replication, *late viral genes* are transcribed. The late proteins are principally viral structural proteins for assembly into new virions; some of these are subject to posttranslational modifications. Each infected cell yields thousands of new virions, which spread to infect other cells.

For most families of DNA viruses, transcription and DNA replication take place in the cell nucleus. Some viruses use the cellular RNA polymerase II and

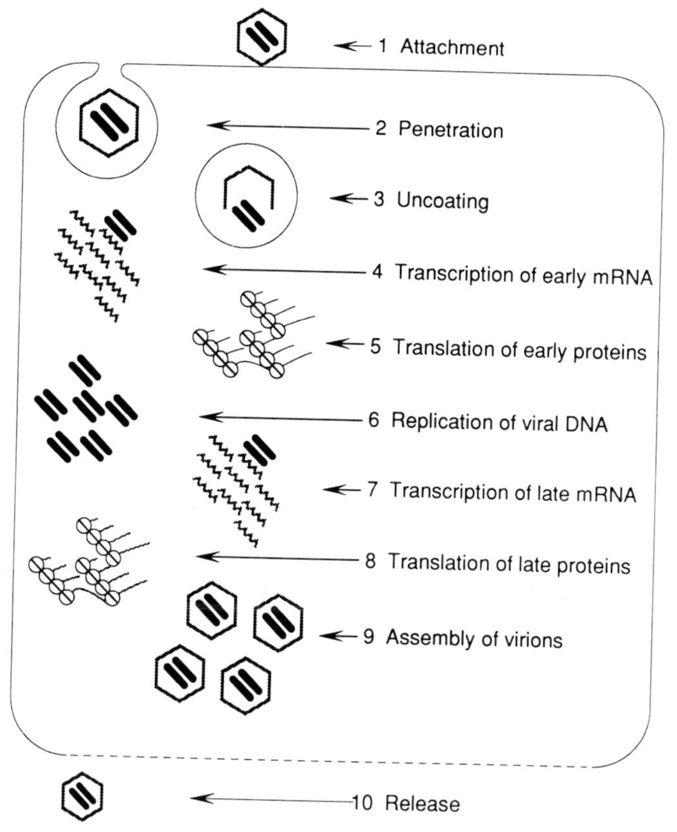

Fig. 3-2 General features of the viral replication cycle, using a nonenveloped DNA virus as a model. No topographic location for any step is implied. One step grades into the next such that, as the cycle progresses, several processes are proceeding simultaneously.

other cellular enzymes, but most have their own genes for a range of other enzymes. In addition, some carry "transforming genes" which induce cellular DNA synthesis, to increase the concentration of cellular enzymes and deoxynucleotides to the levels found only during the S phase of the mitotic cycle.

RNA viruses have the advantage that ribonucleoside triphosphates are available throughout the cell cycle. However, these viruses must encode their own RNA polymerase(s), since cells lack the capacity to copy RNA from an RNA template. Most RNA viruses replicate in the cytoplasm (Table 3-1).

Attachment

To cause infection, virus particles must be able to bind to cells. Ligands on specific molecules on the surface of the virion (*viral attachment proteins*) bind to receptors on the plasma membrane of the cell (Table 3-2). For instance, X-ray crystallography reveals that the receptor for most orthomyxoviruses is the terminal sialic acid on an oligosaccharide side chain of a cellular glycoprotein (or glycolipid), while the ligand is in a cleft at the distal tip of each monomer of

Table 3-2
Examples of Viral Ligands and Cellular Receptors

Virus	Viral protein containing ligand	Target cell	Cell receptor
Epstein-Barr virus	gp350/220	B cell, epithelium	CD21 (C3dg receptor, CR2)
Vaccinia virus	VGF	Various cell types	Epidermal growth factor receptor
Rhinovirus	VP1,3	Nasal epithelium	ICAM-1
Poliovirus	VP1	Intestinal epithelium	Immunoglobulin superfamily
Mouse hepatitis coronavirus	gp180	Liver, macrophages, intestinal epithelium	Carcinoembryonic antigen family[a] (immunoglobulin superfamily)
Human coronavirus	gp180	Respiratory and intestinal epithelium	Human aminopeptidase N (zinc-binding protease)
Parainfluenza virus	H-N glycoprotein	Respiratory epithelium	Gangliosides GD1a, GQ16
Influenza A virus	Hemagglutinin	Respiratory epithelium	Sialic acid-containing glycoprotein
Rotavirus	VP4 and/or VP7	Intestinal epithelium	Acetylated sialic acid on glycoprotein
Reovirus	σ1 (hemagglutinin)	T cell, neurons	β-Adrenergic receptor and/or sialoglycoprotein
Human immunodeficiency virus	gp120	T cell, macrophage	CD4 and/or galactosylcerebroside
Gibbon leukemia virus	gp70	Lymphocyte	Permease

[a]As receptor for mouse hepatitis virus, found only in susceptible strains.

the trimeric viral hemagglutinin glycoprotein. The receptors for a number of other viruses are members of the immunoglobulin superfamily, such as the integrin ICAM-1 (intracellular adhesion molecule-1), which is the major receptor for most rhinoviruses, and CD4, the receptor for the human immunodeficiency viruses. Receptors for other viruses include hormone receptors and permeases. Although there is a degree of specificity about the recognition of particular cellular receptors by particular viruses, quite different viruses, for example, orthomyxoviruses and paramyxoviruses, may utilize the same receptor, and related viruses, for example, human and mouse coronaviruses, or different serotypes of human rhinoviruses may use quite different receptors. Viruses have evolved to make opportunistic use of a variety of membrane glycoproteins as their receptors, but the primary functions of the receptors have nothing to do with viruses.

Recent evidence points to the existence of more than one type of receptor for certain viruses. If situated on different types of cells, distinct receptors could extend the tissue tropism, or even the host range of the virus. There is also evidence suggestive of sequential receptors on the same cell, namely, a "long-range" receptor, via which the virion makes its initial contact, and a "short-range" receptor more intimately associated with the membrane, facilitating entry of the virion by fusion (see Chapter 7).

Uptake (Penetration)

Following attachment, virions can enter cells by one of two main mechanisms, endocytosis or fusion.

Endocytosis

The majority of mammalian cells are continuously engaged in *receptor-mediated endocytosis* for the uptake of macromolecules via specific receptors. Many enveloped and nonenveloped viruses use this essential cell function to initiate infection (Fig 3-3). Attachment to receptors, which cluster at *clathrin-coated pits,* is followed by endocytosis into clathrin-coated vesicles that enter the cytoplasm and, after removal of the clathrin coat, fuse with endosomes (acidic prelysosomal vacuoles). Acidification within the vesicle triggers changes in the capsid protein VP4 of poliovirus, for example, leading to release of RNA from the virion into the cytosol. Likewise, at the acidic pH of the endosomes, the hemagglutinin molecule of influenza virus undergoes a conformational change, which enables fusion to occur between the viral envelope and the endosomal membrane, leading to release of the viral nucleocapsid into the cytoplasm. Many other nonenveloped and enveloped viruses undergo comparable changes.

Fusion with Plasma Membrane

The F (fusion) glycoprotein of paramyxoviruses causes the envelope of these viruses to fuse directly with the plasma membrane of the cell, even at pH 7.

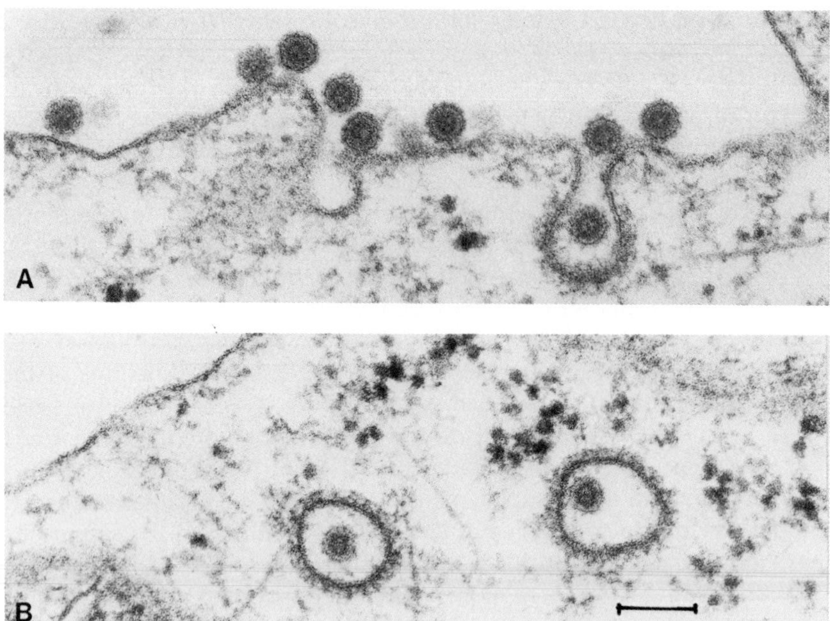

Fig. 3-3 Receptor-mediated endocytosis: penetration by a togavirus (bar, 100nm). (A) Attachment and movement into a clathrin-coated pit. (B) Endocytosis, producing a coated vesicle. [A, From E. Fries and A. Helenius, *Eur. J. Biochem.* **8,** 213 (1979); B, from K. Simons, H. Garoff, and A. Helenius, *Sci. Am.* **246,** 58 (1982), courtesy Dr. A. Helenius.]

This may allow the nucleocapsid to be released into the cytoplasm. A number of other enveloped viruses gain entry in similar fashion.

Uncoating

For viral genes to become available for transcription it is necessary that virions be at least partially uncoated. In the case of enveloped RNA viruses that enter by fusion of their envelope with either the plasma membrane or an endosomal membrane, the nucleocapsid is discharged directly into the cytoplasm, and transcription commences from viral nucleic acid still associated with this structure. With the nonenveloped icosahedral reoviruses only certain capsid proteins are removed, and the viral genome expresses all its functions without ever being released from the core. For most other viruses, however, uncoating proceeds to completion. For some viruses that replicate in the nucleus the later stages of uncoating occur there, rather than in the cytoplasm.

Strategies Of Replication

Replication of most DNA viruses involves mechanisms that are familiar in cell biology: transcription of mRNA from dsDNA and replication of DNA (Fig. 3-2). The situation is quite different for RNA viruses, which are unique in having their genetic information encoded in RNA. RNA viruses with different types of genomes (single-stranded or double-stranded, positive or negative sense, linear or segmented) have necessarily evolved different routes to the production of mRNA. In the case of ssRNA viruses of positive sense, the viral RNA itself functions as messenger, whereas all other types of viral RNA must first be transcribed to mRNA. Since eukaryotic cells contain no RNA-dependent RNA polymerase, negative sense ssRNA viruses and dsRNA viruses must carry an RNA-dependent RNA polymerase in the virion.

Further, eukaryotic cells normally produce "gene-length" mRNA molecules for direct translation into individual proteins, rather than reinitiating translation part way along a polycistronic mRNA. Nuclear DNA viruses use the cellular mechanism of cleavage (and sometimes splicing) of their polycistronic RNA transcripts to yield monocistronic mRNA molecules. RNA viruses, most of which replicate in the cytoplasm and hence do not have access to the RNA processing and splicing enzymes of the nucleus, have developed a remarkable diversity of solutions to the problem. Some RNA viruses have a segmented genome in which each molecule is, in general, a separate gene. Others have a polycistronic genome but produce monocistronic RNA transcripts by termination and reinitiation of transcription. Yet others make use of a nested set of overlapping RNA transcripts, each of which is translated into a single gene product. Finally, some have a polycistronic viral RNA which is translated into a polyprotein that is later cleaved proteolytically to yield the final products.

The diverse strategies followed by viruses of different families for transcription and translation are illustrated diagrammatically in Fig. 3-4 (for DNA viruses) and Fig. 3-5 (for RNA viruses) and described below.

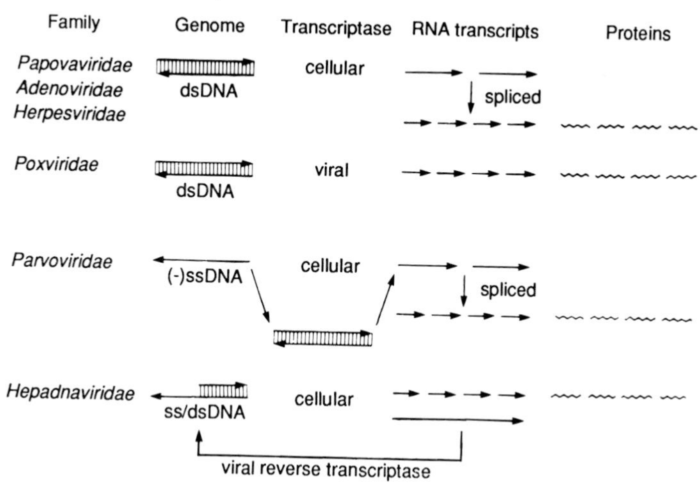

Fig. 3-4 Simplified diagram showing essential features of the transcription and translation of DNA viruses. The sense of each nucleic acid molecule is indicated by an arrow (+ to the right; − to the left). For simplicity, the number of mRNA and protein species for each virus has been arbitrarily shown as four. See text for details.

DNA Viruses

Papovaviruses, Adenoviruses, Herpesviruses

The papovaviruses, adenoviruses, and herpesviruses have in one respect the most straightforward strategy of replication, the viral DNA being transcribed within the nucleus by cellular DNA-dependent RNA polymerase II. There are two or more cycles of transcription, the various *transcription units* (groups of genes under the control of a single promoter) being transcribed in a given temporal sequence. Polycistronic but subgenomic RNA transcripts (corresponding to several genes but less than the whole genome) undergo cleavage and splicing to produce monocistronic mRNAs, introns being removed in the process.

Poxviruses

Poxviruses, which replicate in the cytoplasm, carry their own transcriptase (DNA-dependent RNA polymerase) in the virion. The very large genomes of poxviruses encode numerous other enzymes that make them virtually independent of the cell nucleus. Monocistronic mRNAs are transcribed directly from the viral DNA.

Parvoviruses

The ssDNA of the parvoviruses uses cellular DNA polymerase to synthesize dsDNA, which is then transcribed in the nucleus by cellular DNA-dependent RNA polymerase II and the transcripts are processed by splicing to produce mRNAs.

Hepadnaviruses

Hepadnavirus replication is unique in that it involves reverse transcription of an RNA intermediate to replicate the DNA genome—the mirror image of the process followed by the more extensively studied retroviruses. A supercoiled form of the viral DNA is transcribed in the nucleus by cellular DNA-dependent RNA polymerase II to produce both subgenomic and full-length RNA transcripts. The former serve as mRNA; the latter, known as the pro-genome, migrates to the cytoplasm where it serves as a template for the viral reverse transcriptase, and the minus sense DNA strand produced is in turn the template for synthesis of the dsDNA by the viral DNA polymerase.

RNA Viruses

Picornaviruses, Togaviruses, Flaviviruses, Caliciviruses

The positive sense ssRNA viruses in the families *Picornaviridae, Togaviridae, Flaviviridae,* and *Caliciviridae* require no transcriptase in the virion, since the virion RNA itself functions as mRNA (see Fig. 3-5). The genome of the picornaviruses and flaviviruses, acting as a single polycistronic mRNA, is translated directly into a single *polyprotein* which is subsequently cleaved to give the individual viral polypeptides. One of these proteins is an RNA-dependent RNA polymerase, which replicates the viral genome, transcribing viral RNA

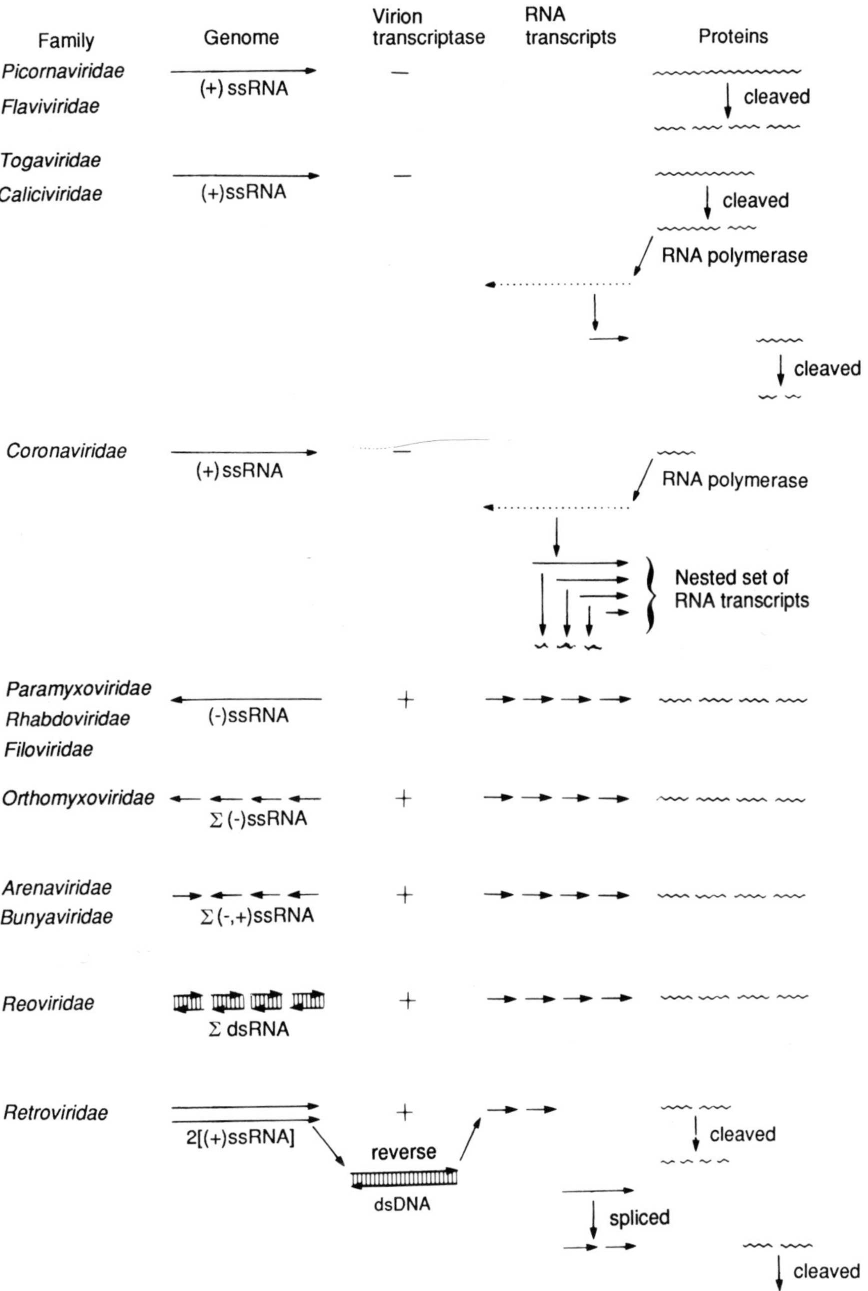

Fig. 3-5 Simplified diagram showing essential features of the replication of RNA viruses. The sense of each nucleic acid molecule is indicated by an arrow (+ to the right; − to the left; −,+ for *Arenaviridae* and *Bunyaviridae* indicates ambisense RNA in one segment). Σ, Segmented genome; 2, diploid genome of *Retroviridae*. For simplicity, the number of mRNA molecules and protein molecules has been arbitrarily shown as four, as have the number of segments in viruses with segmented genomes. See text for details.

into a complementary (minus sense) copy, which in turn serves as a template for the synthesis of plus strand (viral) RNA.

Only about two-thirds of the viral RNA (the 5′ end) of togaviruses is translated. The resulting polyprotein is cleaved into nonstructural proteins, all of which are required for RNA transcription and replication. Viral RNA polymerase makes a full-length minus strand, from which two species of plus strands are copied: full-length virion RNA, destined for encapsidation, and a one-third length RNA, which is colinear with the 3′ terminus of the viral RNA and is translated into a polyprotein from which structural proteins are produced by cleavage. The caliciviruses have not been so extensively studied but also produce both genome-length and subgenomic mRNA species.

Coronaviruses

Initially, part of the plus sense virion RNA of coronaviruses is translated to produce an RNA polymerase, which then synthesizes a genome-length minus strand. From this, a *nested set* of overlapping subgenomic mRNAs with a common 3′-termination site is transcribed. Only the unique 5′-terminal sequence of each successive member of the set of overlapping transcripts is translated.

Paramyxoviruses, Rhabdoviruses, Filoviruses

The minus sense, nonsegmented ssRNA viruses of the families *Paramyxoviridae, Rhabdoviridae,* and *Filoviridae* carry an RNA-dependent RNA polymerase (transcriptase), which transcribes five or more subgenomic plus sense RNAs, each of which serves as a monocistronic mRNA. In contrast, transcription in the replication mode (by the replicase) produces a full-length plus strand which is used as the template for the synthesis of new viral RNA.

Orthomyxoviruses, Bunyaviruses, Arenaviruses

The minus sense RNA viruses of the families *Orthomyxoviridae, Bunyaviridae,* and *Arenaviridae* have segmented genomes, each segment of which is transcribed by a transcriptase carried in the virion to yield an mRNA which is translated into one or more proteins. In the case of the orthomyxoviruses, most of the segments encode single proteins. Furthermore, the ssRNA genome of arenaviruses and certain genera of bunyaviruses is ambisense, that is, part plus sense and part minus sense. The replication strategy of ambisense RNA viruses, like the sense of their genomes, is mixed, with features of both plus sense and minus sense ssRNA viruses.

Reoviruses

Viruses of the family *Reoviridae* have segmented dsRNA genomes. The minus strand of each segment is separately transcribed in the cytoplasm by a virion-associated transcriptase to produce mRNA. These plus sense RNAs also serve as templates for replication. The resulting dsRNA in turn serves as the template for further mRNA transcription.

Retroviruses

In the retroviruses the viral RNA is plus sense, but instead of functioning as mRNA it is transcribed by a viral RNA-dependent DNA polymerase (reverse transcriptase) to produce first an RNA–DNA hybrid molecule, which is in turn converted to dsDNA (by another activity of the same enzyme) and inserted permanently into cellular DNA. This integrated viral DNA (*provirus*) is subsequently transcribed by cellular RNA polymerase II, followed by splicing of the RNA transcript as well as cleavage of the resulting proteins. Some full-length positive sense RNA transcripts associate in pairs to form the diploid genomes of new virions.

Transcription

Having outlined the several contrasting strategies of expression of the viral genome, we are now in a position to describe in more detail the processes of transcription, translation, and replication of viral nucleic acid, beginning with transcription. The viral RNA of most plus sense ssRNA viruses binds directly to ribosomes and is translated in full or in part without the need for any prior transcriptional step. For all other classes of viral genomes, mRNA must be transcribed in order to begin the process of expression of the infecting viral genome. In the case of DNA viruses that replicate in the nucleus, the cellular DNA-dependent RNA polymerase II performs this function. All other viruses require a unique and specific transcriptase that is virus-coded and is an integral component of the virion. Cytoplasmic dsDNA viruses carry a DNA-dependent RNA polymerase, whereas dsRNA viruses have dsRNA-dependent RNA polymerase and minus sense ssRNA viruses carry a ssRNA-dependent RNA polymerase.

Regulation of Transcription from Viral DNA

In 1978 Fiers and colleagues presented the first complete description of the genome of an animal virus (Fig. 3-6). Analysis of the circular dsDNA molecule of the papovavirus SV40 and its transcription program revealed some remarkable facts, many of which can be generalized to other dsDNA viruses. First, the early genes and the late genes are transcribed in opposite directions, from different strands of the DNA. Second, certain genes overlap, so that their protein products have some amino acid sequences in common. Third, some regions of the viral DNA may be read in different reading frames, so that different amino acid sequences are translated from the same nucleotide sequence. Fourth, certain long stretches of the viral DNA consist of introns, which are transcribed but not translated into protein because they are excised from the primary RNA transcript.

For many years it has served us well to think of a one-to-one relationship between a gene and its gene product (protein). Now that we are aware of overlapping reading frames, posttranscriptional cleavage, multiple splicing patterns of RNA transcripts, and posttranslational cleavage of polyproteins, it is often too simplistic to designate a particular nucleotide sequence as a gene encoding a particular protein. It is more appropriate to talk in terms of the *transcription unit,* which is defined as a region of the genome beginning with the transcription initiation site and extending to the transcription termination site (including all introns and exons in between), the expression of which falls

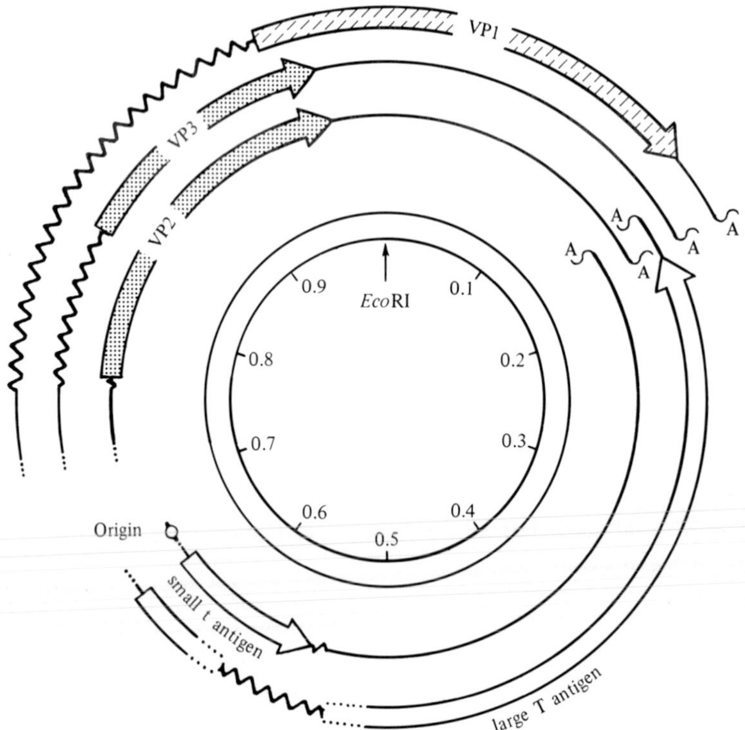

Fig. 3-6 Transcription map of the DNA of the papovavirus SV40. The circular dsDNA is oriented with the *Eco*RI restriction endonuclease cleavage site at zero and the origin of DNA replication (origin) at map position 0.66. The direction of transcription of the early genes is counterclockwise on one DNA strand (open arrows), and that of the late genes is clockwise on the other strand (stippled and shaded arrows). The thin lines indicate regions of the primary RNA transcripts that are not translated into proteins, while the wavy lines indicate regions of the transcripts that are spliced out (introns). The 3′-terminal poly (A) tail of each mRNA is labeled A. The coding regions of the primary transcript are shown as large arrows. The genes for the early proteins small t and large T overlap, as do those for the late proteins, VP1, VP2, and VP3. Large T is coded by two noncontiguous regions of DNA. The amino acid sequence of VP3 corresponds with the C-terminal half of VP2. However, VP1 shares no part of its amino acid sequence with VP2 or VP3, even though the VP1 gene overlaps VP2 and VP3, because its mRNA is transcribed in a different reading frame. [Modified from W. Fiers, R. Contreras, G. Haegemann, G. Rogiers, A. Van de Voorde, H. Van Heuverswyn, J. Van Herreweghe, G. Volckaert, and M. Ysebaert, *Nature (London)* **273,** 113 (1978).]

under the control of a particular promoter. "Simple" transcription units may be defined as those encoding only a single protein, whereas "complex" transcription units code for more than one.

Studies with adenoviruses have elucidated the nature of the mechanisms that regulate the expression of viral genomes, which operate principally, but not exclusively, at the level of transcription. There are several adenovirus transcription units. At different stages of the viral replication cycle, "pre-early," "early," "intermediate," and "late," the various transcription units are transcribed in a given temporal sequence. A product of the early region E1A induces transcription from the other early regions including E1B, but following viral DNA replication there is a 50-fold increase in the rate of transcription from the major late promoter relative to early promoters such as E1B, and a decrease in E1A mRNA levels. A second control operates at the point of termination of transcription. Transcripts that terminate at a particular point early in infection are read through this termination site later in infection to produce a range of longer transcripts with different polyadenylation sites.

Regulation of Transcription from Viral RNA

For RNA viruses, regulation of transcription is generally not as complex as for DNA viruses. In particular, the temporal separation into early genes transcribed before the replication of viral nucleic acid, and late genes thereafter, is not nearly so clear.

Other mechanisms of regulation are required for viruses with nonsegmented minus sense RNA. Once the nucleocapsid is released into the cytoplasm, the RNA polymerase initiates transcription from the 3′ end of the genome. Because there is only a single promoter, one might imagine that only full-length plus strand transcripts could be made. While some full-length plus strands are made as templates for RNA replication, mRNAs corresponding to each gene are also made, in the following fashion. The several genes in the viral RNA are each separated by a consensus sequence that includes termination and start signals as well as a short sequence of U residues which enables the transcriptase to generate a long poly(A) tail for each mRNA by a process of reiterative copying ("stuttering"). The completed mRNA is then released, but the enzyme continues on to transcribe the next gene, and so forth.

Paramyxovirus transcription also involves a process known as "editing." The P gene encodes two proteins, P and V, which share a common N-terminal amino acid sequence but differ completely in their C-terminal sequences because of a shift in the reading frame brought about by the insertion of uncoded G residues into the RNA transcript by transcriptase stuttering.

Regulatory Genes and Responsive Elements

In analyzing viral genomes and RNA transcripts derived from them much attention has been given to identifying the open reading frames in order to derive the amino acid sequence of their gene products. More recently interest has also turned to the untranslated regions of the genome which contain numerous conserved (consensus) sequences, sometimes called *motifs,* which represent responsive elements in the regulatory region and play crucial roles in the expression of the genome. For example, each transcription unit in the viral genome has near its 3′ end an mRNA transcription initiation site (start site), designated as nucleotide +1. Within the hundred or so nucleotides upstream of the start site lies the *promoter,* which up-regulates the transcription of that gene (or genes). Upstream or downstream from the start site there may also be a long sequence with several, maybe repeated, elements known as an *enhancer,* which enhances transcription even further (see Chapter 7). These regulatory regions are activated by the binding of viral or cellular DNA-binding proteins. Several such proteins may bind to adjacent responsive elements in such a way that they also bind to one another or otherwise interact, to facilitate attachment of the viral RNA polymerase. Viral regulatory genes that encode such regulatory proteins may act in *trans* as well as in *cis,* that is, they may *trans*-activate genes residing on a completely different molecule (see Chapters 7 and 35).

A description of the role of one of the six regulatory genes of the human immunodeficiency virus (HIV) will illustrate such regulatory mechanisms. When a DNA copy of the HIV genome is integrated into a chromosome of a resting T cell, it remains latent until a T-cell mitogen or a cytokine induces synthesis of the NF-κB family of DNA-binding proteins. NF-κB then binds to the enhancer present in the integrated HIV provirus, thereby triggering transcription of the six HIV regulatory genes. One of these, *tat*, encodes a protein which binds to a responsive element, TAR, which is present in all HIV mRNAs as well as in the proviral DNA. This greatly augments (*trans*-activates) the transcription of all HIV genes (including *tat* itself) and thereby establishing a positive feedback loop that enables the production of large numbers of progeny virions.

Although the control of HIV transcription is unusually complex because of its complicated replication cycle and requirement for the establishment of latency, HIV contains only nine genes, compared with up to a hundred in the case of some DNA viruses. Thus it may be anticipated that the regulation of expression of other viruses will turn out to be much more complex than had been thought.

Posttranscriptional Processing

Primary RNA transcripts from eukaryotic DNA, and from nucleus-associated viral DNA, are subject to a series of posttranscriptional alterations in the nucleus, known as *processing,* prior to export to the cytoplasm as mRNA. First, a *cap,* consisting of 7-methylguanosine (m^7Gppp), is added to the 5′ terminus of the primary transcript; the cap facilitates the formation of a stable complex with the 40 S ribosomal subunit, which is necessary for the initiation of translation. Second, a sequence of 50–200 adenylate residues is added to the 3′ terminus. This *poly(A) tail* may act as a recognition signal for processing and for transport of mRNA from the nucleus to the cytoplasm, and it may stabilize mRNA against degradation in the cytoplasm. Third, a methyl group is added at the 6 position to about 1% of the adenylate residues throughout the RNA (*methylation*). Fourth, introns are removed from the primary transcript and the exons are linked together in a process known as *splicing.* Splicing is an important mechanism for regulating gene expression in nuclear DNA viruses. A given RNA transcript can have two or more splice sites and be spliced in several alternative ways to produce a variety of mRNA species coding for distinct proteins; both the preferred poly(A) site and the splicing pattern may change in a regulated fashion as infection proceeds.

Special mention should be made of an extraordinary phenomenon known as "cap snatching." The transcriptase of influenza virus, which also carries endonuclease activity, steals the 5′ methylated caps from newly synthesized cellular RNA transcripts in the nucleus and uses them as primers for initiating transcription from the minus sense RNA viral genome.

The rate of degradation of mRNA provides another possible level of regulation. Not only do different mRNA species have different half-lives, but the half-life of a given mRNA species may change as the replication cycle progresses.

Translation

Capped, polyadenylated, and processed monocistronic viral mRNAs bind to ribosomes and are translated into protein in the same fashion as cellular mRNAs. The sequence of events has been closely studied for reovirus. Each monocistronic mRNA molecule binds via its capped 5′ terminus to the 40 S ribosomal subunit, which then moves along the mRNA molecule until stopped at the initiation codon. The 60 S ribosomal subunit then binds, together with methionyl-transfer RNA various initiation factors, after which translation proceeds.

In mammalian cells, mRNA molecules are *monocistronic* (encoding only one protein), and, with few exceptions, translation commences only at the 5′ initiation codon. However, with certain viruses polycistronic mRNA can be translated directly into its several gene products as a result of initiation, or reinitiation, of translation at internal AUG start codons.

Where initiation of translation at an internal AUG is an option, a frameshift can occur. Another mechanism, known as ribosomal frameshifting, occurs fortuitously when a ribosome happens to slip one nucleotide forward or back along an RNA molecule. This phenomenon is exploited by retroviruses to access the reverse transcriptase reading frame within the *gag–pol* mRNA. Thus, considering also the phenomena of RNA splicing and RNA editing described earlier, it can be seen that there are several mechanisms of exploiting overlapping reading frames to maximize the usage of the limited coding potential of the small genomes of viruses.

Most viral proteins undergo posttranslational modifications such as phosphorylation (for nucleic acid binding), fatty acid acylation (for membrane insertion), glycosylation, or proteolytic cleavage (see below). Newly synthesized viral proteins must also be transported to the various sites in the cell where they are needed, for example, back into the nucleus in the case of viruses that replicate there. The sorting signals that direct this traffic are only beginning to be understood, as are the polypeptide chain binding proteins ("molecular chaperones") that regulate folding, translocation, and assembly of oligomers of viral as well as cellular proteins.

Glycosylation of Envelope Proteins

Viruses exploit cellular pathways normally used for the synthesis of membrane-inserted and exported secretory glycoproteins (Fig. 3-7). The programmed addition of sugars occurs sequentially as the protein moves in vesicles progressively from the rough endoplasmic reticulum to the Golgi complex and then to the plasma membrane. The side chains of viral envelope glycoproteins are generally a mixture of simple ("high mannose") and complex oligosaccharides, which are usually N-linked to asparagine but less commonly O-linked to serine or threonine. The precise composition of the oligosac-

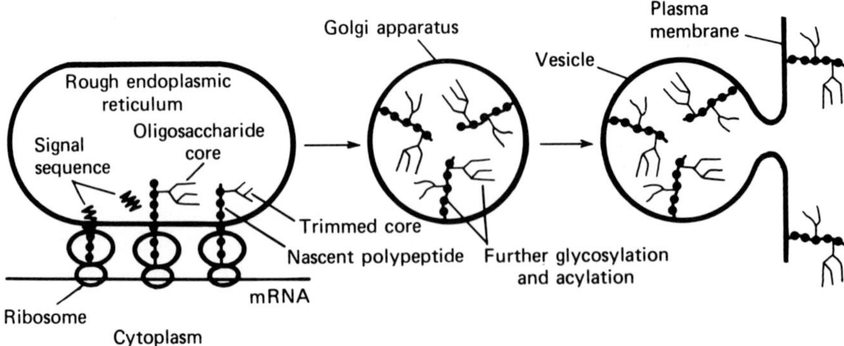

Fig. 3-7 Glycosylation of viral protein. The amino terminus of viral envelope proteins initially contains a sequence of 15–30 hydrophobic amino acids, known as a signal sequence, which facilitates binding of the growing polypeptide chain (dotted) to a receptor site on the cytoplasmic side of the rough endoplasmic reticulum and its passage through the lipid bilayer to the luminal side. Oligosaccharides are then added in N-linkage to certain asparagine residues of the nascent polypeptide by *en bloc* transfer of a mannose-rich "core" of preformed oligosaccharides, and glucose residues are removed by glycosidases ("trimming"). The viral glycoprotein is then transported from the rough endoplasmic reticulum to the Golgi complex. Here the core carbohydrate is further modified by the removal of several mannose residues and the addition of further *N*-acetylglucosamine, galactose, and the terminal sugar, sialic acid or fucose. The completed side chains are a mixture of simple ("high mannose") and complex oligosaccharides. A coated vesicle then transports the completed glycoprotein to the cellular membrane from which the particular virus buds.

charides (glycans) is determined not only by the amino acid sequence and tertiary structure of the proteins concerned, but more importantly by the particular cellular glycosyltransferases prevalent in the type of cell in which the virus happens to be growing at the time.

Posttranslational Cleavage of Proteins

In the case of the plus sense picornaviruses and flaviviruses, the polycistronic viral RNA is translated directly into a single polyprotein which carries proteinase (protease) activity that cleaves the polyprotein at defined recognition sites into smaller proteins. The first cleavage steps are carried out while the polyprotein is still associated with the ribosome. Some of the larger intermediates exist only fleetingly; others are functional for a short period but are subsequently cleaved by additional virus-coded proteases to smaller proteins with alternative functions. Posttranslational cleavage occurs in several other RNA virus families, for example, togaviruses and caliciviruses, in which polyproteins corresponding to large parts of the genome are cleaved. Some viruses encode several different proteases. Most are either trypsin-like (serine or cysteine proteases), pepsin-like (aspartyl proteases), or papain-like (thiol proteases).

Cellular proteases, present in particular organelles such as the Golgi complex or transport vesicles, are also vital to the maturation and assembly of many viruses. For example, cleavage of the hemagglutinin glycoprotein of orthomyxoviruses or the fusion glycoprotein of paramyxoviruses is essential for the production of infectious virions.

Classes of Viral Proteins

Table 3-3 lists the various classes of proteins encoded by the genomes of viruses. In general, the proteins translated from the early transcripts of DNA viruses include enzymes and other proteins required for the replication of viral nucleic acid, as well as proteins that suppress host cell RNA and protein synthesis. The large DNA viruses (poxviruses and herpesviruses) also encode a number of enzymes involved in nucleotide metabolism.

The late viral proteins are translated from late mRNAs, most of which are transcribed from progeny viral nucleic acid molecules. Most of the late proteins are viral structural proteins, and they are often made in considerable excess.

Some viral proteins, including some with other important functions, serve as regulatory proteins, modulating the transcription or translation of cellular genes or of early viral genes. The large DNA viruses also encode numerous additional proteins, sometimes called *virokines*, which do not regulate the viral replication cycle itself but influence the host response to infection (see Chapter 7).

Table 3-3
Categories of Proteins Encoded by Viral Genomes

Structural proteins of the virion[a]
Virion-associated enzymes, especially transcriptase[b]
Nonstructural proteins, mainly enzymes, required for transcription, replication of viral nucleic acid, and cleavage of proteins[c]
Regulatory proteins which control the temporal sequence of expression of the viral genome[d]
Proteins down-regulating expression of cellular genes[e]
Oncogene products (oncoproteins) and inactivators of cellular tumor suppressor proteins[f]
Proteins influencing viral virulence, host range, tissue tropism, etc.[g]
Virokines[g], which act on noninfected cells to modulate the progress of infection in the body as a whole[h]

[a] Comprising capsid and (for some viruses) core and/or envelope.

[b] RNA viruses of plus sense and nuclear DNA viruses do not carry a transcriptase in the virion. Virions of some viruses, e.g., poxviruses, also contain several other enzymes.

[c] DNA and RNA polymerases, helicases, proteases, etc. DNA viruses with large complex genomes, notably poxviruses and herpesviruses, also encode numerous enzymes needed for nucleotide synthesis.

[d] Site-specific DNA-binding proteins (transcription factors) which bind to enhancer sequences in the viral genome, or to another transcription factor. Some may act in *trans* (transactivators).

[e] Usually by inhibiting transcription, sometimes translation.

[f] Upgrade expression of certain cellular genes; may lead to cell transformation and eventually to cancer, as observed with herpesviruses, adenoviruses, papovaviruses, and retroviruses.

[g] Virokines have been recorded so far mainly in the more complex DNA viruses (poxviruses, herpesviruses, adenoviruses) but may be more widespread.

[h] Virokines act mainly by subverting the immune response by inhibiting cytokines, down-regulating MHC expression, blocking the complement cascade, etc. (see Table 7-2).

Replication of Viral Nucleic Acid

Replication of Viral DNA

Different mechanisms of DNA replication are employed by each family of DNA viruses. Because DNA polymerases cannot initiate synthesis of a new DNA strand but only extend synthesis from a short (RNA) primer, one end of the resulting product might be expected to remain single-stranded. Various DNA viruses have evolved different strategies for circumventing this problem. Viruses of some families have a circular DNA genome, others have a linear genome with complementary termini which serve as primers, while yet others have a protein primer covalently attached to each 5′ terminus.

Several virus-coded enzyme activities are generally required for replication of viral DNA: a helicase (with ATPase activity) to unwind the double helix, a helix-destabilizing protein to keep the two separated strands apart until each has been copied, a DNA polymerase to copy each strand from the origin of replication in a 5′ to 3′ direction, an RNase to degrade the RNA primer after it has served its purpose, and a DNA ligase to join the Okazaki fragments together (see below). Often a single large enzyme carries two or more of these activities.

The papovavirus genome, with its associated cellular histones, morphologically and functionally resembles cellular DNA and utilizes host cell enzymes, including DNA polymerase α, for its replication. An early viral protein, large T, binds to sites in the regulatory sequence of the viral genome, thereby initiating DNA replication. Replication of this circular dsDNA commences from a unique palindromic sequence and proceeds simultaneously in both directions. As in the replication of mammalian DNA, both continuous and discontinuous DNA synthesis occurs (of "leading" and "lagging" strands, respectively) at the two growing forks. The discontinuous synthesis of the lagging strand involves repeated synthesis of short oligoribonucleotide primers, which in turn initiate short nascent strands of DNA *(Okazaki fragments),* which are then covalently joined by a DNA ligase to form one of the growing strands.

The replication of adenovirus DNA is quite different. Adenovirus DNA is linear, the 5′ end of each strand being a mirror image of the other (terminally repeated inverted sequences), and each strand is covalently linked to a protein, the precursor of which serves as the primer for adenoviral DNA synthesis. DNA replication proceeds from both ends, continuously but asynchronously, in a 5′ to 3′ direction, using a virus-coded DNA polymerase. DNA replication in adenoviruses does not require the synthesis of Okazaki fragments.

Herpesviruses encode many or all of the "replication proteins" required for DNA replication, including a DNA polymerase, a helicase, a primase, a single-stranded DNA-binding protein, and a protein recognizing the origin of replication. Poxviruses, which replicate entirely within the cytoplasm, are self-sufficient in DNA replication. Hepadnaviruses, like the retroviruses, utilize plus sense ssRNA transcripts as intermediates for the production of DNA by reverse transcription. The ssDNA parvoviruses use 3′ palindromic sequences that form a double-stranded hairpin structure as a primer for cellular DNA polymerase.

Replication of Viral RNA

The replication of RNA is a phenomenon unique to viruses. Transcription of RNA from an RNA template requires an RNA-dependent RNA polymerase, a virus-coded enzyme not found in uninfected cells. The replication of viral RNA requires first the synthesis of complementary RNA, which then serves as a template for making more viral RNA.

Where the viral RNA is of minus sense (orthomyxoviruses, paramyxoviruses, rhabdoviruses, filoviruses, arenaviruses, and bunyaviruses), the complementary RNA will be of plus sense. Whereas most transcripts from such minus sense viral RNA are subgenomic mRNA molecules, some full-length plus strands are also made, in order to serve as templates for viral RNA synthesis (replication). For some viruses there is evidence that the RNA polymerases used for transcription and replication are distinct.

In the case of the plus sense RNA viruses (picornaviruses, caliciviruses, togaviruses, flaviviruses, and coronaviruses) the complementary RNA is minus sense. Several viral RNA molecules can be transcribed simultaneously from a single complementary RNA template, each RNA transcript being the product of a separately bound polymerase molecule. The resulting structure, known as the *replicative intermediate,* is therefore partially double-stranded, with single-stranded tails. Initiation of replication of picornavirus and calicivirus RNA, like that of adenovirus DNA, requires a protein, rather than an oligonucleotide, as primer. This small protein is covalently attached to the 5′ terminus of nascent plus and minus RNA strands, as well as to viral RNA, but not to mRNA. Little is known about what determines whether a given picornavirus plus sense RNA molecule will be directed (1) to a "replication complex," bound to smooth endoplasmic reticulum, where it serves as a template for transcription by RNA-dependent RNA polymerase into minus sense RNA; or (2) to a ribosome, where it serves as mRNA for translation into protein; or (3) to a procapsid, with which it associates to form a virion.

Retroviruses have a genome consisting of plus sense ssRNA. Unlike other RNA viruses, retroviruses replicate via a DNA intermediate. The virion-associated reverse transcriptase, using a transfer RNA (tRNA) molecule as a primer, makes a ssDNA copy. Then, functioning as a ribonuclease, the same enzyme removes the parental RNA molecule from the DNA–RNA hybrid. The free minus sense ssDNA strand is then copied to form a linear dsDNA molecule, which contains an extra copy of sequences necessary for integration and expression, known as the *long terminal repeat (LTR),* at each end. This dsDNA then circularizes and integrates into cellular DNA. Transcription of viral RNA occurs from this integrated (proviral) DNA.

Assembly and Release

Nonenveloped Viruses

All nonenveloped animal viruses have an icosahedral structure. The structural proteins of simple icosahedral viruses associate spontaneously to form capsomers, which undergo self-assembly to form capsids into which viral nucleic acid is packaged. Completion of the virion often involves proteolytic cleavage of one or more species of capsid protein. The best studied example, that of poliovirus, is depicted in Fig. 3-8.

The mechanism of packaging viral nucleic acid into a preassembled empty procapsid has been elucidated for adenovirus. A particular protein binds to a nucleotide sequence at one end of the viral DNA known as the *packaging sequence;* this enables the DNA to enter the procapsid bound to basic core proteins, after which some of the capsid proteins are cleaved to make the mature virion.

Most nonenveloped viruses accumulate within the cytoplasm or nucleus and are released only when the cell eventually lyses.

Maturation and Release of Enveloped Viruses

All mammalian viruses with helical nucleocapsids, as well as some of those with icosahedral nucleocapsids (herpesviruses, retroviruses, flaviviruses, and togaviruses) mature by acquiring an envelope by budding through cellular membranes.

Budding from Plasma Membrane

Most enveloped viruses bud from the plasma membrane. Insertion of the viral glycoprotein(s) into the lipid bilayer occurs by lateral displacement of

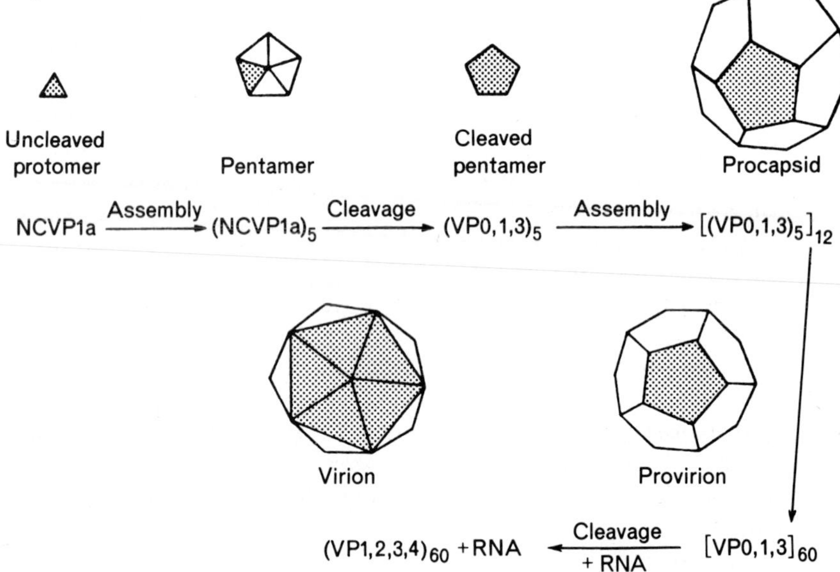

Fig. 3-8 Overview of the assembly of poliovirus, a picornavirus. The capsomer precursor protein (NCVP1a) aggregates to form pentamers; each of the five NCVP1a molecules is then cleaved by viral protease into VPO, VP1, and VP3. Twelve such pentamers aggregate to form a procapsid. A final proteolytic event, which cleaves each VPO molecule into VP2 and VP4, is required for RNA incorporation. The mature virion is a dodecahedron with 60 capsomers, each of which is made up of one molecule each of VP1, VP2, VP3, and VP4. X-ray crystallography shows that the assembling units have extensions that reach across adjacent units to form second and third nearest neighbor relationships.

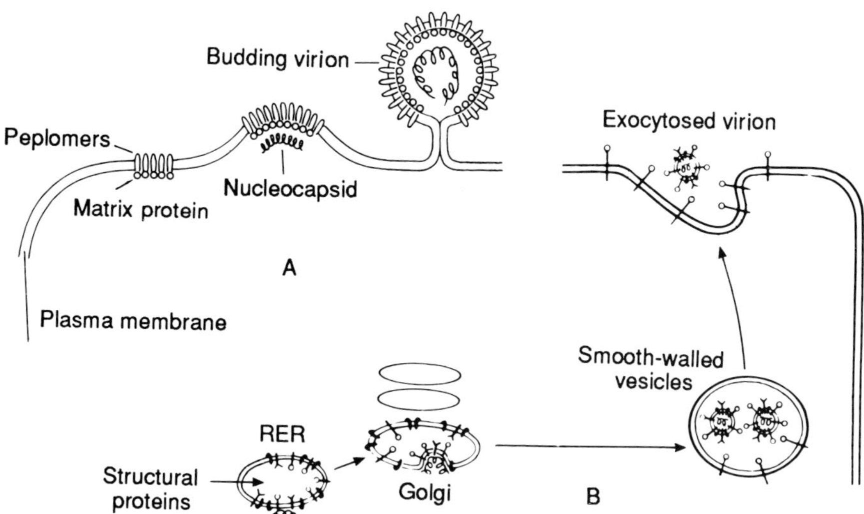

Fig. 3-9 Maturation of enveloped viruses. (A) Viruses whose virion contains a matrix protein (and some viruses which do not) bud through a patch of the plasma membrane in which glycoprotein peplomers have accumulated over the patch of matrix protein. (B) Most enveloped viruses without a matrix protein bud into cytoplasmic vesicles [rough endoplasmic reticulum (RER) or Golgi], then pass through the cytoplasm in smooth-walled vesicles and are released by exocytosis. [B, Modified from K. V. Holmes, *in* "Fields Virology" (B. N. Fields, D. M. Knipe, R. M. Chanock, M. S. Hirsch, J. L. Melnick, T. P. Monath, and B. Roizman, eds.), 2nd Ed., p. 847. Raven, New York, 1990.]

cellular proteins from that patch of membrane (Fig. 3-9A). The monomeric cleaved viral glycoprotein molecules associate into oligomers to form the typical rod-shaped or club-shaped peplomer with a hydrophilic domain projecting from the external surface of the membrane, a hydrophobic transmembrane anchor domain, and a short hydrophilic domain projecting slightly into the cytoplasm. In the case of icosahedral viruses (e.g., togaviruses) each protein molecule of the nucleocapsid binds directly to the cytoplasmic domain of the membrane glycoprotein oligomer, thus molding the envelope around the nucleocapsid. In the more usual case of viruses with helical nucleocapsids, it is the matrix protein which attaches to the cytoplasmic domain of the glycoprotein peplomer; in turn the nucleocapsid protein recognizes the matrix protein and this presumably initiates budding. Release of each enveloped virion does not breach the integrity of the plasma membrane; hence, thousands of virus particles can be shed over a period of several hours or days without significant cell damage (Fig. 3-10). Many but not all viruses that bud from the plasma membrane are noncytopathogenic and may be associated with persistent infections.

Epithelial cells display *polarity,* having an "apical" surface facing the outside world which is separated by a tight junction from the "basolateral" surface. These surfaces are chemically and physiologically distinct. Viruses that are shed to the exterior, for example, influenza virus, tend to bud from the apical surface, whereas others, such as C-type retroviruses, bud through the basolateral membrane (see Chapter 5).

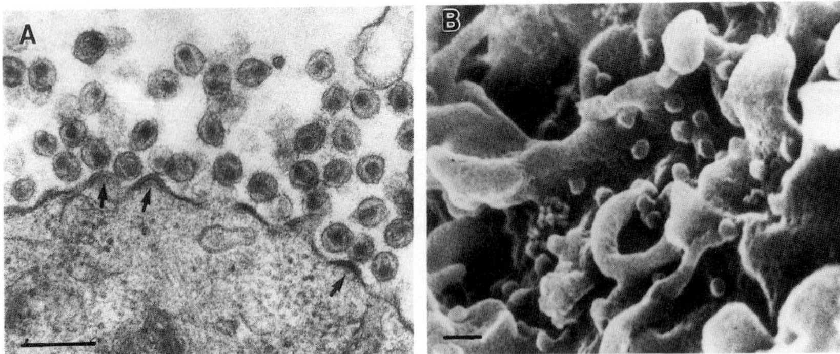

Fig. 3-10 Virions of lentiviruses budding from the plasma membrane (bars, 100 nm). (A) Transmission electron micrograph of feline immunodeficiency virus budding through the plasma membrane of a feline T lymphocyte. Retroviruses have no matrix protein; the arrows indicate accumulations of patches of viral glycoprotein peplomers in the plasma membrane. (B) Scanning electron micrograph of a human immunodeficiency virus budding from the plasma membrane of a human T lymphocyte, from a culture of H9 human T lymphocytes infected *in vitro*. [A, From S. C. E. Friend, C. J. Birch, P. M. Lording, J. A. Marshall, and M. J. Studdert, *Aust. Vet. J.* **67,** 237 (1990), B, courtesy Dr. C. S. Goldsmith.]

Exocytosis

Flaviviruses, coronaviruses, and bunyaviruses mature by budding through membranes of the Golgi complex or rough endoplasmic reticulum; vesicles containing the virus then migrate to the plasma membrane with which they fuse, thereby releasing the virions by *exocytosis* (Fig. 3-9B). Uniquely, the envelope of the herpesviruses is acquired by budding through the inner lamella of the nuclear membrane; the enveloped virions then pass directly from the space between the two lamellae of the nuclear membrane to the exterior of the cell via the cisternae of the endoplasmic reticulum.

Further Reading

Brinton, M. A., and Heinz, F. X., ed. (1990). "New Aspects of Positive-Strand RNA Viruses." American Society for Microbiology, Washington, D.C.

Challberg, M. D., and Kelly, T. J. (1989). Animal virus DNA replication. *Annu. Rev. Biochem.* **58,** 671.

Darnell, J., Lodish, H., and Baltimore, D. (1990). "Molecular Cell Biology," 2nd Ed. Scientific American Books, New York.

Ehrenfeld, E., ed. (1993). Translational Regulation in Virus-Infected Cells. *Semin. Virol.* **4,** 199.

Fields, B. N., Knipe, D. M., Chanock, R. M., Hirsch, M. S., Melnick, J. L., Monath, T. P., and Roizman, B., eds. (1990). "Fields Virology," 2nd Ed. Raven, New York.

Flint, S. J. (1986). Regulation of adenovirus mRNA function. *Adv. Virus Res.* **31,** 169.

Kolakofsky, D., and Mahy, B. W. J., eds. (1989). "Genetics and Pathogenicity of Negative Strand Viruses." Elsevier, Amsterdam.

Krausslich, H. G., and Wimmer, E. (1988). Viral proteinases. *Annu. Rev. Biochem.* **57,** 701.

Levine, A. J. (1992). "Viruses." Scientific American Library, Freeman, New York.

McGeoch, D. J., Randall, R. E., and Elliott, R. M. (1992). Some highlights of virus research in 1991. *J. Gen Virol.* **73,** 2487 (and corresponding reviews annually from 1986 onward).

Marsh, M., and Helenius, A. (1989). Virus entry into cells. *Adv. Virus Res.* **36,** 107.

Perez-Bercoff, R., ed. (1987). "The Molecular Biology of Viral Replication." Plenum, New York.

Peterlin, B. M., ed. (1993). Transcriptional Regulation of Viruses. *Semin. Virol.* **4,** 1.
Rossmann, M. G., and Johnson, J. E. (1989). Icosahedral RNA virus structure. *Annu. Rev. Biochem.* **58,** 533.
Stephens, E. B., and Compans, R. W. (1988). Assembly of viruses at cellular membranes. *Annu. Rev. Microbiol.* **42,** 489.
Strauss, E. G., and Strauss, J. H. (1983). Replication strategies of the single-stranded RNA viruses of vertebrates. *Curr. Top. Microbiol. Immunol.* **105,** 1.
Watson, J. D., Hopkins, N. H., Roberts, J. W., Steitz, J. A., and Weiner, A. M. (1987). "Molecular Biology of the Gene," Vol. 2, 4th Ed. Benjamin-Cummings, Menlo Park, California.

Chapter 4

Viral Genetics and Evolution

Viruses have a greater genetic diversity than any other group of organisms (see Table 1-2). This diversity has been produced by natural selection acting on viral genomes that are continuously changing as a result of mutation, recombination, and reassortment.

Our knowledge of virus genetics was considerably expanded by the development during the 1970s of molecular cloning and nucleotide sequencing methods and the use of monoclonal antibodies, followed in the mid-1980s by the introduction of the polymerase chain reaction. A new genetics has emerged in which the old concept of a one-to-one relationship between a gene and its gene product has been replaced by the realization that the same nucleotide sequence can be used more than once, as a result of a wide variety of previously unexpected biochemical phenomena, such as splicing of RNA transcripts and translation in different reading frames.

Mutation

In every viral infection of an animal or a cell culture, a small number of virus particles replicate, to produce millions of progeny. In such populations, errors in copying the nucleic acid inevitably occur; these are called *mutations*. Many mutations are lethal, because the mutated virus is unable to replicate. Whether a particular nonlethal mutation survives in the genotype depends on whether the resultant change in the gene product is disadvantageous, neutral, or affords the mutant virus some selective advantage. In the laboratory, genetic variants are obtained by subjecting a virus population to some selective condition and isolating a *clone,* that is, a population of viral particles originating from a single virion, usually by growth from a single plaque in a cell monolayer, followed by replaquing.

Types of Mutations

Mutations can be classified either according to the kind of change in the nucleic acid or on the basis of their phenotypic expression.

Classification of Mutants by Changes in the Genome

The most common mutations are single nucleotide substitutions *(point mutations)* or deletions and insertions, which may involve a single nucleotide or more commonly small blocks of nucleotides. Each point mutation has a characteristic frequency of reversion which can be accurately measured. The phenotypic expression of a mutation in one gene may be reversed not only by a back mutation in the substituted nucleotide but, alternatively, by a *suppressor mutation* occurring elsewhere in the same gene, or even in a different gene. For example, some temperature-sensitive mutants of influenza virus developed as potential attenuated live-virus vaccines have reverted to virulence by virtue of an independent suppressor mutation in an apparently unrelated gene, which negates the biological effect of the original mutation. *Deletion mutants* rarely or never revert, so that nonrevertibility is used as a diagnostic criterion of this kind of mutation. Although not mutations, various other sorts of gene rearrangements also occur, especially with DNA viruses and retroviruses, such as duplications, inversions, and incorporation of foreign viral or cellular nucleic acid sequences by recombination, with important biological consequences, as discussed below.

Defective Interfering Mutants

Defective interfering (DI) mutants have been demonstrated to occur naturally in most families of viruses. The properties that define them are that they have a defective genome and thus cannot replicate alone, but can in the presence of a helper virus (usually parental wild-type virus), and that they interfere with the replication of wild-type virus.

All RNA DI particles that have been studied are deletion mutants. In the case of influenza viruses and reoviruses, which have segmented genomes, the defective virions lack one or more of the larger segments and contain instead smaller segments consisting of an incomplete portion of that gene(s). In the case of viruses with a nonsegmented genome, DI particles contain RNA which is shortened—as little as one-third of the original genome may remain in the DI particles of vesicular stomatitis virus. Morphologically, DI particles usually resemble the parental virions, but may be smaller. Sequencing of the RNA reveals simple deletions and a great diversity of structural rearrangements.

Defective interfering particles increase preferentially with serial passage at high multiplicity in cultured cells because their shortened RNA genomes require less time to be replicated, are less often diverted to serve as templates for transcription of mRNA, and have enhanced affinity for the viral replicase, giving them a competitive advantage over their full-length infectious counterparts. These features also explain why on passage the DI particles interfere with the replication of full-length parental RNA with progressively greater efficiency.

The generation of defective genomes of DNA viruses can occur by any of a great variety of modes of DNA rearrangement. For example, papovavirus DI particles usually contain reiterated copies of the genomic origins of replication, which are sometimes interspersed with DNA of host cell origin.

Our knowledge of DI particles derives from studies in cultured cells, but they play a role in some disease conditions. By interference, they may attenuate the lethality of the parental infectious virus, and they may contribute to the pathogenesis of some chronic diseases. However, because their defective and variable nature makes them difficult to detect, much remains to be discovered about their role in disease.

Classification of Mutants by Phenotypic Expression

Mutations that allow the production of progeny virions can also be classified by their phenotypic expression, such as the type of plaque they produce in a cell monolayer (plaque mutants) or their resistance to neutralization by a monoclonal antibody (neutralization *escape mutants*). Mutations affecting antigenic determinants of virion surface proteins may be strongly favored when viruses replicate in the presence of antibody, and they are of importance both in persistent infections, for example, in AIDS, and epidemiologically, as in influenza. Other biologically important mutations are associated with the acquisition of enhanced or reduced virulence for particular host species.

Conditional lethal mutants are produced by a mutation that so affects a virus that it cannot grow under certain conditions determined by the experimenter but can replicate under other, permissive, conditions. Their importance is that a single selective test can be used to obtain mutants in which the mutation may be present in any one of several different genes. The conditional lethal mutants most commonly studied are those whose replication is blocked in certain host cells (*host range mutants*) or at a certain defined temperature (*temperature-sensitive mutants*). With the latter, the selective condition used is the temperature of incubation of infected cells. A point mutation in the genome leads to an amino acid substitution in the translated polypeptide product, which although functional at the *permissive temperature* cannot maintain its correct conformation when the temperature is raised by a few degrees. Temperature-sensitive mutants, and the somewhat similar *cold-adapted mutants,* have been used extensively in attempts to produce attenuated live-virus vaccines.

Mutation Rates

Rates of point mutation in DNA viruses that replicate in the nucleus are probably similar to those in the DNA of eukaryotic cells, since viral DNA replication is subject to the same "proofreading" exonuclease error correction as operates in cells. Errors occur at a rate of 10^{-8} to 10^{-11} per incorporated nucleotide (i.e., per base per replication cycle). Point mutations in the third nucleotide of a codon are often silent, that is, do not result in an altered amino acid, because of redundancy in the genetic code. Some point mutations are lethal because they produce nonfunctional gene products or a stop codon which terminates translation of the message. Viable mutations that are neutral or deleterious in one host may provide a selective advantage in a different host.

The error rate in the replication of viral RNA is much higher than that of viral or cellular DNA, because there is no cellular proofreading mechanism for RNA. For example, the base substitution rate in the 11-kb genome of vesicular stomatitis virus is 10^{-3} to 10^{-4} per base per replication cycle, so that nearly every progeny genome will be different from its parent and from one another in at least one base. This rate of base substitution is about one million times higher than the average rate in eukaryotic DNA. Of course, most of the base substitutions are deleterious and the genomes containing them are lost. However, nonlethal mutations in the genome of RNA viruses accumulate very rapidly. For example, sequence analysis of the genome of two isolates of hepatitis C virus obtained from a chronically infected patient at an interval of 13 years showed that the mutation rate was about 2×10^{-3} base substitutions per genome site per year. The nucleotide changes were unevenly distributed throughout the genome, that is, different genes evolved at different rates. At the population level, an outbreak of human poliomyelitis type 1 in 1978–1979 was traced from the Netherlands to Canada and then to United States. Oligonucleotide mapping of the RNAs obtained from successive isolates of the virus from different people showed that over a period of 13 months of epidemic transmission there were about 100 base changes in a genome of 7441 bases.

It is important to recognize that every virus, as defined by its conventional phenotypic markers, is a genetically complex population that comprises multiple mutants, a minority of which will be dominant under defined conditions of replication. Mutations continue to accumulate in the viral genome during passage in cultured cells, just as they do during replication in the natural host, especially during chronic infections, and in the course of sequential infection of individuals in a population. Nevertheless, in nature some viruses, even some RNA viruses such as measles virus, appear to be relatively stable in their virulence and antigenicity. This appears to be due to a combination of the small size of the inocula involved in aerosol transmission and the multiple selective pressures that operate during the spread of viruses through the bodies of their natural hosts.

Mutagenesis

Spontaneous mutations occur because of chance errors during replication. Mutation frequency can be enhanced by treatment of virions or isolated viral nucleic acid with physical agents such as UV light or X-irradiation or with chemicals such as nitrous acid or nitrosoguanidine. Base analogs, such as 5′-fluorouracil (for RNA viruses) or 5′-bromodeoxyuridine (for DNA viruses), are mutagenic only when virus is grown in their presence because they are incorporated into the viral nucleic acid and produce mutations by miscoding during replication.

Site-Specific Mutagenesis

Instead of relying on chance mutations anywhere in the genome, genetic engineering makes it possible to introduce mutations at any site of interest. Site-specific (or site-directed) mutagenesis enables the experimenter to substitute any selected nucleotide for that in a prescribed position in a DNA molecule (a DNA genome or complementary DNA (cDNA) transcribed from an RNA genome). Several techniques to achieve this are available. For example, the viral DNA is denatured and ssDNA is purified then transferred into an appropriate vector such as bacteriophage M13. A short synthetic oligonucleotide that is homologous to the relevant region except that it contains the desired nucleotide is then annealed to the ssDNA, and the synthesis of this complementary strand primer is completed by DNA polymerase to yield a dsDNA molecule. The hybrid phage DNA is now propagated by transfection of permissive bacteria. Progeny phages containing the mutated genome are selected by hybridization with the oligonucleotide probe, and marker rescue is used to recover the mutated gene in infectious virus. The polymerase chain reaction can be used to verify the location of the genetic lesions.

Until recently, site-directed mutagenesis and other types of genetic engineering were restricted to DNA viruses and plus strand RNA viruses, from which cDNA could be produced by reverse transcription. However, it has now become possible to apply these methods of *reverse genetics* to minus strand viruses, including those with segmented genomes. Site-directed mutagenesis has opened up new research areas, such as dissection of the function of individual genes and the proteins for which they code, or of particular regions of these genes and proteins. At a practical level, mutations can be introduced into particular genes, for example, those contributing to viral virulence, to produce mutants suitable for use as attenuated live-virus vaccines.

Adaptation to Cultured Cells or Laboratory Animals

Major advances in the understanding of viruses involved in human disease usually depend on being able to grow them in cultured cells or in a laboratory animal. Most wild-type viruses initially grow poorly in cell culture or laboratory animals, but they can be "adapted" to their new host by serial passage of high-titer inocula. Such adaptation depends on the spontaneous generation of mutations and progressive selection of the best growing mutant. These host range mutants may contain mutations in any of several genes, often in that encoding the surface protein which binds the somewhat different type of receptor available on the new host cell. For example, influenza virus may acquire the capacity to grow well in the lungs of mice or in cultured cell lines via a point mutation affecting either the ligand of the hemagglutinin (HA) molecule that engages the cell receptor, or the site of proteolytic cleavage of the HA that is essential for uncoating of the endocytosed virion. The mutation may occur within either of these two critical sites, or within a nearby glycosylation sequence (Asn-X-Val/Thr), thereby exposing a site that would otherwise be obstructed by an oligosaccharide side chain.

Some viruses produce clinical signs of disease the first time that they are inoculated into an experimental animal, for instance, eastern equine encephalitis virus in suckling mice. In other cases minimal signs of infection are observed initially, but after serial passage, sometimes prolonged, clinical disease is regularly produced, as, for example, in the adaptation of poliovirus to mice. Likewise, newly isolated viruses at first often fail to grow in certain kinds of cultured cells, but can be adapted by serial passage. Most virologic research is performed with strains of virus adapted to grow rapidly to high yield and to produce plaques or cytopathic changes in continuous cell lines. A frequent by-product of such adaptation to a new experimental host is the coincident attenuation of the virus for its original host; for example, polioviruses, by their adaptation to monkey kidney cell cultures and serial passage in them, were attenuated sufficiently to be used as vaccines.

Genetic Recombination between Viruses

When two different viruses simultaneously infect the same cell, genetic recombination may occur between the newly synthesized nucleic acid molecules; this may take the form of *intramolecular recombination*, *reassortment*, or *reactivation*.

Intramolecular Recombination

Intramolecular recombination involves the exchange of nucleic acid sequences between different but usually closely related viruses during viral replication (Fig. 4-1A). It occurs with all dsDNA viruses, presumably because of strand switching by the viral DNA polymerase. Among RNA viruses, intramolecular recombination has been demonstrated for picornaviruses, coronaviruses, and togaviruses. It may be more widespread, because detection has relied on the recovery of viable recombinants; the use of the polymerase chain reaction following reverse transcription may overcome this lack of sensitivity. For some viruses recombination can also occur between viral and cellular RNA, just as it can between viral and cellular DNA.

There is evidence that western equine encephalitis virus arose as a result of intramolecular recombination between a Sindbis-like virus and eastern equine encephalitis virus. In experimental situations, intramolecular recombination may also occur between viruses belonging to different families; the best example is between SV40 (a papovavirus) and adenoviruses. Both SV40 and adenovirus DNAs occasionally become integrated into cellular DNA, so it is perhaps not surprising to find that when rhesus monkey cells that harbor a persistent SV40 infection are superinfected with an adenovirus, not only does *complementation* occur, the SV40 acting as a helper in an otherwise abortive adenovirus infection, but recombination occurs between SV40 DNA and adenovirus DNA to yield hybrid (recombinant) DNA which is packaged into adenovirus capsids. Integration of viral DNA into cellular DNA by intramolecular recombination occurs in cells transformed by adenoviruses, hepadnaviruses, and polyomaviruses, but not always in cells transformed by papillomaviruses or certain herpesviruses, in which transformation may occur although the viral DNA usually remains episomal (see Chapter 11).

Unlike other RNA viruses, retroviruses have no replicating pool of viral

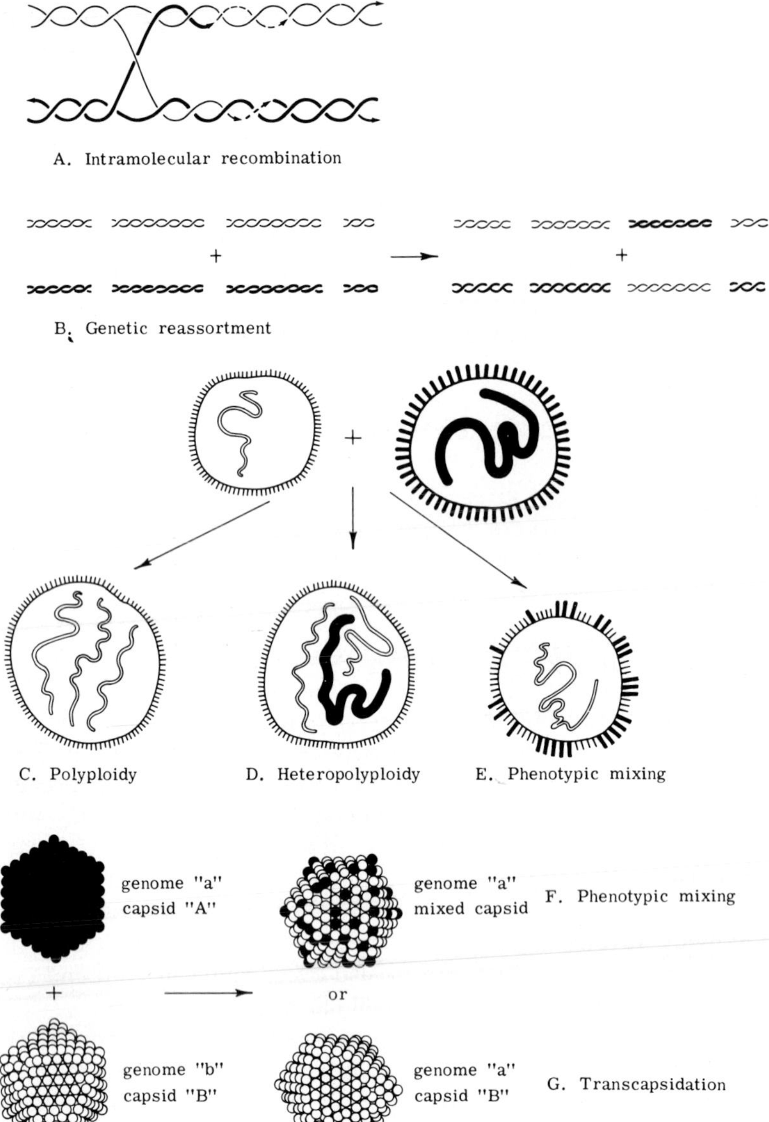

Fig. 4-1 Genetic recombination, polyploidy, phenotypic mixing and transcapsidation. (A) Intramolecular recombination, as in a dsDNA virus. (B) Reassortment of genome fragments, as in reoviruses and orthomyxoviruses. (C) Polyploidy, as seen in unmixed infections with paramyxoviruses. (D) Heteropolyploidy, as may occur in mixed infections with paramyxoviruses and other enveloped RNA viruses. (E–G) Phenotypic mixing: (E) enveloped viruses; (F) viruses with icosahedral capsids; (G) extreme case of transcapsidation or genomic masking.

RNA. Although the genome of retroviruses is plus sense ssRNA, replication does not occur until the genomic RNA is transcribed into DNA by the virion-associated reverse transcriptase and the resultant dsDNA integrated into the DNA of the host cell. However, both minus strand and plus strand recombination occur between the two DNA copies of the diploid genome, as well as between the DNA provirus and cell DNA. In the latter instance, a retrovirus may pick up a *cellular oncogene;* such oncogenes are incorporated into the viral genome to become *viral oncogenes,* which confer the property of rapid oncogenicity on the retrovirus concerned (see Chapter 11).

Reassortment

A variety of recombination called *reassortment* occurs with viruses that have segmented genomes, whether these are ssRNA or dsRNA and consist of 2 (*Arenaviridae*), 3 (*Bunyaviridae*), 8 (influenza A virus), 10 (*Reovirus*), 11 (*Rotavirus*), or 12 (*Coltivirus*) segments (Fig. 4-1B). In a single cell infected with two related viruses within any of these groups, there is an exchange of segments with the production of various stable reassortants. Reassortment occurs in nature and is an important source of genetic variability.

Reactivation

The term *multiplicity reactivation* is applied to the production of infectious virus by a cell infected with two or more virus particles of the same strain, each of which has suffered a lethal mutation in a different gene. Multiplicity reactivation could theoretically lead to the production of infectious virus if animals were to be inoculated with vaccines produced by UV irradiation or treatment with certain chemicals; accordingly these methods of inactivation are not used for vaccine production. *Cross-reactivation* or *marker rescue* are terms used to describe genetic recombination between an infectious virus and an inactivated virus of a related but distinguishable genotype, or a fragment of DNA from such a virus; they are useful techniques for the experimental virologist.

Interactions between Viral Gene Products

All the aforementioned interactions involve physical recombination between the nucleic acid of different viruses, resulting in a permanent heritable alteration in the genomes of the recombinant progeny. In marked contrast, we turn to a totally different class of interactions in which there is no direct association between the parental genomes and no changes in the genomes of the progeny.

Complementation

Complementation is the term used to describe all cases in which a protein encoded by one virus allows a different virus to replicate in doubly infected cells. For example, two temperature-sensitive (*ts*) mutants of the same strain of virus, neither of which is capable of replicating at the nonpermissive temperature, will both replicate in the same cell providing that their mutations are situated in different genes. Thus, as a preliminary step toward genetic mapping, a large panel of random *ts* mutants can be allocated to functional groups (corresponding to separate genes) by testing which pairs can complement one another. It should be noted that, since no genetic recombination is involved, both parents breed true—all the progeny resemble one parent or the other. Complementation can also occur between unrelated viruses, for example, between an adenovirus and adeno-associated virus (a parvovirus) or between SV40 and an adenovirus in monkey cells. In both these examples the first-named serves as a helper virus, providing a gene product that the second virus requires in order to be able to replicate in a cell type that is otherwise nonpermissive for it.

Phenotypic Mixing

Following mixed infection by two viruses that share certain common features, some of the progeny may acquire phenotypic characteristics from both parents, although their genotype remains unchanged. For example, when cells are coinfected with influenza virus and a paramyxovirus, the envelopes of some of the progeny particles contain viral antigens derived from each parent. However, each virion contains the nucleic acid of only one parent, and hence on passage it produces only virions resembling that parent (Fig. 4-1E,F). Phenotypic mixing is an essential part of the life cycle of envelope-defective retroviruses, progeny virions being called *pseudotypes* and having the genome of the defective parental virus but the envelope glycoproteins of the helper retrovirus, in whose company it will always be found.

Experimentally, and in nature with some plant viruses, phenotypic mixing of nonenveloped viruses can take the form of *transcapsidation* (Fig. 4-1G), in which there is partial or usually complete exchange of capsids. For example, poliovirus nucleic acid may be enclosed within a coxsackievirus capsid, or the adenovirus type 7 genome may be enclosed within an adenovirus 2 capsid. Since the viral ligands that govern cell attachment reside in the capsid, transcapsidation can change cell tropism.

Polyploidy

With the exception of the retroviruses, which are diploid, all viruses of vertebrates are haploid, that is, they contain only a single copy of each gene. Even with the retroviruses, diploidy is in no sense comparable to that seen in eukaryotic cells, since both copies of the genome are essentially identical and derived from the same parental virus. Among viruses that mature by budding from the plasma membrane, for example, paramyxoviruses, it is sometimes found that several nucleocapsids (and thus genomes) are enclosed within a single envelope (*polyploidy* or *heteropolyploidy*, Fig. 4-1C,D).

Mapping Viral Genomes

The complete sequence of any viral DNA, or of a DNA copy of any viral RNA, can now be accurately determined. Older techniques for partial characterization of viral DNA or RNA by "mapping" or "fingerprinting" of oligonucleotides produced by enzymatic cleavage are simpler and still useful for identification of viruses with large genomes.

Oligonucleotide Mapping

Several hundred bacterial endonucleases, called *restriction endonucleases,* have been identified and purified from various bacteria. Each recognizes a unique short, *palindromic sequence* of nucleotides (a sequence that reads the same backward as forward), usually four to six nucleotide pairs long. A given restriction endonuclease cleaves the DNA into a precise number of fragments of precise sizes, determined by the location and frequency of the particular palindromic sequence it recognizes. These DNA fragments may be separated by gel electrophoresis. Different viruses, even very closely related strains of the same virus, yield characteristically different restriction endonuclease fragment patterns, sometimes called *fingerprints* or restriction fragment length polymorphisms (RFLPs). These have been invaluable for distinguishing between different species or strains of viruses with large genomes, such as poxviruses or herpesviruses. The order of the fragments can be determined to provide a physical map of the genome. Restriction enzymes can also be used to analyze the molecularly cloned cDNA copies of genes or genomes from RNA viruses, and to locate the specific physical positions of various genetic markers on the viral chromosome.

Before restriction endonuclease mapping and sequencing were widely used, oligonucleotide fingerprinting techniques using T1 ribonuclease provided a fast and powerful procedure for differentiating between different isolates of RNA viruses. Because T1 ribonuclease cuts RNA on the 3′ side of every G residue, very large numbers of short oligonucleotides are produced; hence, adequate separation of the products requires electrophoresis in one dimension followed by chromatography in the other in order to produce a diagnostic fingerprint.

Recombination Maps

Among viruses that undergo intramolecular recombination, the probability of recombination occurring between two markers reflects the distance between them, and recombination frequencies in adjacent intervals are approximately additive. Two-factor crosses are used to determine recombination frequencies between pairs of mutants, and for very close or distant markers three-factor crosses are used to resolve ambiguities.

DNA Sequence Analysis

Much more information can be obtained by determining the complete sequence of nucleotides in a viral genome by either the Maxam–Gilbert method or the dideoxy method of Sanger. *Open reading frames* (ORFs) are translatable sequences starting with the codon for methionine (AUG) and uninterrupted by stop codons (UAA, UAG, UGA). The function of the predicted protein can sometimes be surmised by the similarity of its sequence to that of a viral or cellular protein of known function. Such comparisons are carried out by searching international computer databases of nucleotide and amino acid sequences. It is also possible to find characteristic sequences of amino acids, or *motifs,* that indicate which domains will have particular functions, such as signal sequences for targeting proteins to the endoplasmic reticulum or the plasma membrane, transmembrane sequences, glycosylation sites, and nucleotide binding sites. Short sequence motifs can also be identified which serve as signals in gene expression. The particular methionine codon (AUG) that initiates translation at the beginning of all open reading frames is usually embedded in a consensus sequence GCCGCC/GCCAUGG. Sites of mRNA polyadenylation occur 10–30 bases downstream of the sequence AAUAAA. The start sites for transcription by RNA polymerase II are about 30 base pairs downstream from an A+T-rich sequence, the *TATA box.* And so on.

Recombinant DNA Technology

The discovery of restriction endonucleases and the recognition of other enzymes involved in DNA synthesis (polymerases, ligases, transferases) opened the possibility of deliberately introducing specific foreign DNA sequences into DNA molecules. When the recombinant molecules replicate, there is a corresponding amplification of the foreign DNA. The process is called *molecular cloning.* When the inserted DNA is placed in frame with appropriate upstream and downstream regulatory sequences it is expressed, that is, the polypeptide specified by the foreign DNA is produced. Expression is usually achieved by incorporating the foreign DNA into a bacteriophage or a bacterial plasmid (Fig. 4-2), which serves as a *cloning and expression vector* when introduced into the appropriate prokaryotic or eukaryotic cells. Vectors are available that replicate in bacteria, yeasts, insect and animal cells, and in intact animals. For animal cells, a variety of animal viruses are used as vectors, notably SV40, bovine papilloma virus, retroviruses, and vaccinia virus. The cluster of techniques used is often called recombinant DNA technology or "genetic engineering."

Uses of Genetic Engineering

In addition to the great value of genetic engineering for experimental virology, practical applications to animal viruses include the development of nucleic acid probes for diagnosis by nucleic acid hybridization (see Chapter 12) and novel methods for the production of vaccines, including the use of vaccinia or fowlpox virus as vectors (see Chapter 13). Combined with the polymerase chain reaction and the availability of simple and fast methods of nucleotide sequencing, genetic engineering has also led to studies of animal virus genomes that could not be previously be contemplated. Among the achievements so far are the following:

1. Complete sequencing of the genome of viruses representing all DNA virus families
2. Complete sequencing of cDNA corresponding to the entire genome of viruses representing all RNA virus families
3. Production of labeled nucleic acid probes for viral diagnosis
4. Recognition of the copy number and sequence of viral or proviral DNAs that are integrated into the DNA of transformed cells

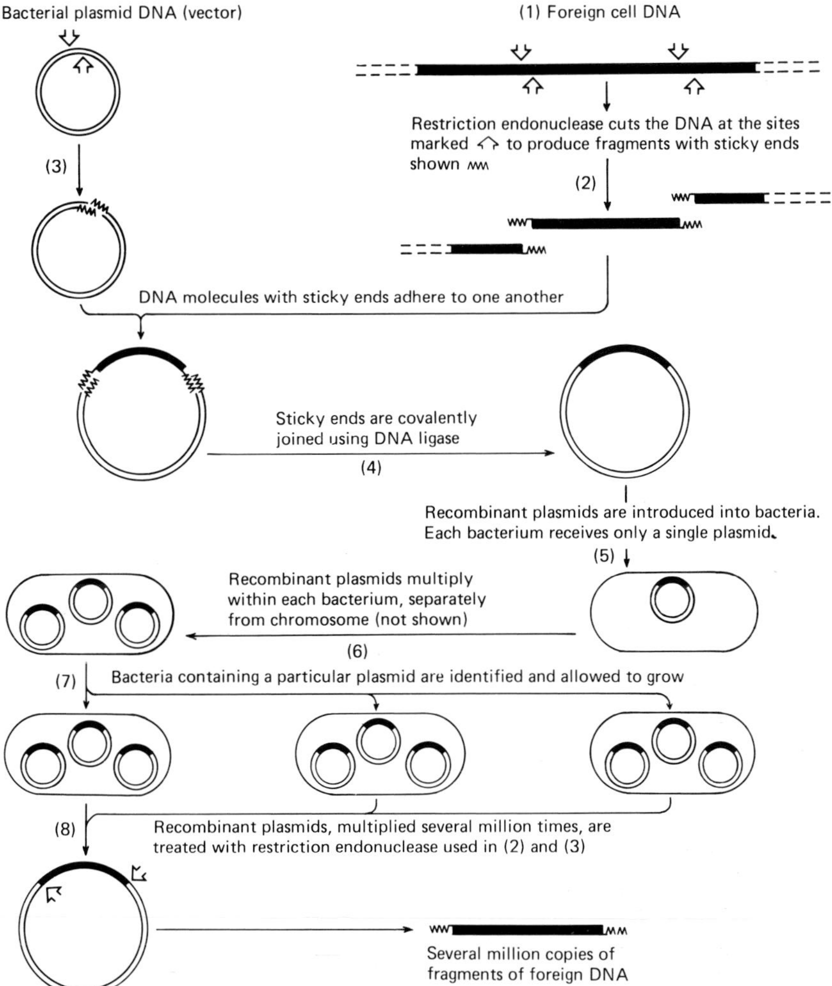

Fig. 4-2 Steps in obtaining recombinant DNA. In parallel, DNA (genome DNA or cDNA from virion RNA or mRNA) from a virus (1) is cut into fragments by a selected restriction endonuclease (2), and the circular DNA molecule of the plasmid vector is cut with the same endonuclease (3). The viral DNA is inserted and ligated into the plasmid DNA, which is thus circularized again (4). The plasmid is then introduced into the host bacterium by transformation (5). Replication of the plasmid as an episome (6) may produce many copies per bacterial cell (for small plasmids), or there may be only one copy (for large plasmids). Bacteria containing the desired plasmid are identified, cloned, and allowed to grow (7). The plasmids are isolated from the bacteria and the viral DNA insert is excised (8) using the same restriction endonuclease employed in steps (2) and (3). In this way a specified gene may be replicated several millionfold. With appropriate genetic engineering including the use of regulatory and termination sequences, the protein product of the inserted gene may be expressed in prokaryotic or eukaryotic cells.

5. Marker rescue by transfection with gene fragments, as a method of genetic mapping and of site-specific mutagenesis
6. Production of proteins coded by specific viral genes using bacterial, yeast, insect, and animal cell expression systems, or by cell-free translation
7. Synthesis of peptides based on DNA sequence data

Genetic Analysis of Viruses that Cannot be Cultured

Growth in cell culture is not essential for the study of viral nucleic acids and proteins, since the introduction of gene cloning and then the polymerase chain reaction (PCR) has made it it possible to produce virtually unlimited quantities of any required nucleic acid. Remarkable progress has been made in the genetic analysis of some noncultivable viruses, such as the papillomaviruses, by using DNA extracted directly from papillomas. A recent milestone in the history of virology was the molecular cloning followed by determination of the complete nucleotide sequence of the hepatitis C virus genome without the virus ever having been seen by electron microscopy, let alone cultured (see Chapter 26).

Transgenic Animals

Transgenic animals provide a new tool for investigating many problems in virology, immunology, and biology in general. In experimental biology transgenic animals have been most commonly produced with mice, either by using retrovirus vectors or more commonly by injecting selected fragments of DNA into one of the two pronuclei of fertilized eggs, some of which, after replacement in foster mothers, develop normally into adults to form the basis of a colony of transgenic animals that are homozygous for the transgene. The technique has enormous potential for enlarging our understanding of viral biology, for it provides insights into the potential role in viral pathogenesis of individual viral gene products in the context of the intact animal.

Evolution of Viruses

Viruses have left no fossil record, hence we must rely on scrutiny of existing viruses for clues to viral origins and evolution. Much can be learned by comparing the nucleotide sequences of the thousands of viral and cellular genes now available in the burgeoning international computer data banks. However, it must be recognized that we are looking through a tiny window of time in the context of the history of life on earth. Although sequencing the genomes of contemporary viruses can shed light on relationships between them and certain cellular genes, we are still in the dark about the origin of viruses. Some scientists regard it as axiomatic that viruses evolved originally from DNA or RNA already present in a cellular organelle or chromosome, or from some form of intracellular parasite such as a bacterium, whereas others picture a primeval "RNA world" in which a form of self-replicating RNA akin to modern viroids predated DNA, proteins, and cells. We also have no idea whether all viruses evolved from a single progenitor, although there is evidence that the plus sense RNA viruses may have.

A computer search reveals that the genomes of virtually all plus sense RNA viruses, whether of animal or plant origin, contain an RNA-dependent RNA polymerase gene with certain conserved motifs, suggesting that this essential enzyme may have been the earliest viral protein. Most families of the larger RNA viruses also carry genes for an RNA helicase and a proteinase activity, both of which display sufficient resemblance to their cellular homologs to indicate that they were originally acquired from cellular nucleic acid by recombination, or vice versa. Indeed, there is much evidence to suggest that among viruses genetic recombination has been a more important evolutionary mechanism than point mutation, and has been responsible for major changes such as the production of the progenitors of all higher level taxa. The theory of "modular evolution" of plus sense RNA viral genomes postulates that, once the cluster of genes essential for genome replication was in place in the prototype virus(es), other genes encoding accessory and structural proteins, less vital but advantageous to the virus, were added to the genome by recombination; subsequently these "modules" or "cassettes" have been acquired from or exchanged with the genome of other, related or unrelated, viruses by genetic recombination or reassortment. Minor changes continue to occur at a very high frequency as a result of point mutations or less frequently nucleotide insertion or deletion. Such mutations do not occur at the same rate in all genes of animal viruses; in influenza A virus, for example, they are more abundant in surface proteins subject to selection by neutralizing antibodies.

Because of the importance of serological methods for diagnostic purposes, clinical virologists place great emphasis on the surface proteins, which distinguish viral strains of relatively recent origin from one another. Phylogenetically, however, the key enzymes concerned with genome replication, as well as gene order, the nature and location of noncoding regulatory sequences, and key features of the viral replication strategy, are more significant in defining a family and its relationship to other families from which it may have diverged millions of years ago. In the so-called *polythetic* approach to taxonomy, viruses are grouped into genera that share a unique set of characters, some but not all of which may be shared with viruses of other genera. These taxonomically useful characters are less variable than some of the other phenotypic characters that are of greater diagnostic, pathogenetic, and epidemiologic importance.

Similar principles and algorithms have been applied to construct evolutionary trees of other viruses. The families of minus sense RNA viruses and of DNA viruses are more heterogeneous than those of the plus sense RNA viruses, and may have arisen from a number of distinct protoviruses. The large DNA viruses such as poxviruses and herpesviruses appear to have captured the genes for numerous cellular enzymes from their host cells at some time in the distant past. The retroviruses share a common genetic feature, the "*gag–pol* replicon" (which encodes a reverse transcriptase) with the "pararetroviruses" (such as hepadnaviruses), as well as with the various classes of noninfectious "retroelements" known as retrotransposons, retroposons, and retrons.

The genetic mechanisms described in this chapter, operating under the pressure of Darwinian selection, have been clearly incriminated in several recent important examples of viral evolution (Table 4-1). Here we shall content ourselves with a description of two viruses that illustrate particularly well the dramatic impact of evolution, even over a relatively short time span. One,

Table 4-1
Examples of Genetic Mechanisms That Have Affected Viral Evolution[a]

Mechanism	Example
Point mutation	Lethal chicken influenza due to a single point mutation
Intramolecular recombination	Western equine encephalitis virus produced by recombination between eastern equine encephalitis virus and a Sindbis-like alphavirus
Genetic reassortment	Pandemic human influenza A subtypes H2N2 (1957) and H3N2 (1968)
Recombination and mutation	Changes in poliovaccine following vaccination
Biased hypermutation[b] (uridine to cytosine transitions)	Evolution of subacute sclerosing panencephalitis virus from measles virus
Genetic rearrangement[c]	Evolution of rubella virus

[a] Based on E. D. Kilbourne, *Curr. Opin. Immunol.* **3**, 518 (1991).
[b] Missense mutations of M gene.
[c] Compared with the alphavirus genome, the order of helicase and NS P3 region is reversed in rubella virus.

myxoma virus, highlights the importance of changes in viral virulence and host resistance. The other, influenza A virus, illustrates how effectively viruses can evolve to avoid the immune response of the host.

Genetic Changes in Virus and Host in Myxomatosis

Myxomatosis, caused by a poxvirus, occurs naturally as a mild infection of rabbits in South America and California (*Sylvilagus* spp.), in which it produces a benign fibroma from which virus is transmitted mechanically by biting insects. However, in laboratory (European) rabbits *(Oryctolagus cuniculus),* myxoma virus causes a lethal infection, a finding that led to its use for biological control of wild European rabbits in Australia.

The wild European rabbit was introduced into Australia in 1859 for sporting purposes and rapidly spread over the southern part of the continent, where it became the major animal pest of the agricultural and pastoral industries. Myxoma virus from South America was successfully introduced into the rabbit population in 1950; when originally liberated the virus produced case–fatality rates of over 99%. This highly virulent virus was readily transmitted by mosquitoes. Farmers undertook "inoculation campaigns" to introduce virulent myxoma virus into wild rabbit populations.

It might have been predicted that the disease and with it the virus would disappear at the end of each summer, owing to the greatly diminished numbers of susceptible rabbits and the greatly lowered opportunity for transmission by mosquitoes during the winter. This must often have occurred in localized areas, but it did not happen over the continent as a whole. The capacity of virus to survive the winter conferred a great selective advantage on viral mutants of reduced lethality, since during this period, when mosquito numbers were low, rabbits infected by such mutants survived in an infectious condition for weeks instead of a few days. Within 3 years such "attenuated" mutants became the dominant strains throughout Australia. Some inoculation campaigns with the virulent virus produced localized highly lethal outbreaks, but in general the viruses that spread through the rabbit populations each year were the "attenuated" strains, which because of the prolonged illness in their hosts provided a greater opportunity for mosquito transmission. Thus the original highly lethal virus was progressively replaced by a heterogeneous collection of strains of lower virulence, but most of them still virulent enough to kill 70–90% of genetically unselected rabbits.

Rabbits that recover from myxomatosis are immune to reinfection. However, since most wild rabbits have a life span of less than 1 year, herd immunity is not so critically important in the epidemiology of myxomatosis as it is in infections of longer lived species. The early appearance of viral strains of lower virulence, which allowed 10% of genetically unselected rabbits to recover, allowed selection for genetically more resistant animals to occur. In areas where repeated outbreaks occurred, the genetic resistance of the rabbits increased steadily such that the case–fatality rate after infection under laboratory conditions with a particular strain of virus fell from 90 to 50% within 7 years. Subsequently, in areas where there were frequent outbreaks of myxomatosis, somewhat more virulent strains of myxoma virus became dominant, because they produced the kind of disease that was best transmitted in populations of genetically resistant rabbits. Thus, the ultimate balance struck between myxoma virus and Australian rabbits involved adaptations of both the virus and host populations, reaching a dynamic equilibrium which finds rabbits greatly reduced compared with premyxomatosis numbers, but still too numerous for the wishes of farmers and conservationists.

Genetic Changes in Influenza A Virus

Human influenza virus was first isolated in 1933. Since that time human influenza viruses have been recovered from all parts of the world, and their antigenic properties have been studied in considerable detail, thus providing an opportunity for observing continuing evolutionary changes.

Influenza A virus occurs in humans, swine, horses, birds, and aquatic mammals. Subtypes are classified according to the two envelope antigens, the hemagglutinin (HA, or H) and neuraminidase (NA, or N). All of the fourteen subtypes of the HA molecule have been found in influenza viruses from birds, three of them also in humans, two in pigs, horses, seals, and whales, and one in mink. The nine NA subtypes show a similar distribution.

The outstanding feature of influenza A virus is the antigenic variability of the envelope glycoproteins, HA and NA, which undergo two types of changes, known as *antigenic drift* and *antigenic shift.* Antigenic drift occurs within a subtype and involves a gradual accumulation of point mutations; those affecting neutralizing epitopes produce strains each antigenically slightly different from its predecessor. In contrast, antigenic shift involves the sudden acquisition of a gene for a completely new HA or NA, giving rise to a novel subtype that spreads rapidly around the world as most or all humans have no immunity to it.

Antigenic Shift

During the past century there have been five pandemics of human influenza, namely in 1890, 1900, 1918, 1957, and 1968. The pandemic at the end of the First World War killed over 20 million people—more than the war itself. In 1957 the H1N1 subtype was suddenly replaced by a new subtype, H2N2, known as "Asian flu" because it originated in China. Within a year over 1 billion people had been infected, but fortunately the mortality was much lower than in 1918, probably because the strain was intrinsically less virulent, although the availability of antibiotics to treat bacterial superinfection undoubtedly saved many lives. In 1968 this subtype was in turn replaced by the "Hong Kong flu" (H3N2). Finally, in 1977 the H1N1 subtype mysteriously reappeared, and since then the two subtypes H3N2 and H1N1 have cocirculated.

The first clear evidence that distinct mechanisms are involved in the processes of antigenic shift and drift came from peptide mapping and partial amino acid sequencing of HA proteins, which demonstrated relatively close relationships between strains within each of the three human subtypes (H1, up until 1957; H2, 1957–1968; H3, 1968 to the present) but major differences between subtypes, indicating that a sharp discontinuity in the evolutionary pattern had occurred with the emergence of H2 in 1957 and H3 in 1968. With the advent of nucleic acid sequencing it became feasible to determine the complete nucleotide sequence of all eight genes of many strains of influenza viruses isolated from several species of animals and birds. Phylogenetic analysis of the data now indicates that all of the influenza viruses of mammals, including humans, originated from the avian influenza gene pool, which itself presumably evolved from a common ancestral avian influenza virus. In 1957, three of the eight genes (HA, NA, and PB1) of the prevalent human H1N1 subtype were replaced by Eurasian avian influenza genes to produce the human H2N2 subtype; then, in 1968 the human H2N2 subtype acquired new HA and PB1 genes from another avian influenza virus of the Eurasian lineage to produce the human H3N2 subtype. Moreover, retrospective serological studies indicate that the 1890 human pandemic subtype was H2N8, the 1900 subtype H3N8, and the 1918 subtype H1N1, suggesting a pattern of recycling of the three human HA subtypes (H1, H2, H3).

Influenza A viruses from birds grow very poorly in humans, and vice versa; indeed, reassortants containing avian internal genes have been tested as experimental vaccines because of their avirulence and their inability to spread from human to human. However, both avian and human influenza virus can replicate in pigs, and genetic reassortment between them can be demonstrated experimentally in that host. An attractive current hypothesis is that antigenic shift in nature occurs when the prevailing human strain of influenza A virus and an avian influenza virus concurrently infect a pig, which serves as a "mixing vessel"; only a reassortant containing mainly genes derived from the human virus but the HA gene derived from the avian virus is able to infect a human and possibly to initiate a pandemic. While such a combination of circumstances may not occur often, it must be appreciated that in rural Southeast Asia, the most densely populated area of the world, hundreds of millions of people live and work in close contact with domesticated pigs and ducks. It is no coincidence that the last two antigenic shifts to produce major pandemics (H1 to H2 in 1957, and H2 to H3 in 1968) emanated from China.

Not all episodes of antigenic shift are attributable to genetic reassortment with avian influenza virus. The strain of H1N1 that reappeared in 1977 closely resembled a human H1N1 strain prevalent around 1950. People old enough to have had prior experience of the H1 subtype before it was replaced by H2N2 in 1957 generally still displayed a considerable degree of immunity to it when it reappeared. How H1N1 survived during this long interval is a complete mystery. It is not inconceivable that it may have escaped from a laboratory in which it had been stored frozen since around 1950. Nor is it clear why the H1 subtype was supplanted by H2 in 1957 (and H2 by H3 in 1968) yet H1N1 and H3N2 have continued to cocirculate since 1977. Genetic reassortants between the human H1N1 and H3N2 subtypes have occasionally been isolated from humans since then.

Antigenic Drift

After antigenic shift introduces a new pandemic subtype of influenza A virus, antigenic drift begins. Point mutations occur at random in all segments of the viral genome, a proportion of which result in nonlethal amino acid changes in the corresponding proteins. Many of these changes will be deleterious or neutral, but substitutions in antigenic sites on the HA molecule will offer a survival advantage in the presence of neutralizing antibody. Thus amino acid substitutions accumulate in the HA molecule of naturally occurring isolates at the rate of about 1% per annum.

Following the derivation of the three-dimensional structure of the influenza HA molecule by X-ray crystallography (Fig. 4-3) it became possible to map the precise location of all the amino acid changes. Most of the changes in field strains arising in nature by antigenic drift are situated on prominent regions of the exposed surface of the HA molecule in the general vicinity of the receptor-binding pocket.

When monoclonal antibodies are used experimentally to select *escape mutants,* each mutant is generally found to contain only a single amino acid substitution, usually involving a change in charge or side-chain length, or sometimes creating an additional glycosylation site. Clearly, this single substitution is sufficient to prevent that particular monoclonal antibody from binding to its epitope and neutralizing the virion by sterically hindering attachment to the host cell, or by other more poorly understood mechanisms. However, there are dozens of overlapping epitopes which tend to cluster into five major antigenic sites (Fig. 4-3), and most infected humans produce neutralizing antibodies directed at several or all of these sites. Hence, if a variant arising under natural circumstances in an immune or partially immune human is to escape neutralization by the spectrum of anti-HA antibodies present in that individual, it would presumably need to contain mutations in more than one antigenic site. Such multiple mutations would arise only very rarely in a given individual but could accumulate sequentially in successive individuals. All known natural strains arising within the H3 subtype by antigenic drift between 1968 and 1988 (as defined by lack of neutralization by convalescent ferret antisera against previous strains) contained at least four different amino acid changes, located in at least two of the five antigenic sites, always including the immunodominant sites A and B.

Attempts have been made to predict the direction of antigenic drift, in order to prepare appropriate vaccines in advance. Certain amino acid residues

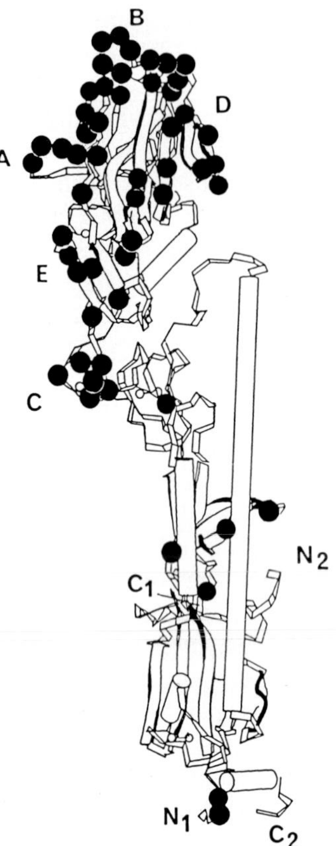

Fig. 4-3 Diagram of one of the three identical monomers of the hemagglutinin molecule of the A/Hong Kong/1968 (H3N2) subtype of influenza A virus, as determined by X-ray crystallography. The monomer is composed of two glycosylated polypeptide chains, HA_1 and HA_2, the amino and carboxyl termini of which are shown as N_1 (N_2) and C_1 (C_2) near the proximal end of the molecule, which is inserted into the viral envelope. The receptor-binding pocket can be seen in the center of the distal tip of the "head" of the molecule, surrounded by the two immunodominant antigenic sites, A and B, and close to the other antigenic sites, C, D, and E. The black dots mark the location of all the amino acid positions that have altered during natural antigenic drift in the H3N2 subtype over the period 1968–1986. Certain amino acid positions in sites A and B have changed a number of times during the 19 years. It can be seen that most of the substitutions have occurred in epitopes sufficiently close to the receptor-binding site for a bound antibody molecule (which is approximately the same size as HA) to sterically hinder attachment of the HA molecule to its receptor on the host cell. [From J. J. Skehel and D. C. Wiley, *in* "RNA Genetics" (E. Domingo, J. J. Holland, and P. Ahlquist, eds.), Vol. 3, p. 142. CRC Press, Boca Raton, Florida, 1988.]

tend to change repeatedly in the course of natural antigenic drift, or selection by monoclonal antibodies *in vitro,* and it may be concluded that they are key residues within immunodominant epitopes. However, there is no reproducible order or direction of the changes; indeed, two separate lineages have arisen during natural drift within the H3N2 subtype of influenza A, and two within influenza B.

Further Reading

Coen, D. M. (1990). Molecular genetics of animal viruses. *In* "Fields Virology" (B. N. Fields, D. M. Knipe, R. M. Chanock, M. S. Hirsch, J. L. Melnick, T. P. Monath, and B. Roizman, eds.), 2nd Ed., p. 123. Raven, New York.

Darnell, J., Lodish, H., and Baltimore, D. (1990). "Molecular Cell Biology," 2nd Ed., p. 189. Scientific American Books, New York.

Domingo, E., Holland, J. J., and Ahlquist, P., eds. (1988). "RNA Genetics," Vols. 1–3. CRC Press, Boca Raton, Florida.

Fenner, F., and Ross, J. (1994). Myxomatosis. *In* "The European Rabbit. The History and Biology of a Successful Colonizer," p. 205. Oxford Univ. Press, Oxford.

Garcia-Sastre, A., and Palese, P. (1993). Genetic manipulation of negative strand RNA viral genomes. *Annu. Rev. Microbiol.* **47,** 765.

Goedell, D. V., ed. (1990). Gene Expression Technology. "Methods in Enzymology," Vol. 185. Academic Press, San Diego.

Holland, J. J. (1990). Defective viral genomes. *In* "Fields Virology" (B. N. Fields, D. M. Knipe, R. M. Chanock, M. S. Hirsch, J. L. Melnick, T. P. Monath, and B. Roizman, eds.), 2nd Ed., p. 151. Raven, New York.

Holland, J. J., ed. (1992). Genetic Diversity of RNA Viruses. *Curr. Top. Microbiol. Immunol.* **176,** 1.

Jarvis, T. C., and Kirkegaard, K. (1991). The polymerase in its labyrinth: Mechanisms and implications of RNA recombination. *Trends Genet.* **7,** 186.

Klenk, H.-D., and Rott, R. (1989). The molecular biology of influenza virus pathogenicity. *Adv. Virus Res.* **34,** 247.

Koonin, E. V., ed. (1992). Evolution of Viral Genomes. *Semin. Virol.* **3,** 311.

Kurstak, E., Marusyk, R. G., Murphy, F. A., and van Regenmortel, M. H. V., ed. (1990). Virus Variability, Epidemiology and Control. *Appl. Virol. Res.* **2.**

Minor, P. D. (1992). The molecular biology of polioviruses. *J. Gen. Virol.* **73,** 3065.

Notkins, A. L., and Oldstone, M. B. A., Ed. (1989). Transgenic mice: Expression of viral genes. "Concepts in Viral Pathogenesis III," p. 158. Springer-Verlag, New York.

Pringle, C. R. (1990). The genetics of viruses. *In* "Topley and Wilson's Principles of Bacteriology, Virology and Immunity" (L. H. Collier and M. C. Timbury, ed.), 8th Ed., Vol. 4, p. 69. Arnold, London.

Ramig, R. F. (1990). Principles of animal virus genetics. *In* "Fields Virology" (B. N. Fields, D. M. Knipe, R. M. Chanock, M. S. Hirsch, J. L. Melnick, T. P. Monath, and B. Roizman, eds.), 2nd Ed., p. 95. Raven, New York.

Rice, C. M., ed. (1992). Animal Virus Expression Vectors. *Semin. Virol.* **3,** 235.

Roux, L., Simon, A. E., and Holland, J. J. (1991). Effects of defective interfering viruses on virus replication and pathogenesis *in vitro* and *in vivo*. *Adv. Virus Res.* **40,** 181.

Sambrook, J., Fritsch, E. F., and Maniatis, T. (1989). "Molecular Cloning: A Laboratory Manual," 2nd Ed. Cold Spring Harbor Laboratory, Cold Spring Harbor, New York.

Smith, D. B., and Inglis, S. C. (1987). The mutation rate and variability of eukaryotic viruses: An analytical review. *J. Gen. Virol.* **68,** 2729.

Steinhauser, D. A., and Holland, J. J. (1987). Rapid evolution of RNA viruses. *Annu. Rev. Microbiol.* **41,** 409.

Chapter 5

Virus-Induced Changes in Cells

The "typical" virus–cell interaction described in Chapter 3 was a *productive* infection of a *permissive* cell by a *cytocidal (lytic)* virus, that is, resulting in the production of infectious virions and the death of the host cell. It is important to appreciate that some cytocidal virus–cell encounters are *nonproductive (abortive)* and that some productive infections are *noncytocidal (nonlytic)* (Table 5-1).

Types of Virus–Cell Interactions

First, even when an infectious virion enters a susceptible cell, more often than not the encounter will be nonproductive—for reasons not fully understood. Second, some types of cells are *nonpermissive* for particular types of virus. The viral replication cycle may be blocked at any point from attachment (owing to the absence of an appropriate receptor) through to the final stage of assembly and release (e.g., owing to the absence of an appropriate cellular protease to cleave an envelope glycoprotein and thus render the progeny virions infectious). If there is a defect in the viral genome, abortive replication will occur even within a fully permissive cell. Two particular examples, discussed in Chapter 4, are the deletion mutants known as defective interfering (DI) mutants and the point mutants known as conditional lethal mutants. Often a wild-type virus that replicates perfectly well in a particular type of host cell finds another cell type nonpermissive; in such cases it is a moot point whether

Table 5-1
Types of Virus–Cell Interactions

Type of infection	Effects on cell	Production of infectious virions	Examples
Cytocidal (lytic)	Morphological changes in cells (cytopathic effects); inhibition of protein, RNA and DNA synthesis; cell death	+	Alphaherpesviruses enteroviruses, reoviruses
Persistent, productive	No cytopathic effect; little metabolic disturbance; cells continue to divide; may be loss of the special functions of some differentiated cells.	+	Arenaviruses, rabies virus, most retroviruses
Persistent, nonproductive	Usually nil	−, but virus may be induced[a]	Measles virus in brain
Transformation	Alteration in cell morphology; cells can be passaged indefinitely; may produce tumors when transplanted to experimental animals	−, oncogenic DNA viruses +, oncogenic retroviruses	Polyomaviruses, adenoviruses Mouse, chicken leukosis and sarcoma viruses

[a] By cocultivation, irradiation, or chemical mutagens.

to allocate the blame to a defect in the cell or the virus—it is the combination that is unsatisfactory. The special case of conditionally defective (satellite) viruses (e.g., hepatitis D virus) that can replicate only in the presence of a helper virus (hepatitis B virus) which supplies an essential gene product was also described in Chapter 4.

Finally, the most important of all nonproductive virus–cell interactions are those particular types of persistent infections known as latent infections, in which one or more complete or defective DNA copies of the viral genome are maintained indefinitely in the cell, either integrated into a host cell chromosome or in the form of a cytoplasmic episome, but are not fully expressed. The cell survives, indeed may divide repeatedly, but no virions are produced unless or until the cell is induced to do so by an appropriate stimulus (see Chapter 10).

Regardless of whether the virus–cell encounter gives rise to infectious progeny, the infection may be lytic or nonlytic. Some viruses, such as arenaviruses and retroviruses, replicate normally without killing the cell; they do not shut down host cell protein, RNA, or DNA synthesis, and virions are released by budding through the plasma membrane. Although some of these viruses may produce subtle changes in certain functions of the host cell that are not essential for its survival ("luxury" functions), the infected cell may continue to divide and new virions continue to be synthesized indefinitely. The extreme case is *transformation* by *oncogenic viruses;* here, one or more copies of the DNA viral genome (or of a cDNA copy of an RNA viral genome) is retained indefinitely inside the cell which is itself not killed but is permanently altered (transformed), sometimes to a state of malignancy (cancer). Certain oncogenic viruses, the retroviruses, often maintain a productive noncytocidal infection in malignant cells; with others, notably the DNA oncogenic viruses, productive infection is incompatible with cell survival, and cancer develops in cells in which only certain viral genes are expressed (see Chapter 11).

Cytopathic Effects of Virus Infections

Cytocidal viruses kill the cells in which they replicate. When a monolayer of cultured cells is infected with a small number of virions, their progeny will be released from the infected cells and will spread through the medium to infect distant (as well as adjacent) cells, so that eventually all cells in the culture will be involved. The resulting cell damage is known as the *cytopathic effect (CPE)* of the virus, and the responsible virus is said to be *cytopathogenic*. The CPE can often be observed by low-power light microscopy in unstained cell cultures (Figs. 5-1 and 5-2A–C); fixation and staining of the cell monolayer may reveal further diagnostic details, notably *inclusion bodies* and *syncytia*. The nature and speed of development of the cytopathic effect are characteristic of the particular virus involved and therefore represent important criteria for the preliminary identification of clinical isolates (see Chapter 12).

Inclusion Bodies

A characteristic morphological change in cells infected by certain viruses is the formation of *inclusion bodies* (or *inclusions*) which are recognized by light microscopy following fixation and staining (Fig. 5-3). Depending on the virus, inclusion bodies may be single or multiple, large or small, intranuclear or intracytoplasmic, round or irregular in shape, and acidophilic (stained by eosin) or basophilic (stained by hematoxylin).

The most striking viral inclusion bodies are the intracytoplasmic inclusions found in cells infected with poxviruses, paramyxoviruses, reoviruses,

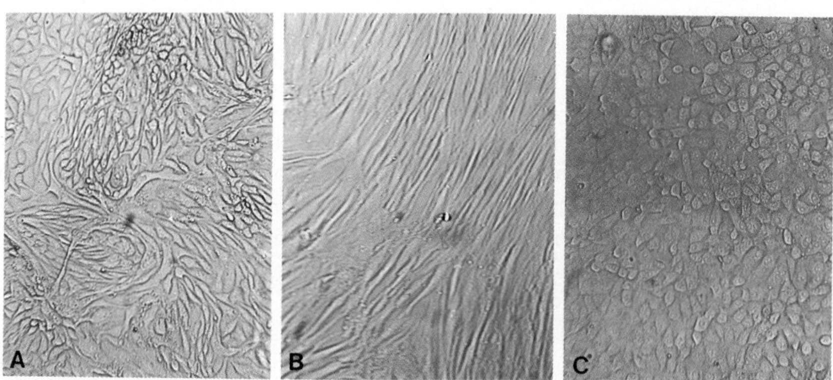

Fig. 5-1 Unstained confluent monolayers of the three main types of cultured cells, as they appear by low-power light microscopy, through the wall of the glass or plastic vessel in which the cells are cultured. Magnification: ×60. (A) Primary monkey kidney epithelium: a mixed population of mainly epithelial cells taken freshly from the body. (B) Diploid strain of human fetal fibroblasts. (C) Continuous line of malignant epithelial cells. (Courtesy I. Jack.)

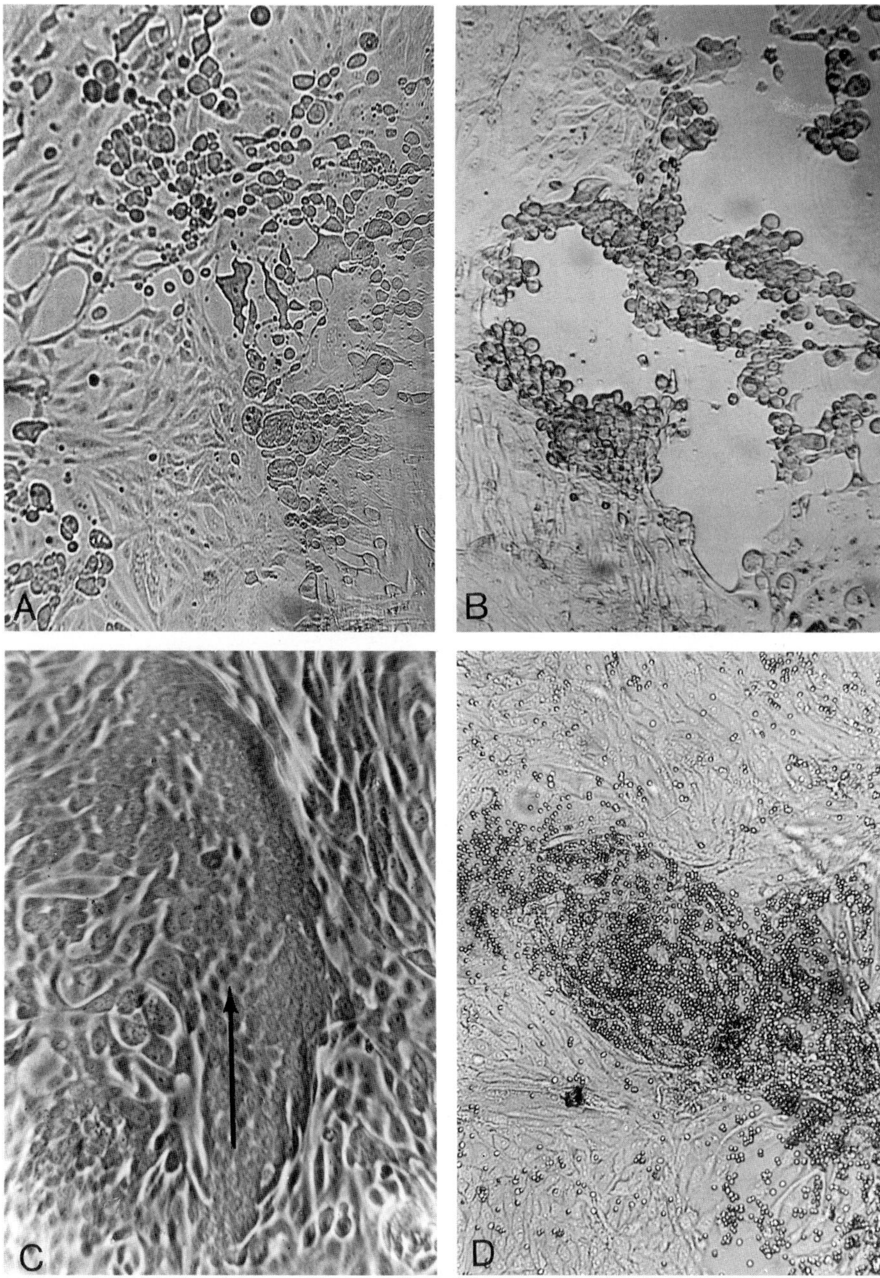

Fig. 5-2 Cytopathic effects produced by different viruses. The cell monolayers are shown as they would normally be viewed in the laboratory, unfixed and unstained. Magnification: ×60. (A) Enterovirus: rapid rounding of cells, progressing to complete cell lysis. (B) Herpesvirus: focal areas of swollen rounded cells. (C) Paramyxovirus: focal areas of cells are fused to form syncytia. (D) Hemadsorption: erythrocytes adsorb to infected cells that incorporate hemagglutinin into the plasma membrane. (Courtesy I. Jack.)

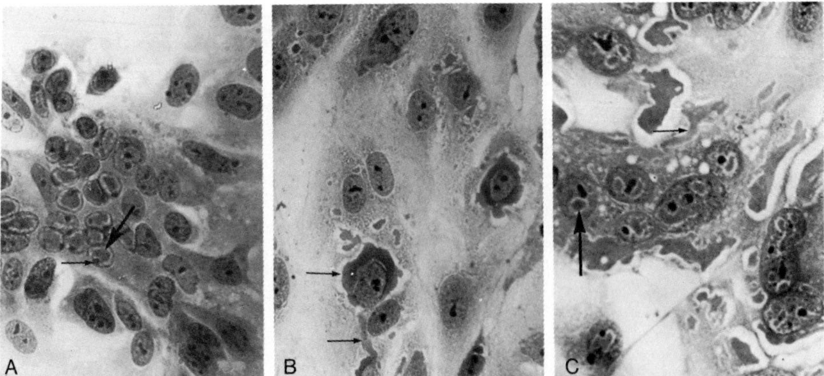

Fig. 5-3 Types of viral inclusion bodies (hematoxylin and eosin stain). Magnification: ×200. (A) Intranuclear inclusions and syncytium (herpesvirus). Small arrow, nucleolus; large arrow, inclusion body. Note margination of condensed nuclear chromatin separated from inclusion body by a halo. (B) Intracytoplasmic inclusions (reovirus). Arrows indicate inclusion bodies, mainly in perinuclear locations. (C) Intranuclear and intracytoplasmic inclusions and syncytia (measles virus). Small arrow, intracytoplasmic inclusion body; large arrow, intranuclear inclusion body. (Courtesy I. Jack.)

and rabies virus and the intranuclear inclusion bodies found in cells infected with herpesviruses, adenoviruses, and parvoviruses. Some viruses, for example, measles virus and cytomegalovirus, may produce both nuclear and cytoplasmic inclusion bodies in the same cell. Many inclusions have been shown to be accumulations of viral structural components; for example, the intracytoplasmic inclusions in cells infected with rabies virus or in "inclusion body encephalitis" (subacute sclerosing encephalopathy, caused by measles virus) are masses of viral nucleocapsids. The basophilic intracytoplasmic inclusions invariably found in cells infected with poxviruses are sites of viral synthesis (viral "factories") made up of masses of viral protein and nucleic acid. In a few instances, for example, adenoviruses in the nucleus and reoviruses in the cytoplasm, inclusion bodies represent crystalline aggregates of virions. Other inclusion bodies, such as those found in the nucleus of cells infected with herpesviruses, are the result of late degenerative changes, as well as condensation and margination of chromatin; these nuclear inclusions are made more obvious by a clear unstained halo which is a shrinkage artifact produced by fixation (Fig. 5-3A).

Effects of Viruses on Plasma Membrane

Polarity of Budding

In the course of their replication, viruses belonging to several families cause changes in the plasma membrane of the cell by insertion of viral glycoproteins. As discussed in Chapter 3, viral glycoproteins are not inserted at random in the plasma membrane. Viruses that mature at the apical surface of epithelial cells are shed into the environment, whereas those maturing at the basolateral surface move to other sites in the body, sometimes entering the bloodstream and establishing systemic infection. For example, rabies virus is shed from the apical surface of salivary gland epithelium into the saliva of a rabid dog, whereas retroviruses bud through the basolateral surface and pass directly from cell to cell or become disseminated. Wild-type Sendai virus, which causes a localized respiratory infection, buds apically, whereas a pantropic variant buds basolaterally; nucleotide sequencing of such pairs should shed light on the "post-codes" that direct the migration of viral and cellular glycoproteins, which are sorted in the trans-Golgi network into distinct transport vesicles for direct delivery to the apical or basolateral surface. Current work suggests that the membrane anchor region represents an important targeting signal.

Cytolysis by Immunologic Mechanisms

Virus-coded antigens in the plasma membrane constitute a target for specific immune mechanisms, both humoral and cellular, which may result in lysis of the cell before significant numbers of new virions are produced, thus slowing the progress of infection and hastening recovery (see Chapter 8). In some cases the immune response may precipitate immunopathologic disease (see Chapter 9). Some viral antigens incorporated in the cell membrane, found in cells that are transformed by viruses, behave as *tumor-specific transplantation antigens* (see Chapter 11).

Cell Fusion

A conspicuous feature of infection of cell monolayers by lentiviruses, paramyxoviruses, some herpesviruses and some other viruses is the production of *syncytia* (see Fig. 5-2C and 5-3C), which result from the fusion of the infected cell with neighboring infected or uninfected cells. Such multinucleate syncytia may also be seen in the tissues of persons infected with these viruses and may represent an important mechanism of spread which avoids exposure of virions to neutralizing antibodies and also allows infection to be transmitted by subviral entities such as nucleocapsids or even viral nucleic acid.

At high multiplicity of infection, paramyxoviruses may cause rapid fusion of cultured cells without any requirement for replication, simply as a result of the action of the fusion (F) protein of input virions attaching to the plasma membrane. Cell biologists have used this phenomenon to produce functional *heterokaryons* by fusing different types of cells. For example, in the pioneering experiments by Milstein and Kohler that produced the first monoclonal antibodies, parainfluenza virus inactivated by irradiation with ultraviolet light was used to produce *hybridoma* cells by fusion of antibody-producing B lymphocytes with myeloma cells.

Hemadsorption

Cells in monolayer culture infected with orthomyxoviruses, paramyxoviruses, or togaviruses, all of which bud from the plasma membrane, acquire the ability to adsorb erythrocytes. This phenomenon, known as *hemadsorption* (see Fig. 5-2D), is due to the incorporation into the plasma membrane of viral glycoproteins assembled into peplomers, and it becomes demonstrable quite early in the replication cycle of noncytocidal or cytocidal viruses. On the envelope of the virion, the same glycoprotein peplomers are responsible for *hemagglutination*, that is, agglutination of erythrocytes by virions. Although hemadsorption and hemagglutination are not known to play a role in the pathogenesis of viral diseases, both techniques are used extensively in their laboratory diagnosis (see Chapter 12).

Other Morphological Changes in Virus-Infected Cells

As seen at the higher resolution provided by the electron microscope, the specific and nonspecific changes in virus-infected cells are dramatic and varied. Early changes in cell structure often involve proliferation of various cell membranes; for example, herpesviruses cause increased synthesis of the nuclear membrane, flaviviruses cause proliferation of endoplasmic reticulum, picornaviruses and caliciviruses cause a distinctive proliferation of microvesicles in the cytoplasm, and many retroviruses cause peculiar fusions of cytoplasmic membranes. Infection by many viruses also leads to a disruption of cytoskeletal fiber systems by depolymerization of microfilaments and/or microtubules, despite the fact that the cytoplasmic cytoskeleton and the nuclear matrix, which provide the physical site for many metabolic activities of the cell, are also used for the subcellular compartmentalization of viral replicative processes. Later in the course of infection, many lytic viruses cause nuclear, organelle, and cytoplasmic rarefaction and/or condensation, with terminal loss of host cell membrane integrity. In many cases the inevitability of cell death is obvious, but in other cases host cell functional loss is subtle and cannot be associated easily with particular ultrastructural pathologic changes. In nonlytic infections most functional losses cannot easily be attributed to damage that is morphologically evident. Specific examples reflecting the range of host cell changes occurring in virus-infected cells are included in many of the chapters in Part II.

In addition to changes directly attributable to viral replication, most virus-infected cells also show nonspecific changes, very much like those induced by physical or chemical insults. The most common early and potentially reversible change is what pathologists call "cloudy swelling;" this change is associated with increasing permeability of the plasma membrane. Electron microscopic study of such cells reveals diffuse swelling of the nucleus, distention of the endoplasmic reticulum and mitochondria, and rarefaction of the cytoplasm. Later in the course of many viral infections the nucleus becomes condensed and shrunken, and cytoplasmic density increases. Cell destruction can be the consequence of further loss of osmotic integrity and leakage of lysosomal enzymes into the cytoplasm. This progression, overall, is called by pathologists "the common terminal pathway to cell death."

Mechanisms of Cell Damage

So many biochemical changes occur in cells infected with cytocidal viruses that the death of the cell cannot readily be ascribed to one particular event; rather it may be the final result of the cumulative action of several biochemical insults. Cell damage can occur even without replication of the virus, for example, when late stages of the expression of the viral genome are blocked experimentally, or in certain abortive infections.

Shutdown of Cellular Protein Synthesis

Most cytocidal viruses code for proteins that shut down the synthesis of cellular proteins. The shutdown is particularly rapid and profound in infections of cultured cells by picornaviruses and some poxviruses and herpesviruses. With some other viruses (e.g., adenoviruses), the shutdown occurs later and is more gradual, whereas with noncytocidal viruses such as arenaviruses and retroviruses there is no shutdown and no cell death.

The mechanisms are varied, and not all are clearly understood. In cases where the inhibition of cellular protein synthesis develops gradually and late in the replication cycle, it may possibly be due to competition for ribosome subunits by the large excess of viral mRNA. Even when viral mRNA is not in excess, the shutdown may provide a selective advantage by allowing the viral message to bind to ribosomes and initiate translation, as has been described with reovirus. An adenovirus early protein inhibits the transport of processed cellular mRNAs from nucleus to cytoplasm, whereas certain herpesviruses bring about selective degradation of cellular mRNA. A special case is the inactivation by a picornavirus protease of the cellular cap-binding protein that is required for the binding of cellular mRNAs to ribosomes; picornavirus RNA is peculiar in having no m^7Gppp cap and being translated quite satisfactorily without it.

Shutdown of Cellular Nucleic Acid Synthesis

Some viruses reduce transcription of cellular mRNA, but the mechanisms are not well understood. Inhibition of cellular DNA synthesis is common, except with those DNA viruses that replicate in the nucleus. This is an inevitable consequence of inhibition of protein synthesis, but more specific mechanisms have been described for certain viruses; these include degradation of cellular DNA by a poxvirus DNase and displacement of cellular DNA from its normal site of replication, seen with herpesviruses.

Cytopathic Effects of Viral Proteins

Large amounts of various viral components accumulate in the cell late in the replication cycle. Some of these, particularly certain capsid proteins (e.g., adenovirus penton and fiber proteins), are toxic to cells. In addition, as discussed above, viral proteins that are inserted into the plasma membrane may cause cell fusion, as well as providing a target for the immune response *in vivo*. Insertion of viral proteins into the plasma membrane can also change membrane permeability, leading directly to loss of osmotic integrity, cell swelling, and death.

Noncytocidal Infections

Noncytocidal viruses usually do not kill the cells in which they replicate. On the contrary, they may produce persistent infection, in which the infected cells produce and release virions but overall cellular metabolism is little affected, with the infected cells continuing to grow and divide. This type of cell–virus interaction is found in cells infected with several kinds of RNA viruses: arenaviruses, retroviruses, and some paramyxoviruses, for example, in all of which virions are released by budding from the plasma membrane. Although such virus-yielding cells may grow and divide in culture for long periods, there are slow, progressive changes that with some exceptions (e.g., some retroviruses) ultimately lead to cell death. In the body, cell replacement occurs so rapidly in most organs and tissues that the slow fall-out of cells due to persistent infection, at least in the short term, may have no effect on overall function. However, persistently infected differentiated cells may lose their capacity to carry out specialized functions, and neurons, once destroyed, are not replaced. Also, antigenic changes produced in the plasma membrane may provoke immune responses which can rapidly lead to destruction of the persistently infected cells and often nearby uninfected cells (see Chapter 9).

Effects on Functions of Specialized Cells

Although they do not immediately kill cells, infections with noncytocidal viruses often interfere with the specialized functions of differentiated cells. For example, lymphocytic choriomeningitis (LCM) virus replicating in somatotropic cells of the pituitary gland of the persistently infected mouse lowers the production of the mRNA for growth hormone in the infected cells, thus impeding the growth and development of the animal. Similarly, LCM virus replicating in β cells of the islets of Langerhans in the pancreas can induce hyperglycemia in the mouse, not dissimilar to insulin-dependent diabetes in humans. β-Adrenergic receptors and opiate receptors are impaired in brain cells persistently infected with measles virus. Viruses that infect lymphocytes may induce a generalized immunosuppression. Rhinovirus infection of the nasal epithelium results in cilial stasis and later in the destruction of cilia, although the cells are often not killed. This effect can be demonstrated in organ culture (Fig. 5-4), and it is important in lowering the resistance of the respiratory tract to secondary bacterial infection.

Interferons

Viral *interference* is said to occur when a virus-infected cell population resists superinfection with the same or a different species of virus. The interfering virus does not necessarily have to replicate to induce interference, and the ability of the challenge virus to replicate may be completely or only partially inhibited. Two main mechanisms have been demonstrated: (1) interference mediated by defective interfering mutants, operating only against the homologous virus (see Chapter 4), and (2) interference mediated by *interferon.*

Properties of Interferons

In 1957 Isaacs and Lindenmann reported that cells of the chorioallantois of embryonated hen's eggs infected with influenza virus released into the medium a nonviral protein, "interferon," which protected uninfected cells against

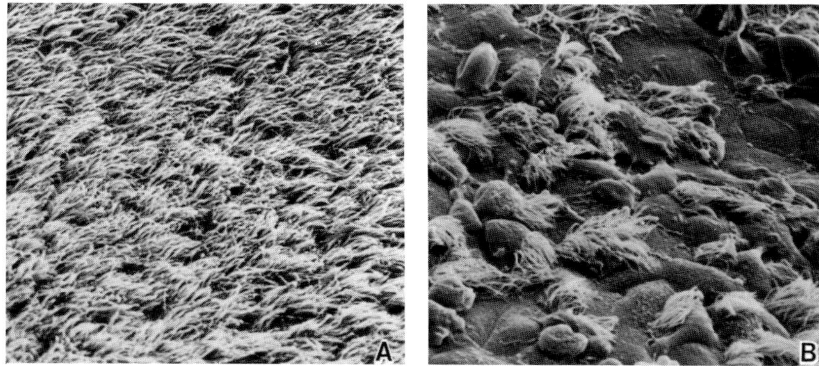

Fig. 5-4 Effect of rhinovirus on bovine tracheal epithelium grown *in vitro* as an explant (organ culture), as shown by scanning electron microscopy. (A) Normal appearance of ciliated cells. (B) Six days after infection many cells are rounded up or detached. [From S. E. Reed and A. Boyde, *Infect, Immun.* **6,** 68 (1972).]

infection with the same or unrelated viruses. This discovery raised hopes that such interferons could be used as safe, nontoxic broad-spectrum antiviral chemotherapeutic agents. Despite an enormous amount of work on this group of proteins since then, the availability of interferon as a therapeutic agent for the treatment of virus diseases remains an unfulfilled dream. The discovery in the early 1980s that interferon had anticancer effects stimulated further research which showed that the interferons were typical members of a large family of normal cellular regulatory proteins called *cytokines*. We now know that there are about 20 human interferons, falling into three chemically distinct types, known as interferon α (which occurs as over a dozen subtypes), interferon β and interferon γ (Table 5-2).

The last decade has seen the discovery and cloning by recombinant DNA technology of the genes for several human interferons, and the corresponding proteins have been purified. Interferons β and γ, but generally not interferon α, are glycosylated. Each subunit has an M_r of about 20,000; interferon α

Table 5-2
Properties of Human Interferons

Property	Interferon α	Interferon β	Interferon γ
Principal source	Leukocytes	Fibroblasts, epithelial cells	T lymphocytes, NK cells
Inducing agent	Virus infection	Virus infection	Antigen (or mitogen)
Number of subtypes	About 20	1	1
Glycosylation	No (most subtypes)	Yes	Yes
Functional form	Monomer	Dimer	Tetramer
Principal activity	Antiviral	Antiviral	Immunomodulation
Mechanism of action	Inhibits protein synthesis	Inhibits protein synthesis	Enhances MHC antigens; activates cytotoxic T cells, macrophages, and NK cells

is active in monomeric form, whereas interferon β occurs mainly as a dimer and interferon γ as a tetramer.

Interferons α and β are not made constitutively in significant amounts, but their synthesis is induced by virus infection—any virus, but especially RNA viruses, multiplying in almost any type of cell, in any vertebrate species. Interferon γ is made only by T lymphocytes (and natural killer, or NK, cells), and only following antigen-specific or mitogenic stimulation; it is a *lymphokine*, with immunoregulatory functions. Some interferons, especially β and γ, display a certain degree of host species specificity; for instance, mouse interferons are ineffective in humans, and vice versa. However, there is little or no viral specificity, in that interferon α, β, or γ induced by, say, a paramyxovirus is fully effective against a togavirus, although certain cloned interferon subtypes may be much more effective against some viruses than against others, for reasons that will become apparent below.

Antiviral Action of Interferons

Following its induction by viral infection, interferon is released from the infected cell and binds to a specific receptor on the plasma membrane of other cells. There appears to be one receptor for interferon α (which can also be utilized by interferon β) and another for interferon γ. Binding of interferon α to its receptor triggers a tyrosine kinase to phosphorylate three proteins which then assemble to form a large DNA-binding transcription factor, which in turn binds to a 14-base pair nucleotide consensus sequence present in the 5′-regulatory region of most interferon-inducible genes. This up-regulates the expression of over 20 cellular genes. Many of the induced *interferon-regulated proteins (IRPs)* directly or indirectly inhibit the replication of virus, each in a different way. Binding of interferon γ to its receptor triggers the phosphorylation of a different transcription factor which recognizes a different regulatory sequence controlling a different set of genes. It has recently become apparent that, although the interferon-treated cell becomes resistant to most viruses, individual interferon-induced proteins are effective only against a limited range of viruses. Three well-studied examples will suffice to illustrate the diversity of biochemical mechanisms involved in interferon-mediated interference.

P1/eIF-2α Kinase

The P1/eIF-2α kinase, which is made constitutively at low levels in untreated cells, is up-regulated by interferon α, β, or γ. Following binding of dsRNA, which is an intermediate or by-product formed in the course of RNA virus replication, this protein kinase phosphorylates its own P1 subunit, thus activating the enzyme to phosphorylate the α subunit of the eukaryotic protein synthesis initiation factor eIF-2, inactivating it. Because eIF-2 is required to initiate synthesis of all polypeptides, interferon-induced P1/eIF-2α kinase can inhibit synthesis of all proteins of any virus.

However, many viruses have developed strategies for circumventing this universal antiviral defense mechanism. For example, adenoviruses encode low-M_r RNAs which bind to P1/eIF-2α kinase, preventing its activation by dsRNA. Reoviruses, which might be expected to be exceptionally susceptible to interferons because they have a dsRNA genome, are not, because their capsid protein σ3 binds much more strongly to its own dsRNA than does P1/eIF-2α kinase. Similarly, the dsRNA-binding vaccinia virus-coded protein p25 inhibits this kinase. The *tat* gene product of the human immunodeficiency virus down-regulates the synthesis of P1/eIF-2α kinase, whereas poliovirus infection appears to degrade the enzyme. Influenza virus infection activates a cellular regulator that inhibits induction of the kinase.

2-5A Synthetase

The interferon-induced enzyme 2-5A synthetase, which also requires dsRNA for its induction, catalyzes the synthesis from ATP of an unusual family of short-lived oligonucleotides known as $(2'-5')\text{pppA(pA)}_n$, or 2-5A for short. In turn, 2-5A activates a cellular endonuclease, RNase L, which destroys mRNA, thereby inhibiting host cell protein synthesis. The 2-5A synthetase/RNase L system is known to be effective against picornaviruses, but its wider relevance has yet to be established.

Mx Protein

A team of Swiss virologists has conducted elegant studies on a family of GTP-binding proteins known as Mx proteins which are induced by interferon α and β and determine the susceptibility of mice to influenza virus infection. Only strains of mice carrying the Mx gene survived challenge with influenza virus; congenic mice with a deletion mutation in the Mx gene died. Dual fluorescent antibody staining of infected tissues revealed that the circle of uninfected cells surrounding a zone of virus-infected cells was synthesizing large amounts of Mx protein. Interferon itself, in the absence of virus infection, similarly induced production of the Mx protein, *in vitro* or *in vivo*. Transfection of cultured Mx^- cell lines with the Mx gene converted them permanently to a state of resistance to influenza virus. Finally, transgenic mice produced by transfecting the Mx gene into Mx^-, influenza-susceptible mice were shown to be resistant to challenge with influenza virus (but not other viruses) and to make the Mx protein following infection with influenza virus or the administration of interferon.

Thus it can be seen that the degree of sensitivity of any particular virus to interferon is influenced by a complex interplay of factors determined by the type of interferon, virus, and host cell. Some host species carry genes that encode an IRP with activity against only one particular species of virus. Further, individual viruses have developed a wide variety of mechanisms for circumventing the antiviral actions of the many IRPs. This intricate web of biochemical pathways represents an instructive illustration of the role of evolution in the never-ending battle for survival between host and parasite.

Implicit in this discussion is that interferons play a significant role in recovery from viral infections. Formal proof of this assertion is provided in Chapter 7.

Interferons as Cytokines

Interferons were discovered as antiviral agents, defined accordingly, and generally regarded as such by virologists for many years. However, it is now recognized that, in addition, interferons exert a wide range of other effects on cells in their role as cytokines. The most important of these additional functions is immunomodulation. Interferon γ, a lymphokine produced almost exclusively by T lymphocytes, is over 100 times more potent than interferons α/β as an immunoregulatory agent and influences the immune response in a wide variety of ways (Table 5-2). Most importantly, interferon secreted by T cells following exposure to virus-infected cells activates T lymphocytes, macrophages, and NK cells in the vicinity to develop their cytotoxic potential. The contributions of interferons and other cytokines to the immune response to viral infections is discussed in more detail in Chapter 8.

Further Reading

Baron, S., Coppenhaver, D. H., Dianzani, F., Fleischman, W. R., Hughes, T. K., Klimpel, G. R., Niesel, D. W., Stanton, G. J., and Tyring, S. K. (1992). "Interferon: Principles and Medical Applications." Univ. of Texas Medical Branch, Galveston.

Carrasco, L. (1987). "Mechanisms of Viral Toxicity in Animal Cells." CRC Press, Boca Raton, Florida.

de Maeyer, E., and de Maeyer-Guignard, J. (1988). "Interferons and Other Regulatory Cytokines." Wiley, New York.

Fraenkel-Conrat, H., and Wagner, R. R., eds. (1984). Viral Cytopathology. "Comprehensive Virology," Vol. 19. Plenum, New York.

Kaariainen, L., and Ranki, M. (1984). Inhibition of cell functions by RNA-virus infections. *Annu. Rev. Microbiol.* **38,** 91.

Kozak, M. (1986). Regulation of protein synthesis in virus-infected animal cells. *Adv. Virus Res.* **31,** 229.

Moller, G., ed. (1987). Gamma Interferon. *Immunol. Rev.* **97.**

Oldstone, M. B. A. (1989). Viral alteration of cell function. *Sci. Am.* **261,** 34.

Pestka, S., Langer, J. A., Zoon, K. C., and Samuel, C. E. (1987). Interferons and their actions. *Annu. Rev. Biochem.* **56,** 727.

Russell, W. C., and Almond, J. W., eds. (1987). Molecular Basis of Virus Disease. *Soc. Gen. Microbiol. Symp.* **40.**

Samuel, C. E. (1991). Antiviral actions of interferon: Interferon-regulated cellular proteins and their surprisingly selective antiviral activities. *Virology* **183,** 1.

Schneider, R. J., and Shenk, T. (1987). Impact of virus infection on host cell protein synthesis. *Annu. Rev. Biochem.* **56,** 317.

Schrom, M., and Bablanian, R. (1981). Altered cellular morphology resulting from cytocidal virus infection. *Arch. Virol.* **70,** 173.

Staeheli, P. (1990). Interferon-induced proteins and the antiviral state. *Adv. Virus Res.* **38,** 147.

Tucker, S. P., and Compans, R. W. (1993). Virus infection of polarized epithelial cells. *Adv. Virus. Res.* **42,** 187.

Chapter 6

Mechanisms of Infection and Spread of Viruses through the Body

To cause infection, viruses, like other infectious agents, must gain entry to the body, multiply, and spread, either locally or systemically, and in generalized infections localize in the appropriate target organ. To be maintained in nature, infectious virions must be shed into the environment, or taken up by an arthropod vector or a needle, or passed congenitally.

Routes of Entry

To infect its host, a virus must first attach to and infect cells of one of the body surfaces, unless they are bypassed by parenteral inoculation via a wound, needle, or bite of an arthropod or vertebrate. The human body can be represented diagrammatically as a set of surfaces, at each of which a sheet of epithelial cells separates host tissue from the outside world (Fig. 6-1). The outer surface proper, the skin, has a relatively impermeable, dry, outer layer of dead cells. In the alimentary canal and in the respiratory tract the surface lining consists of one or more layers of living cells. The urogenital tract, where urine and sexual products are secreted and released into the environment, constitutes another discontinuity in the protective covering of skin, and in the eye the skin is replaced by a transparent layer of living cells to form the conjunctiva.

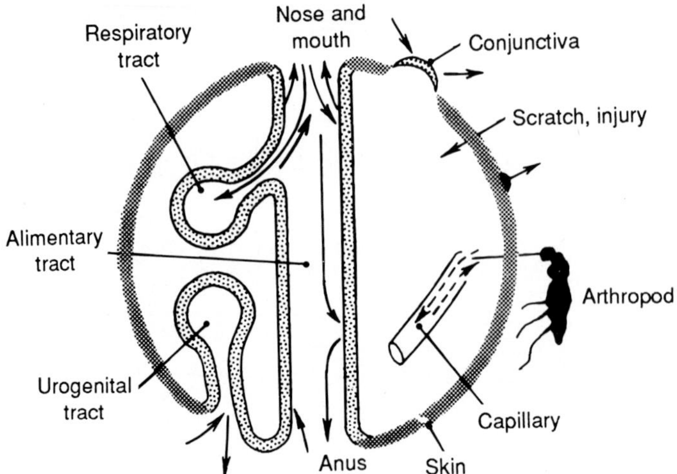

Fig. 6-1 Surfaces of the body in relation to the entry and shedding of viruses. (Modified from C. A. Mims and D. O. White, "Viral Pathogenesis and Immunology." Blackwell, Oxford, 1984.)

Respiratory Tract

Although lined by cells that are susceptible to infection by many viruses, the respiratory tract is ordinarily protected by effective cleansing mechanisms (Fig. 6-2). A mucus blanket and ciliated cells line the nasal cavity and most of the lower respiratory tract. Inhaled virus particles deposited on this surface are trapped in mucus, carried by ciliary action from the nasal cavity and airways to the pharynx, and then swallowed or coughed out. Particles 10 μm or more in diameter are usually deposited on the nasal mucosa over the turbinate bones, which project into the nasal cavity and act as baffle plates. Particles 5–10 μm in diameter may be carried to the trachea and bronchioles, where they are usually trapped in the mucus blanket. Particles of 5 μm or less are usually inhaled directly into the lungs, and some may reach the alveoli, where virions may infect alveolar epithelial cells or be destroyed by alveolar macrophages.

Despite these protective mechanisms, the respiratory tract is, overall, the most important entry site of viruses into the body (Table 6-1). All viruses that infect the host via the respiratory tract probably do so by attaching to specific receptors on epithelial cells. Following respiratory infection, many viruses remain localized (e.g., rhinoviruses, parainfluenza, and influenza viruses), whereas others become systemic (e.g., measles, chickenpox, and rubella viruses).

Alimentary Tract

Many viruses are acquired by ingestion. They may either be swallowed or infect cells in the oropharynx and then be carried to the intestinal tract. The esophagus is rarely infected, probably because of its tough stratified squamous epithelium and the rapid passage of swallowed material over its sur-

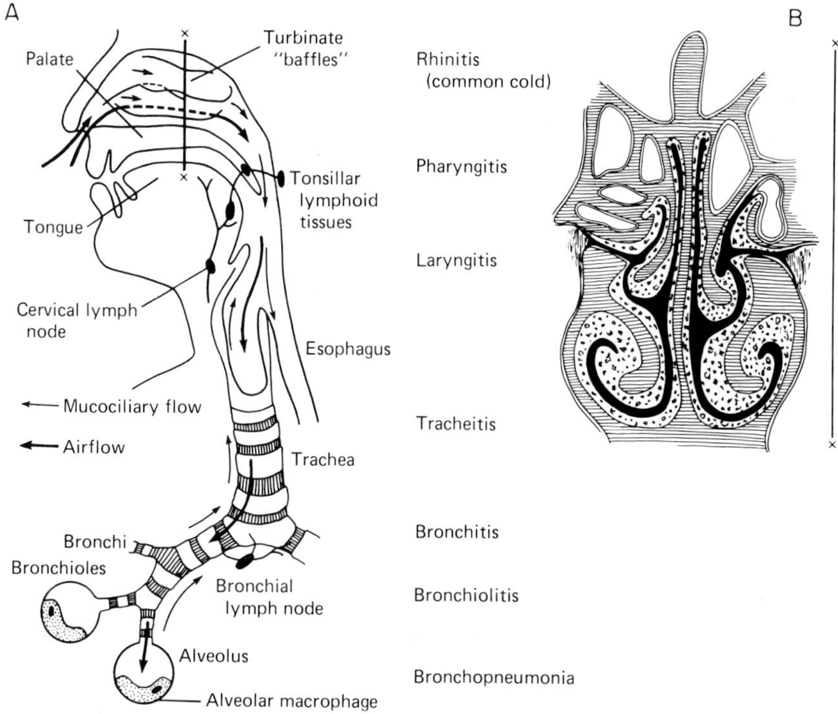

Fig. 6-2 (A) Pathways of infection and mechanical protective mechanisms in the respiratory tract. At right are listed clinical syndromes produced by infection at various levels of the respiratory tract. (B) Enlarged section of turbinates showing the narrow and complicated pathway of inspired air, and thus the ease with which slight cellular swelling "blocks the nose." (From C. A. Mims and D. O. White, "Viral Pathogenesis and Immunology." Blackwell, Oxford, 1984.)

Table 6-1
Viruses That Initiate Infection of Humans via the Respiratory Tract

With production of local respiratory symptoms	
Picornaviridae	Rhinoviruses, some enteroviruses
Coronaviridae	Most types
Paramyxoviridae	Parainfluenza viruses, respiratory syncytial virus
Orthomyxoviridae	Influenza virus
Adenoviridae	Most types
Producing generalized disease, usually without initial respiratory symptoms	
Paramyxoviridae	Mumps, measles viruses
Togaviridae	Rubella virus
Herpesviridae	Varicella virus
Picornaviridae	Some enteroviruses
Papovaviridae	Polyomaviruses
Bunyaviridae	Hantaan virus
Arenaviridae	South American hemorrhagic fever viruses
Poxviridae	Variola virus (smallpox is now extinct)

face. The intestinal tract is partially protected by mucus, which may contain specific secretory antibodies (IgA), but the constant movement of the contents provides opportunities for virions to attach to specific receptors. Virions may also be taken up by specialized M cells that overlie Peyer's patches in the ileum, from which they are passed to adjacent mononuclear cells in which they may replicate.

There are other protective mechanisms in the intestinal tract; from the stomach downward, acid, bile, and proteolytic enzymes may inactivate viruses. In general, viruses that cause intestinal infection, such as enteroviruses, rotaviruses, and caliciviruses (Table 6-2), are acid- and bile-resistant. Rotaviruses and caliciviruses are now recognized as the major causes of viral diarrhea, whereas the great majority of intestinal infections by enteroviruses and adenoviruses are asymptomatic. Some of the enteroviruses (e.g., polioviruses), and hepatitis A and E viruses, are important causes of generalized infection but do not produce signs referable to the intestinal tract.

The emergence of AIDS has drawn attention to the importance of the rectum as a route of infection with HIV and other sexually transmitted agents.

Skin

The skin is the largest organ in the body, and since its outer layer consists of keratinized cells it provides a tough and usually impermeable barrier to the entry of viruses. However, after entry through minor abrasions or by artificial puncture, some viruses replicate in the skin to produce local lesions, for example, papillomaviruses and poxviruses (Table 6-3).

The most efficient way by which viruses are introduced through the skin is via the bite of an arthropod vector, such as a mosquito, tick, or sandfly. Such insects may be mechanical vectors (e.g., for tanapox), but most viruses

Table 6-2
Viruses That Initiate Infection of Humans via the Alimentary Tract

Via mouth or oropharynx	
Herpesviridae	Herpes simplex virus, Epstein-Barr virus, cytomegalovirus, HHV-6
Via intestinal tract	
Producing enteritis	
Reoviridae	Rotaviruses
Caliciviridae	Norwalk and related viruses
Adenoviridae	Some adenoviruses
Producing generalized disease, usually without alimentary symptoms	
Picornaviridae	Many enteroviruses including polioviruses
	Hepatitis A virus
Caliciviridae	Hepatitis E virus
Usually symptomless	
Adenoviridae	Some adenoviruses
Picornaviridae	Some enteroviruses
Reoviridae	Reoviruses

Table 6-3
Viruses That Initiate Infection of Humans via the Skin, Genital Tract, or Eye

Route	Family	Species
Skin		
Minor trauma	*Papovaviridae*	Many types of *Papillomavirus*
	Poxviridae	Molluscum contagiosum, cowpox, orf, milkers' nodes viruses
	Herpesviridae	Herpes simplex viruses
	Hepadnaviridae	Hepatitis B virus
Anthropod bite		
Mechanical	*Poxviridae*	Tanapoxvirus
Biological	*Togaviridae*	All species of *Alphavirus*
	Flaviviridae	All species of *Flavivirus*
	Bunyaviridae	La Crosse, sandfly fever, Rift Valley fever viruses
	Reoviridae	Colorado tick fever virus
Animal bite	*Rhabdoviridae*	Rabies virus
	Herpesviridae	Herpes B virus
Injection	*Hepadnaviridae*	Hepatitis B virus
	Flaviviridae	Hepatitis C virus
	Retroviridae	HIV, HTLV
	Herpesviridae	Cytomegalovirus, Epstein-Barr virus
	Filoviridae	Ebola virus
Genital tract	*Papovaviridae*	Genital types of *Papillomavirus*
	Herpesviridae	Herpes simplex viruses
	Retroviridae	HIV, HTLV
	Hepadnaviridae	Hepatitis B virus
	Flaviviridae	Hepatitis C virus
Conjunctiva	*Adenoviridae*	Several types
	Picornaviridae	Enterovirus 70

introduced in this way replicate in the vector. Viruses that are transmitted by and replicate in arthropod vectors are called *arboviruses.* Infection can be acquired through the bite of an animal, as in rabies. Finally, introduction of a virus by skin penetration may be *iatrogenic,* that is, as a result of human intervention, such as transmission of hepatitis B and C viruses and HIV by contaminated needles or blood transfusion. Generalized infection of the skin, producing an exanthem such as is found in measles, chickenpox, rubella, and several arbovirus diseases, is due to viral dissemination via the bloodstream.

Infection by Other Routes

The genital tract is the route of entry of several important pathogens. Herpes simplex viruses and papillomaviruses produce lesions on the genitalia and perineum. Many others, for example HIV, HTLV, and hepatitis B and C viruses, do not produce local lesions but are sexually transmitted.

The conjunctiva, although much less resistant to viral invasion than the skin, is constantly cleansed by the flow of secretion (tears) and is wiped by the eyelids. It is a rare route of entry, for example, for some adenoviruses and a few enteroviruses.

Mechanisms of Spread in the Body

Viruses may remain localized to the body surface through which they entered: skin, respiratory tract, intestine, genital tract, or conjunctiva. Alternatively, they may cause generalized infections, which are usually associated with viremia and subsequent localization in particular organs.

Local Spread on Epithelial Surfaces

Many viruses replicate in epithelial cells at the site of entry, produce a localized or spreading infection in the epithelium, and are then shed directly into the environment. Infection within the host spreads by sequential infection of neighboring cells. Papillomaviruses initiate infection in the basal layer of the epidermis, but maturation of virions occurs only in cells that have become keratinized as they move toward the skin surface. Since this is a slow process, taking several weeks, warts have a long incubation period. Many poxviruses produce infection via the skin, but in addition to spreading from cell to cell, there is usually also local subepithelial and lymphatic spread. In infection with vaccinia virus a few epidermal cells are infected by scarification and virus spreads locally from cell to cell, primarily in the epidermis, before spreading to the local lymph nodes. The poxviruses that cause molluscum contagiosum, orf, and tanapox remain localized in the skin and produce lumps.

Viruses that enter the body via the intestinal or respiratory tract can spread readily in the layer of fluid that can transport virions over the moist epithelial surfaces; consequently, such infections usually have a short incubation period. After infections of the respiratory tract by paramyxoviruses and influenza virus, or the intestinal tract by rotaviruses, there is little or no invasion beyond the epithelium. Although these viruses usually enter lymphatics and thus have the potential to spread, they do not appear to replicate in the deeper tissues, possibly because the appropriate virus receptors or other permissive cellular factors, such as cleavage-activating proteases, are restricted to epithelial cells, or because the temperature of deeper tissues is higher than the optimal temperature for viral replication.

Restriction of infection to an epithelial surface cannot be equated with lack of severity of clinical disease. Large areas of intestinal epithelium may be damaged by rotaviruses, for example, causing severe diarrhea. The severity of localized infections of the respiratory tract depends on their location; infections of the upper respiratory tract may produce rhinitis but few other signs; infection of the bronchioles or alveoli produces more severe respiratory distress and may predispose to secondary bacterial invasion.

Subepithelial Invasion and Lymphatic Spread

After traversing the epithelium and its basement membrane to reach the subepithelial tissues, virions are immediately exposed to tissue macrophages and can enter the lymphatics that form a network beneath the skin and all mucosal epithelia (Fig. 6-3). Virions that enter lymphatics are carried to local lymph nodes. As they enter, they are exposed to macrophages lining margin-

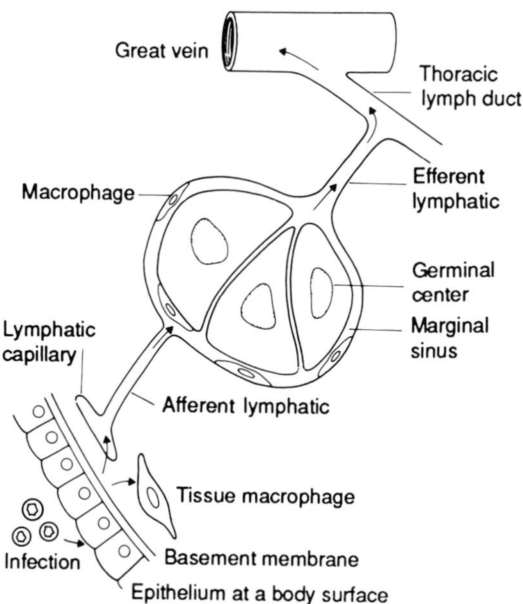

Fig. 6-3 Subepithelial invasion and lymphatic spread of viruses. (From C. A. Mims and D. O. White, "Viral Pathogenesis and Immunology." Blackwell, Oxford, 1984.)

al sinuses and may be engulfed. Virions may be inactivated and processed and their component antigens presented by macrophages and dendritic cells to adjacent lymphocytes in such a way that an immune response is initiated (see Chapter 8). Some viruses, however, replicate in cells of the monocyte/macrophage lineage; others infect lymphocytes. Some virions may pass straight through lymph nodes to enter the bloodstream. Monocytes and lymphocytes circulate through the body, and there is also a constant movement of lymphocytes directly from the blood into the lymph nodes, and in the opposite direction.

There is often a local inflammatory response, the extent of which depends on the extent of tissue damage. Local blood vessels are dilated and rendered more permeable, so that monocytes and lymphocytes, lymphokines, immunoglobulins, and complement components can be delivered directly to the site of infection, with a consequent increase in host resistance, especially after the immune response has been initiated.

Spread by the Bloodstream: Viremia

The blood is the most effective and rapid vehicle for the spread of virus through the body. Once a virus has reached the bloodstream, usually via the lymphatic system (Fig. 6-3), it can localize in any part of the body within minutes. The first entry of virus into the blood is called *primary viremia*. This early viremia may be clinically silent, known to have taken place only because of the invasion of distant organs. Further replication in these sites leads to the sustained liberation of much higher concentrations of virus, producing a *sec-*

ondary viremia (Fig. 6-4), which can in turn lead to the establishment of infection in yet other parts of the body.

In the blood, virions may be free in the plasma or may be associated with leukocytes, platelets, or erythrocytes. Viruses carried in leukocytes, generally lymphocytes or monocytes, are not cleared as readily or in the same way as viruses circulating free in the plasma; being protected from antibodies and other plasma components they can be carried to distant tissues. Monocyte-associated viremia is a feature of measles and many herpesvirus infections, for example. Rarely, as in Rift Valley fever and Colorado tick fever, virions may be associated with erythrocytes. Certain mouse leukemia viruses infect megakaryocytes; the circulating platelets derived from them are infected but do not appear to be important in the pathogenesis of viral infections. Neutrophils have a very short life span and powerful antimicrobial mechanisms; they are rarely infected, although they may contain phagocytosed virions. Hepadnaviruses, togaviruses, flaviviruses, and the enteroviruses that cause viremia circulate free in the plasma.

Virions circulating in the plasma encounter many kinds of cells, but two kinds play a special role in determining their subsequent fate: macrophages and vascular endothelial cells.

Role of Macrophages

Macrophages are very efficient phagocytes and are present in all compartments of the body, in alveoli, subepithelial tissues, sinusoids of the lymph nodes, free in plasma, and above all in the sinusoids of the liver, spleen, and bone marrow. Together with dendritic cells and B lymphocytes, macrophages are antigen-processing and antigen-presenting cells and therefore play a pivotal role in initiation of the primary immune response (see Chapter 8). The antiviral action of macrophages depends on the age and genetics of the host and the site of their origin in the body; indeed, even in a given site there are subpopulations of macrophages that differ in susceptibility. Their state of activation is also important. The kinds of interactions that may occur between macrophages and virions are described in relation to those found in the sinusoids of the liver, the Kupffer cells, in Fig. 6-5.

Differences in virus–macrophage interactions may account for differences in the virulence of virus strains and differences in host resistance. Already efficient phagocytes, macrophages have Fc receptors and C_3 receptors in their plasma membrane, which enhance their ability to ingest virions when these are coated with antibody or complement. If macrophages are susceptible, however, as with dengue viruses, this can lead to infection rather than inactivation, and the antibody may enhance rather than prevent infection.

Role of Vascular Endothelial Cells

The vascular endothelium with its basement membrane and tight cell junctions constitutes the blood–tissue interface, and for particles such as virions often a barrier. Parenchymal invasion by circulating virions depends on localization in the endothelial cells of capillaries and venules, where blood flow is slowest and the barrier is thinnest. Virions may move passively between or through endothelial cells and basement membrane, or they may infect endothelial cells and "grow" through this barrier. This subject has been most intensively studied in relation to viral invasion of the central nervous

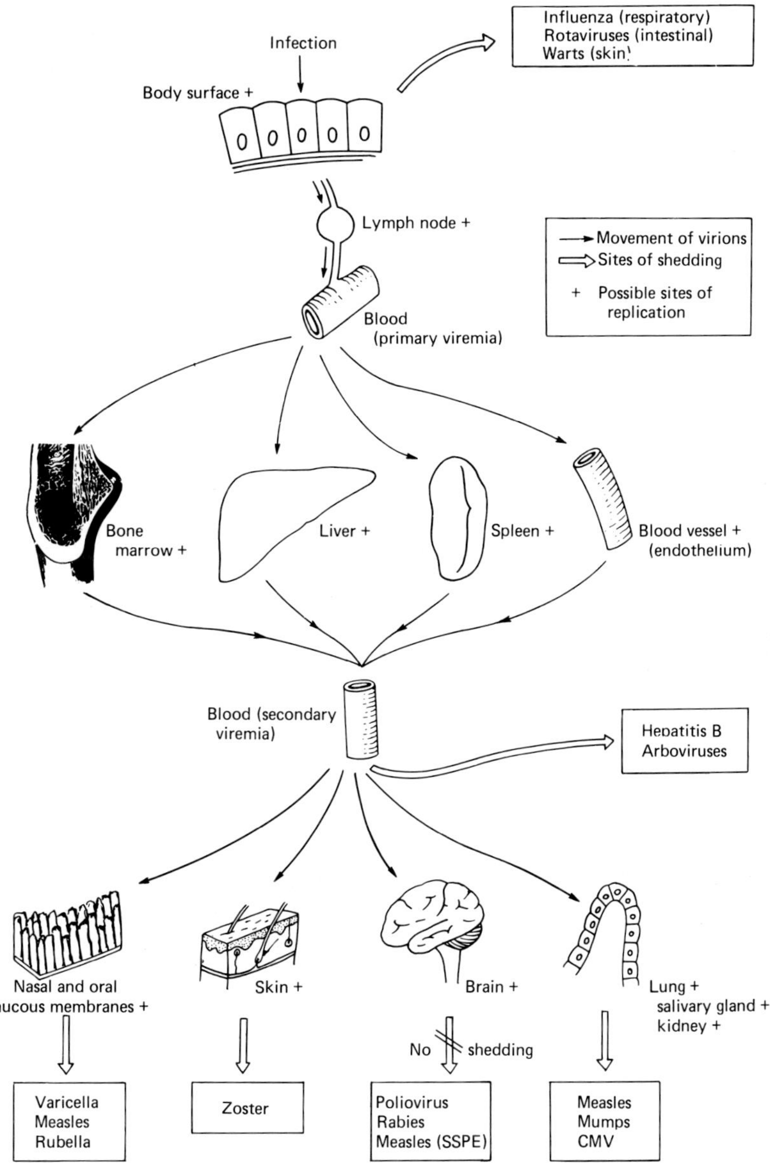

Fig. 6-4 Spread of virions through the body in human viral infections, indicating sites of replication and important routes of shedding of various viruses. (Modified from C. A. Mims and D. O. White, "Viral Pathogenesis and Immunology." Blackwell, Oxford, 1984.)

system (see below), but it also applies to secondary invasion of the skin, pulmonary epithelium, salivary gland epithelium, intestinal epithelium, kidney, and placenta.

Maintenance of Viremia

Neurotropic viruses reach the central nervous system only when the viremia is of adequate magnitude and duration, something that is also a prerequisite if blood-sucking arthropods are to be infected. Because virions circulat-

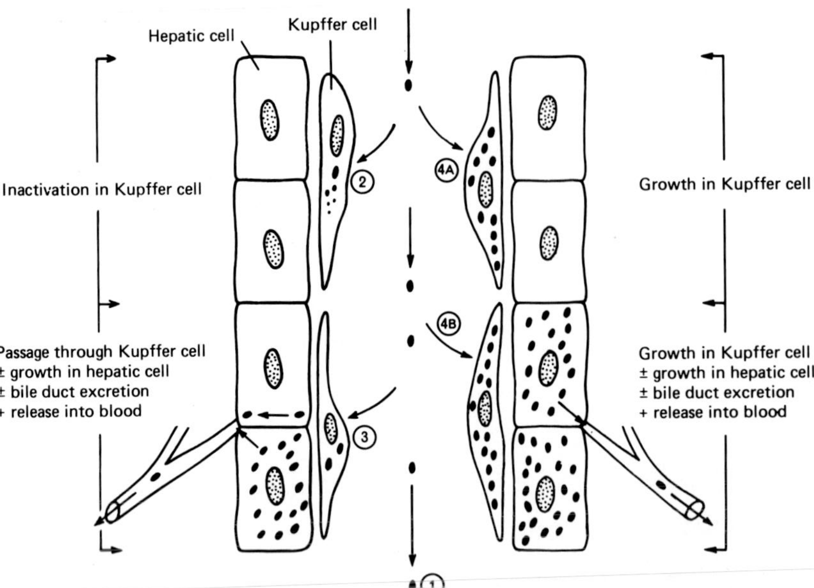

Fig. 6-5 Types of interactions between viruses and macrophages, exemplified by the Kupffer cells lining a sinusoid in the liver. (1) Macrophages may fail to phagocytose virions; for example, in Venezuelan equine encephalitis virus infection this is an important factor favoring prolonged viremia. (2) Virions may be phagocytosed and destroyed. Because the macrophage system is so efficient, viremia with such viruses can be maintained only if virions enter the blood as fast as they are removed. (3) Virions may be phagocytosed and then passively transferred to adjacent cells (hepatocytes in the liver). If, like Rift Valley fever or hepatitis B viruses, the virus replicates in these cells it can cause clinical hepatitis, and the virus produced in the liver can produce a high level of viremia. (4) Virions may be phagocytosed by macrophages and then replicate in them. With some viruses, such as lactate dehydrogenase virus in mice, only macrophages are infected (4A), and progeny virions enhance the viremia, which reaches an extremely high level. More commonly (4B), as in yellow fever, virus replicates in both macrophages and hepatic cells, producing severe hepatitis. (Modified from C. A. Mims and D. O. White, "Viral Pathogenesis and Immunology." Blackwell, Oxford, 1984.)

ing in the blood are continuously removed by macrophages, viremia can be maintained only if there is a continuing introduction of virus into the blood from infected tissues, or if there is impairment of the macrophages. Circulating leukocytes can themselves constitute a site for viral replication, but viremia is usually maintained by infection of the parenchymal cells of organs like the liver, spleen, lymph nodes, and bone marrow. In some infections the viremia is partly maintained by infection of endothelial cells. Striated and smooth muscle cells may be an important site of replication of some enteroviruses, togaviruses, and rhabdoviruses; virions are transferred to the blood via the lymph.

Invasion of Skin

As well as being a site of initial infection, the skin may be invaded via the bloodstream, producing erythema and often a generalized rash. The individual lesions in generalized rashes are described as macules, papules, vesicles, or pustules. A lasting local dilation of subpapillary dermal blood vessels produces a macule, which becomes a papule if there is also edema and infiltration of cells into the area. Primary involvement of the epidermis or separation of epidermis from dermis by fluid pressure results in vesiculation. Erosion or sloughing of the epithelium results in ulceration and scabbing, but prior to ulceration a vesicle may be converted to a pustule by polymorphonuclear cell infiltration. More severe involvement of the dermal vessels may lead to petechial or hemorrhagic rashes, although coagulation defects and thrombocytopenia may also be important in the genesis of such lesions.

Invasion of the Central Nervous System

Because of the critical physiologic importance of the central nervous system and its vulnerability to damage by any process that harms neurons directly or via increased intracranial pressure, viral invasion of the central nervous system is always a serious matter. Viruses can spread from the blood to the brain either (1) after localizing in blood vessels in meninges and choroid plexus, with invasion of the neurons then occurring from the cerebrospinal fluid, or (2) more directly after localizing in blood vessels of the brain and spinal cord (Fig. 6-6). Although the cerebral capillaries represent a morphological blood–brain barrier, most viruses that invade the central nervous system cross these vessels. Some viruses infect the vascular endothelial cells prior to infection of the cells of the brain parenchyma; others appear to be transported across the capillary walls without endothelial cell infection. Rarely, virus may be carried across capillary walls into the brain parenchyma via infected leukocytes. Subsequent spread in the central nervous system can take place via the cerebrospinal fluid or by sequential infection of neural cells.

Some enteroviruses which cause meningitis, rather than encephalitis, may traverse the blood–cerebrospinal fluid junction in the meninges or may grow in the epithelium of the choroid plexus. In such cases virions are found in the cerebrospinal fluid.

The other important route of infection of the central nervous system is via the peripheral nerves, as seen, for example, in rabies, varicella, and herpes simplex. Viruses may pass either (1) centripetally from the body surface to the sensory ganglia or (2) centrifugally from the ganglia to the skin, as in the reactivation of herpes simplex or varicella (as zoster). The rate of travel is quite slow, at up to 10 mm per hour. Herpesvirus capsids travel to the central nervous system in axon cytoplasm, and while doing so also sequentially infect the Schwann cells of the nerve sheath. Rabies virus travels to the central nervous system in axon cytoplasm without infecting cells of the nerve sheath. Following an animal bite, the virus enters the axon cytoplasm from motor axon terminals at neuromuscular junctions; less commonly, after exposure to rabies virus aerosols (as among speleologists in some parts of the world), it passes up the olfactory nerve.

Lytic infections of neurons, whether due to poliovirus, flaviviruses, togaviruses, or herpesviruses, are characterized by the three histologic hallmarks of encephalitis: neuronal necrosis, phagocytosis of neurons by phagocytic cells (neuronophagia), and perivascular infiltration of mononuclear cells (perivascular cuffing), the latter an immune response. The cause of clini-

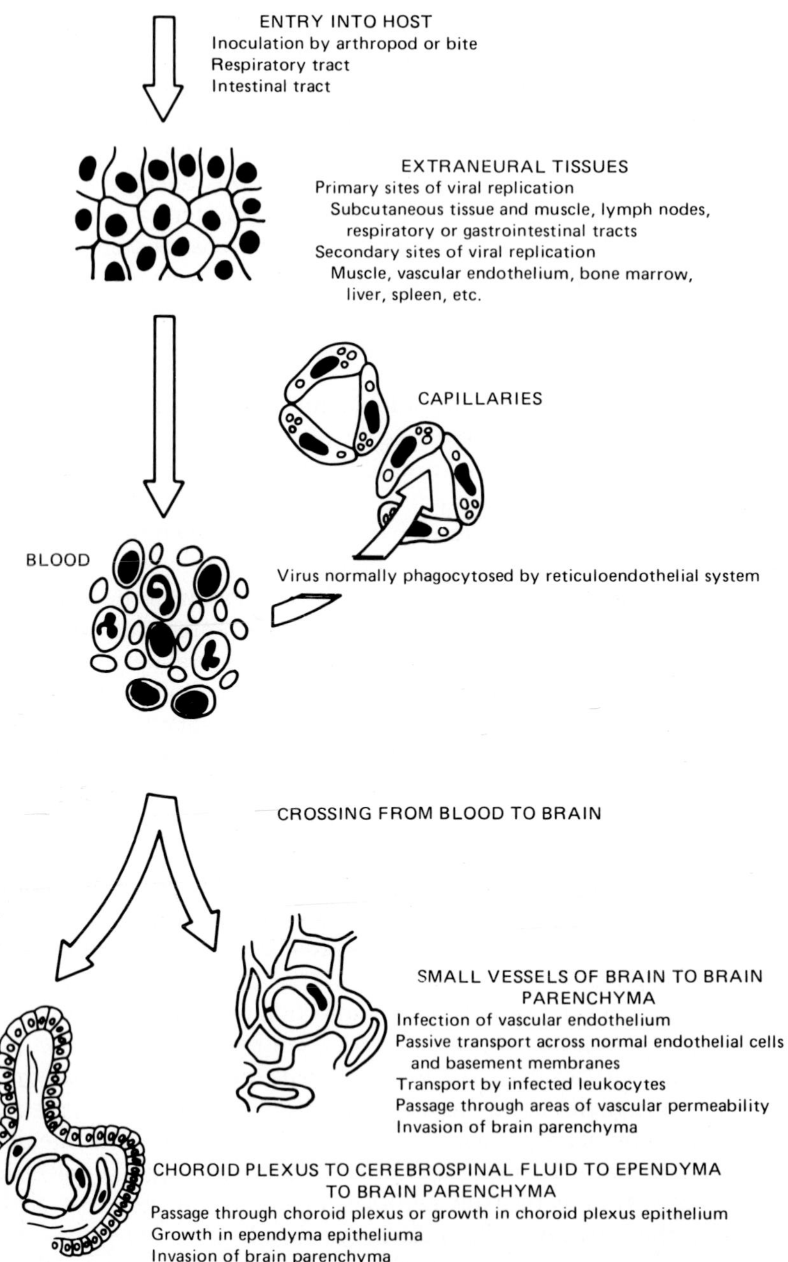

Fig. 6-6 Steps in the hematogenous spread of virus into the central nervous system. (From R. T. Johnson, "Viral Infections of the Nervous System." Raven, New York, 1982.)

cal neurologic signs in other central nervous system infections is more obscure. Rabies virus infection is noncytocidal; it evokes little of the inflammatory reaction or cell necrosis found in other encephalitides, yet it is lethal for most mammalian species. The extensive central nervous system infection of mice congenitally infected with the noncytolytic lymphocytic choriomeningitis virus, readily demonstrable by fluorescent antibody staining, has no recognizable deleterious effect. Characteristic pathologic changes are produced by some of the viruses or viruslike agents (prions) that cause slowly progressive diseases of the central nervous system, discussed in Chapter 10. In scrapie of sheep, for example, there is slow neuronal degeneration and vacuolization; in visna (a chronic lentivirus infection of sheep), changes in glial cell membranes lead to demyelination.

Postinfectious encephalitis is most commonly seen after measles (about 1 per 1000 cases), more rarely after rubella and varicella, and it used to be seen in 1–2 per 100,000 primary vaccinations against smallpox. The pathologic picture is predominantly demyelination without neuronal degeneration, changes unlike those produced by the direct action of viruses on the central nervous system. Allied with the failure to recover virus from the brain of fatal cases, this has led to the view that postinfectious encephalitis is probably an autoimmune disease.

Invasion of Other Organs

Almost any organ may be infected via the bloodstream with one or another kind of virus, but most viruses have well-defined organ and tissue tropisms. The clinical importance of infection of various organs and tissues depends in part on their role in the economy of the body. Thus invasion of the liver, causing severe hepatitis, as in yellow fever and infections with the hepatitis viruses, may be a life-threatening situation, with the additional possibility in the case of hepatitis B and C of the establishment of a chronic carrier state that may eventually result in hepatocellular carcinoma. The critical importance of such organs as the brain, heart, and lung is equally self-evident. Thus the most dangerous viral infections tend to be those that cause encephalitis, pneumonia, carditis, hepatitis, or hemorrhagic fever.

Infection of the testis or accessory sexual organs may lead to excretion of virus in the semen and the risk of transmission during sexual activities. In like manner, localization of virus in the salivary glands (e.g., in mumps), mammary glands, kidney tubules, and lungs leads to excretion in the saliva, milk, urine, and respiratory secretions. Viral infection of the pancreas appears to be one cause of early onset diabetes. Some togaviruses and coxsackieviruses infect muscle cells, while infection of synovial cells by rubella virus or Ross River virus produces arthritis.

Infection of the Fetus

Most viral infections of the mother have no harmful effect on the fetus, but some blood-borne viruses cross the placenta to reach the fetal circulation, sometimes after establishing foci of infection in the placenta. Severe cytolytic infections of the fetus cause fetal death and abortion, a result that was common in smallpox, for example. More important, paradoxically, are the teratogenic effects of less lethal viruses like rubella virus and cytomegalovirus (Table 6-4).

Since Gregg's initial observations in 1941, it has been recognized that maternal rubella contracted in the early months of pregnancy often leads to

Table 6-4
Some Congenital Viral Infections

Syndrome	Virus	Fetus
Fetal death and abortion	Variola virus (extinct)	Human
	Parvovirus B19	Human
Congenital defects	Cytomegalovirus	Human
	Rubella virus	Human
Inapparent, with lifelong carrier state	Lymphocytic choriomeningitis virus	Mouse
Inapparent, with integrated viral genome	Murine leukemia virus	Mouse
	Avian leukosis virus	Chicken

congenital abnormalities in the baby. A variety of abnormalities occur, of which the most severe are deafness, blindness, and congenital heart and brain defects. These defects may not be recognized until after the birth of an apparently healthy baby, or they may be associated with severe neonatal disease—hepatosplenomegaly, purpura, and jaundice—to comprise the "congenital rubella syndrome." Little is known of the pathogenesis of congenital abnormalities in rubella. Classic immunologic tolerance does not develop; children who have contracted rubella *in utero* display high titers of neutralizing antibodies throughout their lives, but there may be some diminution in cell-mediated immunity. Rubella virus is relatively noncytocidal; few inflammatory or necrotic changes are found in infected fetuses. The retarded growth and developmental abnormalities may be due to slowing of cell division leading to the reduced numbers of cells observed in many of their organs. Clones of persistently infected cells might be unaffected by the maternal antibody that develops during the first weeks after the maternal infection, even though such antibody could limit fetal viremia.

Cytomegalic inclusion disease of the newborn results from infection acquired congenitally from mothers suffering an inapparent cytomegalovirus infection during pregnancy. The important clinical features in neonates include hepatosplenomegaly, thrombocytopenic purpura, hepatitis and jaundice, microcephaly, and mental retardation.

Apart from infection of the fetus via the placenta, germ-line transmission of retroviruses, as integrated provirus, occurs commonly in many species of animals but rarely in humans. More important in human medicine because of its epidemiological implications (see Chapter 14) is transovarial transmission of arboviruses, notably bunyaviruses and some flaviviruses, in mosquitoes and ticks.

Virus Shedding

Shedding of infectious virions is crucial to the maintenance of infection in populations (see Chapter 14). Exit usually occurs from one of the body openings or surfaces that are involved in the entry of viruses. With localized infections the same body openings are involved in both entry and exit (see

Fig. 6-1); in generalized infections a greater variety of modes of shedding is recognized (see Fig. 6-4), and some viruses are shed from multiple sites, for example, hepatitis B virus, HIV, or cytomegalovirus in semen, cervical secretions, milk, and saliva. The amount of virus shed in an excretion or secretion is important in relation to transmission. Very low concentrations may be irrelevant unless very large volumes of infected material are transferred; on the other hand, some viruses occur in such high concentrations that a minute quantity of material, for example, less than 5 μl, can transmit infection.

Respiratory and Oropharyngeal Secretions

Many different viruses that cause localized disease of the respiratory tract are shed in mucus or saliva expelled from the respiratory tract during coughing, sneezing, and talking. Viruses are also shed from the respiratory tract in several systemic infections, such as measles, chickenpox, and rubella. A few viruses, for example, the herpesviruses, cytomegalovirus, and EB virus, are shed into the oral cavity, often from infected salivary glands, or from the lung or nasal mucosa, and are transmitted by salivary exchange in kissing and other social activities.

Feces

Enteric viruses are shed in the feces, and the more voluminous the fluid output the greater is the environmental contamination they cause. They are in general more resistant to inactivation by environmental conditions than are the enveloped respiratory viruses, especially when suspended in water, and such viruses can persist for some time outside the body.

Skin

The skin is an important source of virus in diseases in which transmission is by direct contact via small abrasions, for example, molluscum contagiosum, warts, and genital herpes. Several poxviruses may be spread from animals to humans, and sometimes from humans to animals, by contact with skin lesions, for example, the viruses of coxpox, vaccinia, orf, and pseudocowpox. Although skin lesions are produced in several generalized diseases, virus is not shed from the maculopapular skin lesions of measles, nor from the rashes associated with picornavirus, togavirus, or flavivirus infections. Herpesvirus infections, on the other hand, produce vesicular lesions in which virus is plentiful in the fluid of the lesions. Even here, however, virus shed in saliva and aerosols is much more important, as far as transmission is concerned, than that shed via the skin lesions.

Other Routes of Shedding

Urine

Viruria is lifelong in arenavirus infections of rodents and constitutes the principal mode of contamination of the environment by these viruses. However, although a number of human viruses, for example, mumps virus and cytomegaloviruses, replicate in tubular epithelial cells in the kidney and are shed in the urine, this is not a major mode of transmission from human to human.

Milk

Several species of viruses, for example, cytomegalovirus, are excreted in milk, which may serve as a route of transmission to the newborn infant.

Genital Secretions

Many viruses can be found in semen or vaginal secretions. Viruses shed from the genital tract depend on mucosal contact for successful transmission. They include HIV, herpes simplex type 2 virus, and some papillomaviruses. Several other viruses, such as other herpesviruses, hepatitis B and C viruses, and HTLV, are now known to be readily transmissible sexually.

Blood

Viremia is a most important vehicle, not only of viral spread within the host, but also for transmission between hosts. Blood is the usual source from which arthropods acquire viruses by inserting their proboscis into a capillary, and blood may also be the route of transfer of viruses to the ovum or fetus. Hepatitis B, C, and D viruses, HIV, and HTLV were once commonly spread by blood transfusion, and today are often transmitted between intravenous drug users via contaminated needles.

No Shedding

Many sites of viral replication are a "dead end" from the point of view of transmission to other hosts; for example, no virus is shed from the brain or other organs that do not communicate with the outside world. One might question how this could benefit the long-term survival of the virus in nature, but the stepwise augmentation of the titer of virus by replication in cells located in internal organs may be an important prerequisite for shedding from another site, or for infection of blood-sucking arthropods. In animals such as mice and chickens, many retroviruses are not shed, but are transmitted directly in the germ plasm, or by infection of the egg or developing embryo.

Further Reading

Johnson, R. T. (1982). "Viral Infections of the Nervous System." Raven, New York.

Mims, C. A. (1989). The pathogenetic basis of viral tropism. *Am. J. Pathol.* **135,** 447.

Mims, C. A., and White, D. O. (1984). "Viral Pathogenesis and Immunology." Blackwell, Oxford.

Notkins, A. L., and Oldstone, M. B. A., eds. (1984, 1986, 1989). "Concepts in Viral Pathogenesis", Vols. 1–3. Springer-Verlag, New York.

Smith, H., and Sweet, C. (1990). The pathogenicity of viruses. *In* "Topley and Wilson's Principles of Bacteriology, Virology and Immunology" (L. H. Collier and M. C. Timbury, eds.), 8th Ed., Vol. 4, p. 105. Williams & Wilkins, Baltimore, Maryland.

Tyrrell, D. A. J. (1983). How do viruses invade mucous surfaces? *Philos. Trans. R. Soc. London, B* **303,** 75.

Chapter 7

Determinants of Viral Virulence and Host Resistance

Infection is not synonymous with disease. Many infections are *subclinical* (synonyms, *asymptomatic* or *inapparent*), while others produce disease of varying degrees of severity. The outcome of any virus–host encounter is the product of the virulence of the infecting virus, on the one hand, and the susceptibility of the host, on the other. Virus strains may differ greatly in their virulence, that is, their capacity to produce disease or death. Conversely, within a human population infected with a particular strain of virus there are often striking differences in the levels of resistance of different individuals.

The determinants of viral virulence are usually multigenic, and the determinants of host susceptibility/resistance are also multifactorial. Within a susceptible species, the resistance of individual animals varies not only with the genetic constitution of the host (which may affect, among other things, their capacity to mount a rapid immune response), but also with age, nutritional status, stress, and many other factors. Together, these genetic and physiologic factors determine what is called the "nonspecific," "natural," or "innate" resistance of the host, in contrast to acquired, immunologically specific resistance to reinfection that results from the operation of the immune response, as described in the next chapter.

Viral Virulence and Host Resistance

The word *virulence* is used in this book as a measure of *pathogenicity*, that is, the ability of a virus to cause disease in the infected host animal, rather than as a measure of infectiousness or transmissibility, a different property that is discussed in Chapter 14. The virulence of a particular strain of virus administered by a particular route to a particular strain of a laboratory animal can be measured by determining the dose of virus required to cause death in 50% of animals (median lethal dose, LD_{50}). For example, in the susceptible BALB/c strain of mouse, the LD_{50} of a virulent strain of ectromelia virus was 5 virions, compared with 5000 for a moderately attenuated strain and about 1 million for a highly attenuated strain.

In nature, virulence may vary over a wide range, from strains of viruses that almost always cause inapparent infections to others usually causing disease, and finally those that commonly cause death. Meaningful comparison of the virulence of viruses requires that other factors such as the infecting dose and host resistance are equal; these conditions are impossible to meet in heterogeneous outbred populations such as humans except by randomization of large numbers of subjects, hence are usually assayed in inbred strains of mice, which is feasible only for those human viruses that grow in mice.

Susceptibility to infection or disease, or its reciprocal, resistance, can be measured in experimental animals by keeping the strain of virus constant and determining the ratio of the dose that causes infection in 50% of individuals (median infectious dose, ID_{50}) to the LD_{50}. Thus with a virulent strain of ectromelia virus tested in highly susceptible BALB/c mice, the ID_{50} was 1–2 virions and the LD_{50} about 5, whereas for the resistant C57BL strain the ID_{50} was the same but the LD_{50} was 1 million. The severity of an infection therefore depends on the interplay between the virulence of the virus and the resistance of the host. One can regard an acute infection as a race between the ability of the virus to replicate, spread in the body, and cause disease, versus the ability of the host to curtail and control these events. An individual with a high degree of innate resistance may not be seriously affected even by a virulent strain of virus, whereas a relatively avirulent virus can endanger the life of an immunodeficient host.

Variability in the response of individuals to infection is regularly observed during epidemics. For example, during an outbreak of influenza some people (mainly old persons or those with preexisting respiratory disease) may die, while others will suffer a brief attack but quickly recover. The great majority of arbovirus and enterovirus infections are subclinical; for every individual who develops encephalitis during an arbovirus epidemic, or paralytic poliomyelitis during a poliovirus outbreak, dozens more will have no symptoms, the only evidence of infection being a sharp rise in antibody titer. The dose of infecting virus may play a part in determining these differences, but genetic and physiologic factors in the host are probably more important. A unique opportunity to observe the wide variation in innate resistance of healthy young adults of the same age and sex, receiving an identical dose of virus by the same route, arose in 1942 when more than 45,000 U.S. military personnel were inoculated with yellow fever virus vaccine which, in retrospect, turned out to have been contaminated with live hepatitis B virus. Fortunately only 2% developed hepatitis, with incubation periods varying from 10 to 20 weeks, and the disease varied from mild (in 1%) to severe (in 0.1%).

Genetic Determinants of Viral Virulence

Unraveling the genetic basis of virulence has long been one of the major goals of animal virology, and also one of the most difficult to achieve, since many genes, both viral and host, are involved in the outcome of each infection. With advances in molecular genetics it has been possible to dissect the problem in a more precise way, using techniques such as recombinant DNA technology, genetic reassortment, site-specific mutagenesis, and transgenic animals.

Of necessity, most experimental work has been carried out with laboratory animals. The most detailed studies have been those conducted with retroviruses and oncogenic DNA viruses to determine the genetic basis of cellular transformation and oncogenicity (see Chapter 11). Experiments with herpesviruses are beginning to reveal the genetic basis of latency with these viruses (see Chapter 10). With viruses causing acute infections, those with segmented genomes have provided a more easily manipulated experimental model, since each segment of the genome of influenza viruses and reoviruses, for example, is in most cases equivalent to one gene, and reassortants can be readily obtained. Study of a number of reassortants involving different genome segments enables the functions that relate to virulence to be assigned to particular genes. Using a different approach, a detailed understanding of the basis of virulence at the molecular level has been obtained with polioviruses, where it has been possible to compare the sequences of the genomes of wild-type viruses with those of avirulent vaccine strains and then with virulent revertants that arise from time to time in vaccinees or their contacts. Investigations with vaccinia virus have revealed the complexity of the armamentarium of gene products directed against various components of innate or acquired host resistance. Some examples are outlined below to illustrate how molecular biology has opened up the new field of *molecular pathogenesis.*

Most viral genes encode proteins that are essential for viral replication, notably those required for viral entry into the principal host cells, replication of the viral genome, and assembly/release of new virions (see Table 3-3). Additional genes have evolved to maximize the yield of virions by down-regulating expression of cellular genes and up-regulating expression of particular viral genes (and sometimes of particular cellular genes) at appropriate stages of the replication cycle. Recently we have come to recognize a third category of viral genes, which are irrelevant to intracellular replication of virus per se, but which optimize the spread and the ultimate titer of virus in the body as a whole, principally by suppressing various arms of the immune response. These gene products are called *virokines.*

Tropism

All viruses exhibit some degree of host and tissue specificity (*tropism*). The first requirement for infection of a cell is a correspondence between viral attachment molecules (ligands) and cellular receptors (see Table 3-2 and below). Following penetration and uncoating, viral replication may be dependent on the activity of regulatory elements in the viral genome: enhancers, promoters, and transcriptional activators, as well as factors that govern the permissiveness of the cell for complete viral replication (see Chapter 3). Some of these regulatory elements are restricted to certain types of cells or tissues and function as additional important determinants of viral tropism.

Viral Enhancers, Promoters, and Transcription Factors

Enhancers are a class of gene activators which increase the efficiency of transcription of genes under the control of a particular *promoter.* Unlike promoters, which are situated immediately upstream from the transcription site for a particular cluster of genes and are typically about 100 base pairs in length, enhancers can be situated either upstream or downstream from the RNA start site, can exist in either orientation in the DNA, are typically 50–100 base pairs in length, and are often repeated in tandem. These nucleotide sequences often contain, within about 100 nucleotides, a number of motifs representing binding sites for different proteins. Thus, several cellular or viral *site-specific DNA-binding proteins* (*transcription factors,* sometimes known as *transactivating proteins*) may cluster together along a given enhancer, thereby bringing about, in a fashion that is not yet entirely clear, augmentation of binding of DNA-dependent RNA polymerase II to the promoter. Alternatively, viral or cellular proteins can interact not with the enhancer sequence itself, but with an inactive transcription factor, thus activating it. Since many of the transcription factors affecting particular enhancer sequences in viral genomes as well as cellular genomes are restricted to particular cells, tissues, or host species, they can determine the tropism of viruses.

Enhancer regions have been defined in the genomes of retroviruses, several herpesviruses, and hepatitis B virus, and in all cases they appear to influence the tropism of the relevant viruses by regulating the expression of viral genes in specific types of cells. The DNA of a dermatotropic type of papillomavirus contains an enhancer that is specifically active only in keratinocytes, indeed only in a subset of these cells. Further evidence for the tissue specificity of papovavirus enhancers comes from studies of transgenic mice containing the early gene region of JC polyomavirus, a common human virus which very occasionally causes a neurologic disease, progressive multifocal leukoencephalopathy; the offspring of transgenic mice bearing early genes from the JC virus develop a neurologic disease characterized pathologically by dysfunction of myelin-producing oligodendroglia, which mimics the naturally occurring disease.

Virokines

Virokines comprise a major new class of virus-coded proteins that are not required for viral replication *in vitro,* but influence viral pathogenesis *in vivo* by sabotaging the body's innate resistance or adaptive immune response to infection. Many virokines closely mimic normal cellular molecules that are crucial to the immune response, such as cytokines and cytokine receptors. For instance, there are three virokines encoded by poxviruses which closely resemble soluble forms (lacking the membrane anchor) of the receptors for the cytokines tumor necrosis factor (TNF), interferon γ (IFN-γ), and interleukin-1 (IL-1). So far, virokines seem to be confined to the large DNA viruses—poxviruses, herpesviruses, adenoviruses—and to the retroviruses, but the number of genes involved is considerable. More than a quarter of all the genes of vaccinia virus can be deleted without impeding its replication in cell cultures although, as expected, these deletion mutants are attenuated *in vivo*.

Currently only a minority of the genes for virokines have been identified, but their diversity already indicates that we may have only scratched the surface in this exciting new area of molecular pathogenesis. Table 7-1 lists some virokines, which can be grouped as follows: (1) inhibitors of T-cell cytotoxicity, by binding nascent class I MHC protein; (2) inhibitors of cytokines, such as IL-1, IFN-γ, and TNF; (3) inhibitors of the complement cascade; (4) inhibitors of antibody-mediated cytolysis; and (5) cytokine mimics, for example, IL-10.

Models of Viral Virulence

Influenza Virus

Experiments with influenza virus in mice have confirmed the view advanced by the pioneer in the field, F. M. Burnet, that virulence is multigenic,

Table 7-1
Virokines

Virus	Virokine	Host function inhibited	Mechanism
Poxviridae			
Vaccinia virus	B15R	Cytokine	IL-1β receptor homolog binds IL-1β
Vaccinia virus	VCP	Complement	Binds C4b, inhibiting classical pathway
Vaccinia virus	38K serpin	Blood coagulation	Serpin inhibits serine protease
Vaccinia virus	VGF	Nil?	Growth factor stimulates proliferation of surrounding uninfected cells
Cowpox virus	crmA	Cytokine	Serpin prevents protease cleavage of IL-1β
Myxoma virus	?	Cytokine	IFN-γ receptor homolog binds IFN-γ
Fibroma virus	T2	Cytokine	TNF receptor homolog binds TNF
Herpesviridae			
Herpes simplex virus	gE-gI	Antibody-mediated cytolysis	Binds Fc of IgG, preventing complement-mediated cytolysis
Herpes simplex virus	C-1	Complement	Binds C3b, inhibiting alternative and classical pathways
Cytomegalovirus	UL18	Cytotoxic T cells	Homolog of class I MHC binds β-microglobulin, preventing translocation to the cell surface
EB virus	BCRF1	Cytokine, cytotoxic T cells	IL-10 homolog inhibits cytokine synthesis, suppressing Tc cells and stimulating B cells
Adenoviridae			
Adenovirus	E3-19K	Cytoxic T cells	Binds to class I MHC, preventing translocation to cell surface
Adenovirus	Three early proteins	Cytokine	Protects infected cells from cytolysis by TNF
Retroviridae			
HIV	p15E	Protein kinase C	Inhibits signal transduction

that is, that in general, and for influenza virus in particular, no one gene determines virulence. Recent studies show that one essential requirement for virulence in the chicken and neurovirulence in the mouse is that the hemagglutinin protein (HA) must be cleaved. In nonpathogenic strains of avian influenza virus, the HA1 and HA2 polypeptide chains of the hemagglutinin are linked by a single arginine, whereas in virulent strains the linker is a sequence of several basic amino acids and is more readily cleaved by the available cellular protease(s). Virulent strains thus contain HA which is activated (cleaved) in a wide spectrum of different types of cells; hence, they can replicate and spread throughout the host. Nonpathogenic strains, on the other hand, soon reach a barrier of cells which lack appropriate HA-cleaving enzymes.

Perhaps the most striking demonstration of how a minor single nucleotide change in the HA gene can make all the difference between relative avirulence and high virulence was the finding that a single amino acid substitution in the HA protein of influenza A, subtype H5, in 1983 led to a devastating outbreak of fowl plague on poultry farms around Pennsylvania that caused losses amounting to over $60,000,000. The point mutation abolished a glycosylation site, thus exposing to proteolysis the cleavage site previously concealed by an oligosaccharide side chain. However, virulence is only partly explained by this factor, for when reassortants were made between highly virulent and avirulent strains of avian influenza virus, exchange of any one of the eight RNA segments modified the virulence. Studies with a variety of reassortants showed that, for each reassortant, an optimal combination of genes (the optimal "gene constellation") was selected which favored survival in nature and determined virulence.

Reovirus

Detailed analyses of the virulence of reoviruses for mice have been carried out with reassortant viruses. The fact that the protein product of each of the ten genome segments has been isolated and characterized has allowed determination of their functions and effects on virulence (Table 7-2).

Four genes (S1, M2, L2, and S4) encode the four polypeptides that are found on the outer capsid (Fig. 7-1). Each plays a role in determining viru-

Table 7-2
Genetic Analysis of Virulence of Reoviruses; Based on Studies with Mutants and Reassortants

Gene	Protein	Properties	Significance in infections of mice
S1	$\sigma 1$	Reovirus 3: binds to neurons	Neural spread, encephalitis
		Reovirus 1: binds to ependymal cells	Hydrocephalus
		Binds to T and B cells	Infection of lymphocytes
		Recognized by T cells and neutralizing antibodies	Protective antigen
		Reovirus 1: binds to enterocytes Reovirus 3: does not bind to enterocytes	Reovirus 1 more pathogenic than reovirus 3 after oral infection
M2	$\mu 1C$	Reovirus 3: inactivated by intestinal proteases	Nonvirulent by mouth
		Reovirus 1: not inactivated by intestinal proteases	Virulent by mouth
L2	$\lambda 2$	Role in RNA transcription	Affects level of replication in gut, hence role in transmissibility
S4	$\sigma 3$	Inhibits RNA and protein synthesis	Role in persistent infection

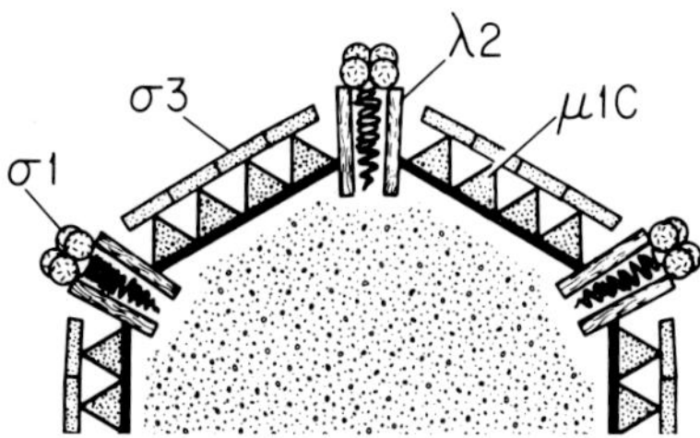

Fig. 7-1 Schematic diagram of part of the reovirus outer capsid, showing the location of the polypeptides that play a major role in virulence. The σ1 protein is located at the vertices of the icosahedron and consists of two components: a globular dimer at the surface, which is responsible for hemagglutination and cell attachment, and an α-helical region which anchors the hemagglutinin by interaction with the λ2 spike protein. The polypeptides μ1C and σ3 are associated with one another, in the ratio of one molecule of μ1C to two molecules of σ3, on the surfaces of the icosahedral capsid. [Modified from L. A. Schiff and B. N. Fields, *in* "Fields Virology" (B. N. Fields, D. M. Knipe, R. M. Chanock, M. S. Hirsch, J. L. Melnick, T. P. Monath, and B. Roizman, eds.), 2nd Ed., p. 1277. Raven, New York, 1990.]

lence. Gene S1 specifies the hemagglutinin (protein σ1), which is located on the vertices of the icosahedron and is responsible for cellular and tissue tropism. With reovirus 3, but not reovirus 1, the σ1 protein is responsible for binding to neurons, whose sequential infection leads to fatal encephalitis, and it is also responsible for viral spread from footpad to the spinal cord via peripheral nerves. Gene M2 specifies polypeptide μ1C, which determines sensitivity of the virion to chymotrypsin and hence affects the capacity of the virus to grow in the intestine. Thus μ1C of reovirus 3 is protease sensitive, and reovirus 3 is avirulent by mouth; μ1C of reovirus 1 is protease resistant, and the virus is infectious by mouth. A reassortant with the M2 gene from reovirus 1 and the S1 gene from reovirus 3 was infectious orally and caused fatal encephalitis.

Gene L2 specifies the λ2 spike protein, which is primarily a core protein but surrounds σ1; it is a guanylyltransferase and plays a role in RNA transcription. The ability of reovirus to reach high titers in the gut and be efficiently transmitted via feces is associated with the λ2 spike protein. Gene S4 specifies polypeptide σ3, which inhibits cellular protein and RNA synthesis in infected cells; mutations in the S4 gene play a role in establishing persistent infection in cultured cells.

Poliovirus

The genome of poliovirus is a single molecule of ssRNA, and the entire nucleotide sequence of several strains has been determined. Because vaccine strains are required to be infectious by mouth but not cause central nervous system disease, the virulence of polioviruses is narrowly defined as the ability to replicate and cause lesions after introduction into the central nervous system of primates.

Comparison of the nucleotide sequence of the poliovirus type 1 vaccine strain with that of the virulent parental strain (Mahoney) from which Sabin derived it shows that there are 55 substitutions in the 7441 bases, scattered along the entire genome. Twenty-one of these substitutions resulted in amino acid changes, involving several of the viral proteins. Studies with recombinants showed that mutations in the VP1 capsid protein gene and other genes, as well as in the 5′ noncoding region, contribute to neurovirulence in primates. On the other hand, with poliovirus type 3 vaccine strain, which reverts to virulence with a frequency at least 10 times that of type 1 vaccine, there are only 10 nucleotides different from those of the parental strain, of which only 3 result in amino acid changes, 1 in each of three viral structural proteins.

The two most important mutations with respect to neurovirulence are a substitution of a single nucleotide at position 472 in the 5′ noncoding region and an amino acid substitution in VP3 which represents a *ts* mutation, destabilizing the capsid. The importance of the 5′ noncoding region in neurovirulence is intriguing; since position 472 also tends to be the first reversion to occur when the vaccine grows in the human gut, it is possible that the wild-type nucleotide at this position allows optimal intestinal replication and that neurovirulence is a consequence of the resultant higher titer of virus in the bloodstream.

Vaccinia and Cowpox Viruses

Poxviruses are much larger and more complex than any viruses yet discussed, and the genetic control of virulence is correspondingly more complex. Interest in the virulence of poxviruses has been stimulated by proposals to use vaccinia virus as a vector for human vaccines. Recent work has shown that many of the large number of genes encoding virokines that influence virulence have effects on the defense mechanisms of the host (see Table 7-1).

Deletion analysis of vaccinia virus has shown that 56 of a total of 198 open reading frames are not required for replication in cultured cells. Many of these are located at the ends of the genome and affect virulence in animals. The first poxvirus gene to be shown to affect virulence was the gene for vaccinia virus thymidine kinase (TK), which is not required for replication in cultured cells; however, TK^- mutants are much less virulent than the wild type in animals. The envelope of vaccinia virus, which contains at least seven virus-coded polypeptides, is not essential for infection of either cultured cells or animals, but enveloped forms are more virulent because they spread around the body more effectively.

Cowpox virus replicates in a wide range of cells, but three open reading frames have been identified which affect its replication in a nonpermissive cell, namely, Chinese hamster ovary (CHO) cells. In animals these genes are probably important in determining host range and tissue tropisms. The vaccinia growth factor (VGF) has homologies with epidermal growth factor and appears to induce localized hyperplasia around foci of infection, thus providing additional metabolically active cells for viral replication. A 35K protein, which has been called the vaccinia virus complement-binding protein, blocks the classical complement pathway and binds to C4b. Then there are at least three distinct serine protease inhibitor (serpin) genes in vaccinia virus. Serpins exert a control over a number of critical events associated with connective tissue turnover, coagulation, complement activation, and inflammatory reactions. The presence of the 38K gene of cowpox virus, which encodes a serpinlike protein, is associated with the bright red pock produced on the chorioallantoic membrane and higher virulence for a number of experimental animals. It has been suggested that the serpin delays the onset or decreases the magnitude of the inflammatory response.

Genetic Determinants of Host Resistance

Genetic differences in susceptibility are most obvious when different animal species are compared. Common viral infections often tend to be less pathogenic in the natural host species than in certain exotic or introduced species. For instance, myxoma virus produces a small benign fibroma in its natural host, *Sylvilagus brasiliensis,* but an almost invariably fatal generalized infection in European rabbits. Likewise, most zoonoses (diseases caused by viruses of wild or domesticated animals that are occasionally transmitted to humans, such as arenavirus, filovirus, and various arbovirus infections) cause more severe diseases in humans than in the natural hosts.

Accurate genetic data on resistance to infection are difficult to obtain in humans, because genetic, physiologic, and environmental differences are generally confounded. With inbred strains of mice, however, it has been possible to study the genetics of resistance to viral infection in some detail. Having identified a susceptible (S) and a resistant (R) strain, the LD_{50} assay is repeated in $(S \times R)F_1$ and F_2 backcrossed mice to determine whether a single gene is responsible, and whether susceptibility or resistance is dominant. Then recombinant or congenic strains can be constructed, with a common genetic background differing only in the gene in question. In this way, susceptibility or resistance of an inbred strain of mouse has sometimes been assigned to a single gene, although later work often reveals one or more additional genes which may influence susceptibility/resistance in quite different ways. For example, the response of mice to murine leukemia viruses is influenced by over a dozen known genes, whereas the response to murine cytomegalovirus is influenced by one gene that maps to the MHC region and a second gene that is not MHC-linked but determines the titer of virus found in the spleen.

Immune Response Genes

Immunologic responses are determined and controlled by a large number of different genes usually located in clusters, but on different chromosomes. A variety of inherited defects in these genes is recognized. These range from the absence of a single immunoglobulin class such as IgA, the commonest primary defect in humans, to agammaglobulinemia, where the number of functional B cells is greatly reduced or absent. Even more serious, for many viral diseases, are congenital deficiencies in cell-mediated immunity, as in thymic dysplasia and the Swiss syndrome. These "experiments of nature" are instructive in teaching us which arms of the immune response are crucial in infections with various viruses. All the congenital immunodeficiency syndromes listed above profoundly impede recovery from numerous viral infections. However, there is one important example of a rare X-linked recessive immunodeficiency syndrome characterized by failure to mount adequate T-cell-mediated immune responses to one particular virus, EB virus; the result is inexorable proliferation of B lymphocytes, terminating in death.

Immune responsiveness to particular antigens differs greatly from one person to another, being under the control of specific *immune response* (*Ir*) genes. There are many of these antigen-specific immunoregulatory genes, most of them situated in the *major histocompatibility complex* (*MHC*), known as the I region of the H2 complex in the mouse and the D/DR region of the HLA complex in humans. These genes encode the class II MHC proteins on the surface of cells of the macrophage lineage and B lymphocytes that are needed to present immunogenic peptides to the receptors carried by helper T cells which help B cells to make antibody (see Chapter 8). Nonresponsiveness of a given host to a given virus is attributable to lack of any class II MHC molecules capable of binding any of the important epitopes of that virus with satisfactory affinity. Even when a viral epitope is satisfactorily presented by antigen-presenting cells, there will be no response if that host does not contain a clone of T cells with a receptor capable of recognizing that particular epitope. Such "holes" in the T-cell receptor repertoire occur with both MHC class II-restricted $CD4^+$ (helper) T cells and MHC class I-restricted $CD8^+$ (cytotoxic) T cells.

Susceptibility of mice to infection with some viruses, for example, cytomegaloviruses, retroviruses, and lymphocytic choriomeningitis virus, has been shown to be linked to particular MHC genotypes. Many reports of such associations in humans have appeared in the literature, for example, affecting susceptibility/resistance to AIDS or seroconversion following hepatitis B vaccination, but the correlations are generally not as dramatic as in the case of, for example, the association of ankylosing spondylitis with HLA B27.

Other Genes Influencing Susceptibility/Resistance to Individual Viruses

Most genes affecting susceptibility/resistance to particular viruses do not map to the MHC region of the genome and presumably operate via any of a wide range of nonimmunologic mechanisms. For instance, there are now precedents for believing that many of the dozens of cellular genes turned on by interferons encode proteins that inhibit the replication of particular viruses only, such as the influenza virus-specific Mx protein (see Chapter 5). In addition, the host cell must carry genes encoding a considerable number of proteins required for the replication of specific viruses; the most obvious are the various receptors needed for entry of different viruses. The capacity to grow in macrophages, rather than being destroyed by them, is a major determinant of success for viruses; a single host gene may determine the permissiveness or otherwise of macrophages.

Macrophages

Macrophages play a central role as determinants of resistance, in part because of their role in the immune response and in part because of their intrinsic susceptibility to infection, which is independent of antibodies or the action of lymphokines, although it is often influenced by those immune factors. Many viruses replicate in macrophages, and in some diseases they appear to be the only susceptible cells. Often viral replication in macrophages is abortive, but sometimes their apparent insusceptibility is due to endogenous interferon production.

Many years ago it was shown that the susceptibility of different strains of mice to mouse hepatitis virus was associated with macrophage susceptibility, *in vivo* and *in vitro*. Other physiologic changes were also seen. Peripheral blood mononuclear cells from susceptible strains of mice showed rapid and striking production of procoagulant activity following infection with the virus that was not seen in cells from resistant strains. Production of interleukin inhibitors and other immunosuppressive factors by macrophages is important in pathogenesis, and tumor necrosis factor from infected macrophages has both antiviral and pathogenic effects.

Macrophages/monocytes play a major role in the pathogenesis of HIV infections. They appear to be the first cells infected, they are the major reservoirs of the virus during all stages of the infection, and they are probably the major vector for the spread of infection to different tissues within the patient and, because they are the predominant cell in most bodily fluids, including semen and vaginal secretions, between individuals.

Cellular Receptors

The presence or absence of receptors on the plasma membrane is a fundamental determinant of susceptibility (see Table 3-2). In the late 1940s it was shown that pretreatment of mice with neuraminidase (sialidase) intranasally conferred substantial protection against intranasal challenge with influenza virus, albeit transient because of rapid regeneration of the sialic acid receptors. Some 40 years later the three-dimensional structure of the influenza HA–receptor complex was determined by X-ray crystallography—the first ligand–receptor combination to be resolved. Indeed, although the virus may be able to utilize more than one type of cellular glycoprotein or glycolipid as a receptor, the conserved pocket near the tip of the HA molecule that constitutes the ligand engages mainly the terminal sialic acid of an oligosaccharide side chain protruding from the receptor. Moreover, although sialic acid can occur in numerous combinations with other sugars, high-affinity binding of HA occurs only to sialic acid (SA) in $\alpha(2,6)$ linkage to galactose (Gal). Passage of the virus in the presence of naturally occurring soluble sialyl glycoproteins which are inhibitory for the virus selects for a mutant with a single amino acid substitution at residue 226 of HA which displays a greater affinity for SAα2,3 Gal than for SAα2,6 Gal and has acquired an altered host range.

Polioviruses provide an example of the importance of cellular receptors at the host species level. These viruses ordinarily infect only primates; mice and other nonprimates are insusceptible because their cells lack the appropriate receptor (a novel member of the immunoglobulin superfamily). Poliovirus RNA, when introduced artificially into mouse cells *in vivo* or in culture, can undergo a single cycle of replication, but since progeny virions encounter mouse cells lacking receptors, they are unable to initiate a second cycle of replication. Transfection of the gene for the human poliovirus receptor into mouse L cells renders them susceptible to infection with poliovirus. Further, transgenic mice expressing the human receptor in all cells are susceptible, but the fact that viral replication is restricted to neurons and muscle reminds us that whereas the presence of an appropriate receptor is necessary for infection, it is not always sufficient—the target cell may be nonpermissive for other reasons.

Theoretically, any normal cellular membrane component could serve as receptor for some virus or another. The more conserved or ubiquitous the receptor, the wider the host range of the virus that exploits it; for example, rabies virus, which uses sialylated gangliosides as well as the acetylcholine receptor, has a very wide host range.

Now that cDNA encoding viral receptors can be cloned in eukaryotic hosts or in transgenic mice, rapid progress can be expected in this field. Doubtless, alternative receptors will be found for many viruses. These may turn out to be situated in different types of cells, organs, or host species, so accounting for the broad tropism of certain viruses. Alternatively, two or more distinct receptors may serve as optional or even sequential receptors on the same host cell. This has been reported for at least two important viruses, human immunodeficiency virus (HIV) and herpes simplex virus (HSV). HIV uses CD4 receptors preferentially but can also infect types of cells on which CD4 cannot be demonstrated; HSV appears to bind initially to a low-affinity receptor, heparan sulfate, which is charged and protrudes conspicuously from the plasma membrane, then it moves to a second, high-affinity receptor. It has been postulated that second receptors, which could be glycolipids, are closer to the lipid bilayer, facilitating viral entry by fusion.

Physiologic Factors Affecting Resistance

A great variety of physiologic factors affect resistance, the most important being the immune response, which is described in detail in Chapter 8. Little is known of many of the nonspecific factors in resistance, but age, nutrition, certain hormones, and cell differentiation play roles in a variety of viral diseases.

Age

Viral infections tend to be most serious at the extremes of life. The high susceptibility of newborns to a number of viral infections is of considerable concern to pediatricians. It can also be exploited for the laboratory diagnosis of viral diseases. Thus the coxsackieviruses were discovered by the inoculation of suckling mice with fecal extracts, and infant mice are still useful for the isolation of arboviruses.

In laboratory animals the first few weeks of life are a period of very rapid physiologic change. For example, during this time mice pass from a stage of immunologic nonreactivity (to many antigens) to immunologic maturity. This change profoundly affects their reaction to viruses like lymphocytic choriomeningitis virus, which induces a persistent tolerated infection when inoculated into newborn mice, but an immune response in mice infected when over 1 week old. In humans, the umbrella of transplacentally acquired maternal antibody protects the infant against many viruses for the first few months of life. The importance of this cover is shown by the fact that viruses such as herpes simplex or varicella virus can cause lethal disease in infants who are born without maternal antibodies. Rotaviruses and respiratory syncytial viruses are the most striking examples of viruses that cause severe disease only in infants in the first year or so of life.

Older infants and children tend to suffer less severely from many virus infections than do premature infants or adults. For example, varicella virus, usually the cause of an uncomplicated disease in children, may produce severe pneumonia in adults, and mumps is complicated by orchitis much more often in adults than in children; poliovirus, hepatitis virus, and EB virus infections are all much more serious in adults.

Malnutrition

Malnutrition can interfere with any of the mechanisms that act as barriers to the replication or progress of viruses through the body. It has been repeatedly demonstrated that severe nutritional deficiencies will interfere with the generation of antibody and cell-mediated immune responses, with the activity of phagocytes, and with the integrity of skin and mucous membranes. However, often it is impossible to disentangle adverse nutritional effects from other factors found in deprived communities. Moreover, just as malnutrition can exacerbate viral infections, so viral infections can exacerbate malnutrition, especially if severe diarrhea is a feature, thus creating a vicious cycle.

Children with protein deficiency of the kind found in many parts of Africa are highly susceptible to measles. All the epithelial manifestations of the disease are more severe, and secondary bacterial infections cause life-threatening disease of the lower respiratory tract as well as otitis media, conjunctivitis, and sinusitis. The skin rash may be associated with numerous hemorrhages, and there may be extensive intestinal involvement with severe diarrhea, which exacerbates the nutritional deficiency. The case–fatality rate is commonly 10% and may approach 50% during severe famines.

Hormones and Pregnancy

There are few striking differences in the susceptibility of males and females to viral infections, except in the obvious instances of viruses with a predilection for tissues such as testis, ovaries, or mammary glands. Pregnancy significantly increases the likelihood of severe disease following infection with certain viruses, an effect that was very pronounced in smallpox and is also seen in infections with hepatitis viruses, especially hepatitis E virus. Latent herpesvirus infections are often reactivated during pregnancy, contaminating the birth canal and leading to infection of the newborn.

The therapeutic use of corticosteroids exacerbates many viral infections and is contraindicated, notably in herpesvirus infections. The precise mechanism is not understood, but corticosteroids reduce inflammatory and immune responses and depress interferon synthesis. It is also clear that adequate levels of these hormones are vital for the maintenance of normal resistance to infection.

Fever

Almost all viral infections are accompanied by fever. The principal mediator of the febrile response appears to be the cytokine interleukin-1 (previously known as endogenous pyrogen). Interleukin-1 is produced in macrophages and is induced during immune responses. It is found in inflammatory exudates and acts on the temperature-regulating center in the anterior hypothalamus. *In vitro* experiments have shown that antibody production and T-cell proliferation induced by interleukin-1 are greatly increased when cells are cultured at 39°C rather than at 37°C.

Fever profoundly disturbs bodily functions. The increased metabolic rate augments the metabolic activity of phagocytic cells and the rate at which inflammatory responses are induced, both of which might be expected to exert antiviral effects. Exposing rabbits to high environmental temperatures greatly diminishes the severity of myxomatosis, usually a lethal disease; on the other hand, lowered environmental temperature increases the severity of the disease produced by attenuated strains of myxoma virus.

Cell Differentiation

The replication of some viruses is determined by the state of differentiation of the cell. The warts produced by papillomaviruses provide a classic example. Productive infection is not seen in the deeper layers of the epidermal tumor but occurs only when the cells become keratinized as they move to the surface layers. Basal cells contain 50–200 copies of viral DNA, but viral antigens and finally viral particles are produced only as the cells differentiate as they approach the surface of the skin. Other examples involve cells of the immune system. For example, measles virus, which does not replicate in normal (resting) peripheral blood lymphocyte cultures, does so after their activation by mitogen. HIV, integrated as a provirus in a resting T cell, cannot replicate until a cytokine induces synthesis of the NF-κB family of DNA-binding proteins, activating the cell to a permissive state (see Chapter 3).

The stage of the mitotic cycle may affect susceptibility. Autonomously replicating parvoviruses replicate only in cells that are in late S phase. Most vulnerable are the rapidly dividing cells of bone marrow, gut, and the developing fetus. The human parvovirus B19 produces lytic infection of dividing cells, the most sensitive target being the erythroid precursor cells. The arrest of erythrocyte production is not clinically apparent in hematologically normal individuals, but in persons with a shortened red cell survival time this arrest results in the transient profound anemia of aplastic crisis. Infrequently, this virus has also been recovered from stillborn edematous fetuses (hydrops fetalis).

Role of Interferons in Recovery from Viral Infection

Interferons are cellular proteins that are induced in virus-infected cells and were first recognized because they interfered with the replication of viruses, although they are now known to have a variety of other physiologic effects. Their properties and mode of action were described in Chapter 5; here we consider the role of interferons in the body. It is difficult to determine which cell types, or even which tissues and organs, are responsible for most interferon production *in vivo*. Extrapolating from findings with cultured cells, one can probably assume that most cells in the body are capable of producing interferons in response to viral infection. Certainly, interferons can be found in the mucus bathing epithelial surfaces such as the respiratory tract, and interferon is produced by most or all cells of mesenchymal origin. Lymphocytes, especially T cells and NK cells, as well as macrophages, produce large amounts of interferon α and γ, and they are probably the principal source of circulating interferon in viral infections characterized by a viremic stage.

There are data supporting a central role for interferons in the recovery of experimental animals and humans following at least some viral infections. Telling evidence that interferon can indeed be instrumental in deciding the fate of the animal following natural viral infection was provided in the early 1970s by Gresser and colleagues, who showed that mice infected with any of several nonlethal viruses, or with sublethal doses of more virulent viruses, die if anti-interferon serum is administered. More recent support is given by the elegant work on the Mx gene and the susceptibility of mice to influenza virus described in Chapter 5. Perhaps the most persuasive evidence, however, comes from a more recent study with transgenic mice. Mice transfected with the gene for human interferon β displayed enhanced resistance to pseudorabies virus, in proportion to the resulting concentration of circulating interferon. Serum from the transgenic mice also protected nontransgenic mice against the unrelated vesicular stomatitis virus, and this protection was abrogated by anti-interferon serum.

Although it is widely thought that interferons constitute the first line of defense in the process of recovery from viral infections, it would be naive to believe that it is the most important factor in recovery. If this were so, one might expect that a systemic infection with any virus, or immunization with a live vaccine, might protect an individual, for a period at least, against challenge with an unrelated virus, yet this cannot be demonstrated. The evidence is somewhat stronger that infection of the upper respiratory tract with one virus will provide temporary and strictly local protection against others. Perhaps this distinction provides the clue that the direct antiviral effect of interferons is limited in both time and space. The main antiviral role of interferons may be to protect cells in the immediate vicinity of the initial focus of infection for the crucial first few days.

Natural Inhibitors of Attachment of Virions to Cells

Blood, mucus, milk, and other body fluids contain a wide range of substances, some of which can coat particular viruses and impede their attachment to cells. For example, influenza virions can be neutralized by mannose-binding lectins (conglutinin and "mannose-binding protein," MBP) found in the plasma of a number of animal species including humans, and in the lungs as pulmonary surfactant proteins, as well as by sialylated glycoproteins found in plasma and respiratory mucus. Cytomegalovirus is often coated with β_2-microglobulin, as is HIV which also binds HLA-DR (α and β chains) and CD4 found in soluble form in plasma as well as on lymphocytes. Antiviral fatty acids derived from lipids in breast milk have also been described. It is not yet clear whether these observations relate to biologically important phenomena. If they do, they may represent the tip of an iceberg of innate natural defense mechanisms which could, with the advent of recombinant DNA technology, be exploited for chemotherapeutic purposes.

Dual Infections

Viral infections of the respiratory tract often lower resistance to bacterial superinfection, notably in measles and influenza. Experimental studies in mice have shown that *Staphylococcus aureus* produces a protease which increases the virulence of influenza virus, presumably by cleavage of the HA molecule. Rotavirus infection of the murine gut is more severe in the presence of an enterotoxigenic strain of *Escherichia coli.*

Even more important are the secondary infections that complicate infections with immunosuppressive viruses. HIV, for example, so profoundly depletes the $CD4^+$ cell population and damages cells of the monocyte lineage that AIDS patients usually die from any of a number of uncontrollable secondary infections. These may be otherwise rare parasites such as *Pneumocystis carinii* and *Toxoplasma gondii,* or mycobacteria, including atypical mycobacteria and *Mycobacterium avium.* Numerous other organisms, including rare fungi, the yeasts *Candida albicans* and *Cryptococcus neoformans,* and herpesviruses, papillomaviruses, adenoviruses, and hepatitis viruses, are frequently isolated from these unfortunate individuals.

Further Reading

Baron, S., Coppenhaver, D. H., Dianzani, F., Fleischman, W. R., Hughes, T. K., Klimpel, G. R., Niesel, D. W., Stanton, G. J., and Tyring, S. K. (1992). "Interferon: Principles and Medical Applications," Univ. of Texas Medical Branch, Galveston.

Berzofsky, J. A. (1987). *Ir* genes: Antigen-specific genetic regulation of the immune response. *In* "The Antigens" (M. Sela, ed.), Vol. 7, p. 1. Academic Press, New York.

Brinton, M. A., and Nathanson, N. (1981). Genetic determinants of virus susceptibility: Epidemiologic implications of murine models. *Epidemiol. Rev.* **3,** 115.

Buller, R. M. L., and Palumbo, G. J. (1991). Poxvirus pathogenesis. *Microbiol. Rev.* **55,** 80.

Chandra, R. K. (1979). Nutritional deficiency and susceptibility to infection. *Bull. WHO* **57,** 167.

Gooding, L. R. (1992). Virus proteins that counteract host immune defenses. *Cell (Cambridge, Mass.)* **71,** 5.

Klenk, H.-D., and Rott, R. (1989). The molecular biology of influenza virus pathogenicity. *Adv. Virus Res.* **34,** 167.

Kolakofsky, D., and Mahy, B. W. J., eds. (1989). "Genetics and Pathogenicity of Negative-Strand Viruses." Academic Press, Orlando, Florida.

Minor, P. D. (1992). The molecular biology of poliovaccines. *J. Gen. Virol.* **73,** 3065.

Nomoto, A., ed. (1992). Cellular Receptors for Virus Infection. *Semin. Virol.* **3,** 77.

Notkins, A. L., and Oldstone, M. B. A., eds. (1984, 1986, 1989). "Concepts in Viral Pathogenesis," Vols. 1–3. Springer-Verlag, New York.

Russell, W. C., and Almond, J. W., eds. (1987). Molecular Basis of Virus Disease. *Soc. Gen. Microbiol. Symp.* **40.**

Tyler, K. L., and Fields, B. N. (1990). Pathogenesis of viral infections. *In* "Fields Virology" (B. N. Fields, D. M. Knipe, R. M. Chanock, M. S. Hirsch, J. L. Melnick, T. P. Monath, and B. Roizman, eds.), 2nd Ed., p. 213. Raven, New York.

Chapter 8

Immune Response to Viral Infections

In response to the constant threat of invasion by microorganisms and viruses, vertebrates have evolved an elaborate set of defensive measures, called, collectively, the immune system. During the initial encounter with a virus, the immune system of the host recognizes as foreign certain viral macromolecules (proteins, carbohydrates) called *antigens,* which elicit several kinds of responses to eliminate the virus and to prevent reinfection. Cells of the *humoral immune system* (B lymphocytes) respond to an antigenic stimulus by producing and secreting specific immunoglobulins called *antibodies;* cells of the cell-mediated immune system (T lymphocytes) respond by secreting several *cytokines* which regulate the immune response by coordinating the activities of the various types of cells involved. These lymphocytes bear highly specific receptors that enable them to interface with discrete sites on the virion or on viral peptides, known as *antigenic determinants* or *epitopes.* This specific recognition event triggers a wide range of effector processes that attack and remove the invading virus and virus-infected cells. The resulting cascade of cell–cell interactions and cytokine secretion amplifies the immune response to match the scale of the virus infection and, in addition, establishes a long-lived memory that enables the immune system to respond more quickly to any reinfection that may occur later in life.

The immune response terminates many viral infections before much damage has been done, resulting in mild or even subclinical infections. This chapter deals with the role of the immune response in recovery from viral infection and resistance to reinfection. Later chapters address situations where the immune system does not function so effectively: where the immune response is actually harmful, causing tissue damage in vital organs (Chapter 9), or where the virus evades the immune system and establishes a persistent infection (Chapter 10).

Components of the Immune System

The immune system comprises several kinds of lymphocytes as well as cells of the monocyte/macrophage lineage, dendritic cells, and NK cells. Lymphocytes, with their specific surface receptors, are the key to immunologic specificity. Any given T or B lymphocyte possesses receptors with specificity for a particular epitope. When T or B lymphocytes bind antigen they respond by dividing to form an expanded clone of cells (clonal expansion). The B lymphocytes differentiate into plasma cells, which secrete specific antibody. The T lymphocytes secrete soluble factors known as *lymphokines* or *interleukins,* which are representatives of a large family of hormones, known generically as cytokines, that modulate the activities of other cells involved in the immune response. Some of the T and B cells revert to long-lived small lymphocytes responsible for *immunologic memory.* Whereas antibodies and the receptors on B cells recognize epitopes on foreign antigens in their native conformation, T-cell receptors recognize short peptides in association with membrane glycoproteins known as *MHC proteins.*

Antigen-Presenting Cells

Cells of the dendritic cell and monocyte/macrophage lineages play a central role in the immune response to viruses, notably in antigen processing and presentation (Fig. 8-1). As a class, cells with these functions are called *antigen-presenting cells* (*APC*). The most efficient antigen-presenting cells are MHC class II-rich dendritic cells, including Langerhans cells of the skin and the interdigitating dendritic cells of lymph nodes, so named because they interdigitate with $CD4^+$ T cells to which they present antigen. Unlike dendritic cells, macrophages are phagocytic; they express some class II MHC protein while resting, but more following activation. After priming and activation, B lymphocytes also serve as antigen-presenting cells; they are important during the latter stages of the primary response and during the secondary response. To understand the intricacies of antigen processing and presentation one must first know something about the structure and intracellular trafficking of MHC proteins.

The *major histocompatibility complex* (*MHC*) is a genetic locus encoding three MHC class I proteins and at least six MHC class II proteins, each of which occurs in one of many alternative allelic forms. Class I glycoproteins are found in the plasma membrane of most types of cells; class II glycoproteins are confined principally to antigen-presenting cells. At the distal tip of each class of MHC protein there is a groove, usually occupied by a short peptide (Fig. 8-2). Peptides recovered from class I molecules are usually 8 or 9 amino acids long; peptides binding to class II proteins range from 13 to 17 amino acids. Specific amino acids lining the groove of any particular MHC protein limit and determine the particular range of peptides that can occupy

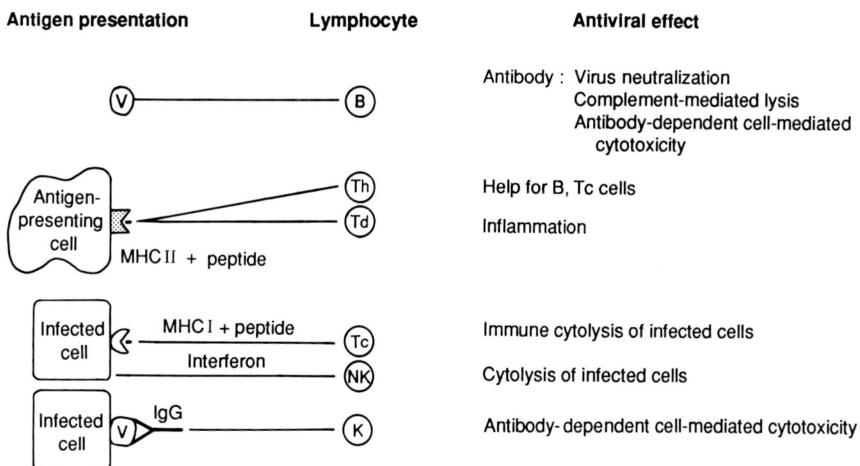

Fig. 8-1 Cells involved in antiviral immune responses. V, Virus or viral antigen; MHC I, MHC II, MHC proteins of class I and II, respectively, with viral peptide bound in groove; Th, helper, Td, delayed hypersensitivity, Tc, cytotoxic T lymphocytes; NK, natural killer cell; K, killer cell. Antigen-presenting cells process virus or viral proteins via the endosomal pathway, and the peptides associate with class II MHC proteins. Some viral protein molecules produced during viral replication undergo proteolysis and the resulting peptides associate with class I MHC proteins through the cytosolic pathway. (Modified from C. A. Mims and D. O. White "Viral Pathogenesis and Immunology." Blackwell, Oxford, 1984.)

it. The peptide–MHC complex is in turn recognized, with a considerably higher degree of specificity, by the *T-cell receptor* (*TCR*) of the appropriate clone of T cells. During ontogeny, positive selection of developing T cells in the thymus by "self" MHC molecules results in mature T cells that can recognize foreign peptides only if they are in the groove of "self" MHC protein, not foreign MHC molecules. This phenomenon is known as *MHC restriction*.

Virus or viral proteins taken up by antigen-presenting cells pass progressively through early endosomes to late (acidic) endosomes and prelysosomes, where they are degraded by proteolytic enzymes (the *endosomal pathway*). Certain of the resulting viral peptides are able to bind to newly synthesized class II MHC molecules which they encounter in these acidic endosomes on their way from the Golgi complex to the plasma membrane. In virus-infected cells, on the other hand, endogenously synthesized viral protein molecules are occasionally degraded in the cytosol, and the resulting peptides are translocated by special transporter molecules into the endoplasmic reticulum, where they assemble with class I MHC molecules to form a stable complex, which is then exported to the cell surface (the *cytosolic pathway*). Thus, in general, peptides derived from endogenously synthesized viral proteins in infected cells associate with MHC class I protein and the complex is recognized by $CD8^+$ T cells, thus eliciting a Tc cell response, whereas peptides derived from exogenous viral proteins endocytosed by antigen-presenting cells associate with class II MHC protein and the complex is recognized by $CD4^+$ T cells, leading to a Th cell response.

Although there is extensive polymorphism of MHC genes between people, any individual has only a limited number of different MHC proteins, and

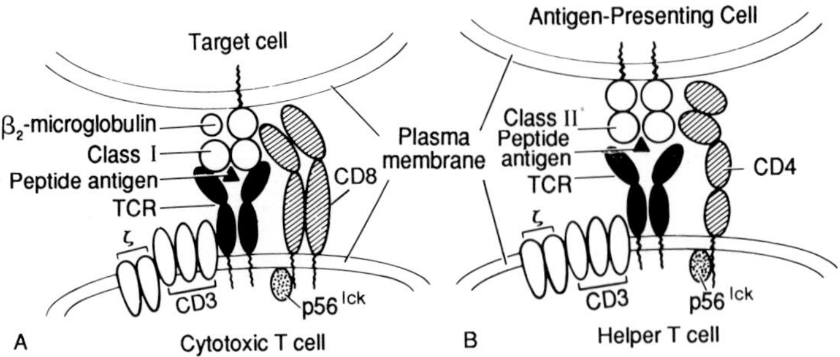

Fig. 8-2 Model for the interaction between MHC proteins, T-cell receptor, and CD4 or CD8. (A) CD8 on a cytotoxic T cell binds to a class I MHC protein on a target cell and interacts with a T-cell receptor (TCR) molecule that is binding both to the same class I MHC protein and to the foreign peptide it is presenting. (B) CD4 on a helper T cell binds to a class II MHC protein on an antigen-presenting cell and interacts with a TCR molecule that is binding both to the class II MHC protein and to the peptide it is presenting. The TCR is shown (heavy shading) as a heterodimer of two polypeptide chains, α and β, each with a constant domain spanning the membrane and a variable region containing the antigen-binding groove; the TCR is associated with the accessory molecule CD3, a complex of three polypeptides (γ, δ, ε) plus a homodimer of ζ, which is thought to serve as a signal transducer for the TCR, being phosphorylated by the tyrosine kinase $p56^{lck}$ (stippled), which is associated with the accessory molecules CD4 and CD8 (striped). The class I MHC protein is shown as a polypeptide chain with three extracellular domains, associated with β_2-microglobulin; the class II MHC is depicted as comprising two polypeptide chains, each having two extracellular domains. Additional pairs of complementary adhesion molecules of different types (not shown) contribute to establishing and stabilizing close contact between the T cell and antigen-presenting cell (including B cells). [From J. R. Parnes, *in* "Encyclopedia of Human Biology" (R. Dulbecco, ed.), Vol. 2, p. 225. Academic Press, San Diego, 1991.]

any given antigenic peptide binds only to certain MHC molecules. If certain peptide–MHC complexes are important in eliciting a protective immune response to a serious viral infection, persons lacking suitable MHC proteins will be genetically more susceptible to that disease. Equally crucial is whether the T-cell repertoire of that individual includes a clone of lymphocytes bearing receptors for that particular MHC–peptide complex.

T Lymphocytes

There are two principal classes of lymphocytes: T lymphocytes, so named because of their dependence on the thymus for their maturation from pluripotent hemopoietic stem cells, and B lymphocytes, derived from the bone marrow. These differ not only in their different antigen receptors, but also in surface markers and function. Functionally, T lymphocytes are classified into four subsets: *helper* (*Th*) and *suppressor* (*Ts*) lymphocytes are regulator cells; *cytotoxic* (*Tc*) and *delayed hypersensitivity* (*Td*) lymphocytes are effector cells (Table 8-1). Close examination of T-cell clones indicates that a single cell type can discharge more than one of these functions and secrete a range of different lymphokines, but it may recognize a given determinant only in association with a particular class of MHC molecule on the surface of the cell.

Table 8-1
Subsets of T Lymphocytes

Subset	Marker	MHC restriction	Function
Helper (Th)	CD4	Class II	Th_1: inflammation Th_2: help for B or Tc cells
Cytotoxic (Tc)	CD8	Class I, usually	Cytolysis of virus-infected cells
Suppressor (Ts)	CD8	?	Suppression of B, Th, or Tc cells

Helper T Cells

The Th cells carry a surface marker known as CD4. They recognize viral peptides in association with class II MHC protein, usually on the surface of an antigen-presenting cell. They then secrete cytokines which activate the Th cells themselves and subsequently activate other types of lymphocytes, such as B cells, helping them to produce antibody, or Tc cells, helping them to acquire cytotoxicity.

Cytotoxic T Cells

The Tc cells carry the CD8 surface marker and generally recognize viral peptides in association with class I MHC molecules. Following activation, they lyse cells which have that particular viral peptide bound to that MHC class I protein on their surface.

Delayed Hypersensitivity T Cells

The Td cells are generally $CD4^+$ (but may be $CD8^+$) and recognize peptides in association with class II (or sometimes class I) MHC protein. The Td cells secrete a variety of lymphokines which set up an inflammatory response and greatly augment the immune response by attracting both monocytes/macrophages and other T cells to the site, and also activating them to proliferate, differentiate, and secrete additional cytokines themselves. The Td cells are generally considered to be a subpopulation of Th cells.

Two major classes of $CD4^+$ T cells have been described. The Th_1 cells ("inflammatory" T cells) are defined as typically secreting the cytokines IL-2, IFN-γ, and TNF-β [plus granulocyte–macrophage colony-stimulating factor (GM-CSF) and IL-3], mediating delayed hypersensitivity (DTH) *in vivo,* and promoting IgG_{2a} production, whereas Th_2 ("helper" T cells) are defined as typically secreting IL-4, IL-5, and IL-6 (plus GM-CSF and IL-3), providing help but not DTH, and promoting a switch to the IgG_1 isotype. However, individual $CD4^+$ T-cell clones vary widely in the particular combinations of cytokines they produce; the two dominant patterns described above tend to emerge in chronic persisting infections or after long-term culture.

Suppressor T Cells

The Ts cells are the least well-characterized class of T cells. Functionally, there is no doubt that certain populations of $CD4^-$, $CD8^+$, I-J^+ T cells can be demonstrated to down-regulate other T-cell- and/or B-cell-mediated immune responses. However, it has proved difficult to clone the T cells that display this property, and some immunologists remain skeptical about the existence of a unique Ts cell type with a unique suppressor function. Currently, it appears likely that T cells may, under particular circumstances, suppress various arms of the immune response in a variety of ways, for example, (1) by direct cognate interaction with an effector B or Tc cell, (2) indirectly, by suppression of a helper T cell, or (3) via antigen-specific soluble factors or antigen-nonspecific cytokines.

The effector response of T cells is generally transient; for instance, in certain acute infections Tc and Td activities peak about 1 week after the onset of a viral infection and disappear by 2–3 weeks. It is not yet clear whether this is attributable to the destruction of infected cells with consequential removal of the antigenic stimulus, or whether it is a function of Ts cells.

γ/δ T Lymphocytes

An entirely different class of T cells with a different type of receptor composed of polypeptide chains designated γ and δ (rather than the conventional α and β chains) is found principally in epithelia such as the skin, gut, and lungs. The γ/δ T lymphocytes display a relatively limited immunologic repertoire, reflecting a highly restricted V (variable) gene usage. There is emerging evidence that these T cells are sometimes involved in the immune response to viral infections, but it is too early to generalize about their role or their importance.

Cytokines

Cytokines are hormones which regulate the proliferation, differentiation, and/or maturation of nearby cells. Many are produced by T lymphocytes (lymphokines) or monocytes/macrophages (monokines) and serve to regulate the immune response by coordinating the activities of the various cell types involved. Thus, although cytokines are not antigen-specific, their production and actions are often antigen-driven.

Cytokines affect cells in the immediate vicinity, particularly at cell–cell interfaces, where directional secretion may occur and very low concentrations may be effective. Target cells carry receptors for particular cytokines. A single cytokine may exert a multiplicity of biological effects, often acting on more than one type of cell. Moreover, different cytokines may exert similar effects, though perhaps via distinct postreceptor signal transduction pathways, resulting in synergism.

Cytokines up-regulate or down-regulate the target cell, and different cytokines can antagonize one another. Typically, a cytokine secreted by a particular type of cell activates another type of cell to secrete a different cytokine or to express receptors for a particular cytokine, and so on in a sort of chain reaction. Because of the intricacy of the cytokine cascade it is rarely possible to attribute a given biological event *in vivo* to a single cytokine.

Cytokines (Table 8-2) can influence viral pathogenesis in a number of ways: (1) augmentation of the immune response, for example, of cytotoxic T cells by tumor necrosis factor or by interferon γ which up-regulates MHC expression; (2) regulation of the immune response, for example, antibody isotype switching by interleukin-4, -5, -6, or interferon γ; (3) suppression of

Table 8-2
Cytokines and Their Sources, Targets, and Effects[a,b,c]

Cytokine	Principal source	Principal targets/effects
IL-1α,β	Monocytes/macrophages	Proliferation of T cells, IL-2 receptor expression, antibody, fever
IL-2	T cells	Proliferation and differentiation of T cells
IL-3	T cells	Hematopoiesis, stem cells and mast cells
IL-4	Th cells, mast cells	Proliferation and differentiation of B, T, M; switch from IgM to IgG_1, IgE
IL-5	Th cells	Proliferation and differentiation of B, E; switch to IgA
IL-6	T cells, macrophages, other	Proliferation and differentiation of lymphocytes; induces antibody; fever
IL-7	Bone marrow stromal cells	Proliferation of pre-B and pre-T cells
IL-8	Various	Chemotaxis, neutrophils and T cells
IL-10	T cells, B cells, monocytes	Immunosuppression; inhibits APC, T cells, and cytokine production
IL-12	Macrophages, B cells	Proliferation and differentiation of T cells
TNF-α,β	Monocytes, other	Antiviral; proliferation and differentiation of T, B, M, N, F; fever, cachexia
TGF-β	Various	Inhibits proliferation of T, B, and stem cells
IFN-α,β	Leukocytes, other	Antiviral; fever
IFN-γ	Th cells	Antiviral; activation of Tc, M, NK; IgM to IgG_{2a} switch; upregulates MHC and Fc receptors
GM-CSF	T, M, F, endothelium	Hematopoiesis, granulocytes, monocytes

[a] Cytokines are pleiotropic in their actions.
[b] Only certain major activities of the best studied cytokines are listed in this condensed summary.
[c] Abbreviations: IL, interleukin; TNF, tumor necrosis factor; TGF, transforming growth factor; IFN, interferon; GM-CSF, granulocyte–macrophage colony-stimulating factor; T, T lymphocyte; B, B lymphocyte; M, monocyte/macrophage; NK, natural killer cell; E, eosinophil; F, fibroblast; APC, antigen-presenting cell.

the immune response, for example, inhibition of the synthesis of interferon γ by interleukin-10; (4) inhibition of viral replication by interferons; and (5) upregulation of viral gene expression, for example, TNF-α and interleukin-6, binding to their receptors on T cells, induce the synthesis of NF-κB, which in turn binds to the regulatory region of the integrated HIV provirus and promotes HIV gene transcription.

B Lymphocytes

Some of the pluripotent hemopoietic stem cells originating from fetal liver and later from bone marrow differentiate into B lymphocytes, which are characterized by the presence on their surface of specific antigen-binding receptors, plus receptors for complement (C3) and receptors for the Fc portion of immunoglobulin. During ontogeny, complex DNA rearrangements occur involving recombination between hundreds of inherited V (variable) immunoglobulin gene segments to yield potentially millions of combinations. Each individual B lymphocyte and its progeny express only one of these immunoglobulin genes; hence, by the time an animal is born there is a vast number of B lymphocytes expressing different monomeric IgM molecules as surface receptor proteins. Such cells have three possible fates: they may react with a self-antigen and usually (but not always) be eliminated; they may react with a foreign antigen, undergo clonal expansion, and secrete antibody to that antigen; or, like other short-lived effector cells, they soon die. In contrast to those of T cells, the receptors of B cells are antibody-like molecules that recognize antigens in their native state, rather than as peptides bound to MHC protein, and hence B cells interact directly with viral proteins or virions. When the particular clones of B cells bearing receptors complementary to any one of the several epitopes of an antigen bind that antigen, they respond, after receiving the appropriate signals from Th cells, by division and differentiation into antibody-secreting plasma cells.

Each plasma cell secretes antibody of only a single specificity, corresponding to the particular V (variable) domains of the immunoglobulin (Ig) receptor it expresses. Initially this antibody is of the IgM class, but a gene translocation then brings about a class switch by associating the V gene segments with a different H (heavy) chain constant domain. Various cytokines also play an important role in isotype switching (Table 8-2). Thus, after a few days IgG, IgA, and sometimes IgE antibodies of the same specificity begin to dominate the immune response. Early in the immune response, when large amounts of antigen are present, there is an opportunity for antigen-reactive B cells to be triggered even if their receptors fit the epitope with relatively poor *affinity;* the result is the production of antibody which binds the antigen with low *avidity.* Later on, when only small amounts of antigen remain, B cells that have evolved by hypermutation in their V_H genes to produce receptors that bind the antigen with high affinity are selected (*affinity maturation*), and hence the avidity of the antibody secreted increases correspondingly.

Immunologic Memory

Following priming by antigen and clonal expansion of lymphocytes, a population of long-lived *memory cells* arises which persists indefinitely. Memory T cells are characterized by particular surface markers (notably CD45RO) and homing molecules (*adhesins*) that are associated with a distinct recirculation pathway. When reexposed to the same antigen, even many years later, memory T cells respond more rapidly and more vigorously than in the primary encounter. Memory B cells, on reexposure to antigen, also display an *anamnestic (secondary) response,* with production of larger amounts of specific antibody.

Little is known about the mechanism of the longevity of immunologic memory in T or B lymphocytes in the absence of demonstrable chronic infection. Possibly the cells are periodically restimulated by the original antigen retained indefinitely as antigen–antibody complexes on follicular dendritic cells in lymphoid follicles, or by surrogate antigen in the form of either fortuitously cross-reactive antigens or anti-idiotypic antibodies. There is also evidence that memory T and B lymphocytes may survive for years without dividing, until restimulated following reinfection.

B Cells as Antigen-Presenting Cells

Memory B cells serve as very efficient antigen-presenting cells. Viral antigen, or the virion itself, binds to the specific immunoglobulin receptors on the B lymphocyte and is endocytosed then cleaved into peptides, which are in turn re-presented on the surface of the B cell. These peptides, which generally represent different epitopes from those recognized (on the same antigen) by the B cell, associate with class II MHC proteins on the B-cell plasma membrane and are presented to the corresponding $CD4^+$ T cells, which respond by secreting cytokines that provide help for the B cell to make antibody. Such *cognate* interaction, involving close physical association of T and B cells, ensures very efficient delivery of "helper factors" (cytokines) from the Th cell to the relevant primed B cell.

Antibodies

The end result of activation and maturation of B cells is the production of antibodies, which react specifically with the epitope identified initially by their receptors. Antibodies fall into four main classes: two monomers, IgG and IgE, and two polymers, IgM and IgA. All immunoglobulins of a particular class have a similar structure, but they differ widely in the amino acid sequences comprising the antigen-binding site, which determines the specificity for a given antigenic determinant.

The commonest immunoglobulin found in serum, IgG, consists of two "heavy" and two "light" chains, and each chain consists of a "constant" and a "variable" domain. The chains are held together by disulfide bonds. Papain cleavage separates the molecules into two identical *Fab fragments,* which contain the antigen-binding sites, and an *Fc fragment,* which carries the sites for various effector functions such as complement fixation, attachment to phagocytic cells, and placental transfer.

The immunologic specificity of an antibody molecule is determined by its ability to bind specifically to a particular epitope. The binding site is located at the amino-terminal end of the molecule and is composed of certain hypervariable sequences within the "variable" domains of both light and heavy chains. Of the approximately 220 amino acids of the variable domain of a heavy chain–light chain pair, between 15 and 30 appear to make up the binding site. Antibody specificity is a function of both the amino acid sequence at these sites and their three-dimensional configuration.

The major class of antibody in the blood is immunoglobulin G (IgG), which occurs as IgG_1, IgG_2, IgG_3, and IgG_4 subclasses. Following systemic viral infections, IgG continues to be synthesized for many years and is the principal mediator of protection against reinfection. The subclasses of IgG differ in the "constant" region of their heavy chains and consequently in biological properties such as complement fixation and binding to phagocytes.

Immunoglobulin M is a particularly avid class of antibody, being a pentamer of 5 IgG equivalents, with 10 Fab fragments and therefore 10 antigen-binding sites. Because IgM is formed early in the immune response and is later replaced by IgG, specific antibodies of the IgM class are diagnostic of recent (or chronic) infection. The IgM antibodies are also the first to be found in the fetus as it develops immunologic competence in the second half of pregnancy. Because IgM does not cross the placenta from mother to fetus, the presence of IgM antibodies against a particular virus in a newborn infant is diagnostic of intrauterine infection.

Immunoglobulin A is a dimer, with four Fab fragments. Passing through epithelial cells, IgA acquires a "secretory component" to become *secretory IgA* (*sIgA*), which is secreted through mucosae into the respiratory, intestinal, and urogenital tracts. Secretory IgA is more resistant to proteases than other immunoglobulins, and it is the principal immunoglobulin on mucosal surfaces, and in milk and colostrum. For this reason IgA antibodies are important in resistance to infection of the respiratory, intestinal, and urogenital tracts, and IgA antibody responses are much more effectively elicited by oral or respiratory than by systemic administration of antigen, a matter of importance in the design and route of delivery of some vaccines (see Chapter 13).

Antibodies directed against certain epitopes on the surface of virions neutralize infectivity; they may also act as opsonins, facilitating the uptake and destruction of virions by macrophages. In addition, antibody may attach to viral antigens on the surface of infected cells, leading to destruction of the cells following activation of the classical or alternative complement pathways, or by arming and activating Fc receptor-bearing cells such as K cells, polymorphonuclear leukocytes, and macrophages (antibody-dependent cell-mediated cytotoxicity).

Complement

The *complement system* consists of a series of serum components which can be activated to "complement" the immune response (Fig. 8-3). As well as the *classical complement activation pathway,* which is dependent on the presence of an antigen–antibody complex, there is also an *alternative antibody-independent pathway.* Both are important in viral infections.

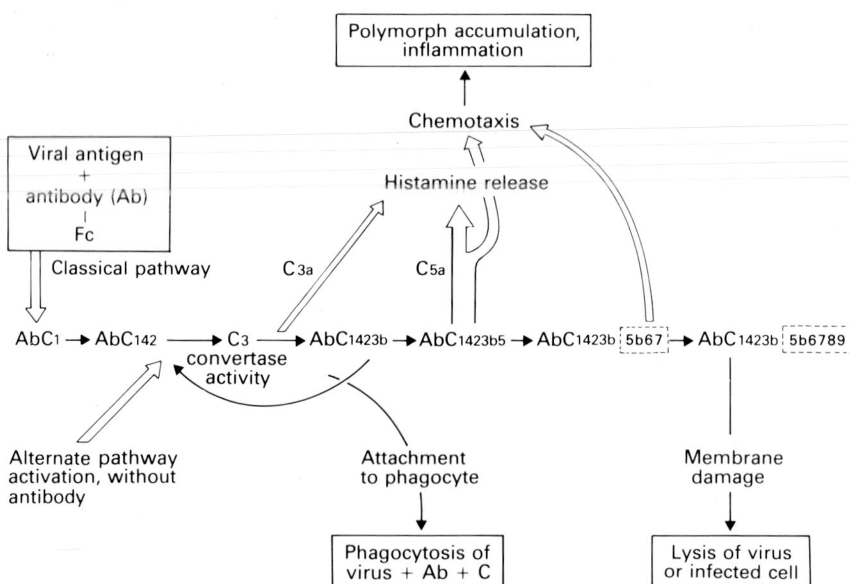

Fig. 8-3 Diagram of the complement activation sequence by the classical and alternative pathways and the antiviral action of complement. The numbers of the complement components are not sequential because they were assigned before the sequence of action was elucidated. (From C. A. Mims and D. O. White "Viral Pathogenesis and Immunology." Blackwell, Oxford, 1984.)

Activation of complement by the classical pathway may lead to the destruction of virions or virus-infected cells, as well as to inflammation. Virions are destroyed as a result of opsonization, enhancement of neutralization, or lysis of the viral envelope. Complement activation following interaction of antibody with viral antigens in tissues leads to inflammation and the accumulation of leukocytes. Activation of complement via the alternative pathway appears to occur mainly after infections with enveloped viruses that mature by budding through the plasma membrane; since it does not require antibody, the alternative pathway can be triggered immediately after viral invasion.

Natural Killer Cells

Natural killer (NK) cells are a heterogeneous group of $CD3^-$, $CD16^+$, $CD56^+$ large granular lymphocytes of uncertain lineage which have the capacity to kill virus-infected cells and tumor cells. The basis for their selectivity for virus-infected cells is not known. They display no immunologic specificity for particular viral antigens, no memory, no MHC restriction, and no dependence on antibody. They may be an important early defense mechanism, since their activity is greatly enhanced within 1 or 2 days of viral infection. Virus-induced activation of NK cells is mediated by interferon, acting synergistically with IL-2, and NK cells themselves secrete several cytokines including IFN-γ and TNF-α.

Immune Responses to Viral Infection

The major features of the immune response to a typical acute viral infection are illustrated in Fig. 8-4. The large boxes highlight three crucially important

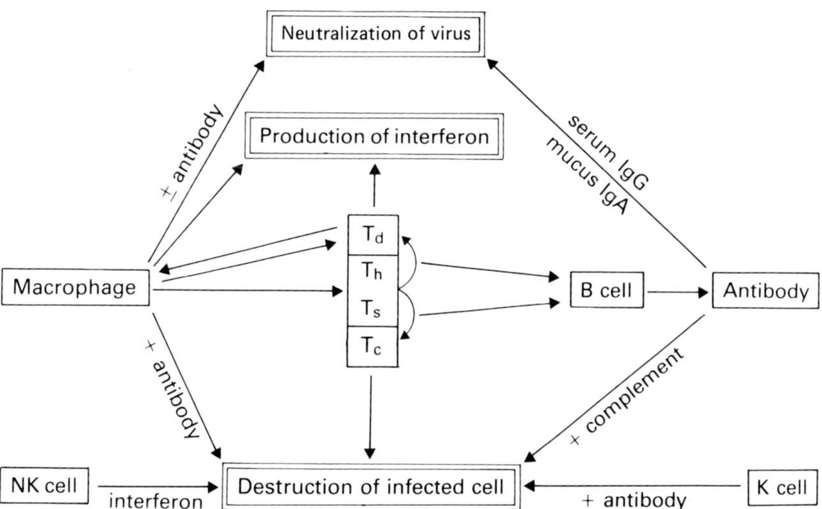

Fig. 8-4 Immune responses to viral infection. For explanation see text. (From C. A. Mims and D. O. White "Viral Pathogenesis and Immunology." Blackwell, Oxford, 1984.)

phenomena which contribute to recovery from infection: (1) destruction of infected cells, (2) production of interferon, and (3) neutralization of the infectivity of virions. The flowchart illustrates, in a simplified fashion, the interactions of the various cell types that participate in these events.

Shortly after infection, some virus particles are phagocytosed by macrophages. Except in the case of certain viruses that are capable of growing in macrophages, the engulfed virions are destroyed. Their proteins are cleaved into short peptides which are presented on the surface of the macrophage in association with class II MHC protein. This combination is recognized by the appropriate clones of $CD4^+$ lymphocytes. The Td (mainly Th_1) lymphocytes respond by clonal proliferation and release of lymphokines, which attract blood monocytes to the site and induce them to proliferate and to differentiate into activated macrophages, the basis of the inflammatory response. The Th_2 lymphocytes respond by secreting a different set of lymphokines that assist the appropriate clones of B cells, following binding of viral antigen, to divide and differentiate into plasma cells. The Tc cells are activated following recognition of viral peptides in association with MHC class I on the surface of infected cells. The Tc response usually peaks at about 1 week after infection, compared with the antibody response which peaks later (2–3 weeks). The NK cell activity is maximal by 2 days, and interferon activity peaks in concert with the peak titer of virus.

Antibody synthesis takes place principally in the spleen, lymph nodes, gut-associated lymphoid tissues (GALT), and bronchus-associated lymphoid tissues (BALT). The spleen and lymph nodes receive viral antigens via the blood or lymphatics and synthesize antibodies, mainly restricted to the IgM class early in the response and IgG subclasses subsequently. On the other hand, the submucosal lymphoid tissues of the respiratory and digestive tracts, such as the tonsils and Peyer's patches, receive antigens directly from overlying epithelial cells, and they make antibodies mainly of the IgA class.

Immune Cytolysis of Virus-Infected Cells

Destruction of infected cells is an essential feature of recovery from viral infections, and it results from any of four different processes, involving cytotoxic T cells, antibody–complement-mediated cytotoxicity, antibody-dependent cell-mediated cytotoxicity, or NK cells. Since some viral proteins, or peptides derived therefrom, appear in the plasma membrane before any virions have been produced, lysis of the cell at this stage brings viral replication to a halt before significant numbers of progeny virions are released.

Cytolysis by cytotoxic T cells occurs by a complex mechanism involving the secretion of perforin, which forms ion channels through the plasma membrane of the target cell. *Antibody–complement-mediated cytotoxicity* is readily demonstrable *in vitro* even at very low concentrations of antibody. The alternative complement activation pathway (see Fig. 8-3) appears to be particularly important in this phenomenon. *Antibody-dependent cell-mediated cytotoxicity* (*ADCC*) is mediated by leukocytes that carry Fc receptors: macrophages, polymorphonuclear leukocytes, and other kinds of killer (K) cells. The NK cells, on the other hand, are activated by interferon, or directly by viral glycoproteins. They demonstrate no immunologic specificity but preferentially lyse virus-infected cells. In addition, in the presence of antibody, macrophages can phagocytose and digest virus-infected cells.

Neutralization of Viral Infectivity

In contrast to T cells, B cells and antibody generally recognize epitopes that are *conformational,* that is, the critical residues that make contact with the antigen-binding site of the antibody molecule are not necessarily contiguous in the primary amino acid sequence but may be brought into close apposition as a result of the folding of the polypeptide chain(s) to produce the native conformation. Such B-cell epitopes are generally located on the surface of the protein, often on prominent protuberances or loops, and generally represent relatively variable regions of the molecule, differing between strains of that species of virus.

Although specific antibody of any class can bind to any accessible epitope on a surface protein of a virion, only those antibodies that bind with reasonably high avidity to particular epitopes on a particular protein of the outer capsid or envelope of the virion are capable of neutralizing viral infectivity. The key protein is usually the one containing the ligand by which the virion attaches to receptors on the host cell. Mutations in critical epitopes on such a protein allow the virus to escape from neutralization by antibody, and the gradual emergence of mutations in a majority of these epitopes leads to the emergence of a novel strain (antigenic drift; see Chapter 4).

Neutralization is not simply a matter of coating the virion with antibody, nor indeed of blocking attachment to the host cell. Except in the presence of such high concentrations of antibody that most or all accessible antigenic sites on the surface of the virion are saturated, neutralized virions may still attach to susceptible cells. In such cases the neutralizing block occurs at some point following adsorption and entry. One hypothesis is that, whereas the virion is normally uncoated intracellularly in a controlled way that preserves its infectivity, a virion–antibody complex may be destroyed by lysosomal enzymes. For example, in the case of picornaviruses, neutralizing antibody appears to distort the capsid, leading to loss of a particular capsid protein and rendering the virion vulnerable to enzymatic attack. With influenza virus, more subtle conformational changes in the hemagglutinin molecule may prevent the fusion event that precedes the release of the nucleocapsid from the viral envelope.

Recovery from Viral Infection

Cell-mediated immunity, antibody, complement, phagocytes, and interferons and other cytokines are all involved in the response to viral infections and may alone or in concert be responsible for recovery, depending on the particular host–virus combination (see Fig. 8-4).

Role of T Lymphocytes

Lymphocytes and macrophages normally predominate in the cellular infiltration of virus-infected tissues; in contrast to bacterial infections, polymorphonuclear leukocytes are not at all plentiful. Depletion of T cells by neonatal thymectomy or antilymphocyte serum treatment increases the susceptibility of experimental animals to most viral infections; for example, T-cell-depleted mice infected with ectromelia virus fail to show the usual inflammatory mononuclear cell infiltration in the liver, develop extensive liver necrosis, and die, in spite of the production of antiviral antibodies and interferon. Virus titers in the liver and spleen of infected mice can be greatly reduced by adoptive transfer of immune T cells taken from recovered donors; this process is class I MHC restricted, implicating Tc cells, and is lifesaving.

Another experimental approach is to ablate completely all immune potential, then add back one or more of the separate components of the immune system. Using mice treated in this way, virus-primed cytotoxic T lymphocytes of defined function and specificity, cloned in culture then transferred to infected animals, have been shown to save the lives of mice infected with lymphocytic choriomeningitis virus, influenza virus, or several other viruses. Generally, greater protection is conferred by $CD8^+$ T cells than by $CD4^+$ T cells. Moreover, transgenic mice lacking $CD8^+$ T cells suffer higher morbidity and mortality than normal mice following virus challenge. Nevertheless, $CD4^+$ T cells have been shown to play a significant role in recovery, as do the cytokines they secrete, notably interferon γ and IL-2.

Although T-cell determinants and B-cell epitopes on surface proteins of viruses sometimes overlap, the immunodominant Tc determinants are often situated on the relatively conserved proteins located in the interior of the virion, or on nonstructural virus-coded proteins that occur only in virus-infected cells. Hence T cell responses are generally of broader specificity than neutralizing antibody responses and display cross-reactivity between strains and serotypes. When the gene encoding a protein that fails to elicit any neutralizing antibody (e.g., the NP, M, or NS protein of influenza virus) is incorporated into the genome of vaccinia virus, the T cells elicited following infection with the construct can adoptively transfer to naive mice complete protection against challenge with influenza virus.

Congenital Immunodeficiencies

The approach least subject to laboratory artifact is simple clinical observation of viral infections in experimental animals or children suffering from primary immunodeficiencies; such studies indicate a key role for T lymphocytes in recovery from generalized viral infections. Animals or humans with severe T-cell deficiencies due to thymic aplasia, lymphoreticular neoplasms, or chemical immunosuppression show increased susceptibility to herpesviruses and to many other viral infections that cannot be controlled by antibody. Perhaps the most informative example is that of measles in infants with thymic aplasia. In these T-cell-deficient infants there is no sign of the usual measles rash but rather an uncontrolled and progressive growth of virus in the respiratory tract, leading to fatal pneumonia. This reveals two aspects of the role of T cells; evidently, in the normal child, the T-cell-mediated immune response controls infection in the lung and plays a vital role in the development of the characteristic skin rash as well.

Role of Natural Killer Cells

The role of NK cells in recovery from viral infection is not yet certain. Athymic mice, which have almost no T cells but normal numbers of NK cells, usually die if infected with viruses that produce generalized viral infections. On the other hand, "beige" mice, which have substantially reduced NK cell activity, or normal mice depleted of NK cells by treatment with NK-specific antibody, show increased replication of some viruses but not others. The NK cells probably represent an innate defense mechanism of particular relevance in the early stage of primary virus infections, but they are less crucial than either T cells or antibody in clearing the infection and play no role in the establishment of immunologic memory.

Role of Antibody

In generalized diseases characterized by a viremia in which virions circulate free in the plasma, circulating antibody plays a significant role in recovery. Unlike those with a T-cell deficit, children with severe primary agammaglobulinemia recover normally from measles or varicella but are about 10,000 times more likely than normal children to develop paralytic disease after vaccination with live attenuated poliovirus vaccine. They have normal cell-mediated immune and interferon responses, normal phagocytic cells and complement, but cannot produce antibody, which is essential if poliovirus spread to the central nervous system via the bloodstream is to be prevented.

Although there is reasonably good evidence that antibody plays a key role in recovery from picornavirus, togavirus, flavivirus, and parvovirus infections, it does not necessarily follow that the antibody is acting solely by neutralizing virions. Indeed it has been shown that certain nonneutralizing monoclonal antibodies can save the lives of mice, presumably by antibody-dependent cell-mediated cytotoxicity or antibody–complement-mediated lysis of infected cells, or by opsonization of virions for macrophages.

Immunity to Reinfection

Whereas a large number of interacting phenomena contribute to recovery from viral infection, the mechanism of acquired immunity to reinfection with the same agent appears to be much simpler. The first line of defense is antibody, which, if acquired by active infection with a virus that causes systemic infections, continues to be synthesized for many years, providing solid protection against reinfection. The degree of acquired immunity generally correlates well with the titer of antibody in the serum. Further, transfer of antibody alone, whether by passive immunization or by maternal transfer from mother to fetus, provides excellent protection in the case of many viral infections. Thus it is reasonable to conclude that antibody is the most influential factor in immunity acquired by natural infection or by vaccination. If the antibody defenses are inadequate, the mechanisms that contribute to recovery are called into play again, the principal differences on this occasion being that the dose of infecting virus is reduced by antibody and that preprimed memory T and B lymphocytes generate a more rapid secondary response.

As a general rule the secretory IgA antibody response is short-lived compared to the serum IgG response. Accordingly, resistance to reinfection with respiratory viruses and some enteric viruses tends to be of limited duration. For example, reinfection with the same serotype of respiratory syncytial virus is not uncommon. Moreover, reinfection at a time of waning immunity favors the selection of neutralization-escape mutants, resulting in the emergence of new strains of viruses such as influenza virus or rhinoviruses by antigenic drift. Because there is little or no cross-protection between antigenically distinct strains of viruses, repeated attacks of respiratory infections occur throughout life.

The immune response to the first infection with a virus can have a dominating influence on subsequent immune responses to antigenically related viruses, in that the second virus often induces a response that is directed mainly against the antigens of the original viral strain. For example, the antibody response to sequential infections with different strains of influenza A virus is largely directed to antigens characterizing the particular strain of virus with which that individual was first infected. This phenomenon, irreverently called "original antigenic sin," is also seen in infections with enteroviruses, reoviruses, paramyxoviruses, and togaviruses. Original antigenic sin has important implications for interpretation of seroepidemiologic data, for understanding immunopathologic phenomena, and particularly for the development of efficacious vaccination strategies.

Passive Immunity

There is abundant evidence for the efficacy of antibody in preventing infection. For example, artificial *passive immunization* (injection of antibodies) temporarily protects against hepatitis A or B, rabies, measles, varicella, and several other viral infections (see Chapter 13). Furthermore, natural passive immunization protects the newborn for the first few months of life against most of the infections that the mother has experienced. In humans this occurs in two ways: (1) maternal antibodies of the IgG class cross the placenta and protect the fetus and the newborn infant during pregnancy and for several months after birth; (2) antibodies of the IgA class are secreted in the mother's milk at a concentration of 1.5 grams per liter (and considerably higher in the early colostrum), conferring protection against enteric infections as long as breast-feeding continues. If the infant encounters viruses when maternal immunity is waning, the virus replicates to only a limited extent, causing no significant disease but stimulating an immune response; thus the infant acquires active immunity while partially protected by maternal immunity. Maternally derived antibody also interferes with active immunization of the newborn and must therefore be taken into account when designing vaccination schedules (see Chapter 13).

Further Reading

Bjorkman, P. J., and Parham, P. (1990). Structure, function and diversity of class I major histocompatibility complex molecules. *Annu. Rev. Biochem.* **59,** 253.

Bloom, B. R., and Oldstone, M. B. A., eds. (1991). Immunity to Infection. *Curr. Opin. Immunol.* **3,** 453.

Braciale, T. J., ed. (1993). Viruses and the Immune System. *Semin. Virol.* **4,** 81.

Brandtzaeg, P. (1989). Overview of the mucosal immune system. *Curr. Top. Microbiol. Immunol.* **146,** 13.

Dimmock, N. J. (1993). Neutralization of animal viruses. *Curr. Top. Microbiol. Immunol.* **183,** 1.

Doherty, P. C., ed. (1992). Viral Immunity. *Semin. Immunol.* **4,** 69.

Doherty, P. C., Allan, W., and Eichelberger, M. (1992). Roles of $\alpha\beta$ and $\gamma\delta$ T cell subsets in viral immunity. *Annu. Rev. Immunol.* **10,** 123.

Germain, R. N., and Margulies, D. H. (1993). The biochemistry and cell biology of antigen processing and presentation. *Annu. Rev. Immunol.* **11,** 403.

Jorgensen, J. L., Reay, P. A., Ehrich, E. W., and Davis, M. M. (1992). Molecular components of T-cell recognition. *Annu. Rev. Immunol.* **10,** 835.

Notkins, A. L., and Oldstone, M. B. A., eds. (1984, 1986, 1989). "Concepts in Viral Pathogenesis," Vols. 1–3. Springer-Verlag, New York.

Paul, W. E., ed. (1993). "Fundamental Immunology," 3rd Ed. Raven, New York.

Roitt, I. M., Brostoff, J., and Male, D. K. (1993). "Immunology," 3rd Ed. Churchill Livingstone, London.

Rothbard, J. B., and Gefter, M. L. (1991). Interactions between immunogenic peptides and MHC proteins. *Annu. Rev. Immunol.* **9,** 527.

Sercarz, E., and Berzofsky, J., eds. (1988). "Immunogenicity of Protein Antigens," Vols. 1 and 2. CRC Press, Boca Raton, Florida.

Steinman, R. M. (1991). The dendritic cell system and its role in immunogenicity. *Annu. Rev. Immunol.* **9,** 271.

Thomas, D. B., ed. (1993). "Viruses and the Cellular Immune Response." Dekker, New York.

Trinchieri, G. (1989). Biology of natural killer cells. *Adv. Immunol.* **47,** 187.

van Regenmortel, M. H. V., and Neurath, A. R., eds. (1985, 1991). "Immunochemistry of Viruses," Vols. 1 and 2. Elsevier, Amsterdam.

Chapter 9

Mechanisms of Disease Production

In the previous four chapters we have analyzed how viruses affect cells, how infection of the body occurs, how viruses spread to various parts of the body, and how the infected person responds to infection by immunologic and other mechanisms. The next three chapters are concerned with the clinical consequences of these interactions, namely, disease. This chapter describes the ways in which viral replication damages tissues and organs, the ways in which the body's own responses may cause damage, and the behavior of viruses in immunocompromised persons; the succeeding two chapters describe persistent infections and viral oncogenesis, respectively.

Viral Damage to Tissues and Organs

The mechanisms by which viruses damage cells were discussed at the cellular and subcellular levels in Chapter 5. Here we apply these concepts at the level of tissues and organs. The severity of disease in humans is not necessarily correlated with the degree of cytopathology produced by the virus *in vitro.* Many viruses that are cytocidal in cultured cells generally do not produce clinical disease; for example, enteroviruses, which cause severe cytopathic effects (CPE) in cultured human cells, usually cause inapparent infections. Conversely, some viruses, such as rabies virus, are noncytocidal *in vitro* but cause a lethal disease. In some organs a great deal of cell and tissue damage can occur without producing disease; for example, a substantial number of liver cells can be destroyed without significant clinical signs. When damage to cells does impair the function of an organ or tissue, this may be of minor importance in muscle or subcutaneous tissue, but of great importance in key organs such as the heart or the brain. Likewise, tissue edema may be unimportant in most sites in the body but may have serious consequences in the brain, because of the resulting increase in intracranial pressure, or in the lung, where it may interfere with gaseous exchange, or in the heart, where it may interfere with conduction.

Direct Damage by Cytocidal Viruses

Sometimes the whole pathologic picture may be explained by the direct damage to cells caused by a highly cytocidal virus. Mice infected intravenously with a large dose of Rift Valley fever virus, for example, developed overwhelming hepatic necrosis within 4 hours of injection and died by 6 hours, because the virions passed quickly through the Kupffer cells to infect the hepatic cells, which were rapidly lysed. In this experimental model, the defense mechanisms of the host were quite unable to cope with the rapid lethal damage to a vital organ. Similarly, the distribution of paralysis in a patient with poliomyelitis is a direct consequence of the distribution of those particular motor neurons in the anterior horn of the spinal cord that are destroyed by this highly cytocidal virus, leaving the muscles supplied by those motor neurons nonfunctional.

Damage to the Epithelium of the Respiratory Tract

Respiratory viruses initially invade and destroy just a few epithelial cells, but they initiate a lesion which can progressively damage the protective layer of mucus and lay bare more and more epithelial cells. As viral replication progresses, large numbers of progeny virions are budded into the lumen of the airway. Early in infection, the beating of cilia, the primary function of which is to cleanse the respiratory tract of inhaled particles, may actually help to move released progeny virus along the airway, thereby spreading the infection. As secretions become more profuse and viscous, the cilial beating becomes less effective and ceases as epithelial cells are destroyed.

In studies of influenza virus infection in experimental animals, the spread of the infection via contiguous expansion from initial foci often does not stop until virtually every columnar epithelial cell at that airway level is infected. The result is complete denuding of large areas of epithelial surface (Fig. 9-1) and the accumulation in the airways of large amounts of transudates, exudates containing inflammatory cells and necrotic epithelial cell debris. Where infection of the epithelium of the nasal passages, trachea, and bronchi proceeds to a fatal outcome, there are usually one or more of three complications: bacterial superinfection (nurtured by the accumulation of fluid and necrotic debris in the airways), infection and destruction of the lung parenchyma and the alveolar epithelium, and/or blockage of airways that are so small in diameter that mucous plugs cannot be opened by forced air movements. Blockage of the airways is of most significance in the newborn. In all of these complications there is hypoxia and a pathophysiologic cascade that leads to acidosis and uncontrollable fluid exudation into airways.

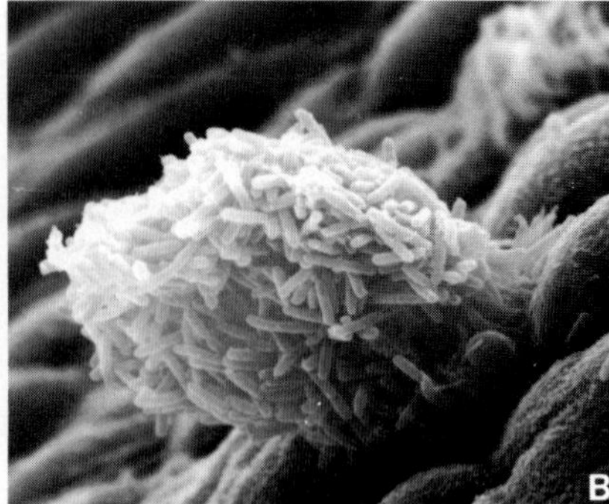

Fig. 9-1 Scanning electron micrographs showing the adherence of *Pseudomonas aeruginosa* to the mouse trachea (bar, 2 μm). (A) Normal mouse trachea, showing a single bacterium (arrow) on a serous cell. (B) Microcolony adhering to desquamating cells in an influenza virus-infected trachea. [From R. Ramphal, P. M. Small, J. W. Shands, Jr., W. Fischlschweiger, and P. A. Small, Jr. *Infect. Immun.* **27,** 614 (1980). Courtesy Dr. P. A. Small, Jr.]

Degeneration of respiratory tract epithelial surfaces during influenza infection is extremely rapid, but so is regeneration. In studies of influenza in ferrets, for example, it has been shown that the development of a complete new columnar epithelial surface via hyperplasia of remaining transitional cells may be complete in a few days. The transitional epithelium and the newly differentiated columnar epithelium that arises from it are resistant to infection, probably by virtue of interferon production and a lack of virus receptors. The role of other host defenses, including soluble factors such as mannose-binding lectins and lung surfactants, as well as macrophages, NK cells, IgA and IgG antibody, and T-cell-mediated immune mechanisms in terminating the infection was discussed in Chapters 7 and 8.

Damage to the Epithelium of the Intestinal Tract

The principal agents causing viral diarrhea in children are the rotaviruses; other viruses that produce diarrhea in children and adults include the caliciviruses, astroviruses, certain adenoviruses, and perhaps coronaviruses. Infection occurs by ingestion, and the incubation period is very short. Rotaviruses infect cells at the tip of the villus and cause marked shortening and occasional fusion of adjacent villi (Fig. 9-2), so that the absorptive surface of the intestine is reduced, resulting in fluid accumulation in the lumen of the gut and diarrhea. Infection generally begins in the proximal part of the small intestine and spreads progressively to the jejunum and ileum and sometimes to the colon. The extent of such spread depends on the initial dose, the virulence of the virus, and the immunologic status of the host. As the infection progresses, the absorptive cells are replaced by immature cuboidal epithelial cells whose absorptive capacity and enzymatic activity are greatly reduced. These cells are relatively resistant to viral infection, so that the disease is often self-limiting if dehydration is not so severe as to be lethal. The rate of recovery is rapid, since the crypt cells are not damaged.

Fluid loss in viral infections of the intestinal tract is mainly a loss of extracellular fluid due to impaired absorption, and osmotic loss due primarily to the presence of undigested lactose in the lumen (in infants), rather than active secretion. As virus destroys the absorptive cells there is a loss of those enzymes responsible for the digestion of disaccharides, and the loss of differentiated cells diminishes glucose carrier, sodium carrier, and Na^+,K^+-ATPase activities. This leads to a loss of sodium, potassium, chloride, bicarbonate, and water, and the development of acidosis. Another cause of acidosis is the increased microbial activity associated with the fermentation of undigested milk. Acidosis can create a K^+ ion exchange across the cellular membrane, affecting cellular functions that maintain the normal potassium concentration.

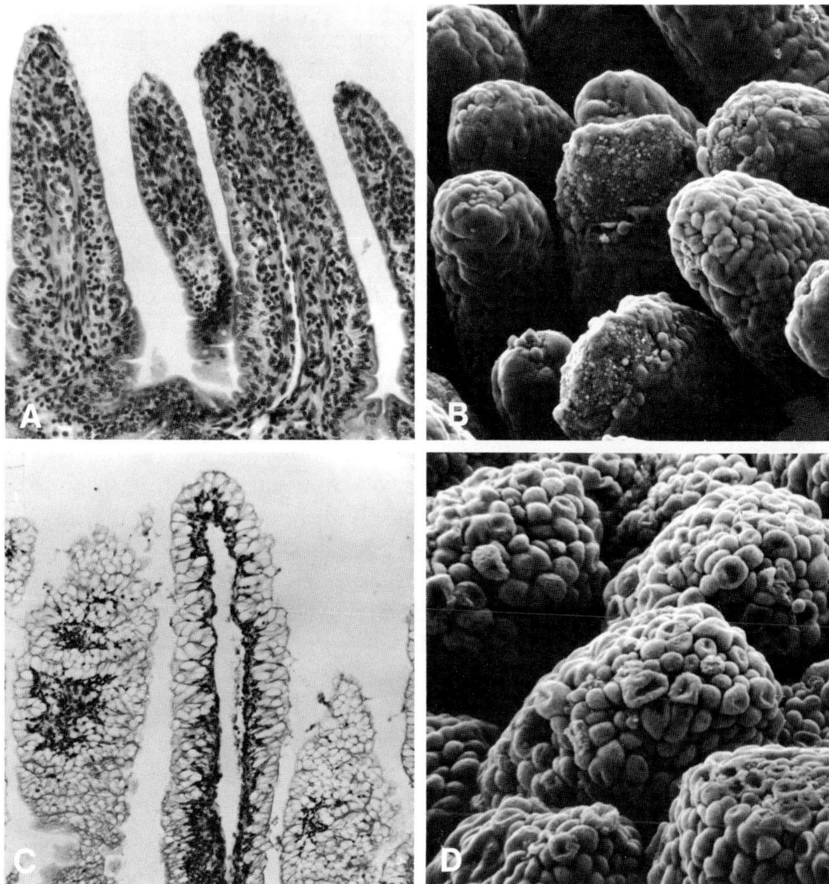

Fig. 9-2 Scanning electron and light micrographs of intestinal tissues from a gnotobiotic calf sacrificed 30 minutes after onset of rotavirus diarrhea. (A) Proximal small intestine with shortened villi and a denuded villus tip (second from right) (hematoxylin and eosin stain; magnification: × 120). (B) Appearance of the same level of intestine as in (A) by scanning electron microscopy, depicting denuded villi (magnification: × 180). (C) Distal small intestine with normal vacuolated epithelial cells and normal villi (hematoxylin and eosin stain; magnification: × 75). (D) Same area as in (C) seen by scanning electron microscopy. Epithelial cells appear round and protruding (magnification: × 210). [From C. A. Mebus, R. G. Wyatt, and A. Z. Kapikian, *Vet. Pathol.* **14,** 273 (1977); and A. Z. Kapikian and R. G. Wyatt, *in* "Textbook of Pediatric Infectious Diseases" (R. D. Feigin and J. D. Cherry, eds.), 3rd Ed., p. 661. Saunders, Philadelphia, Pennsylvania, 1992. Courtesy Dr. A. Z. Kapikian.]

Hypoglycemia owing to decreased intestinal absorption, inhibited glyconeogenesis, and increased glycolysis follow, completing a complex of pathophysiologic changes that, if not promptly corrected by restoration of fluids and electrolytes, results in death.

Epithelial Damage Predisposes to Secondary Bacterial Infection

As well as having direct adverse effects, viral infections often predispose epithelia to secondary bacterial infections, increasing the susceptibility of the respiratory tract, for example, to bacteria that are normal commensals in the nose and throat (see Fig. 9-1). Thus, infections with influenza virus may destroy ciliated epithelia and cause exudation, allowing pneumococci and other bacteria to invade the lungs and cause secondary bacterial pneumonia, which is often the cause of death in elderly people suffering from influenza. Conversely, proteases secreted by bacteria may activate influenza virus infectivity by proteolytic cleavage of the hemagglutinin. Rhinoviruses and respiratory syncytial virus damage the mucosa of the nasopharynx and sinuses, predisposing to bacterial superinfection which commonly leads to purulent rhinitis, pharyngitis, sinusitis, and sometimes otitis media. Similarly, in the intestinal tract, rotavirus infections may lead to an increase in susceptibility to enteropathogenic *Escherichia coli,* and the synergistic effect leads to more severe diarrhea.

Physiologic Changes without Cell Death

In some situations infected cells may show no obvious damage, but, as discussed in Chapter 5, specialized cells may carry out their functions less effectively after infection. For example, lymphocytic choriomeningitis virus infection of hybridoma cells appears harmless, but less antibody is produced by infected than by uninfected cells. In mice, the same virus has no cytopathic effect on cells of the anterior pituitary, but the output of growth hormone is reduced and as a result the mice are runted; likewise, persistent infection of insulin-producing islet cells in the pancreas may result in a lifelong elevation of blood glucose levels (diabetes). Other viruses may indirectly alter the expression of cell surface MHC molecules, leading to destruction of the infected cells by immunologic mechanisms; thus, enhanced class II MHC expression after infection of glial cells by mouse hepatitis virus, perhaps due to the production of interferon γ, may render these cells susceptible to immune cytolysis by Tc cells.

Immunopathology

The immune response is an essential part of the pathogenesis of most virus diseases. Infiltration of lymphocytes and macrophages, with accompanying release of cytokines and inflammation, is a regular feature of viral infection. Such common signs as fever, erythema, edema, and enlargement of lymph nodes have an immunologic basis. In some viral diseases the cardinal manifestations are attributable to the body's immune response. When pathologic changes are ameliorated by immunosuppressive treatment, it can be assumed that immunopathology makes an important contribution to the disease.

For most viral infections it is not known whether immunopathology makes a significant contribution to disease, and, if so, which particular immunologic mechanisms are implicated; nevertheless, it is instructive to speculate about their possible involvement. Immunopathologic mechanisms are traditionally classified into hypersensitivity reactions of types I, II, III, and IV. Although advances in cellular immunology have blurred some of the distinctions, the classification is still convenient.

Type I—Anaphylactic Hypersensitivity

Type I hypersensitivity reactions depend on the interaction of antigens with IgE antibodies attached to the surface of mast cells and basophils via Fc receptors, resulting in the release of histamine, leukotrienes and heparin, and the activation of serotonin and plasma kinins. Except for its possible contribution to some types of rashes and in some acute respiratory infections such as respiratory syncytial virus bronchiolitis, anaphylaxis is probably not particularly important in viral immunopathology.

Type II—Antibody-Dependent Cytotoxic Hypersensitivity

Type II cytolytic reactions occur when antibody, having combined with viral antigen on the surface of infected cells, activates the complement system, leading to cell lysis. Alternatively, binding of antibodies can sensitize virus-infected cells to destruction by Fc receptor-carrying K (killer) cells, polymorphonuclear leukocytes, or macrophages, via antibody-dependent cell-mediated cytotoxicity (ADCC). Although it has been clearly demonstrated that virus-infected cells are readily lysed by all of these mechanisms *in vitro* and there is some evidence that they may be operative in certain herpesvirus infections, the relative importance of antibody-dependent cytotoxic hypersensitivity in viral diseases *in vivo* is unclear.

Type III—Immune Complex-Mediated Hypersensitivity

Antigen–antibody reactions cause inflammation and cell damage by a variety of mechanisms. If the reaction occurs in extravascular tissues there is edema, inflammation, and infiltration of polymorphonuclear leukocytes, which may later be replaced by mononuclear cells. This is a common cause of mild inflammatory reactions. These *immune complex* reactions constitute the classic Arthus response and are of major importance, especially in persistent viral infections. If they occur in the blood, they produce circulating immune complexes, which are found in most viral infections. The fate of the immune complexes depends on the ratio of antibody to antigen. If there is a large excess of antibody, each antigen molecule is covered with antibody and removed by macrophages, which have receptors for the Fc component of the antibody molecule. If the amounts of antigen and antibody are about equal, lattice structures which develop into large aggregates are formed and removed rapidly by the reticuloendothelial system. However, in antigen excess, as occurs in some persistent infections, when viral antigens and virions are continuously released into the blood but the antibody response is weak and the antibodies are of low avidity or nonneutralizing, complexes continue to be deposited in small blood vessels and kidney glomeruli over periods of weeks, months, or even years, leading to impairment of glomerular filtration and eventually to chronic glomerulonephritis.

A classic example is lymphocytic choriomeningitis infection in mice infected *in utero* or as neonates. Viral antigens are present in the blood, and small amounts of nonneutralizing antibody are formed, giving rise to immune complexes which are progressively deposited on renal glomerular membranes; the end result may be glomerulonephritis, uremia, and death. Circulating immune complexes may also be deposited in the walls of the small blood vessels in skin, joints, and choroid plexus, where they attract macrophages and activate complement. Prodromal rashes, which are commonly seen in exanthematous diseases, are probably caused in this way. A more severe manifestation of the deposition of antigen–antibody–complement complexes in capillaries is erythema nodosum (tender red nodules in the skin); when small arteries are involved, as occasionally seen with hepatitis B, periarteritis nodosa results.

In addition to these local effects, by mobilizing soluble mediators antigen–antibody complexes may generate systemic reactions, such as fever, malaise, anorexia, and lassitude, that occur in most viral infections. Little is known about their causes, but fever can be attributed to interleukin-1 and tumor necrosis factor (produced by macrophages), and possibly to interferons. These and other soluble mediators produced by leukocytes, or released from virus-infected cells, may be responsible for other general symptoms as well.

Rarely, systemic immune complex reactions may activate the enzymes of the coagulation cascade, leading to histamine release and increased vascular permeability. Fibrin is deposited in the kidneys, lungs, adrenals, and pituitary gland, causing multiple thromboses with infarcts and scattered hemorrhages, a condition known as *disseminated intravascular coagulation*. This is seen in the hemorrhagic fevers, many of which are zoonoses caused by arenaviruses, bunyaviruses, filoviruses, or flaviviruses; it probably also occurs in other severe systemic diseases.

Type IV—Cell-Mediated Hypersensitivity

Unlike the previous types, the type IV "delayed hypersensitivity" reactions are mediated by cells rather than antibody. They are T-lymphocyte-mediated immune reactions, involving inflammation, lymphocytic infiltration and macrophage accumulation, and activation by lymphokines secreted by Td cells. Such reactions, involving infected endothelial cells of small cutaneous blood vessels, are largely responsible for the maculopapular rash of measles; they also play a role in production of the pustular lesions of smallpox, in which extensive viral replication occurs in the cutaneous epithelium.

The classic model of death due to a cell-mediated immune response is lymphocytic choriomeningitis (LCM) virus infection after primary infection of adult mice by intracerebral inoculation. The virus replicates harmlessly in the meninges, ependyma, and choroid plexus epithelium for about a week, until a Tc-lymphocyte-mediated immune response occurs, causing severe meningitis, cerebral edema, and death. Elsewhere than in the central nervous system, Tc cells help to control infection, but within the rigid confines of the skull these changes are fatal. The death of mice infected in this way can be completely prevented by chemical immunosuppression, by X-irradiation, or by antilymphocyte serum.

Although instructive, intracerebral inoculation of mice with LCM virus is a very unnatural experimental system. When mice are infected with influenza virus by the natural (intranasal) route they develop a lethal pneumonia. In one study it was found that adoptive transfer of influenza virus-primed $CD8^+$ cytotoxic T cells protects mice against intranasal challenge, but $CD4^+$ Th1 cells actually accelerate their demise. In general, although occasionally the cause of immunopathology, cell-mediated immune responses are an important component of the process of recovery from viral infections (see Chapter 8), as becomes evident if they are abrogated by cytotoxic drugs, or are absent, as in some immunodeficiency diseases.

Autoimmunity

Autoantibodies directed against host proteins can be detected quite commonly in viral infections, although usually only transiently and in low titer. In one study, 4% of a large panel of monoclonal antibodies raised against several viruses were found to react with normal tissues. For example, a monoclonal antibody directed against the neutralizing domain of coxsackie B4 virus also reacts against heart muscle; this virus is known to target muscle, including the heart, and to cause myocarditis.

A likely explanation for the widespread cross-reactivity has emerged in the course of computer searches of the international data banks that now contain complete amino acid sequences of thousands of viral and cellular proteins. It transpires that viral proteins share identical or near-identical stretches of 6–10 amino acids with cellular proteins far more frequently than would be predicted by chance. For instance, there is partial homology of amino acid sequence between myelin basic protein (MBP) and several viral proteins, some of which are shown in Table 9-1. Such *molecular mimicry* may be involved in the neurologic disorders associated with the lentivirus infections visna and caprine arthritis–encephalitis, and in the rare occurrence of postvaccinial encephalitis in humans. Inoculation of a neuritogenic epitope in

Table 9-1
Epitope Mimicry between Myelin Basic Protein and Viruses Causing Demyelinating Diseases[a]

Virus	Disease	Epitope mimicry[b]	
Visna	Demyelination	Virus	T G K I P W I L L P G R
		MBP	S G K V P W — L K P G R
Caprine arthritis–encephalitis	Encephalitis	Virus	T G K I P W I L L P G K
		MBP	S G K V P W — L K P G R
Vaccinia	Postinfection encephalitis	Virus	S I N R G F K G V D G R
		MBP	S A H K G F K G V D A Q
Influenza	Guillain-Barré syndrome	Virus	Q L G Q K F E E
		P2[c]	K L G Q E F E E

[a] From P. R. Carnegie and M. A. Lawson, *Today's Life Sci.* **3**(2), 14 (1991).
[b] Single letter amino acid code.
[c] P2, Protein of peripheral nerves.

Table 9-2
Potential Mechanisms of Induction of Autoimmune Disease by Viruses

Molecular mimicry: viral protein elicits humoral and/or cellular immune response cross-reacting with identical or similar epitope fortuitously present on a cellular protein
Polyclonal B-cell activation: B lymphocytes transformed by certain viruses (e.g., Epstein-Barr virus) secrete antibodies against a wide range of antigens, including normal cell proteins
Cytokine induction of MHC antigens: virus induces production of interferon γ and tumor necrosis factor, which induce expression of MHC class II protein on brain cells that do not usually express it (e.g., glial cells), enabling them to present antigens such as myelin to T cells
Exposure of sequestered cellular protein: incorporation into viral envelope or release from virus-infected cell (in altered or precursor form?)
T-cell dysfunction: viral destruction or down-regulation of T cells that normally suppress immune response to self proteins

the P2 protein of peripheral nerve myelin will cause the experimental equivalent of Guillain-Barré syndrome. There is mimicry between this epitope and a sequence in the influenza virus NS2 protein. Ordinarily, this nonstructural protein is removed from influenza virus vaccine during its purification; failure to do this in some batches of swine influenza vaccine used in the crisis program mounted in the United States in 1976–1977 may account for the apparent increase in incidence of the Guillain-Barré syndrome at that time.

A feature of the postinfectious encephalomyelitis that follows a few weeks after about 1 in every 1000 cases of measles (and somewhat less frequently following varicella, mumps, rubella, or vaccinia) is that the virus can rarely be isolated from the brain; however, MBP can be found in the cerebrospinal fluid, and, at least in the case of measles, anti-MBP antibodies and T cells are detectable in the blood. Theoretically, autoimmune damage can continue long after the virus that triggered it has been cleared from the body. Once the cross-reactive viral amino acid sequence has induced a humoral and/or cellular immune response that brings about lysis of normal tissue, cellular protein that is normally sequestered will be released. Providing that the cellular protein is capable of immune recognition by the host, this may precipitate a chain reaction. Implicit in the hypothesis that molecular mimicry is responsible for viral induction of autoimmune disease is the notion that immunologic tolerance has somehow been broken, because the viral epitope has been seen as foreign and has elicited an immune response capable of cross-reacting with a host protein.

Humans suffer from numerous autoimmune diseases, ranging from multiple sclerosis to rheumatoid arthritis. All are incurable, many are common, and most are of unknown etiology. Viruses are a major suspect as triggers for these diseases; however, definitive proof has yet to be produced, and this problem continues to be an important area for research. Current theories for which there is some evidence are summarized in Table 9-2.

Immunosuppression

Because the immune system plays a key role in protection against infections, viral damage to it can exacerbate the severity of disease or predispose to superinfection with other infectious agents. The most dramatic effect of viral suppression of the immune system occurs in infections with the human immunodeficiency virus (HIV), the agent that causes the acquired immunodeficiency syndrome (AIDS). The virus damages the immune system in several different ways. It replicates preferentially in $CD4^+$ T lymphocytes and in cells of the monocyte/macrophage lineage. In the latter, cytopathic effects are slight, but the cell functions of phagocytosis, antigen processing/presentation, and cytokine production may be inhibited. In Th cells, viral cDNA is integrated into the chromosomes, and viral replication occurs only following activation of these lymphocytes by certain cytokines. Death of CD4 T cells can occur by apoptosis, or following fusion with other T cells to form syncytia, or by lysis by $CD8^+$ Tc cells. Deletion of CD4 T cells leads to profound immunosuppression, which can be measured by any of a multitude of indicators of loss of Th_1 (Td) and Th_2 (helper T) cell function, that is, impaired delayed-type hypersensitivity and helper functions, as well as diminished cytokine production. Toward the end of the long incubation period (1 to 10 or more years), the CD4 T-cell count drops below 200 per cubic millimeter and the patient falls victim to a succession of opportunistic infections with a range of other viruses, bacteria, fungi, and protozoa, or to cancer. Death follows inexorably, usually within a year of the development of AIDS.

Infections with certain other viruses may temporarily suppress humoral and/or cell-mediated immune responses. Many viruses are capable of productive or abortive replication in macrophages (see Table 7-3). Cytomegalovirus and Epstein-Barr virus replicate nonproductively in B cells, transforming them and altering their function, and several viruses have been shown to grow productively or abortively in activated T lymphocytes. A good example of the latter is measles virus, which replicates nonproductively in T lymphocytes. It has been known since the work of von Pirquet in 1908 that measles infection depressed skin responses to tuberculin and reactivated latent tuberculosis. Other immunologic abnormalities found during the acute phase of measles include spontaneous proliferation of peripheral blood lymphocytes and increased plasma interferon γ. The depressed delayed-type hypersensitivity response, decreased NK cell activity, increased plasma IgE, and increased soluble IL-2 receptor persist for up to 4 weeks after the onset of the rash. These and other immunologic abnormalities probably account for the susceptibility to secondary infections that causes most of the mortality during outbreaks of measles.

Viral Infections in Immunocompromised Patients

Whereas some viral infections induce immunosuppression, certain noninfectious diseases, such as malignancy, or procedures used for therapy, including chemotherapy, radiation, and organ transplantation, may adversely affect the immune system and thus allow enhanced viral replication. The immunocompromised state may be due to genetic defects in the immune system, such as congenital agammaglobulinemia. Most commonly, however, immune dysfunction is due to some other disease, especially tumors that affect the lymphatic system or infection with the human immunodeficiency virus, or to the chemotherapy or radiotherapy employed to treat tumors or to make organ transplants possible.

Patterns of disease in persons with immune dysfunction depend on the nature of the disorder and the arm of the immune system that is primarily involved in containing (or sometimes exacerbating) the particular viral infection. Thus, many viral infections in immunocompromised subjects follow the disease pattern seen in normal persons. Rarely, if symptoms are largely due to the immune response, the disease may actually be milder in immunocompromised patients. Usually, however, the immunocompromised patient suffers more severe disease, and sometimes relatively innocuous viruses can prove lethal. For example, measles in patients with impaired cell-mediated immunity may produce giant cell pneumonia, sometimes several months after the acute infection, and often with fatal consequences. Immunocompromised individuals are not only threatened by exogenous infections but suffer reactivation of latent viruses, notably the human herpesviruses, as well as adenoviruses and polyomaviruses. These situations are discussed in the chapters dealing with specific viral diseases in Part II of this book.

Further Reading

Fazakerley, J. K., and Buchmeier, M. J. (1993). Pathogenesis of virus-induced demyelination. *Adv. Virus Res.* **42,** 249.

Griffin, D. E. (1991). Immunologic abnormalities accompanying acute and chronic viral infections. *Rev. Infect. Dis.* **13**(Suppl. 1), S129.

Lehmann-Grube, F. (1989). Diseases of the nervous system caused by lymphocytic choriomeningitis virus and other arenaviruses. *In* "Handbook of Clinical Neurology" (R. R. McKendall, ed.), p. 355. Elsevier, Amsterdam.

McChesney, M. B., and Oldstone, M. B. A. (1987). Viruses perturb lymphocyte functions; selected principles characterizing virus-induced immunosuppression. *Annu. Rev. Immunol.* **5,** 279.

Mims, C. A., and White, D. O. (1984). "Viral Pathogenesis and Immunology." Blackwell, Oxford.

Notkins, A. L., and Oldstone, M. B. A., eds. (1984, 1986, 1989). "Concepts in Viral Pathogenesis," Vols. 1, 2, and 3. Springer-Verlag, New York.

Oldstone, M. B. A., ed. (1989). Molecular mimicry. Cross-reactivity between microbes and host proteins as a cause of autoimmunity. *Curr. Top. Microbiol. Immunol.* **145,** 1.

Russell, W. C., and Almond, J. W. (1987). "Molecular Basis of Virus Disease," Fortieth Symp. Soc. Gen. Microbiol., Cambridge Univ. Press, Cambridge.

Webster, A., and Griffiths, P. (1990). Virus infections in immunocompromised patients. *In* "Topley and Wilson's Principles of Bacteriology, Virology and Immunity" (L. H. Collier and M. C. Timbury, eds.), 8th Ed., Vol. 4, p. 151. Arnold, London.

Chapter 10

Persistent Infections

Many viruses, especially those that remain localized in the respiratory tract or the intestinal tract, cause only acute (self-limited) infections, which result rarely in death or more commonly in recovery with elimination of the virus from the body, as was discussed in Chapters 6–9. Others have the capacity to establish infections that persist for years or even for life. Presumably this offers such viruses a long-term evolutionary advantage provided they do not bring about the demise of the host. Most of the viruses we discuss in this chapter have probably satisfied this criterion throughout much of human history, but it is pertinent to note that the balance has been altered for the worse in recent years with the advent of AIDS and the introduction of a range of medical procedures that, on the one hand, rescue congenitally immunodeficient infants and, on the other hand, profoundly suppress the immune system for the purposes of organ transplantation or cancer chemotherapy. Immunosuppression, whether natural or iatrogenic, is the principal trigger for reactivation of normally silent persistent infections.

Establishment and maintenance of a persistent infection implies two obvious prerequisites: avoidance of elimination by the immune system and limitation of expression of the genome (of cytocidal viruses). In the case of a true latent infection in which the genome persists in the absence of any viral replication, there is the further requirement that reactivation of the latent genome to instigate a round of viral replication must occur at some time during the life of the host. Certain other features are not essential to viral persistence. Whether persistence leads to disease is obviously of crucial importance to us, the hosts, but may be of little relevance to the virus whose future in the long haul of evolution depends more critically on shedding into the environment and consequent spread to other hosts.

Persistent viral infections are important for several reasons. (1) They are often of epidemiologic importance, since the carriers of the virus serve as a source of infection for other persons, thereby enabling such viruses to persist in small populations, even if their infectivity is low. (2) They may be reactivated and cause acute episodes of disease. (3) They may lead to immunopathologic disease. (4) They are sometimes associated with neoplasms. (5) They may become established even in vaccinated persons unless the vaccine has elicited a strong memory T-cell response. (6) They cannot be eradicated by antiviral chemotherapy because the latent genome is not susceptible to antiviral agents when not replicating.

Categories of Persistent Infections

There are several alternative routes to persistence. For convenience, persistent infections may be subdivided into four categories:

1. Acute infections with late complications
2. *Latent infections,* in which infectious virus is not demonstrable except when reactivation occurs, and disease is usually absent except perhaps during some or all of such recurrences (see Table 10-1)
3. *Chronic infections,* in which infectious virus is always demonstrable and often shed, and disease may be absent or chronic, or may develop late, often with an immunopathologic or neoplastic basis (see Tables 10-2 and 10-3)
4. *Slow infections,* in which infectious virus gradually increases during a very long preclinical phase, leading to a slowly progressive lethal disease (see Table 10-4)

The key distinctions between the four groups of persistent infections are illustrated diagrammatically in Fig. 10-1. It may be noted that these categories are defined primarily in terms of the extent and continuity of viral replication in the body during the long period of persistence; the presence or absence of shedding and of disease are secondary issues as far as this categorization is concerned. Some persistent infections in all categories are associated with disease, some are not, but the *carrier* status of all such persons makes them a potential source of infection for others.

It is important to recognize that some persistent infections possess features of more than one of these categories, under different circumstances, at different times, or in different cell types. AIDS, for example, presents features of latent, chronic, and slow infections; the viral genome is latent in some cells, while simultaneously being fully expressed in others, and the disease develops only after a very long incubation period. Nevertheless, it is useful to retain the terms so as to focus attention on the vital question, namely, what is the virus doing during its lifelong sojourn in the body?

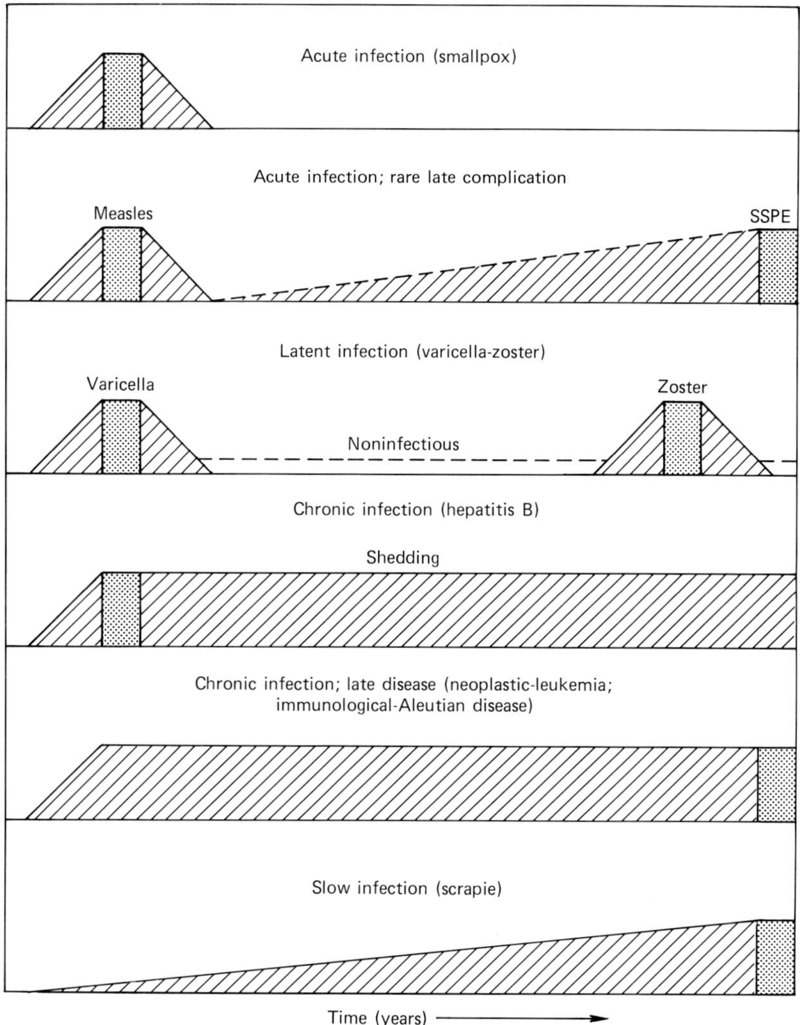

Fig. 10-1 Diagram depicting acute self-limited infection and various kinds of persistent infection, as exemplified by the diseases indicated. The time scale is notional and the duration of various events approximate. Hatching indicates presence of infectious virus; solid line, demonstrable infectious virus; dashed line, virus not readily demonstrable; stippled box, disease episode.

Acute Infections with Rare Late Complications

Subacute Sclerosing Panencephalitis

The paradigm of persistent infection characterized by an acute infection with rare late complications is subacute sclerosing panencephalitis (SSPE), an invariably fatal complication occurring 1–10 years after recovery from measles (Table 10-1). It occurs in only 1 in 300,000 cases and has become very rare indeed following the reduction in measles itself by widespread immunization of children. Nevertheless, it remains of great interest because of the light it sheds on the pathogenesis of persistent infections in the brain.

Patients with SSPE reveal very little measles virus in the brain but exceptionally high titers of neutralizing antibodies in cerebrospinal fluid. The RNA viral genome is detectable in neurons by nucleic acid hybridization, and nucleocapsids are demonstrable by electron microscopy and immunofluorescence. Studies on infected neurons in culture indicate that, in the presence of high-titer neutralizing antibody, antigenic modulation brings about a pronounced inhibition of transcription from the viral genome. Even in the absence of antibody it is clear that there is something peculiar to neurons, and/or their state of differentiation, which down-regulates transcription from the measles viral RNA. Transcription from paramyxovirus genomes characteristically displays polarity, that is, a progressive decline in efficiency of transcription as the polymerase progresses from the 3′ to the 5′ end of the genome. This gradient is much steeper in neurons infected with measles virus, with the result that the production of the envelope proteins, M, F, and H, is reduced and the L protein is hardly made at all. Thus, virtually no infectious virions are assembled, but there is sufficient transcription of the 3′ "housekeeping" genes, N and P/C, to permit viral RNA replication and transfer of nucleocapsids at a very slow rate from cell to cell. Further, during the long incubation period of SSPE, numerous mutations accumulate, seen particularly in the M gene transcripts. All in all, SSPE illustrates beautifully how peculiarities of particular viruses and their host cells may conspire to provide conditions enabling the viral genome to persist—in this case, to the detriment of the host.

Latent Infections

All six human herpesviruses are common and important pathogens which establish lifelong latency and may be reactivated at any time. When the trigger is profound immunosuppression, for example, in AIDS or organ transplantation, reactivation of a latent herpesvirus infection may be lethal. Paradoxically, although all herpesviruses share the common lifestyle of latency, they have evolved divergent strategies for achieving that end. A key distinction is that herpes simplex and varicella–zoster viruses persist in neurons, which are nondividing but long-lived cells, whereas cytomegalovirus, EB virus, and human herpesvirus 6 persist in lymphocytes, which are dividing but short-lived (Table 10-1).

The state of the genome is a central consideration in latency. With some species of viruses the genome must become integrated into a cellular chromosome, whereas for others it survives satisfactorily as a free plasmid (episome) in the cytoplasm or nucleus. Of course, it is essential that any latent viral genome, whatever its physical state or location, be complete if it is ever to be able to program the replication of viable virus following subsequent reactivation (in contrast with the oncogenic potential of many defective viral genomes which may induce cancer following integration even though incapable of further replication). However, the expression of the latent genome is, by defi-

Table 10-1
Examples of Latent Infections: Persistent Infections with Viral Latency and Recrudescence with or without Recurrent Clinical Episodes

Virus	Disease	Target cell		Virus shedding[a]
		During acute attack	During latency	
Herpes simplex virus	1. Primary oral or genital herpes	Epithelial cells	As DNA, in neurons of sensory ganglia	Sporadically in saliva or genital secretions between attacks; plentiful in recurrent herpes vesicles
	2. Recurrent herpes simplex or genitalis	Neurons then epithelium		
Varicella–zoster virus	1. Chickenpox	Widespread	As DNA, in satellite cells of sensory ganglia	From throat and skin lesions; no shedding after recovery from acute disease
	2. Shingles	Neurons then epithelium		From skin lesions; contacts may contract chickenpox
Cytomegalovirus	Usually subclinical except in fetus or immunocompromised	Epithelial cells	As DNA, in salivary glands, kidney epithelium, leukocytes	Sporadically throughout life in saliva and urine, especially during pregnancy
Epstein-Barr virus	1. Glandular faver (mononucleosis) 2. Burkitt's lymphoma, nasopharyngeal carcinoma	Epithelial cells and lymphoid tissue	As DNA, in B cells	In saliva in acute phase
Polyomavirus JC	Progressive multifocal leukoencephalopathy	Unknown (initial), oligodendrocytes (reactivation)	Kidney epithelium	Sporadically in urine, especially during pregnancy

[a] All reactivated by immunosuppression.

nition, repressed, wholly or partially. Generally, only a restricted range of viral genes is transcribed during latency, but of course all are derepressed during reactivation. Often the particular genes that are expressed during latency fulfill a vital role in the maintenance of latency, in a fascinating variety of ways.

Herpes Simplex

Primary infections with herpes simplex virus type 1 (HSV-1) may be subclinical or may produce an acute stomatitis in early childhood or tonsillitis/pharyngitis in adolescence. At intervals of months or years after recovery from the primary infection, the characteristic vesicular lesions reappear, usually on the lips. Herpes simplex virus type 2 (and sometimes type 1) causes comparable initial and recurrent lesions on the male and female genitalia (see Figs. 20–5 and 20–6). During primary infection of the host with herpes simplex virus, viral nucleocapsids are translocated by retrograde axonal flow to the cranial or spinal sensory ganglia, where the viral genome thereafter persists indefinitely. Periodically, reactivation of these latent infections can be triggered by a variety of stimuli, such as "stress," ultraviolet light, fever, nerve injury, or immunosuppression (Fig. 10-2). Clues regarding the mechanism of establishment and maintenance of latency, and of subsequent reactivation, have begun to accumulate.

Following replication in epidermal cells of skin or mucous membrane, some virions access the peripheral nervous system via sensory nerve endings, and the nucleocapsids ascend axons to reach the nucleus of a small minority of the neurons in the corresponding sensory ganglion in the brain stem (e.g., trigeminal ganglion following oral infection) or spinal cord (sacral ganglion following genital infection). Productive infection leads to the destruction of some neurons, but most survive. Studies in mice suggest that these infected neurons are protected by antiviral cytokines secreted by $CD8^+$ T cells that have been activated by exposure to viral peptides presented on the surrounding capsular cells (analogous to microglia) but that cannot lyse infected neurons because neurons do not express significant amounts of MHC class I or II glycoproteins. Interferons reduce expression of HSV "immediate-early" (α) genes, which are absolutely required for the expression of all other HSV genes. Surviving neurons harbor 10–100 copies of the viral genome indefinitely, in the form of nonintegrated, circular, extended concatemers. No virions are made because none of the standard mRNAs are transcribed, hence no viral proteins are synthesized. However, each latently infected neuron produces thousands of molecules of a family of unusual overlapping nonpolyadenylated antisense RNA transcripts known as LATs (latency-associated transcripts). The 2-kb LAT is partially complementary in sequence to the mRNA for a key regulatory protein, which is required for the *trans*activation of all later HSV genes.

Following reactivation by immunosuppression or perhaps by hormonal influences not yet understood, replication of virus is induced in a proportion of the latently infected neurons. Virus is transported down the axon to the periphery where it multiplies once again in epithelial cells in the same general locality as those infected originally. A $CD4^+$ Th_1 cell-mediated inflammatory

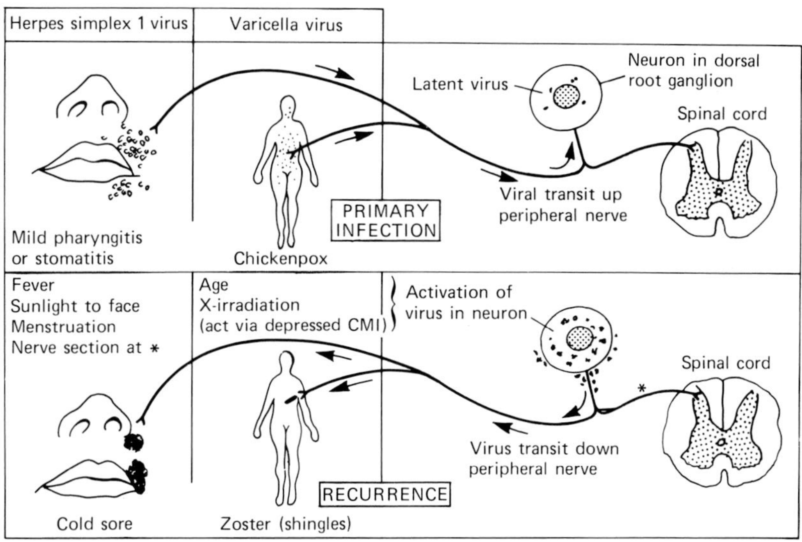

Fig. 10-2 Mechanisms of latency in herpes simplex and zoster. Primary infection occurs in childhood or adolescence; latent virus is located in the cerebral or spinal ganglia. Reactivation of the herpes simplex virus causes recurrent herpes simplex; reactivation of varicella virus causes zoster. (From C. A. Mims and D. O. White, "Viral Pathogenesis and Immunology." Blackwell, Scientific, Oxford, 1984.)

response, perhaps supported by a $CD8^+$ Tc response, quickly leads to the elimination of infected epithelial cells, and preexisting antibody mops up any free virions to bring the recurrence of disease to a halt.

At first glance it may appear improbable that neurons, being nondividing cells, could present any long-term survival advantage to a viral genome. However, neurons do offer certain unique advantages: (1) because they express no MHC antigens they are shielded from lysis by cytotoxic T lymphocytes, and because the latent viral genome expresses no protein they are also a safe haven from lysis by antibody plus complement, or ADCC; (2) because the neuron does not divide there is correspondingly no need for the viral genome to divide to maintain a fixed number of copies per cell; (3) the axon of a sensory neuron, many centimeters long, provides a direct pathway to the periphery, where productive replication in susceptible epidermal cells, even decades later, ensures dissemination of infectious virus to other susceptible hosts.

Herpes Zoster

Herpes zoster, familiarly known as shingles, is characterized by a rash that is usually limited largely to an area of skin innervated by a single sensory ganglion (see Fig. 20-7). It occurs in older people, who have had varicella in childhood. Epidemiologic evidence, supported by serological and virologic studies, including analysis of the genome of virus from both diseases, show that zoster and varicella are two clinical manifestations of the activity of a single virus, varicella–zoster virus (VZV). Zoster results from the reactivation of virus that has remained latent since an attack of varicella, typically many years earlier (see Fig. 10-2). The latency of VZV is more tightly maintained than that of HSV, as indicated by (1) lack of evidence of asymptomatic shedding of virus between attacks, (2) the fact that reactivation is triggered only by declining immunity to the virus, not by fever, ultraviolet light, etc., and (3) the fact that most latent VZV infections never reactivate, and in the remaining 15% only a single recurrence (herpes zoster) occurs and this is largely confined to the elderly. On the other hand, herpes zoster typically involves a whole dermatome, with much more extensive lesions than generally occur with recurrent herpes simplex, and much greater pain, sometimes including protracted neuralgia.

These clinical observations are consistent with the following findings arising from limited investigations of sensory ganglia taken from asymptomatic humans at autopsy. Viral replication cannot be reactivated by explantation or cocultivation of neurons as it can with HSV, nor have neurons been demonstrated to contain the viral genome or any RNA transcripts. However, a tiny proportion of the satellite cells (endothelial and fibroblastic cells resembling glia) which surround the neurons in the ganglion do contain the viral genome as well as several species of RNA transcripts, but no equivalent of the HSV LATs. In contrast, during VZV reactivation, the relevant ganglia of humans or experimental animals display hemorrhagic necrotizing lesions. This has led to the following tentative model. The VZV genome is latent in nonneuronal cells within sensory ganglia. Suppression of T-cell-mediated immunity activates replication of virus which then spreads to involve many of the satellite cells as well as neurons throughout the ganglion, causing severe pain, and lesions in peripheral epidermal cells sometimes covering the whole of the corresponding dermatome.

Epstein-Barr Virus

The mechanism of Epstein-Barr virus (EBV) persistence is completely different. Evolution of the facility for long-term latency of the EBV genome in lymphocytes, which unlike neurons are short-lived dividing cells, has involved the development of two special capabilities: (1) stimulation of the lymphocyte to divide and (2) maintenance of the genome in the form of an episome (plasmid) while also controlling the viral genome copy number in synchrony with the increase in cell number.

In the Third World most infants acquire a subclinical infection with EBV by salivary spread from the mother. In Western countries infection may first occur as a result of amorous interchanges in adolescence, when it may take the form of glandular fever, otherwise known as infectious mononucleosis. This is characterized by a striking proliferation of B lymphocytes which is usually quickly contained by a T-cell-mediated immune response. However, if the patient suffers from any severe underlying congenital or acquired T-cell immunodeficiency, various types of severe EBV-induced progressive lymphoproliferative diseases may prove lethal. Moreover, in certain parts of the world EBV infection in childhood may lead decades later to death from cancer, either a B-cell malignancy known as Burkitt's lymphoma, or nasopharyngeal carcinoma (see Chapter 11).

Virions of EBV replicate productively in epithelial cells of the nasopharynx and salivary glands, but B lymphocytes are generally nonpermissive and constitute the repository for long-term latent infection. The availability of EBV-transformed lymphoblastoid cell lines for biochemical analysis has greatly extended our understanding of the mechanism of EBV latency, in B cells at least. The entire EBV genome persists in the nucleus as multiple copies of a supercoiled covalently closed circular DNA molecule. Only a very small part of the large viral genome is transcribed. A number of the RNA transcripts have unknown function, but others are translated to produce a family of six "Epstein-Barr nuclear antigens" (EBNA-1 to EBNA-6) and a "latent membrane protein" (LMP). EBNA-1 acts on a specific origin of DNA replication, *oriP* (distinct from the origin used for productive infection), and is absolutely required for replication of the EBV plasmid and maintenance of the copy number. EBNA-2 is required for B cell *immortalization,* that is, induction and maintenance of proliferation, which is a key feature of both acute infectious mononucleosis and chronic EBV lymphoproliferative diseases. On the other hand, the membrane protein LMP appears to be implicated in the further step of *transformation,* that is, tumorigenicity resulting from loss of contact inhibition and of response to differentiation signals; this may contribute to the induction of nasopharyngeal carcinoma by EBV. Burkitt's lymphoma cells evade Tc-cell-mediated immune cytolysis by down-regulation of expression of the cell adhesion molecules ICAM-1 and LFA-3 which normally serve as ligands to engage the complementary receptors LFA-1 and CD2 on T cells.

Less is known about reactivation of EBV from the latent state. The switch to replication appears to be triggered by a transactivator known as Zta which transactivates other EBV regulatory genes; this viral gene is presumably repressed during latency.

Cytomegalovirus

Infections with cytomegalovirus (CMV) and EBV are characterized by initial prolonged excretion of virus followed by a state of latency. The CMV infections are normally subclinical, but severe disease occurs in patients with some kind of immunodeficiency, either from AIDS, lymphoreticular disease, or immunosuppression for organ or tissue transplantation. In such patients generalized cytomegalovirus infection may result from reactivation of an endogenous latent infection, or from an exogenous primary infection resulting from the organ graft or from a blood transfusion, reflecting the widespread occurrence of symptom-free carriers in the general population. Indeed, surveys show that at any time about 10% of children under the age of 5 years are excreting CMV in their urine. Transplacental infection of the fetus when a primary CMV infection or reactivation of a latent infection occurs during pregnancy may induce devastating congenital abnormalities in the neonate.

Cytomegalovirus establishes latent infection in salivary glands and kidneys, as well as in monocytes and/or lymphocytes. Virus is shed, intermittently or continuously, particularly into the oropharynx, from which it may be transmitted via saliva, and also into the urine. However, it remains to be proved which cell type(s) comprises the principal reservoir of the viral genome, and whether this persistent infection is one of true latency or whether it is better described as a chronic infection with continuous low-level virus production (see below).

Progressive Multifocal Leukoencephalopathy

Progressive multifocal leukoencephalopathy (PML) is a rare, lethal manifestation of reactivation of an almost universal latent infection with human JC polyomavirus. The acute infection generally occurs in childhood and is usually subclinical, or there may be mild respiratory disease. The virus then persists for life in the kidneys, and perhaps also in the brain or elsewhere, and is shed in urine from time to time, especially during pregnancy or immunosuppression. Almost nothing is known of the mechanism of establishment and maintenance of latency. The PML condition is a demyelinating disease of the brain (see Fig. 18-4), seen only in severely immunocompromised individuals, especially AIDS patients, organ transplant recipients, and those with advanced disseminated lymphoid malignancies. The target cell is the oligodendrocyte, not the neuron.

Chronic Infections

In domestic animals a large number and variety of viral infections fall into the category of chronic persistent infections, which is marked by continuous virus production. Culture of cells from various animal tissues, especially monkey kidney cells, has revealed many such viruses, usually after explantation or by cocultivation with permissive cell lines. There are fewer such infections in humans, but those that do occur cause serious diseases (Tables 10-2 and 10-3).

Hepatitis B

Hepatitis B is the most important chronic viral infection of humans, especially in Asia and Africa, where there are some 250 million carriers. During acute infections the virus replicates in the liver and circulates in the plasma, usually in association with a great excess of smaller particles composed of viral surface antigen (HBsAg). In most infected individuals HBsAg and virions are cleared from the circulation, but in 5–10%, including over 90% of those infected during infancy, a persistent infection is established which can extend for many years, often for life. The carrier state is characterized by continuous production of HBsAg and usually infectious particles, which are plentiful in the bloodstream and less so in semen and saliva, hence the danger of such persons as blood donors and sexual partners. Some carriers develop chronic hepatitis and cirrhosis, and hepatitis B virus (HBV) is an important cause of primary hepatocellular carcinoma.

In acute hepatitis B and in the "high replicative phase" of the chronic carrier state the viral DNA genome exists in hepatocytes as an unintegrated closed circle, and numerous infectious virions are produced. In the subsequent "low replicative phase" of the chronic carrier state the genome is more

Table 10-2
Examples of Chronic Infections: Virus Demonstrable for Long Periods, Sometimes No Disease

Virus	Site of infection	Virus shedding	Disease
Hepatitis B virus	Liver	In plasma[a], genital secretions, saliva	Hepatitis, cirrhosis, hepatocellular carcinoma
Hepatitis C virus	Liver	In plasma[a], genital secretions, saliva	Hepatitis, cirrhosis, carcinoma
Rubella virus[b]	Many organs	Urine, saliva, tears of neonate	Congenital rubella syndrome, teratogenic effects

[a] Much higher titer in plasma, hence higher probability of transmission by transfusion of (unscreened) blood or blood products and needle sharing among injecting drug users.
[b] Congenital infection of fetus during first 16 weeks of pregnancy.

commonly integrated into the chromosomes of the hepatocyte. Transcription from this integrated DNA is restricted largely to subgenomic mRNA, and the major protein translated is HBsAg. No full-length RNA transcripts are produced from the integrated genome, and therefore it cannot replicate as it requires faithful complete RNA copies for the viral reverse transcriptase to utilize as template.

The fact that the number of circulating noninfectious HBsAg particles exceeds the number of virions by a factor of up to 10^5 suggests that this may have evolved as a survival strategy whereby the abundant HBsAg serves as a decoy by absorbing most or all of the neutralizing antibody. Certainly, no free antibody against HBsAg can be detected throughout the many years of the chronic carrier state, but extensive deposition of antigen–antibody complexes in kidneys and arterioles commonly causes "immune complex disease." Cytotoxic $CD8^+$ T lymphocytes, which are principally responsible for clearance of the acute infection by cytolysis of infected hepatocytes, recognize class I MHC associated peptides derived from the two core antigens HBcAg and HBeAg; however, in the low replicative phase of the chronic carrier state, integrated HBV DNA generally produces no RNA transcripts encoding these two antigens, and further, only productively infected hepatocytes express substantial amounts of MHC class I protein. In addition, there is some evidence that chronic carriers display a degree of HBV-specific immunosuppression, mediated by suppressor T cells and/or tolerization of B cells.

Hepatitis C

With the decline in incidence of posttransfusion hepatitis B following effective screening of donated blood, hepatitis C emerged as the commonest cause of post-transfusion jaundice in Western countries. It is also particularly prevalent in injecting drug users. Although clinically milder than hepatitis B, hepatitis C more frequently progresses to chronic hepatitis and vies with alcoholism as the most important cause of liver cirrhosis. Antibody–antigen complexes may circulate in the blood for many years before the virus is eventually eliminated. Little is known of the pathogenetic mechanisms underlying persistence.

Table 10-3
Examples of Chronic Infections, with Virus Always Demonstrable, and with Late Immunopathologic or Neoplastic Disease

Virus/Genus	Host	Site of persistent infection	Late disease[a]
Lymphocytic choriomeningitis virus (*Arenavirus*)	Mouse	Widespread	Glomerulonephritis
Aleutian disease virus (*Parvovirus*)	Mink	Macrophages	Hyperglobulinemia, arteritis, glomerulonephritis
Equine infectious anemia virus (*Lentivirus*)	Horse	Macrophages	Anemia, vasculitis, glomerulonephritis
Human immunodeficiency virus (*Lentivirus*)	Human	Macrophages Lymphocytes	Opportunistic infections due to profound immunodeficiency
Avian leukosis virus[b] (*Retrovirus*)	Chicken	Widespread	Occasionally leukemia

[a] Despite presence of high titers of antibody which fail to eliminate the virus.
[b] Also occurs as latent infection with integrated provirus, transmitted congenitally and not causing disease.

Chronic Infections with Late Disease

A number of interesting diseases of animals fall into the category of chronic infections with late disease. Space does not allow discussion in this book, but some are listed in Table 10-3. In some respects human immunodeficiency virus could be also considered to fit this category, but we have chosen to discuss it as a slow infection (below).

Slow Infections

The term *slow infections* was originally used to describe slowly progressive retroviral diseases found in sheep in Iceland. The term is now used to categorize several viral infections that have a very long preclinical phase (incubation period) leading on to a slowly progressive, invariably lethal disease. The infectious agent can be recovered from infected animals during the preclinical phase as well as after clinical signs have appeared. There are two groups of slow infections (Table 10-4), caused by the lentiviruses and by unclassified agents called *prions* which are postulated to cause the subacute spongiform encephalopathies.

Human Immunodeficiency Viruses/AIDS

Acquired immunodeficiency syndrome (AIDS) is the quintessential persistent infection. It is also the most enigmatic, for it displays features of latent, chronic, and slow infections. At any time far more cells carry the viral genome silently (latency) than are producing new virions. Yet, at all times some cells are manufacturing virus, and this is continuously shed (chronicity), albeit sometimes in small amounts, not only into the bloodstream but also in various bodily secretions. And one can hardly deny that a disease with a mean incubation period (or preclinical stage) of up to a decade is, by definition, a slow infection!

Although first described in 1981 among gay males in the United States, it is now clear that AIDS has been present in Africa for decades and that its customary mode of spread is by conventional heterosexual intercourse. Once infected via genital secretions or blood, the host mounts an immune response but fails to eliminate the virus. A lifelong persistent infection is established in lymphoid organs and the brain, with the virus replicating slowly, principally in $CD4^+$ T lymphocytes (helper T cells) and in macrophages. Small amounts of virus are present continuously or intermittently in the blood and genital secretions; hence, carriers transmit infection to sexual contacts. Throughout a prolonged asymptomatic phase averaging about 10 years, there is a steady decline in the number of $CD4^+$ cells. The patient becomes progressively susceptible to infection with a succession of viruses, bacteria, and fungi, which themselves place heavy demands on the weakened immune system. Cancer may also develop. The cumulative effect of these debilitating opportunistic infections is accompanied by an escalating decline in the $CD4^+$ T cells, and death inevitably follows (see Figs. 35-9 and 35-10).

The human immunodeficiency virus (HIV) belongs to the genus *Lentivirus* (*lenti,* slow) of the family *Retroviridae.* The replication of HIV, like that of all retroviruses (see Chapters 11 and 35), requires reverse transcription of its RNA genome to produce a dsDNA copy (cDNA) which becomes integrated permanently into a chromosome of the host cell as provirus. During the long preclinical phase of HIV infection an increasing proportion of the $CD4^+$ T cells carry HIV cDNA, either integrated or unintegrated, which remains latent until the T cell becomes activated by an appropriate cytokine or by transactivation by another virus (see Chapter 35). Such latently infected T cells express no viral protein and therefore are not susceptible to immune elimination. Moreover, productively infected T cells and macrophages are able to transfer infection to nearby uninfected cells by fusion, involving interaction of the HIV envelope protein with the receptor (CD4) on lymphocytes and macrophages, thus evading neutralizing antibody. In addition, during its long so-

Table 10-4
Slow Infections: Long Preclinical Phase, Slowly Progressive Fatal Disease

Genus or group	Virus or agent	Host	Sites of infection	Disease
Lentivirus	Visna/maedi virus	Sheep	Macrophages, brain and lung	Slowly progressive pneumonia or encephalitis
	Caprine arthritis–encephalitis virus	Goat	Macrophages, brain and joints	Arthritis, encephalitis
	Human immuno-deficiency viruses	Humans	$CD4^+$ T cells, macrophages, brain	Acquired immune deficiency syndrome (AIDS)
Prions	Scrapie agent Kuru agent Creutzfeldt-Jakob agent	Sheep Humans Humans	CNS and lymphoid tissue	Slowly progressive encephalopathy

journ in the body, HIV undergoes an unparalleled degree of genetic drift. The reverse transcriptase is notoriously error-prone, so numerous nucleotide substitutions accumulate in the genome as the years go by. From these mutants two phenotypes tend to be selected. First, escape mutants with amino acid changes in key antigenic determinants recognized by antibody or by T cells (antigenic drift) progressively evade the immune response. Second, there is progressive selection for more virulent mutants with a predilection for $CD4^+$ T cells.

Thus, HIV possesses a remarkable array of weapons that combine to assure its survival in the host: (1) by inducing the formation of syncytia it can spread from cell to cell without encountering antibody; (2) viral cDNA can persist indefinitely intracellularly as a latent infection to be reactivated later; (3) by virtue of the low fidelity of reverse transcription, an inexhaustible supply of different HIV mutants is produced, enabling the virus to alter its cellular tropism, gain in virulence, and evade neutralization by the prevailing antibodies; and most telling of all, (4) the principal pathologic effect of the virus is to suppress the immune system of the host by destruction of helper T cells and macrophages.

Subacute Spongiform Encephalopathies

The term subacute spongiform encephalopathy is used as a generic name for several lethal neurodegenerative diseases that have strikingly similar clinicopathologic features and causative agents, namely, scrapie of sheep and goats, bovine spongiform encephalopathy, mink and feline encephalopathy, wasting disease of deer and elk, and kuru, Creutzfeldt-Jakob disease, Gerstmann–Sträussler–Scheinker syndrome, and fatal familial insomnia in humans. The basic lesion is a progressive vacuolation in neurons and, to a lesser extent, in astrocytes and oligodendrocytes, an extensive astroglial hypertrophy and proliferation, and finally a spongiform change in the gray matter (Fig. 10-3).

The prototype of this group, scrapie, is an infection of sheep, usually transmitted from ewe to lamb. The preclinical phase (incubation period) is very long, up to 3 years, and once signs have appeared the disease progresses slowly but inevitably to paralysis and death. In 1987 an alarming outbreak of a disease very similar to scrapie, bovine spongiform encephalopathy (BSE), emerged in cattle in Great Britain. It is now clear that BSE (colloquially known as "mad cow disease") was due to infection of cattle with scrapie virus via feed that contained bonemeal from sheep carcasses that had been treated in a fashion that failed to destroy the infectivity of the scrapie agent. The use of such cattle feed was prohibited in 1988.

Meanwhile, a number of unusual human diseases meeting the clinical criteria of subacute transmissible spongiform encephalopathy were discovered. The most exotic is kuru (Fig. 10-4). First described in 1957, it was confined to a particular tribe in the remote New Guinea highlands, the women and children of which practiced ritualistic cannibalism of their own deceased relatives. For a time the majority of all deaths in the women of the region were due to kuru. Following the abandonment of this macabre custom, the disease steadily declined in incidence but, because of the exceptionally long incuba-

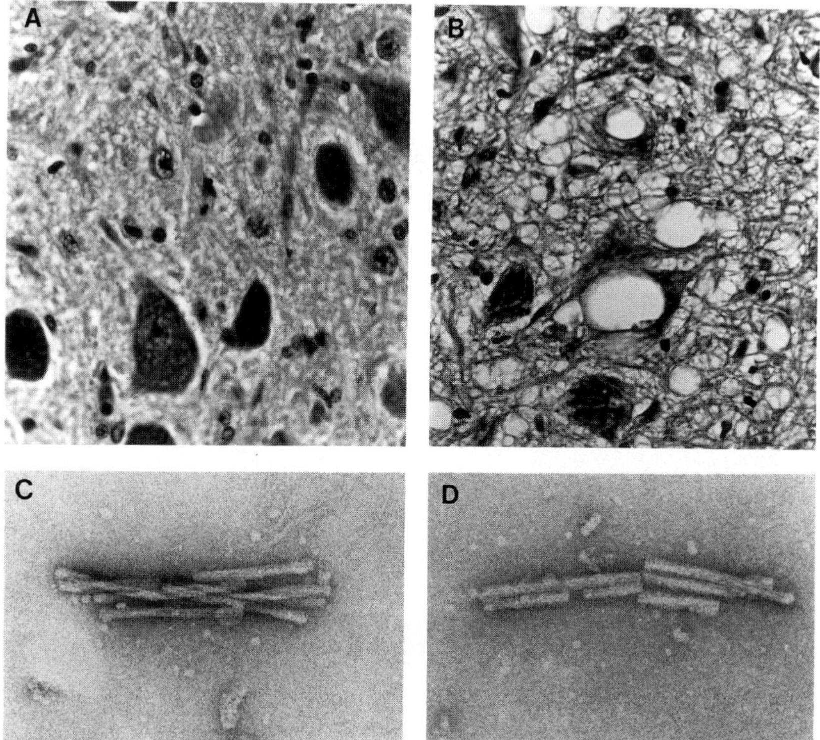

Fig. 10-3 Scrapie. (A) Section of brain of a normal sheep. (B) Section of brain of a sheep that died of scrapie. Note extensive spongiform changes, but no evidence of inflammatory or immune response (hematoxylin and eosin stain; magnification: ×420). (C, D) Electron micrographs of extensively purified prion protein rods from a case of Creutzfeldt-Jakob disease, negatively stained with uranyl formate (bar, 100 nm). (A, B, Courtesy Dr. J. D. Foster; C, D, courtesy Dr. S. B. Prusiner.)

tion period of the disease, occasional cases are still being seen. Gajdusek was able to demonstrate an infectious etiology by serial passage of the agent in chimpanzees, which after a long incubation period died with a disease indistinguishable from the human counterpart.

The importance of Gajdusek's discovery was enhanced when it was demonstrated that a similar or identical agent causes the more widespread sporadic human presenile dementia, Creutzfeldt-Jakob disease (CJD), and that both agents bear a striking resemblance to that of scrapie of sheep. In all human populations that have been adequately studied CJD occurs with a frequency of one in a million persons per year; these sporadic cases do not appear to have been due to infection, although a number of cases of iatrogenic transmission of CJD have been traced to neurosurgery (e.g., dura mater grafts or implantation of cerebral electrodes), corneal transplants, and more recently, administration of human growth hormone derived from pituitary glands. Gerstmann–Sträussler–Scheinker syndrome and fatal familial insomnia are two familial forms of CJD, due to autosomal dominant genes.

Research on this fascinating new group of agents was accelerated by the

Fig. 10-4 Kuru. Child from the Fore region of the New Guinea highlands showing a disorder of gait due to irreversible cerebellar degeneration. (Courtesy of the late Sir Macfarlane Burnet.)

discovery that mice and hamsters could be infected with the scrapie agent, producing an identical disease but with an incubation period of less than 1 year. Using the mouse model it was shown that scrapie could be transmitted sequentially from animal to animal. Early filtration studies suggested that the causative agent was the size of a very small virus. Purification from scrapie-infected brain tissue by density gradient centrifugation revealed that infectivity was associated with filamentous rods called *scrapie-associated fibrils* or *prion rods* (Fig. 10-3). These appear to be artifacts of the purification protocol, although they contain large amounts of infectious scrapie prions. However, the infectious agent is not a conventional virus. Not only do the prion proteins lack nucleic acid, but infectivity is very resistant to inactivation by a variety of physical and chemical treatments that inactivate viruses, including nucleases, UV-irradiation, formalin, mild proteolysis, and even boiling! Moreover, the agent fails to elicit an immune response in the host.

Until recently there have been three theories about the etiologic agents of the subacute spongiform encephalopathies. Some workers believed that the cause was a conventional virus, yet to be discovered. A second view was that the causal agent might be a *virino,* defined as an infectious agent containing nucleic acid insufficient to encode any functional protein but sufficient to serve some regulatory function. The third and now widely accepted hypothesis is that infectivity resides in a *prion,* defined as a small proteinaceous infectious particle. Before speculating on how such an agent could replicate and be infectious, we need to provide more data on the protein itself.

Prions recovered from scrapie brains are oligomers composed of a 33–35 kDa protein designated PrP^{Sc}, standing for the scrapie isoform of the prion protein. This protein has the same amino acid sequence as a protein (PrP^{C}) found in the membrane of normal neurons of the host concerned, but differs in conformation. The genes for the prion proteins of humans reside on the short arm of chromosome 20, and of the mouse on the homologous region of chromosome 2. The genes for the infectious prion proteins of all the subacute spongiform encephalopathies and their laboratory derivatives (from mice and hamsters) have now been sequenced. Changes of host range that in viruses would be ascribed to mutations appear to be due to posttranslational configurational changes in the precursor protein of the new host, with preservation of its amino acid sequence.

CJD can be thought of as the equivalent of a rare somatic mutation, due not to a change in the nucleotide sequence of the prion protein, but to the spontaneous generation of the PrP^{Sc} isoform from the host PrP^{C} by a rare stochastic event, with an incidence of one per million persons per year. Humans suffering from the rare autosomal dominant inherited disease Gerstmann–Sträussler–Scheinker syndrome (GSS) have a single amino acid substitution in the normal cellular prion protein, PrP^{C}, due to a mutation in codon 102, which increases the likelihood of the production of PrP^{Sc}, and therefore of subacute spongiform encephalopathy, about one millionfold. If this point mutation is introduced into the murine PrP^{C} gene and this modified gene is used to produce transgenic mice, the mice spontaneously develop the typical neurologic dysfunction and pathognomonic neuropathology of GSS and scrapie.

If prions are an abnormal isoform of a normal cellular protein, by what mechanism could they reproduce themselves? One hypothesis is that the abnormal prion isoform PrP^{Sc} forms a heterodimer with the normal cellular PrP^{C}, serving as a template which alters the folding of the latter to that of PrP^{Sc}, which in contrast to PrP^{C} is highly resistant to proteolysis. The process would then cascade exponentially, with the conformationally altered cellular molecule in turn serving as a template for modifying the folding of further PrP^{C} molecules as they are produced; these abnormal isoforms accumulate because of their resistance to digestion. In such a fashion a protein molecule might be capable of autocatalytic replication, enabling it to behave like an infectious agent. In the case of the hereditary diseases such as GSS it is assumed that the inherited point mutation itself causes the abnormal conformation which facilitates dimerization to produce the molecule which does the damage.

Strong support for the hypothesis that the causative agent of the subacute spongiform encephalopathies is in fact the scrapie prion protein comes from experiments with transgenic mice in which both copies of the PrP^{C} gene were disrupted. When these homozygous "prion gene knockout" mice were inoculated with mouse scrapie prions the agent failed to replicate (because the mice did not produce PrP^{C}), no pathology was demonstrable, and no disease developed. However, if hamster PrP^{C} genes were introduced into these mice so they produced hamster PrP^{C}, they were fully susceptible to infection with scrapie prions of hamster origin, but not with those of mouse origin.

Pathogenesis of Persistent Infections

The term persistent infections embraces such a wide variety of different conditions that it is not surprising to find that there are several mechanisms whereby the causative viruses bypass the host defenses that eliminate virus in acute infections. The mechanisms include factors related primarily to the virus on the one hand and to the host defenses on the other, although the two kinds of factors interact in some instances (Table 10-5).

Abrogation of the Cytocidal Capacity of the Virus

Restricted Expression of Viral Genes

Obviously a virus cannot persist in a cell it destroys. Therefore, long-term persistence of a potentially cytocidal virus can occur only if the viral genome remains fully or partially silent. Accordingly, latency is maintained only as long as no viral gene with the capacity to kill the cell is expressed. As a rule the few early genes that are transcribed are actually instrumental in the maintenance of latency. Latency is eventually overridden, perhaps following immunosuppression and/or by the action of a cytokine or hormone that derepresses transcription of the whole viral genome, leading to reactivation of viral synthesis. This paradigm is best illustrated by the herpesviruses.

Latency in Nonpermissive, Resting, or Undifferentiated Cells

A particular species of virus may undergo productive replication in one cell type but nonproductive latent infection of another. For example, EB virus may replicate productively in a mucosal epithelial cell but assume the latent state in B lymphocytes; hence, one cell type may serve as a repository which, following reactivation, may seed the other. Even in a given cell type, permissiveness may be determined by the state of cellular differentiation or activation. For instance, papillomaviruses replicate only in fully differentiated epithelial cells. Again, HIV replicates in $CD4^+$ T cells activated by an appropriate cytokine but remains latent in resting $CD4^+$ T cells. Moreover, HIV enjoys a quite different type of association with cells of the monocyte/macrophage lineage, and monocyte-tropic variants tend to be replaced by lymphocyte-tropic variants in the same patient as the years go by.

Noncytocidal Viruses

Arenaviruses and retroviruses are two excellent examples of noncytocidal viruses that establish chronic infections in their rodent hosts without killing the cells in which they replicate and causing little or no damage until certain complications may develop later in life.

Evasion of the Immune Response

Evasion of Neutralization by Antibody

Prions are composed of what is basically a normal cellular protein containing just one or a few amino acid substitutions. Accordingly, they are nonimmunogenic, eliciting neither antibody nor a cellular response; they do not

Table 10-5
Ineffective Immune Responses in Persistent Viral Infections[a]

Phenomenon	Mechanism	Example[b]
Evasion of antibody	Nonimmunogenic prion	Prions
	Blocking by nonneutralizing antibodies	Lymphocytic choriomeningitis virus
	Excess soluble antigen as decoy	Hepatitis B virus
	Suppressor epitopes	HIV
	Antigenic drift	HIV, visna
Reduced antigen display on plasma membrane	Limited viral gene expression	Herpes simplex virus (in neurons)
	"Stripping" by antibody	Measles (SSPE brain)
	Down-regulation of MHC antigen	Adenovirus
	Down-regulation of cell adhesion molecules	Epstein-Barr virus (Burkitt's lymphoma)
	Virus-coded Fc receptor blocks immune lysis	Herpesviruses
Immunosuppression by infection of effector cells	Abrogation of lymphocyte function	HIV ($CD4^+$ T cells)
	Polyclonal B-cell activation	Epstein-Barr virus
	Abrogation of macrophage function	HIV
	Antibody-enhanced infection of macrophage via Fc receptor	Many viruses
Evasion of cytokines	Negation interferon, TNF	Adenovirus
Induction of immunologic tolerance	Congenital infection induces T-cell nonresponsiveness	Congenital rubella and cytomegalovirus
		Lymphocytic choriomeningitis virus
Sequestration in sanctuaries	Blood–brain barrier	HIV (brain)
	Luminal surface of kidneys, glands	Cytomegalovirus, polyomavirus
	Keratinized skin	Papillomavirus

[a] Some of these mechanisms are also operative in nonpersistent infections.
[b] Speculative only; in several instances an association exists but no cause-and-effect relationship has been demonstrated between the immunologic phenomema and the persistent infection listed.

induce interferon production and are not susceptible to its inhibitory effects. This represents the extreme example of immunologic tolerance due to molecular mimicry. There seems to be no mechanism whereby the host can restrict the replication and pathologic effects of these agents.

True viruses are, of course, immunogenic. Some have evolved strategies for evading neutralization by the antibody they elicit.

Cell Fusion Lentiviruses (e.g., HIV), paramyxoviruses (e.g., measles virus), and herpesviruses (e.g., cytomegalovirus) cause adjacent cells to fuse together, enabling the viral genome to spread contiguously from cell to cell without ever being exposed to antibody.

Blocking by Nonneutralizing Antibodies Very high titers of antibody are characteristic of many chronic viral infections, so much so that virus–antibody and antigen–antibody complexes accumulate at the basement membranes of renal glomeruli and other sites, causing a variety of immune complex diseases. Yet the virus is not eliminated. Much of this antibody may be directed against viral proteins or epitopes that are not relevant to neutralization; by binding to the virion nonneutralizing antibodies can block the attachment of neutralizing antibody, by steric hindrance.

Antigen Decoy The chronic carrier state in hepatitis B is marked by the production of a huge excess of noninfectious particles of HBsAg, which may serve as a decoy to mop up neutralizing antibody. Perhaps other chronic infections are also sustained by this strategy.

Immunosuppressive Epitopes The HIV envelope glycoprotein, and the equivalent in other retroviruses, contains a sequence that inhibits T-cell proliferation in response to antigen.

Antigenic Drift Mutations emerge with unusual frequency in retroviruses because their reverse transcriptase is error-prone. During the prolonged incubation period of the slow infections characteristic of lentiviruses such as equine infectious anemia, visna/maedi in sheep, and HIV in humans, there is ample time for antigenic drift to occur in the presence of neutralizing antibody.

Reduced Display of Antigen on Plasma Membrane

An equally dazzling variety of strategies help viruses to persist by diminishing the display of antigen on the cell surface, as seen by T lymphocytes or by antibody.

Limitation of Viral Gene Expression As discussed above, latency is characterized by the production of very few viral proteins, and hence the cell membrane contains little of interest to passing T cells or antibodies. For example, in HSV latency the neuron displays no viral antigen at all. This protects the cell not only against cytotoxic T cells but also against lysis by antibody–complement or by antibody-dependent cell-mediated cytotoxicity.

Antibody-Induced Removal of Antigen from Plasma Membrane Antibody, being at least divalent, can bridge antigen molecules in membranes, bringing about "capping" followed by endocytosis or shedding of the antigen. The phenomenon is readily demonstrable *in vitro* on "carrier cultures" of cells persistently infected with budding viruses and has been postulated to play a role in SSPE, in which antibody also down-regulates transcription of the measles viral genome.

Down-regulation of MHC Antigen Expression Because $CD8^+$ T cells see viral peptides bound to MHC class I antigen (and $CD4^+$ T cells see them in the context of MHC class II), viral persistence is presumably favored by reduction of the concentration of MHC molecules on the cell surface. Adenoviruses encode an early protein that binds to newly synthesized MHC class I antigen, preventing its normal processing thus reducing its cell surface expression.

Down-regulation of Cell Adhesion Molecules As discussed above, Burkitt's lymphoma cells, which carry the EB virus genome, display reduced amounts of the adhesion molecules ICAM-1 and LFA-3, and therefore bind T cells with lower affinity.

Immunosuppression by Infection of Effector Cells

Many viruses can replicate productively or abortively in cells of the reticuloendothelial system, and it is noteworthy that these cells are often implicated in persistent infections. Lymphocytes and monocytes/macrophages represent tempting targets for any virus, in that they move readily throughout the body and can seed virus to any organ, as well as being the key players in the immune response. To render them impotent would ensure persistence. The extreme example of destruction of the body's immune system is provided by HIV, which replicates in $CD4^+$ T lymphocytes and cells of the monocyte/macrophage series. Virtual elimination of helper T cells from the body results in such profound depression of the immune response that the patient dies from intercurrent infections with other opportunistic pathogens, or from cancer.

Abrogation of Lymphocyte Function Numerous other viruses temporarily induce generalized immunosuppression by abrogating the function of one particular arm of the immune response. For example, measles virus suppresses Th_1 cells (DTH). Epstein-Barr virus induces polyclonal activation of B cells which diverts the immune system to irrelevant activity. And so on.

Abrogation of Macrophage Function In many chronic viral infections the virus grows mainly in reticuloendothelial tissue, especially in macrophages. This may impair several key immunologic functions, notably phagocytosis, antigen processing, and presentation to T cells.

Evasion of Cytokines

Interferons are induced by infection with viruses and display a wide range of antiviral as well as immunomodulatory activities (described in Chapters 5 and 7). However, at least some viruses have evolved genes that in one way or another sabotage the specific antiviral action of the key effector molecules known as interferon-regulated proteins (see Chapter 5), whereas others can counter other antiviral cytokines (e.g., an adenovirus gene product partially protects infected cells against tumor necrosis factor).

Induction of Immunologic Tolerance

The probability of an acute infection progressing to chronicity is strongly age-related. In particular, congenital transmission to the neonate, whether transplacental or perinatal, greatly enhances the likelihood. Prolonged chronicity is the rule rather than the exception following congenital infections with hepatitis B virus, rubella virus, cytomegalovirus, parvovirus B19, HTLV-1, or HIV. Presumably this reflects immunologic immaturity/nonresponsiveness. No B-cell tolerance is demonstrable, but there is often a degree of T-cell unresponsiveness to the virus. In the well-studied lymphocytic choriomeningitis (LCM) virus infection of mice, infant mice infected naturally *in utero* do not mount a T-cell-mediated immune response to the virus; the fact that this is reversible indicates that it is due not the deletion of LCM virus-reactive T-cell clones during embryonic life, but rather to clonal anergy or production of suppressor T cells.

Sequestration in Sanctuaries

A striking proportion of persistent infections involve the central nervous system. The brain is insulated from the immune system to some degree by the blood–brain barrier and, further, neurons express very little MHC antigen on their surface, thereby conferring some protection against lysis by cytotoxic T lymphocytes. Herpesviruses, polyomaviruses, and lentiviruses are good examples. HIV is relatively protected not only in the brain but also in the epididymis. Certain other viruses grow in epithelial cells on luminal surfaces,

Table 10-6[a]
Examples of Reactivation of Persistent Viral Infections in Humans

Circumstances	Virus	Features	Clinical severity
Old age	Varicella virus	Rash of shingles	+
Pregnancy	Polyomavirus JC, BK	Viruria	−
	Cytomegalovirus	Replication in cervix	−
	Herpes simplex virus 2	Replication in cervix	Danger to newborn
Immunosuppression by cytotoxic drugs	Herpes simplex virus 1, 2	Vesicular rash	Sometimes severe
	Varicella virus	Vesicular rash (chickenpox or shingles)	++
	Cytomegalovirus	Fever, hepatitis, pneumonitis	Sometimes severe
	Epstein-Barr virus	Increased shedding from throat	−
	Polyomavirus BK, JC	Viruria very common	−
	Hepatitis B virus	Viremia	−
Lymphoid tumours	Varicella virus	Shingles common	+
	Polyomavirus JC	Progressive multifocal leukoencephalopathy	++
AIDS	Cytomegalovirus	Retinitis, colitis, pneumonitis	+
	Polyomavirus JC	Progressive multifocal leukoencephalopathy	++

[a] Based on C. A. Mims and D. O. White, "Viral Pathogenesis and Immunology," Blackwell, Oxford, 1984.

for example, cytomegalovirus in kidney, salivary gland, or mammary gland, and are shed more or less continuously in the corresponding secretions. Most such viruses are not acutely cytopathic, and perhaps because they are released on the luminal borders of cells they avoid cellular immune or inflammatory reactions. Secretory IgA, which is present in the secretions at such sites, does not activate complement, and hence complement-mediated cytolysis or virolysis does not occur. An extreme case of inaccessibility to the immune system is the skin wart; the papillomavirus replicates only in the outer, fully differentiated layers of the dry keratinized lesion.

Reactivation of Persistent Infections

Table 10-6 lists examples of conditions caused by reactivation of persistent viral infections in humans. Further details are given in relevant chapters in Part II of this book.

Further Reading

Ahmed, R., and Stephen, J. G. (1990). Viral persistence. *In* "Fields Virology" (B. N. Fields, D. M. Knipe, R. M. Chanock, M. S. Hirsch, J. L. Melnick, T. M. Monath, and B. Roizman, eds.), 2nd Ed., p. 241. Raven, New York.

Chesebro, B. W., ed. (1991). Transmissible Spongiform Encephalopathies. *Curr. Top. Microbiol. Immunol.* **172.**

Cooper, M. D., ed. (1992). Immunodeficiency. *Curr. Opin. Immunol.* **4,** 461.

Croen, K. D. (1991). Latency of the human herpesviruses. *Annu. Rev. Med.* **42,** 61.

Fauci, A. S., ed. (1993). The Immunology of HIV Infection. *Semin. Immunol.* **5**(3), 147.

Gajdusek, D. C. (1990). Subacute spongiform encephalopathies: Transmissible cerebral amyloidoses caused by unconventional viruses. *In* "Fields Virology" (B. N. Fields, D. M. Knipe, R. M. Chanock, M. S. Hirsch, J. L. Melnick, T. M. Monath, and B. Roizman, eds.), 2nd Ed., p. 2289. Raven, New York.

Gajdusek, D. C. (1992). Genetic control of de novo conversion to infectious amyloids of host precursor proteins: kuru-CJD-scrapie. *In* "Current Topics in Biomedical Research" (R. Kurth and W. K. Schwerdtfeger, eds.), p. 95. Springer-Verlag, Berlin.

Greene, W. C. (1991). The molecular biology of human immunodeficiency virus type 1 infection. *N. Engl. J. Med.* **324,** 308.

Hollinger, F. B., Lemon, S. M., and Margolis, H. S., eds. (1991). "Viral Hepatitis and Liver Disease." Williams & Wilkins, Baltimore, Maryland.

Levy, J. A. (1993). Pathogenesis of human immunodeficiency virus infection. *Microbiol. Rev.* **57,** 183.

Mahy, B. W. J. (1985). Strategies of virus persistence. *Br. Med. Bull.* **41,** 50.

Mims, C. A., and White, D. O. (1984). "Viral Pathogenesis and Immunology." Blackwell, Oxford.

Narayan, O., ed. (1992). The Lentiviruses. *Semin. Virol.* **3**(3), 135.

Notkins, A. L., and Oldstone, M. B. A., eds. (1984, 1986, 1989). "Concepts in Viral Pathogenesis," Vols. 1–3. Springer-Verlag, New York.

Oldstone, M. B. A. (1991). Molecular anatomy of viral persistence. *J. Virol.* **65,** 6381.

Prusiner, S. B. (1991). Molecular biology of prion diseases. *Science* **252,** 1515.

Prusiner, S. B. (1992). Molecular biology and genetics of neurodegenerative diseases caused by prions. *Adv. Virus Res.* **41,** 241.

Prusiner, S. B. (1993). Prion biology and diseases. *The Harvey Lectures* **87,** 85.

Rapp, F., and Cory, J. M. (1988). Mechanisms of persistence in human virus infections. *Microb. Pathog.* **4,** 85.

Roizman, B., and Sears, A. E. (1990). Herpes simplex viruses and their replication. *In* "Fields Virology" (B. N. Fields, D. M. Knipe, R. M. Chanock, M. S. Hirsch, J. L. Melnick, T. M. Monath, and B. Roizman, eds.), 2nd Ed., p. 1824. Raven, New York.

Stevens, J. G. (1989). Human herpesviruses: A consideration of the latent state. *Microbiol. Rev.* **53,** 318.

Chapter 11

Mechanisms of Viral Oncogenesis

Since the mid-1980s there has been a revolution in cell molecular biology which is beginning to shed light on the mechanisms of regulation of cell growth and differentiation, and the disorders of these processes that are expressed as cancer. Although a great deal remains to be learned, we can now say that tumors are initiated by alterations in the expression of one or more genes that regulate cell growth and/or differentiation. The genetic alterations may be caused by chemical or physical agents or by particular viruses, and they appear to involve certain common intracellular molecular pathways. Study of the infections caused by oncogenic viruses, which are currently known to be etiologically involved in about 15% of human cancers, has provided major insights into the nature of these pathways, the elucidation of which is one of the most active fields in virology.

First we must define a few commonly used terms. *Oncology* is the study of tumors. A *benign tumor* is a growth produced by abnormal cell proliferation which remains localized and does not invade adjacent tissue; a *malignant tumor,* in contrast, is usually locally *invasive* and may also be *metastatic,* that is, spread by lymphatic and blood vessels to other parts of the body. Such malignant tumors are often referred to as *cancers.* Malignant tumors of epithelial cell origin are known as *carcinomas,* those arising from cells of mesenchymal ori-

Table 11-1
Viruses Associated with Malignant Tumors in Humans

Virus family	Species of virus	Kind of tumor
Retroviridae	HTLV-1 virus	Adult T-cell leukemia
Papovaviridae	Human papillomaviruses 5, 8[a]	Squamous cell carcinoma[b]
	Human papillomaviruses 16, 18[a]	Genital carcinomas
Hepadnaviridae	Hepatitis B virus	Hepatocellular carcinoma
Herpesviridae	Epstein-Barr virus	Burkitt's lymphoma
		Nasopharyngeal carcinoma
		B-cell lymphomas[c]
Flaviviridae	Hepatitis C virus	Hepatocellular carcinoma?[d]

[a] And several other HPV types less commonly.
[b] Usually as malignant change in the skin condition epidermodysplasia verruciformis in immunocompromised individuals.
[c] In immunosuppressed individuals.
[d] Etiologic role and mechanism not yet firmly established.

gin as *sarcomas,* and those from leukocytes as *lymphomas* (if solid tumors) or *leukemia* (if circulating cells are involved). The process of development of tumors is termed *oncogenesis,* synonyms for which are *tumorigenesis* and *carcinogenesis.* The capacity to study oncogenesis at a molecular level was greatly facilitated when it became possible to induce the essential genetic changes in cultured cells, a process called *cell transformation.*

The discoveries of the viral etiology of avian leukemia by Ellerman and Bang and of avian sarcoma by Rous in 1908 and 1911, respectively, were long regarded as curiosities unlikely to be of any fundamental significance. However, study of these avian viruses and related retroviruses of mice has greatly expanded our understanding of oncogenesis, as well as paving the way for scientists to sequence the genome and unravel the replication cycle of the human immunodeficiency virus with unprecedented speed following its discovery in 1983 (see Chapter 35). Many retroviruses produce tumors in domestic animals, and one has been implicated as a cause of human leukemia. Hepatitis B virus, the herpesvirus Epstein-Barr virus, and several papillomaviruses are also implicated in cancers of humans (Table 11-1). It is noteworthy that the list, if extended to include virus-induced cancers of other animals, embraces four families of DNA viruses but only one family of RNA viruses, namely, the *Retroviridae,* which require an integrated DNA intermediate for their replication.

Oncogenes and Tumor Suppressor Genes

An important element in our present understanding of oncogenesis has come from the discovery of *oncogenes,* which were originally found in retroviruses, where they are collectively referred to as v-*onc* genes. For each of the more than 60 v-*onc* genes so far identified there is a corresponding normal cellular gene, which is referred to as a c-*onc* gene, or as a *protooncogene,* a term which suggests the origin of the v-*onc* genes of retroviruses. The term oncogene is now applied broadly to any genetic element associated with cancer induction, including some cellular genes not known to have a viral homolog and relevant genes of the oncogenic DNA viruses which do not have a cellular homolog. In the normal cell, c-*onc* genes are involved in cellular growth control. The proteins they encode can be assigned to four major classes: growth factors, growth factor receptors, intracellular signal transducers, and nuclear transcription factors (see Table 11-4).

In 1989 a completely different category of cellular genes called *tumor suppressor genes* was discovered which also play an essential regulatory role in normal cells. Their protein products are involved in negative regulation of growth. This regulatory role may be ablated by mutation in the tumor suppressor gene. However, because mutations in genes exerting a negative effect are recessive, both copies must be inactivated for excessive activity of a c-*onc* gene to occur, leading to cancer. Although it currently appears that at least half of all cancers may be associated with altered tumor suppressor genes, these genes may be only indirectly involved in most cancers caused by viruses, at a late stage in the multistep process that leads to full-blown malignancy.

Cell Transformation

The capacity to study oncogenesis at a molecular level was greatly facilitated when it became possible to induce the essential genetic changes in cultured cells, a process called *cell transformation,* which is the *in vitro* equivalent of tumor formation. Transformation by DNA viruses is usually nonproductive (i.e., the transformed cells do not yield infectious progeny virus); transformation by retroviruses is often productive. Viral (or *proviral*) DNA in transformed cells is integrated into the cell DNA, except in the case of papillomavirus and herpesvirus DNAs, which usually remain episomal.

Transformed cells differ in many ways from normal cells (Table 11-2).

Table 11-2
Characteristics of Cells Transformed *in Vitro* by Viruses

1. Viral DNA sequences present, integrated into cellular DNA or as episomes
2. Greater growth potential *in vitro*
   a. Formation of three-dimensional colonies of randomly oriented cells in monolayer culture, usually due to loss of contact inhibition
   b. Capacity to divide indefinitely in serial culture
   c. Higher efficiency of cloning
   d. Capacity to grow in suspension or in semisolid agar (anchorage independence)
   e. Reduced serum requirement for growth
3. Altered cell morphology
4. Altered cell metabolism and membrane changes
5. Chromosomal abnormalities
6. Virus-specified tumor-associated antigens
   a. Some at cell surface behave as tumor-specific transplantation antigens
   b. New intracellular antigens (e.g., T antigens)
7. Capacity to produce malignant neoplasms when inoculated into isologous or severely immunosuppressed animals

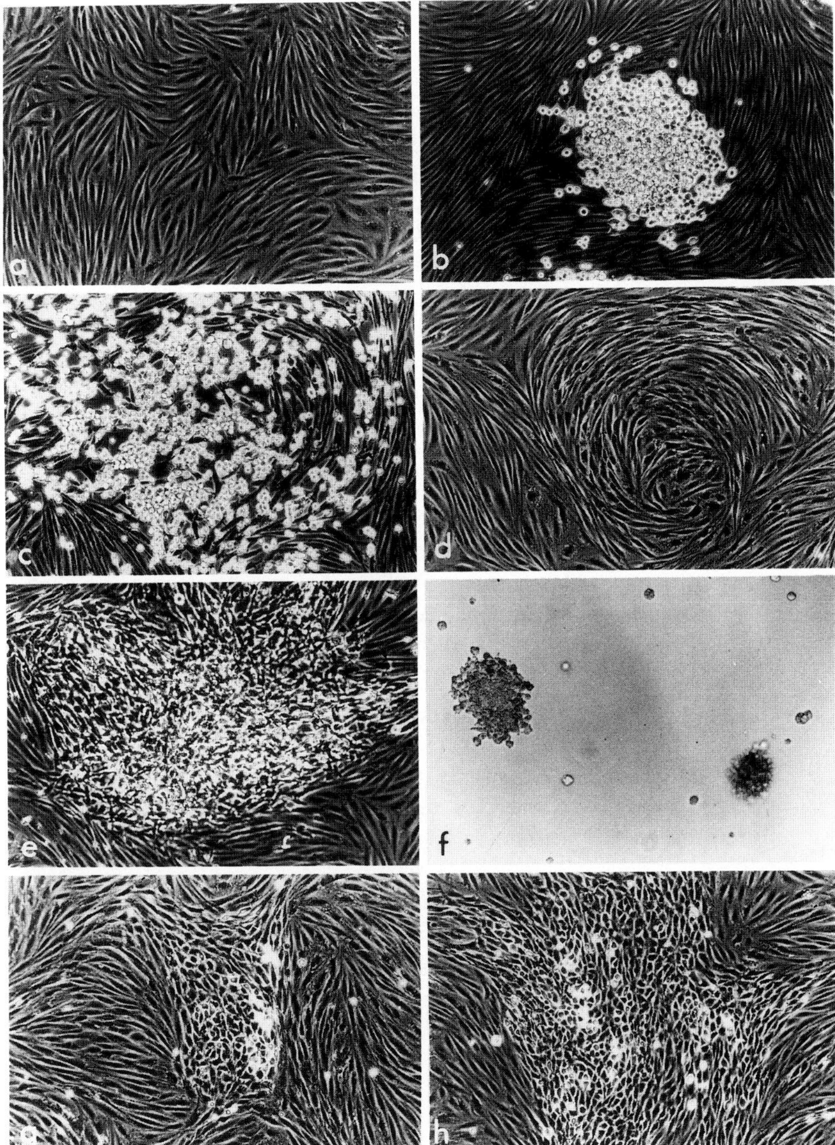

Fig. 11-1 Transformation of a rat fibroblast cell line by different viral oncogenes. (a) Normal F111 cells. (b) Cells transformed by Rous sarcoma virus. (c) Cells transformed by Harvey murine sarcoma virus. (d) Cells transformed by Abelson leukemia virus. (e) Cells transformed by mouse polyoma virus. (f) Polyoma virus-transformed cells in soft agar. (g) Cells transformed by SV40. (h) Cells transformed by simian adenovirus 7. [From T. Benjamin and P. K. Vogt, *in* "Fields Virology" (B. N. Fields, D. M. Knipe, R. M. Chanock, M. S. Hirsch, J. L. Melnick, T. P. Monath, and B. Roizman, eds.), 2nd Ed., p. 322. Raven, New York, 1990. Courtesy Dr. T. Benjamin.]

One of the changes is a loss of control of cell growth (Fig. 11-1); transformed cells acquire a capacity to divide unrestrictedly, which can be demonstrated in a variety of ways, including the capacity to produce tumors in athymic mice, which have defective T-cell immunity but do not support the growth of normal foreign cells. More recently transgenic mice have been used to confirm in whole animals some of the observations of the transformation of cultured cells by cloned genes, either of cellular or viral origin.

Malignant tumors and transformed cells express distinctive antigens, called *tumor-associated antigens.* Cells transformed by nondefective retroviruses also express the full range of viral proteins, and new virions bud from their plasma membranes. In contrast, transformation by DNA viruses usually occurs in cells undergoing nonproductive infection; nevertheless, certain virus-specific antigens are generally demonstrable. Some tumor-associated antigens are located in the plasma membrane, where they constitute potential targets for immunologic attack and are sometimes referred to as *tumor-specific transplantation antigens.*

Tumor Induction by Retroviruses

Although retroviruses of five genera cause leukemias, lymphomas, and/or sarcomas in many different species of animals, in humans only one retrovirus, HTLV-1, has been unequivocally incriminated in oncogenesis, as the cause of a particular form of leukemia. Nevertheless, it is necessary to describe tumor induction by retroviruses in some detail, because such studies are of great importance in understanding mechanisms of oncogenesis.

Categories of Oncogenic Retroviruses

Oncogenic retroviruses are subdivided two ways: according to whether they are replication-competent or replication-defective, and whether they are endogenous or exogenous.

Replication-Competent and Replication-Defective Retroviruses

The genome of a typical replication-competent retrovirus (Fig. 11-2A) consists of two identical copies of a positive sense ssRNA molecule, each of which has three genes: *gag,* encoding the four core proteins, *pol,* encoding the reverse transcriptase, and *env,* encoding the two envelope glycoproteins. A second kind of rapidly oncogenic exogenous retrovirus carries a viral oncogene, generically called v-*onc,* which is responsible for the rapid malignant change of the infected cell. Because the oncogene is usually incorporated into the viral RNA in place of part of one or more normal viral genes (Fig. 11-2C), such viruses are usually defective, dependent on nondefective helper retroviruses for their replication. However, Rous sarcoma virus is atypical in that its genome contains a viral oncogene (v-*src*) in addition to complete copies of the retrovirus genes *gag, pol,* and *env* (Fig. 11-2D), so that it is replication-competent.

Endogenous and Exogenous Retroviruses

A complete DNA copy of the genome (known as the *provirus*) of one, or sometimes more than one, of many retroviruses may be transmitted in the germ-line DNA from parent to offspring and may thus be perpetuated in the DNA of every cell of every individual of certain vertebrate species. Proviral genomes are under the control of cellular regulatory genes and are normally

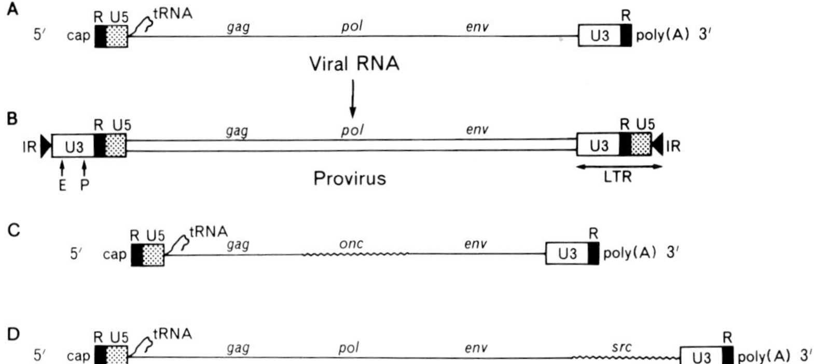

Fig. 11-2 Simplified diagrams of the structure of retrovirus genomes and integrated provirus; in A, C, and D only one of the two identical RNA molecules is shown. (A) Genome of a replication-competent slowly transforming retrovirus. The major coding regions *gag, pol,* and *env* encode the viral proteins. The 5′ terminus is capped and the 3′ terminus is polyadenylated. A short sequence, R, is repeated at both ends of the molecule, while unique sequences, U5 and U3, are located near the 5′ and 3′ termini, immediately proximal to R. There is a 16- to 18-nucleotide sequence adjacent to U5, the primer binding site, which is complementary to the 3′ terminus of a tRNA, which binds to it and acts as a primer for reverse transcriptase. (B) Provirus, a dsDNA, is integrated into cellular DNA. The genome is flanked at each terminus by additional sequences to form the long terminal repeat (LTR). Each LTR comprises U3, R, and U5 plus short inverted repeat sequences (IR) at the distal end. U3 contains the promoter (P) and enhancer (E) sequences, as well as several other sequences with important functions. (C) Genome of a replication-defective rapidly transforming retrovirus. A v-*onc* gene (wavy line) has replaced all of the *pol* and part of the *gag* and *env* genes. (D) Genome of Rous sarcoma virus, a replication-competent rapidly transforming retrovirus. A v-*onc* gene (v-*src*) is present in addition to complete *gag, pol,* and *env* genes.

totally silent in normal animals. Such retroviruses are said to be *endogenous* (Table 11-3). However, expression of proviruses can be induced by various factors such as irradiation, exposure to mutagenic or carcinogenic chemicals, and hormonal or immunologic stimuli, so that virions are synthesized. In contrast, other retroviruses behave as more typical infectious agents, spreading horizontally to contacts, and are said to be *exogenous.*

Most endogenous retroviruses never produce disease, cannot transform cultured cells, and contain no oncogene in the genome. Most exogenous retroviruses, on the other hand, are oncogenic; some characteristically induce leukemia or lymphoma, others sarcoma, and yet others carcinoma, usually displaying a predilection for a particular type of target cell. Exogenous retroviruses can be further subdivided into rapidly oncogenic and slowly oncogenic viruses (Table 11-3). The rapidly oncogenic sarcoma viruses, like Rous sarcoma virus, are the most rapidly acting carcinogens known, causing death in as short a time as 2 weeks after infection in certain host species and rapidly transforming cultured cells *in vitro.* These properties are attributable to the v-*onc* gene which they carry as part of the genome. There are over 60 known v-*onc* genes, the characters of some of which are shown in Table 11-5; most exogenous retroviruses carry only one particular type of v-*onc* gene, for example, v-*src* in the case of Rous sarcoma virus. The genomes of weakly oncogenic ("slowly transforming") viruses contain no v-*onc* gene, but they can induce

Table 11-3
Comparison of Endogenous and Exogenous Retroviruses

Characteristic	Endogenous retroviruses	Exogenous retroviruses	
		Slowly transforming or *cis*-acting retrovirus	Rapidly transforming or transducing retrovirus
Transmission	Vertical (germ line)	Horizontal	Horizontal
Expression	Usually not, but inducible	Yes	Yes
Genome	Complete	Complete	Defective[a]
Replication	Productive	Productive	Requires helper[a]
Oncogene	No	No	Yes
Tumorigenicity	Nil, or rarely leukemia	Leukemia after long incubation period	Sarcoma, leukemia, or carcinoma, after short incubation period
In vitro transformation	No	No	Yes

[a] Except for Rous sarcoma virus.

B-cell, T-cell, or myeloid leukemia with low efficiency and after a much longer incubation period.

Replication of Retroviruses

The complete replication cycle of retroviruses is described in Chapter 35, but certain aspects associated with the integration of the DNA copy of the RNA genome into the cellular DNA need to be described here in order to explain viral oncogenesis. When released in the cytoplasm, the ssRNA genome is converted to dsDNA which is integrated into the chromosomal DNA as provirus (Fig. 11-2A,B). The expression of mRNA from the provirus is under the control of the viral transcriptional regulatory sequences, which include promoter and enhancer elements, both of which are located in the *long terminal repeat (LTR)*. In other ways the proviral DNA behaves as do other chromosomal genes, segregating like other markers and being transmitted via the germ line. The strong promoter in the LTR influences the expression of genes in the vicinity of the insertion site, such that it may increase the expression of downstream cellular genes. Sometimes a cellular gene may be inactivated as a consequence of disruption at the point of insertion of the provirus, and the provirus may also acquire and transduce cellular genes.

Mechanisms of Tumor Production by Retroviruses

Retroviruses produce tumors in one of three ways: (1) the *transducing retroviruses* introduce a v-*onc* gene, which is controlled by viral regulatory sequences present in the LTR of the viral genome, into a chromosome of a normal cell; (2) *cis-activating retroviruses,* which lack a v-*onc* gene, transform cells by integrating close to a c-*onc* gene and thus usurping normal cellular regulation of this gene; and (3) *trans-activating retroviruses* contain a gene that codes for a regulatory protein which may either increase transcription from the viral LTR or interfere with the transcriptional control of specific cellular genes.

Since most transducing retroviruses have lost part of their genome in the course of the original recombination event that led to the acquisition of a c-*onc* gene (now a v-*onc* gene), they are replication-defective and rely for their replication on a helper retrovirus which is always found with them in the animal and in most instances is their progenitor. Transducing retroviruses transform cells rapidly both *in vitro* and *in vivo*. Whereas *cis*-activating retroviruses are replication-competent, induce tumors more slowly, and do not transform cells in culture, *trans*-activating retroviruses either have no oncogenic activity or induce tumors very late, by affecting cellular transcription.

Modes of Action of Transduced Retroviral Oncogenes

The "oncoprotein" products of all retroviral oncogenes appear to act either by interfering with *signal transduction*, that is, the transfer of signals from the membrane to the cell interior, or directly with the regulation of cellular genes. Although the complete pathway via which any given oncoprotein stimulates cell growth is not yet completely comprehended, there is some information on where they fit into the overall picture of growth control of the cell (Fig. 11-3, Table 11-4).

Growth Factors

Only one example of an oncogene encoding a growth factor analog is known, v-*sis*, which codes for one of the two polypeptide chains of platelet-derived growth factor (PDGF). This retroviral v-*onc* gene, under the control of the powerful enhancer/promoter complex in the LTR, induces fibroblasts to manufacture growth factor, which they do not normally express, leading to immortalization of fibroblasts in culture and possibly the induction of malignant tumors.

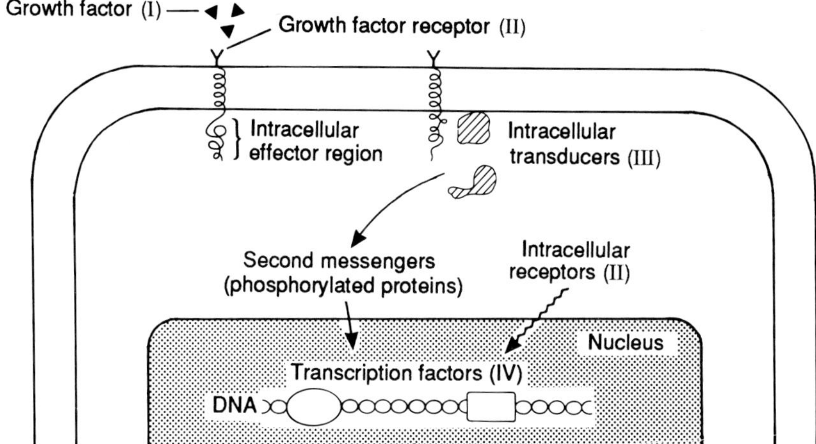

Fig. 11-3 Cellular growth control involves four types of proteins, the genes for which can give rise to oncogenes. They are (I) growth factors, (II) receptors, (III) intracellular transducers, and (IV) intranuclear factors. See Table 11-4 for examples and text for explanation. (Modified from J. E. Darnell, H. Lodish, and D. Baltimore, "Molecular Cell Biology," 2nd Ed., p. 985. Scientific American Books, New York, 1990.)

Table 11-4
Some Retroviral Oncogenes and Functions of Oncoproteins They Encode[a]

Oncogene	Retrovirus	Subcellular location of oncogene product	Nature of oncoprotein
Growth factors			
sis	Simian sarcoma virus	Secreted	Platelet-derived growth factor β chain
Receptors			
Cell surface receptors with protein-tyrosine kinase activity			
erbB	Avian erythroblastosis virus[b]	Plasma membrane	Truncated epidermal growth factor receptor
fms	Feline sarcoma virus	Plasma membrane	Mutated CSF-1 receptor
Intracellular receptors			
erbA	Avian erythroblastosis virus[b]	Nuclear	Thyroxine receptor; activated form prevents differentiation
Intracellular transducers			
Ras proteins			
H-*ras*	Harvey murine sarcoma virus	Plasma membrane	Guanine nucleotide-binding proteins with GTPase activity
Ki-*ras*	Kirsten murine sarcoma virus	Plasma membrane	
Protein-tyrosine kinases			
src	Rous sarcoma virus	Cytoplasm	Protein kinases that phosphorylate tyrosine residues
fps[c]	Fujinami avian sarcoma virus	Cytoplasm	
fes[c]	Feline sarcoma virus	Cytoplasm	
Protein serine/threonine kinases			
mos	Moloney murine sarcoma virus	Cytoplasm	Cytoplasmic serine kinase (cytostatic factor)
Nuclear transcription factors			
myc	Avian myelocytoma virus	Nuclear matrix	Binds to DNA; regulates transcription
myb	Avian myeloblastosis virus	Nuclear matrix	Binds to DNA; regulates transcription
fos	Murine osteosarcoma virus	Nucleus	Transcription factor AP-1
jun	Avian sarcoma virus	Nucleus	Transcription factor AP-1

[a] Oncogenes are designated by three letters in lower case italics; oncogene products (oncoproteins) by the same three letters in roman type with an initial capital. Modified from J. E. Darnell, H. Lodish, and D. Baltimore, "Molecular Cell Biology," 2nd Ed., p. 986. Scientific American Books, New York, 1990.

[b] Transducing retrovirus with two oncogenes.

[c] *fps* and *fes* are the same oncogene derived from avian and feline genomes, respectively.

Growth Factor Receptors and Hormone Receptors

In the normal cell, growth factor receptors bind a particular growth factor and then the receptor sends a growth signal to the cell. For example, the v-*erbB* oncogene product is a truncated epidermal growth factor (EGF) receptor, which lacks most of the extracellular ligand-binding domain but retains the cytoplasmic moiety which possesses tyrosine kinase activity. In the normal cell, this protein kinase becomes activated only after a conformational change in the molecule which follows binding of EGF to its receptor. In the case of the v-*erbB* oncoprotein, however, the enzyme is permanently activated and available to phosphorylate any intracytoplasmic proteins in the vicinity, so initiating the cascade of events culminating in transmission of the appropriate message to the nucleus.

The product of the v-*erbA* gene mimics the intracellular receptor for thyroid hormone, which normally acts by transforming that receptor into a transcription factor. The viral oncoprotein ErbA competes with the natural receptor for the hormone, causing growth without control. In the avian erythroblastosis virus v-*erbA* and v-*erbB* act synergistically to maximize oncogenicity.

Intracellular Transducers

The largest class of oncogenes is derived from genes encoding proteins that act as intracellular transducers, that is, proteins that transmit signals from a receptor to a cellular target. The best understood are the Ha-*ras* and Ki-*ras* oncogenes; their cellular equivalents were the first nonviral oncogenes to be recognized. Most Ras molecules exist in an inactive state in the resting cell, where they bind guanosine diphosphate (GDP). When they receive a physiologic stimulus from a transmembrane receptor they are temporarily activated, leading to the synthesis of guanosine triphosphate (GTP). The *ras* genes acquire transforming properties by mutational changes, mostly point mutations at specific sites, which may stabilize Ras proteins in their active state. This may cause a continuous flow of signal transduction, leading to malignant transformation.

Many oncogenes in this class encode a protein-tyrosine kinase, which differs from the products of the class of oncogenes encoding membrane receptors in that they are cytoplasmic or nuclear proteins, lacking any transmembrane or extracellular domain. Oncogene-encoded kinases recognize and phosphorylate a broad range of target proteins, one or more of which may be crucial to the transformed state.

Nuclear Transcription Factors

By one mechanism or another, the activity of any oncogene must eventually result in a change in the cell nucleus. However, oncogenes of the nuclear transcription factor class encode proteins that directly affect transcription or bind to DNA. The v-*jun* oncoprotein is closely homologous to AP-1, an important transcription factor, and it can bind tightly to another nuclear oncoprotein, Fos. The growth factors and oncoproteins that can induce *jun* are components of mitotic signal chains, and there is evidence that *jun* induces cancer through its role as a transcriptional regulator.

The phosphoprotein genes, of which *myc, myb,* and *fos* are the best known examples, can transactivate other genes and can stimulate DNA replication directly or indirectly. After having been activated by structural and/or regulatory changes, these genes may contribute to tumor development or progression.

Activation of Cellular Oncogenes

There is evidence that c-*onc* genes may themselves be responsible for some transformations. It is not difficult to imagine a tumor arising from overexpression of a c-*onc* gene, or inappropriate expression, for example, in the wrong cell or at the wrong time. Such abnormal c-*onc* transcription may occur in a variety of ways, in several of which oncogenic viruses play a role.

Insertional Mutagenesis

The presence upstream from a c-*onc* gene of an integrated provirus, with its strong promoter and enhancer elements, may greatly amplify the expression of that c-*onc* gene. This is the likely mechanism whereby the weakly oncogenic avian leukosis viruses, which lack a v-*onc* gene, produce tumors. When avian leukosis viruses cause malignancy, the viral genome is generally found to be integrated at a particular location, immediately upstream from a c-*onc* gene. Integrated avian leukosis provirus increases the synthesis of the normal c-*myc* oncogene product 30- to 100-fold. Experimentally, only the LTR need be integrated, and furthermore, by such a mechanism c-*myc* may be expressed in cells in which it is not normally expressed. In some instances a quantitative rather than a qualitative difference in the c-*onc* gene product may be sufficient for oncogenicity.

Transposition

Transposition of c-*onc* genes may result in enhanced expression by bringing them under the control of strong promoter and enhancer elements. For instance, the 8:14 chromosomal translocation that characterizes Burkitt's lymphoma (a tumor of African children associated with Epstein-Barr herpesvirus infection) brings the c-*myc* gene into position just downstream of the strong immunoglobulin promoter. Perhaps v-*onc* genes may sometimes be transposed from their initial site of integration in a similar way.

Gene Amplification

Amplification of oncogenes is a feature of many tumors; for example, a 30-fold increase in the number of c-*ras* gene copies is found in one human cancer cell line, and the c-*myc* gene is amplified in several human tumors. The increase in gene copy number leads to a corresponding increase in the amount of oncogene product, thus producing cancer.

Mutation

Mutation in a c-*onc* gene, for example, c-*ras*, may alter the function of the corresponding oncoprotein. Such mutations can occur either *in situ*, as a result of chemical or physical mutagenesis, or in the course of recombination with retroviral DNA. Given the high error rate of reverse transcription, v-*onc* gene homologs of c-*onc* genes will always carry mutations, and the strongly promoted production of the viral oncoprotein will readily exceed that of the cellular oncoprotein.

Human T-Cell Leukemia Viruses

In 1980 the first human retrovirus was discovered and named the human T-cell lymphotropic virus, type 1 (HTLV-1). Its geographic distribution in Japan, the Caribbean, and central Africa was found to correspond with that of an otherwise uncommon adult T-cell leukemia–lymphoma (ATLL). It is now clear that HTLV-1 is an ancient virus that has been maintained successfully in isolated, relatively inbred racial groups in various parts of the world, being transmitted from mother to baby across the placenta and through breast milk, and between adults by sexual intercourse. Virions are not abundant, being mainly cell-associated, and the efficiency of transmission was apparently too low for long-term survival of the virus in any but such closely knit communities. In modern times, however, its spread has been greatly augmented by sexual promiscuity, blood transfusion, and needle sharing by drug abusers. The virus persists for life, generally causing no disease. However, after a remarkably long incubation period (20–40 years) a very small minority of carriers develop ATLL.

In ATLL, T cells from the tumor carry the HTLV-1 provirus, integrated in a monoclonal fashion, that is, into a particular site within a particular chromosome in every tumor cell within that individual but at a completely different site within every cell of the T lymphoma of another individual. The virus can also be demonstrated to infect and immortalize $CD4^+$ T cells *in vitro*. The mechanism of oncogenesis is of great interest. HTLV-1 fits the definition of a replication-competent exogenous retrovirus, in that it is not genetically inherited in the germ-line DNA but is spread horizontally as a typical infectious agent. Yet its genome carries no oncogene, nor does it transform lymphocytes by insertional mutagenesis, that is, by *cis*-activation of a nearby c-*onc* gene.

The *modus operandi* of HTLV-1 is that of a *trans*-activating retrovirus. In addition to the standard three retroviral genes, *gag, pol,* and *env,* HTLV-1 carries two different regulatory genes known as *tax* and *rex*. Both genes are essential for viral replication, but *tax* also plays a crucial role in transformation of the cell. The product of *tax* is a transcriptional activator protein, which promotes transcription not only from the proviral LTR but also from the regulatory sequences of cellular genes. By acting on the enhancer for the gene encoding the T-cell growth factor interleukin-2 (IL-2), and also on that for the IL-2 receptor gene, the Tax protein establishes an autocrine loop promoting lymphocyte proliferation. The Tax protein also up-regulates two cellular oncogenes, *fos* and *PDGF*, thereby initiating a cascade effect. Mice transgenic for *tax* develop tumors. In a sense, *tax* resembles a viral oncogene. It certainly initiates a chain of events that can lead to cancer. One key question yet to be adequately answered is why, then, does it take up to 40 years for malignancy to become manifest *in vivo*, and why in such a small proportion of those known to be carrying the virus? Is immunosurveillance so effective, or is there an unknown cofactor required for the full flowering of malignancy?

Tumor Induction by DNA Viruses

Although retroviruses are the most important oncogenic viruses in animals, certain DNA viruses (see Table 11-1) are more important as known causes of cancers in humans. DNA tumor viruses interact with cells in one of two ways: (1) productive infection, in which the virus completes its replication cycle, resulting in cell lysis, or (2) nonproductive infection, in which the virus transforms the cell without completing its replication cycle. During such nonproductive infection, the viral genome or a truncated version of it is integrated into the cellular DNA, or alternatively the complete genome persists as an autonomously replicating plasmid (episome). The genome continues to express early gene functions. The molecular basis of oncogenesis by DNA viruses is best understood for polyomaviruses, papillomaviruses, and adenoviruses, all of which contain genes that behave as oncogenes (Table 11-5). These oncogenes appear to act by mechanisms similar to those described for retrovirus oncogenes, primarily in the nucleus, where they alter patterns of gene expression and regulation of cell growth. In every case the relevant genes encode early proteins having a dual role in viral replication and cell transformation. With a few possible exceptions, the oncogenes of DNA viruses have no homologs or direct ancestors (c-*onc* genes) among cellular genes of the host.

The protein products of DNA virus oncogenes are multifunctional, with particular functions related to particular domains of the folded protein, which mimic functions of normal cellular proteins. They interact with host cell proteins at the plasma membrane or within the cytoplasm or nucleus. Polyoma middle T protein (Py-mT), for example, interacts with c-*src*, resulting in increased levels of the protein kinase activity of the Src protein.

Polyomaviruses and Adenoviruses

During the 1960s and 1970s two members of the genus *Polyomavirus*, mouse polyomavirus and simian virus 40 (SV40), as well as certain human adenoviruses (types 12, 18, and 31), were found to induce malignant tumors following inoculation into baby hamsters and other rodents. Although, with the exception of mouse polyomavirus, none of the viruses induces cancer under natural conditions in its natural host, they transform cultured cells of certain other species (see Fig. 11-1) and provide good experimental models for analysis of the biochemical events in cell transformation.

The polyomavirus- or adenovirus-transformed cell does not produce virus. Viral DNA is integrated at multiple sites on the chromosomes of the cell. Most of the integrated viral genomes are complete in the case of the polyomaviruses, but defective in the case of the adenoviruses. Only certain early viral genes are transcribed, albeit at an unusually high rate. By analogy with retrovirus genes, they are called oncogenes (see Table 11-5). Their products, demonstrable by immunofluorescence, used to be known as *tumor (T) antigens*. A great deal is now known about the role of these proteins in transformation. For example, Py-mT of polyoma virus (like the product of the v-*ras* gene of retroviruses) seems to bring about the change in cell morphology, and

Table 11-5
Oncogenes of Adenoviruses and Papovaviruses and Their Products

Virus	Oncogene	Product: Function	Product: Location
Adenoviruses	E1A	Regulates transcription	Nucleus
	E1B	?	Nucleus, membranes
Papillomaviruses	E5	Signaling	Nuclear, membrane
	E6	Transcription/replication	Nucleus, cytoplasm
	E7	Transcription/replication	Nucleus, cytoplasm
Mouse polyomavirus	Py-t	Regulates phosphate activity	Cytoplasm (nucleus)
	Py-mT	Binds and regulates product of c-*src* and related kinases	Plasma membrane
	Py-t	Transcription/replication	Nucleus
Simian virus 40	SV-t	Regulates phosphatase activity	Cytoplasm, nucleus
	SV-T	Initiates DNA synthesis, regulates transcription	Plasma membrane, nucleus

enables the cells to grow in suspension in semisolid agar medium as well as on solid substrates (anchorage independence), whereas Py-T, like the product of the v-*myc* gene of retroviruses, is responsible for the reduction in dependence of the cells on serum and enhances their life span in culture.

Virus can be rescued from polyomavirus-transformed cells, that is, induced to replicate, by irradiation, treatment with certain mutagenic chemicals, or cocultivation with certain types of permissive cells. This cannot be achieved with adenovirus-transformed cells, as the integrated adenoviral DNA contains substantial deletions.

It should be stressed that integration of viral DNA does not necessarily lead to transformation. Many or most episodes of integration of papovavirus or adenovirus DNA have no recognized biological consequence. Transformation by these viruses in experimental systems is a rare event, requiring that the viral transforming genes be integrated in the location and orientation needed for their expression. Even then, many transformed cells revert (*abortive transformation*). Furthermore, cells displaying the characteristics of transformation (see Table 11-2) do not necessarily produce tumors. This capacity needs to be demonstrated independently by transplantation of cells into athymic or syngeneic mice. Further, certain ostensibly normal cell lines commonly used for *in vitro* transformation assays, such as 3T3 cells, are in fact "premalignant"; moreover, transformation of normal cells to the fully malignant state may require the cooperation of more than a single oncogene, for example, either polyoma Py-T, adenovirus E1A, or retrovirus v-*myc* together with at least one other.

On the other hand, integration of the viral genome is not necessary for transformation. As seen below, herpesvirus genomes generally persist as nonintegrated closed circular plasmids, and in the case of the oncogenic herpesvirus EBV, limited transcription from this episomal DNA suffices to transform the cell to the malignant state.

Papillomaviruses

Most human papillomaviruses (HPV) produce benign papillomas (warts) on the skin and mucous membranes which eventually regress spontaneously. However, certain types of HPV initiate changes that may lead to cancer.

There is strong evidence for an etiologic association of human papillomaviruses with carcinoma of the cervix, which is the commonest cancer in women in the developing world and is responsible for about half a million new cases every year. Epidemiologically, cervical carcinoma is strongly correlated with number of different sexual partners; it has almost never been recorded in virgins. Certain genital HPV types, notably 16 and 18, induce cervical dysplasia which may progress to invasive carcinoma.

The viral genome in the form of an unintegrated, autonomously replicating episome is regularly found in the nuclei of the premalignant cells in cervical dysplasia. In contrast, invasive cervical cancers reveal chromosomally integrated HPV-16 or HPV-18 DNA. Each cell carries at least one, and sometimes up to hundreds, of monoclonal incomplete copies of the HPV genome. Integration disrupts one of the early viral genes, E2, and other genes may be deleted, but the viral oncogenes E6 and E7 remain intact and are expressed efficiently. Transfection of cultured cells with just the HPV-16 E6/E7 genes immortalizes them; cotransfection with the *ras* oncogene allows full expression of malignancy. Further, the proteins encoded by the E6/E7 genes from the highly oncogenic HPV types 16 and 18, but not from nononcogenic HPV types, have been demonstrated to bind to the protein products of the human tumor suppressor genes p53 and Rb. Interestingly, it has also been reported recently that HPV-negative cervical cancers reveal a genetic mutation in the p53 gene. Thus it appears that E6/E7-mediated inactivation of a tumor suppressor gene product may represent an important event, but probably not the only one required for the full expression of malignancy in HPV-induced genital cancer.

Most cutaneous types of HPV cause benign skin warts which do not turn malignant. However, there is a rare hereditary condition known as epidermodysplasia verruciformis in which the child becomes infected with one or more unusual HPV types which produce multiple disseminated red scaly patches on the skin; later, some of the lesions tend to undergo malignant change. These squamous cell carcinomas, but generally not those found in immunocompetent persons, are often found to carry, in episomal form, the genome of HPV type 5 or 8, the E6 gene of which displays greater transforming activity than that of other dermatotropic types. The fact that malignant change occurs much more commonly in warts on exposed areas of skin strongly suggests that ultraviolet light is a cocarcinogen in a succession of events culminating in malignancy.

Hepadnaviruses

All three mammalian hepadnaviruses are strongly associated with naturally occurring hepatocellular carcinoma (HCC) in their native hosts; for example, chronically infected woodchucks almost inevitably develop cancer within 2–3 years. Although primary liver cancer of humans is relatively rare in most developed countries, it ranks as one of the commonest cancers in Eastern Asia and sub-Saharan Africa, where 5–20% of the population are carriers of hepatitis B virus (HBV) and perinatal transmission to offspring ensures perpetuation of the pool of chronic carriers, now totaling 250–300 million. Prospective studies have demonstrated that HCC arises about 100 times more commonly in HBV carriers than in controls. An estimated 250,000 people develop HCC each year, typically those who became infected with HBV early in life and developed chronic hepatitis leading to cirrhosis of the liver and eventually to liver cancer 20–50 years postinfection. Hepatitis B virus is second only to tobacco among the known human carcinogens.

In over 80% of cases of HCC, part of the HBV genome is found to be integrated into cellular chromosomes. Integration generally first occurs during chronic infection, decades before cancer becomes apparent. As HBV encodes no integrase, cellular enzymes such as topoisomerase I are presumably involved in this "illegitimate" recombination. It is not restricted to any particular chromosomal location, such as might up-regulate expression of a particular protooncogene. The integrated DNA usually contains deletions, duplications, inversions, and mutations. Later, further recombinative events occur, producing inverted duplications of flanking cellular sequences, often accompanied by further deletion of viral sequences. This suggests that HBV integration somehow promotes genetic instability in the cell.

The hepadnavirus genome contains no oncogene. However, the oncogenic mammalian hepadnaviruses (but not the nononcogenic avian hepadnavirus) contain a gene, X, which encodes a transactivating protein. Most integrated HBV sequences, although truncated, retain at least part of the X gene, including the X promoter and the viral enhancer. It has been postulated that this could deregulate expression of nearby cellular oncogenes. Certain studies of early monoclonal tumors have demonstrated HBV integration within previously unrecognized protooncogenes, encoding cyclin A and retinoic acid receptor b, both of which play important roles in cellular proliferation and differentiation. Moreover, in a minority of HCC cases in woodchucks the viral enhancer is integrated close to a *myc* protooncogene, inducing its overexpression. However, in most human hepatocellular cancers there is no evidence of dysregulation of any cellular oncogene.

It is now recognized that chronic infection with hepatitis C virus (HCV), an RNA virus with no DNA intermediate in its life cycle and no evidence of integration, is at least as important a precursor to HCC as is hepatitis B virus in certain countries, notably Japan. This discovery has reinforced the view that the role of both viruses in carcinogenesis is indirect. Currently there is no unifying hypothesis other than that the cellular proliferation which enables the liver to regenerate remarkably following destruction of hepatocytes maximizes opportunities for sequential development of chromosomal abnormalities that lead to cancer; this sequence may be initiated and/or promoted by genetic abnormalities arising from long-standing chronic infection with HBV or HCV, and may be aggravated by alcoholism, environmental carcinogens such as aflatoxins, or mutagenic oxidants generated by phagocytic and other cells during chronic inflammation.

Definitive proof of an etiologic role for HBV in liver cancer may have to await the results of long-term vaccine trials which commenced in Africa and Asia in the early 1980s. The potential for preventing up to 250,000 deaths from cancer every year, as well as millions more deaths from cirrhosis and liver failure, is ample justification for universal childhood immunization against hepatitis B.

Herpesviruses

Herpesviruses of the subfamily *Gammaherpesvirinae* are the etiologic agents of lymphomas or carcinomas in hosts ranging from amphibia, through birds, to primates including humans. Marek's disease virus, formerly considered a gammaherpesvirus but now classified as an alphaherpesvirus, transforms T lymphocytes, causing them to proliferate to produce a generalized lymphomatosis in chickens. This widespread lethal disease, which inflicts major losses in the poultry industry, is preventable by vaccination with live attenuated vaccines. A recent finding of great interest, yet to be fully evaluated, is that the genome of tumor-inducing Marek's disease virus, but not that of related vaccine strains, contains retrovirus sequences.

Burkitt's Lymphoma

Burkitt's lymphoma (Fig. 11-4) is a malignant B-cell lymphoma of high prevalence in children in tropical Africa at elevations below 1500 meters where malaria is endemic. Epstein and colleagues succeeded in isolating from cultured tumor cells the herpesvirus now known as Epstein-Barr virus (EBV), which more commonly causes infectious mononucleosis (glandular fever) in the developed world.

The regular association between EBV and Burkitt's lymphoma suggested a cause-and-effect relationship. However, EBV is not associated with most of the sporadic cases of Burkitt's lymphoma found outside Africa or New Guinea, while in the areas of high prevalence it is a ubiquitous virus infecting most children subclinically; hence, it was conceivable that the association is fortuitous. However, a long-term prospective seroepidemiologic survey in Uganda firmly established that severe EBV infection acquired very early in life predisposes to Burkitt's lymphoma. In particular, a high antibody titer against the EBV capsid antigen VCA, arising early in childhood, indicates a high risk of subsequently contracting Burkitt's lymphoma.

The EBV genomic DNA is present in multiple copies in each cell of most African Burkitt's lymphomas, in the form of closed circles of the complete viral DNA molecule, found free as autonomously replicating episomes, but the cells do not produce virus unless induced to do so following cultivation *in vitro*. All the cells express the EBV nuclear antigen, EBNA-1, and most also express the early antigen (EA), the viral capsid antigen (VCA), and the membrane antigen (MA). On the other hand, the two major antigens seen by cytotoxic T cells on the surface of cells acutely infected with EBV, namely, EBNA-2 and an integral membrane protein (LMP), are suppressed in Burkitt's lymphoma cells, thus facilitating escape from immunologic surveillance. Unlike asymptomatic carriers of this ubiquitous virus, children with Burkitt's

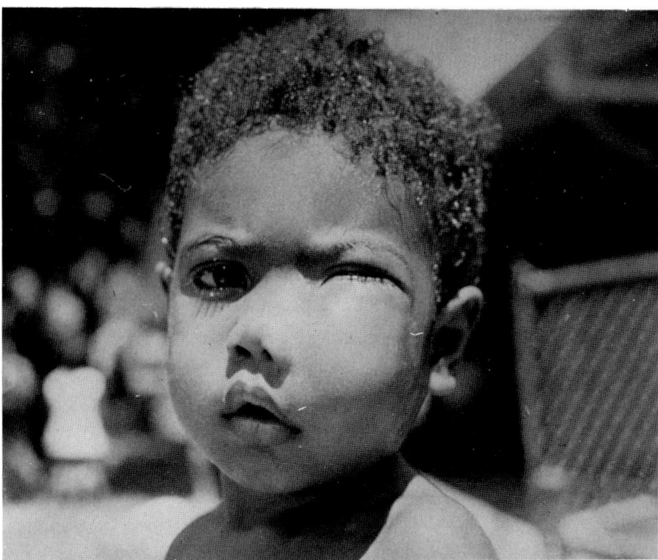

Fig. 11-4 Characteristic facial tumor in a child from Papua New Guinea with Burkitt's lymphoma. (Courtesy Dr. J. Biddulph.)

lymphoma make antibodies to EA and MA, as well as unusually high titers of antibodies against VCA.

The malignant cells also contain a characteristic 8:14 chromosomal translocation. The human c-*myc* oncogene, located on chromosome 8, is transposed to one of three chromosomes that contain genes for immunoglobulin (usually chromosome 14, sometimes 2 or 22), leading to enhanced transcription of c-*myc*. Transgenic mice with the rearranged c-*myc* locus also develop B-cell lymphomas. Some Burkitt's lymphomas also have mutations in the cellular tumor suppressor gene *p53*. Although the details are unclear, Burkitt's lymphoma may develop as a consequence of differentiation arrest due to the translocation, and subsequent growth stimulation resulting from c-*myc* deregulation. In parallel, host defense mechanisms may be weakened by suppression of EBV membrane antigen expression and by blocking programmed cell death mediated by tumor necrosis factor.

Klein proposed that Burkitt's lymphoma develops in three stages: (1) EBV infection arrests B-cell differention and stimulates division, thereby enhancing the probability of chromosomal damage; (2) an environmental cofactor, postulated to be infection with the malaria parasite *Plasmodium falciparum*, impairs the capacity of Tc cells to control this proliferation of EBV-immortalized B cells; (3) chromosomal translocation leads to constitutive activation of the c-*myc* oncogene, resulting in Burkitt's lymphoma.

B-Cell Lymphomas

B-cell lymphomas in immunocompromised patients, rather different from Burkitt's lymphoma, sometimes follow EBV primary infection or reactivation, for example, in AIDS patients or in children with a particular congenital immunodeficiency. Though the molecular pathogenesis of this connection has not been so comprehensively analyzed, the strong direct link with EBV circumstantially strengthens the postulated causality in the case of Burkitt's lymphoma.

Nasopharyngeal Carcinoma

Nasopharyngeal carcinoma (NPC), while uncommon generally, is the commonest cancer in the densely populated regions of southern China and in Cantonese populations who have settled in other parts of the world, as well as in Arctic Inuits (Eskimos) and in East and North Africa. This anaplastic NPC (but not the well-differentiated type more commonly seen in other parts of the world) contains multiple copies of EBV DNA, in episomal form, in every cell. As in the case of Burkitt's lymphoma, the malignant epithelial cells do not synthesize virus except on transplantation into athymic mice or cocultivation *in vitro* with B lymphocytes. Interestingly, EBV normally fails to replicate in or transform cultured epithelial cells, which seem to lack receptors for the virus, but can if EBV or EBV DNA is introduced into epithelial cells artificially. It has been postulated that EBV produced by B lymphocytes trafficking through the copious lymphoid tissue in the pharynx binds to secretory IgA antibody to infect mucosal epithelial cells via the IgA transport pathway. The EBV NPC cells produce EBNA-1 antigen consistently. There is evidence that inappropriate expression of LMP in an undifferentiated epithelial cell may be important in the development of NPC.

The age of primary infection with EBV does not seem to be so critical in NPC, and the interval between infection and development of NPC can range up to 40 years. Reactivation of a latent EBV infection in the nasopharynx, with a consequent rise in anti-VCA antibodies of the IgA class, frequently heralds the development of the tumor. The striking ethnic restriction of EBV NPC, which is retained when Cantonese people emigrate, suggests either a genetic predisposition or a cultural (e.g., dietary) cofactor.

Multistep Oncogenesis

The development of full malignancy requires multiple steps. A potentially neoplastic clone of cells must bypass apoptosis (programmed death), circumvent the need for growth signals from other cells, escape from immunologic surveillance, organize its own blood supply, and possibly metastasize. Thus tumors other than those induced by rapidly transforming retroviruses like Rous sarcoma virus generally do not arise as the result of a single event, but rather by a series of steps leading to progressively greater loss of regulation of cell division. Significantly, the genome of some retroviruses (e.g., avian erythroblastosis virus) carries two different oncogenes (and that of polyomavirus, three), and two or more distinct oncogenes are activated in certain human tumors (e.g., Burkitt's lymphoma). Cotransfection of normal rat embryo fibroblasts with a mutated c-*ras* gene plus the polyomavirus large PyT gene, or with c-*ras* plus the E1A early gene of oncogenic adenoviruses, or with v-*ras* plus v-*myc*, converts them to tumor cells. It should be noted that, whereas v-*ras* and v-*myc* are typical v-*onc* genes, originally of c-*onc* origin, the other two had been assumed to be typical viral genes. Furthermore, a chemical carcinogen can substitute for one of the two v-*onc* genes; following immortalization of cells *in vitro* by treatment with the carcinogen, transfection of a cloned oncogene converts the cloned continuous cell line to a tumor cell line. To achieve full malignancy, mutations in tumor suppressor genes may also be needed.

Such experiments resurrect earlier unifying theories of cancer causation that viewed viruses as analogous to other mutagenic carcinogens, both being capable of initiating a chain of two or more events leading eventually to malignancy. If viruses or oncogenes are considered as cocarcinogens in a chain of genetic events culminating in a tumor, it may be important to determine whether their role is that of initiator or promoter, or either. The most plausible hypothesis may be that (1) c-*onc* genes represent targets for carcinogens (chemicals, radiation, and tumor viruses), and (2) the full expression of malignancy may generally require the mutation or enhanced expression of more than one class of oncogenes, and perhaps also mutations in both copies of critical tumor suppressor genes.

Further Reading

Aaronson, S. A. (1991). Growth factors and cancer. *Science* **254,** 1146.

Benjamin, T., and Vogt, P. K. (1990). Cell transformation by viruses. *In* "Fields Virology" (B. N. Fields, D. M. Knipe, R. M. Chanock, M. S. Hirsch, J. L. Melnick, T. P. Monath, and B. Roizman, eds.), 2nd Ed., p. 317. Raven, New York.

Bishop, J. M. (1991). Molecular themes in oncogenesis. *Cell (Cambridge, Mass.)* **64,** 235.

Blattner, W. A., ed. (1990). "Human Retrovirology: HTLV." Raven, New York.

Cann, A. J., and Chen, I. S. Y. (1990). Human T-cell leukemia virus types I and II. *In* "Fields Virology" (B. N. Fields, D. M. Knipe, R. M. Chanock, M. S. Hirsch, J. L. Melnick, T. P. Monath, and B. Roizman, eds.), 2nd Ed., p. 1501. Raven, New York.

Coffin, J. M. (1990). Retroviridae and their replication. *In* "Fields Virology" (B. N. Fields, D. M. Knipe, R. M. Chanock, M. S. Hirsch, J. L. Melnick, T. P. Monath, and B. Roizman, eds.), 2nd Ed., p. 1437. Raven, New York.

Darnell, J. E., Lodish, H., and Baltimore, D. (1990). "Molecular Cell Biology," 2nd Ed., p. 955. Scientific American Books, New York.

Dulbecco, R. (1990). Oncogenic viruses I: DNA-containing viruses. *In* "Microbiology" (B. D. Davis, R. Dulbecco, H. N. Eisen, and H. S. Ginsberg, eds.), 4th Ed., p. 1103. Lippincott, Philadelphia, Pennsylvania.

Dulbecco, R. (1990). Oncogenic viruses II: RNA-containing viruses (retroviruses). *In* "Microbiology" (B. D. Davis, R. Dulbecco, H. N. Eisen, and H. S. Ginsberg, eds.), 4th Ed., p. 1123. Lippincott, Philadelphia, Pennsylvania.

Gallo, R. C., and Jay, G., eds. (1991). "The Human Retroviruses." Academic Press, New York.

Klein, G., ed. (1981 on). "Advances in Viral Oncology," annual volumes. Raven, New York.

Kung, H.-J., and Vogt, P. K., eds. (1991). Retroviral Insertions and Oncogene Activation. *Curr. Top. Microbiol. Immunol.* **171,** 1.

Rickinson, A. B., ed. (1992). Viruses and Human Cancer. *Semin. Cancer Biol.* **3,** 249.

Schüpbach, J. (1990). "Human Retrovirology." Springer-Verlag, Berlin.

Swanstrom, R., and Vogt, P. K., eds. (1990). Retroviruses: Strategies of Replication. *Curr. Top. Microbiol. Immunol.* **157,** 1.

Teich, N., ed. (1991). Viral Oncogenes, Parts I and II. *Semin. Virol.* **2,** 305.

Tooze, J., ed. (1980). "DNA Tumor Viruses." Cold Spring Harbor Laboratory, Cold Spring Harbor, New York.

Varmus, H. (1988). Retroviruses. *Science* **240,** 1427.

Weinberg, R. A. (1989). "Oncogenes and the Molecular Origins of Cancer." Cold Spring Harbor Laboratory, Cold Spring Harbor, New York.
Weinberg, R. A. (1991). Tumor suppressor genes. *Science* **254,** 1138.
Weiss, R. A., Teich, N., Varmus, H. E., and Coffin, J. M., eds. (1985). "RNA Tumor Viruses," 2nd Ed., Vols. 1 and 2. Cold Spring Harbor Laboratory, Cold Spring Harbor, New York.
zur Hausen, H. (1991). Viruses in human cancers. *Science* **254,** 1167.

Chapter 12

Laboratory Diagnosis of Viral Diseases

Rational Use of the Laboratory

Why bother to establish a definitive laboratory diagnosis of a viral infection when few effective chemotherapeutic agents are available and many viral diseases are clinically obvious or trivial? This is a valid question. The majority of viral infections can be handled without recourse to the laboratory. However, there are several important types of situations where a laboratory diagnosis is needed. These include the following:

1. Diseases may be caused by viruses for which antiviral *chemotherapy* is in fact already available, such as herpesviruses. More importantly, rapid advances in the development of new antiviral agents will surely expand the range of viral diseases for which a precise diagnosis will become necessary. This will be so especially if the new antiviral agents are relatively specific in their antiviral spectra. Furthermore, the emergence of drug-resistant mutants will necessitate drug sensitivity tests following virus isolation.

2. In other diseases the *management of the patient* or the prognosis is influenced by the diagnosis. Specific treatment does not always comprise chemotherapy. For example, the fatal disease rabies is completely preventable by postexposure immunization. Abortion is recommended if rubella is diagnosed in the first trimester of pregnancy. Cesarean section may be the prudent course of action if a woman has primary genital herpes at the time of delivery. Special care and education are required for a baby with congenital

defects attributable to rubella or cytomegalovirus. Immunization at birth is mandatory for a baby of an $HBsAg^{+}$ mother, and so forth. Moreover, establishment of a viral etiology may confer a favorable prognosis and obviate the need for continued antibacterial chemotherapy, for example, in meningitis.

3. Infections may demand *public health measures* to prevent spread to others. For instance, blood banks routinely screen for HIV and hepatitis B and C viruses which may be present in blood donated by symptomless carriers. Herpes simplex type 2 is readily transmissible to sexual partners. Nosocomial infections (e.g., varicella or measles), often in epidemic form, may create havoc in a leukemia ward of a children's hospital, unless hyperimmune IgG is promptly administered to potential contacts following diagnosis of the sentinel case. Documentation of a novel strain of influenza virus may herald the start of a major epidemic against which vulnerable older members of the community should be immunized. Positive identification of a particular arbovirus in a case of encephalitis enables the authorities to promulgate warnings and initiate appropriate antimosquito control measures. Introduction of a dangerous exotic disease demands containment and surveillance, and so on.

4. *Surveillance* of viral infections may determine their significance, natural history, and prevalence in the community, with a view to establishing priorities and means of control, and monitoring and evaluating immunization programs.

5. Continuous surveillance of the community may provide evidence of *new epidemics, new diseases, new viruses,* or *new virus–disease associations.* New viruses and new virus–disease associations continue to be discovered virtually every year. We need only make the point that over 90% of all the human viruses known today were completely unknown at the end of World War II. Opportunities are legion for astute clinicians as well as virologists and epidemiologists to be instrumental in such discoveries.

The traditional approaches to laboratory diagnosis of viral infections have been (1) *isolation of the virus* in cultured cells, followed by identification of the isolate, and (2) detection and measurement of antibody in the patient's serum (*serology*). Although still widely used, both are unacceptably slow to provide an answer; the diagnosis usually arrives too late to exert any influence on treatment, and the attending physician loses faith in the relevance of the laboratory. Therefore, the thrust of developments in diagnostic virology over the past several years has been toward rapid methods that provide a definitive answer in less than 24 hours.

The best methods of diagnosis fulfil the five prerequisites of a truly satisfactory diagnostic assay: $peed, $implicity, $ensitivity, $pecificity, and co$t. Standardized diagnostic reagents of good quality are now commercially available, assays have been miniaturized to conserve reagents, and instruments have been developed to automate the dispensing of solutions into multiwell plastic trays, as well as diluting and rinsing, reading the tests, and computerized analysis and printout of the results. Moreover, a veritable cascade of do-it-yourself kits has flooded onto the market. In the main, these are tests for viral antigen which provide a diagnosis from a single specimen taken directly from the patient during the acute phase of the illness. Solid-phase *enzyme immunoassays (EIAs),* in particular, have revolutionized diagnostic virology and are now the method of choice. More recently, the *polymerase chain reaction (PCR)* is being widely exploited to amplify the number of copies of the viral genome in a clinical specimen to allow them to be identified by hybridization to a labeled *nucleic acid probe.*

Collection, Packaging, and Transport of Specimens

The chance of isolating a virus depends critically on the attention given by the attending physician to the collection of the specimen. Clearly, such a specimen must be taken from the right place at the right time. The right time is as soon as possible after the patient first presents, because virus is usually present in maximum concentration at about the time symptoms first develop, then falls away during the ensuing days. Specimens taken as a last resort when days or weeks of empirically chosen antibacterial chemotherapy have failed are almost invariably useless.

The site from which the specimen is collected will be influenced by the clinical symptoms and signs, together with a knowledge of the pathogenesis of the suspected disease. As a general rule the epithelial surface that constitutes the portal of entry and the primary site of viral replication should be sampled (Table 12-1). The most important specimen, routine in respiratory infections as well as in many generalized infections, is a nasal or throat swab, or in the case of a young child a nasopharyngeal aspirate in which mucus is drawn from the back of the nose and throat into a mucus trap using a vacuum pump (Fig. 12-1). The second important specimen, routine in enteric and many generalized infections, is feces.

Swabs may be taken from the genital tract, from the eye, or from vesicular skin lesions. Some viruses responsible for systemic infections can be isolated from blood leukocytes. Cerebrospinal fluid (CSF) may yield virus in cases of meningitis. Biopsy or autopsy specimens may be taken by needle or knife from any part of the body for virus isolation, or snap-frozen for immunofluorescence. Obviously, tissue taken at autopsy or biopsy for the purpose of virus isolation must not be placed in formalin or any other fixative. In the

Table 12-1
Specimens Appropriate for Laboratory Diagnosis of Various Clinical Syndromes

Syndrome	Specimen
Respiratory	Nasal or throat swab; nasopharyngeal aspirate; sputum
Enteric	Feces
Genital	Genital swab
Eye	Conjunctival (and/or corneal) swab
Skin	Vesicle swab/scraping; biopsy solid lesion
Central nervous system	Cerebrospinal fluid
Generalized	Throat swab[a]; feces[a]; blood leukocytes[a]
Autopsy/biopsy	Relevant organ
Any	Blood for serology[b]

[a] Depending on known or presumed pathogenesis.
[b] Blood allowed to clot then serum kept for assay of antibody.

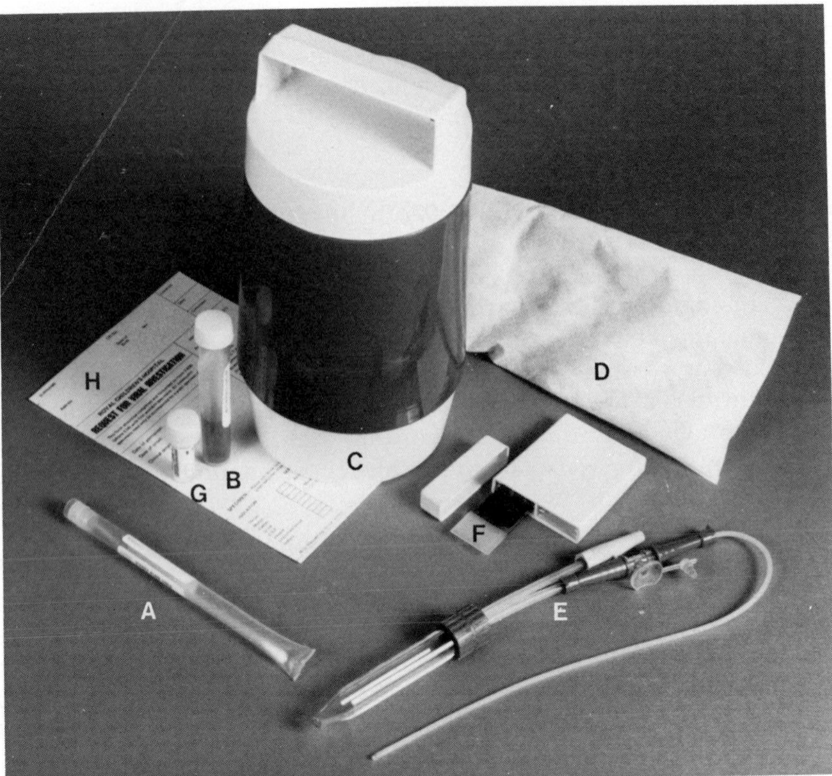

Fig. 12-1 Basic equipment for collection of specimens for virus isolation. (A) Sterile swab. (B) Bottle containing a few milliliters of transport medium (buffered isotonic balanced salt solution plus gelatin and antibiotics). (C) Plastic insulated container. (D) Refrigerant in flexible pack to be wrapped around specimen and placed in flask. (E) Apparatus for aspirating nasopharyngeal mucus by suction. (F) Slides for fixation of aspirated cells. (G) Vial containing anticoagulant for collection of blood for virus isolation from leukocytes and antibody determination on plasma. (H) Requisition form. (Courtesy I. Jack.)

case of many generalized viral diseases it may not be obvious what specimen to take. As a rough working rule it can be said that at a sufficiently early stage in the disease, virus can usually be isolated from a throat swab, or feces, or leukocytes from blood.

Because of the lability of many viruses, specimens intended for virus isolation must always be kept cold and moist. Immediately after collection the swab should be swirled around in a small screw-capped bottle containing virus transport medium. This medium consists of a buffered balanced salt solution, to which has been added protein (e.g., gelatin or albumin) to protect the virus against inactivation, and antibiotics to prevent the multiplication of bacteria and fungi. (If it is at all probable that the specimen will also be used for attempted isolation of bacteria, rickettsiae, chlamydiae, or mycoplasmas, the collection medium must not contain antibiotics—the portion used for virus isolation can be treated with antibiotics later.) The swab stick is then broken off aseptically into the fluid, the cap is tightly fastened and secured with adherent tape to prevent leakage, and the bottle is labeled with the patient's name, date of collection, and nature of specimen, then dispatched immediately to the laboratory, accompanied by a properly completed laboratory submission form which includes an informative clinical history, a provisional diagnosis, and request for a particular test.

If a transit time of more than an hour or so is anticipated the container should be sent refrigerated (but not frozen), with "cold packs" (4°C) or ice in a thermos flask or styrofoam box (Fig. 12-1). International or transcontinental transport of important specimens, particularly in hot weather, generally requires that the container be packed in dry ice (solid CO_2) to maintain the virus in a frozen state. Governmental and International Air Transport Association (IATA) regulations relating to the transport of biological materials require such precautions as double-walled containers with absorbent padding in case of breakage. Permits must be obtained from the appropriate authorities for interstate and international transportation.

Direct Identification of Virus, Viral Antigen, or Viral Genome

The last two decades have witnessed the development of an impressive variety of rapid diagnostic methods based on the detection of virus, viral antigens, or viral nucleic acid in specimens taken directly from the patient without augmentation by cultivation *in vitro*. By and large, these are tests of high sensitivity, capable of detecting the very small concentrations of virus found in patients' feces or throats even at the height of excretion.

Direct Detection of Virions by Electron Microscopy

Perhaps the most obvious method of virus identification is direct visualization of the virion itself. The morphology of most viruses is sufficiently characteristic to assign them to the correct family. Moreover, viruses that have never been cultured may nevertheless be recognized by electron microscopy (EM). During the 1970s EM was the means to the discovery in feces of several new groups of previously noncultivable viruses. The human rotaviruses, caliciviruses, astroviruses, hepatitis A, and previously unknown types of adenoviruses and coronaviruses were all first identified in this way.

The biggest limitation of EM as a diagnostic tool is its low sensitivity. Because of the time required for even a skilled microscopist, occupying a very expensive machine, to scan the grid adequately, specimens need to contain at least 10^7 virions per milliliter to be satisfactory. Such levels are often surpassed in feces and vesicle fluid, but not in respiratory mucus. Feces are first clarified by low-speed centrifugation; the supernatant is then subjected to ultracentrifugation to deposit the virions. Alternatively, salts and water can be removed from a drop of virus suspension hanging from a carbon-coated plastic support film by diffusion into agar, leaving the concentrated virions on the film. The specimen is then negatively stained with phosphotungstate, or sometimes uranyl acetate, and scanned by EM.

Immunoelectron Microscopy

Definitive identification (and further concentration) of virions may be achieved by adding specific antibody to the specimen before depositing the virus–antibody complexes by centrifugation, and negative staining; the virions are observed to be aggregated (Fig. 12-2). To improve visualization, the antibody may be labeled with gold. Solid-phase immunoelectron microscopy (IEM) techniques have been developed in which virus-specific antibody is first bound to the plastic supporting film on the copper grid. Sensitivity can be enhanced by a double layering procedure, whereby staphylococcal protein A (which binds the Fc moiety of IgG) is first bound to the film, then virus-specific antibody, then the sample. It should now be practicable to extend solid-phase IEM to specimens other than feces, urine, serum, and vesicle fluid.

Detection of Antigen by Enzyme Immunoassay

Enzyme immunoassay (EIA), often known affectionately as ELISA (for enzyme-linked immunosorbent assay), has revolutionized diagnostic virology. The EIAs can be designed in different formats to detect antigen or antibody. The exquisite sensitivity of the method enables less than 1 ng of viral antigen per milliliter to be detected in specimens taken directly from the patient.

The methodologic configurations are almost limitless: a wide variety of direct, indirect, and reversed assays has been described. Most are solid-phase EIAs; the "capture" antibody is attached (by simple adsorption or by covalent bonding) to a solid substrate, typically the wells of polystyrene or polyvinyl microtiter plates, so facilitating washing steps. The simplest format is the

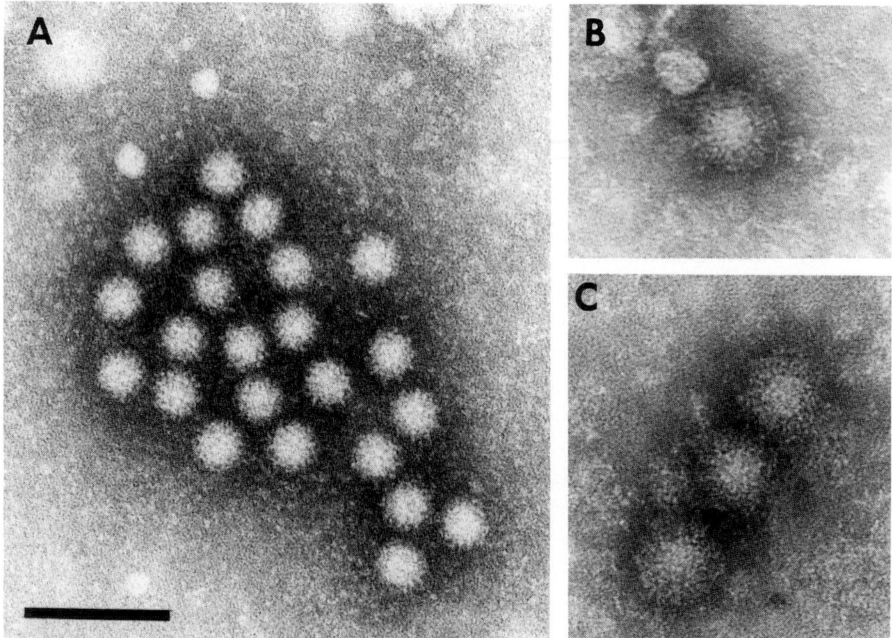

Fig. 12-2 Immunoelectron microscopy. Norwalk agent from the feces of a patient with gastroenteritis; aggregation with convalescent serum containing a low (A) or high (B and C) titer of antibody against the virus. Bar, 100 nm. (Courtesy Dr. A. Z. Kapikian.)

direct EIA (Fig. 12-3, left). Virus and soluble viral antigens from the specimen are allowed to adsorb to the capture antibody. After unbound antigen has been washed away, an enzyme-labeled antiviral antibody (the "detector" antibody) is added. (Various enzymes can be linked to antibody; horseradish peroxidase and alkaline phosphatase are the most commonly used.) After a final washing step, readout is based on the color change that follows addition of an appropriate organic substrate for the particular enzyme. The colored product of the action of the enzyme on the substrate should be clearly recognizable by eye. The test can be made quantitative by serially diluting the antigen to obtain an end point, or by using spectrophotometry to measure the amount of enzyme-conjugated antibody bound to the captured antigen.

A further refinement takes advantage of the extraordinarily high binding affinity of avidin for biotin. The antibody is conjugated to biotin, a reagent of low M_r which gives reproducible labeling and does not alter antigen-binding capacity. The antigen–antibody complex is recognized with high sensitivity simply by adding avidin–labeled enzyme, then substrate (Fig. 12-3, right). Subsequent modifications of EIAs have been aimed at increasing sensitivity even further. High-energy substrates are available which release fluorescent, chemiluminescent, or radioactive products that can be identified in very small amounts.

Indirect immunoassays are widely used because of their somewhat greater sensitivity and avoidance of the need to label each antiviral antibody in the laboratory repertoire. Here, the detector antibody is unlabeled, and a further layer, labeled (species-specific) anti-immunoglobulin, is added as the "indicator" antibody (Fig. 12-4, right); of course, the antiviral antibodies constituting the capture and detector antibodies must be raised in different animal species. Alternatively, labeled staphylococcal protein A, which binds to the Fc moiety

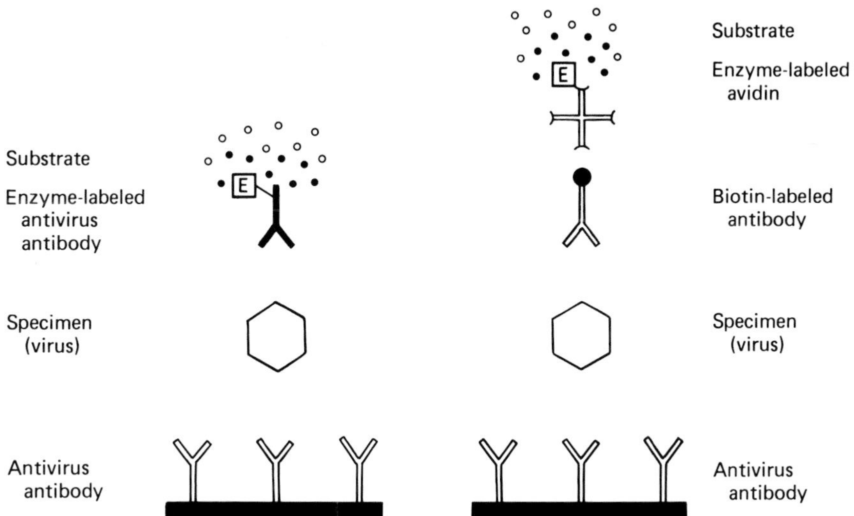

Fig. 12-3 Enzyme immunoassays (EIA or ELISA) for detection of virus and/or viral antigen. *Left:* Direct. *Right:* Avidin–biotin.

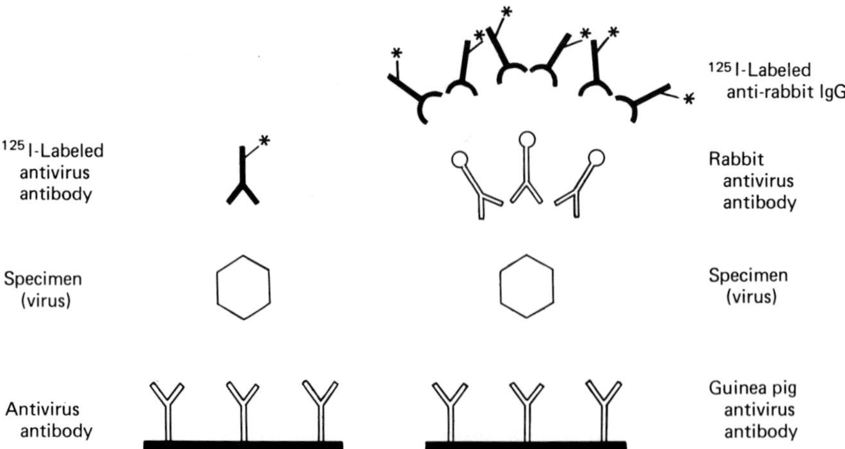

Fig. 12-4 Radioimmunoassays for detection of virus and/or viral antigen. *Left:* Direct. *Right:* Indirect.

of IgG of most mammalian species, can be used as the indicator in indirect immunoassays.

Increasingly, monoclonal antiviral antibodies (MAbs) are being used as capture and/or detector antibodies in immunoassays. Their obvious advantages are that they represent well-characterized, highly pure, monospecific antibody of a defined class, recognizing only a single epitope, are free of "natural" and other extraneous antibodies against host antigens or adventitious agents concurrently infecting the animal, and can be made available in large amounts as reference reagents. It is important to select MAbs of high affinity, but not of such high specificity that some strains of the virus sought might be missed in the assay. Indeed, the specificity of the assay can be predetermined by selecting a MAb directed at an epitope that is either confined to a particular viral serotype or common to all serotypes within a given species or genus.

Radioimmunoassay and Time-Resolved Fluoroimmunoassay

Radioimmunoassay (RIA) predates EIA but is progressively being superseded by it. The only significant difference is that the label is not an enzyme but a radioactive isotope such as ^{125}I, and the bound antibody is measured in a gamma counter (Fig. 12-4). The RIA is a highly sensitive and reliable assay that lends itself well to automation, but the cost of the equipment and the health hazard of working with radioisotopes argue against its use in small laboratories.

Time-resolved fluorescence immunoassay (TR-FIA) is a nonisotopic immunoassay, in which the indicator antibody is labeled with a fluorophore (a europium chelate). Following excitation by light, the fluorophore emits fluorescence of a different wavelength which is measured in a time-lapse fluorometer. The method is as sensitive as RIA and has the advantage of a stable label, but it requires costly equipment that can be contemplated only by laboratories with a large throughput of specimens.

Latex Particle Agglutination

Perhaps the simplest of all immunoassays is the agglutination by antigen of small latex beads previously coated with antiviral antibody. The test can be read by eye within minutes. Not surprisingly, diagnostic kits based on this method have become popular with small laboratories and with medical practitioners. However, they currently suffer from low sensitivity and low specificity. Thus, false negatives occur unless large numbers of virions are present, and therefore this assay for antigen tends to be restricted to examination of feces. False positives, on the other hand, occur quite commonly with fecal specimens in particular. If these problems can be overcome, latex agglutination may develop a better reputation for reliability.

Immunofluorescence

If antibody is labeled with a fluorochrome, the antigen–antibody complex, when excited by light of short wavelength, emits light of a particular longer wavelength, which can be visualized as fluorescence in an ordinary microscope when light of all other wavelengths is filtered out. The sensitivity of the method is generally too low to detect complexes of fluorescent antibody with virions or soluble antigen; hence, the antigen in the test typically takes the form of virus-infected cells. There are two main variants of the technique.

Direct Immunofluorescence

Direct immunofluorescence is conducted as follows. A frozen tissue section, or an acetone-fixed cell smear or monolayer on a coverslip, is exposed to fluorescein-tagged antiviral antibody (Fig. 12-5, left). Unbound antibody is then washed away, and the cells are inspected by light microscopy using a powerful ultraviolet/blue light source. The apple-green light emitted from the specimen is revealed (against a black background) by incorporating filters in the eyepieces that absorb all the blue and ultraviolet incident light (Fig. 12-6).

Indirect Immunofluorescence

Indirect ("sandwich") immunofluorescence differs in that the antiviral antibody is untagged, but fills the role of the meat in the sandwich. It binds to antigen and is itself recognized by fluorescein-conjugated anti-immunoglobulin (Fig. 12-5, right). The high affinity of avidin for biotin can also be exploited in immunofluorescence, by coupling biotin to antibody and fluorescein to avidin.

Though technically demanding, immunofluorescence has proved to be of great value in the early identification of viral antigens in infected cells taken

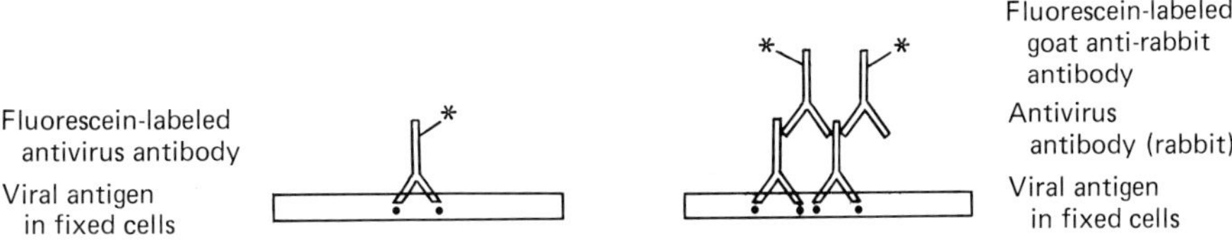

Fig. 12-5 Immunofluorescence. *Left:* Direct. *Right:* Indirect.

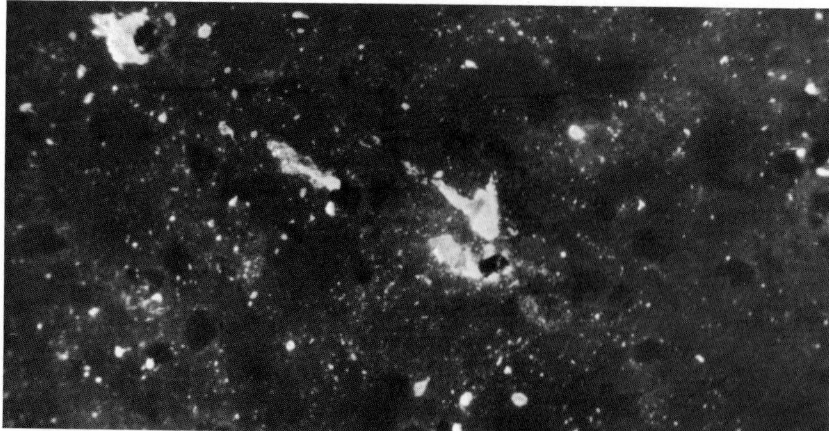

Fig. 12-6 Immunofluorescence for the diagnosis of animals infected with rabies virus. Tissue impressions are made by lightly touching the brain of the suspect animal onto a microscope slide. Following air-drying and acetone fixation the tissue is stained with fluorescein-conjugated anti-rabies virus globulin, washed, mounted, and examined by ultraviolet/blue light microscopy. Rabies virus antigen, present in cytoplasmic inclusions, is identified as apple-green fluorescence against a black background. (Courtesy Dr. F. A. Murphy.)

from patients with diseases known to have a relatively small number of possible etiological agents. There is no difficulty in removing partly detached infected cells from the mucous membrane of the upper respiratory tract, genital tract, eye, or from the skin, simply by swabbing or scraping the infected area with reasonable firmness. Cells are also present in mucus aspirated from the nasopharynx; aspirated cells must be extensively washed by centrifugation to remove mucus before fixation and staining. Respiratory infections with paramyxoviruses, orthomyxoviruses, adenoviruses, and herpesviruses are particularly amenable to rapid diagnosis by immunofluorescence. Immunofluorescence can also be applied to infected tissue, for example, brain biopsies for the diagnosis of such lethal diseases as herpes simplex encephalitis or measles SSPE, or at necropsy, for the verification of rabies in the brain of animals captured after biting humans (Fig. 12-6).

Immunoperoxidase Staining

An alternative method of locating and identifying viral antigen in infected cells is to use antibody coupled to horseradish peroxidase; subsequent addition of hydrogen peroxide together with a benzidine derivative forms a colored insoluble precipitate. The advantages of the method are that the preparations are permanent and only an ordinary light microscope is needed. The principal disadvantage is that endogenous peroxidase, present in a number of tissues, particularly leukocytes, produces false positives, but this problem can be circumvented by meticulous technique and adequate controls.

Nucleic Acid Hybridization and Gene Amplification

The principle of nucleic acid hybridization is that single-stranded DNA will hybridize by hydrogen-bonded base pairing to another single strand of DNA (or RNA) of complementary base sequence. Thus the two strands of the target DNA molecule are first separated by boiling, then, following cooling, allowed to hybridize with a labeled single-stranded DNA or RNA probe present in excess. The reaction can be accomplished in solution, which is useful for determining the kinetics of annealing or the stoichiometry of the reaction, from which can be calculated the percentage homology (similarity) between the two sequences. The conditions set for annealing, especially temperature and ionic strength, determine the degree of discrimination of the test. Under conditions of low *stringency* a number of mismatched base pairs are tolerated, whereas at high stringency such a heteroduplex is unstable.

The other major factor determining the specificity of the test is, of course, the nature of the probe itself. This may correspond in length with the whole viral genome, or a single gene, or a much shorter nucleotide sequence deliberately chosen to represent either a variable or a conserved region of the genome, depending on whether it is intended that the probe be type-specific or more versatile, for example, a probe able to bind to the DNA of any type of human papillomavirus. The oligonucleotide sequence intended as a probe is produced by chemical synthesis or by genetic cloning in a bacterial plasmid or phage.

Traditionally, radioactive isotopes such as ^{32}P and ^{35}S have been used to label nucleic acids or oligonucleotides intended as probes for hybridization tests, with the signal being read by counting in a spectrometer or by autoradiography. The trend is now toward nonradioactive labels. Some of these (e.g., fluorescein or peroxidase) produce a signal directly, whereas others (e.g., biotin or digoxigenin) act indirectly by binding to another labeled compound which then emits the signal. Biotinylated probes can be combined with various types of readouts, for example, an avidin-based EIA. Chemiluminescent substrates, such as luminol, are also being widely exploited. Indeed, the 1990s are witnessing a proliferation of diagnostic kits for various diseases, many of them based on novel labels and/or methods of readout.

Dot-Blot Hybridization

Most of the methods favored today are two-phase systems, generically known as *filter hybridization*. In its simplest format, *dot (blot) hybridization*, DNA or RNA extracted from virus or infected cells is denatured, then spotted directly onto a charged nylon or nitrocellulose membrane filter to which it binds tightly on baking. The single-stranded DNA or RNA probe is then hybridized to the target nucleic acid *in situ* on the membrane, and unbound probe is washed away. The signal generated by the bound probe is measured by autoradiography if the probe is radioactive, or by the formation of a colored precipitate if an enzyme label is used. By choosing RNA as a probe, sensitivity can be improved and the incidence of false positives reduced by treating the filters with RNAse before counting.

In Situ *Hybridization*

In situ hybridization has been widely used by pathologists to screen patients with persistent infections or viral cancers for evidence of integrated or nonintegrated copies of the viral genome. Frozen sections on slides are probed, much as described above, and the intracellular location of viral sequences is revealed by autoradiography or immunoperoxidase cytochemistry.

Southern Blot Hybridization

The introduction of Southern blotting (Fig. 12-7) elevated the technology to a higher level of sophistication. Restriction endonucleases are used to cleave the DNA into shorter oligonucleotides which are then separated by electrophoresis on an agarose or acrylamide gel. After staining with ethidium bromide to reveal the position of the fragments, the gel is treated successively with acidic and basic solutions to depurinate and denature the DNA, which is then transferred by electrophoresis, diffusion, or other means ("blotting") onto a nylon or nitrocellulose membrane filter. Individual bands are revealed by hybridization of a labeled DNA or RNA probe followed by autoradiography or a color development process. "Northern" blotting of RNA follows similar principles but is generally not so sensitive.

Polymerase Chain Reaction

The polymerase chain reaction (PCR) (Fig. 12-8) constitutes the greatest advance in molecular biology since the advent of recombinant DNA technology. It enables a single copy of any gene sequence to be amplified *in vitro* at least a millionfold within a few hours. Thus viral DNA extracted from a very small number of virions or infected cells can be amplified to the point where it

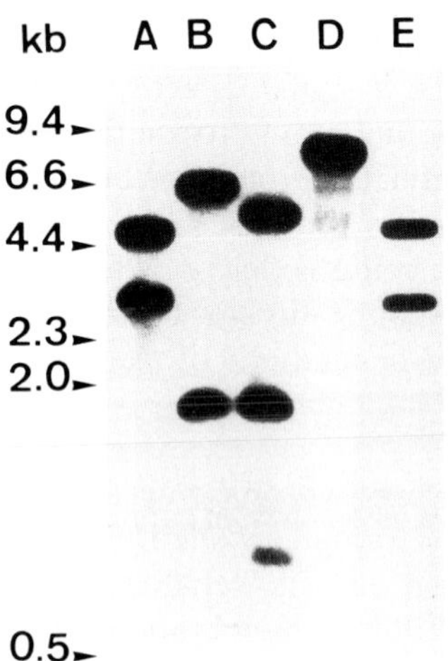

Fig. 12-7 Southern blot hybridization of DNA extracted from a wartlike skin lesion. The DNA was first digested with restriction enzymes *Bgl*II (lane A), *Eco*R1 (lane B), *Pst*I (lane C), *Pvu*II (lane D), and *Bam*HI (lane E), and then the fragments were separated on the basis of size by agarose gel electrophoresis. After transfer to a nylon membrane, the blot was hybridized at high stringency to a probe consisting of ^{32}P-labeled human papillomavirus type 8 (HPV-8) DNA, then washed and exposed to X-ray film. The strong hybridization and the pattern of restriction fragments identify the virus as type 5C (HPV types 5 and 8 are close relatives and are usually associated with skin lesions of people suffering from epidermodysplasia verruciformis). The HPV genome is normally found in cells as a circular plasmid of about 8 kbp, but it is easier to estimate DNA size by gel electrophoresis when it is linear (i.e., after digestion of the plasmid with restriction enzymes). In lane D (*Pvu*II digest), the plasmid is cut only once (to produce the full-length linear form), whereas the other enzymes cut two or more times. The total size of the type 5C genome is 7.7 kb, and using the molecular size markers at left and adding the fragments of each digest, it is clear that the total size is the same for each lane. (Courtesy Drs. H. E. Trowell and M. L. Dyall-Smith.)

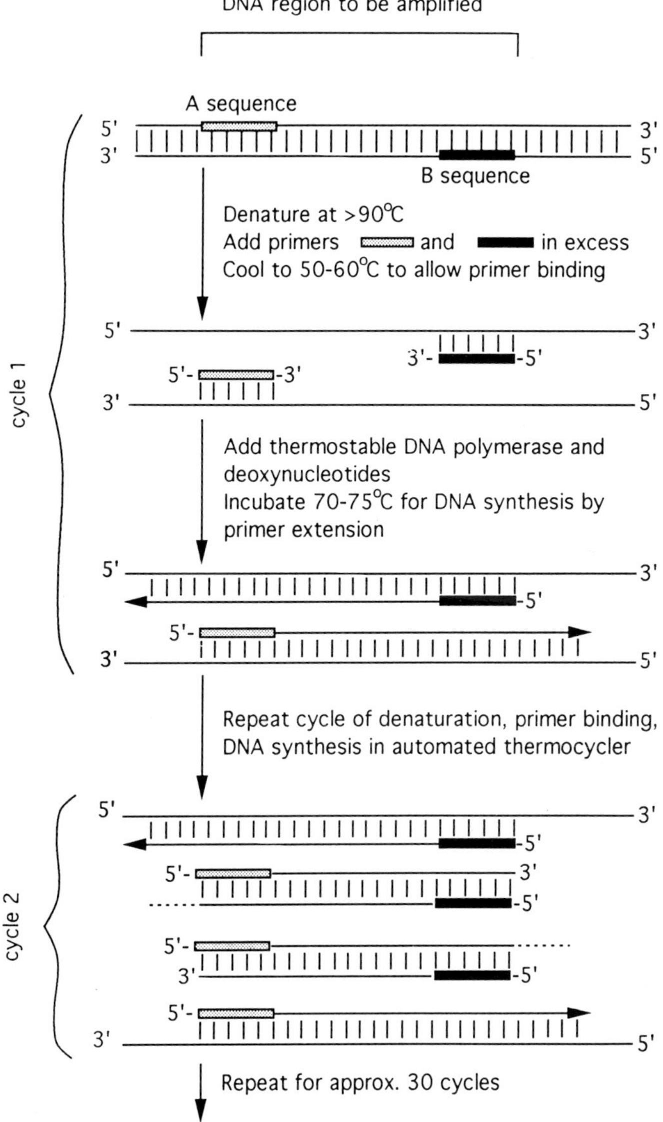

Fig. 12-8 Amplification of part of a DNA sequence by the polymerase chain reaction (PCR). Oligonucleotide primers must first be made according to the sequences at either end of the portion of DNA to be amplified. After the DNA has been denatured by heating, the primers can hybridize to the complementary sequences on the opposite strand. In the presence of heat-resistant DNA polymerase and deoxynucleotide triphosphates, two new copies of the desired region are produced. The cycles of melting, annealing, and extension are repeated rapidly; each time, the amount of target DNA sequence doubles. After the first few cycles virtually all the templates consist of just the short region chosen for amplification. After 30 cycles, taking about 3 hours, the region bounded by the chosen primers has been amplified many millionfold. (Courtesy Drs. I. H. Holmes and R. Strugnell.)

can readily be identified using labeled probes in a hybridization assay. Moreover, the PCR can be modified for the detection of viral RNA, by incorporating a preliminary step in which reverse transcriptase is used to convert the RNA to DNA. It is not necessary or usual to amplify the whole genome, but it is necessary to know at least part of the nucleotide sequence, in order to synthesize two oligonucleotide primers representing the extremities of the region one chooses to amplify.

There are three steps in the process: (1) melting the target DNA at 95°C, (2) cooling to 50°–60°C to allow binding of two oligonucleotide primers, corresponding with short sequences (on opposite strands of the template DNA) which flank the segment of DNA that is to be amplified, and (3) extension from the oligonucleotide primers by DNA polymerase to form two DNA strands that are complementary copies of the original (plus and minus) target strands, that is, the DNA located between and including the two primers. The cycle of melting, primer binding, and primer extension is repeated many times, and the number of DNA copies increases exponentially. Thus after 30 cycles the number of DNA copies, beginning with a single copy of the target sequence, is over one million. Thirty cycles can be completed within 4 hours using a suitable automated thermal cycling device. By using the heat-stable DNA polymerase (*Taq*) of *Thermus aquaticus,* an organism naturally found in hot springs, it is not necessary to replenish the enzyme between cycles. The amplified DNA may be detected by agarose gel electrophoresis or by hybridization with labeled DNA or RNA probes culminating in any of a wide variety of readout systems (Fig. 12-9).

Selection of the most suitable pair of primers is a matter of central importance. They may be chosen to be highly specific for a particular virus strain or, alternatively, to represent consensus sequences within a gene that is conserved within a given family or genus. By probing for a conserved gene such as that for RNA polymerase it is even possible to discover a previously unknown agent. Reactions must be carried out under very carefully controlled conditions of ionic strength, temperature, primer concentration, and nucleotide concentration. Deviations can result in nonspecific amplification. The problem of contamination of samples with extraneous DNA has delayed the transition of PCR from research tool to routine diagnostic method, and great care is required to avoid such contamination. However, a number of technical refinements have been introduced, meticulous attention to which has made PCR a reliable methodology in experienced laboratories.

Isothermal amplification is a new technique that does not require the temperature cycling and accompanying equipment of PCR. Three enzymes involved in the replication of retroviruses are simply mixed with either a DNA or RNA template and DNA primers at a constant temperature, and millionfold amplification is rapidly achieved.

These several developments have so dramatically enhanced the sensitivity and versatility of nucleic acid hybridization that probing for the viral genome may overtake probing for antigen as the diagnostic method of choice in many laboratories. The procedure is invaluable when dealing with (1) viruses that cannot be cultured satisfactorily, (2) specimens that contain predominantly inactivated virus, as a result of prolonged storage or transport, or (3) latent infections in which the viral genome lies dormant and virus is not

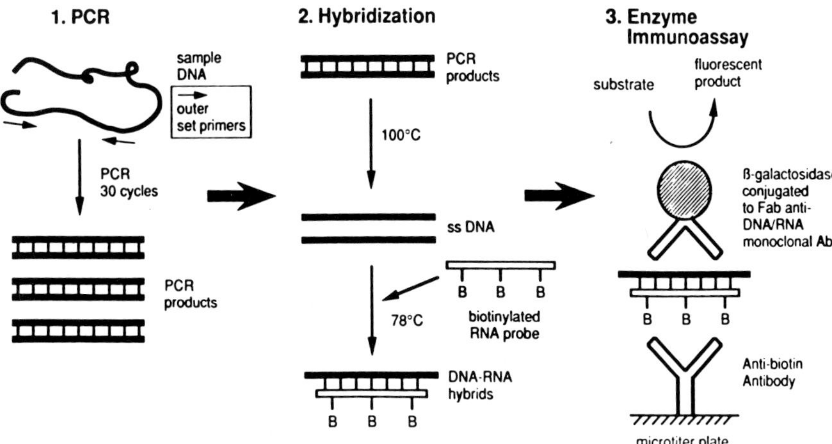

Fig. 12-9 Detection of PCR products using a biotinylated RNA probe and an enzyme immunoassay readout based on capture of labeled DNA–RNA hybrid molecules by antibiotin antibody. [Reproduced with permission from R. H. Yolken, *World J. Microbiol. Biotechnol.* **7**, 150 (1991).]

being made. However, the added sensitivity provided by gene amplification can actually create new and unexpected problems. For example, some studies have reported DNA of carcinogenic types of papillomavirus in such a high proportion of healthy women, the vast majority of whom will never develop carcinoma of the cervix, that the medical relevance of the data is called into question. Quantitative PCR, currently in its infancy, is required to determine the viral load, or at least the genome copy number prior to gene amplification.

Virus Isolation

Despite the explosion of new techniques for same-day diagnosis of viral disease by demonstration of virus, viral antigen, or viral nucleic acid in specimens taken directly from the patient, it is still true to say that few techniques achieve quite the sensitivity of virus isolation in cell culture. Theoretically at least, a single viable virion present in a specimen can be grown in cultured cells, thus expanding the numbers of virions many millionfold to produce enough material to be characterized antigenically. Virus isolation remains the "gold standard" against which newer methods must be compared. Moreover, it is the only technique that can detect the unexpected, that is, identify a totally unforeseen virus, or even discover an entirely new agent. Accordingly, even those laboratories well equipped for rapid diagnosis sometimes also inoculate cell cultures in an attempt to isolate the virus. Culture is the only method of producing a supply of live virus for further examination, such as drug sensitivity testing. Research and reference laboratories, in particular, are always on the lookout for new viruses, which require comprehensive characterization. Moreover, large quantities of virus must be grown up in cultured cells to produce diagnostic antigens and monoclonal antibodies for distribution to other laboratories.

Preparation of Specimens

On arrival in the laboratory the specimen is refrigerated then processed as soon as possible. Before inoculation into cell culture, material is shaken from the swab, or feces are dispersed, using a vortex mixer. Tissue specimens are homogenized in a high-speed blender. Cell debris and bacteria are deposited by low-speed centrifugation, after which the supernatant from particularly "dirty" specimens like feces may be passed through a 0.45-μm syringe-top membrane filter. Some of the original sample and some of the filtrate should be retained at 4°C or frozen (−70°C), at least until the isolation attempt is completed. The remainder of the filtrate is inoculated into cell cultures, and/or rarely into chick embryos or newborn mice.

Inoculation and Maintenance of Cultures

The choice of the most suitable host cell system will depend on the virus one expects to find, in the light of the clinical history of the patient and the source of the specimen. The precise routine differs from one laboratory to another. Primary cultures, established directly from tissue (e.g., primary monkey or human kidney cells), contain a variety of types of differentiated cells, hence support the growth of a broad spectrum of viruses. However, most species of monkeys are now protected and in any case may harbor silent simian viruses, whereas human embryonic tissue has become less readily available in many countries as a result of changes in the law or in the frequency or surgical techniques of abortion. Further, most of the cell types in primary cultures do not survive more than 2 or 3 passages *in vitro,* making primary cultures so expensive and inconvenient that their use in diagnostic virology has diminished greatly in recent years.

On the other hand, continuous heteroploid cell lines, such as HEp-2 or HeLa derived from human cancers, or Vero, LLC-MK2, or BSC-1 from normal monkey kidney, have the great advantage of immortality, hence can be passaged indefinitely *in vitro* (and/or stored frozen in liquid nitrogen for years). Any given cell line is generally capable of supporting the growth of only a limited range of viruses but may represent the ideal *in vitro* host system for a particular virus. Diploid lines (strains) of human embryonic fibroblasts (known as HEF, or HDF for human diploid fibroblasts) offer a useful compromise, in that they can be successfully passaged for a year or so *in vitro* (or stored frozen indefinitely) yet are susceptible to a wide range of different human viruses.

To cover themselves against most contingencies some laboratories have a policy of inoculating a given specimen simultaneously into one type of primary culture, one diploid strain of human fibroblasts, and perhaps, in the case of respiratory specimens, one malignant cell line. However, most laboratories today would tend to rely almost entirely on HEF (HDF) for routine isolations and reserve other cells for particularly fastidious viruses only. For example, primary monkey kidney (PMK) cells are suitable for paramyxoviruses, orthomyxoviruses, and enteroviruses; the RD line of human rhabdomyosarcoma for coxsackie A and echoviruses; Vero or baby hamster kidney (BHK-21) cells for arboviruses; RK-13 or Vero cells for rubella virus; and so on.

Monolayer cultures for viral diagnostic purposes are generally grown in screw-capped glass tubes or vials, sometimes containing a removable coverslip for subsequent staining. Microtiter plates are very convenient for neutralization tests, where large numbers of cultures are required and economy of medium, equipment, and space is important, but the risk of cross-contamination argues against them for virus isolation. The inoculated cultures are held at the temperature and pH of the human body (37°C, pH 7.4), except in the case of viruses of the upper respiratory tract such as influenza viruses, rhinoviruses, and coronaviruses, which grow best at 33°C, the temperature encountered in the nasal mucosa. Cultures may be slowly rotated on a roller drum, or kept stationary. The cultures are inspected two or three times a week for the development of cytopathic effects (CPE).

Recognition of Viral Growth

Although rapidly growing viruses such as poliovirus or herpes simplex virus can be relied on to produce detectable CPE within a day or two, some of the notoriously slow agents such as cytomegalovirus, rubella virus, and some adenoviruses may not produce obvious CPE for 1–4 weeks. By this time, the uninoculated control cultures will often be showing nonspecific degeneration, so robbing the virologist of an appropriate standard of comparison. In such cases it may be necessary to subinoculate the cells and supernatant fluid from the infected culture into fresh monolayers ("blind passage"), after which clear-cut CPE usually appear promptly.

When, at any stage, degenerative changes suggestive of viral multiplication become evident, the virologist has a number of courses open. The CPE is often sufficiently characteristic, even in the living unstained culture viewed *in situ* (see Fig. 5-2), for the trained observer to be able to tender a provisional diagnosis to the physician immediately. Alternatively, it may be instructive to fix and stain the infected monolayer to identify the CPE. Inclusion bodies and multinucleate giant cells (syncytia) can be identified by staining with hematoxylin and eosin or Giemsa stain; if present, these changes are usually sufficiently characteristic to enable the virologist to place the isolate at least within the correct viral family (see Fig. 5-3). Coverslip cultures can also be stained with fluorescent antibody for positive identification. If none of these histologic methods proves to be diagnostic, virus must be extracted from the culture for further examination.

Some viruses are relatively noncytocidal for cultured cells (see Chapter 5 and Table 5-1). Their growth in monolayer culture may sometimes be recognized by means of hemadsorption. Most viruses that hemagglutinate will be amenable to test by hemadsorption; the growth of paramyxoviruses or orthomyxoviruses, for example, is routinely recognized in this way (see Fig. 5-2D).

Animals and Chick Embryos

Many viruses will grow satisfactorily in chick embryos or newborn mice, but neither animal is now commonly used in the diagnostic laboratory because cell culture is generally the simpler option. The use of mice can be limited to the isolation of arboviruses, rabies virus, and some of the group A coxsackieviruses, for which suckling mice less than 24 hours old are injected intracerebrally and/or intraperitoneally, then observed for up to 2 weeks for the development of pathognomonic signs before sacrificing for histologic examination of affected organs (see Fig. 23-5). Chick embryos are still used, together with primate kidney cell cultures, for the isolation of influenza A viruses from humans. Three to four days after amniotic and allantoic inoculation of 10-day-old embryonated hen's eggs, the fluids are tested for hemagglutination (see Chapter 31 for details).

Larger animals are virtually never used for routine isolation of human viruses. Primates, especially chimpanzees, are employed by a small number of well-endowed research laboratories for attempted isolation of human viruses not yet cultivable in any other nonhuman host, for example, hepatitis viruses (Chapters 22, 23, 24, 26), HIV (Chapter 35), and various agents responsible for slow infections of the human brain (Chapter 10), as well as for experimental studies of viral pathogenesis and immunity, and for testing new viral vaccines (Chapter 13).

Identification of Viral Isolates

Antigenic Characterization

A newly isolated virus can usually be provisionally allocated to a particular family on the basis of the clinical findings, the particular type of cell culture yielding the isolate, and the visible result of viral growth (CPE, hemadsorption, etc.). Definitive identification, however, rests on antigenic characterization. By using the new isolate as antigen against a panel of antisera, for example, in an EIA, the virus can first be placed into its correct family or genus. If it is considered important to do so, one can then go on to identify the species or serotype by moving to MAbs, or to one of the more discriminating serologic procedures, such as neutralization or hemagglutination inhibition. This sequential approach is applicable only to families that share a common family antigen and hence is of no help with picornaviruses, for example. The range of available immunologic techniques is now almost embarrassingly wide. Some are best suited to particular families of viruses. Each laboratory makes its own choice of favored procedures, based on considerations of sensitivity, specificity, reproducibility, speed, convenience, and cost.

The simplest way of identifying a newly isolated virus is by fluorescent antibody staining of the infected cell monolayer itself (see Fig. 12-6). This can provide an answer within an hour or so of recognizing the earliest suggestion of CPE. Immunofluorescence is best suited to the identification of viruses with no close relatives, or in epidemic situations when a particular virus is suspected; otherwise replicate cultures must be screened with a range of antisera. Most other immunologic approaches are applied to identification of virions and soluble antigen present in the cell culture supernatant. Today, the EIA technique is probably the most commonly used, though neutralization remains the gold standard for defining and distinguishing serotypes (Fig. 12-10).

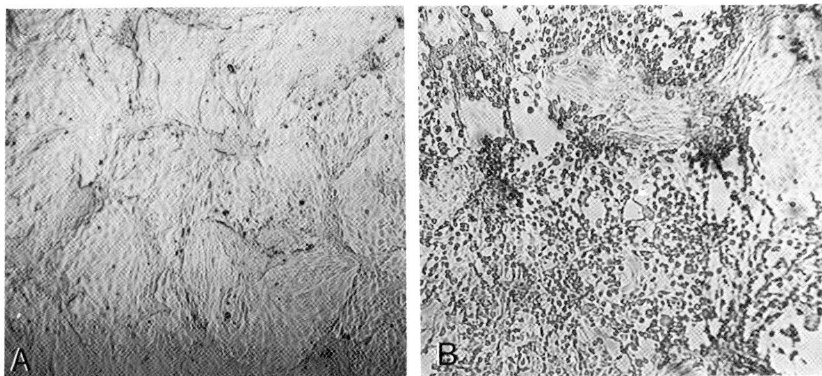

Fig. 12-10 Virus neutralization test. A child developed meningitis during a summer epidemic of coxsackievirus B2. An enterovirus was isolated from the feces. One hundred $TCID_{50}$ units of the virus was incubated at 37°C for 60 minutes with a suitable dilution of "inactivated" (56°C, 30 minutes) anti-coxsackievirus B2 serum (a reference serum previously raised in a rabbit). The mixture was inoculated onto a monolayer of monkey kidney cells in a well of a microculture tray (A). Virus similarly incubated with normal rabbit serum was inoculated into well B. The cultures were incubated at 37°C for several days and inspected daily for CPE. Unstained; magnification: ×23. Interpretation: The infectivity of the virus isolate has been neutralized by anti-coxsackievirus B2 serum (no CPE); the control culture (B) shows typical CPE. (Courtesy I. Jack.)

Characterization of the Viral Genome

For most routine diagnostic purposes it is often not necessary to "type" the isolate antigenically. On the other hand, there are certain situations when epidemiologic information of importance to the public health can be obtained by going even further to characterize very subtle differences between variants, strains, or subtypes within a given serotype. Short of determining the complete nucleotide sequence of the viral genome, there are several simpler methods for deriving sufficient information about the sequence to identify such variants. (1) The PCR, using strategically chosen primers, can be used to amplify a diagnostic region within the genome for partial nucleotide sequencing. (2) A labeled nucleic acid probe, corresponding to the whole genome or a variable part of the genome, may be employed in a hybridization test, under predetermined conditions of stringency, to measure the degree of homology with the isolate. (3) Appropriately chosen restriction endonucleases may be used to cut viral DNA into fragments that can then be separated by agarose gel electrophoresis and stained with ethidium bromide or silver to produce a diagnostic "restriction map." (4) Ribonuclease T1 may be used to cut labeled viral RNA into fragments which can be separated by two-dimensional electrophoresis to yield a unique "oligonucleotide fingerprint" revealed by autoradiography.

Minor degrees of genetic drift, often too subtle to be reflected in antigenic differences, can be picked up in these ways. Reference to a library of known fingerprints enables the investigator to trace the epidemiologic origin of individual isolates with respect to time, geography, and even potentially to a particular person. For example, it is possible to trace the origin of a given strain of herpes simplex virus type 2 in a newborn baby to the nursery of the maternity hospital, or of HSV-2 in an adult to a sexual contact. Other striking examples of the power of these techniques in tracking the origin and course of outbreaks of poliomyelitis are discussed further in Chapter 23.

Interpretation

The isolation and identification of a particular virus from a patient with a given disease are not necessarily meaningful in themselves. Concurrent subclinical infection with a virus unrelated to the illness in question is not uncommon. Koch's postulates are as relevant here as in any other microbiological context, but are not always easy to fulfill. In attempting to interpret the significance of any virus isolation one must be guided by the following considerations. First, the site from which the virus was isolated is important. For example, one would be quite confident about the etiologic significance of rubella virus isolated from any organ of a congenitally affected infant, or of mumps virus isolated from the CSF of a patient with meningitis, because these sites are usually sterile, that is, they have no normal bacterial or viral flora. On the other hand, recovery of an echovirus from the feces, or herpes simplex virus from the throat, may not necessarily be significant, because such viruses are often associated with inapparent infections. Second, interpretation of the significance of the isolation in such instances will be facilitated by isolation of the same virus from several cases of the same illness during an epidemic. Third, knowledge that the virus and the disease in question are often causally associated engenders confidence that the isolate is significant.

Measurement of Serum Antibodies

We have discussed identification of viruses or their antigens, using panels of antibodies of known specificity. Serological techniques may also be employed the other way around, to identify antibody using panels of known antigens. "Paired sera" are taken from the patient, the "acute-phase" serum sample as early as possible in the illness and the "convalescent-phase" sample at least 2 weeks later. Blood is collected in the absence of anticoagulants and given time to clot, then the serum is separated. Acute and convalescent serum samples should be tested simultaneously. For certain tests that measure inhibition of some biological function of virus, for example, virus neutralization or hemagglutination inhibition, the sera must first be "inactivated" by heating at 56°C for 30 minutes and sometimes treated by additional methods, to destroy various types of nonspecific inhibitors of infectivity or hemagglutination, respectively. Prior treatment of the serum is not generally required for assays that simply measure antigen–antibody binding, such as EIA, RIA, Western blot, latex agglutination, immunofluorescence, or immunodiffusion. The paired sera are then titrated for antibodies using any of a wide range of available serologic techniques. Space does not permit detailed description of all of these techniques. The basic principles of those most widely used today are set out in Table 12-2; three of these are illustrated in detail in Figs. 12-11 through 12-13.

For the safety of laboratory workers it is imperative to inactivate the

Table 12-2
Serological Procedures Used in Virology

Technique	Principle
Enzyme immunoassay	Antibody binds to antigen; enzyme-labeled anti-Ig binds to antibody; substrate changes color
Radioimmunoassay	Antibody binds to antigen; radiolabeled anti-Ig binds to antibody
Western blot	Virus disrupted, proteins separated by gel electrophoresis and transferred (blotted) onto nylon membrane; antiserum binds to viral proteins; labeled anti-Ig binds to particular bands; revealed by EIA or autoradiography
Latex particle agglutination	Antibody agglutinates antigen-coated latex particles
Virus neutralization	Antibody neutralizes infectivity of virion; CPE inhibition, plaque reduction, or protection of animals
Hemagglutination inhibition	Antibody inhibits viral hemagglutination
Immunofluorescence	Antibody binds to intracellular antigen; fluorescein-labeled anti-Ig binds; fluoresces by UV microscopy
Immunodiffusion	Antibodies and soluble antigens produce visible lines of precipitate in a gel
Complement fixation	Antigen–antibody complex binds complement, which is thereafter unavailable for lysis of sheep RBC in presence of antibody to RBC

infectivity of any particularly dangerous viruses employed as antigens in serological tests other than neutralization (e.g., arenaviruses, rhabdoviruses, togaviruses, or flaviviruses). This can be done, without destroying antigenicity, by γ-irradiation, or by photodynamic inactivation with ultraviolet light in the presence of psoralen. Increasingly, viral proteins produced by recombinant DNA technology are being employed instead of whole virus as antigen in serological tests.

Sensitivity and Specificity

The two most important parameters of any diagnostic assay are specificity and sensitivity. It is vital to understand the difference. The *sensitivity* of a given test is a measure of the percentage of those with the disease (or infection) in question who are identified as positive by that test. For example, a particular EIA used to screen a population for HIV antibody may display a sensitivity of 98%, that is, of every 100 infected people, 98 will be correctly diagnosed and 2 will be missed (the *false-negative* rate equals 2%). In contrast, the *specificity* of a test is a measure of the percentage of those without the disease (or infection) who yield a negative result. For example, the same EIA for HIV antibody may have a specificity of 97%, that is, of every 100 uninfected people, 97 will be correctly diagnosed as clear but 3 will be incorrectly scored as infected (the *false-positive* rate equals 3%).

Whereas sensitivity and specificity are fixed percentages intrinsic to the particular diagnostic assay, the *predictive value* of an assay is greatly affected by the prevalence of the disease (or infection) involved. Thus, if this EIA is used to screen a high-risk population with a known HIV prevalence of 50% the predictive value of the assay will be high; however, if it is used to screen blood donors with a known HIV prevalence of 0.1%, the great majority of the 3.1% who register as positive will in fact be false positives and will require follow-up with a confirmatory test of much higher specificity. This striking illustration draws attention to the importance of selecting diagnostic assays with a particular objective in mind. A test with high sensitivity is required when the aim is to screen for a serious infection, the diagnosis of which must not be missed; a test with high specificity is required for confirmation that the diagnosis is correct.

The sensitivity of a given immunoassay is really a measure of its ability to pick up small amounts of antibody (or antigen). For instance, EIA, RIA, and neutralization assays generally display substantially higher sensitivity than IEM, immunofluorescence, complement fixation, or immunodiffusion. Improvements in sensitivity are achieved by miniaturization of assays and use of purified reagents and sensitive instrumentation. On the other hand, the specificity of an immunoassay is a measure of its ability to discriminate, thus is influenced mainly by the purity of the key reagent, that is, antigen (when testing for antibody) or antibody (when testing for antigen).

Quite apart from the problem of fortuitous cross-reactions caused by the use of impure reagents, it is crucial to understand that the specificity of any assay can be manipulated to match one's particular objective. For example, because of the presence of antigenic determinants common to several or all serotypes within a given viral genus, the serum of a patient will contain a wide range of antibodies cross-reactive with virions of any one of the serotypes used as antigen in an EIA. To render the test more specific it would be necessary to use genetically cloned type-specific antigen, or even a synthetic peptide corresponding to an epitope known to be unique to a particular serotype. Alternatively, one can select an immunoassay which registers only those particular antibodies that bind to type-specific epitopes on the virus; virus neutralization (Fig. 12-10) and hemagglutination inhibition (Fig. 12-11) are specific because they measure only those antibodies that inhibit infection or hemagglutination, by binding to type-specific epitopes on the particular protein that mediates attachment of the virion to the cell and has therefore been most subject to antigenic change.

Immunoblotting (Western Blotting)

Perhaps the ultimate refinement is to test simultaneously but independently for antibodies against all of the proteins present in a particular virus. There are three key steps to Western blotting (Fig. 12-12). First, purified virus is solubilized with the anionic detergent sodium dodecyl sulfate (SDS) and the constituent proteins separated into discrete bands according to M_r by polyacrylamide gel electrophoresis (SDS-PAGE). Second, the separated proteins are electrophoretically transferred ("blotted") onto a nitrocellulose membrane to immobilize them. Finally, the test serum is allowed to bind to the viral proteins on the membrane, and their presence is demonstrated using a radiolabeled or enzyme-labeled anti-species antibody. Thus, immunoblotting permits demonstration of antibodies to some or all of the proteins of any given virus and can be used not only to discriminate between infection with closely

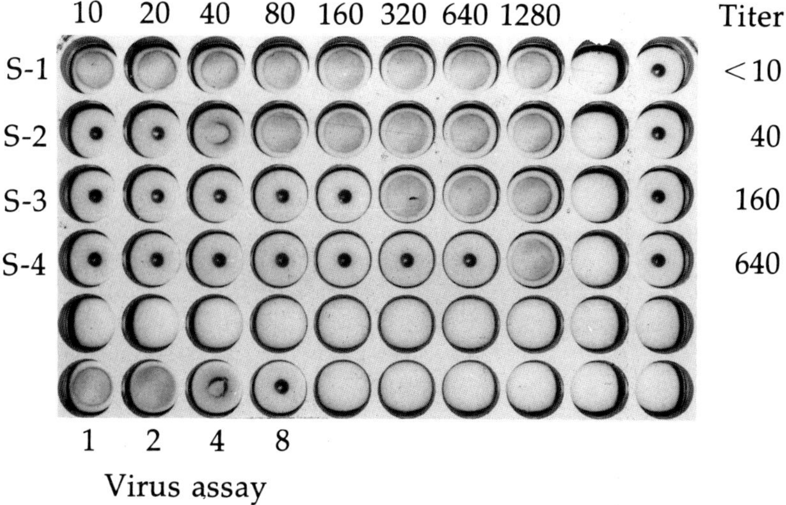

Fig. 12-11 Hemagglutination inhibition test, used for titrating antibodies to the viral hemagglutinin. Titers are expressed as reciprocals of dilutions. In the example illustrated, an individual was immunized against the prevalent strain of influenza virus. Serum samples S-1, S-2, S-3, and S-4 were taken, respectively, before immunization, 1 week and 4 weeks after the first dose of vaccine, and 4 weeks after the second. The sera were treated with periodate and heated at 56°C for 30 minutes to inactivate nonspecific inhibitors of hemagglutination, then diluted in twofold steps from 1/10 to 1/1280. Each cup received four hemagglutinating (HA) units of the relevant strain of influenza virus, and a drop of red blood cells. Where enough antibody is present to coat the virions, hemagglutination has been inhibited; hence the erythrocytes settle to form a button on the bottom of the cup. On the other hand, where insufficient antibody is present, erythrocytes are agglutinated by virus and form a shield. The virus assay (bottom line) indicates that the viral hemagglutinin used gave partial agglutination (the end point) when diluted 1/4. Interpretation: The patient originally had no hemagglutinin-inhibiting antibodies against this particular strain of influenza virus. One injection of vaccine produced some antibody; the second injection provided a useful booster response. (Courtesy I. Jack.)

related viruses sharing certain antigens, but also to monitor the presence of antibodies to different antigens at different stages of infection.

Applications of Serology

A significant (conventionally, fourfold or greater) rise in antibody titer between serum samples is indicative of recent infection. Because of the necessary interval between the two specimens a diagnosis is provided only in retrospect. Nevertheless, there are two particular situations when serology is still the diagnostic method of choice despite the delay involved in waiting for a rising titer to make itself apparent. The first is when it is not practicable to attempt cultivation of viruses that are notoriously difficult to isolate, for example, some togaviruses and bunyaviruses. The second is when the management of the case is not urgent; for example, in a women with a rash during the first 4 months of pregnancy a clear demonstration of a rising antibody titer against rubella virus constitutes a strong indication for abortion.

There are also a number of situations when finding antibody in a single specimen of serum can be diagnostic. The first and most important is when

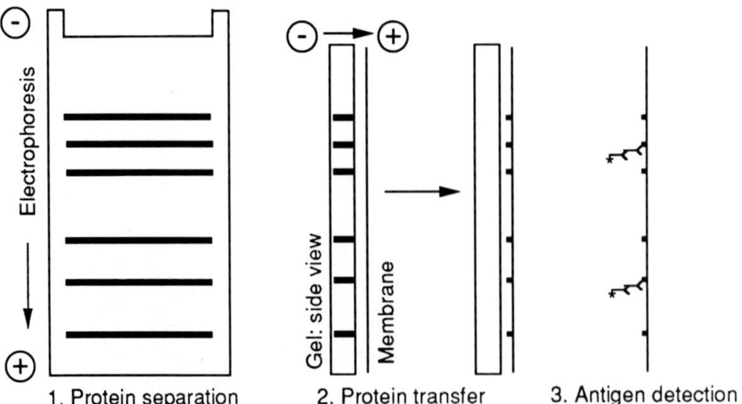

Fig. 12-12 Immunoblotting (Western blotting) for identification of antibodies. (1) Purified virus is digested with the anionic detergent sodium dodecyl sulfate (SDS) and subjected to electrophoresis on a polyacrylamide slab gel (PAGE), which separates the different viral proteins according to M_r. (2) The bands of viral protein are then transferred ("blotted") onto a nitrocellulose membrane by electrophoresis in a different plane to immobilize them. (3) The unoccupied areas of the membrane are blocked ("quenched") by saturation with a suitable protein, then washed, dried, and cut into strips which can be used to test individual patient sera. Each test serum or plasma ("primary" antibody) is then incubated with one strip to enable antibodies to bind to the individual viral proteins. Following rinsing, bound antibody is detected by the addition of enzyme-labeled anti-human immunoglobulin ("secondary" antibody). Following another wash, the bands are revealed by the addition of a substrate chosen to produce an insoluble colored product. (Courtesy of Drs. I. H. Holmes and R. Strugnell.)

specific antibody of the IgM class is found; IgM serology is discussed below. Second, when a seriously ill patient recently returned from abroad is found to have antibodies against an exotic virus, such as Lassa, Marburg, or Ebola virus (all African), it is prudent to assume this agent to be the cause of the illness. Third, in the case of certain persistent infections (e.g., HIV), where it is known that most or all infections progress, the presence of antibody can be used as a reliable diagnostic marker of ongoing infection (which may or may not be responsible for the disease currently under investigation).

Today, the major application of serology is not to diagnose current or recent illness in a patient, but rather to test for immunity. Immunized or unimmunized populations or individuals can be screened for immunity to any given virus. The coverage and efficacy of vaccines can be tested. The susceptibility of close contacts of an individual suffering from infection with a potentially dangerous virus can be determined with a view to protecting at-risk nonimmune individuals by segregation or immunization, or at least establishing a baseline against which to monitor them subsequently.

Immunoglobulin M Class-Specific Antibody Assays

A rapid diagnosis can be made on the basis of a single acute-phase serum by demonstrating virus-specific antibody of the IgM class. Because IgM antibodies appear early after infection but drop to low levels within 1–2 months and generally disappear altogether within 3 months, they are diagnostic of recent (or chronic) infection. Moreover, if found in a newborn baby, they are diagnostic of intrauterine infection, because maternal IgM, unlike IgG, does not cross the placenta. All of the immunoassays described above can readily be rendered IgM class-specific; EIA, RIA, and immunofluorescence are the most generally useful. Typical indirect RIAs for virus-specific IgM are depicted in Fig. 12-13.

A particular problem with IgM assays is interference by the so-called *rheumatoid factor* (RF), which is antibody, mainly of the IgM class, directed against the constant region (Fc) of normal IgG. Though first described in rheumatoid arthritis and other autoimmune diseases, RF is in fact prevalent in many infectious diseases, and it is found in the majority of congenitally infected neonates. RF produces false positives in IgM immunoassays, because it binds to antiviral (as well as normal) IgG in human serum, forming IgM–IgG complexes, which in turn bind the anti-human IgM employed as detector antibody in the assay format depicted in the left-hand side of Fig. 12-13.

To minimize the impact of RF, it is advisable to employ a *reverse* (or *capture*) *IgM assay,* in the simplest form of which monoclonal anti-human IgM is used as the capture antibody and labeled virus as the detector/indicator. More commonly, unlabeled virus is the detector, then labeled monoclonal antiviral IgG, or perhaps better still $F(ab')_2$, is added as indicator (Fig. 12-13, right).

A less common source of false-positive IgM assays is cross-reaction between closely related viruses; for example, infection with one particular coxsackievirus might recall a low-level secondary IgM response against another coxsackievirus serotype with which the patient had been infected earlier in life. False-negative IgM results are occasionally encountered in infants, in some respiratory infections, and during reactivation of latent herpesviruses. Provided that these several traps for the unwary are negotiated with due caution, IgM capture assays have an important place in the armamentarium of rapid diagnostic methods. Class-specific immunoassays can also be designed

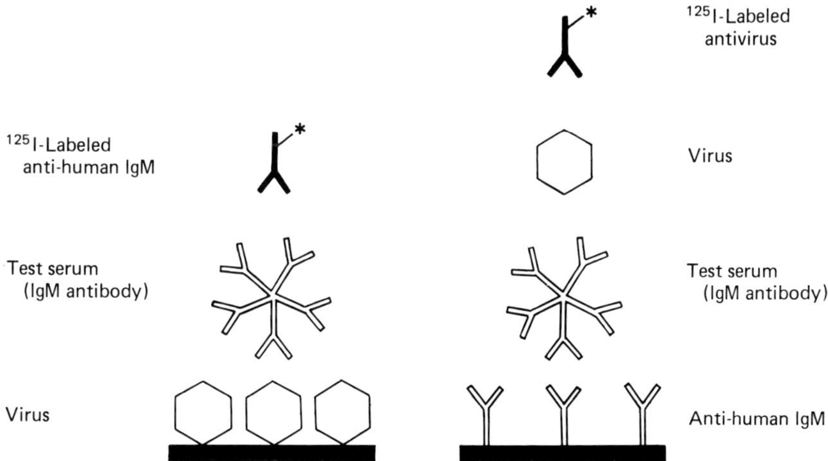

Fig. 12-13 Radioimmunoassays for detection of specific antiviral antibodies of the IgM class. *Left:* Indirect. *Right:* Reverse.

Table 12-3
Advantages and Disadvantages of Various Diagnostic Methods

Diagnostic method	Advantages	Disadvantages/problems
Virus isolation	Permits further study of agent Usually highly sensitive Readily available	Slow, time-consuming, can be difficult Useless for nonviable virus Selection of cell type, etc., may be critical
Direct observation by electron microscopy	Rapid Detects viruses that cannot be isolated Detects nonviable virus	Relatively insensitive Limited to a few viral infections
Serological identification of virus or antigen, e.g., EIA	Rapid and sensitive Provides information on serotypes Readily available, often as diagnostic kits	Not applicable to all viruses Interpretation may be difficult
Nucleic acid probes (with or without gene amplication by PCR)	Rapid Very sensitive (especially after PCR) Potentially applicable to all viruses	May not be readily available Risk of DNA contamination in PCR
Antibody conversion (acute and convalescent sera)	Useful in relating cases to a disease outbreak	Slow, late (retrospective) Interpretation may be difficult
IgM serology	Rapid	False positives may occur

to measure specific antiviral antibodies of the IgA, IgE, or IgG class, or of any given IgG subclass.

A summary of the major strengths and limitations of the several alternative approaches to the diagnosis of viral infections is given in Table 12-3.

Laboratory Safety

Although it has truthfully been observed that virology is one of the less hazardous human occupations, compared with building, mining, or driving a car, it is also true that many cases of serious illness and several hundred deaths from laboratory-acquired infection have been recorded over the years, particularly from togaviruses, flaviviruses, arenaviruses, filoviruses, and other Biosafety Level 4 pathogens. Potentially hazardous procedures are set out in Table 12-4. In addition to specimens taken for detection of virus or antigen, "normal" serum, which may contain HIV or hepatitis B or C virus, has to be treated with caution at all times. The safest procedure is to regard all specimens as potentially infectious.

Precautions to avoid laboratory hazards consist essentially of good laboratory technique. Rigorous aseptic technique must be practiced. Mouth pipetting is banned. Laboratory coats must be worn at all times, and gloves must be used for handling potentially infectious materials including patient specimens. Various classes of biological safety cabinets providing increasing levels of containment are available for procedures of various degrees of biohazard.

Table 12-4
Laboratory Hazards

Hazard	Cause
Aerosol	Homogenization (e.g., of tissue in blender)
	Centrifugation
	Ultrasonic vibration
	Broken glassware
	Pipetting
Ingestion	Mouth pipetting
	Eating or smoking in laboratory
	Inadequate washing/disinfection of hands
Skin penetration	Needle stick
	Hand cut by broken glassware
	Leaking container contaminating hands
	Pathologist handling infected organs
	Splash into eye
	Animal bite

Biosafety Level 4 viruses can be handled only in maximum-containment laboratories (see Fig. 32-2). Careful attention must be given to sterilization, where possible by autoclaving, of all potentially infectious waste as well as equipment. Special arrangements for the disposal of "sharps" are essential. Spills are cleaned up with an appropriate chemical disinfectant. Personnel should be immunized against such diseases as hepatitis B and poliomyelitis (as well as against exotic agents in the special laboratories handling them). Limitations should be placed on the type of work undertaken by pregnant or immunosuppressed employees. The U.S. Centers for Disease Control in conjunction with the National Institutes of Health have issued detailed guidelines on appropriate laboratory procedures and containment facilities for working with viruses (and other microorganisms), classified according to the level of hazard each presents.

Further Reading

Arnheim, N., and Erlich, H. (1992). Polymerase chain reaction strategy. *Annu. Rev. Biochem.* **61,** 131.

Balows, A., Hausler, W. J., Herrmann, K. L., Isenberg, H. D., and Shadomy, H. J., eds. (1991). "Manual of Clinical Microbiology," 5th Ed. American Society for Microbiology, Washington, D.C.

Centers for Disease Control/National Institutes for Health. (1988). "Biosafety in Microbiological and Biomedical Laboratories," 2nd Ed. Publication NIH 88-8395. U.S. Government Printing Office, Washington, D.C.

Freshney, R. I. (1988). "Culture of Animal Cells: A Manual of Basic Techniques," 2nd Ed. Liss/Wiley, New York.

Gardner, P. S., and McQuillin, J. (1980). "Rapid Virus Diagnosis: Application of Immunofluorescence," 2nd Ed. Butterworth, London.

Hierholzer, J. C. (1991). Rapid diagnosis of viral infection. *In* "Rapid Methods and Automation in Microbiology and Immunology" (A. Vaheri and A. Balows, eds.), p. 556. Springer-Verlag, New York.

Innis, M. A., Gelfand, D. H., Sninsky, J. J., and White, T. J., eds. (1990). "PCR Protocols: A Guide to Methods and Applications." Academic Press, San Diego.

Johnson, F. B. (1990). Transport of viral specimens. *Clin. Microbiol. Rev.* **3,** 120.

Kurstak, E. (1985). Progress in enzyme immunoassays: Production of reagents, experimental design, and interpretation. *Bull. WHO* **63,** 793.

Lennette, E. H., Halonen, P., and Murphy, F. A., eds. (1988). "Laboratory Diagnosis of Infectious Diseases, Volume II: Viral, Rickettsial and Chlamydial Diseases." Springer-Verlag, New York.

Palmer, E. L., and Martin, M. L. (1988). "Electron Microscopy in Viral Diagnosis." CRC Press, Boca Raton, Florida.

Sambrook, J., Fritsch, E. F., and Maniatis, T. (1989). *In vitro* amplification of DNA by the polymerase chain reaction. *In* "Molecular Cloning: A Laboratory Manual," 2nd Ed., Chapter 14, p. 14-2. Cold Spring Harbor Laboratory, Cold Spring Harbor, New York.

Schmidt, N. J., and Emmons, R. W., eds. (1989). "Diagnostic Procedures for Viral, Rickettsial and Chlamydial Infections," 6th Ed. American Public Health Association, Washington, D.C.

Specter, S., and Lancz, G. J., eds. (1991). "Clinical Virology Manual," 2nd Ed. Elsevier, Amsterdam and New York.

Wilcek, M., and Bayer, E. A., eds. (1990). Avidin–Biotin Technology. "Methods in Enzymology," Vol. 184. Academic Press, San Diego.

Yolken, R. H. (1982). Enzyme immunoassays for the detection of infectious antigens in body fluids: Current limitations and future prospects. *Rev. Infect. Dis.* **4,** 35.

Chapter 13

Immunization against Viral Diseases

Immunization is the most generally applicable way of preventing infectious disease. The control of so many important viral diseases by immunization is arguably the outstanding medical achievement of the twentieth century, recognized by the award of several Nobel prizes.

Traditionally, there have been two major strategies for the production of viral vaccines, one employing live avirulent virus and the other employing chemically inactivated virus. Recombinant DNA technology has now opened up an exciting range of additional options (Table 13-1).

Live-Virus Vaccines

The most successful viral vaccines are live avirulent mutants. They have been instrumental in dramatically reducing the incidence of several important diseases of childhood (see Fig. 15-1) and for eradicating smallpox (see Chapter 15). The key to their success is the fact that the live virus multiplies in the recipient, eliciting a lasting immune response but causing little or no disease. In effect, a live vaccine produces a subclinical infection, nature's own way of immunizing.

Table 13-1
Approaches to Designing Viral Vaccines

Live-virus vaccines
Virus attenuated in virulence by serial passage in cultured cells
Cold-adapted mutants and reassortants
Virus attenuated by gene deletion or site-directed mutagenesis
Antigens expressed via recombinant live viral or bacterial vectors
Nonreplicating antigens
Inactivated virions
Purified proteins
Proteins from genes cloned in prokaryotic or eukaryotic cells
Synthetic peptides
Others
Anti-idiotypic antibodies
Viral DNA

Attenuated Live-Virus Vaccines

Most of the live-virus vaccines in common usage today have been derived empirically by serial passage in cultured cells. Adaptation of virus to more vigorous growth in cultured cells is fortuitously accompanied by progressive loss of virulence for the natural host. Avirulence is demonstrated initially in a convenient laboratory model, often a mouse, then a primate, before being confirmed by clinical trials in human volunteers. Because of the requirement that the vaccine must not be so attenuated that it fails to replicate satisfactorily *in vivo,* it is sometimes necessary to compromise with a strain that does in fact induce trivial symptoms in a few of the recipients.

During dozens of passages in cultured cells these host range mutants accumulate numerous point mutations; it is generally not known which of these are responsible for attenuation. For most viruses, several genes contribute to virulence in different ways. Moreover, the avirulence of attenuated vaccines has generally not yet been characterized in terms of their pathogenesis in the vaccinee. In the case of some experimental vaccines administered by the respiratory route, multiplication of the attenuated virus is severely restricted and confined to the upper respiratory tract. On the other hand, the oral poliovaccine replicates in intestinal cells but has lost the capacity to infect the critical target cells in the spinal cord.

Despite the outstanding success of empirically derived live vaccines, a vigorous program of research is aimed at replacing what some see as "genetic roulette" with a more calculated approach to the design of live viruses of reduced virulence. These approaches, which have already yielded several promising experimental vaccines, are discussed below.

Temperature-Sensitive and Cold-Adapted Mutants

The observation that *temperature-sensitive (ts) mutants* (unable to replicate satisfactorily at temperatures much higher than normal, e.g., 40°C) generally display reduced virulence suggested that they might make satisfactory live vaccines. Unfortunately, even vaccines containing more than one *ts* mutation displayed a disturbing tendency to revert toward virulence during replication in humans.

Attention then moved to *cold-adapted (ca) mutants*, derived by adaptation of virus to grow at suboptimal temperatures. The rationale is that such a mutant might provide a safer vaccine for intranasal administration, in that it would replicate well at the lower temperature of the nose (33°C) but not at the temperature of the more vulnerable lung. Cold-adapted influenza vaccines containing mutations in almost every gene do not revert to virulence. They are used as "master strains" into which genes for novel hemagglutinin (HA) and/or neuraminidase (NA) proteins can be introduced by reassortment, but they are not as immunogenic as had been hoped.

Deletion Mutants and Site-Directed Mutagenesis

The problem of back mutation could be circumvented by deleting nonessential genes that contribute to virulence. The large DNA viruses, in particular, carry a certain amount of genetic information that is not absolutely essential, at least for replication in cultured cells. Genetic surgery has been used to construct deletion mutants of certain herpesviruses, one of which (pseudorabies of swine) is now in use in veterinary practice.

Site-directed mutagenesis now permits the introduction of prescribed nucleotide substitutions at will (see Chapter 4). When more becomes known about particular genes influential in virulence it will be possible not only to delete or modify these genes but also to construct vaccines with any designated nucleotide sequence. Indeed, the day may come when licensing authorities demand that new vaccines be fully defined genetically, that is, that the complete nucleotide sequence of the vaccine virus be known and perhaps even stipulated.

Live Recombinant Viruses or Bacteria as Antiviral Vaccines

It is ironic that, no sooner had the World Health Organization (WHO) recommended the abandonment of vaccination following the eradication of smallpox, than an exciting new use was discovered for vaccinia virus. Recombinant DNA technology has opened a novel approach to vaccination which could prove to be of widespread applicability. The concept is to insert the gene for the protective surface protein of any chosen virus into the genome of an avirulent virus, which can then be administered as a live vaccine. Cells in which the vector virus replicates *in vivo* will produce this foreign protein and the body will mount both a humoral and a T-cell-mediated immune response to it.

In the original construct the hepatitis B surface antigen (HBsAg) gene, flanked by the nonessential vaccinia gene for thymidine kinase (TK) and its promoter, was inserted into a bacterial plasmid (Fig. 13-1). Mammalian cells infected with vaccinia virus were then transfected with this chimeric plasmid. Recombination occurred between the vaccinia DNA and the plasmid DNA. Selection for virus with a TK^- phenotype enabled vaccinia virions containing the HBsAg gene to be recovered. This construct and analogous vaccinia recombinants incorporating genes from a range of other viruses have been

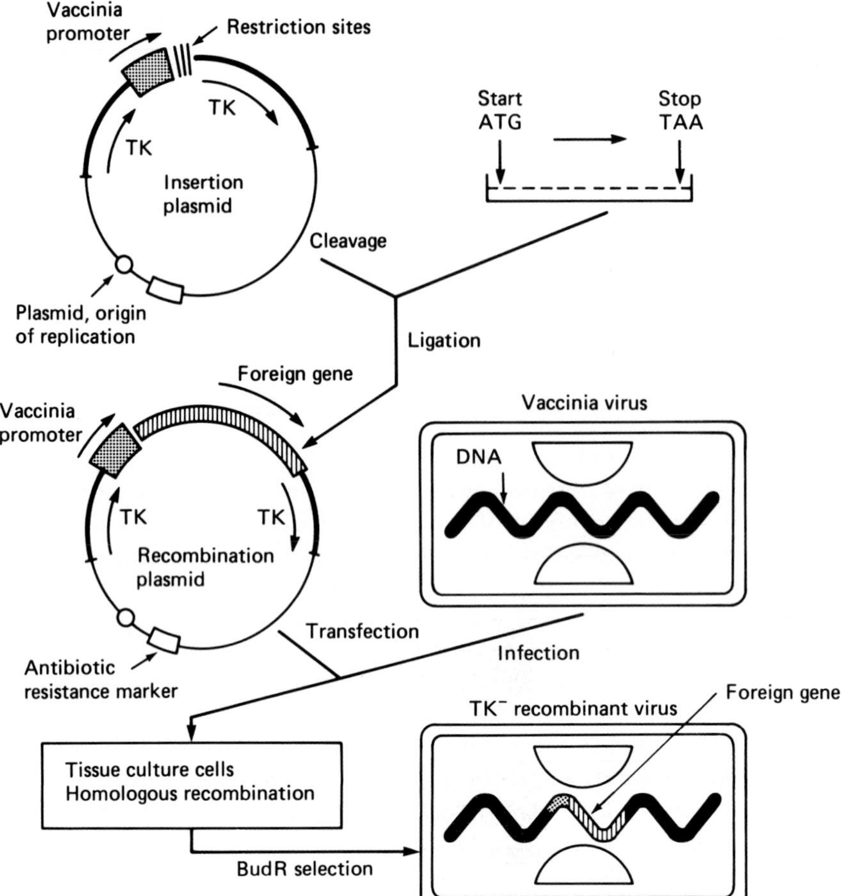

Fig. 13-1 Method of construction of a recombinant vaccinia virus carrying a selected gene from another virus. TK, Thymidine kinase gene of vaccinia virus; BudR, bromodeoxyuridine. (Courtesy Dr. B. Moss.)

demonstrated to protect animals against the diseases they cause. For example, a vaccinia–rabies recombinant protects foxes and raccoons against rabies when incorporated in bait dropped from the air. Because the large vaccinia genome can accommodate at least a dozen foreign genes and still be packaged satisfactorily within the virion, it would theoretically be possible to construct a single recombinant virus as a vaccine capable of protecting against most of the common viral (and nonviral) infectious diseases of childhood. The potential benefits, particularly to the Third World, are self-evident.

DNA viruses of the families *Poxviridae, Herpesviridae,* and *Adenoviridae* can all be used as vectors, but vaccinia virus offers obvious advantages (Table 13-2). The major problem with vaccinia virus is that with unmodified smallpox vaccine there were always minor side effects, and in about 1 in 100,000 persons there were severe complications, with a substantially higher incidence in severely immunocompromised individuals. This problem has in principle been overcome by making deletion mutants of vaccinia virus that produce

Table 13-2
Live Vaccinia Virus as Vector for Cloned Viral Genes

Advantages
Large genome can accommodate several foreign inserts
Production inexpensive
Relatively stable, even in tropics
Delivery (multiple puncture) by nonmedical personnel
Cell-mediated immunity also elicited
Disadvantages
Rare serious side effects with regular vaccinia virus vaccine[a]
Revaccination with vaccinia vector within 5 years may be unreliable

[a] Attenuated vectors are now available; most have not yet been tested in large numbers of people, to determine the possible occurence of rare severe complications.

trivial lesions but still satisfactorily express incorporated foreign genes. It has also been found that birdpox viruses, which undergo restricted replication in mammalian cells, can express foreign genes effectively. In both systems, the incorporation of genes encoding certain cytokines may greatly enhance the immune response. Recombinant poxviruses produced by these approaches are currently undergoing clinical trials.

Live Bacteria as Expression Vectors

Recombinant DNA technology also allows the expression of a viral epitope on the surface of a bacterium. The general approach is to insert the DNA encoding a protective viral epitope into a region of the genome of a bacterium that encodes a prominent surface domain on a protein normally situated on the exterior of the organism. Provided that the added loop protruding from the resultant hybrid protein does not seriously interfere with its transport, packaging, stability, or function, that organism will present the passenger epitope to the immune system of the host. Enteric bacteria which multiply naturally in the gut would seem to be ideal expression vectors for presenting protective epitopes of virulent enteric viruses to the gut-associated lymphoid tissue. The main candidates currently under development as potential vehicles are attenuated (vaccine) strains of *Salmonella typhi, S. typhimurium, Escherichia coli,* and BCG, the world's most widely used live bacterial vaccine.

Inactivated Virus and Virus Subunit Vaccines

Inactivated Virus Vaccines

Inactivated ("killed") vaccines are made from virulent virus by chemically destroying its infectivity while retaining its immunogenicity. The traditional inactivating agent was formaldehyde, but this is being supplanted by β-propiolactone and ethylenimines; one of the advantages of β-propiolactone is that it is completely hydrolyzed within hours to nontoxic products. Being noninfectious, such vaccines are generally safe but need to be injected in large amounts to elicit an antibody response commensurate with that attainable by a much smaller dose of live-virus vaccine. Normally, even the primary course comprises two or three injections, and further ("booster") doses may be required at intervals over the succeeding years to revive waning immunity.

Purified Protein Vaccines

Large doses of inactivated virions often produce a febrile response as well as local reaction at the injection site, especially in young children. It seems logical to remove all nonessential components of the virion and inoculate only the relevant immunogen, namely, the particular surface (envelope or outer capsid) protein against which neutralizing antibodies are directed (Table 13-3). For example, the HA and NA glycoproteins can be extracted from influenza virions with detergent and used as a subunit vaccine.

Viral Proteins from Cloned Genes

To produce viral proteins more inexpensively on an industrial scale, recombinant DNA technology is now being widely exploited to clone the appropriate genes in prokaryotic or eukaryotic cells. Particular advantages of this approach are that safe, noninfectious vaccines can be made against viruses which cannot be cultured satisfactorily *in vitro,* for example, HIV or hepatitis B, or for which it has proved difficult to develop safe, effective live-virus vaccines.

Once the critical structural protein conferring protection has been identified, its gene (or, in the case of an RNA virus, a cDNA copy of the gene) may be cloned in bacteria, yeasts, insect cells, or mammalian cells. The yield of functional viral protein from bacteria is often poor, for a variety of reasons. Yeast cells have some capacity to glycosylate and secrete proteins, and the industrial technology for the large-scale cultivation of yeasts is well developed. The current hepatitis B vaccine, produced in *Saccharomyces cerevisiae,*

Table 13-3
Defined Antigen Vaccines

Approaches
Purified protein
Protein "cloned" by recombinant DNA technology (in yeasts, insect, or mammalian cells)
Synthetic peptide
Advantages
Production and quality control simple
Nontoxic, nonallergenic
Feasible even if virus cannot be cultured
Safer if virus is highly virulent (e.g., HIV), persistent (e.g., herpesvirus), or carcinogenic (e.g., hepatitis B virus)
Disadvantages
Less immunogenic than whole virus
Requires adjuvant
Requires booster injections
May fail to elicit cell-mediated immunity

was the first genetically engineered vaccine to be licensed for use in humans. Mammalian cells present advantages over cells from lower eukaryotes in that they possess the machinery for posttranslational processing of mammalian viral proteins including glycosylation and secretion. A wide variety of mammalian DNA virus vectors have been constructed to facilitate or prevent viral replication as desired, or to regulate gene expression. A fascinating alternative is the use of insect viruses of the baculovirus family growing in cultured moth cells (or caterpillars!). The promoter for the gene encoding the nonessential viral polyhedrin protein is so strong that the product of a foreign gene inserted within the polyhedrin gene may comprise up to half of all the protein the infected cells make.

Synthetic Vaccines

Techniques have been developed for locating and defining epitopes on viral proteins, and it is possible to synthesize peptides corresponding to these antigenic domains (Table 13-4). Such synthetic peptides have been shown to elicit neutralizing antibodies against HIV and a few other viruses, but in general have been disappointing. The main reason for this is that, in the native antigen, most of the epitopes recognized by antibodies are not continuous but assembled, that is, not a linear array of contiguous amino acids but an assemblage of amino acids which, although separated in the primary sequence, are brought into close apposition by the folding of the polypeptide chain(s). In contrast, the epitopes recognized by T lymphocytes are short, linear peptides (bound to MHC protein). Further, some of these "T-cell epitopes" are conserved between strains of viruses and therefore elicit a cross-reactive T-cell response. Thus, attention is moving toward the construction of artificial heteropolymers of T-cell epitopes and B-cell epitopes, perhaps coupled to a peptide facilitating fusion with cell membranes to enhance uptake. Such constructs might prime T cells to respond more vigorously when boosted with an inactivated whole-virus vaccine, or on natural challenge.

Table 13-4
Synthetic Peptides as Potential Vaccines

Advantages
Short defined amino acid sequence representing protective epitope
Safe, nontoxic, stable
T-cell epitopes are naturally presented in the form of peptides
Artificial constructs may be engineered to contain B-cell and T-cell epitopes, or epitopes of one or more proteins
Disadvantages
Poorly immunogenic; adjuvant, liposome, iscom, or polymer needed
Most B-cell epitopes are assembled (discontinuous)
May be too specific, not protecting against natural variants
Single-epitope vaccine will readily select mutants
No response in those lacking appropriate class II MHC antigen

DNA Vaccines

An even more revolutionary technique for vaccination may be in the wings. To the surprise of immunologists, it has been found that intramuscular injection of cloned viral DNA, in a plasmid with suitable promoters, can produce long-lasting antibody and cell-mediated immune responses to the protein(s) encoded by that DNA. Likewise, the immunomodulatory effects of various cytokines can be produced by the intramuscular injection of the relevant DNA. The potential of this technique as a method of vaccination is being vigorously explored by pharmaceutical companies, although all concerned realize that much research and development lie ahead if it is ever to become a practical proposition.

It will be necessary, for example, to obtain detailed information on such critical aspects as the location (especially elsewhere than in muscle cells) and fate of the injected plasmids, their persistence, the mechanism of expression, including the processing of peptides through cytosolic and endosomal pathways and presentation via class I and II MHC antigens to Tc and Th lymphocytes, the continuity of priming of various classes of memory T cells, the mechanism of protection *in vivo* in the short and long term, the mass of DNA required in humans as opposed to mice, and the optimum methods of presentation (packaging) and protection of the injected DNA. Above all, licensing authorities will need to be convinced that there is no realistic possibility of integration of the foreign nucleic acid into the host genome which might lead to cancer.

Anti-Idiotypic Antibodies

The antigen-binding site of the antibody produced by each B-cell clone contains a unique amino acid sequence known as its *idiotypic determinant* or *idiotype (Id)*. Antibodies can be raised against this idiotype; they are known as *anti-idiotypic antibodies (anti-Id)*. Because the original antigen binds to the same variable region of the antibody molecule as does anti-Id, they might be expected to have similar conformations. If so, anti-Id could be employed as a surrogate antigen, that is, as a vaccine to elicit an antiviral immune response.

There are now a number of examples of anti-idiotypic antibodies that elicit specific antiviral B-cell and/or T-cell responses. It is still uncertain whether this points the way to a novel vaccine strategy.

Methods for Enhancing Immunogenicity

The immunogenicity of inactivated vaccines, and especially of purified protein vaccines and synthetic peptides, usually needs to be enhanced in some way. This may be achieved by mixing the antigen with an *adjuvant* or incorporating it into *liposomes* or into an *immunostimulating complex (iscom)*.

Adjuvants

Adjuvants are substances that, when mixed with vaccines, potentiate the immune response, humoral and/or cellular, so that a lesser quantity of antigen and/or fewer doses will suffice. Adjuvants differ greatly in their chemistry and in their modes of action, which may include the following: (1) prolongation of release of antigen; (2) activation of macrophages, leading to secretion of cytokines and attraction of lymphocytes; (3) mitogenicity for lymphocytes. Alum is the only adjuvant currently licensed for use in humans, and has been widely used for years, but it is not particularly effective. Mineral oil adjuvants are used in animals but are too reactogenic to be acceptable in humans. There is a clear requirement for better adjuvants, preferably chemically defined and of known mode of action. For instance, there are now a number of experimental formulations based on muramyl dipeptide, which can also be chemically coupled to synthetic antigens or incorporated into liposomes.

Liposomes and Iscoms

Liposomes consist of artificial lipid membrane spheres into which proteins can be incorporated. When purified viral envelope proteins are used, the resulting "virosomes" (or "immunosomes") somewhat resemble the original envelope of the virion. This enables one not only to reconstitute virus envelope-like structures lacking nucleic acid and other viral components, but also to select nonpyrogenic lipids and to incorporate substances with adjuvant activity. Although liposomes have not fully lived up to expectations as immunogens, other types of membrane-bound micelles (i.e., aggregates of protein molecules) can fully restore the immunogenicity lost when viral glycoprotein is removed from its original milieu.

When viral envelope glycoproteins or synthetic peptides are mixed with cholesterol plus a glycoside known as Quil A, a spherical cagelike structure 40 nm in diameter is formed. These *iscoms* (immunostimulating complexes) have been shown experimentally to display significantly enhanced immunogenicity but have not yet been developed commercially as vaccines.

Comparison of Different Classes of Vaccines

The relative advantages and disadvantages of live-virus vaccines compared with inactivated or subunit vaccines are summarized in Table 13-5 and discussed below.

Immunologic Considerations

Naturally acquired immunity to reinfection is virtually lifelong in the case of most of the viruses that reach their target organ(s) via systemic (viremic) spread. This solid immunity is attributable to antibody of the IgG class, which successfully neutralizes the incoming challenge virus. Immunity to those respiratory and enteric viruses whose pathogenic effects are manifested mainly at the site of entry is attributable mainly to antibodies of the IgA class and tends to be of shorter duration. Thus the principal objective of artificial immunization by vaccine is to elicit a high titer of neutralizing antibodies of the appropriate class, directed against the relevant epitopes on the surface of the

Table 13-5
Advantages and Limitations of Live and Inactivated Vaccines

Property	Live	Inactivated
Route of administration	Natural or injection	Injection
Dose of virus; cost	Low	High
Number of doses	Single, generally	Multiple
Need for adjuvant	No	Yes
Duration of immunity	Many years	Generally less than live vaccines
Antibody response	IgG; IgA (mucosal route)	IgG
Cell-mediated response	Good	Uncertain
Heat lability	Yes	No
Interference	Oral poliovaccine only	No
Side effects	Occasional, mild	Occasional, local
Reversion to virulence	Rarely; oral poliovaccine only	No

virion, in the hope of preventing initiation of infection by neutralization of the challenge virus.

This ideal is not always realizable, but it may not matter if the progress of the infection can be curtailed sufficiently to allow time for the emergence of the quite different set of immunologic mechanisms that contribute to recovery from viral infection. Provided that enough memory T and B cells are still present at the time of challenge, anamnestic T-cell- and B-cell-mediated responses will be mounted without undue delay. In particular, recovery will be accelerated by cytotoxic T-cell-mediated cytolysis of infected cells. Such "memory-dependent immunity" may be particularly important in diseases with a relatively long incubation period.

It has proved difficult to produce effective vaccines against three classes of viral diseases: respiratory infections, sexually transmitted diseases, and persistent infections. Mucosal immunity, mediated by IgA, is important in both respiratory and sexually transmitted diseases; there is evidence that vaccination by the convenient oral route may generate satisfactory mucosal immunity in the respiratory and genital tracts, as well as in the intestinal tract, as a result of lymphocyte trafficking between different compartments of the "common mucosal pathway." Antigenic drift and shift pose yet another problem, circumvention of which requires constant revision of the antigenic composition of vaccines such as those used to control influenza. Special difficulties also attend vaccination against viruses known to establish persistent infections, such as herpesviruses and retroviruses; a vaccine must be outstandingly effective if it is to prevent not only the primary disease but also the establishment of lifelong latency.

Subclinical infection is, by and large, extremely effective, inducing lifelong immunity following systemic infection. Live avirulent vaccines, preferably but not necessarily delivered via the natural route, are obviously the nearest approach to this ideal. The track record of live vaccines against major human diseases such as smallpox, yellow fever, poliomyelitis, and measles is clearly superior to that of most of the inactivated vaccines devised so far. Why is this so?

First, one must consider the quantitative advantage of a live immunogen which replicates many millionfold following delivery. Second, whether delivered via the natural route or not, live virus will be presented to the various arms of the immune system in a natural way. Not only will virions be processed and presented via the endosomal pathway to MHC class II restricted helper T cells, but peptides derived from newly synthesized viral proteins in infected cells will be presented via the cytosolic pathway to class I restricted cytotoxic T cells. Thus live vaccines are generally found to be much more effective in eliciting cell-mediated immunity. Furthermore, live vaccines may elicit a broader immunologic response, because many cytotoxic and helper T cells are directed to conserved epitopes on internal or nonstructural proteins that are shared between different strains of virus. There is also the possibility that live vaccine virus persists in some form for years or that larger amounts of relevant antigens are sequestered, on follicular dendritic cells or elsewhere, for longer than with other types of vaccines. It does appear that memory T and B lymphocytes are more effectively recruited by live than by inactivated vaccines, though very little is known about the optimal conditions of antigen presentation for eliciting B- or T-cell memory.

The efficacy of an immunogen depends not only on the dose of protein delivered but also on the form in which that protein is presented to the immune system. For example, immunogen in certain forms may elicit a stronger response in some classes of T cells than others, and it may be crucial to avoid, say, a suppressor T-cell response at the expense of a helper or cytotoxic response. An excessive DTH response or IgE response might also be detrimental under some circumstances. There have been some unforeseen immunopathologic consequences of immunization with certain inactivated respiratory vaccines, for example.

Special problems attend the choice of submolecular immunogens such as synthetic peptides. Because each individual in an outbred population carries a unique set of MHC molecules, any particular T-cell epitope is unlikely to be recognized by all humans. Therefore, if synthetic peptides or other submolecular constructs are to be used as vaccines, it will be necessary to ensure that they include at least one T-cell epitope recognized by a class I molecule and one recognized by a class II molecule present in most people, as well as at least one B-cell epitope that is immunodominant in a large majority of the population and that elicits a protective antibody response.

Vaccine Safety and Efficacy

To be acceptable, a vaccine must be both safe and efficacious. Licensing authorities have become extremely vigilant and have insisted on rigorous safety tests since residual live virulent virus in certain batches of inactivated vaccines caused a number of tragedies in pioneering days. Live attenuated vaccines present a different set of challenges that must be met before any particular product is released onto the market; these are discussed below.

Contaminating Viruses

Because vaccine viruses are grown in cells derived from humans or animals, there is always a possibility that a vaccine will be contaminated with another virus, derived from those cells or from the medium (especially the serum) in which the cells are cultured. For example, primary monkey kidney cell cultures, once widely use for the manufacture of poliovaccines, have, at one time or another, yielded over 75 simian viruses, some of which are pathogenic for humans. This danger has led to a swing away from primary cell cultures toward well-characterized continuous cell lines which can be subjected to comprehensive screening for endogenous agents before being certified as safe, then stored frozen for many years as seed lots.

Underattenuation

Some excellent human viral vaccines in routine use, such as rubella and measles vaccines, produce some symptoms—in effect, a very mild case of the disease—in a minority of recipients. Attempts to attenuate virulence further by additional passages in cultured cells have been accompanied by a decline in the capacity of the virus to replicate in humans, with a corresponding loss of immunogenicity. Such trivial side effects as do occur with current human viral vaccines are of no real consequence and do not prove to be a significant disincentive to immunization, provided that parents of vaccinated children are adequately informed in advance.

Genetic Instability

A different problem occurs in the case of vaccine strains with an inherent tendency to revert toward virulence during replication in the recipient. The only example in a human vaccine in general usage is oral poliovaccine (OPV). Exceedingly rarely (less than once in every million vaccinees), "vaccine-associated" poliomyelitis is seen, either in a congenitally immunodeficient baby or, even more rarely, in an unvaccinated parent to whom the virulent revertant has spread. For this reason, inactivated poliovaccine is preferred in known immunocompromised individuals.

Heat Lability

Live vaccines are vulnerable to inactivation by high ambient temperatures, which presents a particular problem in the tropics. Because most tropical countries also have underdeveloped health services, difficulties are encountered in maintaining the "cold chain" from manufacturer to the point of delivery, namely, the child in some remote rural village. To some extent the problem has been alleviated by the addition of stabilizing agents to the vaccines, and by packaging them in freeze-dried form for reconstitution immediately before administration. In other cases simple portable refrigerators have been developed and placed in the field.

Interference

Live vaccines delivered by mouth or nose depend for their efficacy on replication in the enteric and respiratory tract, respectively. Interference can occur between different live viruses contained in vaccines delivered via the natural route (e.g., between the three serotypes of poliovirus in OPV if their concentrations are not appropriately balanced), or between the vaccine virus and itinerant enteric or respiratory viruses that happen to be growing in the vaccinee at the time. The latter is the reason why OPV is routinely adminis-

Table 13-6
Viral Vaccines Recommended for Human Use[a,b]

Disease	Vaccine strain	Cell substrate	Attenuation	Inactivation	Route
Yellow fever	17D	Chick embryo	+	−	Subcutaneous
Poliomyelitis	Sabin 1, 2, 3	HEF	+	−	Oral
Measles	Schwartz	CEF	+	−	Subcutaneous
Rubella	RA27/3	HEF	+	−	Subcutaneous
Mumps	Jeryl Lynn	CEF	+	−	Subcutaneous
Rabies	Pasteur	HEF	−	BPL	Intramuscular
Influenza	A/H3N2, B, A/H1N1	Chick embryo	−	BPL or formalin or HANA[b] subunits	Intramuscular
Hepatitis A	HM175	HEF	−	Formalin	Intramuscular
Hepatitis B	Hepatitis B	—	−	Cloned HBsAg[c]	Intramuscular

[a] A wise variety of different viral strains and cell substrates are used in different countries; the selection listed is not comprehensive.
[b] BPL, β-Propiolactone; CEF, chick embryo fibroblast cultures; HEF, diploid strain of human embryonic fibroblasts; HANA, hemagglutinin/neuraminidase.
[c] Hepatitis B surface antigen purified after gene cloning in yeast.

tered on three occasions, separated by at least 2 months. Interference is not a problem with systemically administered polyvalent vaccines, for example, the live measles–mumps–rubella (MMR) vaccine.

Characteristics of the major human viral vaccines in general usage around the world are set out in Table 13-6. Others are in use against viruses of restricted geographic distribution, such as Japanese encephalitis. There are many more vaccines in various stages of development or clinical trial, for example, against rotavirus, HIV, several herpesviruses, several respiratory viruses, and so on.

The subjects of vaccination policy, passive immunization, and eradication of diseases by vaccination are discussed in Chapter 15 together with a typical schedule for immunization of infants against common viral diseases (Table 15-1). Detailed discussion of current immunization practices and/or vaccine development prospects for individual viral diseases is provided in each chapter of Part II of this book.

Further Reading

Ada, G. L. (1993). Vaccines. *In* "Fundamental Immunology" (W. E. Paul, ed.), 3rd Ed., p. 1309. Raven, New York.

Binns, M. M., and Smith, G. L., eds. (1992). "Recombinant Poxviruses." CRC Press, Boca Raton, Florida.

Bittle, J. L., and Murphy, F. A., eds. (1989). Vaccine Biotechnology. *Adv. Vet. Sci. Comp. Med.* **33,** 1.

Brown, F., ed. (1990). Modern Approaches to Vaccines. *Semin. Virol.* **1,** 1.

Brown, F., ed. (1991). Impact of biotechnology on protection against infectious diseases. *World J. Microbiol. Biotechnol.* **7,** 105.

Brown, F., Chanock, R. M., Ginsberg, H. S., and Lerner, R. A., eds. (1990–1992). "Vaccines 90," "Vaccines 91," "Vaccines 92." Cold Spring Harbor Laboratory, Cold Spring Harbor, New York.

Dimmock, N. J., Griffiths, P. D., and Madeley, C. R. (1990). "Control of Virus Diseases," Society for General Microbiology Symposium 45. Cambridge Univ. Press, Cambridge.

Kurstak, E., Marusyk, R. G., Murphy, F. A., and van Regenmortel, M. H. V., eds. (1988). New Vaccines and Chemotherapy. *Appl. Virol. Res.* **1.**

Moss, B. (1991). Vaccinia virus: A tool for research and vaccine development. *Science* **252,** 1662.

Murphy, B. R., and Chanock, R. M. (1990). Immunization against viruses. *In* "Fields Virology" (B. N. Fields, D. M. Knipe, R. M. Chanock, M. S. Hirsch, J. L. Melnick, T. P. Monath, and B. Roizman, eds.), 2nd Ed., p. 469. Raven, New York.

Ogra, P. L., and Garofalo, R. (1990). Secretory antibody response to viral vaccines. *Prog. Med. Virol.* **37,** 156.

Schödel, F. (1992). Prospects for oral vaccination using recombinant bacteria expressing viral epitopes. *Adv. Virus Res.* **41,** 409.

Ulmer, J. B., Donnelly, J. J., Parker, S. E., Rhodes, G. H., Felgner, P. L., Dwarki, V. J., Gromkowski, S. H., Deck, R. R., DeWitt, C. M., Friedman, A., Hawe, L. A., Leander, K. R., Martinez, D., Perry, H. C., Shiver, J. W., Montgomery, D. L., and Liu, M. A. (1993). Heterologous protection against influenza by injection of DNA encoding a viral protein. *Science* **259,** 1745.

van Regenmortel, M. H. V., and Neurath, A. R., eds. (1990). "Immunochemistry of Viruses, Volume II: The Basis for Serodiagnosis and Vaccines." Elsevier, Amsterdam.

WHO Report. (1990). Potential use of live viral and bacterial vectors for vaccines. *Vaccine* **8,** 425.

Woodrow, G. C., and Levine, M. M., eds. (1990). "New Generation Vaccines." Dekker, New York.

Chapter 14

Epidemiology of Viral Infections

Epidemiology is the study of the determinants, dynamics, and distribution of diseases in populations. The risk of infection and/or disease in a human population is determined by characteristics of the virus, of individual hosts, and of the host population, such as innate and acquired resistance, and behavioral, environmental, and ecologic factors that affect virus transmission from one person to another. Epidemiology, which is part of the science of population biology, attempts to meld these factors into a unified whole.

By introducing quantitative measurements of disease trends, epidemiology has come to have a major role in improving our understanding of the nature of diseases and in alerting and directing disease control activities. Epidemiologic study is also effective in clarifying the role of viruses in the etiology of diseases, understanding the interaction of viruses with environmental determinants of disease, determining factors affecting host susceptibility, unraveling modes of transmission, and testing of vaccines and drugs on a large scale.

Computations and Data Used by Epidemiologists

Calculations of Rates

The comparison of disease experience in different populations is expressed in the form of rates, the purpose of the multiplier (10^n = 1000, 100,000, 1,000,000, etc.) being to produce a rate that is a manageable whole number. Two rates are widely used: the *incidence rate* and the *prevalence rate;* another statistic of importance is the death rate. In all cases the denominator (total number of persons at risk) may be as general as the total population in a state or country or as specific as the population known to be susceptible or at risk (e.g., the number of persons in a specified population who lack antibodies to the virus of interest). In each situation it is imperative to be clear about the nature of the denominator—indeed, epidemiology has been called "the science of the denominator." All these rates may be affected by various attributes that distinguish one person from another: age, sex, genetic constitution, immune status, nutrition, and various behavioral parameters. The most widely applicable attribute is age, which may encompass, and can therefore be confounded by, immune status as well as various physiologic variables.

Incidence

The incidence, or *attack rate,* is a measure of frequency over time, for example, monthly or annual incidence, and is especially useful for acute diseases of short duration. For acute infections three parameters determine the incidence of infection (or disease): the proportion of susceptible persons, the

$$\text{Incidence rate} = \frac{\text{Number of cases} \times 10^{n}}{\text{population at risk}} \text{ in a specified period of time}$$

proportion of susceptibles that are infected, and the proportion of infected persons who suffer disease. The proportion susceptible to a specific virus reflects their past history of exposure to that virus and the duration of immunity. The proportion infected during a year or a season may vary considerably, depending on factors such as number and density, season, and for arbovirus infections the vector population. Of those infected, only some develop overt disease. The ratio of clinical to subclinical (inapparent) infections varies greatly with different viruses (see Table 14-1). For example, in measles almost 100% of infections are clinically apparent, whereas less than 1% of those infected with an encephalitogenic arbovirus or with poliovirus develop encephalitis or poliomyelitis, respectively.

The *secondary attack rate,* applied to comparable relatively closed groups like households or classrooms, is a useful measure of the "infectiousness" of viruses transmitted by aerosol or droplet spread. It is defined as the number of persons in contact with the primary or *index case* who become infected or ill within the maximum incubation period, as a percentage of the total number of susceptible persons exposed to infection.

Prevalence

It is difficult to measure the incidence of chronic diseases, especially where the onset is insidious, and for such diseases it is customary to determine the prevalence, that is, the ratio, at a particular point in time, of the number of cases currently present in a population divided by the population. Prevalence is a snapshot of the frequency that prevails at a given time, and it

$$\text{Prevalence rate} = \frac{\text{Number of cases} \times 10^{n}}{\text{population at risk}} \text{ at a particular time}$$

is thus a function of both incidence and duration of the disease. *Seroprevalence* relates to the occurrence of antibody to a particular virus in a population, and because neutralizing antibodies often last for many years, or even for life, seroprevalence rates usually represent the cumulative experience of a population.

Death Rates

Deaths from a disease can be categorized in two ways: the *cause-specific mortality rate* (the number of deaths from the disease in a given year, divided by the total population at midyear), usually expressed per 100,000; or the *case–fatality rate* (the percentage of persons with a particular disease who die from the disease).

Terms Used in Epidemiology

An infectious disease is characterized as *endemic* when there are multiple or continuous chains of transmission resulting in continuous occurrence of disease in the population of a limited region. *Epidemics* are peaks in disease incidence that exceed the endemic baseline or expected rate of disease (Fig. 14-1). The size of the peak required to constitute an epidemic is arbitrary and is related to the background endemic rate, the morbidity rate (frequency of illness), and the anxiety that the disease arouses because of its severity; for example, a few cases of encephalitis are regarded as an epidemic, whereas a few cases of influenza are not. A *pandemic* is a worldwide epidemic, such as the pandemics caused by the Asian and Hong Kong strains of influenza A virus in 1957 and 1968.

The *incubation period* is the interval between infection and the onset of symptoms. In many diseases, such as measles and chickenpox, persons become infectious for others a day or so before they become sick themselves. Infected persons shed virus and remain infectious for a variable period, the *period of infectivity,* which tends to be relatively short in acute infections and much longer in chronic or slow infections, in which virus may be shed in

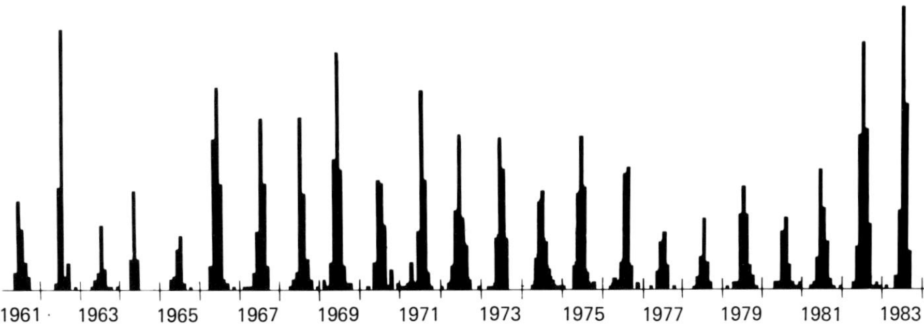

Fig. 14-1 Epidemic occurrence of respiratory syncytial virus infections, as seen in the data on monthly admissions to Fairfield Hospital for Infectious Diseases, Melbourne, Australia, between 1961 and 1983. Epidemics of varying magnitude occur each winter (June to August), with significant disease occurring mainly in infants. (Data courtesy A. A. Ferris, F. Lewis, N. Lehmann, and I. D. Gust.)

small amounts, even from asymptomatic carriers, for years. Table 14-1 sets out these parameters for some common human viral diseases.

Sources of Data

The collection of accurate data about the occurrence of disease is more difficult than the computation of the rates just described. Data on the denominator (population) are readily available, but it is difficult to obtain accurate information on the number of cases. Where such information is regarded as essential for public health purposes, infectious diseases are made "notifiable" by law, so that all cases should be reported to the relevant public health authority. In practice, physicians tend not to be sufficiently conscientious about reporting, and in some cases are not even consulted by the patient. To help overcome this problem, many public health authorities enlist the assistance of selected general practitioners to establish a network of "sentinel practices," and of hospital and public health diagnostic laboratories to provide integrated infor-

Table 14-1
Epidemiologic Features of Common Human Viral Diseases

Disease	Mode of transmission	Incubation Period[a] (days)	Period of infectivity[b]	Clinical:subclinical ratio[c]
Influenza	Respiratory[d]	1–2	Short	Moderate
Common cold	Respiratory[d]	1–3	Short	Moderate
Bronchiolitis	Respiratory[d]	3–5	Short	Moderate
Dengue	Mosquito bite	5–8	Short	Moderate
Herpes simplex	Saliva, sexual	5–8	Long	Moderate
Enteroviruses	Enteric, respiratory	6–12	Long	Low
Poliomyelitis	Enteric	5–20	Long	Low
Measles	Respiratory[d]	9–12	Moderate	High
Smallpox	Respiratory[d]	12–14	Moderate	High
Chickenpox	Respiratory[d]	13–17	Moderate	Moderate
Mumps	Respiratory[d], saliva	16–20	Moderate	Moderate
Rubella	Respiratory, congenital	17–20	Moderate	Moderate
Mononucleosis	Saliva, parenteral	30–50	Long	Low
Hepatitis A	Enteric	15–40	Long	Low[e]
Hepatitis B	Parenteral, sexual, perinatal	50–150	Very long	Low[e]
Rabies	Animal bite	30–100	Nil	High
Warts	Contact, sexual	50–150	Long	High
AIDS	Sexual, parenteral, congenital	1–10 years	Very long	High[f]

[a] Until first appearance of prodromal symptoms. Diagnostic signs, e.g., rash or paralysis, may not appear until 2–4 days later.
[b] Most viral diseases are highly transmissible for a few days before symptoms appear, as well as after. Long, >10 days; short, <5 days.
[c] High, >90%; low, <10%.
[d] Also by contact.
[e] In Third World, but moderate in industrialized countries, where infection usually occurs at a later age.
[f] Eventually, after a long incubaton period.

mation about clinical cases, virus identifications, and serology. To disseminate information and maintain the interest of the practitioners and the laboratories, local data on infectious diseases and reports on interesting findings elsewhere in the world are distributed in the form of newsletters, such as the *Morbidity and Mortality Weekly Report* (MMWR) in the United States, and the *Weekly Epidemiological Record* of the World Health Organization.

Special arrangements for surveillance are used in focused campaigns, such as those designed to eradicate diseases (see Chapter 15), and public health authorities, such as the Centers for Disease Control in the United States, respond promptly to the occurrence of an unusual outbreak of disease by setting up a special "task force" of appropriately experienced epidemiologists and microbiologists to investigate the problem. In a number of instances the first recognition of a new disease or the isolation of a previously unknown virus has resulted from such investigations, for example, the disease AIDS and Lassa, Marburg, and Norwalk viruses, and in 1993, the pulmonary syndrome hantavirus in the United States.

Seroepidemiology

Traditional surveillance is based on the reporting of clinical disease. However, examination of sera for antibodies gives a more accurate index of the prevalence of a particular virus. By detecting antibodies to particular viruses in various age groups it is possible to determine how effectively viruses have spread, or how long it has been since the last appearance of nonendemic viruses. Correlation of serological tests with clinical observations also makes it possible to determine the ratio of clinical to subclinical infections.

Seroepidemiology is extremely useful in public health operations and research. Because of the expense of collecting and properly storing sera, advantage is taken of a wide range of sources, such as planned population surveys, entrance examinations for military and other personnel, blood banks, hospitals, and public health laboratories. Such sera can be used to determine the prevalence of particular infections, to evaluate eradication and immunization programs, and to assess the impact, dynamics, and geographic distribution of new, emerging, and reemerging viruses. For example, serological surveys were used to determine the prevalence and geographic distribution of hepatitis B and C viruses, Lassa virus, Ebola virus, and the human immunodeficiency viruses and, using stored sera, to estimate how long they had been present in various populations. It is regrettable that exaggerated concerns about the possibility of breaches of privacy have led to the imposition of constraints on the use of anonymous stored sera as an epidemiologic research resource.

Molecular Epidemiology

The term molecular epidemiology has been applied to the use of molecular biological methods for the epidemiologic investigation of viral diseases. Many of the techniques described in Chapter 12 can be used. With DNA viruses (e.g., herpes simplex virus), restriction endonuclease mapping can be used for the identification of field isolates with a precision that surpasses serological methods, whereas oligonucleotide fingerprinting of ribonuclease digests is similarly useful for characterizing RNA viruses. With viruses that have segmented genomes, such as rotaviruses, polyacrylamide gel electrophoresis of genomic RNA provides valuable supplementation to serological typing. The polymerase chain reaction can be used to amplify very low copy numbers of a particular viral DNA or RNA in diagnostic samples prior to detection by nucleic acid hybridization. Nucleotide sequencing is used to distinguish vaccine strains from wild strains of poliovirus and also to determine the geographic origins of wild strains of polioviruses and of dengue viruses.

A sophisticated application of molecular biology to epidemiology was that developed by the Centers for Disease Control to monitor polioviruses in the United States and subsequently in the project of the Pan American Health Organization to eradicate poliomyelitis from the Americas. Kew and colleagues selected a region of the viral genome, involving part of the VP1 open reading frame, which displays numerous differences between the three poliovirus serotypes, and then sequenced this region in hundreds of field isolates. In this way they were able to define, within each of the three serotypes, a number of "genotypes," arbitrarily defined as "a group of polioviruses having no more than 15% genomic divergence within the VP1/2A region."

Like other RNA viruses, polioviruses mutate with high frequency during their replication, although many of the nucleotide substitutions are silent because they occur in the third base of a codon and are translated into an unchanged amino acid. These and other mutations that do not decrease the "fitness" of the genotype accumulate in the genome leading to a process of genetic drift, allowing chains of transmission to be deduced. Unlike the situation with influenza viruses, where novel antigenic strains resulting from antigenic drift rapidly spread around the world, genetic drift among polioviruses (which is not accompanied by antigenic drift) tends to remain localized to particular endemic regions. Hence the origin of outbreaks of poliovirus in nonendemic countries can be inferred by a study of their genotypes. As might be expected, most introductions into Europe in recent years have come from the eastern Mediterranean region, whereas every type 1 poliovirus isolated in the United States since 1970 was introduced from Mexico (until 1979) or Central America (until 1989). However, it came as a surprise to find that the 1980–1981 outbreak among the Palestinian population of the Gaza Strip and the West Bank was traced to a probable importation from the Andean region of South America.

Types of Epidemiologic Investigations

Investigation of Causation

Cross-Sectional, Case-Control, and Cohort Studies

Epidemiologic methods used to determine the incidence and prevalence of infectious diseases, the relationships between cause and effect, and the evaluation of risk factors include the *cross-sectional study*, the *case–control study*, and the *cohort study*. A cross-sectional study can be carried out relatively quickly and provides data on the prevalence of particular diseases in the population. A case–control study, the most common kind of investigation, starts after the disease has occurred and attempts to identify the cause; it is thus a *retrospective study,* going back in time to determine causative events. Although it does not require the creation of new data or records, a case–control study does require careful selection of the control group, matched to the test group so as to avoid bias. The advantages of the retrospective study are that it lends itself to quick analysis, is relatively inexpensive to carry out, and is the only practicable method of investigating rare occurrences.

In cohort studies, also called *prospective studies,* investigation starts with a presumed cause, and a population exposed to the causative virus is followed into the future to identify correlated resulting effects. This type of study requires the creation of new data and records, and careful selection of the control group to be as similar as possible to the exposed group, except for the absence of contact with the presumed causative virus. It does not lend itself to quick results as groups must be followed until disease is observed, which makes such studies expensive. However, when cohort studies are successful, proof of the cause–effect relationship is often incontrovertible.

The discovery of the causation of congenital defects by rubella virus provides examples of both retrospective and prospective studies. Dr. N. M. Gregg, an ophthalmologist working in Sydney, Australia, was struck by the large number of cases of congenital cataract he saw in 1940–1941, and by the fact that many of the children also had cardiac defects. By interviewing the mothers he found that the great majority of them had experienced rubella early in the related pregnancy. His hypothesis that there was a causative relation between maternal rubella and congenital defects quickly received support from other retrospective studies, and prospective studies were then organized. Groups of pregnant women were sought who had experienced an acute exanthematous disease during pregnancy, and the subsequent occurrence of congenital defects in their children was compared with that in women who had not experienced such infections. Gregg's predictions were confirmed and the parameters defined more precisely.

Human Volunteer Studies

Epidemiologic aspects of several specifically human diseases that have not been reproduced in other animals have been studied in human volunteers; for example, early work with yellow fever, hepatitis viruses, rhinoviruses, and a range of other respiratory viruses involved human volunteer studies. Many major discoveries that have led to the control of viral diseases were possible only with the use of human volunteers. An absolute requirement is that the investigators obtain informed consent from the subjects or, in the case of minors, from their parents. It is essential in such work that careful consideration be given to any short- or long-term risks that may be involved, including the possibility of transferring other agents that may be present in the inoculum as contaminants.

Proving a Causal Relationship between Virus and Disease

One of the great landmarks in the scientific study of infectious diseases was the development of the Henle–Koch postulates of causation. They were originally drawn up with bacteria and protozoa in mind, not viruses, and were revised in 1937 by Rivers, who developed another set of criteria to cover viruses. With the advent of tissue culture for viral diagnosis many new viruses were discovered—"viruses in search of disease"—and Huebner further revised the Koch and Rivers postulates. Later the problem arose of determining whether viruses were causally involved in various chronic diseases and in cancers, a question that is still of major concern to medical virologists. Because the relevant disease cannot be reproduced by inoculation of experimental animals, scientists have to evaluate the probability of "guilt by association," a difficult procedure that relies on an epidemiologic approach. Two tools that are of central importance in assisting the epidemiologist are immunologic investigations and demonstration of the presence of the viral genome in tumor cells by the use of nucleic acid probes or the polymerase chain reaction.

The immunologic criteria were first formulated by Evans in an assessment of the relationship of Epstein-Barr (EB) virus to infectious mononucleosis, at a time when there was no method of isolation of the virus. Subsequently the same approach was applied to investigating the role of EB virus in Burkitt's lymphoma. A large prospective study carried out on 45,000 children in an area of high incidence of Burkitt's lymphoma in Africa showed that (1) EB virus infection preceded development of the tumors by 7–54 months; (2) exceptionally high EB virus antibody titers often preceded the appearance of tumors; and (3) antibody titers to other viruses were not elevated. In addition, it was demonstrated that the EB virus genome is always present in the cells of Burkitt's lymphomas in African children, and that a malignant lymphoma can be induced in certain primates with EB virus or EB virus-infected lymphocytes (see Chapter 11). There is thus a very strong causal association between EB virus and Burkitt's lymphoma in African children.

Vaccine and Drug Trials

The immunogenicity, potency, safety, and efficacy of vaccines are first studied in laboratory animals, followed by small-scale closed trials in normal adults to determine safety (Phase 1 trials) and the immune response (Phase 2 trials), followed by large-scale field trials in a place where efficacy can be tested (Phase 3 trials). Sometimes these trials can be combined. The Phase 3 trials employ epidemiologic methods to analyze the results, rather like those of the cohort study just described. Perhaps the best-known vaccine trial was the famous "Francis field trial" of the Salk inactivated poliovirus vaccine, carried out in 1954, but similar trials were necessary for live-virus poliovirus vaccines, for yellow fever vaccines before this, and for measles, mumps, and rubella vaccines since. There is no alternative way to evaluate new vaccines, and the design of field trials has now been developed so that they yield maximum information with minimum risk and cost. Even with this system, however, a serious problem may be recognized only after a vaccine has passed into commercial use. Similar considerations apply to trials of new antiviral chemotherapeutic agents (see Chapter 16).

Another kind of epidemiologic investigation that can provide etiologic information and data on the value of vaccines or therapeutic agents is the long-term study of a closed population, such as "Junior Village" in Washington, or the long-term family studies that have been carried out in several cities in the United States. Because of the present advanced state of diagnostic virology, such studies now yield a much greater array of valuable data than was possible a few years ago, but they are very expensive and require long-term dedication of both personnel and money. Conducted with careful attention to the ethical problems involved, these studies have greatly augmented our understanding of viral infections and accelerated progress toward the control of several of them. When used for the estimation of the value of vaccines or therapeutic agents, long-term population studies have the exceptional advantage that they include all of the variables occurring in a natural population.

Population studies may also be used for the early detection of the first appearance of virus in a region. Such *sentinel studies* are widely used for assessing the seasonal prevalence of arbovirus infections; for example, sentinel chickens are used for the early detection of eastern equine encephalitis and St. Louis encephalitis viruses in the southern United States and for the detection of Murray Valley encephalitis virus in Australia.

Mathematical Modeling

From the time of William Farr, who studied infectious diseases in the 1840s, mathematicians have been interested in "epidemic curves" and secular trends in the incidence of infectious diseases. With the development of mathematical modeling using the computer there has been a resurgence of interest in the dynamics of infectious diseases within populations. Since modeling involves predictions about future occurrences of diseases, models carry a degree of uncertainty; it is sometimes said that "for every model there is an equal and opposite model." However, models are being developed to predict patterns of disease transmission, critical population sizes required to support the continuous transmission of viruses with short and long incubation periods, the dynamics of endemicity of viruses that establish persistent infection, and the important variables in age-dependent viral pathogenicity.

Computer modeling has provided useful insights into the effects of different vaccination regimes and different levels of acceptance of vaccination against rubella and measles on important complications of these two diseases, congenital rubella and subacute sclerosing panencephalitis, respectively. One effect of vaccination programs is to raise the average age at which unimmunized individuals contract the infections against which that vaccine protects. Up to a certain level, increasing herd immunity may actually increase the risk in especially vulnerable (older) individuals, by increasing the age of first exposure to the virus. Thus, unless the acceptance rate is high, vaccination of infants against rubella may paradoxically increase the risk of the most important complication of the natural disease, namely, congenital rubella. A second conclusion from these studies is that because of the cyclic incidence of the viral exanthemata, with intervals between peaks that increase as vaccination coverage improves, the evaluation of vaccination programs that stop short of countrywide elimination must be carried out over a prolonged period of time.

Virus Transmission

Viruses survive in nature only if they can be transmitted from one host to another, whether of the same or another species. Transmission cycles require virus entry into the body, replication, and shedding with subsequent spread to another host. Molecular and cellular aspects of entry and shedding were described in Chapters 3 and 6; here we mention only those aspects that are relevant to epidemiology.

Virus transmission (Fig. 14-2) may be *horizontal* or *vertical.* Vertical transmission describes transmission from mother to offspring. However, most transmission is horizontal, that is, between individuals within the population at risk, and can occur via direct contact, indirect contact, or a common vehicle, or may be airborne, vector-borne or iatrogenic. Some viruses are transmitted in nature via several modes, others exclusively via one mode.

Horizontal Transmission

Direct contact transmission involves actual physical contact between an infected and a susceptible person, by hand or body contact, including kissing and various types of sexual contact, such as occurs in transmission of some herpesviruses, papillomaviruses, and poxviruses.

Indirect contact transmission occurs via *fomites,* such as shared eating utensils, children's toys, handkerchiefs, towels, bed linen, improperly sterilized surgical equipment, or shared syringes and needles; transmission of many enteric, respiratory, and hepatitis viruses is by indirect contact.

Common vehicle transmission includes fecal contamination of food and water supplies (fecal–oral transmission, e.g., of hepatitis A and E viruses, rotaviruses, and caliciviruses).

Airborne transmission, resulting in infection of the respiratory tract, occurs via droplets and droplet nuclei (aerosols) emitted from infected persons during talking, coughing, or sneezing (e.g., influenza). Large droplets settle quickly, but microdroplets evaporate forming droplet nuclei (<5 μm in diameter) which remain suspended in the air for extended periods. Droplets may travel only a meter or so; droplet nuclei may travel longer distances, but enveloped respiratory viruses do not survive for more than a few hours, being sensitive to inactivation at ambient temperatures. Environmental sources of airborne transmission include virus-contaminated dust, thought to be the source of arenavirus infections.

Arthropod-borne transmission involves the bite of arthropod vectors (e.g., mosquitoes transmit dengue, yellow fever, and some encephalitis viruses, ticks transmit other encephalitis viruses, and sandflies transmit sandfly fever; see Table 14-6).

Other terms are used to describe transmission by mechanisms that embrace more than one of the above routes. **Iatrogenic transmission** occurs as a direct result of some activity of the attending doctor, nurse, or other person in the course of caring for patients, usually via nonsterile equipment, multiple-use syringes, blood transfusions, the use of human blood products, or inadequate handwashing. **Nosocomial transmission** occurs while a person is in a

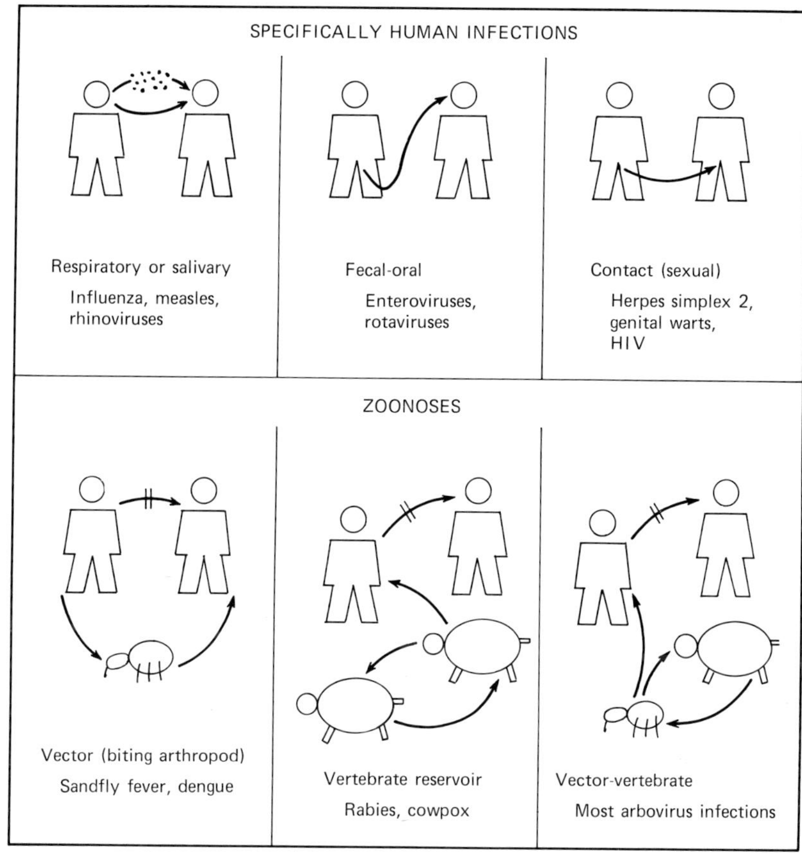

Fig. 14-2 Modes of transmission of human viral diseases. (Modified from C. A. Mims, "The Pathogenesis of Infectious Disease," 2nd Ed. Academic Press, London, 1982.)

hospital or clinic. The lethal Ebola virus outbreaks in Zaire and the Sudan in 1976 were classic examples of iatrogenic nosocomial infections; more common examples of nosocomial virus infections are the occurrence of chickenpox, influenza, and respiratory syncytial virus cross-infections in hospital settings. Hepatitis B and C viruses, and to a lesser extent HIV, can also be transmitted by doctors, dentists, acupuncturists, tattooists, etc., but are more of a risk to attending staff and laboratory personnel, via needle stick and similar injuries.

Zoonotic Transmission

Because most viruses are host-specific, most human viral infections are maintained in nature within the human population. However, there are a number of viruses that spread naturally between several different species of animals, for example, rabies and the arboviral encephalitides. The term *zoonosis* is used to describe infections that are transmissible from animals to humans (see Fig. 14-2). The zoonoses, whether involving domestic or wild animal reservoirs, usually occur only under conditions where humans are engaged in activities involving close contact with animals (Table 14-2), or if the viruses are transmitted by arthropods (see Table 14-6).

Table 14-2
Non-Arthropod-Borne Viral Zoonoses

Virus family	Virus	Reservoir host	Mode of transmission to humans
Herpesviridae	Herpes B	Monkey	Animal bite
Poxviridae	Cowpox	Rodents, cats, cattle	Contact, through abrasions
	Monkeypox	Squirrel, monkeys	
	Pseudocowpox	Cattle	
	Orf	Sheep, goats	
Rhabdoviridae	Rabies	Various mammals	Animal bite, scratch, respiratory
	Vesicular stomatitis	Cattle	Contact with secretions[a]
Filoviridae	Ebola, Marburg	?Monkeys	Contact; iatrogenic (injection)[b]
Orthomyxoviridae	Influenza A[c]	Birds, pigs	Respiratory
Bunyaviridae	Hantaan	Rodents	Contact with rodent urine
Arenaviridae	Lymphocytic choriomeningitis, Junin, Machupo, Lassa	Rodents	Contact with rodent urine

[a] May also be arthropod-borne.
[b] Also human-to-human spread.
[c] Usually maintained by human-to-human spread; zoonotic infections occur only rarely, but reassortants between human and avian influenza viruses (perhaps arising during coinfection of pigs) may result in human pandemics due to antigenic shift.

Vertical Transmission

The term vertical transmission is usually used to describe infection that is transferred from mother to embryo/fetus/newborn prior to, during, or shortly after parturition, although some authorities prefer to restrict the term to situations where infection occurs before birth. Certain retroviruses of birds and mammals are vertically transmitted via the integration of proviral DNA directly into the DNA of the germ line of the fertilized egg. Rubella virus and cytomegalovirus are transmitted to the fetus via the placenta, whereas others, such as herpes simplex type 2 virus, are transmitted during passage through the birth canal. Yet others, such as hepatitis B virus, may be transmitted perinatally or postnatally via saliva, milk, or other secretions. Vertical transmission of a virus may be lethal to the fetus, and the cause of abortion, or it may be associated with congenital disease or cause congenital abnormalities.

Mechanisms of Survival of Viruses in Nature

Perpetuation of a virus in nature depends on the maintenance of serial infections, that is, a chain of transmission; the occurrence of disease is neither required nor necessarily advantageous. Indeed, although clinical cases may be somewhat more productive sources of infectious virus than inapparent infections, the latter are generally more numerous and do not restrict the movement of infectious individuals, and thus provide a major mechanism of viral dissemination. As our knowledge of the different features of the pathogenesis, species susceptibility, routes of transmission, and resistance of viruses to the environment has increased, epidemiologists have been able to recognize four different mechanisms (or cycles) by which viruses maintain serial transmission in their host(s): acute self-limiting infections, persistent infections, vertical transmission, and arthropod transmission.

Most viruses have a principal mechanism for survival, but if this mechanism is interrupted, for example, by a sudden decline in population of the host species due to another disease or a short-term climate change, other mechanisms, previously less apparent, may emerge. This should be remembered when relating the epidemiology of a specific disease to particular mechanisms of survival, as proposed in Table 14-3.

An appreciation of the mechanisms for viral perpetuation is valuable in designing and implementing control programs. For example, knowledge that variola virus caused an acute self-limiting infection in which the vast majority produced clinical disease, and that it had no animal host, was important in the successful eradication of smallpox.

Acute Self-Limiting Infections

The majority of human viral infections fall into the category of acute self-limiting infections. They lack the survival advantages of persistence, vertical transmission, or multiple host species, with or without arthropod transmission. Although optimum transmissibility is crucial, the perpetuation of viruses that cause systemic infections with lifelong immunity is possible only in

Table 14-3
Factors Influencing Survival of Viruses in Nature[a]

Virus family	Example	Survival factor
Parvoviridae	Human parvovirus	Large population; virus stable
Papovaviridae	Human papillomaviruses	Persistent in chronic lesions; virus stable
Adenoviridae	Human adenoviruses	Persistent infection; virus stable
Herpesviridae	Herpes simplex virus	Persistent infection; recurrent infectivity
Poxviridae	Molluscum contagiosum virus	Persistent in chronic lesions; virus stable
Picornaviridae	Human enteroviruses	Large population; ?antigenic drift
Caliciviridae	Norwalk virus	Large population; virus stable
Togaviridae	Ross River virus	Zoonosis; arthropod-borne
	Rubella virus	Large population; virus persistent after congenital infection
Flaviviridae	Dengue viruses	Large population[b]; arthropod-borne
Coronaviridae	Human coronaviruses	Large population
Paramyxoviridae	Mumps virus	Large population
Rhabdoviridae	Rabies virus	Zoonosis; dead-end infection in humans
Orthomyxoviridae	Influenza A virus	Large population; antigenic drift and shift
Bunyaviridae	Rift Valley fever virus	Zoonosis; arthropod-borne
	Hantaan virus	Zoonosis; persistent in rodents
Arenaviridae	Lymphocytic choriomeningitis virus	Zoonosis; persistent in rodents
Reoviridae	Human rotaviruses	Large population; ?antigenic shift
Retroviridae	Human immunodeficiency virus	Persistent; recurrent infectivity; vertical transmission

[a] Unless they produce persistent infections, all nonzoonotic viruses require large human populations; perpetuation is also favored by heat stability of the virion.
[b] In rural Aftrica, may also persist in small human populations, as a zoonotic disease, with monkeys as reservoir hosts.

large, relatively dense populations. Viruses that cause superficial mucosal infections with short-lived immunity may survive in somewhat smaller populations, and their capacity to survive may be enhanced by antigenic drift (see below).

Transmissibility

Several factors contribute to optimal transmission of viruses. Some relate to properties of the virion itself, others to the extent and nature of shedding from the body, and others to social interactions. Enveloped viruses infecting mucosal surfaces bud from the apical surface of epithelial cells to maximize shedding into the outside world. Obviously, human-to-human transmission will be enhanced by shedding of high titers of virus containing a high proportion of infectious virions. Respiratory viruses tend to be shed over a relatively brief period (a few days) but are expelled in high concentration as an aerosol generated by explosive sneezing or coughing, thus ensuring transmission to close contacts. Enteric viruses are also shed in large numbers but usually for a longer period (a week or more) in feces, which may contaminate hands, fomites, food, and water. Enveloped respiratory viruses are relatively labile, especially during summer or in the tropics year-round. In contrast, most enteric viruses are nonenveloped and may survive for several days or weeks in water or dust, or on fomites, as may poxviruses, adenoviruses, papillomaviruses, and hepatitis viruses.

Seasonality

Many viral infections show pronounced seasonal variations in incidence. In temperate climates, arbovirus infections transmitted by mosquitoes or sandflies occur mainly during the summer months, when vectors are most numerous and active. Infections transmitted by ticks occur most commonly during the spring and early summer months. More interesting, but also more difficult to explain, are the variations in seasonal incidence of infections in which humans are the only host animals.

Table 14-4 shows the season of maximal incidence of several human respiratory, enteric, and generalized infections. In temperate climates most respiratory infections are most prevalent in winter or to a lesser extent in spring or autumn. Annual winter outbreaks of severe respiratory syncytial virus infections in infants are a feature of communities in temperate climates, the major impact of each epidemic lasting for only 2 or 3 months (see Fig. 14-1); epidemics of influenza also occur almost exclusively in the winter, but they vary greatly in extent from year to year (see Fig. 31-2). Many of the exanthemata of childhood transmitted by the respiratory route peak in the spring. Among enteric viruses, infections with enteroviruses (like most enteric bacterial infections) are maximal in the summer, but caliciviruses show no regular seasonal patterns; rotaviruses tend to be more prevalent in winter. Infections with the herpesviruses: HSV, cytomegalovirus, and EB virus, which are transmitted by intimate contact with saliva and other bodily secretions, show no seasonal variations in incidence, nor do other sexually transmitted diseases. The patterns shown in Table 14-4 are found in both the northern and southern hemispheres. Different factors probably affect seasonality in the tropics, where wet and dry seasons tend to replace summer and winter. The peak incidence of

Table 14-4
Season of Maximal Incidence of Specifically Human Viral Infections in Temperate Climates[a]

Type of infection	Winter	Spring	Summer	Autumn	None in particular
Respiratory					
Adenoviruses		+			
Rhinoviruses		+		+	
Influenza	+				
Coronaviruses	+				
Respiratory syncytial virus	+				
Parainfluenza 1 and 2				+	
Parainfluenza 3					+
Enteric					
Enteroviruses			+		
Rotaviruses	+				
Caliciviruses					+
Generalized					
Rubella		+			
Measles		+			
Mumps		+			
Varicella		+			
Hepatitis B					+
Herpes simplex 1 and 2					+
Cytomegalovirus					+
Epstein-Barr virus					+
Most arboviruses			+		

[a] Seasonality is often different in tropical climates, whre there is little temperature fluctuation between summer and winter, but the occurrence of some infectious diseases is influenced by "wet" and "dry" seasons.

measles and chickenpox is late in the dry season, with an abrupt fall when the rainy season begins, whereas influenza and rhinovirus infections reach a peak during the rainy season.

Both biological and sociological factors may play a role in these seasonal variations. Measles, influenza, and vaccinia viruses survive in air better at low rather than high humidity, whereas polioviruses, rhinoviruses, and adenoviruses survive better at high humidity. All survive better, in aerosols, at lower temperatures. These situations correspond with conditions prevalent during the seasons when infections due to these viruses are most prevalent. It has also been suggested that there may be seasonal changes in the susceptibility of the host, perhaps associated with changes in nasal and oropharyngeal mucous membranes, such as drying as a result of smoke, central heating, or air conditioning.

Seasonal differences in social activities may also markedly influence the opportunities for transmission of viruses, especially by the respiratory route. Although experience in the Arctic and Antarctic show that cold weather alone is not enough to cause "colds" and other respiratory infections, the crowding into restricted areas and ill-ventilated vehicles and buildings that occurs in temperate climates in winter promotes the exchange of respiratory viruses. In places subject to monsoonal rains, the onset of the rains early in summer is accompanied by greatly reduced movement of people, both in daily affairs and to fairs and festivals. While this may reduce the opportunity for exchange of viruses with those from other villages, confinement to smoke-filled huts maximizes the opportunity for transfer of respiratory viruses within family groups.

Critical Community Size

Population size and density play a role in perpetuation that depends in the main on whether the virus produces acute self-limiting or persistent infections. In general, survival of viruses that produce acute self-limiting infections requires that the susceptible host population should be large and relatively dense. Such viruses may disappear from a population because they exhaust the potential supply of susceptible hosts as they acquire immunity to reinfection. Persistent viruses, on the other hand, may survive in very small populations, sometimes by spanning generations. Depending on the duration of immunity and the pattern of virus shedding, the *critical community size* varies considerably with different viruses. The principle can be exemplified by two common infections of children, measles and chickenpox.

Measles is a cosmopolitan disease that is characteristic in this respect of the generalized viral infections of childhood, like rubella, mumps, and poliomyelitis. Persistence of the virus in a community depends on a continuous supply of susceptible subjects. With an incubation period of about 12 days, maximum viral excretion for the next 6 days, and solid immunity to reinfection, between 20 and 30 susceptible individuals would need to be infected in series to maintain transmission for a year. Because, for a variety of reasons, nothing like such precise one-to-one transmission occurs, many more than 30 susceptible persons are needed to maintain endemicity. Analyses of the incidence of measles in large cities and in island communities have shown that a population of about 500,000 persons is needed to ensure a large enough annual input of new susceptibles, provided by the annual incidence of births, to maintain measles indefinitely as an endemic disease.

Because infection depends on close contact, the duration of epidemics of measles is correlated inversely with population density. If a population is dispersed over a large area, the rate of spread is reduced and the epidemic will last longer, so that the number of susceptible persons needed to maintain endemicity is reduced. On the other hand, in such a situation a break in the transmission cycle is much more likely. If a large proportion of the population is initially susceptible, the intensity of the epidemic builds very quickly. Attack rates were almost 100% in an epidemic of measles in southern Greenland in 1951, which ran through the entire population in about 40 days before running out of susceptibles and disappearing completely. Such *virgin-soil epidemics* in isolated communities may have devastating consequences, due rather to lack of medical care and the disruption of social life than to a higher level of genetic susceptibility or abnormal virulence of the virus.

The peak age incidence of measles depends on local conditions of population density and the chance of exposure. In large urban communities, before the days of vaccination, epidemics occurred every 2–3 years, exhausting most of the currently available susceptible persons, and epidemics on a continental scale used to occur annually in the United States. Although newborns provide the input of susceptibles each year, the age distribution of cases in unvaccinated communities is primarily that of children just entering school, with a peak of secondary cases (family contacts) about 2 years old. The cyclic nature of measles outbreaks is determined by several variables, including the build-up of susceptibles, introduction of the virus, and environmental conditions that promote viral spread. Both the seasonality of infectivity (Table 14-4 and the occurrence of school holidays affect the epidemic pattern. Following the widespread introduction of immunization programs the epidemiology of measles has changed dramatically (see Chapter 28).

Although chickenpox is also an acute exanthema in which infection is followed by lifelong immunity to reinfection, it requires a dramatically smaller critical community size for indefinite persistence of the disease: less than 1000, compared with 500,000 for measles. This is explained by the fact that varicella virus, after being latent for decades, may be reactivated and cause zoster (see Chapter 10). Although zoster is not as infectious as chickenpox itself (secondary attack rates of 15%, compared with 70% for varicella), it can, in turn, produce chickenpox in susceptible children and grandchildren.

Effects of Immunity

Immunity acquired from prior infection or from vaccination plays a vital role in the epidemiology of viral diseases. In most generalized infections, acquired immunity, attributable largely to circulating IgG antibody, appears to be lifelong. This occurs even in the absence of repeated subclinical infections, as evidenced by studies of measles and poliomyelitis in isolated populations. In a paper on measles in the Faroe Islands written in 1847, Panum demonstrated that the attack rate was almost 100% among exposed susceptibles, but that immunity conferred by an attack of measles experienced during an epidemic 65 years earlier remained solid in spite of no further introduction of the virus to the islands in the interim.

The situation is different with viral infections that are localized to mucosal surfaces such as the respiratory tract, since mucosal immunity is relatively short-lived. A large number of serotypes of rhinoviruses and a few serotypes of coronaviruses and enteroviruses can produce superficial infections of the mucous membrane of the upper respiratory tract. The seemingly endless succession of common colds suffered by urban communities reflects a series of minor epidemics, each caused by a different serotype of one of these viruses. Protection against reinfection is due mainly to antibodies in the nasal secretions, primarily IgA. Although short-lived type-specific immunity does occur, there is no intertypic cross-immunity, hence the convalescent individual is still susceptible to all other rhinoviruses and coronaviruses. Most persons contract between two and four colds each year.

In urban communities young children appear to be particularly important as the persons who introduce rhinoviruses into families, mainly because they bring back viruses from school and from neighbors' children and partly because they often shed larger amounts of virus than adults. A feature of rhinovirus infection that had not been observed previously emerged from family studies, namely, prolonged shedding of virus for up to 3 weeks, long after the acute symptoms have subsided. The epidemiology of coronaviruses, another important cause of colds, is rather similar; however, there are only a few serotypes, and they tend to produce colds in winter, and in adults rather than children.

Epidemiologic observations on isolated human communities illustrate the need for a constant supply of susceptible subjects or antigenically novel viral serotypes to maintain respiratory diseases in nature, and the importance of repeated (often subclinical) infections in maintaining herd immunity. Explorers, for example, are notably free of respiratory illness during their sojourn in the Arctic and Antarctica, despite the freezing weather, but invariably contract severe colds when they again establish contact with other humans.

Antigenic Drift and Shift

In influenza, respiratory tract immunity due to IgA selects for mutants displaying antigenic variation in the hemagglutinin (antigenic drift), leading to the emergence of new strains that can infect individuals immune to infection with previous strains of that subtype, as discussed in Chapter 4. Antigenic drift may also account for the evolution of the numerous antigenic types of rhinoviruses, enteroviruses, and other viruses that cause superficial infections confined largely to mucosal surfaces.

The more radical changes known as antigenic shift, and attributable to genetic reassortment in influenza A virus, occur much less frequently but constitute a major antigenic change that may cause widespread epidemics in a world population with no immunity (Chapters 4 and 31). As genetic reassortment can also occur in rotaviruses and there are rotaviruses affecting several animal hosts, antigenic shift may also occur with these viruses, although this has yet to be demonstrated under natural conditions.

Persistent Infections

Persistent viral infections (see Chapter 10), whether or not associated with episodes of clinical disease, play an important role in the perpetuation of many viruses. Infection with members of certain viral families is characteristically associated with continuous or intermittent shedding (Table 14-5). In the extreme case, shedding by a persistently infected person can reintroduce

Table 14-5
Viral Infections of Humans Associated with Persistent Viral Excretion

Family	Species	Comments
Herpesviridae	Epstein-Barr virus, cytomegalovirus, HHV6	Intermittent shedding in saliva, genital secretions, milk
	Herpes simplex virus types 1 and 2	Recurrent excretion in saliva, genital secretions, for years
	Varicella virus	Recurs as zoster many years later
Adenoviridae	All adenoviruses	Intermittent excretion from throat and/or feces
Papovaviridae	BK, JC polyomaviruses	Excreted in urine
Hepadnaviridae	Hepatitis B virus	Persistent viremia; may occur in semen, saliva
Arenaviridae	All arenaviruses	In rodents, intermittent shedding in urine
Togaviridae	Rubella virus	Persistent only in congenitally infected children; excreted in urine
Flaviviridae	Hepatitis C virus	Persistent viremia; may occur in semen
Retroviridae	Human immunodeficiency viruses	Persistent viremia, excretion in genital secretions for years

virus into a population in which all members have been born since the last clinically apparent episode of disease (an immunologically naive population.). This transmission pattern is important for the survival of herpesviruses in small populations (see above for varicella–zoster).

Persistence of infection, production of disease, and transmission of virus are not necessarily linked. Thus persistent arenavirus infections have little adverse effects on their rodent reservoir hosts but are transmitted very efficiently. On the other hand, the persistence of viruses in the central nervous system, as with measles virus in subacute sclerosing panencephalitis, is of no epidemiologic significance, since no infectious virus is shed, but the infection is ultimately lethal. It is reasonable to postulate that viruses of families such as the *Herpesviridae* and *Retroviridae,* so well adapted to lifelong persistence and transmission within small isolated populations, were among the important specifically human viruses found in our hominid ancestors and man the hunter-gatherer.

Vertical Transmission

Transmission of virus from the mother to the embryo, fetus, or newborn, as described above, can be important in virus survival in nature; all arenaviruses, several herpesviruses and retroviruses, and some togaviruses may be transmitted in this way (see Table 6-4). Indeed, when the consequence of vertical transmission is lifelong persistent infection with continuing or recurrent shedding, as in the case of arenavirus infection of rodents, the long-term survival of the virus is assured on both counts. Transmission in the perinatal period is also an important mode of transmission for some viruses, notably hepatitis B virus, several herpesviruses, and the retroviruses HTLV and HIV. Vertical transmission in arthropod vectors is an important mode of perpetuation of some arthropod-borne viruses.

Arthropod Transmission

Arthropod transmission is ecologically the most complex of the four modes by which viruses are perpetuated. Several arthropod-borne diseases are discussed in appropriate chapters of Part II, but it is useful to consider here some common features useful in understanding their epidemiology and control. Over 500 arboviruses are known, of which around 80 are capable of infecting humans, although only some 20 of these cause important diseases (Table 14-6). A few of the human-pathogenic arboviruses are tick-borne, but most are transmitted by insects, mainly mosquitoes and some sandflies (*Phlebotomus* spp.) and midges (*Culicoides* spp.).

Arthropod transmission involves replication of the virus in the arthropod vector, which acquires virus by feeding on the blood of a viremic animal or person. Replication of the ingested virus, initially in the insect gut, and its spread to the salivary gland take several days (the *extrinsic incubation period*); the interval varies with different viruses and is influenced by ambient temperature. (The *intrinsic incubation period* is the incubation period in the vertebrate host.) Virions in the salivary secretions of the vector are injected into new vertebrate hosts during blood meals.

Table 14-6
Major Arthropod-Borne Viral Zoonoses

Virus family	Virus genus	Disease	Reservoir hosts	Arthropod vector
Togaviridae	*Alphavirus*	Chikungunya fever[a]	Monkeys, humans	
		Eastern equine encephalitis	Birds	
		Western equine encephalitis	Birds	Mosquitoes
		Venezuelan equine encephalitis[a]	Mammals, horses	
		Ross River polyarthritis[a]	Mammals	
Flaviviridae	*Flavivirus*	Japanese encephalitis	Birds, pigs	
		St. Louis encephalitis	Birds	
		West Nile fever	Birds	Mosquitoes
		Murray Valley encephalitis	Birds	
		Yellow fever[b]	Monkeys, humans	
		Dengue[b]	Humans, monkeys	
		Kyasanur Forest disease	Mammals	Ticks
		Tick-borne encephalitis	Mammals, birds	
Bunyaviridae	*Phlebovirus*	Rift Valley fever	Mammals	Mosquitoes
		Sandfly fever[a]	Mammals	Sandflies
	Nairovirus	Crimean–Congo hemorrhagic fever	Mammals	Ticks
	Bunyavirus	California encephalitis	Mammals	
		La Crosse encephalitis	Mammals	Mosquitoes
		Tahyna virus infection	Mammals	
		Oropouche fever	?Mammals	Mosquitoes, midges
Reoviridae	*Coltivirus*	Colorado tick fever	Mammals	Ticks

[a] In certain episodes, transmitted from person to person, by insects.
[b] Usually transmitted from person to person, by mosquitoes.

Arthropod transmission provides a way for a virus to cross species barriers, since the same arthropod may bite birds, reptiles, and mammals that rarely or never come into close contact in nature. Vertebrate reservoir hosts are usually wild mammals or birds which generally sustain subclinical infections. Humans are rarely involved in primary transmission cycles, although urban yellow fever and dengue provide important exceptions to this generalization.

Transmission of some arboviruses from one vertebrate host to another can also occur by mechanisms not involving an arthropod at all. Thus, in central Europe a variety of small rodents and ticks are reservoir hosts of tick-borne encephalitis virus. Goats, cows, and sheep are incidental hosts and are subclinically infected by tick bites; however, they excrete virus in their milk, and newborn animals may be infected by drinking infected milk. Humans may be infected either by being bitten by a tick or by drinking milk from an infected goat (Fig. 14-3).

Although some arboviruses can be maintained by transovarial transmission in arthropods, vertebrates represent important amplifier hosts, and the typical *maintenance cycle (enzootic cycle)* of an arbovirus is based on regular alternation between the invertebrate and vertebrate hosts. In contrast to the arthropod, which carries the virus throughout its short life, infected vertebrates usually recover rapidly, eliminate the virus, and develop a lasting immunity to reinfection. To be an efficient reservoir host, the vertebrate must be abundant and have a rapid turnover rate, and after infection it must maintain a high level of viremia for an adequate period. In turn, the invertebrate, to be an efficient vector, must have a sufficiently low threshold of infection by that virus, must then carry and shed the virus in salivary secretions for life, and not be adversely affected by it, and must have a distribution, flight range, longevity, and biting habits well adapted to the habits and habitat of its principal vertebrate host. Under these circumstances, the virus flourishes

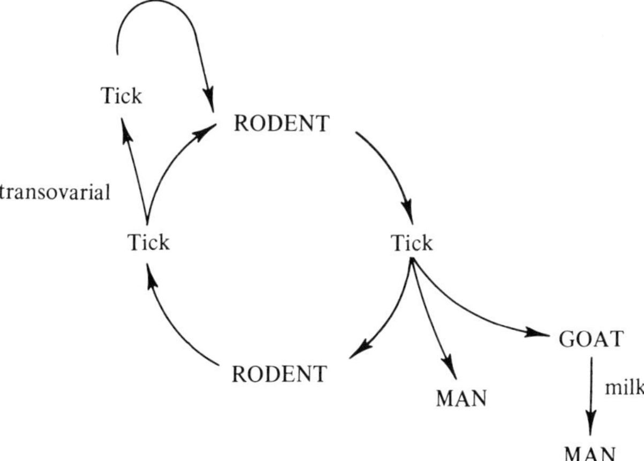

Fig. 14-3 Central European tick-borne flavivirus infection. Congenital transmission in the tick occurs on a scale sufficient to maintain the virus in nature. The virus is transmitted by ticks to many other animals, including cows, goats, and humans. Humans can also become infected by drinking milk from an infected goat.

indefinitely in a state of peaceful coexistence with vertebrate and invertebrate hosts.

Disease occurs when unusual vertebrate hosts become involved. When humans live in regions where a particular arbovirus is enzootic they are vulnerable to infection, and a proportion of those infected may suffer severe, even lethal, disease. Visitors such as tourists, soldiers, or forest workers are even more vulnerable as, unlike the indigenous population, they will not have acquired immunity from subclinical infection in childhood. In tropical countries where the arthropod vector is plentiful year-round, the risk is always present and human disease is endemic (e.g., jungle yellow fever). In regions subject to monsoonal rains, epidemics of mosquito-borne diseases may occur toward the end of the wet season, for example, Japanese encephalitis in parts of Southeast Asia. In some temperate countries, and particularly in arid areas, human epidemics of mosquito-borne arbovirus disease occur following periods of exceptionally heavy rain.

When arthropods are active, arboviruses replicate alternately in vertebrate and invertebrate hosts. A puzzle that has concerned many investigators has been to understand what happens to the viruses during the winter months in temperate climates, when the arthropod vectors are inactive. An important mechanism for overwintering is *transovarial transmission* from one generation of arthropods to the next. With arthropods such as ticks that have several larval stages, this is necessarily associated with *transstadial transmission*. Transovarial infection occurs in most tick-borne arbovirus infections and is often sufficient to ensure survival of the virus independently of a cycle in vertebrates; as far as virus survival is concerned, vertebrate infection is only important in amplifying the population of infected ticks.

Transovarial transmission also occurs with some mosquito-borne bunyaviruses and flaviviruses. For example, some bunyaviruses are found in high northern latitudes where the mosquito breeding season is too short to allow virus survival by horizontal transmission cycles alone; many of the first mosquitoes to emerge each summer carry virus as a result of transovarial transmission, and the pool of virus is rapidly amplified by horizontal transmission in mosquito–mammal–mosquito cycles.

Vertical transmission in arthropods may not explain all arbovirus overwintering, but other possibilities are still unproved or speculative. For example, hibernating vertebrates have been thought to play a role in overwintering. In cold climates, bats and some small rodents, as well as snakes and frogs, hibernate during the winter months. Their low body temperature has been thought to favor persistent infection, with recrudescent viremia occurring when normal body temperature returns in the spring. Although demonstrated in the laboratory, this mechanism has never been proved to occur in nature.

Examples of the complexity of the life cycles of arboviruses are given in Chapters 25, 26, and 33. Many ecologic changes produced by human activities disturb natural arbovirus life cycles and have been incriminated in the geographic spread or increased prevalence of the diseases they cause: (1) population movements and human intrusion into new arthropod habitats, notably tropical forests; (2) deforestation, with development of new forest–farmland margins and exposure of humans to new arthropods; (3) irrigation, especially primitive irrigation systems, which pay no attention to arthropod control; (4) uncontrolled urbanization, with vector populations breeding in accumulations of water and sewage; (5) increased long-distance air travel, with potential for carriage of arthropod vectors and of persons incubating diseases such as dengue and yellow fever; (6) new routing of long-distance bird migrations brought about by new man-made water impoundments.

Further Reading

Anderson, R. M., and May, R. M. (1991). "Infectious Diseases of Humans: Dynamics and Control." Oxford Univ. Press, Oxford.

Beran, G. W. (1994). "Handbook of Zoonoses, 2nd Ed., Section B. Viral." CRC Press, Boca Raton, Florida.

Berg, E. (1987). "Methods of Recovering Viruses from the Environment." CRC Press, Boca Raton, Florida.

Black, F. L., and Singer, B. (1987). Elaboration versus simplification in refining mathematical models of infectious disease. *Annu. Rev. Microbiol.* **41,** 677.

Burnet, F. M., and White, D. O. (1972). "Natural History of Infectious Disease," 4th Ed. Cambridge Univ. Press, Cambridge.

de-Thé, G. (1979). The epidemiology of Burkitt's lymphoma: Evidence for a causal association with Epstein-Barr virus. *Epidemiol. Rev.* **1,** 32.

Evans, A. S., ed. (1989). "Viral Infections of Humans: Epidemiology and Control," 3rd Ed. Plenum Medical, New York.

Fenner, F., Henderson, D. A., Arita, I., Jezek, Z., and Ladnyi, I. D. (1988). "Smallpox and Its Eradication." World Health Organization, Geneva.

Fox, J. P., and Hall, C. E. (1980). "Viruses in Families." PSG Publ., Littleton, Massachusetts.

Kew, O. M., Nottay, B. K., Rico-Hesse, R., and Pallansch, M. A. (1990). Molecular epidemiology of wild poliovirus transmission. *Appl. Virol. Res.* **2,** 199.

Kramer, M. S. (1988). "Clinical Epidemiology and Biostatistics." Springer-Verlag, New York.

Mims, C. A. (1981). Vertical transmission of viruses. *Microbiol. Rev.* **41,** 267.

Nathanson, N. (1990). Epidemiology. *In* "Fields Virology" (B. N. Fields, D. M. Knipe, R. M. Chanock, M. S. Hirsch, J. L. Melnick, T. P. Monath, and B. Roizman, eds.), 2nd Ed., p. 267. Raven, New York.

Panum, P. L. (1847, reprinted 1940). "Observations Made during the Epidemic of Measles on the Faroe Islands in the Year 1846." American Publ., New York.

Yorke, J. A., Nathanson, N., Pianigiani, G., and Martin, J. (1979). Seasonality and the requirements for perpetuation and eradication of viruses in populations. *Am. J. Epidemiol.* **109,** 103.

Chapter 15

Prevention, Control, and Eradication of Viral Diseases

Nowhere in medicine is the adage "Prevention is better than cure" more appropriate than in viral diseases, for there is no effective treatment for most viral infections whereas there are several methods that may be applicable to their prevention or control. The ultimate step, which eliminates the need for control, is global eradication, which was achieved for smallpox in 1977. Within any country, control measures operate at various levels. Exotic diseases may be excluded by quarantine, a procedure that has more relevance nowadays in veterinary than in human medicine. Hygiene and sanitation are important methods of controlling enteric infections, and vector control may be useful for arbovirus diseases. However, the most generally useful control measure is vaccination.

Quarantine

Originally introduced in the fifteenth century for the control of plague, quarantine of shipping was used by the English colonists in North America in 1647 to try to prevent the entry of yellow fever and smallpox. Quarantine proved very effective in keeping Australia free of endemic smallpox, and in delaying the entry of pandemic influenza into that country in 1919. However, with the onset of air travel and the consequent arrival of passengers before the end of the incubation period, quarantine became much less effective. It was replaced, for smallpox, by the widespread requirement that international travelers had to have a valid certificate of smallpox vaccination, but this is no longer necessary. Currently, a similar provision operates for travelers who come from or pass through countries where yellow fever is endemic.

Hygiene and Sanitation

Hygiene and sanitation have had a profound effect on the incidence of enteric infections, both viral and bacterial. Viruses that infect the intestinal tract are shed in feces, and in many human communities recycling of feces back into the mouth following fecal contamination of food or water is common. A more voluminous and more fluid output (diarrhea) increases the environmental contamination. Hands contaminated at the time of defecation and inadequately washed may transfer viruses directly or indirectly to food, which is a particular problem if it occurs among those responsible for the preparation of meals to be eaten by others. In many densely populated parts of the world there are no reticulated sewerage systems, and sewage may seep into wells, streams, or other drinking water supplies, particularly after heavy rains. Explosive outbreaks of hepatitis E, poliomyelitis, or gastroenteritis occur from time to time even in sewered areas when sewerage mains burst or overflow to contaminate drinking water supplies.

Raw sewage contains 10^3–10^6 infectious virus particles per liter (typically 10^3–10^4 per liter in Western countries), mostly enteroviruses, caliciviruses, adenoviruses, and rotaviruses. Titers drop 100-fold (typically to 10–100 pfu per liter) following treatment in a modern activated sludge plant, because virions adsorb to the solid waste which sediments as sludge. The primary sludge is generally subjected to anaerobic digestion, which reduces the titer of virus significantly. Some countries require that the treated sludge be inactivated by pasteurization prior to being discharged into rivers and lakes or being utilized as land fill or fertilizer in agriculture.

In countries where wastewater has to be recycled for drinking and other domestic purposes, the treated effluent is further treated by coagulation with alum or ferric chloride, adsorption with activated carbon, and finally chlorination. Evidence that by-products of chlorination are toxic to fish and may be carcinogenic for humans has encouraged several countries to turn instead to ozonation. Ozone is a very effective oxidative disinfectant, for viruses as well as bacteria, provided that most of the organic matter, to which viruses adsorb, is removed first.

There is also a good case for chemical disinfection of recycled wastewater used for nondrinking purposes, such as agricultural irrigation by sprinklers, public fountains, and industrial cooling towers, because such procedures disseminate viruses in aerosols and contaminate vegetables. However, the lability of viruses to heat, desiccation, and ultraviolet light ensures that the virions remaining in wastewater from which most of the solids have been removed will be inactivated without intervention within a few weeks or months, depending on environmental conditions. Even during a cold northern winter, the number of viable enteroviruses in standing water drops by about 1 log per month; during a hot dry summer the rate of decay is as high as

2 logs per week. Hence storage of the final effluent in an oxidative lagoon for 1–2 months is an inexpensive and effective way of inactivating viruses.

Hygienic measures obviously have less effect on the incidence of respiratory infections, although washing of hands contaminated with respiratory secretions from a nose or handkerchief is important in minimizing the risk of transfer of many respiratory viruses. Attempts to achieve "air sanitation" by filtration and/or ultraviolet irradiation in public buildings have proved to have only a marginal effect, although these measures are an important feature of the biosafety cabinets that are widely used in virus laboratories. Respiratory viral infections are probably more common now than they have ever been, because of growing populations and the constant and extensive movement of people within cities, from rural to urban areas, and internationally. For human respiratory viruses, the population of the world now constitutes a single ecosystem, although seasonal differences between the northern and southern hemispheres affect the incidence of particular diseases, like influenza, at any particular time.

Vector Control

Because the control of their vertebrate reservoir hosts is usually difficult to achieve, control of arbovirus infections may be approached by attacking the arthropod vectors, minimizing the opportunities for exposure to them, or enhancing human resistance by vaccination. Approaches to vector control are well illustrated by examining the methods advocated by the World Health Organization (WHO) with respect to mosquitoes, namely, elimination of breeding sites and destruction of mosquitoes or their larvae. The flight range of many vector mosquitoes is so limited that much can be achieved by concentrating on the immediate vicinity of human settlements, particularly in the case of anthropophilic species such as *Aedes aegypti*. Any still water constitutes a potential breeding site. Swamps and ditches can be drained, and water-collecting refuse such as discarded tires, tin cans, and plastic containers should be destroyed. Larvicidal chemicals are placed in domestic water jars, and kerosene or diesel oil layered on the surface of nonpotable water. Biological control of mosquito larvae by fish or microorganisms, such as certain bacterial spores, is likely to be more widely exploited in the future. However, the growth of cities in the developing countries, largely by the growth of urban slums, has greatly accentuated the difficulties of control of domesticated *Aedes* mosquitoes, which is now less effective than it was 30 years ago. For arbovirus infections other than urban dengue and yellow fever the vectors breed over too wide a geographic area to make vector control feasible except on a local scale, usually in the face of a threatened epidemic.

The use of insecticides is a controversial issue, because there are environmental objections and mosquitoes eventually develop resistance. Policy should be decided on the basis of a risk–benefit analysis on a situation-by-situation basis. Some countries have based their arbovirus control programs on aerial insecticide spraying, but most retain this approach for emergency control in the event of an epidemic, aimed at rapid reduction of the adult female mosquito population. Organophosphorus insecticides such as malathion or fenitrothion are delivered as an ultra-low-volume (short-acting) aerosol generated by spray machines mounted on backpacks, trucks, or low-flying aircraft. Spraying of the luggage bays and passenger cabins of aircraft with insecticides reduces the chances of intercontinental transfer of exotic arthropods, whether infected or noninfected.

Avoidance of exposure to the bite of arthropods is another obvious precautionary measure. Personal protection against mosquito bites can be achieved by the use of screens on doors and windows, nets over beds, protective clothing, especially at dusk, and repellants.

Change of Lifestyle

Certain lifestyles that have become common in Western countries during the last few decades, such as the sharing of needles during the intravenous administration of addictive drugs and promiscuous male homosexuality, are associated with increased risks of infection with a variety of agents, notably hepatitis B and C viruses and HIV. Observations on the incidence of gonorrhea in developed countries suggest that fear of AIDS has increased the use of condoms and somewhat reduced the incidence of promiscuous male homosexual practices, but in general changes in lifestyle designed to reduce the incidence of disease are difficult to achieve, as is evident with cigarette smoking and alcoholism.

Immunization

Each of the foregoing methods of control of viral diseases is focused on reducing the chances of infection. The most generally effective method of control, immunization, is directed primarily at making the individual resistant to infection, although by reducing the incidence of cases it can also substantially reduce the chances of infection. Immunization may be active, that is, induction of an immune response by administration of antigen (vaccine), or passive, that is, administration of antibody (immune serum or immunoglobulin).

As outlined in Chapter 13 and discussed at length in the chapters of Part II, there are now effective vaccines for many common viral diseases. They are especially effective in diseases with a viremic phase, such as poliomyelitis, yellow fever, and the acute exanthemata. It has proved much more difficult to immunize effectively against infections of the alimentary or respiratory tracts. The dramatic success of vaccination programs in reducing the incidence of poliomyelitis, measles, mumps, and rubella in the United States over the period 1950 to 1980 is illustrated in Fig. 15-1.

Vaccination Policy

The principal aim of vaccination is to protect the vaccinated individual, although vaccination on a wide enough scale will also enhance *herd immunity* to such an extent that transmission can be restricted, or even arrested altogether, in a given community or country. Even in the case of a dreaded disease such

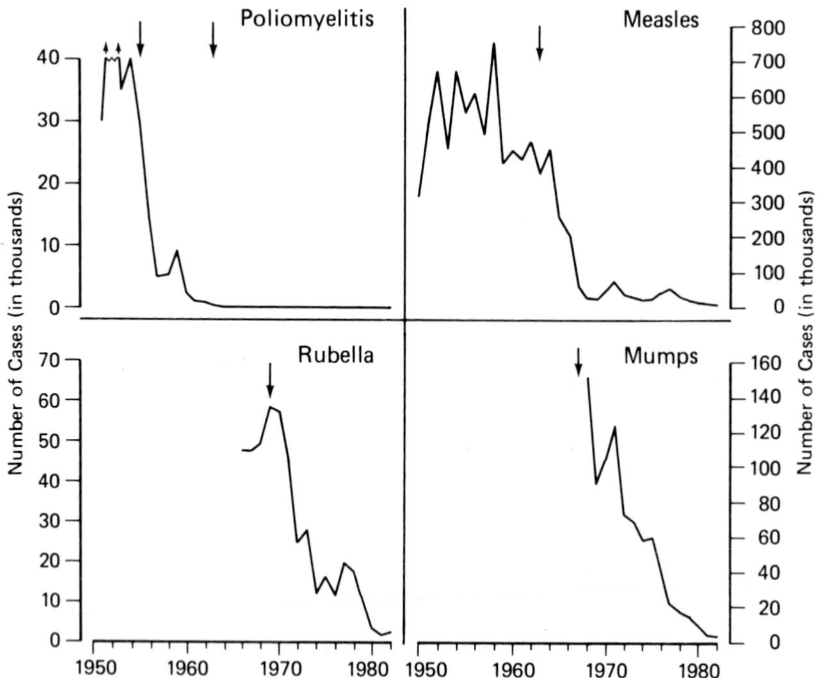

Fig. 15-1 Fall in the incidence of poliomyelitis, measles, rubella, and mumps in the United States following the introduction of vaccination against the diseases (arrows). Inactivated poliovirus vaccine was introduced in 1954, live vaccines in 1963. The other three, for measles (introduced in 1963), mumps (1967), and rubella (1969), are all live vaccines. (Compiled from data kindly supplied by the U.S. Centers for Disease Control.)

as poliomyelitis, it is difficult to maintain enthusiasm for a program of universal immunization after the disease has become very rare. Consequent complacency has resulted in a degree of resurgence of poliomyelitis and measles in a number of countries with strong immunization programs. Continuation of routine vaccination after the threat of the disease has almost vanished is doubly essential because the absence of wild virus in the population has left unvaccinated people uniquely susceptible, by removing the protective effect of subclinical infections. For these reasons it is essential for all countries to maintain highly organized and resolute health services, which need to pay particular attention to unimmunized pockets, such as urban ghettos, immigrants, and certain religious minorities.

Acceptability of a vaccine by a community is governed by a complex equation, balancing efficacy against safety, fear of disease against fear of needles and side effects, and complacency and inertia against the persuasive powers of the health services. If the disease is lethal or debilitating, both the people and the vaccine-licensing authorities will accept a risk of even quite serious consequences of vaccination in a tiny minority of recipients. If, on the other hand, the disease is perceived as trivial, no side effects will be countenanced. Where more than one satisfactory vaccine is available, considerations such as cost and ease of administration tip the balance.

A significant impediment to comprehensive vaccine coverage of the community is the unnecessarily complicated immunization schedules officially recommended by some government health authorities. Most of the currently available vaccines, bacterial and viral, are aimed at preventing diseases the risks of which are greatest in infancy; hence, these are given during the first 6 months of life for oral poliovaccine (and for diphtheria, pertussis, and tuberculosis) or after maternal antibody has disappeared in the case of the live vaccines for measles, mumps, and rubella (Table 15-1). Polyvalent vaccines, such as that available for measles–mumps–rubella (MMR vaccine), have a major practical advantage in minimizing the number of visits that the mother must make to the clinic. Characteristics of the major human viral vaccines in common use are shown in Table 15-2. Many more are in various stages of development or clinical trial, or are used only in the particular geographic locations where such diseases occur (e.g., various arboviral diseases). The vaccines used in particular diseases are discussed in the relevant chapters of Part II.

Passive Immunization

Instead of actively immunizing with viral vaccines it is possible to confer short-term protection by the intramuscular inoculation of antibody, either as immune serum or as immune (serum) globulin. Human immunoglobulin is

Table 15-1
Schedules for Immunization against Human Viral Diseases[a]

Vaccine	Primary course	Subsequent doses
Live vaccines		
Poliomyelitis	2, 4, and 15 months[b]	School entry (5 years)
Measles	12–15 months[c,d]	4–6 years or 11–12 years
Rubella	12–15 months[d]	11–12 years
Mumps	12–15 months[d]	11–12 years
Yellow fever	Before travel to endemic area	Booster after 10 years
Inactivated vaccines		
Influenza	Autumn annually[e]	Annual booster
Rabies	Preexposure: 0, 7, 28 days[f] Postexposure: 0, 3, 7, 14, 28, 90 days[h]	Every 2 years[g]
Hepatitis A	3 doses before travel to endemic area	Booster after 10 years
Hepatitis B	When at risk,[i] then 1 and 6 months later[j]	Booster after 10 years

[a] Schedules vary from country to country. This table is to be taken only as a guide.
[b] Two or three doses spaced two months apart, commencing between 2 and 6 months of age, conveniently timed to coincide with diphtheria–pertussis–tetanus (DPT) vaccine. The third dose is perhaps better delayed to 15 months, at the time of MMR vaccine.
[c] Given shortly after first birthday in most developed countries, but at 9 months in developing countries where measles death rate is high in the second 6 months of life.
[d] Usually as combined measles–mumps–rubella (MMR) vaccine.
[e] Vulnerable groups only, especially the aged and chronic cardiopulmonary invalids.
[f] Veterinarians, animal handlers, etc.
[g] If in rabies-endemic country or at high risk.
[h] Plus rabies immune globulin.
[i] Currently, vulnerable groups only, e.g., family contacts and babies of carriers, staff of hemodialysis units, blood banks, laboratories, hospitals, institutions for mentally retarded, etc., immunosuppressed persons, drug addicts; but eventually all infants worldwide.
[j] In countries with a high HBsAg carrier rate all infants should be vaccinated.

Table 15-2
Viral Vaccines Recommended for Use in Humans[a,b]

Disease	Vaccine strain	Cell substrate	Attenuation	Inactivation	Route
Yellow fever	17D	Chick embryo	+	–	Subcutaneous
Poliomyelitis	Sabin 1, 2, 3	HEF	+	–	Oral
Measles	Schwarz	CEF	+	–	Subcutaneous
Rubella	RA27/3	HEF	+	–	Subcutaneous
Mumps	Jeryl Lynn	CEF	+	–	Subcutaneous
Rabies	Pasteur	HEF	–	BPL	Intramuscular
Influenza	A/H1N1, A/H3N2, B	Chick embryo	–	BPL, formalin, or HANA subunits	Intramuscular
Hepatitis B		Recombinant DNA in yeast	–	Nil, HBsAg	Intramuscular

[a] A wide variety of different viral strains and cell substrates are used in different countries; the selection listed is not comprehensive.
[b] BPL, β-Propiolactone; CEF, chick embryo fibroblast cultures; HEF, diploid strain on human embryo fibroblasts; HBsAg, hepatitis B surface antigen purified after gene cloning in yeast; HANA, mixture of purified hemagglutinin and neuraminidase spikes.

preferred, because heterologous protein may provoke serum sickness or anaphylaxis. Pooled normal human immunoglobulin contains reasonably high titers of antibody against all the common viruses that cause systemic diseases in humans, but specific high-titer immunoglobulin can also be collected from individuals known to have recently recovered from a particular infection, for example, herpes zoster. Passive immunization should be regarded as an emergency procedure for the protection of unimmunized individuals exposed to special risk; it is an important prophylactic measure against hepatitis A (in travelers to developing countries), hepatitis B (in newborn babies of infected mothers or in unimmunized laboratory or health workers following a needle stick or comparable accident), rabies (following a bite from a potentially rabid animal), measles (in unimmunized close contacts of a patient), or varicella (to protect newborn babies of mothers with chickenpox at the time of delivery).

Specific antibody can also occasionally be used as therapy for an established viral disease; for example, immune plasma reduces the mortality from Lassa fever. A wider role for antibody in treatment and postexposure prophylaxis is being reexamined in the light of the availability of monoclonal antibodies of high specificity and high titer.

The Expanded Immunization Programme

In the industrialized countries immunization is carried out reasonably effectively by the public health authorities and private medical practitioners. However, in Third World countries immunization used to be available only to the small wealthy elite; there were neither the health services, the political will, nor the funds to provide for the poor majority. To capitalize on the health infrastructure that had been developed in Third World countries to support the Intensified Smallpox Eradication Programme, the World Health Organization in 1977 established the Expanded Programme on Immunization (EPI), with the specific goal of immunizing the world's children against six diseases for which there were satisfactory vaccines at that time: diphtheria, measles, poliomyelitis, tetanus, tuberculosis, and whooping cough. In 1985 the WHO program was greatly strengthened by the participation of the United Nations Children's Fund (UNICEF) as the provider of vaccines and augmentation of the funding. In 1990 the United Nations organized a World Summit for Children which endorsed the goals of the EPI and established the Children's Vaccine Initiative. As well as improving coverage with the present EPI vaccines, which is now about 80% overall, it is envisaged that over the next decade several additional vaccines may be added to the Programme, in certain cases only in countries where there is a special risk. These are vaccines for yellow fever, hepatitis A, hepatitis B, and Japanese encephalitis (for which satisfactory vaccines have already been licensed), and rotavirus, respiratory syncytial virus, and dengue (for which vaccines are under development), plus the bacterial vaccines for pneumococcus, meningococcus, and *Haemophilus influenzae* B.

Eradication

Control, whether by vaccination alone or by vaccination plus the various other methods aimed at lessening the chances of infection, is an ongoing process, which must be maintained indefinitely. If a disease could be eradicated, so that the causative agent was no longer present anywhere in the world, except possibly in microbiologically and militarily secure laboratories, control of that disease would no longer be required. The term *elimination* is used for the interruption of transmission within a country or region, a situation that does not exclude the importation of infection from outside; *eradication* means elimination from the world.

So far, global eradication has been achieved for only one disease, smallpox, the last naturally occurring case of which was reported in Somalia in October 1977. Smallpox eradication was achieved by an intensive effort that involved a high level of international cooperation and utilized a potent and very stable vaccine that was easy to administer. However, mass vaccination alone could not have achieved eradication of the disease from the densely populated tropical countries where it remained endemic in the 1970s, because it was impossible to achieve the necessary very high level of vaccine coverage. The effective strategy was to combine vaccination with *surveillance and containment*, by which cases were actively sought out, isolated, and their contacts vaccinated, first in the household and then at increasing distances from the index case.

The global smallpox eradication campaign was a highly cost-effective operation. The expenditure by the World Health Organization between 1967 and 1979 was $81 million, to which could be added about $32 million in bilateral aid contributions and some $200 million in expenditures by the endemic countries involved in the campaign. Against this expenditure of about $313 million over the 11 years of the campaign could be set an *annual* global expenditure of about $1000 million for vaccination, airport inspections, etc., made necessary by the existence of smallpox. This equation takes no account of the deaths, misery, and costs of smallpox itself, or of the complications of vaccination.

The achievement of global eradication of smallpox gave rise to discussions as to whether any other diseases could be eradicated worldwide. International conferences were held to discuss two other viral diseases, namely, measles and poliomyelitis. The biological characteristics of the three diseases that affect the ease of eradication are set out in Table 15-3. These diseases share several essential characteristics: (1) no animal reservoir, (2) lack of recurrent infectivity, (3) one or few stable serotypes, and (4) an effective vaccine. However, some features that were very important in the eradication of smallpox from tropical countries, notably the lack of infectivity in the prodromal stage, which made surveillance and containment possible, are lacking in measles, whereas the preponderance of subclinical infections in poliomyelitis renders its eradication more difficult. In addition to the biological properties of smallpox being particularly favorable, there were strong financial incentives for the industrialized countries to promote the global eradication of smallpox, because of the costs associated with vaccination of international travelers, port inspections, etc. However, in contrast to their limited participation in the smallpox eradication campaign, UNICEF and other major sources of international funds have not only provided solid financial support for the EPI since 1985, but, heartened by the excellent results obtained with the reduction of poliomyelitis, have expressed great interest in the eradication of this disease.

In 1988 the WHO, UNICEF, and some other organizations agreed on joint activity to facilitate immunization against poliomyelitis, and in May 1988 the World Health Assembly voted to commit the WHO to global eradication of

Table 15-3
Comparison of Features Influencing the Feasibility of Eradication of Measles and Poliomyelitis, Compared with Smallpox, in Which All Features Were Favorable

	Smallpox	Measles	Poliomyelitis
Biological features			
Reservoir host in wildlife	No	No	No
Persistent infection occurs	No	Yes[a]	No
Number of serotypes	1	1	3
Antigenically stable	Yes	Yes	Yes
Vaccine			
Effective	Yes	Yes[b]	Yes
Cold chain necessary	No	Yes	Yes
Number of doses	1	2	4
Infectivity in prodromal stage	No	Yes	Yes
Subclinical cases occur	No	No	Yes
Early containment possible	Yes	No	No
Sociopolitical features			
Country-wide elimination achieved	Yes[c]	No	Yes[d]
Financial incentive for assistance	Strong	Weak	Weak[e]
Records of vaccination required	No—scar	Yes	Yes

[a] As subacute sclerosing panencephalitis, but since no shedding occurs in this disease it is epidemiologically irrelevant.
[b] Vaccination is ineffective in the presence of maternal antibody.
[c] Before the Intensified Smallpox Eradication Programme commenced, in many countries.
[d] Before global eradication proposed, in several countries.
[e] But since 1985 considerable help provided by UNICEF, the World Bank, and others.

poliomyelitis by the year 2000. Regional offices of the WHO have declared their aims to reach this goal in their regions earlier than this: the Pan American Health Organization (PAHO) for the Americas by 1990 and the Western Pacific Region by 1995. The PAHO program has progressed very well, utilizing novel techniques for surveillance (flaccid paralysis notification) and for vaccination (national vaccination days, mopping up vaccination around recognized cases). As this book goes to print, in March 1994, no cases of paralytic poliomyelitis due to a wild poliovirus had been reported in the Americas since August 1991.

Because cases become infectious before the subject becomes ill, it is not possible to control measles by vaccination supplemented by surveillance and containment, but rather only by attaining very high levels of effective immunization, estimated to be 96% in the United States. Such a level is very difficult to achieve, because it is almost impossible to reach this proportion of the population with a vaccine that is invariably potent. In addition, maternal antibody inhibits infection by the standard Schwarz vaccine until some time between 9 and 12 months after birth; hence, vaccination is not recommended before the infant is 12 months old. However, in developing countries where measles is still a common disease, many infections occur in children aged 9–12 months old. It has also become apparent that in the United States (and presumably in other countries where the basic immunity provided by vaccination is not being boosted by later subclinical infections), vaccination in infancy does not provide certain protection throughout life. It is therefore recommended that there should be a booster inoculation of vaccine at school entry. Elimination of measles could be achieved in countries with a good health service and the political will to do it, but global eradication appears to be a distant and, in the face of the increasing world population and the increasing poverty of many developing countries, probably an unattainable goal.

Further Reading

Anderson, R. M., and May, R. M. (1991). "Infectious Diseases of Humans: Dynamics and Control." Oxford Univ. Press, Oxford.

Benenson, A. S., ed. (1990). "Control of Communicable Diseases in Man," 15th Ed. American Public Health Association, Washington, D.C.

Evans, A. S., ed. (1989). "Viral Infections of Humans: Epidemiology and Control," 3rd Ed. Plenum, New York.

Fenner, F. (1986). The eradication of infectious diseases. *S. Afr. Med. J. "Festschrift Suppl."* 11 October 1986, p. 35.

Fenner, F., Henderson, D. A., Arita, I., Jezek, Z., and Ladnyi, I. D. (1988). "Smallpox and Its Eradication." World Health Organization, Geneva.

Hovi, T. (1991). Remaining problems before eradication of poliomyelitis can be accomplished. *Prog. Med. Virol.* **38,** 69.

Kurstak, ed. (1993). "Control of Viral Diseases," 2nd Ed. Dekker, New York.

Melnick, J. L., ed. (1984). Enteric Viruses in Water. *Monogr. Virol.* **15.**

Pennington, J. E., ed. (1986). Properties and characteristics of a new immunoglobulin G intravenous preparation. *Rev. Infect. Dis.* **8** (Suppl. 4), S371.

World Health Organization. (1991). Safe use of pesticides: Fourteenth report of the WHO Expert Committee on Vector Biology and Control. *WHO Tech. Rep. Ser.* **813.**

Chapter 16

Chemotherapy of Viral Diseases

The hundreds of antibiotics now available to fight bacteria have no activity against viruses. The only circumstances in which it may be appropriate to prescribe antibacterial antibiotics in viral infections are as follows: (1) to prevent or treat serious superinfection of a viral disease (e.g., bacterial pneumonia complicating influenza, or *Pneumocystis carinii* in AIDS); or (2) to play safe pending the laboratory identification of the etiologic agent in the case of serious illnesses where there is real possibility of a bacterial cause, as in meningitis or pneumonia. In general, however, the widespread practice of prescribing antibiotics as a knee-jerk response to any infection is bad medicine.

Why has our hunt for antiviral agents yielded such a meagre harvest so far? The explanation is not hard to find. Being obligate intracellular parasites, viruses are absolutely dependent on the metabolic pathways of the host cell for their replication. Hence, most agents that block the replication of viruses are lethal to the cell. In recent years, however, we have come to know much more about the biochemistry of viral replication. This has led to a more rational approach to the search for antiviral chemotherapeutic agents. Two developments, the discovery of acyclovir and the production of interferon by recombinant DNA technology, provided the pharmaceutical industry with a great boost. The HIV pandemic has accelerated and diversified research in antiviral chemotherapy, and at least some antiviral agents have become standard weapons in the armamentarium of clinical practice.

Strategy for Development of Antiviral Agents

We now recognize a considerable number of steps in the viral replication cycle that represent potential targets for selective attack. Any process that is more essential to the replication of the virus than to the survival of the cell is theoretically vulnerable. Examples include the following: (1) attachment of the virion to its cellular receptor followed by its entry and uncoating; (2) transcription by the viral transcriptase from the viral genome to produce viral mRNA (or cDNA, in the case of the retroviruses); (3) regulation of the viral transcription program by viral regulatory genes; (4) translation of viral mRNA into protein; (5) posttranslational cleavage of proteins by virus-coded proteases; (6) replication of viral DNA or RNA by a virus-coded DNA polymerase or RNA-dependent RNA polymerase; (7) assembly/maturation of the virion. Table 16-1 provides examples of classes of antiviral agents known to target these key steps in the viral replication cycle. Many of the agents are still experimental, some still at the stage of laboratory investigation, and others are undergoing clinical trial.

A logical approach to the discovery of new antiviral chemotherapeutic agents is to isolate or synthesize substances that might be predicted to serve as a substrate or as an inhibitor of a known virus-coded enzyme, such as a transcriptase, replicase, or protease. A further refinement of this approach is well illustrated by the nucleoside analog acycloguanosine (acyclovir), an inhibitor of the herpesvirus DNA polymerase required for replication of viral DNA. Acyclovir is in fact an inactive prodrug which requires another

Table 16-1
Targets for Antiviral Chemotherapy

Process	Target	Agent
Attachment/uncoating	Ligand on virion	Receptor analogs, disoxaril
Transcription of viral genome	Viral transcriptase	Transcriptase inhibitors, antisense oligonucleotides
Reverse transcription	Reverse transcriptase (RT)	Zidovudine,[a] nonnucleoside RT inhibitors
Regulation of transcription	Regulatory proteins or their binding sites	HIV *tat* inhibitors
Processing of RNA transcripts	Various	Ribavirin[a]
Translation	mRNA	Interferons,[a] antisense oligonucleotides
Posttranslational cleavage	Viral protease	Protease inhibitors
Replication of DNA	Viral DNA polymerase	Acyclovir,[a] other nucleoside analogs[a]
Replication of RNA	Viral replicase	Replicase inhibitors
Assembly of the virion	Membrane protein (ion channel)	Rimantadine,[a] protease inhibitors

[a] Already licensed for human use.

herpesvirus-coded enzyme, thymidine kinase, to phosphorylate it to its active form. Because the viral enzyme occurs only in infected cells, such prodrugs are nontoxic for uninfected cells.

Having found an agent which, in its native state or following modification by a virus-coded enzyme, displays a degree of specificity for a viral enzyme, synthetic pathway, or any other process integral to the viral replication cycle, analogs (congeners) of the prototype are then synthesized with a view to enhancing activity and/or selectivity. Now that any viral gene can be cloned and expressed by recombinant DNA technology, purified viral enzymes such as the reverse transcriptase and the protease of HIV can be assayed directly *in vitro*, in the presence and absence of competitive inhibitors. Techniques such as cryoenzymology, X-ray crystallography, and nuclear magnetic resonance can be harnessed to elucidate structure–function relationships.

X-ray crystallography has opened up a major new approach to the search for antiviral agents. Now that the three-dimensional structure of the whole surface of isometric virions such as picornaviruses is known (see frontispiece), it is feasible to characterize the receptor-binding site (the ligand) on the critical capsid protein in atomic detail. Complexes of viral proteins with purified cell receptors or receptor mimics can be crystallized and examined directly. The receptor-binding site on the virion has generally turned out to be a "canyon," cleft, or depression on the external surface of the protein. The next step is to analyze the structure of the viral protein when bound to a compound known to neutralize infectivity, thereby confirming the identity of the receptor-binding site and supplying vital information on the nature of the interaction between the two. Further data are provided by mapping the position of the particular amino acid residues found to be substituted in resistant mutants of virus selected by growth in the presence of the antiviral compound, or by using site-specific mutagenesis to identify which residues are crucial. This information can then be exploited to design better synthetic drugs, using computer modeling to optimize the fit and the binding energy of the drug–virus interaction. Clearly, this exciting new approach lends itself best to the development of drugs that act by binding directly to the capsid (or envelope) of the virion itself, thereby blocking the early steps in the replication cycle, namely, attachment, penetration, or uncoating.

Another approach has recently emerged from the discovery of regulatory genes in the genome of HIV and other viruses. Potentially, the replication of such viruses could be blocked by agents that bind either to the protein product of such a regulatory gene or to the recognition site in the regulatory region of the viral genome with which that protein normally interacts, such as the TAR region to which the product of the HIV *tat* gene binds.

The critical test of any putative chemotherapeutic agent is, of course, inhibition of viral replication. In the presence of dilutions of the agent, the multiplication of suitable indicator viruses in cultured cells is measured by reduction in the yield of virions or of some convenient viral marker. Toxicity of the drug for human cells may be measured crudely by the cytopathic effect (in the absence of virus) or, more sensitively, by reduction in cell plating efficiency or cell doubling time. The therapeutic index may be defined as the ratio of the minimum cell-toxic dose to minimum virus-inhibitory dose. In general, only those agents displaying a therapeutic index of at least 10 and preferably 100–1000 are worth pursuing further. The literature is cluttered with worthless reports of "antiviral" agents that are lethal for cells. Indeed, many agents that look promising in cell culture also fall by the wayside at the next hurdle, namely, the experimental animal.

Before embarking on human clinical trials, the pharmacology and toxicology of the drug must be thoroughly investigated in experimental animals, including primates. Ideally, the drug should be water-soluble, chemically and metabolically stable, moderately apolar, and taken into cells satisfactorily. Pharmacokinetic studies, first in animals and then in humans, address such questions as the mechanism and rate of absorption following various routes of administration, tissue distribution, metabolism, detoxification, and excretion of the drug. Tests for acute toxicity encompass comprehensive clinical surveillance of all the body systems, biochemical tests (e.g., for liver and kidney function), hematologic examination, tests for immunosuppression, and so on. Longer term investigations screen for chronic toxicity, allergenicity, mutagenicity, carcinogenicity, and teratogenicity.

Clinical Application

Methods of Delivery

The route of administration of an antiviral agent is a prime consideration in assessing its general acceptability. The oral route is naturally by far the most convenient for the patient. Nasal drops or sprays may be acceptable for upper respiratory infections but can be irritating, whereas continuous delivery of aerosols through a face mask or oxygen tent is generally appropriate only for very sick hospitalized patients. Topical preparations (creams, ointments, etc.) are satisfactory for superficial infections of skin, genitalia, or eye, provided they are relatively localized; penetration of drugs through the skin can be enhanced by mixing with substances such as polyethylene glycol. Parenteral administration is the only option in the case of some drugs and may, in any case, be required for serious systemic infections; intravenous infusion of course necessitates hospitalization.

Currently, many experimental drugs have to be used in very high, potentially toxic concentrations because of poor solubility or poor penetration into cells. Delivery of antiviral concentrations of compounds into cells can sometimes be achieved by incorporating the drug into liposomes or by conjugating it to a hydrophobic membrane anchor. Sophisticated chemistry may also be required to modify potential antivirals, such as synthetic peptides or oligonucleotides, which are otherwise rapidly degraded intra- or extracellularly. In the future we may also see antiviral drugs conjugated to antiviral antibody, or incorporated into liposomes coated with such antibody, to direct them to virus-infected cells.

Strategies to Minimize Emergence of Drug-Resistant Mutants

It is already clear that mutants resistant to many of the available antiviral drugs readily emerge *in vitro* and *in vivo*, especially during long-term therapy of chronic infections and in immunocompromised patients. Often, resistance is the result of a single point mutation in the gene encoding the particular viral protein that is the target of the drug. Stepwise increases in the degree of resistance may occur as further nucleotide substitutions accumulate, often in a particular order. Clinical isolates may be tested for drug sensitivity by growth in cultured cells in the presence of serial dilutions of the agent. For virus/drug combinations where resistance is regularly associated with particular mutations, it may be feasible to develop a PCR that differentiates resistant from sensitive isolates without the need for culture; for example, in cases of HIV infections treated with zidovudine (AZT), only the critical region of the reverse transcriptase gene needs to be sequenced.

To minimize the emergence of drug-resistant viral mutants we need to capitalize on the lessons learned in handling the problem of antibiotic resistance in bacteria. Antiviral agents should be used only when absolutely necessary, but administered in adequate dosage. Certain lifesaving drugs may need to be retained for designated diseases only, and/or as replacement therapy following the emergence of resistance to the standard drug. Combined therapy, preferably using agents with distinct modes of action, minimizes the probability of emergence of resistant mutants. Furthermore, by allowing one or both drugs to be given in lower dosage, combined therapy can reduce the incidence of toxic side effects. In certain instances, particular drug combinations display synergism, as has been observed when interferon α is combined with acyclovir or zidovudine, for example.

Clinical Priorities

Diseases against which no satisfactory vaccine is available, including those with a large number of different etiologic agents, are prime targets for antiviral chemotherapy. The common cold is an admirable example on both counts, but there are so many serotypes that chemotherapeutic agents, to be sufficiently broad spectrum, will need to be directed at molecules (or ligands) that are conserved across the genus *Rhinovirus*. Other respiratory infections, gastroenteritis, hepatitis, and infectious mononucleosis must also be high on the list of priorities. Effective chemotherapy is also needed to treat reactivation of latent infections such as herpes simplex and zoster, even though the latent infection itself will not be eliminated. Reactivation of herpesvirus infections is a particular problem in immunocompromised individuals, such as AIDS patients or transplant recipients. Chronic infections, for example, hepatitis B and C, AIDS, congenital rubella, or cytomegalovirus infections, may be particularly amenable to antiviral chemotherapy, as might some other long drawn out diseases of currently unknown etiology such as certain cancers, autoimmune diseases, and degenerative conditions of the brain, should any of these turn out to be of viral causation. Finally, we must not forget less common but lethal viral diseases, such as encephalitis, rabies, and the hemorrhagic fevers, for which successful chemotherapy would be lifesaving.

Chemoprophylaxis may also have a role, not only in the prevention of complications, such as orchitis or meningoencephalitis in mumps, but also in limiting the spread of diseases like hepatitis, mononucleosis, influenza, measles, or rubella to unimmunized family contacts. A special case is AIDS, which has such a long incubation period that development of the disease might be delayed, or conceivably prevented, by long-term chemoprophylaxis in individuals found to have seroconverted to HIV.

Interferons

Ever since their discovery in 1957, the interferons seemed to offer advantages that would ensure their future as ideal antiviral agents. They are natural cellular products of viral infection and display a broad spectrum of activity against essentially all viruses (see Chapters 5, 7, and 8). Early clinical trials were conducted with inadequate amounts of semipurified interferons which had been produced by treating cultured human leukocytes or fibroblasts with a paramyxovirus or with a synthetic double-stranded RNA. However, in 1980 a major breakthrough was achieved when the gene for a human interferon α (IFN-α) was cloned and expressed in *Escherichia coli*. Since then, the genes for all known subtypes of human IFN-α as well as IFN-β and IFN-γ have been cloned in prokaryotic and/or eukaryotic cells. The yields so obtained are vastly greater than those from leukocyte, lymphoblastoid, or fibroblast cultures, and the cost of production has declined substantially.

Interferons are not effective by mouth, therefore are injected. IFN-α is much more active *in vivo* than IFN-β or IFN-γ, probably because the latter do not achieve or maintain the required blood levels after intramuscular administration. Toxic side effects are regularly observed and may be marked with doses in excess of 10^7 units per day, even when highly purified cloned IFN subtypes are employed. Fever regularly occurs at high dosage but lasts only a day or so. Severe fatigue is the most debilitating symptom and may be accompanied by malaise, anorexia, myalgia, headache, nausea, vomiting, weight loss, erythema and tenderness at the injection site, partial alopecia (reversible), dry mouth, reversible peripheral sensory neuropathy, or signs referable to the central nervous system. Various indicators of myelosuppression (granulocytopenia, thrombocytopenia, and leukopenia) and abnormal liver function tests, both reversible on cessation of therapy, are regularly observed if high-dose interferon administration is prolonged.

Although the advent of less expensive interferons produced by recombinant DNA technology made it possible to treat patients with the sort of dosage that is required to produce clinically beneficial effects, it must be said that the successes are still rather modest. For example, some hepatitis C carriers respond to prolonged treatment with interferon α, but most relapse following withdrawal of the drug (Chapter 26). Genital warts have been successfully treated, and juvenile laryngeal papillomatosis, a severe condition calling for repeated surgical removal following recurrences, can be arrested by local injection of interferon; however, the tumors reappear when therapy is withdrawn (Chapter 18).

In the late 1970s reports of the partial regression of several types of cancers following interferon treatment precipitated a flurry of excitement, not only in cancer circles, but even on Wall Street, where the stocks of the newly emerging recombinant DNA based biotechnology companies skyrocketed overnight. However, subsequent more carefully controlled trials have given much less encouraging results. Only a minority of patients with only certain types of cancers, such as hairy cell leukemia and chronic myelocytic leukemia, respond favorably to vigorous interferon therapy, with temporary remission or partial regression. In these situations interferons may be acting not as antivirals but as cytokines exerting immunomodulatory effects. The euphoria of the late 1970s has been supplanted by a more cautious, balanced belief that interferons, while no panacea, may nevertheless come to occupy a limited place in both antiviral and anticancer chemotherapy. The future may lie in synergistic combinations of interferons with other types of antiviral or anticancer agents.

Inhibitors of Viral DNA Polymerase

Many of the successful antiviral agents described to date are nucleoside analogs, most of which are restricted in their antiviral activity to the herpesviruses. The early prototypes, such as adenine arabinoside, were relatively undiscriminating inhibitors of both cellular and viral DNA synthesis which produced toxic side effects, directed especially at dividing cells in the bone marrow and gastrointestinal tract.

Acycloguanosine (Acyclovir) and Homologs

A major breakthrough in antiviral chemotherapy occurred in 1977 when Elion and colleagues developed a prodrug that depends on a viral enzyme to convert it to its active form. Acycloguanosine, now commonly known as acyclovir, is a guanine derivative with an acyclic side chain, the full chemical name being 9-(2-hydroxyethoxymethyl)guanine (Fig. 16-1). Its unique advantage over earlier nucleoside derivatives is that the herpesvirus-encoded enzyme, thymidine kinase (TK), which has broader specificity than cellular TK, is required to phosphorylate acycloguanosine intracellularly to acycloguanosine monophosphate (ACG-P); a cellular GMP kinase then completes the phosphorylation to the active agent, acycloguanosine triphosphate (ACG-PPP) (Fig. 16-2). Further, ACG-PPP inhibits the herpesvirus-encoded DNA polymerase at least 10 times more effectively than it does cellular DNA polymerase α. It acts as both inhibitor and substrate of the viral enzyme, competing with GTP and being incorporated into DNA, leading to chain termination because acyclovir lacks the 3′-hydroxyl group required for chain elongation. Since activation of the prodrug needs the viral TK, acyclovir is essentially nontoxic to uninfected cells but is powerfully inhibitory to viral DNA synthesis in infected cells, giving it much greater selectivity than the earlier nucleoside analogs.

Herpes simplex viruses types 1 and 2 (HSV-1 and -2) are both very susceptible to acyclovir; varicella–zoster virus (VZV) is susceptible at somewhat higher concentrations of the drug. Other human herpesviruses, which do not possess a gene coding for TK, are susceptible to acycloguanosine only at much greater doses; this results from limited production of ACG-P by cellular GMP kinase. The relative sensitivity of different herpesviruses seems to de-

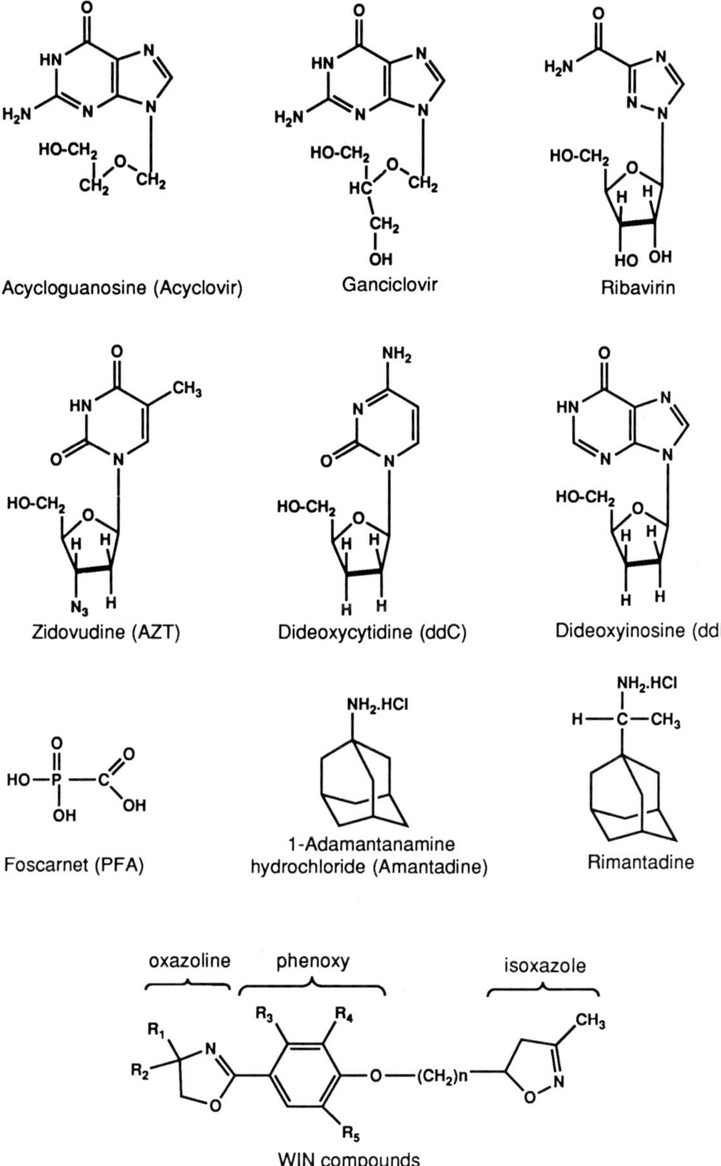

Fig. 16-1 Some antiviral chemotherapeutic agents.

pend on a rather complex interplay of at least three variables: (1) the efficiency of the virus-coded TK (if any) in converting acyclovir to ACG-P; (2) the efficiency of cellular kinases in converting this intermediate to ACG-PPP; and (3) the susceptibility of the viral DNA polymerase to ACG-PPP. The use of acyclovir for treatment of various herpesvirus diseases is discussed in Chapter 20.

Acyclovir (Zovirax) may be delivered orally, by slow intravenous infusion, or topically as an aqueous cream. As anticipated from *in vitro* studies, the

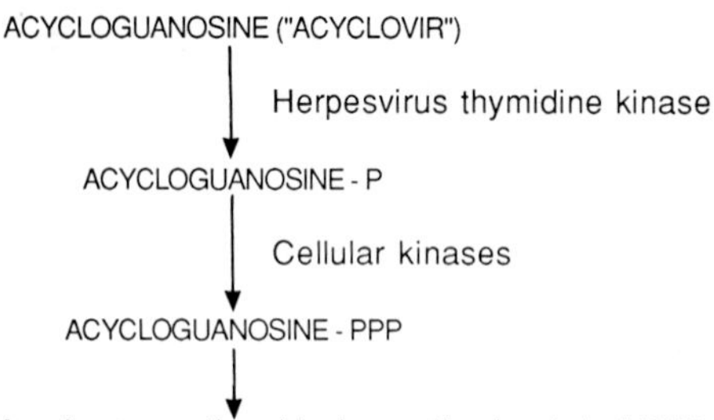

Fig. 16-2 Mechanism of inhibition of herpesvirus replication by acycloguanosine.

drug is essentially nontoxic. Acyclovir levels must be carefully monitored in patients with dehydration or renal impairment, as the drug, which is excreted unchanged through the kidneys, is rather insoluble, and crystalluria may occur.

Acyclovir-resistant mutants of HSV can be recovered *in vivo* and in cell culture. The mutation is usually located in the gene coding for the viral thymidine kinase, but more rarely is seen in the DNA polymerase gene. There are two kinds of TK mutants: (1) those failing to produce appreciable levels of TK and (2) those in which the enzyme is produced but has an altered substrate specificity such that it can no longer satisfactorily phosphorylate acyclovir. The former (TK^- mutants), the most common, may contain a mutation, deletion, or insertion leading to premature termination of translation or the production of a nonfunctional enzyme, whereas the latter (TK^a mutants) result from a point mutation causing a more subtle alteration in substrate specificity so that the enzyme no longer phosphorylates acyclovir. The TK^- mutants, while able to establish latent infection of ganglia, have a reduced ability to reactivate but can still induce severe disease in immunocompromised hosts.

Certain derivatives of acyclovir display greater activity against varicella-zoster and/or herpes simplex virus, in cultured cells and mice. Clinical investigations have begun on a number of such analogs. For example, valaciclovir is better absorbed orally than is acyclovir, and is rapidly converted to acyclovir *in vivo*. The TK^- mutants resistant to acyclovir often display cross-resistance to related nucleoside analogs.

Ganciclovir

A derivative of acyclovir, 9-(1,3-dihydroxy-2-propoxy)methylguanine (DHPG), known as ganciclovir (Fig. 16-1), is the first drug to offer satisfactory therapy for cytomegalovirus (CMV) infections. CMV encodes an enzyme, not yet characterized, which phosphorylates ganciclovir to the monophosphate; further phosphorylation by cellular kinases yields the active triphosphate, which inhibits the viral DNA polymerase. Resistance in some mutants maps to the phosphorylation gene, in others to the DNA polymerase gene.

Ganciclovir has been used principally to treat severe CMV infections such as retinitis, colitis, and pneumonia in AIDS patients and in transplant recipients. It is not effective by mouth. Given intravenously for some weeks, ganciclovir may produce a temporary remission in a proportion of cases, but unfortunately the condition generally recurs following its withdrawal. The drug is very toxic, severe neutropenia and thrombocytopenia being common side effects, so this compound should be reserved for life-threatening CMV infections in immunocompromised individuals.

Ribavirin

A rather different nucleoside analog, 1-β-D-ribofuranosyl-1,2,4-triazole-3-carboxamide, known as ribavirin (Fig. 16-1), was first synthesized in 1972. Despite extensive investigation, it has still not been licensed for general use in many countries because of unpersuasive evidence of antiviral activity in humans as well as indications of toxicity. At first sight, the drug would appear to have potential, as it inhibits the growth of a wide spectrum of RNA and DNA viruses in cultured cells and experimental animals, by what appears to be a multipoint mechanism of action. However, this early promise has not been matched by a comparable degree of efficacy in humans. Further, following oral administration at the usual dosage of about 1 gram per day, a substantial minority of recipients develop a reversible anemia with increased reticulocyte numbers and elevated serum bilirubin levels, while immunosuppressive and teratogenic effects have been demonstrated in animals. The fact that ribavirin monophosphate inhibits the cellular enzyme IMP dehydrogenase, decreasing the pool of GTP, as well as inhibiting guanylyltransferase-mediated 5′-capping of mRNA, suggest that it may be acting on cellular pathways that are somewhat more critical to the virus than to the cell.

Oral or intravenous ribavirin has been claimed to reduce mortality from infections with the exotic Lassa and Hantaan viruses. Ribavirin has also found a niche in the treatment of severe respiratory syncytial viral infections in infants when delivered as an aerosol; a nebulizer is required to generate a small-particle aerosol which is then administered via a mask or oxygen tent for 3–6 days. The value of ribavirin aerosol in the management of severe influenza and parainfluenza infections has yet to be established.

Trisodium Phosphonoformate (PFA, Foscarnet)

The search for inhibitors of enzymes that catalyze the transcription or replication of viral DNA or RNA need not be restricted to the nucleoside analogs. For example, trisodium phosphonoformate, known also as phosphonoformic acid (PFA) or foscarnet (Fig. 16-1), inhibits the DNA polymerase of herpesviruses and hepatitis B, as well as the reverse transcriptase of HIV, by blocking the pyrophosphate binding site on the enzyme. Resistance maps to the DNA polymerase gene. Foscarnet in the form a cream has been claimed to accelerate healing of recurrent facial or genital herpes lesions, and given systemically it can halt the progression of cytomegalovirus infections in immunocompromised patients. The drug also displays some activity against hepatitis B and HIV *in vivo*.

Although foscarnet displays some selectivity in that it inhibits cellular DNA polymerase α only at higher concentrations than required to inhibit viral DNA polymerase, it accumulates in bone and is too toxic for the kidneys to be advocated for infections that are not life-threatening.

Inhibitors of Reverse Transcriptase

Zidovudine and Homologs

The advent of AIDS and of human cancers caused by other retroviruses quickened interest in the search for inhibitors of the crucial retroviral enzyme, reverse transcriptase (RT). Reverse transcriptase is a complex molecule carrying at least four different enzyme activities: reverse transcriptase, DNA polymerase, ribonuclease H, and ribonuclease D. The first compound to display sufficient antiviral activity *in vivo* to be licensed for human use was 3′-azido-2′,3′-dideoxythymidine, otherwise known as azidothymidine, AZT, or zidovudine (Fig. 16-1).

AZT is phosphorylated by cellular kinases to AZT triphosphate (AZT-PPP), which exerts its antiviral effect against HIV by the following mechanisms. AZT-PPP inhibits HIV reverse transcriptase, being accepted by the enzyme in preference to TTP. AZT-PPP binds to RT approximately 100 times more efficiently than it does to the cellular DNA polymerase α. AZT-PPP is incorporated into the growing HIV DNA chain, leading to premature chain termination. In addition, AZT monophosphate (AZT-P) competes successfully for the enzyme thymidylate kinase, resulting in depletion of the intracellular pool of TTP. Clearly, since the HIV provirus persists indefinitely in nondividing cells, it cannot possibly be expected that zidovudine administered to already infected people could do more than suppress replication; it could never eliminate the viral genome from the body.

Zidovudine is rapidly absorbed following oral administration and rather rapidly metabolized by hepatic glucuronidation (half-life about 1 hour) so that the drug needs to be given 2 or 3 times daily. Very ill hospitalized patients may be treated by continuous intravenous infusion. Side effects of zidovudine are frequent. The most important result from the toxicity of the drug for bone marrow, namely, macrocytic anemia (often requiring red cell transfusions) and granulocytopenia (neutropenia). Headache, nausea, and insomnia are common, and many patients develop a myopathy resulting in (reversible) wasting of proximal muscle groups. Severe side effects may demand suspension of treatment, reduction in the daily dosage, combined therapy with another nucleoside analog or interferon α or another cytokine displaying synergism, or replacement or alternation with another drug.

Early clinical trials in 1986 indicated that after several weeks of treatment with zidovudine advanced AIDS patients survived longer, experienced fewer and less severe opportunistic infections, temporarily regained helper T cells and DTH capability, displayed reduced levels of HIV core antigen (p24) in the blood, and generally felt better, as measured on the Karnofsky scale of well-being. However, progress of the disease is not arrested, and life expectancy is generally extended for only a year or two. The tendency now is to institute therapy much earlier, even prophylactically in seropositive individuals before symptoms become apparent, with a view to continuing medication for life. This raises major problems of (1) long-term toxicity and (2) emergence of drug-resistant mutants.

Resistance to zidovudine arises regardless of the regime employed. Mutations tend to occur in particular codons of the RT gene, resistance increasing as mutations accumulate over many months. Following withdrawal of the drug, AZT-resistant mutants tend to be replaced by susceptible virus. Cloned wild-type and mutant RT can be used to measure the binding affinity of various drugs, while the recent elucidation of the structure of RT by X-ray crystallography should enable further inhibitors to be designed to act not only against the RT domain but also against the ribonuclease H and other domains of the enzyme complex.

Other dideoxynucleosides, such as dideoxycytidine (ddC) and dideoxyinosine (ddI), do not show cross-resistance with AZT, nor do several different types of nonnucleoside inhibitors, such as the "TIBO" compounds that block HIV RT activities in different ways. Hence, the probability of emergence of drug resistance can be minimized by (1) combined chemotherapy or (2) alternating courses of different drugs, for example, AZT with ddC or interferon α. This regimen also alleviates the problem that the principal serious side effect of prolonged AZT usage is bone marrow toxicity, whereas ddI and ddC tend rather to cause peripheral neuropathy. Further details about the chemoprophylaxis and chemotherapy of AIDS with AZT and other drugs are given in Chapter 35.

Ion Channel Blockers

Some 30 years ago a simple three-ringed symmetrical amine known as 1-aminoadamantane hydrochloride, or more commonly as amantadine, was synthesized and shown to inhibit the replication of influenza A viruses. The principal target of amantadine is now recognized to be the protein M2, which is a minor component of the influenza viral envelope that is thought to play a key role in stabilizing the viral hemagglutinin (HA). M2 forms a tetrameric transmembrane ion channel which serves to reduce the transmembrane pH gradient in acidic endosomes. Such endosomes are involved both in the uncoating of the incoming virion and in the transport of newly synthesized HA from the trans-Golgi cisternae to the plasma membrane. Amantadine acts at these two distinct steps in the replication cycle as an ion channel blocker. First, by raising the pH of the endosome, amantadine prevents the pH 5-mediated conformational change in the HA molecule required for fusion of viral envelope with endosomal membrane to release the ribonucleoprotein of the infecting virion. Later in the replication cycle, by disturbing the ionic environment within the exocytic pathway, amantadine prevents the assumption of the correct conformation of newly synthesized HA destined for incorporation into the envelope of budding virions.

Therapeutically, amantadine is only marginally effective against influenza A virus (and not at all against influenza B), but it has been reported to reduce the severity of symptoms in about 50% of cases if given early. Administered prophylactically, however, it can reduce the incidence of clinical influenza quite significantly (50–90% in various trials). Because, in practice, this demands the ingestion of 200 mg daily for 1–2 months from the commencement of an influenza epidemic in the community, it is important to consider the possible complications of such a prolonged chemoprophylactic regime.

Side effects commonly occur with amantadine. They relate mainly to the central nervous system (loss of concentration, insomnia, nervousness, lightheadedness, drowsiness, anxiety, confusion) but are generally reversible and mild, unless the recommended dose is exceeded or the patient has a history of mental illness or renal disease.

Rimantadine, or α-methyl-1-adamantane-methylamine hydrochloride (Fig. 16-1), is a very similar drug which has been extensively used in Russia for several years. Because side effects are less frequent and less severe than with amantadine and efficacy is at least as good, rimantadine appears to be the drug of choice. Like amantadine, rimantadine is given orally, but either drug can be delivered to hospitalized patients by aerosol spray, perhaps in conjunction with ribavirin. A major potential problem is that resistant mutants arise rapidly to amantadine and rimantadine and are transmissible to contacts. As expected the mutants contain amino acid substitutions in M2.

Although these compounds are relatively specific for influenza A virus, they may represent the prototype of a class of compounds that may be found to act as ion channel blockers against many other viruses with similar low M_r hydrophobic transmembrane proteins, such as the vpu protein of HIV, the 1A protein of respiratory syncytial virus, or the LMP protein of Epstein-Barr virus.

Blocking Attachment or Uncoating of Virion

One theoretical option for antiviral therapy is to inhibit the very first step in the viral replication cycle, namely, attachment of the virion to its specific receptor on the plasma membrane of the host cell. This could be accomplished by substances designed to mimic either the cell receptor or the viral ligand. For example, attachment of HIV to its receptor (CD4) can be blocked by soluble CD4 (which binds to the virion) or by a synthetic peptide corresponding to the ligand on the HIV envelope glycoprotein gp120 (which binds to the cell receptor). A major problem with ligand mimics is that, by saturating the cell receptor, they will presumably interfere with the normal physiologic function of that membrane glycoprotein. Receptor mimics may be safe but would need to be demonstrated not to elicit an autoimmune response.

X-Ray crystallography has provided detailed information on the binding site of a wide range of antiviral agents that block uncoating of picornaviruses. Many of the studies to date have used human rhinovirus type 14 (HRV-14) as a model. Most of the drugs, in spite of their diversity of chemical structure, bind to the same site on HRV-14, namely, a hydrophobic pocket which lies immediately beneath the floor of the canyon that comprises the ligand (receptor-binding site) on the viral capsid protein VP1. Hydrophobic interactions result in deformation of the canyon floor which may inhibit attachment of the virion to its cell receptor but, more importantly, inhibits uncoating of the virion. This is thought to occur by locking VP1 into a position that prevents the disassembly of the virion which normally occurs in the acidic environment of the endosome. When administered prophylactically, but not therapeutically, some such antivirals have been claimed to reduce the symptoms of the common colds induced by certain sensitive rhinovirus serotypes but not others. While this is just a start, it does engender some optimism that this new approach to antiviral drug design may eventually provide us with an answer to the question most commonly addressed to virologists, namely, "when will you produce a cure for the common cold?"

Similar research based on a combination of X-ray crystallography and computer modeling is currently being directed at the influenza viruses. The three-dimensional structure of both envelope glycoproteins is known. Sialyloligosaccharides mimicking the receptor for the viral hemagglutinin might be expected to block attachment of virion to host cell, whereas compounds binding the enzyme active site of the viral neuraminidase might inhibit release of new virions from infected cells.

Inhibitors of Viral Proteases

Cleavage of viral proteins by proteinases (proteases) is required at several stages in the viral replication cycle: activation of some viral enzymes, post-translational cleavage of the polyprotein product of polycistronic mRNA, activation of envelope fusion glycoproteins, and maturation of the virion. As many of the proteases are virus-coded, it should be possible to find or devise agents that specifically inhibit the viral protease without interfering with essential cellular proteases. The three-dimensional structure of the HIV aspartyl protease has been solved by X-ray crystallography, and several pharmaceutical companies are competing to produce different types of protease inhibitors, some of which are already undergoing clinical trials. One approach has been to synthesize peptides corresponding to the sequence on the *gag–pol* polyprotein that represents the cleavage site for the HIV protease; such peptides, or close analogs of the native sequence, bind to the active site of the enzyme and inhibit its activity. Unfortunately, they are often of low solubility, have low oral bioavailability, and are rapidly degraded in serum. Hence, following the precedent of the very effective class of oral antihypertensive drugs that inhibit the protease, angiotensin-converting enzyme, attempts are being made to produce a stable molecule that mimics the conformation of the peptide but without amide bonds.

Virus-Specific Oligonucleotides

Theoretically, short synthetic "antisense" oligodeoxynucleotides, complementary in sequence to viral mRNA, might be able to inhibit viral gene expression. For example, hybridization to viral mRNA might prevent the splicing, transport, or translation of that mRNA, or render it susceptible to degradation by RNase H. On the other hand, hybridization to viral DNA or cDNA might block transcription or replication, or block the attachment of DNA-binding regulatory proteins. Short single-stranded DNA sequences of this sort have displayed antiviral activity against HIV, HSV, and influenza viruses in cultured cells, but there are major problems with the specificity, stability, and uptake of oligodeoxynucleotides. Statistically, the genome of the average human contains no sequence precisely complementary to any random oligonucleotide of 18 or more bases. In practice, however, significant binding occurs even if there are one or two mismatches, and nonspecific binding to proteins has also been demonstrated. Moreover, oligonucleotides are rapidly degraded by extracellular and intracellular nucleases unless the backbone of the molecule is modified to circumvent this problem. Third, the uptake of oligonucleotides into cells is very inefficient; attempts are being made to facilitate entry, for example, by coupling to a hydrophobic peptide or lipid.

An avant-garde variation on this theme is to produce transgenic plants or animals that constitutively make antisense RNA or a ribozyme (RNA with RNase activity specific for a particular nucleotide sequence). Some such transgenic plants and animals display resistance to the virus in question. Clearly, transgenic humans are a fantasy, but the current commercial interest in creating plants and animals that are resistant to microbial disease may well benefit research on antisense oligonucleotides as antiviral agents in humans.

Inhibitors of Regulatory Proteins

The majority of the genes of HIV are regulatory genes, the sole or principal function of which is to control the expression of other genes. For example, the protein product of the *tat* gene binds to a specific responsive element called TAR which is present in both the integrated HIV cDNA and all HIV mRNAs. This results in augmentation of the expression of all the HIV genes, including *tat* itself, thereby constituting a positive feedback loop that enables production of large numbers of progeny virions. Clearly, an agent capable of binding either to the Tat protein or to the TAR nucleotide sequence would be expected to be a most effective inhibitor of HIV replication. Several such agents have been demonstrated to reduce HIV production from chronically infected cells in culture, and the least toxic of them seem likely to undergo clinical trials in the near future. On the reasonable assumption that many or most other viruses will be found to carry regulatory genes, this novel approach to antiviral chemotherapy has considerable appeal.

Further Reading

Chapman, M. S., Giranda, V. L., and Rossmann, M. G. (1990). The structures of human rhinovirus and Mengo virus: Relevance to function and drug design. *Semin. Virol.* **1,** 413.

Darby, G., and Larder, B. (1991). Resistance of herpesviruses and human immunodeficiency virus to antiviral drugs. *Adv. Gene Technol.* **2,** 177.

de Clercq, E. (1993). Antiviral agents: Characteristic activity spectrum depending on the molecular target with which they interact. *Adv. Virus Res.* **42,** 1.

de Clercq, E., and Jeffries, D. J., eds. (1993). "Antiviral Chemotherapy." Wiley, London.

Dimmock, N. J., ed. (1990). "Control of Virus Diseases," Society of General Microbiology, Symposium 45. Cambridge Univ. Press, London.

Field, H. J. (1988). "Antiviral Agents: The Development and Assessment of Antiviral Chemotherapy." CRC Press, Boca Raton, Florida.

Freifeld, A. G., and Ostrove, J. M. (1991). Resistance of viruses to antiviral drugs. *Annu. Rev. Med.* **42,** 247.

Galasso, G. J., Whitley, R. J., and Merigan, T. C., eds. (1990). "Antiviral Agents and Viral Diseases of Man," 3rd Ed. Raven, New York.

Lentz, T. L. (1990). The recognition event between virus and host cell receptor: A target for antiviral agents. *J. Gen. Virol.* **71,** 751.

McKinlay, M. A., Pevear, D. C., and Rossmann, M. G. (1992). Treatment of the picornavirus common cold by inhibitors of viral uncoating and attachment. *Annu. Rev. Microbiol.* **46,** 635.

Marsden, H. S., ed. (1992). Antiviral Therapies. *Semin. Virol.* **3**(1), 1.

Mills, J., and Corey, L., eds. (1993). "Antiviral Chemotherapy," 3rd Ed. Elsevier, New York.

Oxford, J., and Oberg, B. (1985). "Conquest of Viral Diseases." Elsevier, Amsterdam.

Sen, G. C., and Ransohoff, R. M. (1993). Interferon-induced antiviral actions and their regulation. *Adv. Virus Res.* **42,** 57.

Staeheli, P. (1990). Interferon induced proteins and the antiviral state. *Adv. Virus Res.* **38,** 147.

Part II

Viruses of Humans

Chapter 17

Parvoviridae

The smallest of all human viruses, the *Parvoviridae* contain so little genetic information in their unique single-stranded DNA molecule that they are able to replicate only in dividing cells (or, in the case of members of the *Dependovirus* genus, in the presence of a helper virus). This requirement for dividing cells accounts for their predilection for bone marrow, gut, and the developing fetus. Parvoviruses of cats, dogs, and mink cause panleukopenia and enteritis, while a rat parvovirus causes congenital malformation of the fetus. The human parvovirus B19, originally discovered fortuitously in the serum of asymptomatic blood donors, has subsequently turned out to be associated not only with a very common exanthematous disease of children, but also with aplastic crises in patients with chronic hemolytic anemia, as well as with hydrops fetalis. Another human parvovirus has been reported to cause gastroenteritis.

Properties of *Parvoviridae*

Only 20–25 nm in diameter, the *Parvoviridae* (Table 17-1) consist of a simple icosahedral shell surrounding a linear single-stranded DNA molecule of very limited coding potential (5 kb for human parvovirus B19 and 4.7 kb for human dependoviruses). The three-dimensional atomic structure of canine parvovirus has been solved by X-ray crystallography, revealing a capsid composed of one major and two minor polypeptides arranged to form 60 protein subunits. The virus is very stable, resisting 60°C for some hours and variation from pH 3 to 9.

The ssDNA parvovirus genome is of negative polarity, but some virions (up to 50% of them in the case of the genus *Dependovirus*) package a positive

Table 17-1
Properties of *Parvoviridae*

Three mammalian genera: *Parvovirus, Erythrovirus,* and *Dependovirus*
Icosahedral virion, 20–25 nm, capsid composed of 60 protein subunits
Linear minus sense ssDNA genome, 5 kb, palindromic hairpins at each end
Unique mechanism of DNA replication
Replicate in nucleus of cycling cells, using cellular enzymes
Relatively stable to heat (60°C) and pH (3–9)
Dependovirus usually requires helper virus; persists by integration

strand instead. The genomes of several human and animal parvoviruses have been sequenced, and all display long terminal palindromic sequences enabling each end of the molecule to fold back on itself to form a hairpin structure.

There are two genera of parvoviruses that contain viruses of humans. The genus *Dependovirus* was so named because the only known representatives, the human "adeno-associated viruses" (AAV), are usually found in association with an adenovirus which serves as a helper in their replication (Fig. 17-1A). The dependoviruses were, therefore, assumed to be defective and to be absolutely dependent on a helper adenovirus (or herpesvirus) for their replication, but this is now recognized to be not entirely true (see below). The genus *Parvovirus* (Fig. 17-1B) contains several viruses of animals and some that may cause diarrhea in humans; parvovirus B19 has now been allocated to a separate genus, *Erythrovirus*.

Viral Replication

Unlike those double-stranded DNA viruses (e.g., polyomavirus) that induce resting host cells to enter the S phase, the tiny ssDNA parvoviruses lack this capacity and hence can replicate only in dividing cells. Parvoviruses replicate in the nucleus; transcription and replication of the genome occur there, the nonstructural proteins accumulate there, and virions assemble there. There are no enzymes in the virion itself; a cellular DNA polymerase is used to transcribe the viral ssDNA into dsDNA, which then serves as the template for transcription of mRNA by cellular DNA-dependent RNA polymerase II. Nonstructural proteins are encoded by the left side of the genome and structural proteins by the right. Alternative splicing patterns give rise to several mRNA species which are translated into a greater number of different proteins than the limited coding potential of the short genome might suggest. The splicing program differs from one parvovirus to another. Certain nonstructural proteins serve to *trans*activate the viral promoter(s) and to downregulate transcription from certain cellular promoters.

The mechanism of replication of the genome is extraordinary. The palindromic 3′-terminal sequence serves as a self-primer for initiation of synthesis of a plus sense DNA strand to give a double-stranded replicative intermediate from which progeny minus strands are in turn transcribed, again using the hairpin structure as a primer. The detection in infected bone marrow cells of a dimeric form of the replicative intermediate, in which each of

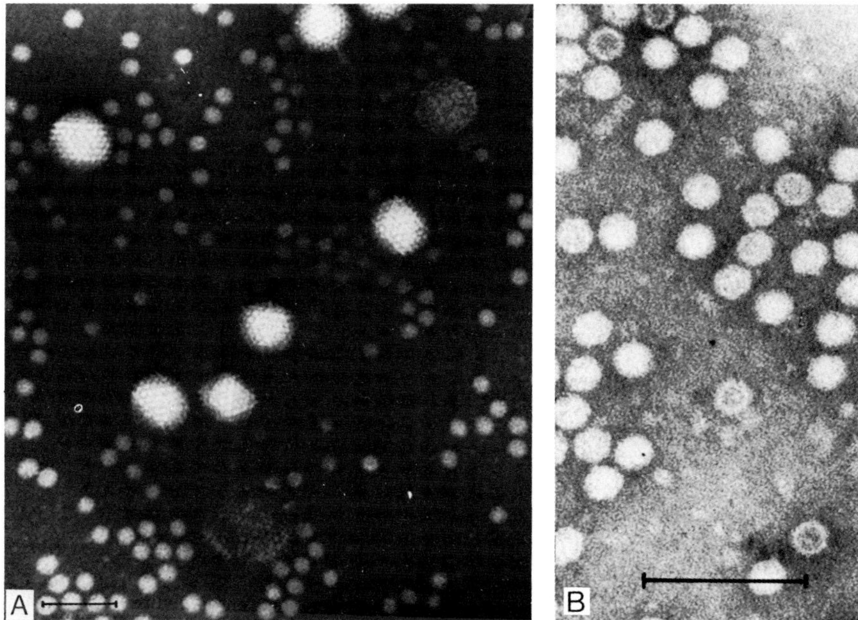

Fig. 17-1 *Parvoviridae.* Negatively stained preparations. (A) Human *Dependovirus* (adeno-associated virus) particles together with helper adenovirus. (B) *Parvovirus.* Bars, 100 nm. (A, Courtesy Dr. H. D. Mayor; B, courtesy Dr. E. L. Palmer.)

the two hydrogen-bonded strands consists of a complete plus strand and a complete minus strand covalently linked together to form a concatemer, has led to a model postulating that the growing strand replicates back on itself to produce a tetrameric form from which two complete plus strands and two complete minus strands are generated by endonuclease cleavage (Fig. 17-2).

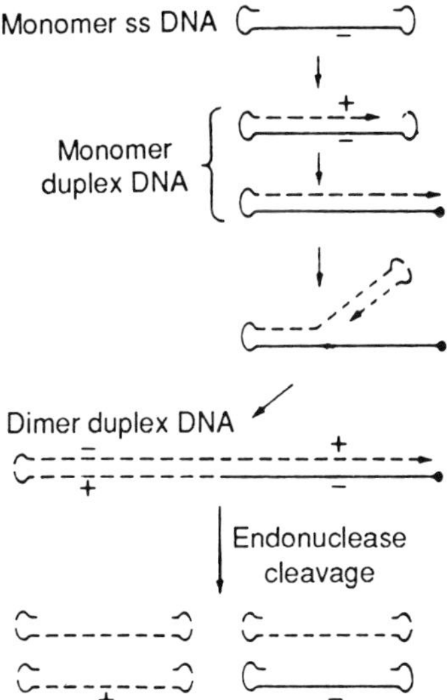

Fig. 17-2 Parvovirus DNA replication. See text for details of the postulated mechanism. [Modified from K. Ozawa, G. Kurtzman, and N. Young, *Science* **233,** 883 (1986).]

Parvovirus B19

A parvovirus designated B19, now allocated to the genus *Erythrovirus*, and originally isolated from the blood of healthy donors, has turned out to be the cause of a range of quite distinct clinical syndromes (Table 17-2). The commonest, erythema infectiosum, is a mild self-limited condition seen in normal children and adults. The rarer but more serious manifestations of infection occur only in patients with (1) an underlying congenital or acquired immunodeficiency, or (2) a requirement for accelerated erythropoiesis, for example, in chronic hemolytic anemia, or (3) pregnancy, in which both (1) and (2) also apply to some degree.

Pathogenesis and Immunity

Our understanding of the pathogenesis of parvovirus B19 infection was illuminated by a study in human volunteers—something that can be contemplated with equanimity only in the case of a relatively harmless virus such as this (Fig. 17-3). Just over a week after intranasal inoculation of seronegative volunteers, a short-lived but high-level viremia reached its peak and virus was shed for a few days from the throat. By about the tenth day, no erythroid precursors could be detected in bone marrow, nor reticulocytes in the blood; hemoglobin declined only very slightly, and there were no symptoms of anemia. Clinically, the volunteers displayed a biphasic illness. The first episode comprised fever, malaise, myalgia, and chills occurring around days 8–11, corresponding with the peak levels of virus in the bloodstream and the destruction of erythroblasts in the bone marrow. In contrast, the rash and arthralgia occurred between days 17 and 24, that is, after the viremia had disappeared but at a time when IgM antibodies had peaked and IgG had begun to rise, consistent with the hypothesis that the rash and arthritis are mediated by immune complexes, as in the case of Aleutian mink disease, another parvovirus infection.

These conclusions are supported by laboratory studies on chronic hemolytic anemia patients undergoing a naturally occurring parvovirus B19-induced transient aplastic crisis. Erythroblasts then reticulocytes vanish just a few days after viremia reaches its peak. In these patients, however, the aplastic crisis is profound because the average life of their circulating red cells is only 15–20 days (compared with 120 days in a normal individual) and, despite the efforts of the bone marrow to compensate with increased erythrocyte production, the hemoglobin level is already low.

Table 17-2
Human Infections Associated with *Parvoviridae*

Genus	Virus	Disease
Erythrovirus	B19	Erythema infectiosum (fifth disease) Arthritis (especially in young women) Aplastic crisis in chronic hemolytic anemia Chronic anemia in immunodeficiency syndromes Hydrops fetalis
Dependovirus	AAV (1–5)	Nil

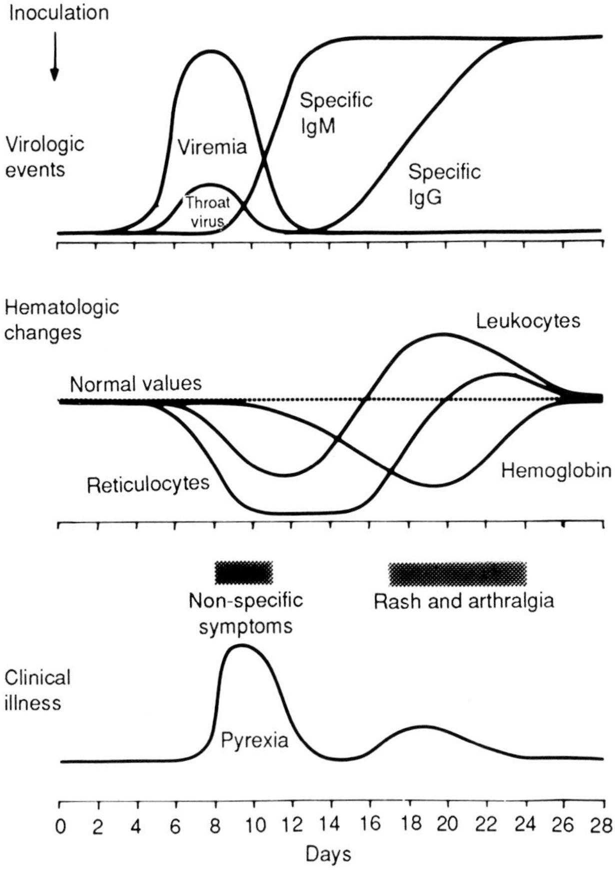

Fig. 17-3 Schematic representation of the virologic, hematologic, and clinical events in parvovirus B19 infection. (From A. J. Zuckerman, J. E. Banatvala, and J. R. Pattison, eds., "Principles and Practice of Clinical Virology," 2nd Ed. Wiley, Chichester, 1990.)

The selectivity of B19 for erythrocyte progenitors has been demonstrated directly in cultures of bone marrow cells; the formation of erythroid but not myeloid colonies is inhibited. The basis of the transient thrombocytopenia, lymphopenia, and neutropenia seen in some patients has yet to be established. However, there is evidence of inhibition of megakaryocyte colony formation *in vitro* without virus replication but with expression of a nonstructural viral protein which may be cytotoxic.

The immunologic basis of viral persistence and disease progression in chronic arthritis, and in chronic anemia in the immunocompromised as well as in hydrops fetalis, merits further investigation. It is also possible that other types of cells are permissive or semipermissive for parvovirus infection.

Clinical Features

Erythema Infectiosum

Erythema infectiosum, once known as "fifth disease," is an innocuous contagious exanthem of childhood which has been well known to pediatricians for over a century though its cause remained a mystery until the 1980s. An erythematous rubella-like rash on the face gives the child strikingly flushed cheeks. The rash, which also involves the limbs and trunk, fades rapidly within a day or two to develop the appearance of fine lace (Fig. 17-4). Though fleeting, the rash may reappear during the next few weeks or months following such stimuli as bathing or exposure to sunlight. As with rubella, arthralgia is seen occasionally in children and is a regular feature in adults, especially women, with a predilection for the peripheral joints of the hands, wrists, knees, and ankles. Indeed, polyarthralgia, sometimes without a rash, is often the dominant feature and may smolder on for weeks or months. Approximately 25–50% of infections are asymptomatic.

Transient Aplastic Crisis

Transient aplastic crisis is a temporary but potentially life-threatening complication of various forms of chronic hemolytic anemia, such as sickle cell anemia, thalassemia, or hereditary spherocytosis. It is now clear that the great majority of these episodes are directly attributable to infection with parvovirus B19. The patient presents with the pallor, weakness, and lethargy characteristic of severe anemia and is found to have suffered a sudden drop in hemoglobin associated with almost total disappearance of erythrocyte precursors from the bone marrow and of reticulocytes from the blood. There is usually no rash. Recovery generally occurs spontaneously within a week but is expedited by blood transfusion, which is sometimes lifesaving.

Other Complications of Parvovirus B19 Infection

Chronic anemia in immunodeficient patients has been observed when parvovirus B19 infects patients with acute leukemia on chemotherapy, AIDS patients, bone marrow transplant recipients, or children with congenital immune deficiency states. *Hydrops fetalis* is marked by generalized edema, which is presumed to result from severe anemia and congestive cardiac failure in the fetus. Current evidence suggests that a proportion of fetuses dying from this rather rare condition are delivered by women who became infected with parvovirus B19 a few weeks beforehand.

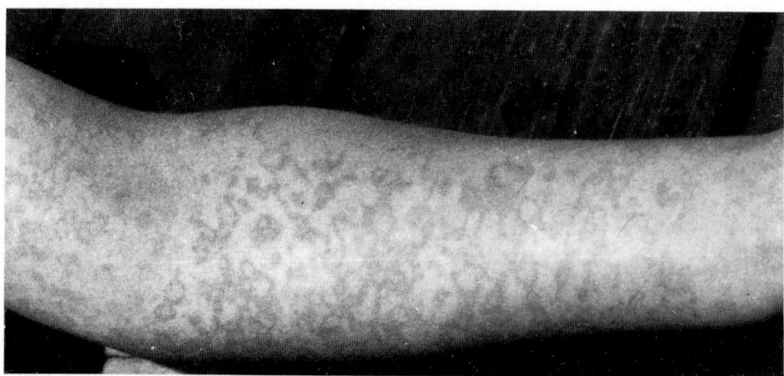

Fig. 17-4 Lacy appearance of the rash of erythema infectiosum, due to infection with parvovirus B19. (Courtesy Dr. M. Bucens.)

Laboratory Diagnosis

Parvovirus B19 has been cultured in cells from human bone marrow or fetal liver in the presence of erythropoietin and interleukin-3, but the system is not yet practicable for everyday diagnostic use. The virus has also been grown in a human megakaryocytic leukemia cell line, MB-02, in the presence of the cytokine GM-CSF, following erythroid differentiation as a result of treatment with erythropoietin.

Currently, the cornerstone of diagnosis of acute parvovirus infections is the demonstration of antibodies of the IgM class (or of a significant rise in IgG antibodies) by enzyme immunoassay (EIA) or radio immunoassay (RIA), using antigen produced by molecular cloning in mammalian or insect cells. The best way to detect virus is by nucleic acid hybridization or PCR for viral DNA, or EIA for viral antigen, in acute-phase serum, especially in aplastic crisis.

Histologically, giant pronormoblasts are found in bone marrow cells. The infected erythroid precursor cells display characteristic large eosinophilic intranuclear inclusions with surrounding margination of the nuclear chromatin. Electron microscopy reveals crystalline arrays of virions in the nucleus. *In situ* hybridization using cloned parvovirus DNA is useful for demonstrating the presence of the genome in acute or chronic infections (including hydrops fetalis).

Epidemiology

Parvovirus B19 is ubiquitous, common, and highly contagious. It is present year-round with a tendency to produce spring epidemics among schoolchildren 4–10 years old. Because parvovirus B19 is readily transmitted by respiratory secretions and close contact, attack rates among susceptible individuals during such epidemics within day-care centers and schools are about 25%, and may reach 50% in household contacts. Over 50% of adults are seropositive and immune. Transmission probably occurs during the incubation period of erythema infectiosum, the patient being no longer infectious by the time the rash appears (Fig. 17-3). In contrast, chronic hemolytic anemia patients are infectious for up to 1 week after onset of an aplastic crisis, and immunocompromised patients with chronic B19 anemia may excrete virus for months or years.

A second route of transmission is transplacental. About 30% of primary maternal infections lead to infection of the fetus, but they usually do no harm. Prospective studies suggest that less than 10% of primary maternal infections cause fetal death and rarely if ever congenital malformations. Transmission can also occur via blood transfusion, but so rarely that routine screening of blood is not justified. Factor VIII administered to hemophiliacs is a greater problem as the virus is sufficiently heat-stable to survive in clotting factor concentrates.

Treatment and Control

Erythema infectiosum requires no treatment, but aplastic crises in chronic hemolytic anemia can be life-threatening, often requiring supportive care and sometimes blood transfusion. Intravenous administration of normal human immunoglobulin has been shown to be beneficial in the treatment of severe persistent anemia in immunocompromised patients.

There is a substantial risk of transmission of parvovirus B19 to susceptible children or staff in schools and day-care centers, or hospitals during nosocomial outbreaks or nursing of chronic hemolytic anemia patients with an aplastic crisis. Hence, pregnant nonimmune women, immunocompromised individuals, or those with chronic hemolytic anemia should be appraised of the potential risks.

Enteric Parvoviruses

A confusing variety of "small round viruses" have been visualized by electron microscopy in feces from normal and ill people. Among those resembling parvovirus, some were from people with gastroenteritis acquired in a common-source outbreak, such as consumption of raw shellfish. None of these "fecal" parvoviruses has been shown to bear any serological relationship to B19 (or AAV), although nucleotide sequencing of their DNA reveals striking homology with B19, particularly in the left side of the genome which encodes the nonstructural proteins. The relationship between the two and the clinical importance of fecal parvoviruses need to be established.

Dependoviruses

The five serotypes of adeno-associated virus (AAV) found in humans were so named because they were first isolated from the throat or feces of humans concurrently infected with an adenovirus (Fig. 17-1A). It was believed that AAV were defective viruses requiring an adenovirus as a helper virus, hence the generic name *Dependovirus*. Indeed it was clearly demonstrated that, in the absence of its helper, a tandemly repeated double-stranded form of the AAV genome becomes integrated into the cellular genome where it persists indefinitely until rescued by subsequent superinfection with adenovirus; moreover, various adenovirus (or herpesvirus) "early" proteins transactivate AAV gene expression and provide other functions that facilitate the replication of AAV by modifying certain cellular activities. Very probably, this is how AAV, despite its genetic limitations, survives in nature.

Now, however, it is apparent that the so-called dependoviruses are not exclusively dependent on a helper virus, in that they are also capable of replicating autonomously in cells treated with any one of several types of chemical agents which either synchronize cell division or otherwise provide suitable conditions for the growth of these extremely fastidious agents. Thus, the dependoviruses may not be as different from the "autonomous" parvoviruses as used to be thought.

There is no evidence that the dependoviruses cause any disease. They have, however, approached the ultimate in successful parasitism.

Further Reading

Anderson, L. J. (1990). Human parvoviruses. *J. Infect. Dis.* **161,** 603.

Anderson, M. J. (1987). Parvoviruses as agents of human disease. *Prog. Med. Virol.* **34,** 55.

Berns, K. I. (1990). Parvovirus replication. *Microbiol. Rev.* **54,** 316.

Berns, K. I. (1990). Parvoviridae and their replication. *In* "Fields Virology" (B. N. Fields, D. M. Knipe, R. M. Chanock, M. S. Hirsch, J. L. Melnick, T. P. Monath, and B. Roizman, eds.), 2nd Ed. p. 1743. Raven, New York.

Cotmore, S. F., and Tattersall, P. (1987). The autonomously replicating parvoviruses of vertebrates. *Adv. Virus Res.* **33,** 91.

Pattison, J. R., ed. (1988). "Parvoviruses and Human Disease." CRC Press, Boca Raton, Florida.

Tijssen, P., ed. (1990). "Handbook of Parvoviruses," Vols. 1 and 2. CRC Press, Boca Raton, Florida.

Young, N. (1988). Hematologic and hematopoietic consequences of B19 parvovirus infection. *Semin. Hematol.* **25,** 159.

Chapter 18

Papovaviridae

The main interest in papovaviruses lies in the probability that they cause cancer in humans, as certain members certainly do in other animal species. It has long been known that viruses of the genus *Papillomavirus* cause skin warts. More recently, however, it has become clear that papillomaviruses are a frequent cause of sexually transmitted disease and are strongly linked to carcinoma of the cervix, one of the commonest cancers of women. Certain members of the genus *Polyomavirus*, on the other hand, infect most people subclinically, persist for life, and are frequently reactivated by immunosuppression.

Properties of *Papovaviridae*

The family *Papovaviridae* contains two genera, *Papillomavirus* and *Polyomavirus*, the former having a larger virion and a larger genome (see Table 18-1). The virion is a small naked icosahedron (*Polyomavirus*, 45 nm; *Papillomavirus*, 55 nm) with 72 capsomers (Fig. 18-1). The genome consists of a covalently closed supercoiled circular dsDNA molecule of 5 kbp (*Polyomavirus*) or 8 kbp (*Papillomavirus*), shown in Fig. 18-2.

Papillomaviruses

Host species-specific papillomaviruses have been found in many animals and birds. Most cause benign papillomas in the skin or mucous membranes. There are numerous human papillomaviruses (HPV), most displaying a predilection for a particular site in the body. Some have oncogenic potential.

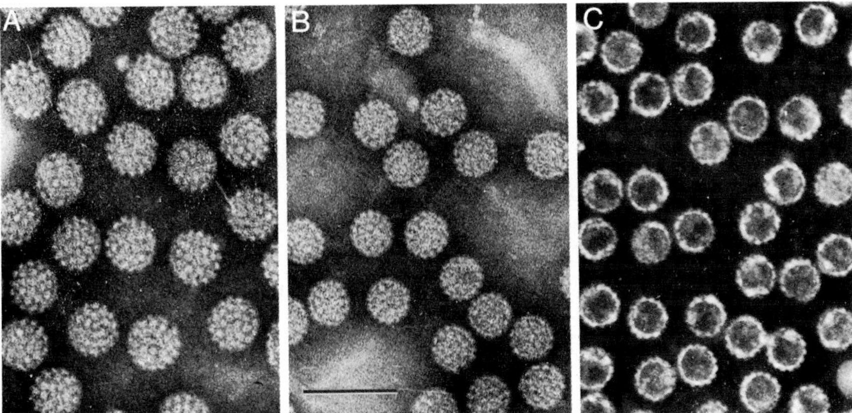

Fig. 18-1 *Papovaviridae*. (A) *Papillomavirus*. (B) *Polyomavirus*. (C) *Polyomavirus*, empty virions. Bar, 100 nm. (Courtesy Dr. E. A. Follett.)

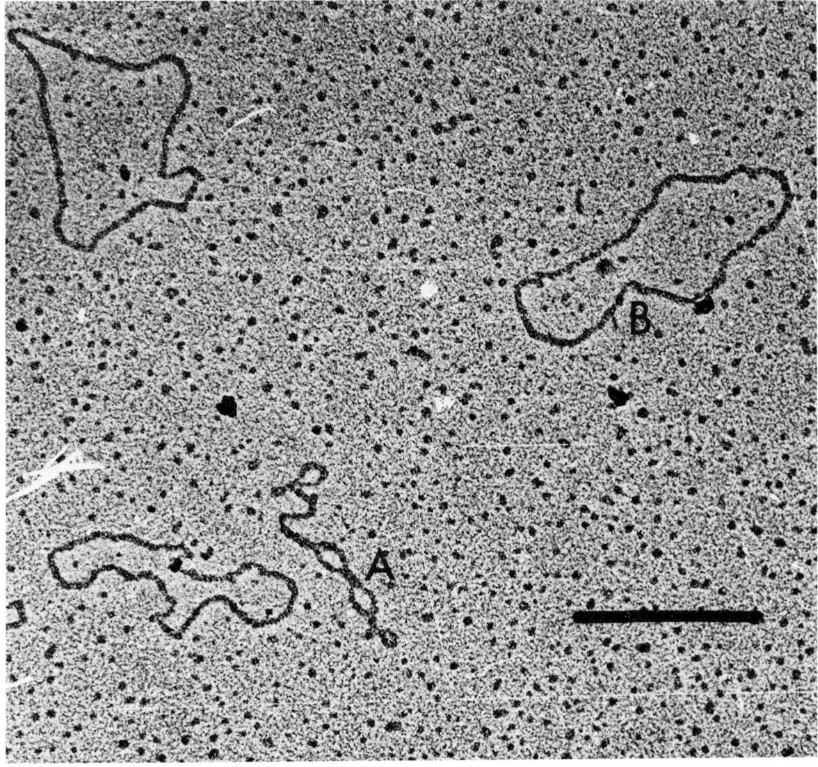

Fig. 18-2 DNA molecules extracted from the papovavirus SV40 (bar, 0.5 μm). Molecules of SV40 exist in two major forms. When the DNA is isolated from the virus particles, most occurs in the configuration shown in (A) as double-stranded closed-circular molecules containing superhelical twists. If one of the DNA strands is broken, the superhelical twists are relieved and the molecule assumes a relaxed circular configuration (B). (Courtesy Dr. P. Sharp.)

Properties of Papillomaviruses

Paradoxically, we already know a great deal about the DNA of papillomaviruses despite our inability to grow them in conventional cell cultures. The small circular dsDNA genome readily lends itself to cloning, thus ample quantities are available for sequencing, for development of diagnostic probes, and for exploration of the putative role of papillomaviruses as carcinogens. Incidentally, papillomavirus DNA itself is proving to be a cloning vector of great potential (see Chapter 4); bovine papillomavirus type 1 (BPV-1) DNA in particular has been exploited as a eukaryotic cloning vector for the manufacture of products such as interferons. The genomes of several human papillomaviruses have been sequenced and found to contain three distinct regions: (1) an upstream regulatory region that controls transcription and replication; (2) at least seven early genes encoding proteins required for transcription, DNA replication, and cell transformation; and (3) two late genes encoding the major and minor capsid proteins. All the open reading frames are situated on one strand.

Classification

As human papillomaviruses have been grown in cultured cells only recently and with great difficulty, neutralization tests have been unavailable, hence classification is currently based on differences in the genome itself; the types so defined are "genotypes" rather than serotypes. Conventionally, an HPV strain has been accepted as a new type if its DNA shows less than 50% duplex formation with all known types in a liquid reassociation kinetics reaction under stringent conditions. Progressively, however, this is likely to be replaced by comparison of complete nucleotide sequences. At the time of writing over 60 types of HPV are recognized, and the total is rising rapidly. These types have been assigned to about a dozen groups reflecting the degree of homology of their nucleotide sequences.

Viral Replication

Following entry and uncoating, the viral genome migrates to the nucleus where transcription, DNA replication, and virion assembly occur. The early genes are copied from a single promoter and the transcripts subjected to differential splicing to generate the mRNAs for the seven early proteins. These include regulatory proteins, some with *trans*activating properties which derepress the genes for certain cellular enzymes and stimulate cellular DNA synthesis. Replication of the viral genome is initiated by binding of nonstructural proteins E1 and E2 to the unique origin of replication on the viral DNA. In the replicating cells comprising the basal layer of epidermis, only early HPV genes are expressed, but this suffices to augment cellular proliferation (hyperplasia); the viral genome replicates only very slowly as an autonomous plasmid in the nuclei of these cells. However, in the terminally differentiated cells that comprise the outer layers of the epithelium, full expression of the viral genome and vegetative DNA synthesis occur. Late genes are transcribed from a second promoter; these encode the structural proteins, L1 and L2, which, after various posttranslational modifications, are directed to the nucleus via their nuclear localization signals, to be assembled into virions (Table 18-1).

Table 18-1
Properties of *Papovaviridae*

Two genera: *Polyomavirus* and *Papillomavirus*; potentially oncogenic
Nonenveloped icosahedral virion, 45[a] or 55[b] nm in diameter
Circular supercoiled dsDNA genome, 5[a] or 8[b] kbp
Transcription of overlapping open reading frames, splicing, DNA replicaton all occur in nucleus, using cellular enzymes; early gene products induce cellular enzymes and transform cell; papillomavirus late (capsid) proteins made only in terminally differentiated epithelial cells
Genome persists, integrated[a] or as episome[b]

[a] *Polyomavirus*.
[b] *Papillomavirus*.

Pathogenesis and Immunity

Papillomaviruses are not only host species-specific but also greatly restricted in their tissue tropism, multiplying only in epithelial cells of the skin or certain mucous membranes, with different HPV types displaying preferences for different sites in the body. Broadly speaking, they fall into two groups: cutaneous types (infecting skin) and mucosal types (infecting the genital tract and sometimes the respiratory tract, oral cavity, or conjunctiva). Furthermore, production of virions is absolutely dependent on the state of differentiation in the various layers of squamous epithelium.

Warts have a long incubation period of up to 2 years. Virus penetrates the skin through an abrasion and infects the basal cell layer. In these relatively undifferentiated replicating cells, only early viral genes are expressed and limited viral DNA replication maintains the genome as a stable nuclear plasmid (episome). Expression of early viral genes stimulates proliferation of the basal cells; the resulting hyperplasia leads to acanthosis (thickening) and generally to a protruding papilloma, after an incubation period of several months. Capsid proteins are synthesized and virions produced only in the terminally differentiated keratinocytes in the outer layers of the epithelium which are producing keratin but no longer dividing. Hyperkeratosis is a prominent feature of skin warts. Histologically, the stratum granulosum is characterized by the presence of koilocytes: large cells with cytoplasmic vacuolation and nuclear distortion.

Warts tend to disappear, usually synchronously, within a couple of years. This sudden regression is not determined by antibody titer and is generally ascribed to the T-cell-mediated immune response, but it is not clear what triggers it or, conversely, what delays it for so long.

Although the papillomas induced by HPV on the external genitalia are fundamentally similar to those described above, cervical lesions display some important differences. The virus enters during sexual intercourse, possibly via a minor abrasion in the vicinity of the squamocolumnar border, where cells are proliferating. After an incubation period of one or more (average 3) months a flat condyloma develops. Infections with certain HPV types may progress over a period of years through the various stages of cervical intraepithelial neoplasia (CIN) to invasive squamous carcinoma. Papanicolaou smears of cervical dysplasia reveal the characteristic koilocytes. The HPV genome may persist for years, as a nonintegrated nuclear episome, not only within the lesion but also in histologically normal mucous membrane for up to a few centimeters around it. In contrast, cervical carcinomas harbor the genome of HPV, most commonly type 16, 18, 31, 33, or 35, in the form of incomplete copies of the viral DNA integrated at random sites within host cell chromosomes.

Because HPV cannot yet be grown in conventional monolayer cell cultures our knowledge of the mechanism of carcinogenesis has been acquired principally from *in vitro* transformation of cells transfected with molecularly cloned HPV DNA, or from studies in athymic or transgenic mice. The oncogenicity of the high-risk genital HPV types 16 and 18 has been ascribed to oncogenes E6 and E7 which are expressed from integrated defective HPV DNA in over 90% of all cervical carcinomas and encode proteins that bind to normal cellular tumor suppressor gene products. Full expression of malignancy may require the subsequent activation of cellular oncogenes as a result of destabilization of the host DNA. This along with the relationship of dermatotropic types 5 and 8 to squamous cell carcinomas of skin is discussed in Chapter 11.

Clinical Features

Certain HPV types display a predilection for the skin and others for mucous membranes, and this is a convenient way of classifying the several clinical presentations of infection (Table 18-2). Certain lesions caused by particular types have the potential to progress to cancer, especially in the cervix.

Genital Infections

Many infections of the genital tract are subclinical, but there are several important clinical presentations. *Condyloma acuminatum, anogenital warts,* and *exophytic warts* are the names given to the large moist pedunculated excres-

Table 18-2
Diseases Caused by Human Papillomaviruses

Site	Clinical presentation	Types[a]
Genital tract	Condyloma acuminatum	**6, 11,** 42, 43, 44, 55, and others
	Genital malignancies	**16, 18,** 31, 33, 35, 39, 45, 51, 52, 56
Respiratory tract	Respiratory papillomas	**6, 11**
Mouth	Focal epithelial hyperplasia	**13, 32**
	Oral papillomas	**6,** 7, **11,** 16, 32
Skin	Plantar wart	**1,** 2, 4
	Common wart	**2, 4,** and others
	Flat wart	**3, 10,** 28, 41
	Butchers' warts	**7**
	Epidermodysplasia verruciformis[b]	**5, 8,** 9, 12, 14, 15, 17, 19–25, 36, 46, 47

[a] Common types in bold type.
[b] Types **5, 8,** and less commonly 17, 20, and 47 are the principal types so far associated with malignant change in epidermodysplasia verruciformis.

cences of soft papillomas found on the external genitalia, perineum, vaginal introitus, penis, or anus, caused commonly by HPV types 6 and 11 (Fig. 18-3). *Condyloma planum (flat wart)* is the more usual presentation in the cervix, with types 6 and 11 again being the commonest. *Cervical carcinoma* may develop 20–50 years after infection with certain mucosal HPV types. There is a slow progression through three stages of lesions variously called *cervical dysplasia* or *cervical intraepithelial neoplasia* (CIN). The flat condyloma is sometimes classified as CIN-1; CIN-3 is often designated *carcinoma in situ.* The fully malignant, invasive carcinoma is usually but not always of the squamous type. Over 90% of all cancers of the cervix contain HPV DNA, usually of type 16 or 18. The same types are associated with carcinomas of the vulva, vagina, penis, and anus.

Respiratory Papillomatosis

Much more rarely the genital HPV types infect the respiratory tract, usually in young children but also sometimes in young adults. Classically, the lesions begin in the larynx, but they grow copiously, obstructing the airway, and tend to spread also to other parts of the respiratory tract. Malignant change occurred occasionally in the days when children with laryngeal papillomatosis were treated with irradiation. Papillomas caused by these genital types are also seen occasionally on the conjunctiva.

Oral Infections

Focal epithelial hyperplasia, associated almost exclusively with HPV types 13 and 32, is a condition characterized by multiple nodular lesions in the mouth and is particularly prevalent in Eskimos and in South and Central American Indians. *Oral papillomas* of the more conventional kind can be caused by the

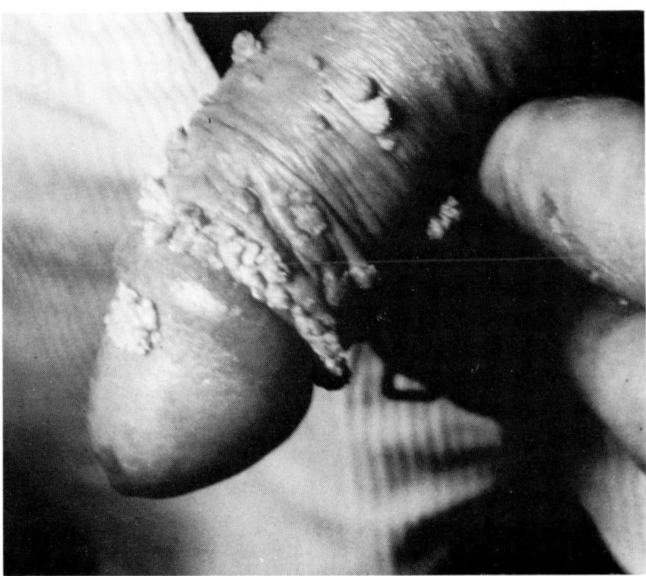

Fig. 18-3 Genital warts (condyloma acuminata). (Courtesy Dr. D. Bradford.)

sexually transmitted types 6, 11, and 16. Common warts on the lips are usually type 2. HPV-16 DNA has also been reported in a minority of oral carcinomas.

Skin Warts

Perhaps 10% of all schoolchildren and young adults experience a crop of warts on the skin, probably acquired in the course of body-contact recreational activities. Generally, warts regress within 2 years. *Common warts,* seen on prominent regions subject to abrasion such as hands and knees, are raised papillomas with a rough surface, often caused by type 2 or 4. Type 7 is the agent of "butchers' warts," an occupational disease of meat-handlers. *Flat warts,* sometimes called *plane warts,* are smaller, flatter, smoother, and more numerous, seen especially on the arms, face, and knees of youngsters. Types 3 and 10 are often involved. *Plantar warts* are painful deep endophytic warts found on the weight-bearing regions of the heel and sole of the foot; palmar warts are similar. Type 1 is the major etiological agent.

Epidermodysplasia verruciformis is a rare condition seen in people with a particular autosomal recessive hereditary cell-mediated immunodeficiency which has not yet been fully defined. The infection, acquired in childhood but persisting for life, is characterized by numerous warts widely disseminated over the skin. The lesions are of two varieties: (1) flat warts, commonly caused by types 3 and 10, as in normal children and (2) reddish-brown macular scaly patches from which can be isolated any of nearly 20 rare HPV types found almost exclusively in these patients (presumably they must cause subclinical infections in normal people). *Squamous cell carcinoma* (SCC) arises in about one-third of all epidermodysplasia verruciformis patients, after many years, in one or often several of the macular lesions situated on areas of the skin exposed to sunlight, which is obviously therefore a critical cofactor in the genesis of this malignancy. The tumors, usually carrying the genome of HPV type 5 or 8, are often slow-growing *in situ* carcinomas but may be invasive SCCs which metastasize. HPV-5 or HPV-8 DNA has also been detected in SCCs on exposed areas of the skin of a small number of chronically immunosuppressed recipients of renal allografts as well as from some apparently immunocompetent individuals, but an etiologic association has yet to be proved.

Laboratory Diagnosis

Clinically, the diagnosis of skin warts and condyloma acuminata on the external genitalia generally poses no great problems, but histology is required whenever malignancy is a possibility. Cervical condylomas, visualized by colposcopy, are often flat and indistinguishable from cervical intraepithelial neoplasia. The Papanicolaou smear is widely used for routine screening of women for premalignant and malignant changes in the cervix.

Although human papillomavirus has recently been successfully grown in organ cultures of floating rafts of infected skin treated with the phorbol ester TPA to increase keratinocyte differentiation, isolation of virus is not an appropriate approach to laboratory diagnosis of papillomavirus infections. Because only terminally differentiated nonreplicating cells of squamous epithelium are fully permissive for viral replication, virions and capsid proteins are demonstrable only in the outer keratinized layers of skin warts. However, the HPV genome can persist for years in basal cells, as an episome in benign papillomas or premalignant cervical dysplasia, or integrated in cancer cells. Thus, to be confident of not overlooking HPV infections, the only generally applicable diagnostic approach is to probe for viral DNA or for particular viral genes (see Chapter 12 for detailed discussion). Nucleic acid hybridization is currently employed mainly for the investigation of condyloma and cervical carcinoma.

Nucleic Acid Hybridization

The DNAs of most HPV types have been cloned and can be labeled with a radioactive isotope, such as ^{32}P, ^{35}S, or ^{3}H, or with a nonradioactive label such as biotin. Furthermore, exon-specific probes can be constructed to search for individual HPV genes or for the corresponding mRNAs, for example, DNA probes specific for the L1 ORF (encoding the major capsid protein) of the common HPV types, or for the oncogenes E6 and E7. Often, the first approach will be to screen for conserved papillomavirus DNA sequences by using as a probe either a single type or an appropriate cocktail of a few distantly related types, under "relaxed" (low stringency) conditions. Determination of the type requires the use of particular HPV probes, at high stringency.

Southern blot hybridization has served as the gold standard for identification of HPV DNA in human tissue; it offers the advantages of high specificity and high sensitivity, being capable of detecting one copy of the HPV genome per cell under conditions of high stringency (see Fig. 12-7). *Dot-blot hybridization* is simpler, but its sensitivity is lower. *In situ hybridization* has been widely employed by pathologists to screen numbers of gynecologic specimens, including cervical "touch smears," frozen sections, or even formalin-fixed sections, on slides; sensitivity is low, but the method presents the advantage that the autoradiograph (or immunoperoxidase cytochemistry) reveals the topographic location of the viral genome in particular layers of the epidermis or the tumor.

Gene amplification by the polymerase chain reaction (PCR) has emerged as by far the most sensitive method of detecting low numbers of genome copies and can detect low-level latent infection in clinically and histologically normal tissue, as well as the larger copy numbers present in papillomas, dysplasias, and carcinomas, with a sensitivity several orders of magnitude higher than Southern blotting. However, false positives resulting from contamination or the use of unsuitable primers plagued many of the early PCR studies of genital HPV infections, flooding the literature with misleading results. The PCR amplification should be conducted by experts and interpreted with caution. We need to discover the implications of asymptomatic persistence of HPV DNA in low copy number in healthy tissue. Although identification of a carcinogenic HPV type has prognostic value, it is important to remember that the great majority of women infected with HPV types 16 or 18 will never develop cervical cancer. Telltale histologic change in a Papanicolaou smear is still a better indicator of impending malignant change than is identification of a highly oncogenic type of HPV.

Epidemiology

Transmission of skin warts occurs mainly in school-age children via direct contact through abrasions, with the possibility of subsequent spread by scratching (autoinoculation). Plantar warts are readily picked up from the wet floors of public swimming pools and bathrooms. Genital warts, on the other hand, are spread by sexual intercourse; not surprisingly their incidence skyrocketed in parallel with the sexual revolution of the late 1960s and 1970s. Some cases of oral or respiratory papillomatosis may also be sexual in origin, but most such infections in young children are assumed to be acquired during passage of the baby through an infected birth canal. Fortunately, the efficiency of transmission in this manner must be very low, considering the rarity of laryngeal papillomatosis vis-à-vis the high frequency of cervical infection.

Treatment and Prevention

The fact that skin warts regress spontaneously encourages the perpetuation of mythical cures ranging from hypnosis to Tom Sawyer's infallible dead-cat-in-the-cemetery-at-midnight cure. A comparable rate of success can be assured by letting nature take its course. However, skin warts can be removed by cryotherapy or caustic chemicals, laryngeal papillomas by laser, external genital warts by cryotherapy, podophyllin, laser, or diathermy, and cervical dysplasia by laser or diathermy; invasive carcinoma requires surgery. As discussed in Chapter 16, it has been reported that interferon α or β injected intramuscularly and/or into the lesion itself, may be effective in causing genital or laryngeal papillomas to regress temporarily, but recent studies do not support these claims.

Experimental vaccines have been produced by recombinant DNA technology. For example, infection with live vaccinia recombinants incorporating the genes for the HPV-16 capsid proteins L1 and L2 yields particles resembling empty HPV capsids, which might be expected to induce synthesis of neutralizing antibodies. Similar vaccinia recombinants incorporating the HPV-16 E6/E7 genes are undergoing human trials in the hope that they may be used therapeutically to boost the cytotoxic T-cell response to E6/E7 peptides presented on the surface of cervical intraepithelial neoplasia or invasive carcinoma cells.

Polyomaviruses

The prototype after which the genus *Polyomavirus* was named is the polyoma virus of mice. Though causing only harmless inapparent infections in mice when spread by natural routes, the virus induces many different types of malignant tumors ("polyomas") when artificially injected into infant rodents, such as hamsters. Another *Polyomavirus*, simian virus 40 (SV40), infects monkeys subclinically but also induces tumors after inoculation into baby rodents. During the 1960s and 1970s these two viruses became the principal models for the biochemical investigation of virus-induced malignancy. The newly discovered techniques of molecular biology were brought to bear on the expression of the integrated viral genome and associated cellular changes in cultured fibroblasts transformed by these viruses *in vitro* (see Chapters 3 and 11, and Figs. 3-6 and 11-1). Subsequently two human polyomaviruses, designated BK and JC, were discovered. BK virus was recovered from the urine of a renal transplant recipient, and JC virus from the brain of a patient with a rare demyelinating condition, progressive multifocal leukoencephalopathy (PML). Like SV40, BK and JC are oncogenic in newborn hamsters and transform mammalian cells *in vitro,* but there is no evidence that they cause human cancer. Both are ubiquitous in humans, producing inapparent infections that persist for many years in the urinary tract, and may be reactivated by immunosuppression.

BK Polyomavirus

BK virus infects most children before the age of 10, often subclinically but sometimes associated with mild upper respiratory symptoms, suggesting that transmission may occur via the respiratory route. The viral genome persists for life in the kidney without any apparent ill effects. Reactivation occurs during the last trimester of about 3% of pregnancies, causing asymptomatic shedding of virus intermittently in urine. Following immunosuppression, for example, in kidney transplantation, reactivation is demonstrable in about a third of all patients, and urinary shedding continues for days to months, with no detectable loss of renal function.

JC Polyomavirus

JC virus has a similar natural history, although primary infection may occur somewhat later in childhood and only about 75% of the population has antibody. Again, lifelong persistence is established in the kidney, and virus is shed in urine sporadically throughout life and more frequently during pregnancy or immunosuppression. Unlike BK virus, however, JC virus causes a lethal disease, progressive multifocal leukoencephalopathy (PML). PML is a rare subacute demyelinating disease of the CNS which is seen mainly as a complication of advanced disseminated malignant conditions such as Hodgkin's disease or chronic lymphocytic leukemia, but also in primary or secondary immunodeficiency syndromes, especially AIDS, or following immunosuppression for organ transplantation (Fig. 18-4). The target cell is the oligodendrocyte, in which the virus undergoes a lytic productive infection; neurons are unaffected. Histologically, the disease is characterized by multiple foci of demyelination in the brain, accompanied by proliferation of giant bizarre astrocytes. The surrounding oligodendrocytes are enlarged, with swollen nuclei occupied by a prominent inclusion body, which in fact contains a crystalline aggregate of thousands of virions.

The disease PML represents a reactivation of long-standing persistent infection of the kidney, and perhaps also of the brain. Previously rare and confined largely to the elderly, it is now seen principally in AIDS patients. Indeed, the HIV-1 Tat protein has been shown to *trans*activate transcription of the late JC viral genes.

Two types of JC virus have recently been defined. JC virus isolated from

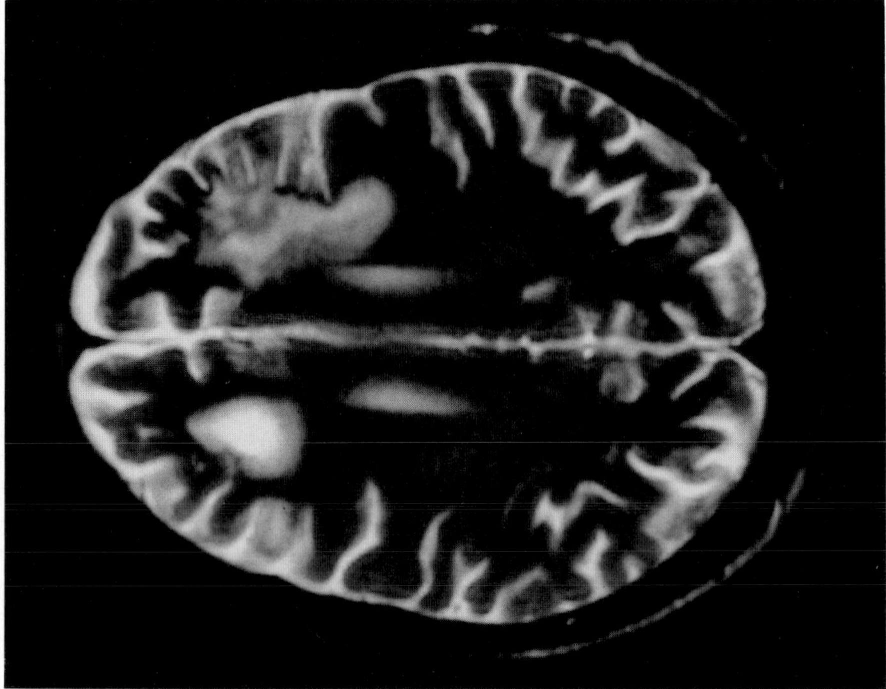

Fig. 18-4 Magnetic resonance image scan showing lesions of progressive multifocal leukoencephalopathy (PML) in the frontal lobes of a patient with AIDS. (Courtesy of Dr. R. L. Doherty, Fairfield Hospital for Infectious Diseases, Melbourne.)

the brain of PML patients usually differs from the "archetype" found in the urine of asymptomatic carriers by extensive deletions and duplications in the nucleotide sequences within the promoter/enhancer region of the genome.

Laboratory Diagnosis

BK virus can be isolated from urine in cultured human diploid fibroblasts, or JC virus from urine or brain in human fetal glial cells, and the two viruses can be distinguished by hemagglutination inhibition. Much simpler, however, is direct detection of antigen in urine by enzyme immunoassay, or direct detection of the viral genome by PCR and nucleic acid hybridization (e.g., Southern blotting). Following brain biopsy or autopsy, JC viral DNA can be demonstrated by *in situ* hybridization, JC antigens by immunofluorescence, and virions by electron microscopy.

Further Reading

Arthur, R. R., and Shah, K. V. (1989). The occurrence and significance of papovaviruses BK and JC in the urine. *Prog. Med. Virol.* **36,** 42.

Galloway, D. A., and McDougall, J. K. (1989). Human papillomaviruses and carcinomas. *Adv. Virus Res.* **37,** 126.

Howley, P. M. (1990). Papillomavirinae and their replication. *In* "Fields Virology" (B. N. Fields, D. M. Knipe, R. M. Chanock, M. S. Hirsch, J. L. Melnick, T. P. Monath, and B. Roizman, eds.), 2nd Ed., p. 1625. Raven, New York.

Howley, P. M., and Broker, T. R., eds. (1990). Papillomaviruses. *UCLA Symp. Mol. Cell. Biol.* **124.**

Klein, G., ed. (1989). *Adv. Viral Oncol.* **8.**

Koutsky, L. A., Galloway, D. A., and Holmes, K. K. (1988). Epidemiology of genital human papillomavirus infection. *Epidemiol. Rev.* **10,** 122.

Munoz, N., Bosch, F. X., and Jensen, O. M., eds. (1989). "Human Papillomaviruses and Cervical Cancer." Oxford Univ. Press, Oxford.

Pfister, H., ed. (1990). "Papillomaviruses and Human Cancer." CRC Press, Boca Raton, Florida.

Reeves, W. C., Rawls, W. E., and Brinton, L. A. (1989). Epidemiology of genital papillomaviruses and cervical cancer. *Rev. Infec. Dis.* **11,** 426.

Roman, A., and Fife, K. H. (1989). Human papillomaviruses: Are we ready to type? *Clin. Microbiol. Rev.* **2,** 166.

Salzman, N. P., ed. (1986). "The Papovaviridae, Volume 1: The Polyomaviruses." Plenum, New York.

Salzman, N. P., and Howley, P. M., eds. (1987). "The Papovaviridae, Volume 2: The Papillomaviruses." Plenum, New York.

Shah, K. V. (1990). Polyomaviruses. *In* "Fields Virology" (B. N. Fields, D. M, Knipe. R. M. Chanock, M. S. Hirsch, J. L. Melnick, T. P. Monath, and B. Roizman, eds.), 2nd Ed., p. 1609. Raven, New York.

Shah, K. V., and Howley, P. M. (1990). Papillomaviruses. *In* "Fields Virology" (B. N. Fields, D. M. Knipe, R. M. Chanock, M. S. Hirsch, J. L. Melnick, T. P. Monath, and B. Roizman, eds.), 2nd Ed., p. 1651. Raven, New York.

Syrjänen, K., Gissmann, L., and Ross, L. G., eds. (1987). "Papillomaviruses and Human Disease." Springer-Verlag, Berlin and New York.

zur Hausen, H. (1991). Human papillomaviruses in the pathogenesis of anogenital cancer. *Virology* **184,** 9.

zur Hausen, H. (1994). Papillomaviruses. *Annu. Rev. Microbiol.* **48,** in press.

Chapter 19

Adenoviridae

In 1953 Rowe and colleagues, having observed that certain explant cultures of human adenoids spontaneously degenerated, isolated a new infectious agent which became known as "adenovirus." Before long, it became evident not only that adenoviruses may persist for years as latent infections of lymphoid tissues, but also that they are a significant cause of disease in the respiratory tract and the eye. Subsequently, certain serotypes were found to infect the genitourinary tract, whereas more recently discovered adenoviruses have been associated with gastroenteritis or with infections of immunocompromised patients (e.g., in AIDS).

Following the discovery that certain human adenoviruses produce malignant tumors in baby rodents, some of the world's most capable molecular biologists turned their attention to the biochemistry of adenovirus replication and oncogenesis. Although it later became clear that adenoviruses play no role in human cancer, the spin-off from this research had a major impact on our understanding of the expression of mammalian as well as viral genes. To cite just three examples, adenovirologists discovered the phenomenon of splicing of RNA transcripts, were the first to describe the role of proteins as primers for initiation of DNA replication, and pioneered techniques for elucidating the regulation of transcription programs in mammalian cells.

Properties of *Adenoviridae*

The family *Adenoviridae* is defined by the properties listed in Table 19-1. The virion is a perfect icosahedron (Fig. 19-1B and 1-1A). The outer capsid, about

Table 19-1
Properties of *Adenoviridae*

Genus *Mastadenovirus* comprises all 47 human serotypes
Icosahedral virion, 80–90 nm, 252 capsomers, 12 fibers at vertices, 12 structural proteins
Linear, dsDNA genome, 36–38 kbp, inverted terminal repeats, protein primer at each 5′ terminus
Complex program of transcription from seven early, intermediate, and late promoters; splicing
Transcription, DNA replication, and virion assembly occur in nucleus

80 nm in diameter, is composed of two main types of capsomers: 240 "hexons" make up the 20 triangular faces of the icosahedron, while 12 "pentons" form the 12 vertices (Fig. 1-2A). From each penton protrudes a "fiber," giving the virion the appearance of a communications satellite (Fig. 19-1A). The genome, which is associated with an inner protein core, consists of a single linear molecule of dsDNA, 36–38 kbp in size for mammalian adenoviruses,

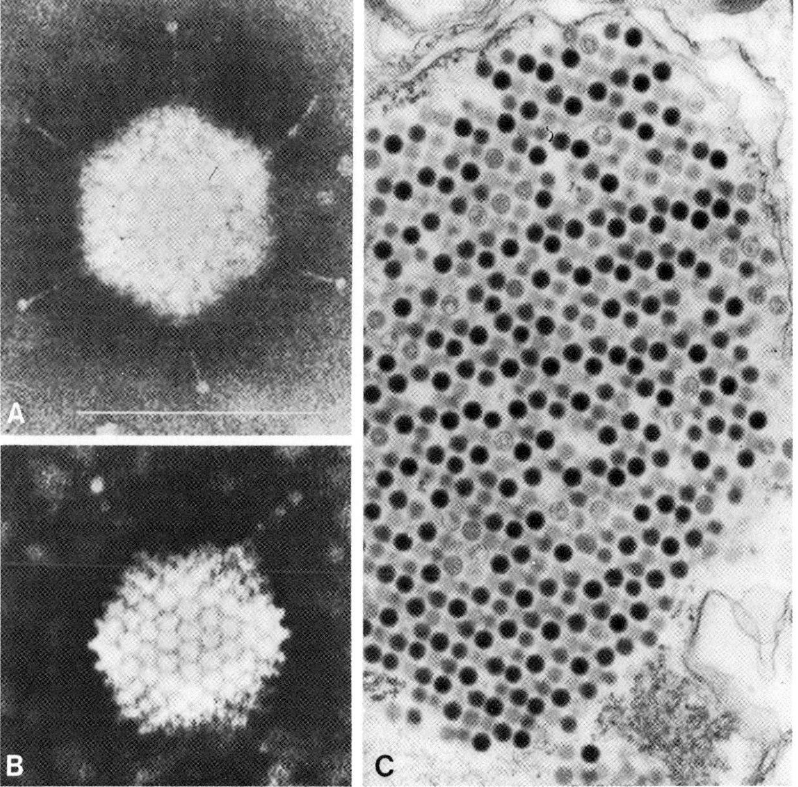

Fig. 19-1 *Adenoviridae.* (A, B) Negatively stained preparations. (A) Virion showing fibers projecting from the vertices. (B) Virion showing icosahedral array of capsomers. Capsomers at the vertices (pentons) are surrounded by five nearest neighbors, all the others (hexons) by six. (C) Section showing crystalline array of mature virions in the nucleus of a human fibroblast cell. Bar, 100nm. [A and B, From R. C. Valentine and H. G. Periera, *J. Mol. Biol.* **13,** 13 (1965); C, courtesy Dr. A. K. Harrison.]

with inverted terminal repetitions. The DNA, in association with a 55K protein which is covalently linked to each 5′ terminus, is infectious when transfected into susceptible cells.

Classification

Mammalian and avian adenoviruses comprise two distinct genera, designated *Mastadenovirus* and *Aviadenovirus* respectively. In turn, the genus *Mastadenovirus* comprises numerous adenovirus serotypes specific for particular mammalian species. Currently, 47 serotypes of human adenovirus (sometimes designated h-Ad1 to h-Ad47) are recognized. They are assigned to six subgenera (A–F) on the basis of various biochemical and serological criteria which generally match up quite well with previous assignments on the basis of certain biological characteristics, notably oncogenicity and hemagglutination (Table 19-2). Members of subgenus A were found to be highly oncogenic for baby rodents and subgenus B less so, and the six subgenera are also compatible with the earlier subdivision based on agglutination of rat or monkey red cells and with the degree of homology of the genomes.

Designation as a distinct serotype is based on a serological difference (of >16-fold) in reciprocal neutralization assays. The fiber protein is responsible for hemagglutination and is type-specific, whereas the hexon protein carries genus-specific, subgenus-specific, intertype-specific, and type-specific epitopes and, like the fiber, elicits neutralizing antibodies. Numerous isolates from AIDS patients have proved difficult to type, as they share epitopes with the fiber and/or hexon proteins of one or more distinct serotypes. Nucleic acid sequencing should reveal whether these so-called intermediate strains have arisen by mutation or by recombination.

Viral Replication

Adenoviruses bind to the host cell receptor via the fiber and enter the cell by endocytosis. The outer capsid is then removed, and the core comprising the viral genome with its associated histones enters the nucleus where mRNA transcription, viral DNA replication, and assembly of virions occur. In the nucleus the genome is transcribed by cellular RNA polymerase II according to a complex program involving both DNA strands (see Chapter 3). Products of the E1A region are required to up-regulate transcription of other regions of

Table 19-2
Classification of Human Adenoviruses

Subgenus	Serotypes
A	12, 18, 31
B	3, 7, 11, 14, 16, 21, 34, 35
C	1, 2, 5, 6
D	8–10, 13, 15, 17, 19, 20, 22–30, 32, 33, 36–39, 42–47
E	4
F	40, 41

the genome, including negating the inhibition of viral promoters by cellular repressors. RNA transcribed in a precisely controlled temporal order from nine separate promoters is spliced in several alternative ways, again according to a complex but ordered program. About a dozen nonstructural early proteins are translated before the genome can be replicated. Viral DNA replication, using a viral protein as primer, a virus-coded DNA polymerase and DNA-binding protein, plus cellular factors, proceeds from both ends by a strand displacement mechanism described in Chapter 3. Following DNA replication, late transcription units are expressed, giving rise to structural proteins, which are made in considerable excess. Virions are assembled in the nucleus, where they form crystalline aggregates (Fig. 19-1C). Shutdown of cellular DNA synthesis occurs relatively early in the viral replication cycle, whereas cellular RNA and protein synthesis decline progressively during the second half of the cycle.

Pathogenesis and Immunity

Adenoviruses multiply initially in the pharynx, conjunctiva, or small intestine, and they generally do not spread beyond the draining cervical, preauricular, or mesenteric lymph nodes. As the disease process remains relatively localized, the incubation period is short (5–8 days). Most of the enteric infections and some of the respiratory infections are subclinical. Generalized infections are occasionally seen, especially in immunocompromised patients, and deaths do occur, particularly from type 7 which is the most pathogenic human adenovirus. At autopsy, lungs, brain, kidney, liver, and other organs reveal the characteristic basophilic nuclear inclusions.

Infection with the common endemic types 1, 2, and 5 persists asymptomatically for years in the tonsils and adenoids of a child, and virus is shed continuously in the feces for many months after the initial infection, then intermittently for years thereafter. The mechanism of this persistence is uncertain; perhaps viral replication is held in check by the antibody synthesized by these lymphoid organs. Fluctuation in shedding indicates that latent adenovirus infections can be reactivated. For example, infection with *Bordetella pertussis* can do so, and measles can actually be followed by adenovirus pneumonia. Subgenus B viruses are frequently shed in the urine of immunocompromised persons; types 34 and 35 were both originally isolated from renal transplant recipients and can be recovered quite commonly from AIDS patients. In addition, a wide variety of other serotypes, particularly from subgenus D, have been recovered from the feces of AIDS patients. Many of these serotypes are new and/or rare, and some appear to be genetic recombinants. Presumably these novel serotypes and intermediate strains arose in this cohort as a result of a combination of factors, principally sexual promiscuity, fecal–oral spread, and immunosuppression. The common serotypes of subgenus C, which are characterized by prolonged latency in lymphoid tissue, can also be reactivated in AIDS patients and are not uncommonly recovered from the blood. Indeed, as a general rule an isolate of an adenovirus from an AIDS patient is likely to be subgenus B if it comes from urine, C from blood, and D from feces (Table 19-3).

In contrast to most respiratory viral infections, adenovirus infections lead to lasting immunity to reinfection with the same serotype, perhaps because of the extent of involvement with lymphoid cells in the alimentary tract and the regional lymph nodes. Maternal antibody generally protects infants under the age of 6 months against severe lower respiratory disease.

Recent research has revealed that several adenovirus genes encode proteins that diminish the host response to infection. For example, one small protein encoded by a gene within the E3 transcription unit binds to the nascent heavy chain of the class I MHC antigen, preventing its transport to the cell surface and thereby minimizing the presentation of adenovirus peptides to cytotoxic T lymphocytes. A second E3 gene product protects infected cells against lysis by tumor necrosis factor, whereas a third binds to the epidermal growth factor receptor, causing its internalization and degradation. Doubtless additional gene products that modulate the host response to infection in a wide variety of unexpected ways will be discovered in the near future.

Clinical Syndromes

Only about half of the known serotypes of human adenoviruses have been causally linked to disease (Table 19-3). Adenoviruses 1–8 are by far the commonest species worldwide and are responsible for most cases of adenovirus-induced disease. Some 5% of acute respiratory illnesses in children under the age of 5 years (but <1% of those in adults) have been ascribed to adenoviruses. The recently cultivated enteric serotypes 40 and 41 have been claimed to cause up to 10% of infantile gastroenteritis. Several of the lower numbered serotypes as well as types 8, 19, and 37 are major causes of eye infections, whereas types 19 and 37 also cause genital infections.

Respiratory Infections

Pharyngitis is seen particularly in young children. The infant presents with a cough, nasal congestion, and fever, the throat is inflamed, and there is often an exudative tonsillitis or rarely a pertussis-like illness. Adenoviruses 1–7 are usually responsible for these common sporadic infections, which are relatively trivial except when otitis media or pneumonia supervenes.

Acute respiratory disease (ARD) is the name given to a syndrome characterized by fever, pharyngitis, cervical adenitis, cough, and malaise which occurs in epidemic form when military recruits assemble in camps. Types 4 and 7 are most often responsible.

Pneumonia, often severe and occasionally fatal, may develop in young children infected with any of the common serotypes, but particularly types 7 and 3. In some of the colder parts of the world, such as northern China or Canada, adenoviruses are an important cause of pneumonia in infants under the age of 2 years. The case–fatality rate can be quite high, and long-term impairment of pulmonary function can accompany the development of obliterative bronchiolitis or bronchiectasis. ARD in military recruits also occasionally progresses to a pneumonitis. Of increasing importance in this era of organ transplantation and AIDS are the severe infections, including pneumonia, seen in immunosuppressed patients.

Table 19-3
Diseases Caused by Human Adenoviruses

Disease	Age	Common serotypes[a]	Major subgenus	Major source
Respiratory infections				
Pharyngitis	Young children	**1, 2,** 3, **5,** 6, 7	B, C	Throat
Acute respiratory disease	Military recruits	3, **4, 7,** 14, 21	B, E	Throat
Pneumonia	Young children	1, 2, **3,** 4, 5, **7,** 21	B, C	Throat
	Military recruits	**4, 7**	B, E	Throat
Ocular infections				
Pharyngoconjunctival fever	Children	1, 2, **3, 4,** 6, **7**	B, C, E	Throat, eye
Epidemic keratoconjunctivitis	Any age	**8,** 19, **37**	D	Eye
Genitourinary infections				
Cervicitis, urethritis	Adults	19, **37**	D	Genital secretions
Hemorrhagic cystitis	Young children	**11,** 21	B	Urine
Enteric infections				
Gastroenteritis	Young children	31, **40, 41**	A, F	Feces
Infections in immunocompromised individuals				
Encephalitis, pneumonia,	Any age, including AIDS patients	7, **11, 34, 35**	B	Urine, lung
Gastroenteritis	AIDS patients	Many D including 43–47	D	Feces
Generalized	AIDS patients	2, 5	C	Blood

[a] Only the commonly occurring serotypes are listed; those most commonly associated with particular syndromes are in bold type.

Ocular Infections

Pharyngoconjunctival fever tends to occur in outbreaks, for example, at children's summer camps where "swimming pool conjunctivitis" may occur with or without pharyngitis, fever, and malaise. Adenoviruses 3, 4, and 7 are commonly responsible. Type 4 has caused a number of nosocomial outbreaks of conjunctivitis or pharyngoconjunctival fever among hospital staff.

Epidemic keratoconjunctivitis is a more severe eye infection, commencing as a follicular conjunctivitis and progressing to involve the cornea (keratitis). Originally reported in industrial workers exposed to dust and trauma, the disease was once known as "shipyard eye." Highly contagious and often occurring in epidemic form, the infection is caused by members of subgenus D. Type 8 was the principal agent until 1973 when type 19 temporarily took over until, in 1976, type 37 suddenly appeared, spread worldwide, and today remains (with type 8) the predominant cause of epidemic keratoconjunctivitis. During epidemics, nosocomial spread can readily occur in eye clinics.

Genitourinary Infections

Cervicitis and urethritis are common manifestations of venereal infection with type 37, which was first identified in prostitutes. Cystitis, seen mainly in young boys, is caused by type 11 and more rarely by type 21. In its acute hemorrhagic form cystitis is characterized by hematuria as well as dysuria and frequency of micturition. Adenoviruses commonly establish asymptomatic persistent infection of the kidney and may be shed in the urine for months or years. This is observed particularly in immunocompromised individuals, such as renal transplant recipients.

Enteric Infections

Gastroenteritis in infants is commonly caused by two recently discovered adenovirus serotypes, 40 and 41. These enteric adenoviruses, previously visualized by electron microscopy in feces but regarded as uncultivable, can now be grown in cultured cells. Their recovery from outbreaks of gastroenteritis, including in day-care centers, and significantly more frequently from symptomatic patients than from controls, confirms that they do indeed cause the disease. However, many other adenoviruses that replicate in the intestine or in the throat are excreted asymptomatically in the feces for weeks or months, hence carefully controlled studies are required before assigning them an etiologic role in gastroenteritis.

Infections in Immunocompromised Patients

Adenoviruses may cause life-threatening infections in at least three groups of immunocompromised patients. In children with severe combined immune deficiency disease the common serotypes of subgenera A, B, and C can cause serious conditions such as pneumonia or meningoencephalitis. Kidney, liver, or bone marrow transplant recipients, and patients with AIDS, often shed subgenus B serotypes such as 11, 34, or 35 in urine for prolonged periods, and AIDS patients also excrete the recently described subgenus D serotypes (43–47) in feces.

Laboratory Diagnosis

Depending on the clinical presentation, appropriate specimens for diagnosis include feces; pharyngeal swab, nasopharyngeal aspirate, transtracheal aspirate, or bronchial lavage; conjunctival swab, corneal scraping, or tears; genital secretions; urine; and tissues from biopsy (e.g., of liver or spleen) or autopsy (e.g., lung or brain).

Enzyme immunoassay (EIA) is emerging as the diagnostic method of choice for the detection of soluble viral antigen in feces or nasopharyngeal secretions. A monoclonal antibody (MAb) to a hexon epitope common to all adenovirus serotypes (or polyclonal serum) suffices to identify the family; then, if desired, a type-specific MAb can be used to identify the particular adenovirus concerned. Reliable reagents are now available commercially, and the EIA should give 90–95% specificity and 70–90% sensitivity compared with the more laborious process of isolation of the virus in cultured cells. Even the so-called noncultivable adenoviruses can be identified by this method, or by immunoelectron microscopy.

Immunofluorescence can be employed to demonstrate adenoviral antigen in cells from the respiratory tract, eye, urine, or biopsy or autopsy material, following low-speed centrifugation of the specimen and then fixation of the pelleted cells.

Virus isolation is still the approach of most diagnostic and reference laboratories. Propagation of adenoviruses in cultured cells is time-consuming because many serotypes are very slow-growing. Human malignant cell lines such as HeLa, HEp-2, KB, or A-549 cells, or diploid human embryonic fibroblasts (HDF) derived, for example, from lung or tonsil, are the substrates of choice. The fastidious enteric adenoviruses 40 and 41 have only recently been cultivated *in vitro* and require special cell lines such as Graham-293 or special conditions (e.g., low-serum medium). The common adenoviruses (types 1–7) generally produce cytopathic effects (CPE) within 1–2 weeks; the cells become swollen, rounded, and refractile, cluster together like a bunch of grapes, and reveal characteristic basophilic intranuclear inclusions after staining (Fig. 19-2). Other serotypes, however, especially those of subgenera A and D, grow very slowly, and CPE, often nonclassical, may not become evident for a month. Confirmation of the isolate as an adenovirus can be made by immunofluorescence on the (fixed) cell monolayer. Appropriate type-specific antisera, or MAbs directed to type-specific epitopes on the fiber, can then be chosen to type the isolate by hemagglutination inhibition and/or neutralization.

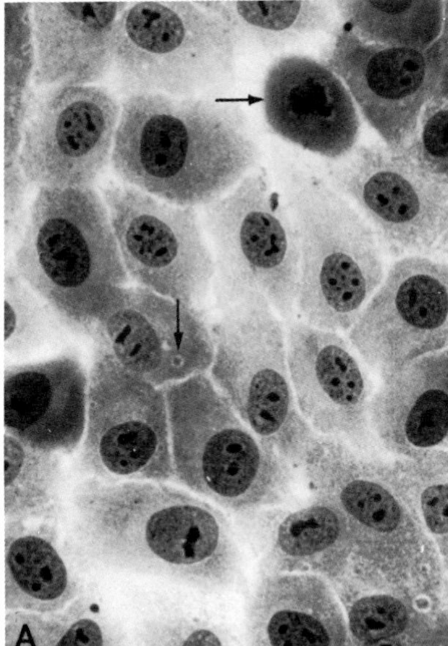

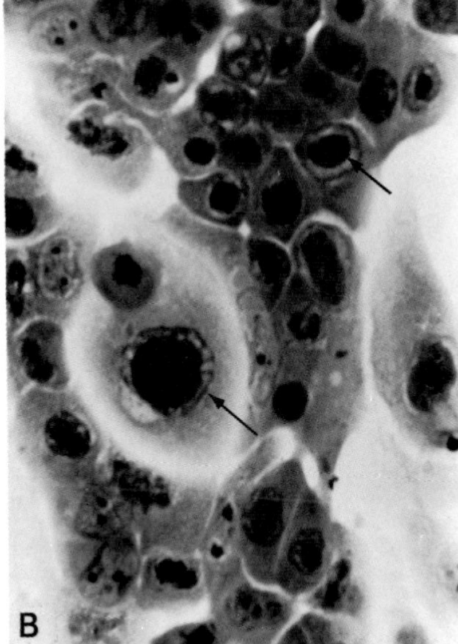

Fig. 19-2 Cytopathic effects induced by adenoviruses (hematoxylin and eosin stain; magnification: ×400). (A) Normal monolayer of HEp-2 cells. Horizontal arrow marks cell in mitosis and the vertical arrow marks phagocytosed cell debris, not to be confused with viral inclusion bodies. (B) Cytopathic effects induced by adenovirus in HEp-2 cells. Note distended cells containing basophilic intranuclear inclusions (arrows), which consist of masses of virions (see Fig. 19-1C). Threads of chromatin sometimes radiate from the nuclear inclusions to the periphery of the nucleus (Courtesy I. Jack.)

Epidemiology

Although mainly associated with disease of the respiratory tract and eye, and often transmitted by respiratory droplets or contact, adenoviruses probably spread principally via the enteric (fecal–oral) route. Long-term family studies have demonstrated that, following infection of young children, very large numbers of adenovirus particles are shed in feces (10^{11} virions per gram) over a period of several months and succeed in infecting about half of all susceptible members of the same family. Doubtless this explains the endemicity of the common members of subgenus C (1, 2, 5) and the fact that most children acquire one or more of these viruses by the age of 2 years. About half of such infections are subclinical, the others presenting as pharyngitis or pharygoconjunctival fever. The enteric adenoviruses of subgenus A and many of those of subgenus D are generally isolated from feces, whereas those of subgenus F may spread exclusively via the fecal–oral route. Adenoviruses 40 and 41 have been clearly demonstrated to cause gastroenteritis in children year-round, with occasional outbreaks, for example, in schools or hospitals. Adenoviruses may be responsible for about 10% of cases of gastroenteritis in children.

Respiratory spread, via droplets or contact, occurs particularly in the case of the so-called epidemic serotypes of subgenera B and E, which are responsible for outbreaks of pharyngoconjunctival fever in children (types 3, 4, 7) or ARD in military recruits (4, 7) in late winter and spring. Altogether adenoviruses have been calculated to cause 5–10% of respiratory viral infections.

Eye infections may be acquired in several ways, including transfer of respiratory secretions on fingers following infection via the respiratory or alimentary routes. However, two important direct routes of entry to the eye are well established. Outbreaks of "swimming pool conjunctivitis" (with or without pharyngoconjunctival fever) often occur in children during the summer. In addition, a number of major outbreaks of epidemic keratoconjunctivitis have been traced to the surgeries of particular ophthalmologists or hospitals whose aseptic technique leaves something to be desired. These iatrogenic infections are attributable to contaminated towels, ophthalmic solutions, and instruments such as tonometers. Hand-to-eye transfer is also particularly important.

Adenoviruses of subgenus B are also commonly excreted in urine, whereas subgenus D types 19 and 37 can presumably be transmitted venereally (as well as by contact, in eye infections) because they cause genital infections of males and females.

Overall, fewer than half of the 47 types of adenoviruses currently known have been unequivocally demonstrated to cause any disease. This applies particularly to the 29 members of subgenus D. Serotypes 1–8 comprise about 90% of isolates worldwide.

Control

Fecal–oral spread of adenoviruses within families can be reduced by personal hygiene. Chlorination of swimming pools, drinking water, and wastewater largely removes the risk of outbreaks from these communal sources. Prevention of contact spread of eye infections by ophthalmologists and nurses in eye clinics demands that special attention be paid to early identification and segregation of epidemic keratoconjunctivitis patients, hand washing, separate paper towels and ophthalmic solutions, and adequate disinfection of equipment. Similar precautions should also be taken to minimize the possibility of nosocomial outbreaks in wards where a patient with a severe adenovirus infection is being nursed.

Vaccine

The regularity of outbreaks of ARD in U.S. military recruits prompted the development in the 1960s of a vaccine for protection of this particular population. The approach was novel. Live virulent virus is enclosed in a gelatin-coated capsule and given by mouth. In this way the virus bypasses the throat, in which it would normally cause disease, but is released in the intestine, where it grows without producing disease. Because of lymphocyte recirculation, such vaccines induce mucosal immunity in the respiratory tract as well as in the intestinal tract. When the two important serotypes, 4 and 7, cultured in human fibroblasts, are combined in such a live vaccine, they grow in the gut with little or no mutual interference and produce highly effective immunity to challenge. This experience has led to intensive research on the possible use of recombinant adenoviruses for protection against a range of infections in which mucosal immunity plays an important role.

Further Reading

Doerffler, W., ed. (1984). The Molecular Biology of Adenoviruses, Vols. 1–3. *Curr. Top. Microbiol. Immunol.* **109, 110,** and **111.**

Doerffler, W., ed. (1986). "Adenovirus DNA: The Viral Genome and Its Expression." Nijhoff, Boston.

Flint, S. J. (1986). Regulation of adenovirus mRNA function. *Adv. Virus Res.* **31,** 169.

Fox, J. F., Hall, C. E., and Cooney, M. K. (1977). The Seattle Virus Watch. VII. Observations of adenovirus infections. *Am. J. Epidemiol.* **105,** 362.

Ginsberg, H. S., ed. (1984). "The Adenoviruses." Plenum, New York.

Hierholzer, J. C. (1989). Adenoviruses. *In* "Diagnostic Procedures for Viral, Rickettsial and Chlamydial Infections" (N. J. Schmidt and R. W. Emmons, eds.), 6th Ed., p. 219. American Public Health Association, Washington, D.C.

Hierholzer, J. C., Wigand, R., Anderson, L. J., Adrian, T., and Gold, J. W. M. (1988). Adenoviruses from patients with AIDS: A plethora of serotypes and a description of five new serotypes of subgenus D (types 43–47). *J. Infect. Dis.* **158,** 804.

Horwitz, M. S. (1990). Adenoviridae and their replication. *In* "Fields Virology" (B. N. Fields, D. M. Knipe, R. M. Chanock, M. S. Hirsch, J. L. Melnick, T. P. Monath, and B. Roizman, eds.), 2nd Ed., p. 1679. Raven, New York.

Nevins, J. R. (1987). Regulation of early adenovirus gene expression. *Microbiol. Rev.* **51,** 419.

Schmitz, H., Wigand, R., and Heinrich, W. (1983). Worldwide epidemiology of human adenovirus infections. *Am. J. Epidemiol.* **117,** 455.

Waddell, G. (1988). Adenoviridae: The adenoviruses. *In* "Laboratory Diagnosis of Infectious Diseases: Principles and Practice. Volume II. Viral, Rickettsial and Chlamydial Diseases" (E. H. Lennette, P. Halonen, and F. A. Murphy, eds.), p. 284. Springer-Verlag, Berlin and New York.

Waddell, G., Allard, A., Johansson, M., Svensson, L., and Uhnoo, I. (1987). Enteric adenoviruses. *Ciba Found. Symp.* **128,** 63.

Warren, D., Nelson, K. E., Farrar, J. A., Hurwitz, E., Hierholzer, J. C., Ford, E., and Anderson, L. J. (1989). A large outbreak of epidemic keratoconjunctivitis: Problems in controlling nosocomial spread. *J. Infect. Dis.* **160,** 938.

Chapter 20

Herpesviridae

All herpesviruses have the capacity to persist in their hosts indefinitely, in the form of an episome in the nucleus of the cells that harbor them. Virtually every vertebrate species that has been carefully searched is found to support at least one host-specific herpesvirus which has evolved with that host species for millennia. Sometimes, as in humans, host-specific herpesviruses of different subfamilies occupy distinct ecologic niches, noncompetitively, in particular types of cells within a given individual (Table 20-1). Varicella (chickenpox) and herpes simplex viruses establish latent infections in neurons. On reactivation, the varicella virus precipitates an attack of herpes zoster (shingles), whereas herpes simplex type 1 typically causes recurrent attacks of labial herpes; herpes simplex type 2 is mainly responsible for genital herpes. Cytomegalovirus, Epstein-Barr (EB) virus, and human herpesvirus 6 (HHV-6) persist in lymphocytes. With the control of rubella by immunization, cytomegalovirus is now the major infectious cause of mental retardation and other congenital defects. EB virus is the etiologic agent of infectious mononucleosis and is associated with two types of human cancer, a carcinoma and a lymphoma. HHV-6 causes a common exanthem in children.

The importance of these six human pathogens is increasing as a result of developments in modern medicine and changing sexual practices. Herpesviruses are frequently reactivated in AIDS and following immunosuppressive therapy for organ transplantation or cancer. Under these circumstances, or in infections of the fetus or newborn infant, lethal disseminated disease may occur. Although most herpesviruses pose unsolved problems for vaccine development, some of them are amenable to antiviral chemotherapy.

Table 20-1
Herpesviruses of Humans

Subfamily/genus	Official name	Vernacular name	Biological properties
Alphaherpesvirinae			
Simplexvirus	Human herpesvirus 1	Herpes simplex virus 1	Fast-growing, cytolytic Latent in neurons
	Human herpesvirus 2	Herpes simplex virus 2	
	Cercopithecine herpesvirus 1	Simian herpes B virus	
Varicellovirus	Human herpesvirus 3	Varicella–zoster virus	
Betaherpesvirinae			
Cytomegalovirus	Human herpesvirus 5	Cytomegalovirus	Slow-growing, cytomegalic Latent in salivary glands, kidneys
Roseolovirus	Human herpesvirus 6		Latent in macrophages, lymphocytes
Gammaherpesvirinae			
Lymphocryptovirus	Human herpesvirus 4	Epstein-Barr (EB) virus	Lymphoproliferative Latent in B lymphocytes

Properties of *Herpesviridae*

The herpesvirus virion comprises four concentric layers: an inner *core*, surrounded by an icosahedral *capsid*, then an amorphous *tegument*, and finally an *envelope* (Table 20-2). The DNA genome is wound like a ball of wool and is associated with a protein core with the shape of a torus or doughnut which appears to be suspended by fibrils anchored to the inner side of the surrounding capsid. The capsid is an icosahedron, 100 nm in diameter, composed of 162 hollow capsomers: 150 hexamers and 12 pentamers (Fig. 20-1B). Surrounding the capsid is a layer of globular material, known as the tegument, which is enclosed by a typical lipoprotein envelope with numerous small glycoprotein peplomers (Fig. 20-1A). The envelope is fragile and the virion somewhat pleomorphic, ranging in diameter from 120 to 200 nm.

Table 20-2
Properties of Herpesviruses

Spherical enveloped virions, 120–200 nm (usually about 150 nm) in diameter
Icosahedral capsid, 100–110 nm, with 162 capsomers; surrounded by amorphous tegument, then lipid envelope containing numerous different glycoproteins, some forming peplomers
Linear dsDNA genome with reiterated sequences characteristic of the genus, 125–229 kbp; associated with toroid protein core
Replicates in nucleus, with sequential transcription and translation of immediate early (α), early (β), and late (γ) genes producing α, β, and γ proteins, respectively, the earlier of which regulate transcription from later genes
DNA replication and encapsidation occur in nucleus; envelope is acquired by budding through nuclear membrane
Productive infection in permissive cells is cytocidal; intranuclear inclusions, sometimes cytomegalic cells or syncytia
Establish latent infections, with genome persisting in the nucleus of neurons or lymphocytes; only a small subset of genes is expressed; reactivation triggers replication and recurrent or continuous shedding of infectious virus

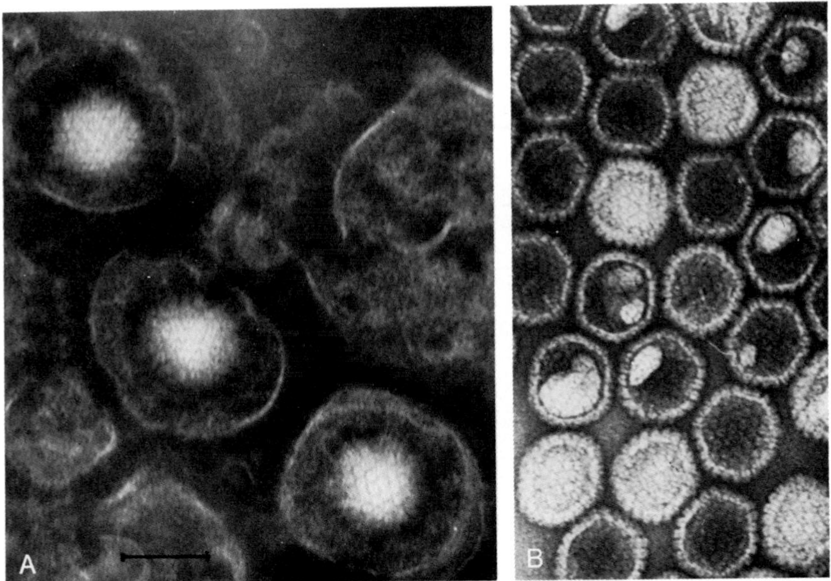

Fig. 20-1 *Herpesviridae.* Negatively stained preparations of the prototype herpesvirus, herpes simplex virus type 1. (A) Enveloped particles. (B) Icosahedral capsids with 162 capsomers. Bar, 100 nm. (Courtesy Dr. E. L. Palmer.)

The virion contains over 30 proteins, of which about 6 are present in the nucleocapsid, 10-20 in the tegument, and 10 in the envelope; a smaller number are associated with the DNA in the core. The envelope proteins are mainly glycoproteins, most but not all of which project as peplomers. Antigenic relationships are complex. There are some shared antigens within the family, but different species have distinct envelope glycoproteins.

The herpesvirus genome consists of a linear dsDNA molecule which is infectious under appropriate experimental conditions. There is a remarkable degree of variation in the composition, size, and structure of herpesvirus DNAs. Physical maps based on ordering of restriction endonuclease fragments of viral DNA (restriction maps) are useful for epidemiologic analysis. The genomes of most of the human herpesviruses have been sequenced, an impressive achievement since these, together with those of the poxviruses, are by far the largest human viral genomes, encoding from about 70 to around 200 proteins. The genomes of the alphaherpesviruses appear to be colinear, that is, the presence and order of the individual genes are similar. Herpesviruses genomes display some unusual features (Fig. 20-2). Reiterated DNA sequences generally occur at both ends and in some viruses also internally, dividing the genome into two unique sequences, designated large (U_L) and small (U_S). When these reiterated sequences are inverted in their orientation, the unique L and S components can invert, relative to one another, during replication, giving rise to two or four different isomers of the genome, present in equimolar proportions. Further, intragenomic and intergenomic recombinational events can alter the number of any particular reiterated sequence, creating *polymorphism.*

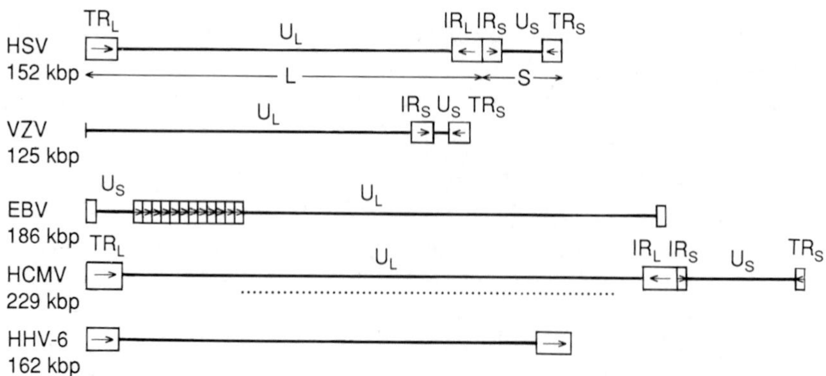

Fig. 20-2 Genome structure of human herpesviruses. Most human herpesvirus genomes comprise two regions designated long (L) and short (S). Terminal repeat (TR) and internal repeat (IR) sequences may bracket unique sequences (U_L, U_S) of both L and S (HSV) or only S regions (VZV). Repeat sequences are shown as boxes and are inverted as indicated by the direction of the arrows. The repeat sequences allow the DNA they bracket to invert relative to the rest of the genome such that where both U_L and U_S are bracketed by repeat sequences, four isomers are made and packaged in equimolar amounts into virions. Where only S is bracketed by repeat sequences (VZV) two equimolar isomers are made. The genome of Epstein-Barr virus contains terminal repeat and major internal repeat (MIR) sequences in a variable number of copies. Unique sequences (U_L and U_S) are demarcated by these repeat families. Near the extremities of U_L are two regions, D_R and D_L, whose sequences are almost identical. The genome of human cytomegalovirus has a structure similar to that of HSV in the layout of repeats and unique sequences; it occurs in four isomers but is much larger. The genome of HHV-6 is still being investigated but appears to consist of a single unique sequence flanked by a pair of large direct repeats. [Based on D. J. McGeoch, *Semin. Virol.* **3**, 402 (1992).]

Classification

Subdivision of the family into three subfamilies was originally based on biological properties (see Table 20-1). The subfamily *Alphaherpesvirinae* includes herpes simplex virus types 1 and 2 and varicella–zoster virus. They all grow rapidly, lyse infected cells, and establish latent infections in sensory nerve ganglia. The subfamily *Betaherpesvirinae* includes human cytomegalovirus and HHV-6. Their replication cycle is slow and produces large, often multinucleate cells (cytomegalia). The viral genome remains latent in lymphoreticular tissue, secretory glands, kidneys, and other tissues. The subfamily *Gammaherpesvirinae* contains the Epstein-Barr virus. It replicates in lymphoid cells and may also be cytocidal for epithelial cells. Latency is frequently demonstrable in lymphoid tissue. As the genomes of an increasing number of herpesviruses are sequenced, herpesvirus taxonomy will progressively be based on the conservation of particular genes and gene clusters, the gene order, and the arrangement of the terminal sequences involved in packaging of the genome.

Viral Replication

Herpesvirus replication has been most extensively studied with herpes simplex virus (HSV); betaherpesviruses and gammaherpesviruses replicate more slowly and exhibit certain significant differences but generally follow a similar pattern. Unlike certain other DNA viruses such as papovaviruses and parvoviruses which stimulate the cellular DNA synthetic machinery, herpesviruses themselves encode most of the enzymes they require to increase the pool of deoxynucleotides and to replicate viral DNA. This facility is vital for viral replication in resting cells such as neurons, which throughout most of the life of the host never make DNA and do not divide. Interestingly, about half of the 73 genes of herpes simplex virus are not essential for viral replication in cultured cells, and it is likely that a similar ratio applies in other herpesviruses; presumably many of these additional genes encode regulatory proteins and virokines which optimize growth, dissemination, and pathogenicity *in vivo* by such devices as extending tissue tropism, establishing and maintaining latency, and suppressing the host immune response.

The HSV virion attaches via its envelope glycoprotein gC to the heparan sulfate moiety of cellular proteoglycans, then may form a firmer association between its gD glycoprotein and a second, unknown cellular receptor. Entry into the cytoplasm requires viral glycoproteins gB, gD, and gH and occurs by pH-independent fusion of the virion envelope with the plasma membrane. Tegument proteins are released, one of which shuts down cellular protein synthesis, and the capsid is transported along the cytoskeleton to a nuclear pore, where viral DNA is released, enters the nucleus, and circularizes.

Viral gene expression is tightly regulated, with three classes of mRNA, α, β, and γ, being transcribed in strictly ordered sequence by the cellular RNA polymerase II (Fig. 20-3). Another of the released tegument proteins transactivates transcription of the five "immediate early" (α) genes. This viral protein associates with two cellular proteins to form a multiprotein complex that specifically recognizes a nucleotide sequence in the promoter region of the viral DNA, triggering transcription by the cellular polymerase. The α mRNAs are transported to the cytoplasm and translated to the several α proteins,

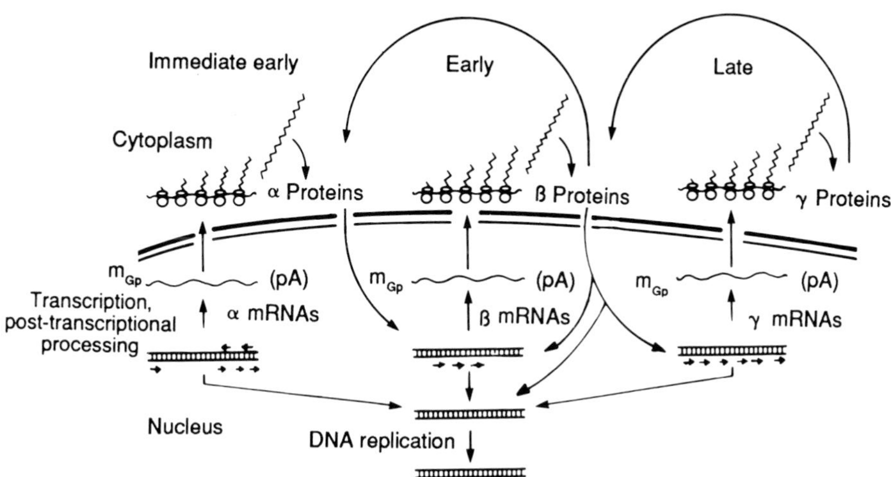

Fig. 20-3 Diagram representing transcription, translation, and DNA replication of herpes simplex virus (see text). Transcription and posttranscriptional processing occur in the nucleus, and translation in the cytoplasm. Some of the α and β proteins are involved in further transcription, and some β proteins participate in DNA replication. (Courtesy Dr. B. Roizman.)

which are regulatory proteins that control the expression of all later genes. One α protein initiates transcription of the "early" (β) genes. The β proteins are enzymes required to increase the pool of nucleotides (e.g., thymidine kinase, ribonucleotide reductase) and others needed for viral DNA replication (e.g., a DNA polymerase, primase–helicase, topoisomerase, single-strand and double-strand DNA-binding proteins).

The viral genome probably replicates by a rolling circle mechanism. Following DNA replication, certain β proteins induce the program of transcription to switch again, and the resulting "late" (γ) mRNAs are translated into the γ proteins, most of which are structural proteins required for morphogenesis of the virion. Capsid proteins assemble to form empty capsids in the nucleus. Unit-length viral DNA cleaved from newly synthesized DNA concatemers is packaged to produce nucleocapsids, which then associate with patches of nuclear membrane to which tegument and glycosylated envelope proteins have bound. This triggers envelopment by budding through the nuclear membrane. Enveloped virions accumulate in endoplasmic reticulum, and the mature virions are released by exocytosis. Virus-specific proteins are also found in the plasma membrane, where they are involved in cell fusion, may act as Fc receptors, and are presumed to be targets for immune cytolysis.

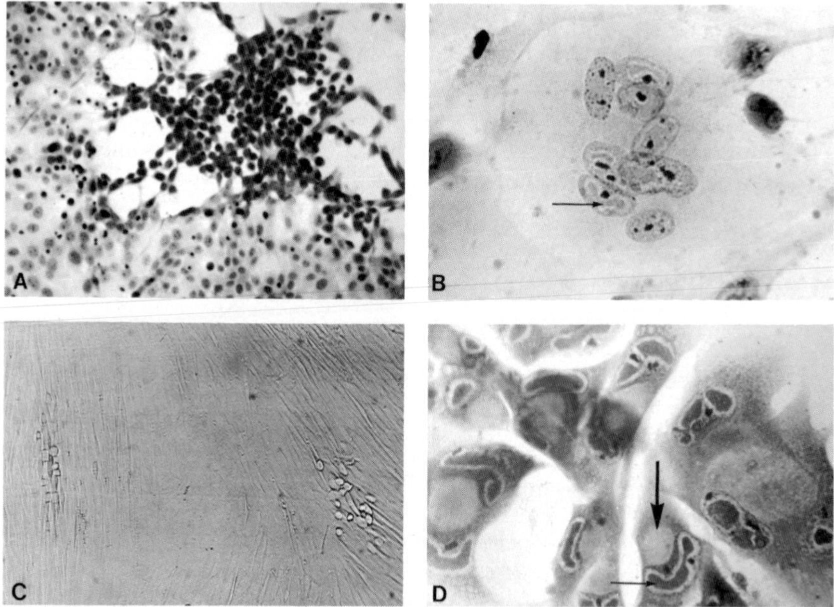

Fig. 20-4 Cytopathic effects induced by herpesviruses. (A) Herpes simplex virus in HEp-2 cells showing early focal cytopathology (hematoxylin and eosin stain; magnification: ×40). (B) Varicella virus in human kidney cells (hematoxylin and eosin stain; magnification: ×150), showing multinucleated giant cell containing acidophilic intranuclear inclusions (arrow). (C) Human cytomegalovirus in human fibroblasts (unstained; magnification: ×25), showing two foci of slowly developing cytopathology. (D) Human cytomegalovirus in human fibroblasts (hematoxylin and eosin stain; magnification ×150), showing multinucleate giant cells with acidophilic inclusions in the nuclei (small arrow) and cytoplasm (large arrow), the latter being characteristically large and round. (Courtesy I. Jack.)

Such productive infections (as opposed to latent infections) are lytic, as a result of virus-induced shutdown of host protein and nucleic acid synthesis. Major changes are obvious microscopically, notably margination and pulverization of chromatin and the formation of large eosinophilic intranuclear inclusion bodies, which are characteristic of herpesvirus infections and can usually be found both in herpesvirus-infected tissues and in appropriately stained cell cultures (Fig. 20-4).

Herpes Simplex Viruses

Pathogenesis and Immunity

Although preexisting neutralizing antibody directed against envelope glycoproteins, notably gB and gD, may successfully prevent primary infection and limit spread of herpes simplex virus from epithelial cells to nerve endings, cell-mediated immunity is the key to recovery from primary infection and maintenance of latency. At the site of epidermal infection viral antigens are presented on dendritic cells and macrophages to $CD4^+$ Th_1 lymphocytes, which initiate viral clearance by secreting cytokines such as interferon γ (IFN-γ) that recruit and activate macrophages and natural killer (NK) cells. $CD4^+$ and $CD8^+$ T cells, as well as NK cells and antibody-dependent cell-mediated cytotoxicity (ADCC), lyse infected cells. The overt epithelial infection is cleared, but some virus ascends the local sensory neurons by retrograde axonal transport and establishes lifelong latency in the corresponding spinal or cerebral ganglion (see Figs. 10-1 and 10-2). The mechanism of establishment, maintenance, and reactivation of latency was discussed in Chapter 10. Experimental studies in animal models, as well as the clinical observation that immunocompromised humans are much more prone to severe HSV infections and to reactivation, make it clear that $CD8^+$ T-cell-mediated immunity is the key to recovery from primary infection. HSV-specific $CD8^+$ T cells may also suppress the full expression of HSV DNA during the establishment of latency in sensory ganglia. We do not yet understand what is the common link between the apparently disparate events known to trigger reactivation (immunosuppression, "stress," trauma, ultraviolet irradiation, fever, etc.).

Recurrences of disease are typically less severe than primary disease, and the frequency and severity diminish with time. The two HSV types display a degree of selectivity in their tissue tropism. HSV-2 replicates to a higher titer than does HSV-1 in genital mucosa, is more likely to lead to encephalitis and severe mental impairment in neonates, is twice as likely as HSV-1 to establish reactivatable latent infection, and is subject to recur almost 10 times as frequently; the converse applies to orolabial infections, where HSV-1 predominates. There is also evidence of extraganglionic latency at the site of primary infection, for example, genital tract or cornea, but the nature and biological significance of this are unknown. The severity of primary HSV infection is influenced by three major factors: (1) age, premature infants being particularly vulnerable, (2) site, systemic and brain infections being much more serious than infections confined to epithelial surfaces, and (3) immunocompetence, T-cell-mediated immunity being crucial in the control of infection.

Two of the HSV glycoproteins, gE and gI, form a receptor for the Fc domain of IgG. This Fc receptor is found on the surface of both virions and infected cells and can protect both against immunologic attack, by steric hindrance resulting from binding of normal IgG, or from "bipolar bridging" of HSV antibody which can attach to gE/gI by its Fc end and simultaneously to another HSV glycoprotein via one Fab arm. Moreover, gC is a receptor for the C3b component of complement and may protect infected cells from antibody–complement-mediated cytolysis.

Clinical Features

In considering the several clinical presentations it is important to distinguish between *primary* and *recurrent* infections (Table 20-3). Primary infections with HSV are generally inapparent, but when clinically manifest they tend to be more severe than are recurrences in the same dermatome. Since immunity to exogenous reinfection is long-lasting, nearly all second infections with the same HSV type are reactivations of an endogenous latent infection. However, because cross-immunity is only partial, reinfection with the heterologous serotype can occur (e.g., genital herpes caused by HSV-2 in an HSV-1 immune person); such cases are called "initial disease, nonprimary infection" and are usually mild.

Oropharyngeal Herpes Simplex

Primary infection with HSV-1 most commonly involves the mouth and/or throat (Fig. 20-5A,C). In young children the classic clinical presentation is gingivostomatitis. The mouth and gums become covered with vesicles which soon rupture to form ulcers. Though febrile, irritable, and in obvious pain with bleeding guns, the child recovers uneventfully. In adults, primary infection more commonly presents as a pharyngitis or tonsillitis.

Following recovery from primary oropharyngeal infection the individual

Table 20-3
Diseases Produced by Herpes Simplex Viruses

Disease	Primary (P) or recurrent (R)	Age	Frequency	Severity	Type
Gingivostomatitis	P	Young children	Common	Mild	1
Pharyngotonsillitis	P	Adults	Common	Mild	1 > 2
Herpes labialis	R	Any	Common	Mild	1 > 2
Genital herpes	P, R	>15 years	Common	Mild/moderate	2 > 1
Keratoconjunctivitis	P, R	Any	Common	Mild/moderate	1
Skin infection[a]	P, R	Any	Rare	Mild/moderate	1, 2[b]
Encephalitis	P, R	Any	Rare	Severe[c]	1 > 2[d]
Neonatal herpes	P	Newborn	Rare	Severe[c]	2 > 1
Disseminated herpes	P, R	Any	Rare	Severe[c]	1 > 2

[a] Including herpes simplex virus infection of burns, eczema herpeticum, etc.
[b] Skin above waist, 1 > 2; below waist, 2 > 1; arms, either 1 or 2.
[c] Often fatal.
[d] HSV-2 in neonates.

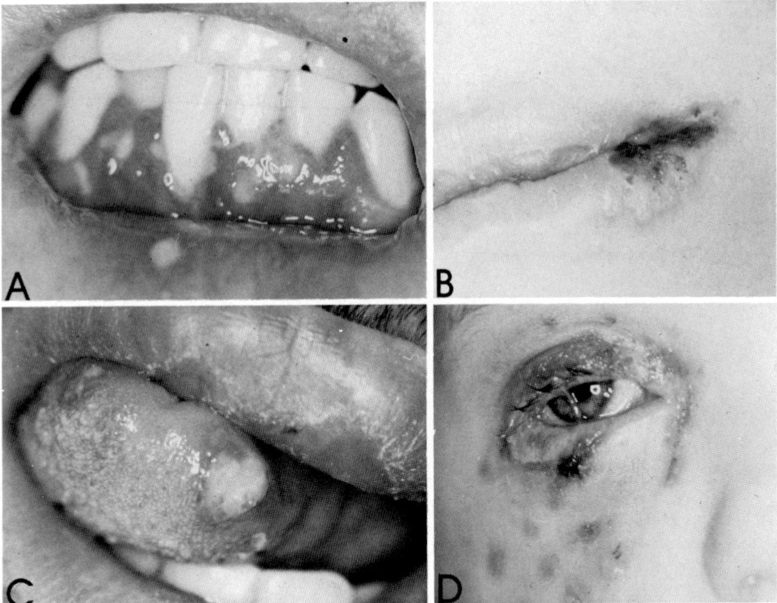

Fig. 20-5 Herpes simplex. Note vesicles, which rupture to become ulcers, on (A) gums, (B) lips, (C) tongue, and (D) eye. (Courtesy Dr. J. Forbes.)

retains HSV DNA in the trigeminal ganglion for life and has at least a 50% chance of suffering recurrent attacks of herpes labialis (otherwise known as herpes facialis, herpes simplex, fever blisters, or cold sores) from time to time throughout the remainder of life. A brief prodromal period of hyperesthesia heralds the development of a cluster of vesicles, generally around the mucocutaneous junction on the lips (Fig. 20-5B).

Genital Herpes

About 80% of primary genital herpes is caused by HSV-2, but an increasing minority is attributable to HSV-1, possibly as a result of orogenital sexual practices. As a sexually transmitted disease it is seen mainly in young adults, but it is occasionally encountered following accidental inoculation at any age, or in young girls who have been victims of sexual abuse. Though the majority of primary infections are subclinical, disease often occurs and is occasionally severe, particularly in females. Ulcerating vesicular lesions develop on the vulva, vagina, cervix, urethra, and/or perineum in the female, penis in the male (Fig. 20-6), or rectum and perianal region in male homosexuals. Local manifestations are pain, itching, redness, swelling, discharge, dysuria, and inguinal lymphadenopathy; systemic symptoms, notably fever and malaise, are often quite marked, especially in females. Spread may occur to the central nervous system, causing mild meningitis in about 10% of all cases. Initial disease is less severe in those with immunity resulting from previous infection with HSV-1 or previous subclinical infection with HSV-2.

The vast majority of recurrent genital herpes is caused by HSV-2; recurrences are also more frequent than following oropharyngeal HSV-1 infection.

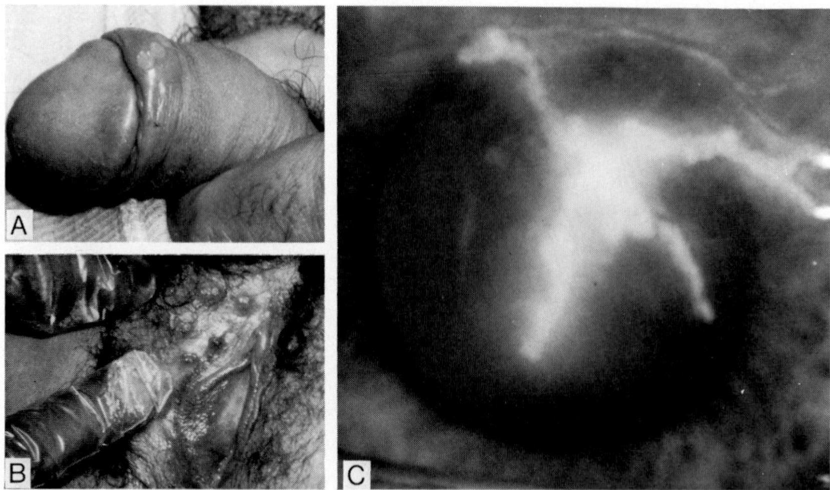

Fig. 20-6 Herpes simplex. HSV-2 genital herpes in the male (A) and female (B). (C) Dendritic ulcers on the cornea of the eye, caused by HSV-1 (stained with fluorescein). (A, B, Courtesy Dr. D. Bradford; C, courtesy Dr. H. MacLean.)

Although recurrences are less severe than was the primary attack, the resultant pain, sexual frustration, and sense of guilt can have psychological as well as sexual consequences.

Keratoconjunctivitis

Primary infection of the eye (Figs. 20-5D and 20-6C) may occur *de novo* or may result from autoinoculation. Involvement of the cornea (keratitis) often leads to a characteristic "dendritic ulcer" which may progress to involve the stroma beneath. Immunopathologic mechanisms including autoimmunity are thought to be involved. The corneal scarring that results from repeated HSV infections may lead to blindness; such recurrences are usually unilateral.

Skin Infections

Primary and recurrent HSV infections may also involve any region of the skin. Rarely this occurs by direct traumatic contact, for example, in wrestlers or rugby players—"herpes gladiatorum"! An occupational hazard for dentists and nurses is herpetic paronychia ("whitlow"). Much more dangerous, disseminated skin infections with HSV can complicate burns or eczema.

Encephalitis

Although encephalitis is a rare manifestation of HSV infection, this virus is nevertheless the commonest sporadic (nonepidemic) cause of this disease. The virus may spread to the brain during primary or recurrent HSV infection, but vesicles are usually not present on the body surface. HSV-1 is usually responsible except in the case of neonates. The temporal lobes are principally involved. Untreated, the case–fatality rate is 70%, and the majority of survivors suffer permanent neurologic sequelae. HSV is also a significant cause of meningitis and has more rarely been associated with a range of other neurologic conditions.

Neonatal Herpes

Herpes neonatorum is a serious disease acquired by babies, usually from their mothers during delivery. The majority are HSV-2 infections acquired during passage through an infected birth canal, but some may be acquired postnatally, and a very few prenatally by viremic transmission across the placenta or by ascending infection from the cervix. If virus is present in the maternal genital tract at the time of delivery the risk of neonatal herpes ranges from 3–4% in recurrent infection to 30–40% in primary infection; neonatal herpes simplex infection occurs in 1 in 5000–10,000 live births.

Neonatal herpes may present as (1) disseminated disease, with a case–fatality rate of 80%, most of the survivors being left with permanent neurologic or ocular sequelae; (2) encephalitis, with high mortality; or (3) disease localized to mucocutaneous surfaces such as skin, eye, and mouth (which may, however, progress to disseminated disease if not treated promptly).

Disseminated Herpes in Compromised Hosts

Patients particularly at risk of potentially lethal disseminated HSV infections are those who are already compromised by (1) congenital immunodeficiency disorders or malignancy, (2) immunosuppression (e.g., for organ transplantation), (3) severe malnutrition with or without concomitant measles, or (4) severe burns, eczema, or certain other skin conditions.

Laboratory Diagnosis

Some of the clinical presentations of HSV, such as recurrent herpes labialis, are so characteristic that laboratory confirmation is not required. Others are not nearly so clear-cut, for example, if the lesions are atypical or if vesicles are not visible at all, as in encephalitis, keratoconjunctivitis, or herpes genitalis infections that are confined to the cervix. Specimens include vesicle fluid, cerebrospinal fluid (CSF), or swabs or scrapings from the genital tract, throat, eye, or skin, as appropriate. Rising concern about herpes genitalis has led to a situation where genital swabs are now the most common specimen received by many virus diagnostic laboratories. Isolation in cell culture is the method of choice, providing that the specimen is taken early, placed in an appropriate transport medium, kept on ice, and transferred to the laboratory without delay. Human fibroblasts or Vero cells are both very sensitive, but HSV replicates rapidly in any mammalian cell line. Distinctive foci of swollen, rounded cells appear within 1–5 days (Fig. 20-4A). The diagnosis can be confirmed within 24 hours by immunofluorescent staining of the infected cell culture. Differentiation of HSV-1 from HSV-2 is not usually relevant to the management of the case but may be required for epidemiologic studies; it simply involves selection of appropriate monoclonal antibodies (MAbs) that distinguish the two.

Speed is important in situations where the patient would benefit from early commencement of antiviral therapy. Currently the most reliable of the more rapid diagnostic alternatives is demonstration of HSV antigen in cells scraped from lesions, genital tract, throat, or cornea, or brain biopsy, by immunofluorescence or immunoperoxidase staining (using type-specific MAbs if desired). Enzyme immunoassays (EIAs) on CSF or detergent-solubilized cells/mucus offer an alternative. Rapid diagnosis of encephalitis is particularly difficult, yet important if appropriate chemotherapy is to be commenced promptly. Brain biopsy has been used in the past but is unnecessarily invasive. The polymerase chain reaction (PCR) is now the method of choice for detection of HSV DNA in CSF. EIA or radioimmunossay (RIA) can also be used to demonstrate anti-HSV IgM in CSF, as well as an abnormally high ratio of total HSV antibody in CSF to antibody in serum. In most other HSV infections, however, serology is not widely used, except for epidemiologic studies. Most HSV antibodies react with both serotypes, but type-specific antigens (e.g., baculovirus-cloned external domain of the gG glycoproteins of HSV-1 and HSV-2) are now available for EIA or "immunodot" screening assays.

Epidemiology

Herpes simplex viruses spread principally by close person-to-person contact with lesions or mucosal secretions. HSV-1 is shed principally in saliva, which may be communicated to others directly, for example, by kissing, or indirectly, via contaminated hands, eating utensils, etc. Autoinoculation by fingers may occasionally spread virus to the eye or genital tract. HSV-2, on the other hand, is transmitted mainly, but not exclusively, by sexual intercourse. Neonates can be infected during passage through an infected birth canal, or rarely transplacentally.

The probability and age of acquisition of HSV are closely linked to socioeconomic circumstances. In the developing world and in the poorer communities within developed nations, most children first become infected with HSV-1 within a few years after losing the "umbrella" of maternal antibody, and over 80% are seropositive by adolescence. Doubtless this reflects the influence of such factors as overcrowding, poor hygiene, and patterns of human contact including social mores. In contrast, the majority of children in more affluent communities escape infection until adolescence, when there is a second peak of infection as a result of salivary exchange via kissing; the prevalence in this cohort may never climb beyond 40–50%. A similar situation obtains with HSV-2; in the United States about one in every four adults is seropositive, but the prevalence is four times higher in blacks than in whites. Because almost all the transmission of HSV-2 is sexual, most initial HSV-2 infections in blacks are subclinical, as they occur in adolescents and young adults with preexisting antibody against HSV-1, whereas those in whites tend to be more severe and to recur more frequently thereafter. Shedding in genital secretions (typically HSV-2) or saliva (typically HSV-1) occurs sporadically and is not limited to recurrences with overt symptoms; for example, studies indicate that HSV-1 can be isolated from the saliva of up to 20% of children and 1–5% of adults at any given time. At least 20% of people suffer from recurrent herpes labialis and perhaps 5% from recurrent genital herpes.

Control

The very substantial risk of contracting genital herpes from a partner who is a known carrier can be reduced by diligent use of condoms irrespective of whether lesions are present. The risk of herpetic paronychia in dentists, and in nurses handling oral catheters, can be eliminated by wearing gloves. Patients with active herpes simplex virus infections, particularly heavy shedders such as babies with neonatal herpes or eczema herpeticum, should be isolated from patients with immunosuppressive disorders.

The strategy of risk assessment in pregnant women and the place of delivery by cesarean section in the prevention of neonatal herpes are both controversial. Until recently many countries advocated a series of cultures of genital specimens taken at least weekly through the final 4–6 weeks of gestation from "high risk" women, usually defined as those with a clinical history of genital herpes. However, this costly procedure has been shown to be a poor predictor of viral shedding at the time of labor and therefore of fetal risk, and it is no longer advocated except when primary genital herpes is diagnosed or suspected. Currently, the preferred policy is to await the onset of labor, then carefully examine the cervix and vulva and collect swabs for testing. Culture should be undertaken, as the most reliable indicator of infectious virus. Sensitive rapid diagnostic methods cannot be recommended for general application until they become more reliable, readily available, and cost-effective. If genital lesions are present at labor, the baby should be delivered by cesarean section as soon as possible and not more than 4 hours after the membranes have ruptured, because of the risk to the fetus of ascending infection. If no lesions are present vaginal delivery is appropriate; however, should the vaginal swabs subsequently prove to be HSV positive, evidence of neonatal infection should be sought by taking samples from the baby's eyes, nose, throat, umbilicus, and anus.

Chemotherapy

The drug of choice for treatment of alphaherpesvirus infections is acycloguanosine (acyclovir), the mechanism of action of which was discussed in Chapter 16. Acyclovir in an ointment formulation is used topically for the treatment of ophthalmic herpes simplex. Intravenous acyclovir (10 mg/kg 3 times daily) should be commenced promptly and continued for 2–3 weeks in the case of the life-threatening HSV diseases, encephalitis, neonatal herpes, and disseminated infection in immunocompromised patients. Oral acyclovir (200 mg 5 times daily for 10 days) is appropriate treatment for primary herpes genitalis and may be of some value for unusually severe primary orofacial herpes. Any benefit of treating recurrent herpes genitalis or herpes labialis is marginal. Acyclovir has been used cautiously during pregnancy for maternal infection and in selected cases late in pregnancy when it is unlikely to carry any risk of fetal damage. To date, fetal abnormality has not been associated with the use of acyclovir during pregnancy, though a theoretical risk remains.

A case can be made for prolonged oral prophylaxis (200 mg 2 or 3 times daily) to suppress recurrences in patients with a history of unusually severe and/or frequent attacks of herpes genitalis. Because this strategy invites the emergence of drug-resistant mutants, those on long-term prophylaxis should be monitored for this, as well as for reduced creatinine clearance as an index of impaired kidney function. Chemoprophylaxis is also justified in patients receiving an organ transplant, or in immunocompromised patients with any clinically apparent mucocutaneous HSV infection, or for the prevention of neonatal herpes in a baby delivered vaginally from a mother with proven primary herpes genitalis at the time of delivery. In none of these situations is acyclovir a panacea, and it does not prevent the establishment of latency; at best it can be claimed that early administration of the drug will restrict the progression of the disease and ameliorate symptoms.

Some of the newer nucleoside analogs now undergoing clinical trials may prove to be superior to acycloguanosine. Older antiviral agents such as adenine arabinoside (vidarabine) and phosphonoformate (foscarnet) are too toxic to be recommended for general use, but the latter can be used to treat acyclovir-resistant life-threatening HSV infections, for example, in AIDS patients (see Chapter 16).

Vaccines

The development of herpesvirus vaccines has proved to be difficult. Prevention of the primary mucocutaneous infection will require a strong T-cell memory response and perhaps also a high titer of type-specific mucosal IgA antibody. Prevention of establishment of latency may be impossible because virus gains access to nerve ganglia not by viremic spread but by retrograde axonal transport. Nevertheless, research is proceeding on a number of fronts, particularly with a view to the prevention of genital and neonatal herpes; delivery of such a vaccine could be delayed until puberty, when a significant proportion of recipients would already have been primed by natural infection with HSV-1. Various HSV glycoproteins elicit neutralizing antibodies, but the most important appears to be gD. Several of the experimental HSV vaccines at various stages of development or testing consist of genetically cloned gD (or gD plus gB) associated with an adjuvant, whereas others are live vaccinia/HSV-gD recombinants, or live attenuated HSV mutants, genetically engineered by deletion of particular genes to be nonneurovirulent or replication-defective mutants.

Varicella–Zoster Virus

A single herpesvirus known as varicella–zoster virus (VZV) is responsible for two almost universal human diseases: varicella (chickenpox), one of the exanthemata of childhood, and herpes zoster (shingles), a disabling disease of aged persons and immunocompromised patients.

Pathogenesis and Immunity

Varicella virus enters by inhalation and replicates initially in the mucosa of the respiratory tract and oropharynx. Its progression during the 10 to 20-day incubation period (typically 14 days) is presumed to be comparable to that seen in other generalized exanthemata. Dissemination occurs via lymphatics and the bloodstream, and virus multiplies in mononuclear leukocytes and capillary endothelial cells. Eventually the rash results from multiplication of virus in epithelial cells of the skin; prickle cells show ballooning and intranuclear inclusions, and virus is plentiful in the characteristic vesicles. At this time, virus is presumed to ascend the axons of various sensory nerves to localize in sensory ganglia, where it becomes latent for life until reactivated by immunosuppression, as discussed in Chapter 10. The primary infection is generally rapidly controlled by T-cell-mediated immunity. Varicella usually induces prolonged immunity, but subclinical reinfections may occur more frequently than once thought, and mild disease is sometimes seen, particularly in immunocompromised individuals.

Herpes zoster occurs when varicella virus in a sensory ganglion is reactivated and descends the sensory nerve within the axon. In contrast with herpes simplex, VZV latency is tight; herpes zoster occurs in only 10–20% of those previously infected with VZV, it is mainly confined to the elderly, and more than a single attack is exceptional. Zoster annually affects about 1% of persons aged 50–60 years, with the incidence climbing rapidly thereafter to the point where most 80 year olds have suffered an attack. The condition is particularly common in patients suffering from Hodgkin's disease, lymphatic leukemia, or other malignancies, or following treatment with immunosuppressive drugs or irradiation of or injury to the spine. A decline in the level of cell-mediated immunity is thought to precipitate the attack of herpes zoster, and the protracted course of the disease in older or immunocompromised patients is presumably due to their weaker cell-mediated immune response.

Clinical Features

Varicella

The rash of chickenpox appears suddenly, with or without prodromal fever and malaise. Erupting first on the trunk, then spreading centrifugally to the head and limbs, crops of vesicles progress successively to pustules then scabs. Though painless, the lesions are very itchy, tempting the child to scratch, which may lead to secondary bacterial infection and permanent scarring. Painful ulcerating vesicles also occur on mucous membranes such as the mouth and vulva. The disease tends to be more severe in adults, in whom potentially life-threatening varicella pneumonia occurs quite commonly (see Fig. 36-3B).

Neurologic complications are uncommon but potentially serious. In about 1 case in 1000, encephalitis develops a few days after the appearance of the rash; this can be lethal, particularly in adults. Rarer neurologic developments include the Guillain-Barré syndrome and Reye's syndrome (see Chapter 36).

Varicella is a particularly dangerous disease in immunocompromised persons or in nonimmune neonates. Children with deficient cell-mediated immunity, whether congenital or induced by malignancy (e.g., leukemia), anticancer therapy, or steroid therapy, are especially vulnerable. Not only may the skin manifestations be necrotizing and hemorrhagic, but the disease becomes disseminated, involving many organs including the lungs, liver, and brain.

If women have not been infected as children, varicella tends to be more serious in pregnancy and may affect the fetus. Infections occurring in the few days immediately before or after parturition in a nonimmune woman can be particularly dangerous, since the baby has no maternal antibody to protect it and may die from disseminated varicella. Very rarely, maternal infection in the first half of pregnancy has been associated with congenital malformations in the fetus (cutaneous scarring, limb hypoplasia, and eye abnormalities).

Herpes Zoster

Herpes zoster results from reactivation of virus which has remained latent in one or more sensory ganglia following an attack of chickenpox many years earlier. Vesicles are generally unilateral and confined to the area of skin innervated by a particular sensory ganglion (*zoster,* girdle), usually on the trunk or on the face involving the eye (Fig. 20-7); scattered lesions outside the dermatome primarily affected may also occur. The accompanying pain is often very severe for up to a few weeks, but postherpetic neuralgia, which occurs in half of all patients over 60 years of age, may persist for many months. Motor paralysis and encephalomyelitis are rare complications. Disseminated (visceral) zoster is sometimes seen in cancer patients or those otherwise immunocompromised.

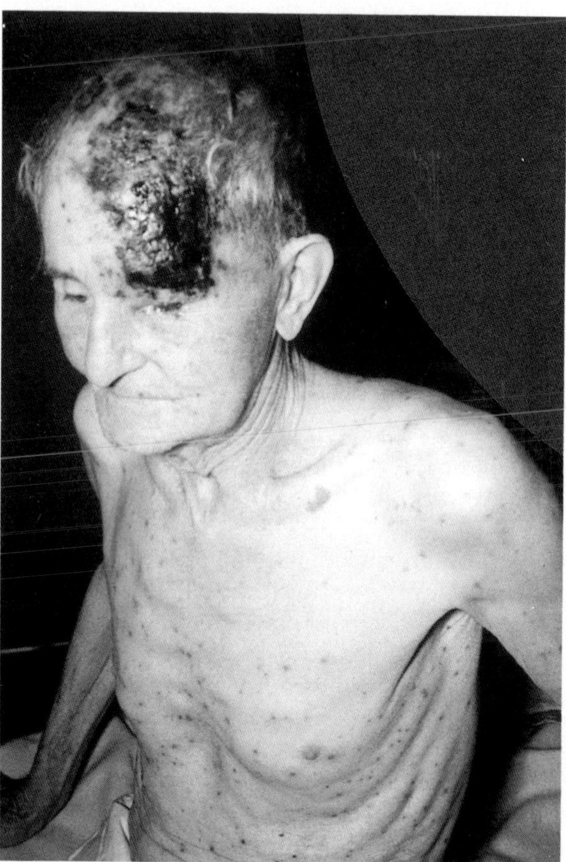

Fig. 20-7 Herpes zoster. Note unilateral distribution of the lesions on the dermatome supplied by the ophthalmic division of the trigeminal nerve. In addition there are scattered lesions elsewhere on the skin (disseminated zoster), not an uncommon complication of the disease. (Courtesy Dr. J. Forbes.)

Laboratory Diagnosis

The clinical picture of both varicella and herpes zoster is so distinctive that the laboratory is rarely called on for assistance. Smears from the base of early skin lesions, or sections from organs taken at autopsy, may be fixed and stained to reveal the characteristic intranuclear inclusions within multinucleated giant cells, but positive identification using monoclonal fluorescent antibody is the rapid diagnostic technique of choice. Alternatively, EIA can be used to demonstrate VZV antigens in vesicle fluid. Finally, PCR can be used to amplify DNA extracted from virions in vesicle fluid, for detection by nucleic acid hybridization.

The virus can be isolated from early vesicle fluid in cultures of human embryonic lung fibroblasts; however, virus tends to remain cell-associated, very little being released, and hence the cytopathic effect (CPE) develops slowly in distinct foci over a period of 2 or more weeks (Fig. 20-4B). VZV antigen can be demonstrated in nuclear inclusions by immunofluorescence before the end of the first week.

Recent infection can also be confirmed by detecting a rising titer of antibody, or by IgM serology preferably using EIA. Immune status, for example, of potentially vulnerable leukemic children following contact, can be determined rapidly using a latex agglutination test. Specific tests for cell-mediated immunity are also used for research purposes and vaccine assessment.

Epidemiology and Control

Varicella occurs throughout the year but is most prevalent during late winter and spring. Epidemics occur among groups of susceptible children, for example, in schools or children's hospitals. Most children become infected during their first years at school. Spread probably occurs via airborne respiratory droplets generated from vesicles on oropharyngeal mucosa, as well as by contact with skin lesions or fomites. Children are highly contagious and should be excluded from school for as long as moist vesicles are present on the skin; 1 week is normally sufficient.

Passive immunization with zoster immune globulin (ZIG), obtained originally from convalescent zoster patients, has an important place in the prevention of varicella. The ZIG should be administered to nonimmune pregnant women who have come into close contact with a case of varicella within the preceding 3 days but is ineffectual if further delayed. If a pregnant woman has actually contracted varicella within the few days before (or very soon after) delivery, the probability of disseminated disease in the baby may be reduced by injecting both the mother (immediately) and the baby (at birth) with ZIG. Administration of ZIG is also indicated for immunocompromised patients who become exposed to the risk of infection; a typical crisis situation would be the occurrence of a case of varicella in the leukemia ward of a children's hospital.

Chemotherapy

Varicella in the normal child can generally be managed by prevention of itching, scratching, and secondary bacterial infection. However, varicella pneumonitis requires vigorous treatment with intravenous acyclovir (10 mg/kg 3 times daily for 7–10 days), as does herpes zoster when it involves the eye. Acyclovir may be given orally (800 mg 5 times daily for 7 days) to accelerate healing in other severe cases of zoster, provided it is commenced promptly.

Vaccine

Japanese and American workers have produced a live attenuated vaccine from the Oka strain of VZV, derived by serial passage in cultured human and guinea pig fibroblasts. A single injection protects about 90% of recipients for several years at least. Protection is somewhat lower in immunocompromised children, who constitute the principal target group for such a vaccine. The vaccine induces fever and a few skin papules, occasionally in normal children but much more frequently in immunocompromised children. For instance, a significant minority of children with leukemia, or on steroid therapy (e.g., for nephrotic syndrome), developed mild varicella following vaccination. The attenuated vaccine often establishes latent infection in dorsal ganglia and may lead to zoster in the years ahead, but there is evidence that such reactivation may occur less frequently than following natural varicella infection. At the time of writing, the Oka vaccine is licensed for general use in normal children in Japan and for use in immunocompromised children in several other countries.

Cytomegalovirus

Throughout much of the world cytomegalovirus (CMV) infection is acquired subclinically during childhood, but in some of the more affluent communities it tends to be delayed until an age when it is capable of doing considerably more damage. In particular, primary infections during pregnancy can lead to severe congenital abnormalities in the fetus. Iatrogenic infections may follow blood transfusion, or immunosuppression, for example, for organ transplantation, and CMV is a major cause of blindness or death in AIDS patients. Of all the herpesviruses, CMV is the one responsible for most morbidity and mortality in immunocompromised hosts.

Pathogenesis, Immunity, and Epidemiology

Once infected with CMV, an individual carries the virus for life and may shed it intermittently in saliva, urine, semen, cervical secretions, and/or breast milk. Up to 10% of people may be found to be shedding virus at any time, especially young children. The intermittent nature of CMV shedding and the fluctuations observed in antibody levels suggest that asymptomatic exacerbations occur on several occasions throughout life. For example, reactivation occurs during pregnancy, rising markedly as term approaches. Hormonal factors may be at work here, but immunosuppression is generally the most powerful trigger. CMV is one of the commonest causes of death in AIDS patients as well as in recipients of grafts, especially following bone marrow transplantation. The virus can be isolated from over 90% of patients profoundly immunosuppressed for organ or tissue transplantation. Such infections generally involve reactivation of a latent (or low-level chronic) infection that has been lying dormant in cells of either the donor or the recipient of the graft.

Relatively little is known about the pathogenesis of CMV infection and the mechanism of latency. During the viremia observed in acute infection, whether primary or reactivated, virus can be recovered from monocytes, polymorphs, and to a lesser extent T lymphocytes. That these and other cells are potentially permissive has been confirmed by *in vitro* cultivation of CMV in monocytes, endothelial cells, vascular smooth muscle cells, and some $CD8^+$ T cells, but not B cells. However, it is almost impossible to reactivate CMV by cocultivation of leukocytes from healthy carriers with susceptible fibroblasts *in vitro*. PCR or *in situ* hybridization studies reveal that only about 1% of peripheral blood mononuclear cells from carriers contain the viral genome and that only the major immediate-early gene (IE1) is transcribed and translated. There is evidence that the viral genome may persist principally in endothelial cells, stromal cells, and/or ductal epithelial cells in salivary glands and renal tubules, from which virus is shed into saliva and urine, respectively. In summary, it is not yet clear which cell types constitute the principal reservoir of the viral genome, nor whether persistence is maintained by a continuous low-level chronic productive infection or by true latency in which episome copy numbers are maintained but expression of most genes is restricted until derepressed by immunosuppression.

Cell-mediated immunity appears to be principally responsible for controlling CMV. Studies of murine CMV in mice and of human CMV in humans have shown that NK cells are important early in infection, and that $CD8^+$ T lymphocytes directed at the major immediate-early protein, IE1, confer protection. Neutralizing antibodies directed mainly against the envelope glycoprotein gB, and to a lesser extent gH, may contribute to protection, but exogenous reinfection can occur. It is uncertain whether this is generally with a different strain; more than one strain has been isolated concurrently from a single individual.

Although CMV infection elicits a CMV-specific cell-mediated immune response, it can also be generally immunosuppressive, thereby predisposing to secondary infection with bacterial or fungal agents. It is unclear whether this should be ascribed to functional impairment of antigen-presenting cells or of T cells as a result of productive or abortive replication.

A number of characteristics of infection by CMV may contribute to its ability to persist in the body by evading the immune system. First, the virus multiplies extremely slowly and can spread contiguously from cell to cell by fusion, so escaping neutralization by antibody. Second, CMV infection reduces cell surface expression of class I MHC proteins, thus perhaps protecting infected cells to some degree from lysis by cytotoxic lymphocytes. A CMV protein, UL18, which displays homology with the heavy chain of MHC class I proteins, can interact with β_2-microglobulin, thereby not only interfering with T-cell recognition but also coating the free virion and protecting it from antibody. Further, like other herpesviruses, CMV encodes a protein with the functional characteristics of an Fc receptor, which protects the plasma membrane of the infected cell against immune attack, as described above for HSV.

Transplacental infection with CMV is now the commonest viral cause of prenatal damage to the fetus. Approximately 1% of all babies become infected *in utero*. Whereas the majority of maternal infections are endogenous recurrences (reactivation) and are generally uneventful, most cases of overt cytomegalic inclusion disease result from primary infections occurring during the first 6 months of pregnancy. Hence this syndrome tends to be a disease of affluence, as over 50% of women in developed countries, compared with less than 10% in Third World countries, are still seronegative as they enter the child-bearing years. Primary infection during the first 6 months of pregnancy carries a 30–40% risk of prenatal infection and 10–15% risk of clinical abnormalities in the neonate. The risk of fetal infection following recurrent CMV infection during pregnancy is 0.5–1%, with a low risk of abnormality. Higher titers of virus are transmitted to babies from mothers with primary than with recurrent infection. The preexisting immunity present in the latter confers considerable protection against disease in the fetus, and major deficits other than unilateral deafness are rare in babies infected prenatally as a result of reactivation of maternal CMV.

A minority of babies with clinical abnormalities at delivery are stillborn or die shortly after birth. Autopsy may reveal fibrosis and calcification in the brain and liver. Typical cytomegalic cells may be found in numerous organs; indeed, they may be found in salivary glands or renal tubules of many normal children (Fig. 20-8A). Progressive damage may occur not only throughout pregnancy but also after birth. The affected infant synthesizes specific IgM, and immune complexes are plentiful, but CMV-specific and nonspecific cell-mediated immune responses are markedly depressed.

Postnatal or natal (intrapartum) infection is commoner than prenatal infection and may occur via at least three different routes. Because most maternal infections occurring at this time are not primary but reactivated infections, the neonate is at least partially protected by preexisting maternal IgG antibodies and therefore only rarely develops disease. Approximately 10% or more of women shed CMV from the cervix at the time of delivery; some of the babies who become infected subclinically may acquire the infection during delivery. Second, 10–20% of nursing mothers shed CMV in their milk, and their infants have a 50% chance of becoming infected via breast-feeding; such infections are subclinical and perhaps the most common mode of transmission of CMV in the neonatal period. Later, oropharyngeal secretions are believed to constitute the principal vehicle of transmission, not only in childhood but again in adolescence, via direct contact or contamination of hands, eating utensils, etc. Kissing and sexual contact no doubt account for the sudden increase in CMV seropositivity from 10–15% to 30–50% between the ages of 15 and 30 in countries such as the United States, though it is difficult to determine unequivocally the relative importance of these two routes. CMV is also shed intermittently in cervical secretions and in semen, hence sexual transmission may be significant; for example, CMV infection is almost universal among promiscuous male homosexuals.

The two remaining mechanisms of acquiring CMV, blood transfusion and organ transplantation, constitute special cases of iatrogenic infection. Almost all those who receive multiple transfusions of large volumes of blood develop CMV infection at some stage. Most such episodes are subclinical, but the mononucleosis syndrome is not uncommon. More serious manifestations of

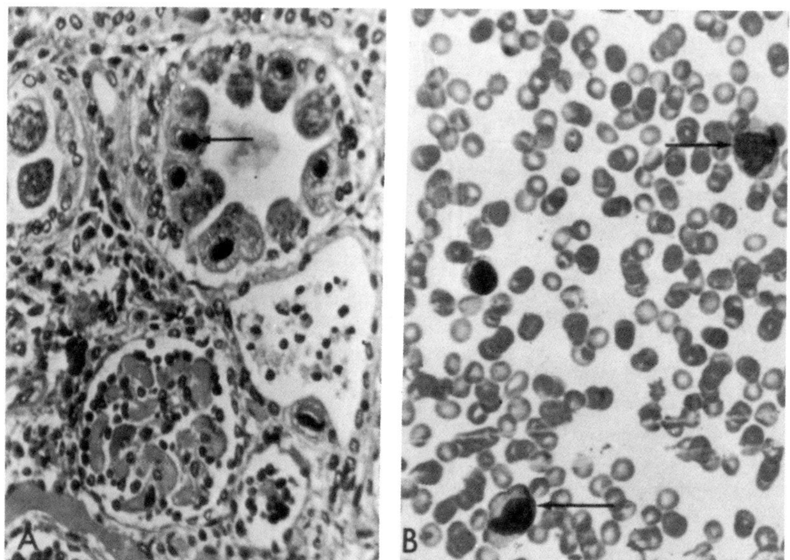

Fig. 20-8 Histopathology of some diseases caused by herpesviruses. (A) Cytomegalic inclusion disease. Section of kidney (hematoxylin and eosin stain; magnification: ×250). Arrow indicates one of several "cytomegalic" cells inside a renal tubule. The typical cytomegalic cell is greatly swollen, with an enlarged nucleus distended by a huge inclusion which is separated by a nonstaining halo from the nuclear membrane, giving the cell the appearance of an owl's eye. (B) Infectious mononucleosis (glandular fever). Smear of peripheral blood (Leishman stain; magnification: ×400). Note large "atypical lymphocytes" (arrows). (Courtesy I. Jack.)

primary infection may occur following transfusion of seropositive (infected) blood into seronegative premature infants, pregnant women, or immunocompromised patients. CMV infection of recipients of kidney, heart, liver, or bone marrow transplants may also be of exogenous origin, introduced via the donated organ or via accompanying blood transfusions. Such an exogenous infection may be primary or may be a reinfection with a different CMV strain. On the other hand, the profound immunosuppression demanded for organ transplantation, or indeed for other purposes such as for cancer therapy, is quite sufficient to reactivate previous infection. Primary CMV infections, in particular, are serious and often lethal in immunocompromised patients. They are also often associated with rejection of the organ graft, for example, glomerulopathy in a transplanted kidney. Reactivation of latent CMV is also one of the commonest opportunistic infections leading to death in AIDS patients.

The relationship between the circumstances of transmission of CMV and the commoner clinical outcomes is summarized in Table 20-4.

Clinical Features

Prenatal Infection (Cytomegalic Inclusion Disease)

Cytomegalovirus infection during pregnancy is now the major viral cause of congenital abnormalities in the newborn. Approximately 0.5–2% of babies are born with asymptomatic CMV infection, and 1 in 2000 has signs of cyto-

Table 20-4
Cytomegalovirus Infections

Age or immunocompetence	Route	Disease caused by primary infection
Prenatal	Transplacental	Encephalitis, hepatitis, thrombocytopenia; long-term sequelae include brain damage, nerve deafness, retinopathy
Perinatal	Cervical secretions, breast milk, saliva	Nil
	Blood transfusion	Pneumonitis, disseminated disease
Any age	Saliva or sexual intercourse	Mononucleosis, mild hepatitis
	Blood transfusion	Mononucleosis
Immunocompromised[a]	Saliva, sex, organ graft	Pneumonia, hepatitis, retinitis, encephalitis, myelitis, gastrointestinal disease

[a] Diseases shown occur less commonly after reactivation of a latent infection.

megalic inclusion disease (CID) (Fig. 20-9). The classic syndrome is not always seen in its entirety. The infant is usually small, with petechial hemorrhages, jaundice, hepatosplenomegaly, microcephaly, encephalitis, and sometimes choriorentinitis and/or inguinal hernia. The abnormalities of the brain and eyes are associated with mental retardation, cerebral palsy, impairment of hearing, and rarely impairment of sight. Many of these infants require special care for life. Socially and educationally important intellectual or perceptual deficits such as hearing loss, subnormal IQ, epilepsy, and behavioral problems may not become apparent until as late as 2–4 years after birth.

Mononucleosis

Most infections acquired after birth, by whatever route, are subclinical. Not uncommonly, however, a syndrome resembling EB virus mononucleosis but milder is seen, particularly in young adults and in recipients of blood transfusions. Typically, the patient presents with prolonged fever and on examination is found to have splenomegaly, abnormal liver function tests, and lymphocytosis, often with "atypical lymphocytes" such as those observed with EB virus infections (see below). In contrast to EB virus mononucleosis, however, pharyngitis and lymphadenopathy are uncommon, and heterophile antibody is absent.

Other Clinical Presentations

Cytomegalovirus infections may be widely disseminated and may present in a wide variety of more serious forms, particularly in immunocompromised patients (e.g., AIDS patients or recipients of organ transplants) or in premature infants following blood transfusion. Almost any organ may be seriously affected. The most important presentations are interstitial pneumonia, hepatitis, chorioretinitis, arthritis, carditis, chronic gastrointestinal infection, and various CNS diseases, especially encephalitis, Guillain-Barré syndrome, and transverse myelitis.

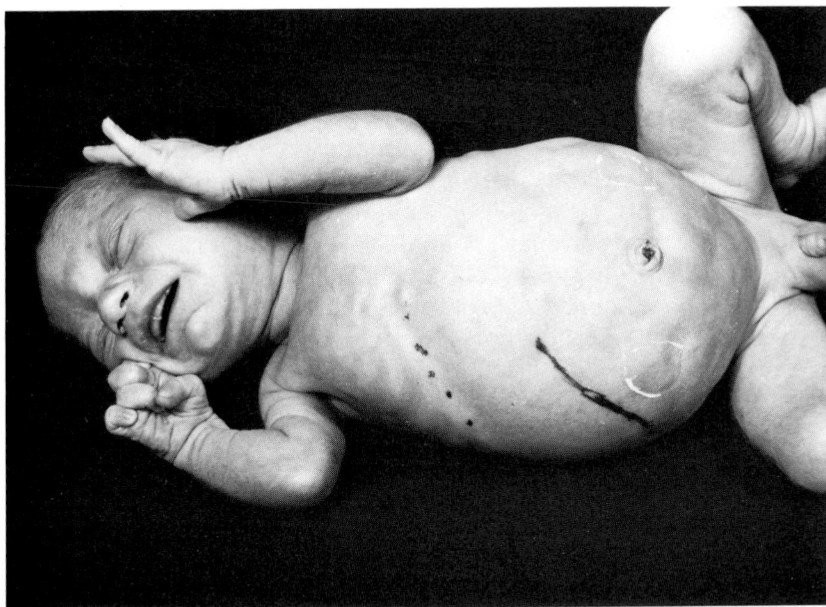

Fig. 20-9 Congenital cytomegalic inclusion disease. Features are retardation of growth, microcephaly, thrombocytopenia, and hepatosplenomegaly. The edge of the liver has been marked. (Courtesy Dr. K. Hayes.)

Laboratory Diagnosis

Laboratory confirmation of cytomegalic inclusion disease (CID) in a newborn infant is of great importance as it affects the medical and educational management of the child and facilitates parental counseling. Further, life-threatening CMV infections in immunocompromised persons are becoming increasingly common, and early diagnosis is needed to instigate appropriate antiviral therapy. In particular clinical circumstances it may be important not only to detect CMV but to discern whether the patient is undergoing an active acute infection rather than a latent or chronic low-grade infection, and if so, whether that acute episode is a primary infection of exogenous origin or an endogenous reactivation of a persistent infection in that individual. Because virus can be detected intermittently in urine or saliva of asymptomatic carriers, these specimens are suitable only for epidemiologic surveys or for detection of virus (or viral DNA or viral antigen) in newborn infants with suspected CID. The more relevant specimen in other clinical situations is blood leukocytes, because cell-associated viremia, especially if high-titer, correlates with invasive disease. Quantitative assays for virus load are likely to be developed and used to assess the need for prompt antiviral therapy. Other specimens appropriate to particular clinical presentations include bronchoalveolar lavage and various organs taken at biopsy or autopsy.

Cytomegalovirus is isolated in cultured human fibroblasts; sensitivity can be increased by sedimenting the specimen onto the cells by low-speed centrifugation. The virus replicates very slowly, a single cycle having a latent period of 36–48 hours. Moreover, new virus tends to remain cell-associated, with spread occurring mainly to adjacent cells; hence CPE may not become evident for up to a month, but foci of swollen refractile cells with cytoplasmic granules are detectable (Fig. 20-4C). When stained, these cells are found to contain a number of nuclei with pathognomonic large amphophilic "skeinlike" nuclear inclusions and smooth round acidophilic cytoplasmic masses (Fig. 20-4D). As early as 24–36 hours after inoculation the monolayer can be stained by the immunofluorescence (IF) or immunoperoxidase techniques, using a monoclonal antibody against an immediate-early nuclear antigen; obviously this expedites the diagnosis but is not as sensitive as awaiting the development of CPE. There is only a single serotype of human CMV, but different strains can be distinguished by kinetic neutralization or restriction endonuclease mapping. The latter technique can be employed to identify the source of virus.

Amplification assays involving the PCR have been developed for the detection of CMV DNA and are at least as sensitive as virus culture, but much faster. The readout can be anything from simple ethidium bromide staining of the product following separation by electrophoresis on an agarose gel, to dot-blot or other appropriate methods of DNA hybridization (see Chapter 12). If meticulously conducted, PCR promises to be more reproducible than detection of antigenemia by immunofluorescence or immunoperoxidase staining of peripheral blood leukocytes with monoclonal antibody.

Discrimination of primary infection from endogenous reactivation is not easy in the absence of a recent negative prebleed. CMV-specific IgM is the only available indicator and can be detected by EIA using genetically cloned matrix phosphoprotein pp150 as antigen. As IgM does not cross the placenta, finding CMV-specific IgM in a newborn baby is diagnostic of prenatal infection.

Blood and organ donors can be screened for latent CMV infection, and intended recipients screened for susceptibility, by testing for antibody against the immunodominant pp150 antigen, using a rapid latex agglutination test or a more reliable EIA.

Control

Iatrogenic infection via transfusion or organ transplantation can be reduced by screening donor and recipient for evidence of CMV infection. The presence of antibody in the intended recipient indicates a significant degree of immunity, whereas antibody in the donor provides a warning that virus, or reactivatable viral episomes, may be transmitted to the recipient. The potentially dangerous combination is averted if seronegative recipients are given organs or blood taken only from seronegative donors. The amount of testing involved is not cost-effective for routine blood transfusions but should be considered when the recipients are premature infants, pregnant women, or immunocompromised individuals. Removal of the leukocytes by filtration of donor blood effectively prevents transmission of CMV.

Following laboratory diagnosis of primary CMV infection in pregnancy, it is possible to diagnose prenatal CMV infection in the fetus by amniocentesis followed by PCR and/or isolation of virus from amniotic fluid.

Chemotherapy

Ganciclovir (see Chapter 16 for details) is the drug of choice for the treatment of severe CMV infections such as pneumonia, chorioretinitis, or colitis in AIDS patients and recipients of organ grafts. It may also be used prophylactically, for example, promptly following detection of virus in the blood or in bronchoalveolar lavage fluid of bone marrow transplant recipients. When administered intravenously for up to 3 months ganciclovir often displays hematologic toxicity, and resistant mutants can emerge. Foscarnet is even more toxic but is a useful second-line drug when CMV develops resistance to ganciclovir (see Chapter 16).

Vaccines

Live attenuated CMV vaccine strains have been developed by serial passage in human fetal fibroblasts. Although the vaccines appear to confer some protection against disease they do not protect against infection, and therefore presumably latent infection can still develop. The question of the safety of live CMV vaccines must also be addressed, particularly in relation to adequacy of attenuation, establishment of persistence, subsequent reactivation, and possible oncogenicity. Alternative products that have proved to be at least partially protective in mice challenged with murine CMV include genetically cloned vaccines comprising the gB envelope glycoprotein, which elicits neutralizing antibody, and live vaccinia virus incorporating the nucleotide sequence encoding an epitope from the immediate-early protein IE1, which elicits a cytotoxic T-cell response; neither is close to human application.

Human Herpesvirus 6

Human herpesvirus 6 (HHV-6) was discovered in human lymphocytes in 1986. Since then it has been shown to be ubiquitous, infecting most children worldwide in the first year or two of life. The virus causes a generally harmless febrile illness sometimes associated with a rash which has been known since the early twentieth century as exanthem subitum, roseola infantum, or sixth disease.

Properties of Human Herpesvirus 6

The biological properties of HHV-6 resemble those of CMV, and sequencing of the genome (Fig. 20-2) has confirmed a taxonomic relationship. The genome occurs as a single isomer consisting of a unique 142-kb segment flanked by a direct repeat sequence of 10–13 kb in single copy at each end. HHV-6 is now classified as a species within a newly defined genus, *Roseolovirus*, within the subfamily *Betaherpesvirinae*. There are two variants of HHV-6, HHV-6A and HHV-6B, distinguishable of restriction endonuclease mapping and reactivity with subsets of virus-specific monoclonal antibodies. Group A strains have been isolated mainly from adults, whereas most exanthem subitum isolates have been group B.

Pathogenesis

The principal target cell of HHV-6 appears to be the dividing $CD4^+$ T lymphocyte. Infected T cells show ballooning, often with more than one nucleus and nuclear and/or cytoplasmic inclusions, before eventually dying. Macrophages are persistently infected and may comprise an important reservoir. Transformed B lymphocytes, NK cells, megakaryocytes, glial cells, fibroblasts, and epithelial cells have also been reported to support productive replication of certain HHV-6 strains in culture.

The virus persists in the body, but the mechanism of latency is not known. HHV-6A is readily demonstrable by *in situ* hybridization and immunofluorescence in salivary glands, and is regularly isolated from saliva, suggesting that salivary glands may represent a major reservoir and saliva the main route of transmission. Reactivation is precipitated by immunosuppression, as used for bone marrow or organ transplantation, or in AIDS patients. This has been demonstrated mainly by a rise in anti-HHV-6 IgG titer, indicative of either exogenous infection or reactivation of an endogenous latent infection; reactivation has not been correlated with any particular symptoms or with renal graft rejection even though HHV-6 is often found in the kidney and in urine of asymptomatic immunocompetent people. There is evidence that HHV-6 up-regulates HIV expression in $CD4^+$ T cells.

Clinical Features

A recent large U.S. prospective study found that a remarkable 14% of febrile infants under 2 years of age yielded HHV-6 from their blood, although most did not show the classic rash of exanthem subitum (roseola). The commonest presentation was a high fever, with irritability and malaise, lymphadenopathy (particularly involving suboccipital nodes), and mild injection of the pharynx and tympanic membranes; a rash was seen in about 10% of the infants. Encephalopathy may rarely occur, and there have been isolated reports of other possible clinical associations, such as hepatitis; however, causal relationships have not been demonstrated except with exanthem subitum.

Laboratory Diagnosis

Virus can be isolated from peripheral blood mononuclear cells of roseola patients in the early febrile stage of the illness, or from saliva of adults intermittently throughout life, by centrifugation-enhanced infection of phytohemagglutinin-activated peripheral blood leukocytes or of T-cell lines. HHV-6 antigen can be identified in infected cells by immunofluorescence using an appropriate MAb. EIA and immunoblot assay have been successfully applied to detection of both antigen and antibodies. The PCR is also available for detection of the genome in productive or latent infection.

Epidemiology and Control

Serological surveys indicate that almost all children have become infected by 2–3 years of age, strongly suggesting intrafamilial spread, probably from the mother shortly after maternal antibodies have declined in the infant. The high incidence of shedding of virus in saliva points to this as the likely, but not necessarily the only, natural route of transmission.

Chemotherapy

As expected from its similarity to CMV, HHV-6 is susceptible to ganciclovir and foscarnet but relatively resistant to acyclovir.

Epstein-Barr Virus

Pathogenesis

Epstein-Barr virus (EBV) replicates in epithelial cells of the nasopharynx and salivary glands, especially the parotid, lysing them and releasing infectious virions into saliva. The B lymphocytes infiltrating the lymphatic tissue of the infected oropharyngeal mucosa may in turn become infected but are generally not permissive for virus production. EBV binds via its envelope glycoprotein gp350/gp220 to the CR2 complement receptor, CD21, and enters the B cell, where it normally fails to replicate but establishes lifelong latency (see Chapter 10 for detailed description). The genome persists as a plasmid, relatively few genes of which are expressed, but at least one of the viral gene products, EBNA-2, immortalizes the B cell. Each resulting B-cell clone produces B-cell growth factor and secretes its own characteristic monoclonal antibody. The "heterophile" antibodies that result from this polyclonal B-cell activation represent just one manifestation of the immunologic chaos that characterizes the acute phase of infectious mononucleosis. There is also a general depression of cell-mediated immunity with an increase in suppressor T cells. Meanwhile, clones of $CD8^+$ cytotoxic T lymphocytes that recognize class I MHC bound peptides derived from EBV proteins EBNA-2 to EBNA-6 and latent membrane protein (LMP) become activated to proliferate and lyse B cells and oropharyngeal epithelial cells expressing that protein on their surface. It is these cytotoxic T lymphoblasts, not the infected B cells, that comprise the pathognomonic "atypical lymphocytes" that characterize infectious mononucleosis (Fig. 20-8B). In individuals with congenital or acquired T-cell immunodeficiencies, the virus fails to be cleared and B-cell lymphomas or other lethal conditions overwhelm the patient.

Strains of EBV fall into two classes, based mainly on differences in sequence of the EBNA-2 gene. Type A, which is the most common in Europe and the United States, is regularly isolated from B lymphocytes, whereas type B is only recovered from B cells of immunocompromised individuals and otherwise appears to be confined to mucosal epithelia and secretions therefrom.

Clinical Features

Infectious Mononucleosis (Glandular Fever)

In young children EBV infections are asymptomatic or very mild. However, when infection is delayed until adolescence, as happens in developed countries, the result is often infectious mononucleosis. Following a long incubation period (4–7 weeks), the disease begins insidiously with headache, malaise, and fatigue. The clinical presentation is protean, but three regular features are fever, pharyngitis, and generalized lymphadenopathy. The fever is high and fluctuating, and the pharyngitis is characterized by a white or grey malodorous exudate covering the tonsils and may occasionally be so severe as to obstruct respiration. The spleen is often enlarged and liver function tests abnormal. Occasionally a rash may appear, especially following treatment with ampicillin; the reason for this strange association is unknown. The disease usually lasts for 2–3 weeks, but convalescence may be very protracted.

An extraordinary range of further developments can occasionally occur. Neurologic complications include the Guillain-Barré syndrome, Bell's palsy, meningoencephalitis, and transverse myelitis. Other complications include hemolytic anemia, thrombocytopenia, carditis, nephritis, or pneumonia.

Infection in Immunocompromised Hosts

A variety of syndromes associated with uncontrolled progression of infection occur in individuals with congenital or acquired inability to mount an adequate immune response to EBV. For example, a rare fatal polyclonal B-cell proliferative syndrome caused by EBV infection occurs in families with an X-linked recessive immunodeficiency associated with a reduced ability to synthesize interferon γ *(X-linked lymphoproliferative syndrome).* There is an inexorable expansion of virus-infected B cells and suppression of normal bone marrow cells, and about half of the boys die within a month from sepsis or hemorrhage; the remainder develop dysgammaglobulinemia or die from malignant B-cell lymphomas.

Much more common is *progressive lymphoproliferative disease,* seen in transplant recipients, immunodeficient children, or AIDS patients. In these immunocompromised individuals, the absence of cell-mediated immunity permits unrestrained replication of EBV, acquired exogenously or reactivated from the latent state. Some of the infections present as mononucleosis, but others are atypical, presenting, for example, as pneumonitis or hepatitis. Infants with AIDS develop a lymphocytic interstitial pneumonitis; adults with AIDS may develop EBV-associated lymphoproliferative conditions of various kinds.

Burkitt's Lymphoma and Nasopharyngeal Carcinoma

The reader is referred to Chapter 11 for a discussion of the highly malignant neoplasms Burkitt's lymphoma and nasopharyngeal carcinoma and their relationship to EBV.

Laboratory Diagnosis

The clinical picture of infectious mononucleosis is so variable that laboratory confirmation is required. This rests on (1) differential white blood cell counts; (2) heterophile antibodies, and (3) EBV-specific antibodies. By the second week of the illness, white blood cells total 10,000–20,000 per cubic millimeter or even higher. Lymphocytes plus monocytes account for 60–80% of this number. Of these, at least 10%, and generally more than 25%, are "atypical lymphocytes" (large pleomorphic blasts with deeply basophilic vacuolated cytoplasm and lobulated nuclei, Fig. 20-8B), which persist for 2 weeks to several months.

Many years ago Paul and Bunnell made the empirical observation that sera from glandular fever patients agglutinate sheep erythrocytes. It is now clear that the agglutinins are just one of the heterophil IgM antibodies elicited by EBV infection. The Paul–Bunnell test is still a valid diagnostic aid, however. Sera are first absorbed with guinea pig kidney cells to remove Forssman antibody (and serum-sickness antibodies, which are rare these days). Commercially available horse erythrocyte slide agglutination kits are convenient, although heterophile antibodies are made *in vivo* for only a month, results are often negative in children, and neither sensitivity nor specificity is high. Another assay for heterophile antibodies is the ox red cell hemolysin test, which requires no serum absorption and is specific but remains positive for only a month or two.

The EBV-specific antibodies offer a more reliable indicator of infection now that EIAs have become available to replace the cumbersome immunofluorescence readout that was previously employed. IgM antibody against the EBV capsid antigen VCA develops to high titer early in the illness and declines rapidly over the next 3 months or so; it therefore represents a good index of primary infection. However, because IgM assays tend to be plagued by false positives and negatives, the trend is toward using a panel of genetically cloned EBV antigens to screen for IgG antibodies and considering the resulting profile. Because antibodies to EBNA first appear a month or more after onset of disease and then persist, a rising titer is diagnostic. (It should be noted, however, that antibody to EBNA-1 is specifically missing from many immunocompromised patients with severe chronic active infection.) On the other hand, antibodies against the early antigen EA-D are diagnostic of acute, reactivated, or chronic active infection, as they decline rapidly and are not detectable in asymptomatic carriers. IgG (or total) antibodies against VCA represent the most convenient measure of past infection and immune status.

Isolation of virus is rarely used as a diagnostic procedure because no known cell line is fully permissive for EBV. The only method for "growing" EBV *in vitro* is to inoculate infected oropharyngeal secretions or peripheral blood leukocytes onto umbilical cord lymphocytes and to demonstrate the immortalization of the latter to produce a lymphoblastoid cell line that can be stained successfully with fluoresceinated monoclonal antibody against EBNA. The other problem with virus isolation (or use of PCR to detect viral DNA in infected cells) for routine diagnosis of EBV-induced disease is that such a high proportion of asymptomatic individuals carry the latent genome and/or shed virus for life.

Epidemiology

Following an attack of mononucleosis, virus is found in saliva for several months, and more sensitive assays reveal that most EBV-seropositive people secrete the virus in lower titer thereafter, continuously or intermittently, probably for life. Chronic shedding is highest in less developed countries, in young children, in early pregnancy, and in immunocompromised patients. EBV is ubiquitous in most developing countries. Almost all infants become infected in the first year or two of life, probably by salivary exchange, contamination of eating utensils, and perhaps also by respiratory aerosols. At this age almost all infections are subclinical; glandular fever is virtually unknown in the Third World. By way of contrast, in countries with higher standards of living many persons reach adolescence before first encountering the virus. The intimate osculatory contact involved in kissing is the principal means of transmission; hence, mononucleosis is a disease of 15–25 year olds. Some 80% of people eventually acquire infection and are permanently immune.

As with CMV, conventional respiratory transmission of EBV by droplets does not appear to be significant. For example, casual roommates of mononucleosis patients are not at increased risk; closer contact seems to be required. By analogy with CMV it may be reasonable to speculate that EBV can also be transmitted by sexual intercourse. Male homosexuals have a very high rate of seropositivity. Blood transfusions can also rarely transmit the virus.

Control

Chemotherapy

Unfortunately none of the present armory of nucleoside analogs has been unequivocally shown to have any effect on EBV infections.

Vaccines

If a successful vaccine could be developed against EBV it may be feasible to prevent not only mononucleosis but also nasopharyngeal carcinoma (see Chapter 11). However, formidable barriers will delay progress. A conventional live or inactivated vaccine is a long way off since the virus has yet to be grown satisfactorily in cultured cells. Research has commenced on "subunit" and recombinant vaccines based on the major membrane glycoprotein complex gp350/gp220, and live vaccinia or attenuated varicella virus constructs containing the corresponding gene.

Herpes B Virus

Macaque monkeys suffer from infection with an alphaherpesvirus known variously as herpes B virus, herpesvirus simiae, or cercopithecine herpesvirus 1. Its natural history is very like that of HSV-1 infection in humans. A number of fatal cases of ascending paralysis and encephalitis in animal handlers have followed the bite of a monkey, and herpes B virus is a continuing risk to personnel working with primates, as in zoos or laboratories.

Further Reading

Adler, S. P. (1990). New insights into human cytomegaloviral infections. *Prog. Med. Virol.* **37,** 136.
Caserta, M. T., and Hall, C. B. (1993). Human herpesvirus-6. *Annu. Rev. Med.* **44,** 377.
Chou, S. (1992). Cytomegalovirus infection. *Curr. Opin. Infect. Dis.* **5,** 427.

Further Reading

Corey, L., and Spear, P. G. (1986). Infections with herpes simplex viruses. *N. Engl. J. Me*. 686 and 749.

Croen, K. D. (1991). Latency of the human herpesviruses. *Annu. Rev. Med.* **42,** 61.

Croen, K. D., and Straus, S. E. (1991). Varicella–zoster virus latency. *Annu. Rev. Microbiol.* **45,** 265.

Davison, A. J. (1991). Varicella–zoster virus. *J. Gen. Virol.* **72,** 475.

Davison, A. J., ed. (1993). HSV and other alphaherpesviruses. *Semin. Virol.* **4,** 123.

Demmler, G. J. (1991). Summary of a workshop on surveillance for congenital cytomegalovirus disease. *Rev. Infect. Dis.* **13,** 315.

Epstein, M. A., and Achong, B. G., eds. (1986). "The Epstein-Barr Virus: Recent Advances." Heinemann, London.

Evans, A. S., ed. (1989). "Viral Infections of Humans: Epidemiology and Control," 3rd Ed., Chapters 8, 10, 14, and 25–28. Plenum, New York.

Fields, B. N., Knipe, D. M., Chanock, R. M., Hirsch, M. S., Melnick, J. L., Monath, T. P., and Roizman, B., eds. (1990). "Fields Virology," 2nd Ed., Chapters 64–73, pp. 1787–2054. Raven, New York.

Galasso, G. J., Whitley, R. J., and Merigan, T. C., ed. (1990). "Antiviral Agents and Viral Diseases of Man," 3rd Ed., Chapters 6–8, 13, 14, 16, 17, and 19. Raven, New York.

Gershon, A. (1987). Live attenuated varicella vaccine. *Annu. Rev. Med.* **38,** 41.

Ho, M. (1991). "Cytomegaloviruses: Biology and Infection," 2nd Ed. Plenum, New York.

Hyman, R. W. (1987). "Natural History of Varicella–Zoster Virus." CRC Press, Boca Raton, Florida.

Jones, J. F., and Straus, S. E. (1991). Chronic Epstein-Barr virus infection. *Annu. Rev. Med.* **42,** 195.

Landini, M. P. (1993). New approaches and perspectives in cytomegalovirus diagnosis. *Prog. Med. Virol.* **40,** 157.

Lopez, C., Mori, R., Roizman, B., and Whitley, R. W. (1990). "Immunobiology and Prophylaxis of Human Herpesvirus Infections." Plenum, New York.

McDougall, J. D., ed. (1990). Cytomegaloviruses. *Curr. Top. Microbiol. Immunol.* **154.**

McGeoch, D. J. (1989). The genomes of human herpesviruses; contents, relationships, and evolution. *Annu. Rev. Microbiol.* **43,** 235.

Meyers, J. D. (1991). Prevention and treatment of cytomegalovirus infections. *Annu. Rev. Med.* **42,** 179.

Nahmias, A. J., Dowdle, W. R., and Schinazi, R. F., eds. (1986). "The Human Herpesviruses." Elsevier, New York.

Okano, M., Thiele, G. M., Davis, J. R., Grierson, H. L., and Purtilo, D. T. (1988). Epstein-Barr virus and human diseases: Recent advances in diagnosis. *Clin. Microbiol. Rev.* **1,** 300.

Okano, M., Matsumoto, S., Osato, T., Sakiyama, Y., Thiele, G. M., and Purtilo, D. T. (1991). Severe chronic active Epstein-Barr virus infection syndrome. *Clin. Microbiol. Rev.* **4,** 129.

Ostrove, J. M. (1990). Molecular biology of varicella–zoster virus. *Adv. Virus Res.* **38,** 45.

Pellett, P. E., Black, J. B., and Yamamoto, M. (1992). Human herpesvirus 6—the virus, and the search for its role as a human pathogen. *Adv. Virus Res.* **41,** 1.

Roizman, B., ed. (1982–1985). "The Herpesviruses," Vols. 1–5. Plenum, New York.

Roizman, B. (1991). Herpesviruses. *In* "Encyclopedia of Human Biology" (R. Dulbecco, ed.), Vol. 4, p. 187. Academic Press, Orlando, Florida.

Roizman, B., and Sears, A. E. (1987). An inquiry into the mechanisms of herpes simplex virus latency. *Annu. Rev. Microbiol.* **41,** 543.

Roizman, B., and Sears, A. E. (1990). Herpes simplex viruses and their replication. *In* "Fields Virology" (B. N. Fields, D. M. Knipe, R. M. Chanock, M. S. Hirsch, J. L. Melnick, T. P. Monath, and Roizman, B., eds.), 2nd Ed., p. 1795. Raven, New York.

Roizman, B., Desrosiers, R. C., Fleckenstein, B., Lopez, C., Minson, A. C., and Studdert, M. J. (1992). The family *Herpesviridae:* An update. *Arch. Virol.* **123,** 425.

Rouse, B. T., ed. (1992). Herpes Simplex Virus: Pathogenesis, Immunobiology and Control. *Curr. Top. Microbiol. Immunol.* **179.**

Rubin, R. H., ed. (1990). Cytomegalovirus Infections: Epidemiology, Diagnosis, and Treatment Strategies. *Rev. Infect. Dis.* **12** *(Suppl. 7).*

Schlossberg, D., ed. (1989). "Infectious Mononucleosis," 2nd Ed. Springer-Verlag, Berlin.

Stevens, J. G. (1989). Human herpesviruses: A consideration of the latent state. *Microbiol. Rev.* **53,** 318.

Straus, S. E. (1988). The chronic mononucleosis syndrome. *J. Infect. Dis.* **157,** 405.

Straus, S. E. (1992). Acute progressive Epstein-Barr virus infections. *Annu. Rev. Med.* **43,** 437.

Chapter 21

Poxviridae

The family *Poxviridae* is divided into two subfamilies, *Chordopoxvirinae* (poxviruses of vertebrates) and *Entomopoxvirinae* (poxviruses of insects); only the former are of importance in medicine. The subfamily *Chordopoxvirinae* contains eight genera, distinguished on the basis of genetic, antigenic, and morphologic differences. Several poxviruses causes diseases in humans: smallpox (now extinct), vaccinia (including a strain called buffalopox virus), monkeypox, molluscum contagiosum, cowpox, milker's nodes, orf, and tanapox (Table 21-1). Smallpox and molluscum contagiosum are specifically human diseases; the others are zoonoses.

All diseases caused by poxviruses are associated with skin lesions, which may be localized or may be part of a generalized rash, as in smallpox and human monkeypox. Smallpox, once one of the great plagues of mankind, which has played an important role in human history and a central role in the development of virology, was eradicated from the world in 1977 (see Chapter 15) and is not further discussed.

Properties of *Poxviridae*

The poxviruses are the largest and most complex of all viruses. Figure 21-1A,C,D illustrates the structure of the brick-shaped virion of vaccinia virus, which is characteristic of that of all the poxviruses affecting humans except those belonging to the genus *Parapoxvirus*, shown in Fig. 21-1B. There is no nucleocapsid conforming to either of the two types of symmetry found in

Table 21-1
Diseases Produced in Humans by Poxviruses

Genus	Disease	Clinical features
Orthopoxvirus	Smallpox (now extinct)	
	Variola major	Generalized infection with pustular rash; case–fatality rate 10–25%
	Variola minor	Generalized infection with pustular rash; case–fatality rate <1%
	Vaccination (vaccinia)	Local pustule, slight malaise
	Complications (rare)	Postvaccinial encephalitis; high mortality Progressive vaccinia; high mortality Eczema vaccinatum; low mortality Autoinoculation and generalized vaccinia; nonlethal
	Monkeypox	Generalized with pustular rash; case–fatality rate 15%
	Cowpox	Localized ulcerating infection of skin, usually acquired from cows or cats
Parapoxvirus	Milker's nodes	Trivial localized nodular infection of hands acquired from cows
	Orf	Localized papulovesicular lesion of skin acquired from sheep
Molluscipoxvirus	Molluscum contagiosum	Multiple benign nodules in skin
Yatapoxvirus	Yabapox	Localized skin tumors acquired from monkeys (rare)
	Tanapox	Localized skin lesions probably from arthropod bites; common in parts of Africa

most other viruses; hence, the virion is sometimes called a "complex" virion. An outer membrane of tubular-shaped lipoprotein subunits, arranged rather irregularly, encloses a dumbbell-shaped core and two "lateral bodies" of unknown nature. The core contains the viral DNA and associated proteins. Especially in particles released naturally from cells, rather than by cellular disruption, there is an envelope (Fig. 21-1D) which contains cellular lipids and several virus-specified polypeptides.

The nucleic acid is dsDNA (Table 21-2), varying in size from 130 kbp for parapoxviruses to 220 kbp for cowpox virus; the DNA of vaccinia virus (191,636 bp) and variola virus (186, 102bp) have been completely sequenced. Restriction endonuclease maps of the genome provide the definitive criterion for the allocation of strains to a particular species of the genus *Orthopoxvirus* (e.g., cowpox virus, which may infect many different species of animals); species of *Parapoxvirus* cannot be so readily grouped in this way.

There are over 100 different polypeptides in the virion. The core proteins include a transcriptase and several other enzymes, and numerous antigens are recognizable by immunodiffusion. The lipoprotein outer membrane of the virion is synthesized *de novo,* not derived by budding from cellular membranes; the envelope, when present, is derived from membranes of the Golgi apparatus but contains several virus-specific polypeptides. Most of the pro-

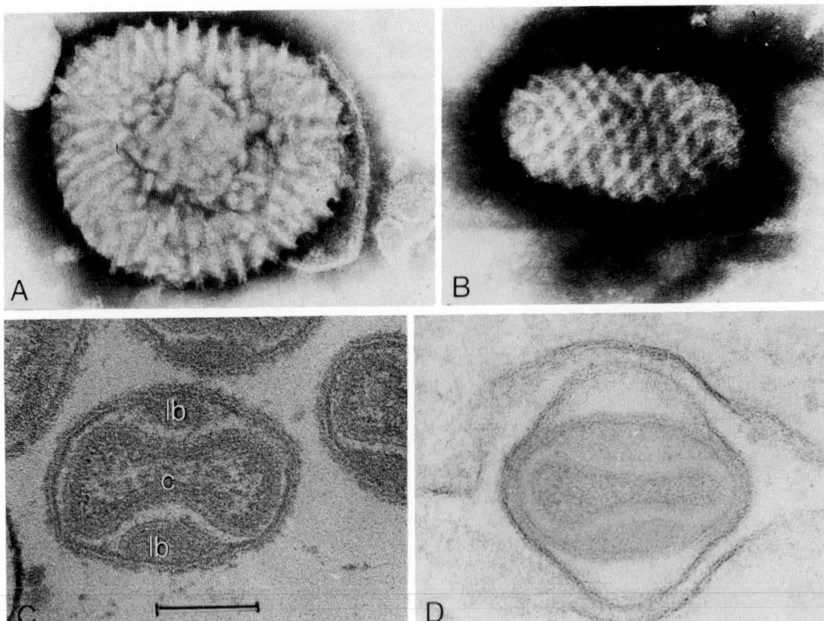

Fig. 21-1 *Poxviridae.* (A) Negatively stained vaccinia virion, showing surface structure of rodlets or tubules characteristic of the outer membrane of the genera *Orthopoxvirus, Molluscipoxvirus,* and *Yatapoxvirus.* (B) Negatively stained orf virion, showing characteristic surface structure of the outer membrane of the genus *Parapoxvirus.* (C) Thin section of vaccinia virion in its narrow aspect, showing the biconcave core (c) and the two lateral bodies (lb). (D) Thin section of mature extracellular vaccinia virion lying between two cells. The virion is enclosed by an envelope originating from altered Golgi membranes. bar 100 nm. [A, D, From S. Dales, *J. Cell Biol.* **18,** 51 (1963); B, from J. Nagington, A. A. Newton, and R. W. Horne, *Virology* **23,** 461 (1964); C, from B. G. T. Pogo and S. Dales, *Proc. Natl. Acad. Sci. U.S.A.* **63,** 820 (1969).]

teins are common to all members of any one genus, although each species is characterized by certain specific polypeptides, whereas a few others appear to be shared by all poxviruses of vertebrates. There is extensive cross-neutralization and cross-protection between viruses belonging to the same genus, but none between viruses of different genera. Genetic recombination

Table 21-2
Properties of Poxviruses of Vertebrates

Eight genera; members of genera *Orthopoxvirus, Parapoxvirus, Yatapoxvirus,* and *Molluscipoxvirus* infect humans
Most genera: virion brick-shaped with rounded corners, 250 × 200 × 200 nm, irregular arrangement of tubules on outer membrane; *Parapoxvirus*: virion ovoid, 260 × 160 nm, with regular spiral arrangement of tubule on outer membrane
Complex structure with core, lateral bodies, outer membrane, and sometimes envelope
Linear dsDNA, 130 kbp (*Parapoxvirus*), 170–250 kbp (*Orthopoxvirus*)
Transcriptase, transcription factor, poly(A) polymerase, capping enzyme, methylating enzymes in virion
Cytoplasmic replication; enveloped particles released by exocytosis, nonenveloped particles released by cell lysis

occurs readily between viruses of the same genus, but rarely between those of different genera.

Poxviruses are resistant to ambient temperatures and may survive many years in dried scabs. Orthopoxviruses are ether-resistant, but parapoxviruses are ether-sensitive.

Viral Replication

Replication of poxviruses occurs in the cytoplasm and can be demonstrated in enucleate cells. To achieve this total independence from the cell nucleus, poxviruses, unlike other DNA viruses, have evolved to encode dozens of enzymes required for transcription and replication of the viral genome, several of which must be carried in the virion itself. After fusion of the virion with the plasma membrane or via endocytosis, the viral core is released into the cytoplasm (Fig. 21-2).

Transcription is initiated by the viral transcriptase, and a transcription factor, capping and methylating enzymes, and a poly(A) polymerase also carried in the core of the virion enable functional capped and polyadenylated mRNAs to be produced, without splicing, within minutes after infection. The polypeptides produced by translation of these mRNAs complete the uncoating of the core, and transcription of about 100 "early" genes, distributed throughout the genome, occurs before viral DNA synthesis begins. Early proteins include DNA polymerase, thymidine kinase, and several other enzymes required for replication of the genome. Poxvirus DNA replication involves the production of concatemeric forms as intermediates, but details of the mechanism are still unknown.

With the onset of DNA replication there is a dramatic shift in gene expression. Transcription of "intermediate" and "late" genes is controlled by binding of specific viral proteins to characteristic promoter sequences. Virion assembly occurs in circumscribed areas of the cytoplasm ("viral factories").

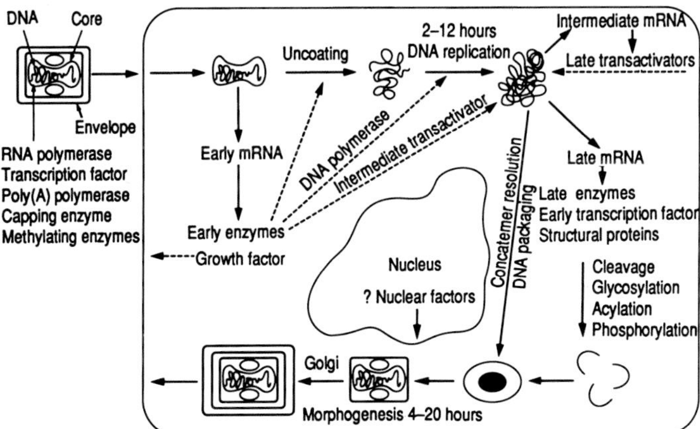

Fig. 21-2 Diagram illustrating the replication cycle of vaccinia virus. See text for details. [From B. Moss, *Science* **252,** 1662 (1991).]

Spherical immature particles can be visualized by electron microscopy; the outer bilayer becomes the outer membrane of the virion, and the core and lateral bodies differentiate within it. This outer membrane is not derived by budding from cellular membrane but is synthesized *de novo*. Some of the mature particles move to the vicinity of the Golgi complex, acquire an envelope, and are released from the cell by exocytosis. However, most particles are not enveloped and are released later by cell disruption. Both enveloped and nonenveloped particles are infectious, but enveloped particles are more rapidly taken up by cells and appear to be important in the spread of virions through the body.

Several poxvirus genes code for proteins that are secreted from infected cells and affect the response of the host to infection. Among these virokines is a homolog of epidermal growth factor, a complement regulatory protein, proteins conferring resistance to interferon, and yet others suppressing the immune response by inhibiting certain cytokines (see Table 7-1). Now that the complete sequence of vaccinia virus is known, we can anticipate the discovery of many additional genes affecting the host response to infection.

Pathogenesis and Immunity

All poxvirus infections are associated with lesions of the skin, which may be localized or generalized. The lesions associated with many diseases are pustular, but lesions due to molluscum contagiosum virus, parapoxvirus, and yatapoxvirus are proliferative. Generalized poxvirus infections have a stage of leukocyte-associated viremia, which leads to localization in the skin and to a lesser extent in internal organs. Immunity to such infections is prolonged. However, in some localized poxvirus infections, notably those produced by parapoxviruses, immunity is short-lived and reinfection is common.

Laboratory Diagnosis

The morphology of the virions is so characteristic that electron microscopy is used to identify them in negatively stained vesicle fluid or biopsy material taken directly from skin lesions. All orthopoxvirus virions have the same appearance, which is shared by the virions of molluscum contagiosum and tanapox; milker's nodes and orf viruses can be distinguished by the distinctive appearance of the parapoxvirus virion.

The usual method of isolation of orthopoxviruses is by inoculation of vesicle fluid or biopsy material on the chorioallantoic membrane (CAM) of chick embryos, where discrete lesions known as pocks become visible within a few days. Cowpox virus produces much more hemorrhagic lesions on the CAM than does vaccinia virus. The parapoxviruses, molluscum contagiosum virus, and tanapox virus do not grow on the CAM. Orthopoxviruses grow well in cell culture, parapoxviruses and tanapox virus less readily, and molluscum contagiosum virus has not yet been satisfactorily cultivated.

Identification of the particular species of orthopoxvirus, for example, differentiation between variola and monkeypox viruses or between vaccinia and cowpox viruses, can be made by animal inoculation methods, of which the CAM and rabbit skin systems are the most useful. Each species of orthopoxvirus has a distinctive DNA map demonstrable by digestion with restriction endonucleases.

Human Infections with Orthopoxviruses

Vaccinia

The origin of vaccinia virus is obscure, but it probably evolved from cowpox or smallpox virus. For smallpox vaccination, the virus was inoculated into the superficial layers of the skin of the upper arm by a "multiple puncture" technique. Severe complications occasionally occurred in children with eczema who were mistakenly vaccinated or were infected by contact. Eczema vaccinatum was rarely fatal, especially if treated with vaccinia-immune human gammaglobulin. Other very rare but more serious complications were progressive vaccinia, which occurred only in persons with defective cell-mediated immunity, and postvaccinial encephalitis.

With the eradication of smallpox, routine vaccination of the general public with vaccinia virus ceased to be necessary, and the requirement that international travelers should have a valid vaccination certificate was abolished. Vaccination of military personnel has also ceased in most countries. However, strains of recombinant vaccinia virus as vectors incorporating genes for protective antigens for several different pathogens are being developed for the production of vaccines against several diseases (see Chapter 13), although none is yet in use. Much less virulent strains of vaccinia virus than those used for smallpox vaccination are being developed for use as vectors, and the strains used will be much less likely to produce serious complications than those previously used for smallpox vaccination.

Buffalopox

Buffalopox has occurred in water buffaloes *(Bubalis bubalis)* in Egypt, the Indian subcontinent, and Indonesia and still occurs in India. By restriction mapping, the causative virus has been shown to be vaccinia virus, although most strains differ from laboratory strains of vaccinia virus (and those used for smallpox vaccination in India) in some biological properties. The disease is characterized by pustular lesions on the teats and udders of milking buffaloes; occasionally, especially in calves, a generalized disease is seen. Human infections produce lesions on the hands and face of milkers, who are no longer protected by vaccination against smallpox. The epidemiology is illustrated in Fig. 21-4B.

Human Monkeypox

Human infections with monkeypox virus, a species of *Orthopoxvirus,* were discovered in West and Central Africa in the early 1970s, after smallpox had been eradicated from the region. The signs and symptoms are very like those of smallpox, with a generalized pustular rash, fever, and toxemia (Fig. 21-3).

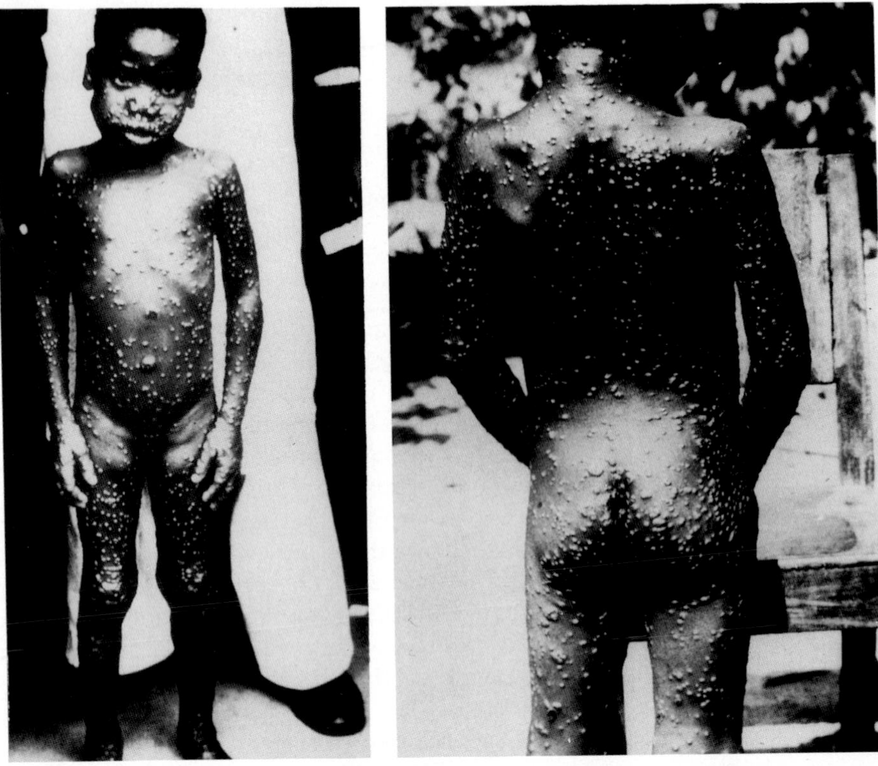

Fig. 21-3 Human monkeypox. Front and rear views of a 7-year-old Zairean girl with monkeypox, on the eighth day after the appearance of the rash, which is indistinguishable in its evolution, appearance, and distribution from the rash of smallpox. The gross enlargement of the superficial lymph nodes seen in human monkeypox was not seen in smallpox. [From J. G. Breman, Kalisa Ruti, M. V. Steniowski, E. Zanotto, A. I. Gromyko, and I. Arita, Human monkeypox, 1970–79. *Bull. WHO* **58,** 165 (1980).]

Human monkeypox occurs as a rare zoonosis in villages in tropical rain forests in West and Central Africa, especially in Zaire. It is probably acquired by direct contact with wild animals killed for food, especially squirrels and monkeys. A few cases of person-to-person transmission occur, but the secondary attack rate is too low for the disease to become established as an endemic human infection. Up to December 1986, when intensive surveillance ceased, only 400 cases had been diagnosed. Vaccination with smallpox vaccine (vaccinia virus) immunizes against monkeypox but is not justified, since the disease is so rare.

Cowpox

Humans can acquire three different poxvirus infections from cows, usually as lesions on the hands after milking: vaccinia (in the days of smallpox vaccination), cowpox (caused by an *Orthopoxvirus*), and milker's nodes (caused by a *Parapoxvirus*). Despite the name, the reservoir hosts of cowpox virus are rodents, from which it occasionally spreads to cats, cows, humans, and zoo animals, including large cats and elephants (Fig. 21-4A).

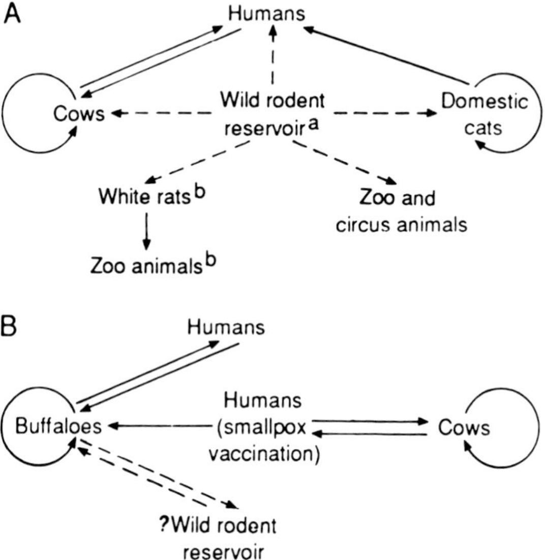

Fig. 21-4 Diagram illustrating the epidemiology of cowpox (A) and buffalopox (B). Solid lines denote known paths of transmission, broken lines presumed or possible paths of transmission. (A) There are probably several different wild rodent hosts of cowpox virus in Europe and adjacent parts of Asia, from which cowpox virus enters the other animals indicated. [a]Natural hosts include gerbils and susliks in Turkmenia, *Rattus norvegicus* in Russia, and probably voles and field mice in Britain. [b]Outbreak in the Moscow Zoo. (B) In the days of smallpox vaccination, vaccinated humans sometimes infected cows and water buffaloes with vaccinia virus. Buffalopox, caused by vaccinia virus, appears to have remained enzootic in several states in India, and it constitutes a continuing source of infection of humans, who in turn may reinfect buffaloes. By analogy with cowpox, it is possible that the true reservoir host of buffalopox (vaccinia) virus is some species of wild rodent, but if so this has not been identified.

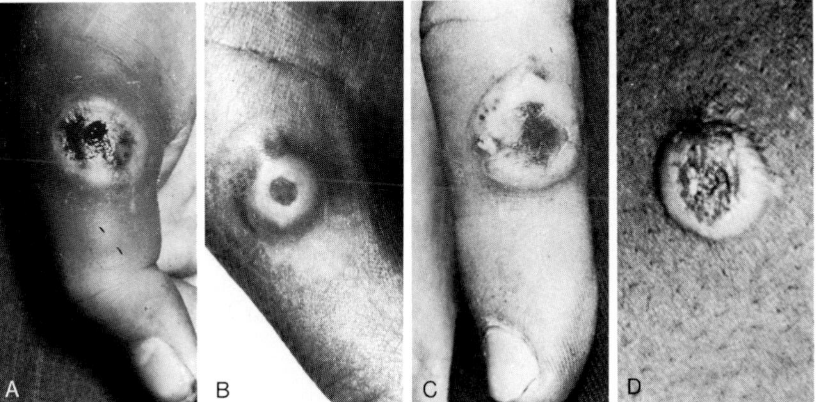

Fig. 21-5 Localized zoonotic infections with poxviruses. Lesions on hands acquired by milking infected cows: (A) cowpox, caused by an orthopoxvirus, and (B) milker's nodes, caused by a parapoxvirus. (C) Orf, a parapoxvirus lesion acquired by handling sheep or goats suffering from contagious pustular dermatitis. (D) Tanapox, lesion on arm of a child in Zaire, probably transmitted mechanically by mosquitoes from an animal reservoir host. (A, B, Courtesy Dr. D. Baxby; C, courtesy Dr. J. Nagington; D, courtesy Dr. Z. Jezek.)

Cowpox virus has been found only in Europe and adjacent parts of the former Soviet Union. It produces ulcers on the teats and the contiguous parts of the udder of cows, and it is spread through herds by the process of milking. Currently, infection with cowpox virus is more commonly seen among domestic cats, from which it is occasionally transmitted to humans. The lesions in humans usually appear on the hands and develop just like primary vaccinia (Fig. 21-5A), although fever and constitutional symptoms can be more severe.

Human Infections with Parapoxviruses

Milker's Nodes

Milker's nodes (Fig. 21-5B) also occur on the hands of humans, derived from lesions on cows' teats. Unlike cowpox, it is primarily a disease of cattle and occurs worldwide. In humans the lesions are small nonulcerating nodules. Immunity following infection of humans does not last long, and second attacks may occur at intervals of a few years. The disease is trivial, and no measures for prevention or treatment are warranted.

Orf

Orf is an old Saxon term applied to the infection of humans with the virus of contagious pustular dermatitis ("scabby mouth") of sheep and goats. The disease of sheep, which occurs all over the world, is found particularly in lambs during spring and summer and consists of a papulovesicular eruption that is usually confined to the lips and surrounding skin. Infection of humans occurs as a single lesion on the hand or forearm (Fig. 21-5C) or occasionally on the face; a slowly developing papule becomes a flat vesicle and eventually heals without scarring. Orf is an occupational disease associated with handling of sheep or goats.

Molluscum Contagiosum

The specifically human disease molluscum contagiosum, caused by the only virus in the genus *Molluscipoxvirus*, consists of multiple discrete modules 2–5 mm in diameter, limited to the epidermis, and occurring anywhere on the body except on the soles and palms. The nodules are pearly white or pink in color and are painless. At the top of each lesion there is an opening through which a small white core can be seen. The disease may last for several months before recovery occurs. Cells in the nodule are greatly hypertrophied and contain large hyaline acidophilic cytoplasmic masses called molluscum bodies. These consist of a spongy matrix divided into cavities in each of which are clustered masses of viral particles that have the same general structure as those of vaccinia virus.

The incubation period in human volunteers varies between 14 and 50 days. Attempts to transmit the infection to experimental animals have failed, and reported growth in cultured human cells has been hard to reproduce. The disease is most commonly seen in children and occurs worldwide, but it is much more common in some localities, for example, parts of Zaire and Papua New Guinea. The virus is transmitted by direct contact, perhaps through minor abrasions and sexually in adults. In developed countries communal swimming pools and gymnasiums may be a source of infection.

Yabapox and Tanapox

Two members of the genus *Yatapoxvirus* occur naturally only in tropical Africa. Yabapoxvirus was discovered because it produced large benign tumors in Asian monkeys kept in a laboratory in West Africa. Subsequently, cases occurred in primate colonies in the United States. Similar lesions have occurred after accidental inoculation of a laboratory attendant handling affected monkeys and in human volunteers.

Tanapox is a relatively common skin infection of humans in parts of Africa extending from eastern Kenya to Zaire. It appears to be spread mechanically by insect bites from a reservoir in wild animals of some unknown species. The skin lesion starts as a papule and progresses to an umbilicated vesicle (Fig. 21-5D), but pustulation never occurs. Occasionally multiple lesions occur. There is usually a febrile illness lasting 3–4 days, sometimes with severe headache, backache, and prostration.

Further Reading

Buller, R. M. L., and Palumbo, G. J. (1991). Poxvirus pathogenesis. *Microbiol. Rev.* **55,** 80.

Dumbell, K., and Richardson, M. (1993). Virological investigations of specimens from buffaloes affected by buffalopox in Maharashtra State, India, between 1985 and 1987. *Arch. Virol.* **128,** 257.

Fenner, F. (1990). Poxviruses. *In* "Fields Virology" (B. N. Fields, D. M. Knipe, R. M. Chanock, M. S. Hirsch, J. L. Melnick, T. P. Monath, and B. Roizman, eds.), 2nd Ed., p. 2113. Raven, New York.

Fenner, F. (1994). Poxviral zoonoses. *In* "Handbook of Zoonoses, 2nd Ed., Section B, Viral" (G. W. Beran, ed.), p. 503. CRC Press, Boca Raton, Florida.

Fenner, F., and Nakano, J. H. (1988). *Poxviridae:* The Poxviruses. *In* "Laboratory Diagnosis of Infectious Diseases, Volume II: Viral, Rickettsial and Chlamydial Diseases" (E. H. Lennette, P. Halonen, and F. A. Murphy, eds.), p. 177. Springer-Verlag, New York.

Fenner, F., Henderson, D. A., Arita, I., Jezek, Z., and Ladnyi, I. D. (1988). "Smallpox and Its Eradication." World Health Organization, Geneva.

Fenner, F., Wittek, R., and Dumbell, K. R. (1989). "The Orthopoxviruses." Academic Press, San Diego.

Jezek, Z., and Fenner, F. (1988). Human monkeypox. *Monogr. Virol.* **17.**

Moss, B. (1992). Molecular biology of poxviruses. *In* "Recombinant Poxviruses" (M. M. Binns and G. L. Smith, eds.), p. 45. CRC Press, Boca Raton, Florida.

Robinson, A. J., and Lyttle, D. J. (1992). Parapoxviruses: Their biology and potential as recombinant vaccines. *In* "Recombinant Poxviruses" (M. M. Binns and G. L. Smith, eds.), p. 285. CRC Press, Boca Raton, Florida.

Turner, P. C., and Moyer, R. W., eds. (1990). Poxviruses. *Curr. Top. Microbiol. Immunol.* **163,** 125.

Chapter 22

Hepadnaviridae and *Deltavirus*

In 1963 Blumberg, a geneticist investigating hereditary factors in the sera of isolated racial groups, discovered an antigen in the serum of an Australian aborigine that reacted with sera from multiply transfused American hemophiliacs. In due course the antigen was demonstrated to be present on the surface of particles with three different morphologic forms (Fig. 22-1) and to be associated with the disease serum hepatitis, now known as hepatitis B. The 22 nm particles of "Australia antigen," subsequently renamed HBsAg, for hepatitis B surface antigen, were found to be noninfectious, but the 42 nm particles were shown to be infectious virions capable of transmitting hepatitis to chimpanzees. The unique characteristics of these viruses led to their classification within a new family, named *Hepadnaviridae* to reflect the association with hepatitis and the DNA genome. The very small genome replicates via a unique mechanism.

Hepatitis B is one of the world's major unconquered diseases. Some 300 million people are chronic carriers of the virus, and a significant minority go on to develop cirrhosis or cancer of the liver from which over 1 million die every year. Although hepatitis B virus (HBV) has yet to be cultivated reproducibly *in vitro,* reliable diagnostic procedures and a much-needed vaccine are available.

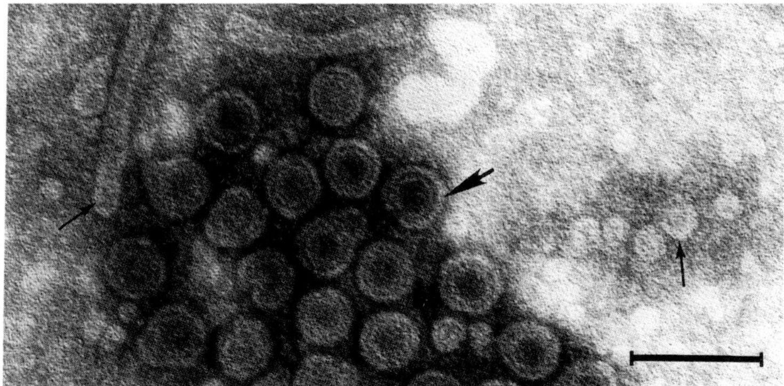

Fig. 22-1 *Hepadnaviridae*. Negatively stained preparation of hepatitis B virions (large arrow) and accompanying HBsAg particles (small arrows). Bar, 100 nm. (Courtesy of Dr. I. D. Gust and J. Marshall.)

Properties of *Hepadnaviridae*

The family *Hepadnaviridae* contains hepatitis viruses specific for humans, woodchucks, ground squirrels, ducks, and herons. We are concerned here only with the agent of human hepatitis B (Table 22-1).

The virion possesses two shells (Fig. 22-1), a 27 nm icosahedral nucleocapsid (core) constructed from 180 capsomers, surrounded by a closely fitting envelope. The virion is relatively heat-stable but labile to acid and to lipid solvents. The genome consists of a 3.2 kbp molecule of circular dsDNA of most unusual structure (Fig. 22-2). The plus strand is incomplete, leaving 15–50% of the molecule single-stranded; the minus strand is complete but contains a discontinuity ("nick") at a unique site. The 5′ termini of the plus and minus strands overlap by about 240 nucleotides and include short direct repeats, DR1 and DR2, producing "cohesive" ends that base-pair to maintain the chromosome in a relaxed circular configuration. A "terminal protein" is

Table 22-1
Properties of Hepatitis B Virus

Spherical enveloped virion, 42 nm, enclosing inner icosahedral 27 nm nucleocapsid (core)
Envelope contains glycoprotein, HBsAg; core contains phosphoprotein, HBcAg, plus polymerase with three enzyme activities: reverse transcriptase, DNA polymerase, RNase H
Circular dsDNA genome, 3.2 kb, cohesive 5′ ends; minus strand nicked, 5′ terminal protein; plus strand incomplete, 5′ RNA primer
Four overlapping open reading frames: *S*, *C*, *P*, and *X*
Genome converted to supercoiled covalently closed circular form and transcribed in nucleus to produce full-length pregenome RNA and subgenomic mRNAs
RNA pregenome in cytoplasmic core particles reverse transcribed to dsDNA; some return to nucleus to augment pool of viral supercoiled DNA
Cores bud through endoplasmic reticulum, acquiring lipid membrane containing HBsAg; noncytocidal
Subtypes differ in allelic pairs of HBsAg determinants (*d* or *y*; *r* or *w*)

covalently attached to the 5′ end of the minus strand, whereas a 5′-capped oligoribonucleotide primer is attached to the 5′ end of the plus strand.

The minus strand contains four open reading frames: *pre-S/S*, *pre-C/C*, *P* (or *POL*), and *X* (Fig. 22-2). The *P* gene, which compromises 80% of the genome and overlaps all the other genes, encodes a polymerase with three distinct enzymatic functions (DNA polymerase, reverse transcriptase, and RNase H) and also encodes the terminal protein primer. Gene *X*, spanning the cohesive ends of the genome, encodes a transactivating protein that up-regulates transcription from all the viral and some cellular promoters. The *C* gene has two initiation sites that divide it into a *pre-C* and a *C* region, producing two distinct proteins, HBeAg and HBcAg, respectively. The *pre-S/S* gene encodes the envelope protein, S, which occurs in three forms: a large (L) protein, translated from the first of the three in-phase initiation codons, is a single polypeptide encoded by the *pre-S1*, *pre-S2*, plus *S* regions of the genome and occurs in the envelope of infectious virions; a middle-sized (M) protein comprises the product of *pre-S2* plus *S;* and finally, the most abundant product is the S protein, the basic constituent of noninfectious HBsAg particles, comprising only the product of the *S* ORF. All three forms are glycosylated, and the *pre-S1* product is myristylated.

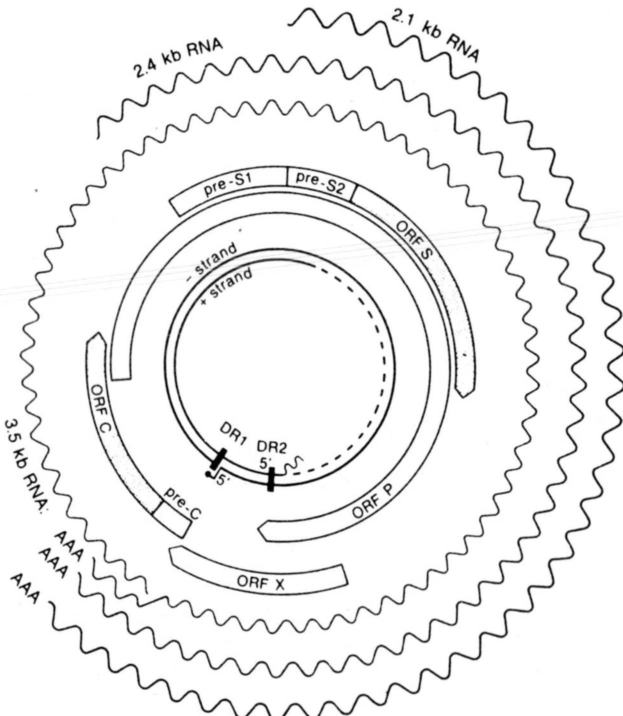

Fig. 22-2 HBV genome. Wavy lines denote viral transcripts; boxes, viral open reading frames (ORFs); arrows, direction of transcription and translation; innermost circles, structure of virion DNA; DR1, DR2, positions of direct repeat sequences involved in the priming steps in viral DNA synthesis. [From D. Ganem and H. E. Varmus, *Annu. Rev. Biochem.* **56,** 651 (1987).]

There are a number of subtypes of hepatitis B virus, defined by various combinations of antigenic determinants present on the HBsAg. All have the same group-specific determinant, *a*, but there are four major subtype-specific determinants, certain pairs of which (*d* and *y*; *r* and *w*) tend to behave as alleles, that is, as mutually exclusive alternatives. The *w* determinant displays considerable heterogeneity. Hence we have subtypes designated *ayw*, ayw_2, ayw_3, ayw_4, *ayr*, adw_4, *adr*, and so on. There are also some more unusual combinations, and additional determinants such as *q* and *x* or *g* have been described, as well as some variants with mutations in any of the determinants. Different subtypes tend to show characteristic geographical distributions, though they often overlap.

Viral Replication

In the absence of a conventional cell culture system for HBV much of our current understanding of the replication of hepadnaviruses comes from studies of woodchuck, ground squirrel, or duck hepatitis viruses, or from the growth of HBV in human hepatocytes either *in vivo* or in primary culture, or following transfection of hepatoma cell lines by various HBV DNA constructs.

The HBV genome is remarkably compact and makes use of overlapping reading frames to produce seven primary translation products from only four ORFs ("genes"): *S*, *C*, *P*, and *X*. Transcription and translation are tightly regulated via the four separate promoters and at least two enhancers plus a glucocorticoid-responsive element. Transcription occurs in the nucleus, whereas replication of the genome takes place in the cytoplasm, inside protein cores that represent intermediates in the morphogenesis of the virion. Replication of the dsDNA genome occurs via a unique mechanism involving the reverse transcription of DNA from an RNA intermediate. Thus hepadnaviruses are sometimes categorized as "retroid" viruses because of the similarity in replication strategy to the retroviruses, although in a sense the two strategies are mirror images of one another. The key difference is that, in the case of the retroviruses, the plus sense ssRNA is packaged as the genome of the virion, whereas in the case of the hepadnaviruses, the ssRNA is the intracellular intermediate in the replication of the dsDNA genome (Fig. 22-3).

The virion attaches to the hepatocyte via a sequence in the pre-S1 protein and enters by receptor-mediated endocytosis. Following removal of the envelope, the nucleocapsid is translocated to the nucleus and the viral genome released. The short (plus) strand of viral DNA is then completed to produce a full-length *relaxed circular (RC)* dsDNA molecule. This in turn is converted to a *covalently closed circular (CCC)* form by removal of the protein primer from the minus strand and of the oligoribonucleotide primer from the plus strand, elimination of the terminal redundancy from the minus strand, and ligation of the two ends of the DNA. This closed circular form becomes twisted to yield what is known as *supercoiled (SC)* DNA, which is the template for transcription by cellular RNA polymerase II. The "minus" strand only is transcribed to give mRNAs of 2.1 and 2.4 kb, plus a 3.4 kb RNA transcript known as the *pregenome* that is actually longer than the genome itself because it contains terminally redundant sequences. Following transport to the cytoplasm, the 3.4 kb species is translated to yield the C (core) antigens and the polymerase, while

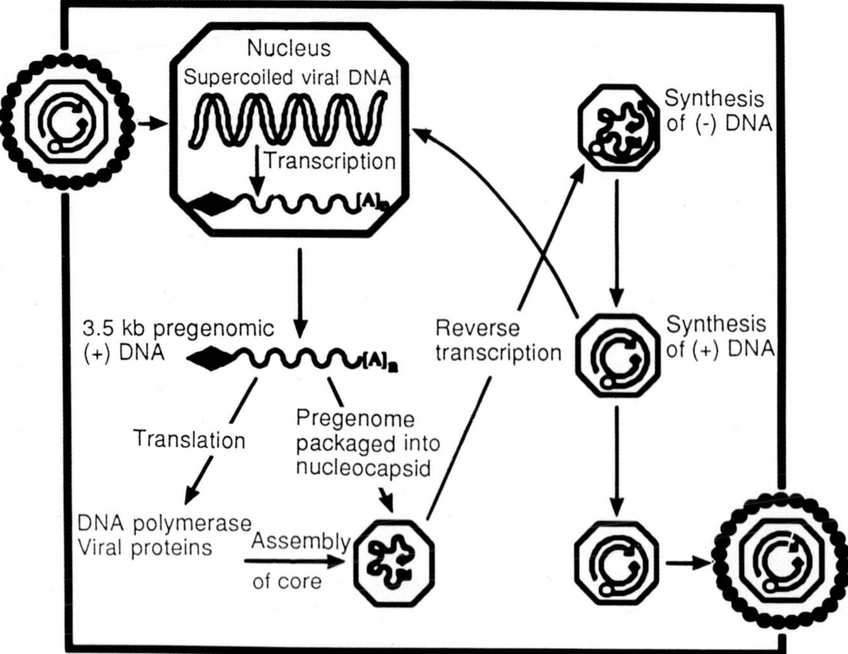

Fig. 22-3 Replication cycle of hepatitis B virus. See text for details. [Modified from G. Civitico, Y. Y. Wang, C. Luscombe, N. Bishop, G. Tachedjian, I. Gust, and S. Locarnini, *J. Med. Virol.* **31**, 90 (1990) Courtesy Dr. S. Locarnini.]

the 2.1 and 2.4 kb transcripts are translated from three different initiation codons to yield the three forms of S (surface) antigens.

Replication of the viral genome occurs via a mechanism absolutely distinct from that of any other DNA virus. The RNA pregenome associates with the polymerase and core protein to form an immature core particle in the cytoplasm. Within this structure the reverse transcriptase, primed by the virus-coded terminal protein, transcribes a complementary (minus) strand of DNA. Meanwhile, the RNase H progressively degrades the RNA template from its 3′ end, leaving only a short 5′ oligoribonucleotide which, following transposition to base-pair with a complementary site on the newly synthesized minus strand, serves as the primer for the DNA polymerase to transcribe a DNA plus strand. Some of the core particles, containing newly synthesized viral DNA, are recycled back into the nucleus to amplify the pool of HBV genomes available for transcription. The remainder are assembled into virions before the plus strand of the genome has been completed. The cores bud through those areas of endoplasmic reticulum into which the L, M, and S antigens have been inserted, thereby acquiring an HBsAg-containing lipid envelope. Vesicles transport the virions to the exterior without cell lysis.

Clinical Features of Hepatitis B

Most HBV infections are subclinical, particularly during childhood, but about one-third of adult infections are icteric. The course of acute viral hepatitis is conventionally divided into three phases: (1) preicteric, (2) icteric, and (3) convalescent. Following a long incubation period of 6–26 weeks in the case of hepatitis B, the *preicteric (prodromal)* phase commences with malaise, lethargy, anorexia, and commonly nausea, vomiting, and pain in the right upper abdominal quandrant. A minority of patients develop at this time a type of serum sickness characterized by mild fever, urticarial rash, and polyarthritis, resembling a benign, fleeting form of acute rheumatoid arthritis. Any time from 2 days to 2 weeks after the prodromal phase begins, the *icteric* phase commences, heralded by dark urine (bilirubinuria) closely followed by pale stools and jaundice. The *convalescent* phase may be long and drawn out, with malaise and fatigue lasting for weeks.

There are a number of possible outcomes (Fig. 22-4). Less than 1% of the icteric cases die of fulminant hepatitis. Most recover uneventfully following complete regeneration of the damaged liver within 2–3 months, but some progress to chronic infection. This may take the form of an asymptomatic carrier state, defined as HBs antigenemia persisting for at least 6 months, or of chronic persistent hepatitis or chronic active hepatitis causing progressive liver damage, which may lead eventually to cirrhosis and/or to primary hepatocellular carcinoma. A proportion of those with chronic persistent or chronic active hepatitis develop manifestations of immune-complex disease, systemic necrotizing vasculitis (polyarteritis nodosa) and membranoproliferative glomerulonephritis being the two most common.

The typical course of the hepatitis B carrier can be divided into high and low replicative phases. In the *high replicative phase* (or *productive phase*), the liver is supporting a large amount of viral replication; HBsAg, viral DNA, and HBeAg are found to very high titer in the peripheral compartment. In contrast, the *low replicative phase* (or *restricted phase*) is characterized by much lower titers of HBsAg, extremely low levels of viral DNA, absence of HBeAg, but the presence of anti-HBe. This switch from HBeAg to anti-HBe, usually associated with a hepatitic flare, is known as the *seroconversion illness*. In a number of carriers, this hepatitic flare is not associated with HBeAg clearance and seroconversion to anti-HBe, and is referred to as an *abortive seroconversion*.

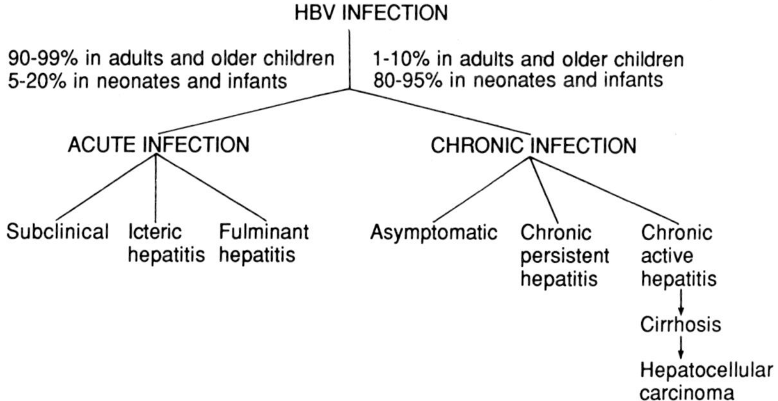

Fig. 22-4 Clinical outcomes of hepatitis B virus infection.

Some chronic carriers of hepatitis B virus can have a number of such abortive seroconversions in a lifetime, resulting in the development of quite severe liver disease. Chronic infection with HBV can result in the development of chronic liver disease characterized by chronic persistent hepatitis, chronic active hepatitis, and cirrhosis. Some carriers have a very poor prognosis, developing rapidly progressive liver disease, whereas other carriers have a much more benign prognosis.

Pathogenesis and Immunity

The tropism of HBV for hepatocytes appears to be determined, in part at least, by *cis*-acting regulatory elements in the HBV genome. The state of differentiation of the liver cells may also be important, and certain viral enhancer elements are inducible by hormones. There is evidence of at least limited viral replication also in bile duct epithelium, pancreatic acinar cells, B lymphocytes, and monocytes. Liver biopsy at the height of acute hepatitis reveals necrosis of liver parenchyma and the accompanying histopathology shown in Fig. 22-5.

By immunofluorescence or immunoelectron microscopy HBcAg is demonstrable in the cytoplasm and nucleus of infected hepatocytes in acute or chronic active hepatitis, whereas HBsAg is seen only in the cytoplasm. By autoradiography, HBV DNA is found in the cytoplasm of productively infected cells but integrated into the chromosomes of chronically infected carriers in the low replicative phase. HBsAg and HBeAg (but not HBcAg), as well as virions, viral DNA, and viral DNA polymerase, are found free in the serum of patients with acute or chronic active hepatitis. In chronic infection, including the symptom-free carrier state, 22 nm HBsAg particles continue to be synthesized and are found in concentrations up to 10^{13} per milliliter in serum for years or even for life; in contrast, infectious virions rarely exceed 10^8 per ml, even in acute infection.

The mechanism of hepatic cytopathology is not fully understood. Viral replication per se is generally noncytocidal for hepatocytes, though transfection experiments in cultured cell lines and transgenic mice indicate that accumulation of unnatural concentrations of HBcAg or pre-S1 protein can be toxic. It is widely assumed that the damage in natural infection is mediated immunologically, principally by class I restricted $CD8^+$ cytotoxic T lymphocytes which predominate in the mononuclear cell infiltration of the liver. Hepatocytes express relatively little class I antigen, but the interferon induced by HBV infection up-regulates class I MHC expression, as well as activating Tc and NK cells and interfering directly with viral replication.

Recovery from HBV infection is attributed principally to $CD8^+$ cytotoxic T lymphocytes recognizing peptides derived from HBcAg and HBeAg. Immunity to reinfection is due to neutralizing antibodies directed exclusively at HBsAg, including the pre-S regions, which are particularly immunogenic and also carry determinants recognized by $CD4^+$ helper T cells. Cross-immunity, arising principally from antibodies directed against the shared *a* determinant, confers protection against heterologous subtypes.

The determinants of persistence (see Chapter 10) are not defined, in spite

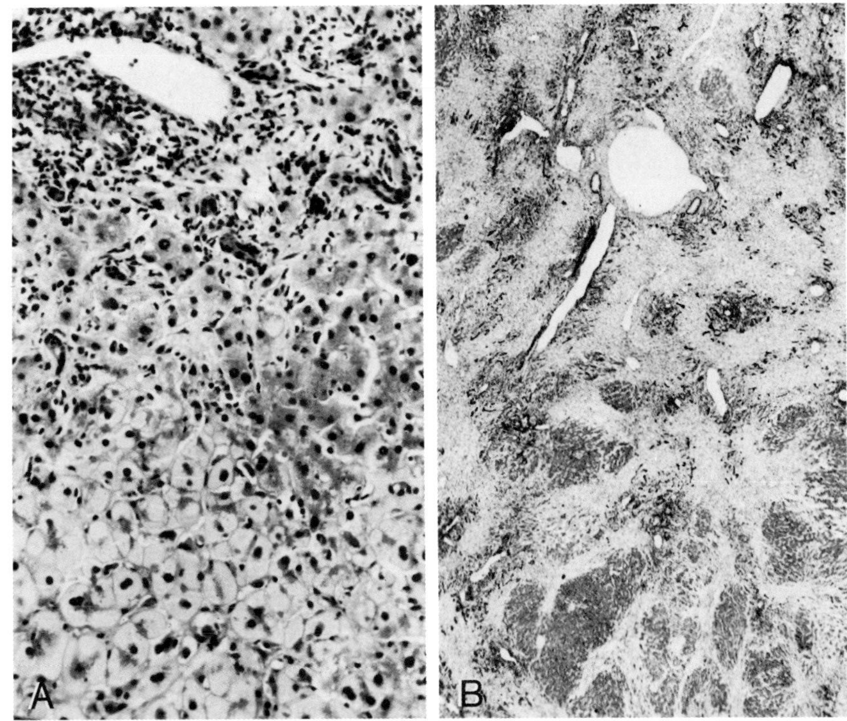

Fig. 22-5 Viral hepatitis. (A) Liver biopsy 2 weeks after onset of acute hepatitis (hematoxylin and eosin stain; magnification: ×125). Note focal necrosis within liver lobule with characteristic ballooning of hepatocytes (bottom left), biliary stasis (center right), and inflammatory cell infiltration around a widened portal tract (top left). (B) Liver at autopsy of a patient who died 25 days after the onset of viral hepatitis (hematoxylin and eosin stain; magnification: ×12). Note massive necrosis of liver parenchyma and subsequent collapse leaving portal tracts with proliferating ductules and nodules of regenerating liver cells. (Courtesy Dr. P. S. Bhathal and I. Jack.)

of many hypotheses. Immunocompetence and age of acquisition of infection are the most obvious parameters. Infants acquiring HBV perinatally from a carrier mother have a greater than 90% probability of themselves becoming a chronic carrier, whereas people first infected as adults have only a 1–5% chance. Although it is not certain that this striking difference relates to immunologic immaturity, rather than dose or route of infection, it is noteworthy that most immunocompromised individuals become carriers, regardless of age.

Recycling of multiple copies of newly synthesized dsDNA genomes from the cytoplasm back into the nucleus is a feature of persistent HBV infection during the high replicative phase, and integration of the HBV genome into hepatocyte chromosomes is observed in chronic carriers, mainly in the low replicative phase. Transcription from integrated DNA is restricted to subgenomic mRNA, with HBsAg being the principal protein produced in long-standing carriers, and even that diminishing with the passing years. The strategy of secreting into the bloodstream up to 100,000 times more antigen in the form of noninfectious 22 nm HBsAg particles than of infectious virions may serve as a decoy to mop up neutralizing antibody, which is generally not demonstrable by standard assays in the presence of HBsAg.

Immune complexes are associated with both the serum sickness often seen during the prodrome of the acute illness and with the polyarteritis nodosa or glomerulonephritis sometimes encountered in chronic hepatitis, and there is evidence that antigen–antibody complexes may induce membranous glomerulonephritis, especially in children. Moreover, the fact that HBsAg carriers fail to respond to vaccination with HBsAg suggests either that their B cells are tolerized or that suppressor T cells are responsible for the inability of chronically infected individuals to mount an immune response capable of rejecting the virus. Furthermore, hepatocytes harboring the viral genome may not present sufficient or appropriate viral peptide to class I restricted Tc cells to render them accessible to elimination by immune cytolysis. In chronic HBV carriers, class I MHC expression on hepatocytes and infiltration of both $CD8^+$ Tc cells and $CD4^+$ Th cells are all lower than during acute infection.

Variants of HBV have been isolated from Mediterranean and Asian patients with chronic liver disease. "Pre-S" mutants with point mutations or deletions in the *pre-S2* region produce higher viremia than is seen with wild-type virus. "Pre-core" mutants, which make no HBeAg, display enhanced aggressiveness and are commonly found in patients with severe chronic active hepatitis or cirrhosis, and even fulminant hepatitis. These cases are characterized by very high levels of HBcAg in hepatocyte nuclei and cytoplasm, unrelenting fibrosis, and a poor prognosis.

The reader is referred to Chapter 11 for a discussion of the pathogenesis and epidemiology of primary hepatocellular carcinoma and its relationship to the hepatitis B carrier state.

Laboratory Diagnosis

Routine biochemical tests of liver function distinguish viral hepatitis from the many nonviral, for example, obstructive or toxic, causes of jaundice. Characteristically, levels of serum transaminases (aminotransferases) are elevated markedly (5- to 100-fold) in acute symptomatic viral hepatitis, whether due to hepatitis A, B, C, D, or E. Alanine aminotransferase (ALT) and aspartate aminotransferase (AST) rise together late in the incubation period to peak about the time jaundice appears; they gradually revert to normal over the ensuing 2 months in an uncomplicated case. Serum bilirubin may rise anything up to 25-fold, depending on the severity of the case, and may of course be close to normal in anicteric viral hepatitis.

To date, HBV has been reproducibly isolated in chimpanzees and certain other simian species, and irreproducibly in organ cultures or primary cultures of human hepatocytes, which are impracticable for routine diagnostic use. Serology forms the basis of the diagnosis of hepatitis B and of the differentiation of the various clinical forms of hepatitis B from one another. Although many types of immunoassays have been successfully applied to HBV, the most widely used and most sensitive have been radioimmunoassay (RIA) and enzyme immunoassay (EIA). Six markers, all found in serum, are of particular diagnostic importance: HBsAg, HBV DNA, HBeAg, antibody to HBsAg (anti-HBs), anti-HBe, and anti-HBc.

Acute infection with hepatitis B virus (Fig. 22-6A) is characterized by the appearance of HBsAg in the blood a month or two after infection, rising to a peak shortly before symptoms develop, then gradually disappearing coinci-

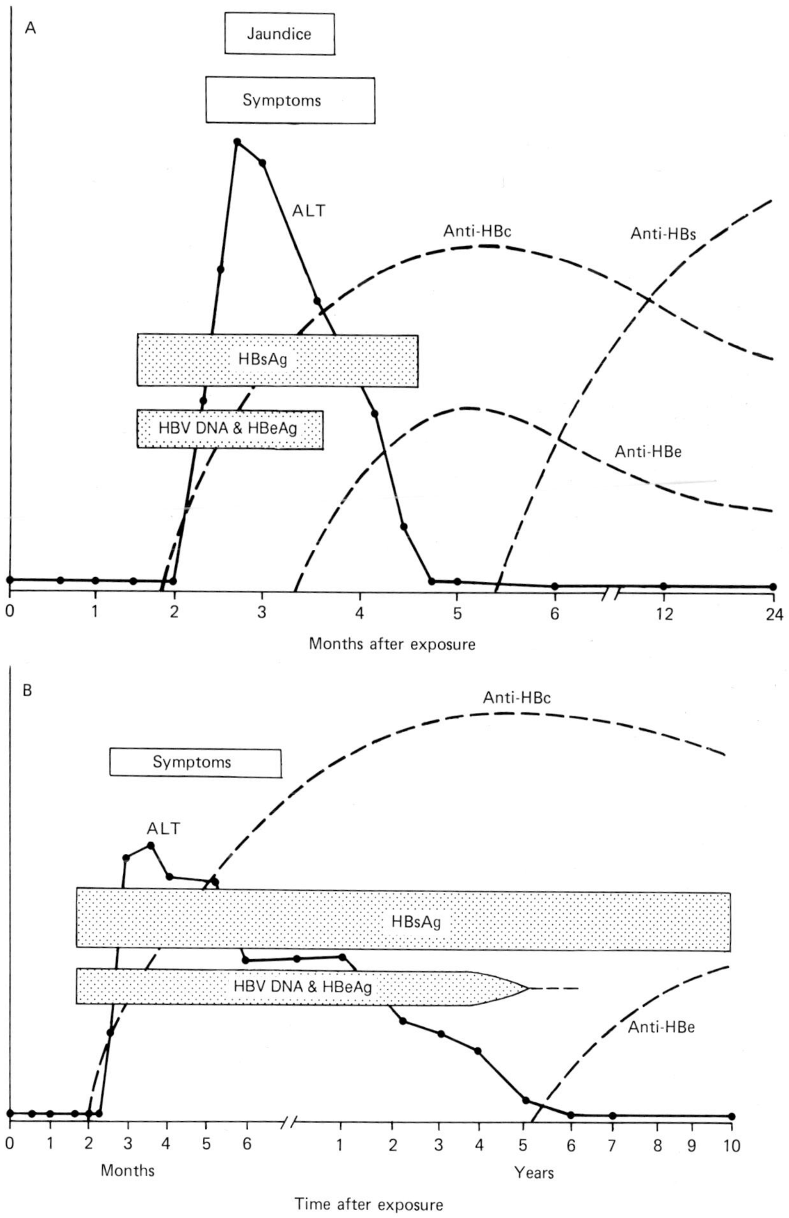

Fig. 22-6 Serological events associated with the typical course of acute type B hepatitis (A) and the development of the chronic hepatitis B virus carrier state (B). HBsAg, hepatitis B surface antigen; HBeAg, hepatitis B e antigen; ALT, alanine aminotransferase; anti-HBs, antibody to HBsAg; anti-HBc, antibody to hepatitis B core antigen; anti-HBe, antibody to HBeAg. [Modified from J. H. Hoofnagle, *Annu. Rev. Med.* **32,** 1 (1981). Copyright 1981 by Annual Reviews Inc.]

dentally with the fall in transaminase levels over the next few months. Viral DNA and HBeAg appear at about the same time as HBsAg but disappear more abruptly when symptoms and enzyme levels peak. The first antibody detected is anti-HBc; it usually appears before symptoms develop, rises rapidly to high titer, and persists indefinitely. Anti-HBs, on the other hand, does not become detectable until HBsAg has been cleared and recovery is complete (usually within a year); indeed, there is sometimes a window in time during which neither HBsAg nor anti-HBs is demonstrable, and anti-HBc is the only positive marker of infection.

Chronic hepatitis B infection (Fig. 22-6B) is characterized by the persistence of HBsAg for at least 6 months, but often for years or even for life. As long as HBs antigenemia persists anti-HBs antibody is usually not found free, but in about 10% of carriers it is complexed in low amounts with HBsAg. Anti-HBc rises to very high titer and persists for life in the normal way. This is the picture with the HBsAg *carrier state.*

Chronic active hepatitis is distinguished from the asymptomatic carrier state by progression of liver damage, as indicated by continuing elevation of serum transaminase levels and histologic evidence on liver biopsy. Persistence of HBV DNA, viral polymerase, HBeAg, and virions implies active viral multiplication, high infectivity, and progressive liver damage, the hallmarks of the high replicative phase. In contrast, anti-HBe, which develops only after HBeAg disappears and enzyme levels have declined, indicates a longer standing carrier state characteristic of the low replicative phase.

Table 22-2 summarizes the patterns of serological markers that characterize the various outcomes of hepatitis B infection. Note that the key markers are HBsAg, anti-HBs, anti-HBc, and HBV DNA; the pattern of these can distinguish most of the important situations. Note also that (1) the single most reliable marker of past or present HBV infection is anti-HBc; (2) persistence of HBV DNA in chronic active hepatitis portends an unfavorable outcome in the long run; and (3) anti-HBs, which is the neutralizing antibody, appears only after HBsAg has vanished, hence is a reliable indicator of recovery and of immunity to reinfection.

Table 22-2
Serological Markers of Hepatitis B Infection

	Serological marker[a]						
Clinical condition	HBsAg	Anti-HBs	Total anti-HBc	IgM anti-HBc	HBeAg[b]	Anti-HBe	Viral DNA
Acute hepatitis	+	−	+	++	+→−	−→+	+
Chronic active hepatitis	+	−	+	+	+	−	+
Asymptomatic carrier state	+[c]	−	+	−[d]	−	+	−[e]
Past infection: immunity	−	+	+	−	−	+→−	−
Past immunization	−	+	−	−	−	−	−

[a] Arrow means transition in due course from one state to the other.
[b] If infected with wild-type HBV, not with pre-core mutant.
[c] Persisting for more than 6 months.
[d] Low titer.
[e] Very low titer.

Epidemiology

Although hepatitis B first came to the attention of the Western world as an iatrogenic disease transmitted accidentally by inoculation of contaminated blood ("serum hepatitis"), it is self-evident that this is not the natural mode of spread. In the populous areas of the developing world of high HBV endemicity (Southeast Asia including China, Indonesia, Philippines, and the Pacific islands, the Middle East, Africa, and the Amazon basin), where the majority of people are seropositive and 8–15% are chronic carriers, most become infected at birth or in early childhood. Some 5–12% of parturient women are HBsAg positive, of whom nearly half are also HBV DNA positive, resulting in efficient perinatal transmission with a 70–90% probability of the infant itself becoming a carrier. As the infant generally becomes HBsAg positive only 1–3 months after birth, it is considered that most perinatal infections result from contamination of the baby with blood during parturition, rather than transplacentally; breast milk or maternal saliva are probably responsible occasionally. However, up to half of all children in medium to high prevalence communities who become carriers acquire the infection from intrafamilial contact with chronically infected siblings or parents secreting virus in oozing skin sores, blood, or saliva; between 1 and 5 years of age the probability of becoming a chronic carrier following infection is of the order of 25%. In adolescents and adults transmission is principally by sexual intercourse; only 1–5% of primary infections acquired at this age progress to chronicity.

The picture is quite different in the developed world, where the carrier rate is generally less than 1% except in ethnic minorities (e.g., Asian immigrants) and in injecting drug users. Perinatal spread is correspondingly less common, and sexual (including homosexual) transmission among adolescents and adults is a significant risk. Percutaneous transmission by iatrogenic invasive procedures represents the most common identifiable mode of spread, with injecting drug users constituting the largest cohort of carriers. Posttransfusion hepatitis B and infection of hemophiliacs by contaminated factor VIII have now almost disappeared as a result of routine screening of blood and organ donors, but hepatitis B can represent a major occupational risk for laboratory workers who are vulnerable to accidental infection by blood spill or needle-stick injury. Less than 1 μl of blood contaminating a syringe or needle can readily transmit hepatitis B from one individual to another. Professionals occupationally at risk include dentists, surgeons, pathologists, mortuary attendants, and technicians and scientists working in serology, hematology, biochemistry, and microbiology laboratories in hospitals or public health institutions, blood banks, or hemodialysis units. However, the availability of an effective vaccine has greatly reduced this occupational risk. Tattooing, acupuncture, and ear-piercing without rigorous sterilization of equipment constitute other potential routes of transmission, as do certain body-contact sports such as wrestling and rugby football.

Control

Blood banks today routinely screen donors for HBsAg using a sensitive enzyme immunoassay or radioimmunoassay. This has almost eliminated posttransfusion hepatitis B. Intravenous drug users should be the target of educational campaigns to reduce the extreme risk of transmission that accompanies needle sharing. Some countries such as Australia have established programs for free distribution of disposable sterile syringes and needles to registered drug users. Sexual transmission is obviously difficult to address, other than by general education advocating moderation, monogamy, screening of sexual partners for HBsAg, and so on. Partners of known carriers should be vaccinated and encouraged to avoid contact with the carrier's blood or other secretions, for example, by using condoms, covering skin sores or abrasions, and eschewing the sharing of toothbrushes, razors, eating utensils, etc. Perinatal transmission to newborn infants of carrier mothers can be minimized by inoculation at birth with both hepatitis B vaccine and hepatitis B immune globulin.

Prevention of infection in health care workers and their patients is based on vaccination and universal precautions in the ward, theater, and laboratory founded on the presumption that any patient may be infectious. Barrier techniques include wearing of gloves, gowns, masks, and eyeglasses to prevent exposure to blood in high-risk situations, such as during invasive procedures, avoidance of mouth-pipetting and of eating or smoking while working, meticulous hand washing routines, careful attention to the disposal of blood and body fluids and to cleaning up blood spills with appropriate chemical disinfectants [such as 2% glutaraldehyde, 0.5% (5000 ppm available chlorine) sodium hypochlorite, or 1–5% formalin], special precautions in disposal of used needles, the use of disposable equipment wherever possible, and appropriate procedures for the sterilization of reusable equipment. These approaches to aseptic technique and sterilization of equipment should apply equally to dentists, acupuncturists, tattooists, etc.

Passive Immunization

Hepatitis B immune globulin, obtained by plasmapheresis of subjects with high titers of anti-HBs, is effective for postexposure prophylaxis of hepatitis B, for example, in unvaccinated people who have been recently exposed to infection with HBsAg positive blood in needle-stick accidents or in newborn infants born to HBsAg positive mothers. However, it is most efficacious when used in conjunction with vaccination (active–passive prophylaxis).

Vaccines

Paradoxically, an effective vaccine against hepatitis B was produced and licensed in 1981 even though the virus has not yet been reliably cultured *in vitro*. Such high concentrations of HBsAg can be found in the sera of human carriers that this prototype vaccine was prepared simply by purifying 22 nm HBsAg particles directly from this source, followed by chemical treatment to inactivate any accompanying HBV or other contaminating virus. This "plasma-derived" vaccine is still being produced and is widely used, particularly in developing countries where the need is greatest, but it has progressively been replaced by the world's first genetically engineered human vaccine.

In the mid-1980s the HBsAg gene, cloned into a plasmid, was used to transfect the yeast *Saccharomyces cerevisiae,* and the nonglycosylated form of HBsAg particle produced was extracted and purified for use as a vaccine. Like the plasma-derived vaccine, the recombinant vaccine is adsorbed with the adjuvant aluminum hydroxide, stored cold but not frozen, and administered by intramuscular injection into the deltoid in a course of three doses, separated by 1 month then 5 months, respectively. There are no side effects other than an occasional (5–20%) sore arm, and protective levels of neutralizing antibody are elicited in over 90% of recipients (95% of neonates). Of the nonresponders, only half seroconvert following a second full course, suggesting the absence of an appropriate immune response gene. Renal dialysis patients and immunodeficient or elderly recipients also respond suboptimally. Protection studies demonstrate that immunity in immunocompetent vaccinees remains solid for a decade or so, and that those who do become infected during this period generally develop an anamnestic anti-HBs response as well as an anti-HBc response and do not progress to chronicity. Nevertheless, longer term studies may reveal that a booster dose of vaccine is desirable after about 10 years.

Recombinant HBsAg vaccines are continually being refined. First, inclusion of the nucleotide sequence encoding pre-S1 and pre-S2 in the cloned gene construct enhances the immunogenicity of the resultant protein particles. Second, HBsAg produced by mammalian cell lines such as CHO are glycosylated normally and thus more closely resemble the natural human product. Third, encapsulation of HBsAg in a polyglycan or other matrix may enhance immunogenicity by creating a depot from which antigen is released slowly or in pulses. Finally, the HBsAg gene has been incorporated into live vaccinia virus or adenovirus recombinants and demonstrated to confer protection.

Vaccination strategies differ from country to country. In populous areas of the world where the HBsAg carrier rate is high the top priority must be to interrupt the chain of perinatal and early horizontal transmission by vaccinating all infants at birth. For convenience, the course of injections should be synchronized with those for diphtheria/pertussis/tetanus and/or with oral poliovaccine, and should become an integral part of the Expanded Programme of Immunization (EPI). Pilot programs have been conducted in several developing countries since the late 1980s. Finance is the major impediment to implementation on a world scale.

The strategy of universal immunization of newborn infants should be the aim of all countries, but current policy in most of the developed world where the HBsAg carrier rate is less than 1% is to target only high-risk groups in the community (Table 22-3). Several of these cohorts are notoriously difficult to access, and the experience in the United States has been that the incidence of HBV has actually increased in spite of strenuous efforts to implement this policy. Moreover, the cost of locating and testing for HBsAg carriage approaches that of the vaccine. As a result, health authorities in the United States and elsewhere have recently been lobbying for funding to support a policy of universal vaccination of all children. A number of manufacturers are developing "pentavalent" vaccines against diphtheria, pertussis, tetanus, *Haemophilus influenzae* type B, and hepatitis B, for administration to infants at

Table 22-3
Priority Candidates for Hepatitis B Vaccine in Developed Countries[a]

Newborn babies born of HBsAg positive mothers[b] and of immigrants from countries of high HBV prevalence[c]
Sexual and close household contacts of HBV carriers
Needle-stick injuries from HBV carriers
Parenteral drug abusers
Homosexually active men
Heterosexually promiscuous persons including prostitutes and those with sexually transmitted disease
Hemophiliacs, hemodialysis patients, or others requiring frequent transfusions or blood products
Clients and staff of homes for the developmentally disabled and other custodial institutions, such as prisons
Health care and public safety workers potentially exposed to human blood[d]
Immigrants and ethnic minorities from countries of high HBV prevalence[c]
International travelers to HBV endemic areas who anticipate close or sexual contact with the local population

[a] Based on Recommendations of the Immunization Practices Advisory Committee, Centers for Disease Control (1990). Protection against viral hepatitis. *Morbidity Mortality Weekly Report* **39**, RR-2.
[b] Identified by screening of pregnant women.
[c] Immigrants and refugees from East Asia, Africa, etc., should ideally be screened for HBV markers; if an HBV carrier is detected, all members of the family (as well as all subsequent newborn infants) should be vaccinated.
[d] Vaccinate students undergoing professional training.

birth, 1–2 months, and 6–18 months of age, in lieu of the currently popular DPT schedule of 2, 4, 6, and 15 months.

Chemotherapy

Theoretically, any of the several enzymatic functions of the HBV polymerase offer tempting targets for chemotherapeutic attack. In practice, however, inhibitors of reverse transcriptase (such as AZT) have proved to be ineffective, as have inhibitors of DNA polymerase (such as acyclovir and adenine arabinoside), with the exception of ganciclovir, especially in combination with phosphonoformate (foscarnet), which has recently been reported to reduce HBV replication in chronic carriers following liver transplantation. The newer nucleoside analogs penciclovir and its oral form, famciclovir, have also been reported to inhibit HBV DNA replication *in vitro* and *in vivo*. An alternative approach is to inhibit the generation and processing of the main transcriptional template, the viral supercoiled DNA, using compounds known to interfere with DNA topology. The DNA gyrase inhibitor nalidixic acid, a well-characterized oral antibacterial antibiotic, blocks viral replication at this level and has displayed efficacy against duck hepatitis virus *in vitro* and *in vivo*.

Interferons (see Chapter 16) have an established place in the treatment of chronic hepatitis B but have been generally disappointing. Their efficacy seems to be largely limited to a particular cohort of patients with chronic active hepatitis in the window between the high and low replicative phases. In this subset, which represents only a small minority of the HBV carriers worldwide, a 6-month course of interferon α, 5–10 million units injected three times a week, brings about a response in up to 50% of cases, but there is often a relapse after cessation of therapy. The naturally occurring human immunomodulatory thymic hormone and interferon inducer thymosin α-1 appears to have a similar effect and is being tested for synergism with other anti-HBV chemotherapeutic agents.

Deltavirus (Hepatitis D)

In 1977 a young Italian physician, Rizzetto, detected a novel antigen, which he called δ (delta), in the nuclei of hepatocytes from particularly severe cases of hepatitis B. Delta antigen was also found inside 36 nm viruslike particles, the "delta agent," the outer coat of which was serologically indistinguishable from HBsAg. It transpired that the δ agent, now known as hepatitis D virus (HDV), is a defective *satellite* virus, found only in association with its helper virus, HBV. The tiny RNA genome of HDV, smaller than that of any known animal virus, encodes its own nucleoprotein (the delta antigen), but the outer capsid of the HDV virion is composed of HBsAg, encoded by the genome of HBV coinfecting the same cell.

Currently classified as the sole member of a new free-standing genus, *Deltavirus,* HDV is absolutely unique among human viruses, and indeed is without precedent among mammalian viruses. Some regard it as a subviral agent falling outside the definition of a virus. It displays features characteristic of several different classes of plant pathogens known as viroids, virusoids, satellite RNAs, and satellite viruses, all of which have RNA genomes resembling that of HDV in certain respects, but some of which encode no coat protein or no proteins at all. The HDV genome, like that of several of these subviral plant pathogens, is a covalently closed circle of single-stranded RNA with self-cleaving *(ribozyme)* and self-ligating activity. Although HDV is currently the only known mammalian example of this strange class of agents, it is reasonable to suppose that other comparable infectious agents of humans await discovery.

Properties of Hepatitis D Virus

The HDV virion is roughly spherical and heterogeneous in size (30–40 nm with a mean of 36–38 nm). The coat is composed of HBsAg, mainly lacking the pre-S regions. The genome is a covalently closed circle of minus sense RNA of only 1.7 kb, with extensive base pairing creating a secondary structure enabling it to fold into an unbranched rodlike structure. There is no sequence similarity (homology) with either HBV DNA or cellular DNA. Currently, the only known gene product is HDAg, which is encoded by the largest ORF on the antigenome. HDAg has at least three functional domains: (1) an RNA-binding domain, which no doubt accounts for its intimate association with the viral genome, (2) a nuclear localization signal, which directs the infecting genome and newly synthesized HDAg to the site of viral transcription and replication, and (3) a leucine zipper, which is assumed to promote interaction between HDAg and HBsAg in assembly of the virion (Table 22-4).

Table 22-4
Properties of Hepatitis D Virus

Satellite virus requiring hepatitis B virus as helper, by HDV/HBV coinfection or by superinfection of HBV carrier
Spherical virion, 36–38 nm, HBsAg coat, HDAg nucleoprotein
Circular minus sense ssRNA genome, 1.7 kb
Transcription and genome replication occur in nucleus using host RNA polymerase II
RNA replication by rolling circle mechanism; ribozyme self-cleavage of multimeric nascent strand, then self-ligation to circularize
RNA editing allows read through of stop codon to yield two versions of HDAg which regulate replication, morphogenesis, and release

Viral Replication

Although HDV can be grown in primary cultures of hepatocytes in the presence of its helper virus HBV, this is too cumbersome a system to produce definitive biochemical data on the details of viral replication. However, a number of clear-cut and surprising findings have come from transfection experiments using either (1) genetically cloned DNA copies of the HDV RNA genome or (2) genomic RNA itself, delivered in liposomes to cells that have been engineered to express HDAg. Replication of the HDV genome itself requires neither hepatocytes, human cells, nor HBV; a wide range of mammalian cells will support HDV RNA replication in the absence of HBV, but the latter is necessary for the production of HDV virions.

Following entry and uncoating, the genome with associated HDAg is transported to the nucleus where it utilizes the host RNA polymerase II to transcribe complementary RNA of two distinct forms: (1) full-length circular plus sense RNA, the "antigenome," which serves as the intermediate in replication of the genome, and (2) a shorter polyadenylated linear transcript that is exported to the cytoplasm and serves as mRNA for translation into HDAg. Replication of the genome occurs in the nucleus by what is known in plant viroids as the double rolling circle mechanism; the template rolls as the antigenome is synthesized, to produce excessively long plus strand linear copies (Fig. 22-7). This multimeric transcript then cleaves itself to yield a unit-length antigenome, one specific nucleotide sequence serving as the substrate (cleavage site) and another as a magnesium-dependent enzyme. A ligase activity then joins the ends of the linear transcript to yield the monomeric closed circular plus sense RNA antigenome. A similar sequence of events then generates new copies of the genome, using the antigenome as a template.

During RNA replication of HDV an extraordinary example of RNA editing is observed; a specific mutation (UAG → UGG) occurs in the termination codon at the end of the ORF for the truncated (small) form of HDAg (P24), enabling the host RNA polymerase II to read through it to yield the mRNA encoding the large form of HDAg (P27). P24 is required for HDV RNA replication whereas P27 inhibits it, hence the switch from production of P24 to P27 suppresses further genome replication and promotes its packaging. Assembly of the genome–HDAg complex with HBsAg in the cytoplasm then produces the complete virion for release from the cell.

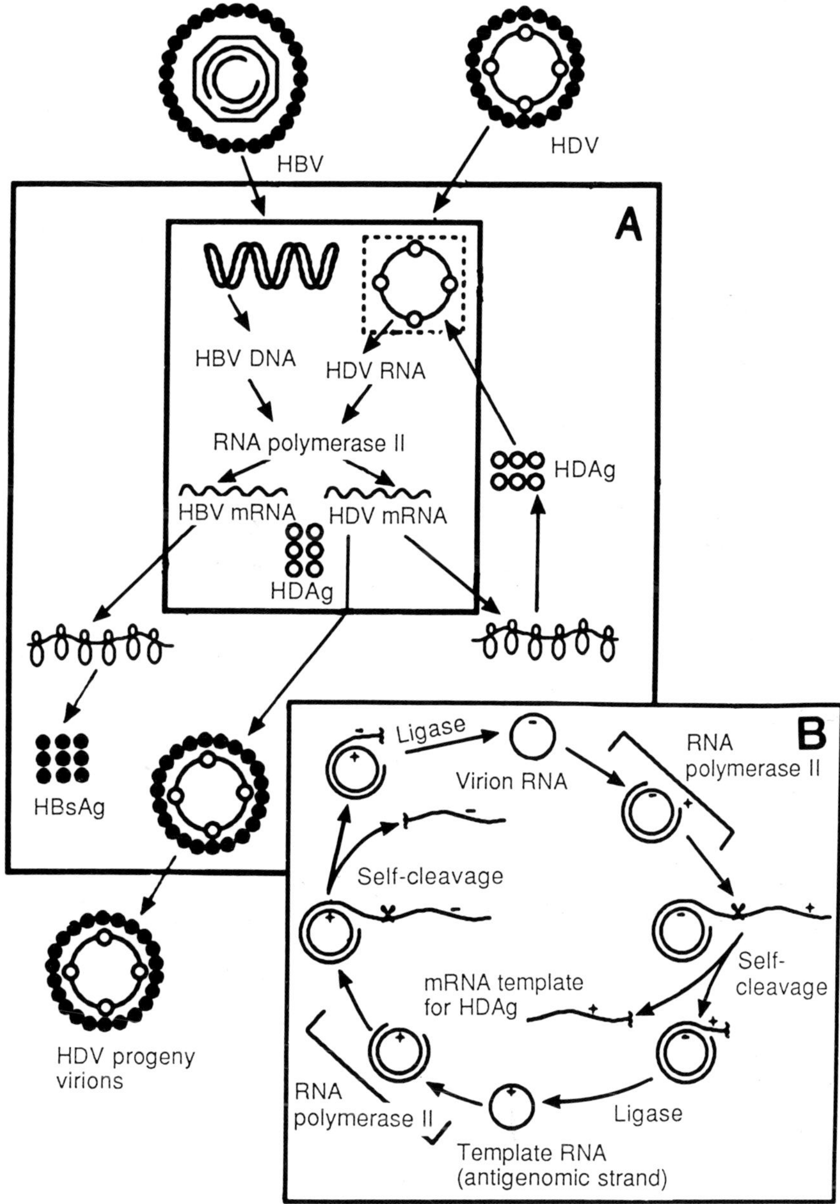

Fig. 22-7 (A) Replication of hepatitis D virus in the presence of hepatitis B virus. (B) Enlargement of the replication cycle of the hepatitis D virus genome. See text for details. (Courtesy Drs. S. A. Locarnini and A. L. Cunningham.)

Pathogenesis and Clinical Presentations

There are two main patterns of HDV infection, coinfection and superinfection. *Coinfection* is defined as simultaneous primary infection of an HBV-susceptible individual with HDV and HBV. It most commonly results from parenteral transmission, for example, in intravenous drug abusers. The incubation period depends on the size of the inoculum of HBV and is characteristic of hepatitis B (6 weeks to 6 months). Both viruses replicate simultaneously or sequentially, giving rise either to a single episode of clinical hepatitis B plus D or in 10–20% of cases to two discrete episodes, namely, hepatitis B followed by hepatitis D, depending on the ratio of HBV to HDV in the inoculum. Coinfections with HDV and HBV can be more severe than the disease caused by HBV alone, and the incidence of fulminant hepatitis higher, but the proportion of acute infections that progress to chronicity is no higher than following HBV infection alone.

Superinfection of an HBsAg carrier by HDV is the commoner occurrence and the more serious. Because large numbers of hepatocytes are already producing HBsAg, HDV is able to replicate without delay; after a relatively short incubation period (3 weeks), viremia of up to 10^{11} virions per milliliter and a severe clinical attack of hepatitis D ensues. Case–fatality rates of up to 20% from fulminant hepatitis have been observed in some centers. Almost all the other cases go on to chronic active hepatitis, and progression to cirrhosis can be rapid. Overall, chronic hepatitis D has a very poor prognosis.

Clinically and histologically the presentation of hepatitis D does not differ significantly from other types of viral hepatitis, except in its severity. A high proportion of all deaths from fulminant hepatitis represent superinfections or coinfections with HDV and HBV, and in developed countries many of the HBsAg carriers who progress to chronic active hepatitis or cirrhosis are infected with HDV.

Little is known of the basis of the liver damage, but the observation that the extent of the cytopathology correlates better with the amount of the small (P24) form of HDAg in liver cell nuclei than with the number of inflammatory cells suggests that P24 is directly cytotoxic for hepatocytes. The P24 form predominates during acute infection, when large numbers of virions are being produced, whereas P27 is found mainly in chronic infection, when much lower titers of virus are present. Following HDV superinfection of an HBsAg carrier the titer of HBsAg in the serum may be depressed during the period of maximum HDV replication, possibly as a result of competition for host RNA polymerase II, which is required by both viruses for transcription, and/or competition for HBsAg (which is required by both viruses, but the number of HDV particles produced often exceeds that of HBV virions by 1000-fold).

Laboratory Diagnosis

Figure 22-8 depicts the time course of the several diagnostic serological markers throughout the progression of HDV/HBV coinfection (Fig. 22-8A) or HDV superinfection of an HBsAg carrier (Fig. 22-8B). Simultaneous coinfection typically produces acute hepatitis B/D which tends to resolve spontaneously. Toward the end of the incubation period, shortly after the appearance of HBsAg and prior to the rise in anti-HBc IgM or transaminases, HDAg can be demonstrated in serum by EIA or immunoblotting, and HDV RNA by hybridization to a radiolabeled RNA probe (dot blot or Northern blot), reflecting the very high titer of infectious HDV particles produced. In most instances, however, HDV is cleared rapidly, and the only demonstrable marker in the later stages of the acute disease may be anti-HD IgM, which can be identified in an EIA, using recombinant HDAg cloned in *Escherichia coli* or in a hepatoma cell

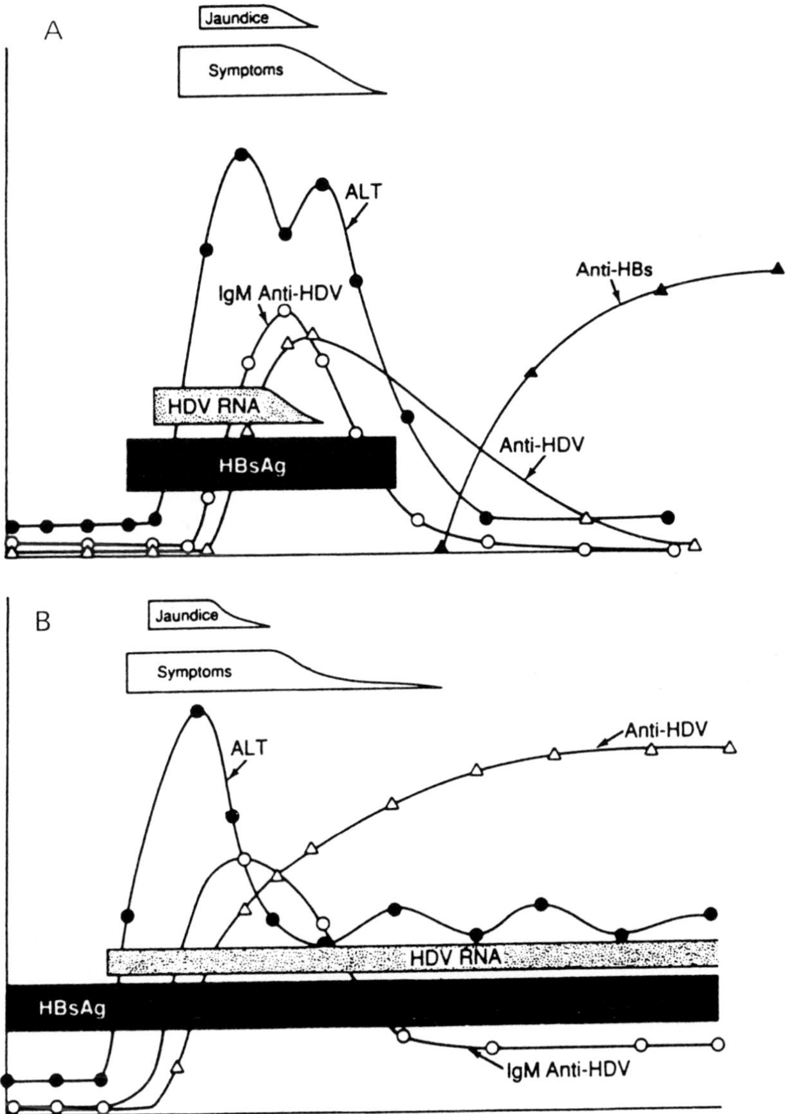

Fig. 22-8 The time course of serological markers in relation to clinical illness in hepatitis D (HDV) infection. (A) Coinfection with HDV plus hepatitis B virus (HBV). (B) Superinfection with HDV of an HBsAg carrier, leading to chronic HDV infection. ALT, Alanine aminotransferase; anti-HBs, antibody to hepatitis B surface antigen; anti-HDV, antibody to hepatitis D antigen. [From J. H. Hoofnagle and A. M. Di Bisceglie, *in* "Antiviral Agents and Viral Diseases of Man" (G. J. Galasso, R. J. Whitely and T. C. Merigan, eds.), 3rd Ed., pp. 445 and 446. Raven, New York, 1990.]

line. Total antibody to HDV is demonstrable for longer, but even the IgG response is usually short-lived following resolution of the infection, so screening for antibody is not a reliable indicator of past HDV infection.

In contrast, superinfection by HDV of an HBsAg carrier generally induces a more severe acute disease, after a shorter incubation period, and progresses to chronic active hepatitis and/or cirrhosis. Acute superinfection is distinguished from acute coinfection by the absence of anti-HBc IgM. In chronic hepatitis delta infection, alanine aminotransferase levels fluctuate but remain elevated, whereas anti-HD IgM as well as total anti-HD and HDV RNA remain demonstrable in the serum for months or years. Table 22-5 summarizes the patterns of serological markers of HDV infection.

Epidemiology

The world distribution of HDV roughly parallels that of its helper virus HBV, but is more patchy; for example, HDV is rare in the high HBV prevalence countries of east Asia, including most parts of China. High HDV prevalence regions include the western Amazon basin, parts of central Africa, Romania, and isolated provinces in China. In these areas over 20% of HBsAg carriers and over 60% of chronic hepatitis cases have markers of HDV infection. For example, a remote tribe of indigenous Venezuelan Indians has been found to suffer a 10% annual HDV infection rate in HBV carriers, with a high case–fatality rate from fulminant hepatitis or from rapidly progressive chronic hepatitis. Prevalence is moderately high in southern and eastern Europe, the Middle East, the central Asian republics, the Indian subcontinent, and Central and South America. Three genotypes of HDV have been described so far: genotype 1 is widely distributed; genotype 2 is a single isolate from Japan; and genotype 3, from South America, is associated with a severe form of hepatitis D characterized by high mortality and a characteristic histologic lesion in the liver called a morula cell.

Transmission of HDV occurs by exactly the same routes as does HBV, namely, parenteral, perinatal, sexual, and contact. In developing countries of high HDV prevalence it is likely that most of the spread occurs horizontally among children by contact with open skin lesions, etc., and among adolescents and young adults by sexual intercourse. Perinatal transmission is less common than with HBV. In developed countries of low prevalence, such as North America, temperate South America, Australia, and western and north-

Table 22-5
Serological Markers of Hepatitis D Infection

	Serological marker[a]					
	HBV		HDV			
Clinical condition	HBsAg	IgM anti-HBc	HDV RNA	HDAg	IgM anti-HD	Total anti-HD
Acute hepatitis (HDV/HBV coinfection)	(+)	+	+	(+)	(+)	(+)
Acute hepatits (HDV superinfection)	+	–	+	(+)	(+)	(+)
Chronic hepatitis (HDV/HBV)	+	–	+	–	+	+

[a] Parentheses indicate that the marker is present at some stage but cannot be guaranteed throughout the acute illness.

ern Europe, the majority of HDV infections are acquired parenterally, particularly by intravenous drug users sharing needles. Prolonged epidemics occurred in the latter cohort during the 1970s, with about half of all HBsAg positive injecting drug users in some Western countries becoming infected. The overall trend more recently has been a decline in the prevalence of HDV in this high-risk group. Sexual transmission to partners frequently occurs, but it is less efficient than sexual transmission of HBV. HDV outbreaks have also been reported in hemophiliacs receiving contaminated clotting factors and in hemodialysis units, as well as among HBV carriers in institutions for the developmentally disabled.

Control

Control of HDV centers on prevention of coinfection with HBV or of superinfection of HBV carriers and hence requires all the measures that apply to the prevention of HBV infection, including vaccination against HBV. Interferon-α, 5 megaunits daily, produces some amelioration of the disease in about half of all chronic hepatitis D patients but needs to be continued indefinitely as relapse invariably occurs following cessation of treatment. Liver transplantation currently offers the only therapeutic option in the management of end-stage liver failure.

Further Reading

Bradley, D. W., ed. (1993). Hepatitis Viruses. *Semin. Virol.* **4(5),** 269.

Centers for Disease Control. (1990). Protection against viral hepatitis: Recommendations of the Immunization Practices Advisory Committee (ACIP). *Morb. Mortal. Wkly. Rep.* **39, RR-2.**

Chisari, F. V., Ferrari, C., and Mondelli, M. U. (1989). Hepatitis B virus structure and biology. *Microb. Pathog.* **6,** 311.

Ganem, D., and Varmus, H. E. (1987). The molecular biology of the hepatitis B viruses. *Annu. Rev. Biochem.* **56,** 651.

Gerin, J. L., Purcell, R. H., and Rizzetto, M., ed. (1991). The Hepatitis Delta Virus. *Prog. Clin. Biol. Res.* **364,** 1.

Hadler, S. C., and Fields, H. A. (1991). Hepatitis delta virus. *In* "Textbook of Human Virology" (R. B. Belshe, ed.), 2nd Ed., p. 749. Mosby Year Book, St. Louis, Missouri.

Hollinger, F. B. (1990). Hepatitis B virus. *In* "Fields Virology" (B. N. Fields, D. M. Knipe, R. M. Chanock, M. S. Hirsch, J. L. Melnick, T. P. Monath, and B. Roizman, eds.), 2nd Ed., p. 2171. Raven, New York.

Hollinger, F. B., Lemon, S. M., and Margolis, H. S., eds. (1991). "Viral Hepatitis and Liver Disease." Williams & Wilkins, Baltimore, Maryland.

Howard, C. R. (1986). The biology of hepadnaviruses. *J. Gen. Virol.* **67,** 1215.

Locarnini, S. A., and Cunningham, A. L. (1993). Clinical treatment of viral hepatitis. *In* "Antiviral Chemotherapy" (E. deClercq and D. J. Jeffries, eds.), Wiley, Chichester.

Locarnini, S. A., and Gust, I. D. (1988). Hepadnaviridae: Hepatitis B virus and the delta virus. *In* "Laboratory Diagnosis of Infectious Diseases: Principles and Practice. Volume II. Viral, Rickettsial and Chlamydial Diseases" (E. H. Lennette, P. Halonen, and F. A. Murphy, eds.), p. 750. Springer-Verlag, Berlin and New York.

Margolis, H. S., Alter, M. J., and Hadler, S. C. (1991). Hepatitis B: Evolving epidemiology and implications for control. *Semin. Liver Dis.* **11,** 84.

Marion, P. L. (1988). Use of animal models to study hepatitis B virus. *Prog. Med. Virol.* **35,** 43.

Mason, W. S., and Seeger, C., eds. (1991). Hepadnaviruses: Molecular Biology and Pathogenesis. *Curr. Top. Microbiol. Immunol.* **168,** 1.

Purcell, R. H., and Gerin, J. L. (1990). Hepatitis delta virus. *In* "Fields Virology" (B. N. Fields, D. M. Knipe, R. M. Chanock, M. S. Hirsch, J. L. Melnick, T. P. Monath, and B. Roizman, eds.), 2nd Ed., p. 2275. Raven, New York.

Purcell, R. H., Hoofnagle, J. H., Ticehurst, J., and Gerin, J. L. (1989). Hepatitis viruses. *In* "Diagnostic Procedures for Viral, Rickettsial and Chlamydial Infections" (N. J. Schmidt and R. W. Emmons, ed.), 6th Ed., p. 957. American Public Health Association, Washington, D.C.

Robinson, W. S. (1990). Hepadnaviridae and their replication. *In* "Fields Virology" (B. N. Fields, D. M. Knipe, R. M. Chanock, M. S. Hirsch, J. L. Melnick, T. P. Monath, and B. Roizman, eds.), 2nd Ed., p. 2137. Raven, New York.

Robinson, W. S. (1991). Liver carcinoma and the role of hepatitis B virus. *In* "Encyclopedia of Human Biology" (R. Dulbecco, ed.), Vol. 4, p. 755. Academic Press, San Diego.

Shikata, T., Purcell, R. H., and Uchida, T., eds. (1991). "Viral Hepatitis C, D and E." Elsevier, Amsterdam.

Taylor, J. M. (1992). The structure and replication of hepatitis delta virus. *Annu. Rev. Microbiol.* **46,** 253.

Tiollais, P., Pourcel, C., and Dejean, A. (1985). The hepatitis B virus. *Nature (London)* **317,** 489.

Zuckerman, A. J., ed. (1988). "Viral Hepatitis and Liver Disease." Alan R. Liss, New York.

Chapter 23

Picornaviridae

The family *Picornaviridae* comprises one of the largest and most important families of viruses. Some 68 enteroviruses including the polioviruses may infect the human enteric tract, hepatitis A virus is a major pathogen, and over 100 rhinoviruses cause common colds.

Picornaviruses have featured in many of the milestones of virology. In 1897, foot-and-mouth disease was the first infection of any animal to be shown to be caused by an agent that could pass through a filter that held back all bacteria—thus the "filterable viruses" were born. In 1949 Enders and colleagues first propagated poliovirus in monolayers of mammalian cells cultivated *in vitro*—thus modern "cell culture" was born. Based on this technologic breakthrough, inactivated and live attenuated poliovaccines developed by Salk and Sabin in the United States in 1954 and 1961–1962 respectively, were licensed. Their use quickly led to the demise of infantile paralysis throughout the developed world; in 1988 the World Health Organization (WHO) resolved to strive for the same result in the developing world by the year 2000. In 1981, poliovirus became the first RNA animal viral genome to be molecularly cloned and sequenced. Then, in 1985 poliovirus type 1 and rhinovirus type 14 were the first human viruses whose three-dimensional structure was solved in atomic detail by X-ray crystallography (see frontispiece), opening new approaches to antiviral chemotherapy and vaccinology. Finally, in 1991 authentic infectious virions were synthesized *in vitro* from poliovirus RNA in a cell-free cytoplasmic extract from uninfected cells—yet another first for poliovirus.

Properties of *Picornaviridae*

The picornavirus virion is a naked icosahedron, only 30 nm in diameter, appearing smooth and round in outline by electron microscopy (Fig. 23-1). The capsid is constructed from 60 protein subunits (*protomers*), each comprising a single molecule of each of four polypeptides, VP1, VP2, VP3, and VP4, derived by cleavage of a single polyprotein (see Fig. 3-8). The detailed structure of the virion, solved by X-ray crystallography, was discussed in Chapter 1 (see Fig. 1-4).

The genome is a single linear molecule of single-stranded RNA of positive polarity, 7–8 kb, polyadenylated at its 3′ end, with a protein, VPg, covalently linked to its 5′ end. Being a single molecule of plus sense, the RNA has messenger function and is infectious (Table 23-1).

Classification

The family *Picornaviridae* is divided into five genera, three of which, *Enterovirus, Rhinovirus,* and *Hepatovirus,* include human pathogens. The enteroviruses and hepatoviruses, which infect via the alimentary tract, are remarkably resistant to the conditions prevailing in the gut, namely, acidic pH, proteolytic enzymes, and bile salts. The rhinoviruses, which multiply in the nose, are acid-labile; they also have a substantially greater buoyant density by cesium chloride equilibrium gradient centrifugation.

The genus *Enterovirus* contains 68 viruses known to infect humans. Not to put too fine a point on it, their nomenclature is a mess (Table 23.2). First, they are known as "types" but should more properly be regarded as species because most of them share no antigens and hence can be distinguished by complement fixation or immunodiffusion as well as neutralization tests. Second, they were historically allocated to three groups (polioviruses, cox-

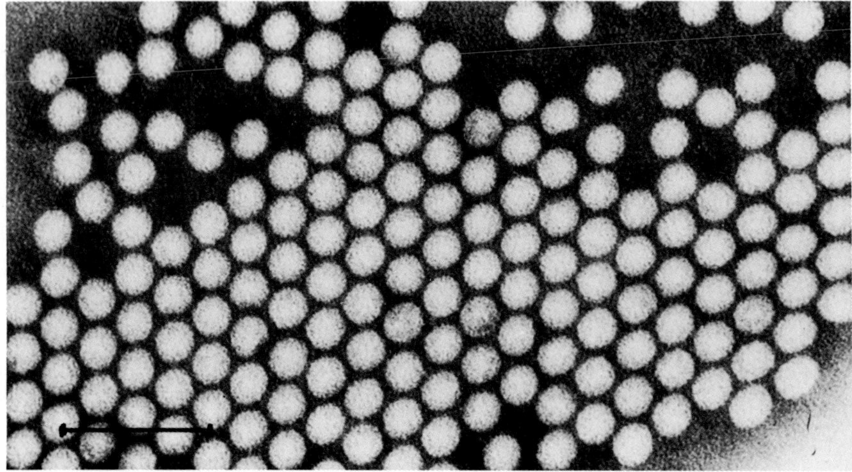

Fig. 23-1 *Picornaviridae.* Negatively stained virions of poliovirus. Bar, 100 nm. (Courtesy Dr. J. J. Esposito.) See frontispiece for a high-resolution picture of the virion of a rhinovirus.

Table 23-1
Properties of Picornaviridae

Genera *Enterovirus* and *Hepatovirus* acid-stable (pH >3), *Rhinovirus* acid-labile (pH <5)
Nonenveloped icosahedral capsid, 30 nm in diameter, 60 protomers
Linear, plus sense ssRNA genome, 7–8 kb, protein primer VPg at 5′ end, polyadenylated at 3′ end, infectious
Virion RNA acts as mRNA and is translated into a polyprotein which then cleaves itself progressively to yield nonstructural and structural proteins
Cytoplasmic replication, cytocidal

sackieviruses, and echoviruses) on the basis of criteria which are now regarded as trivial. Coxsackieviruses are pathogenic for infant mice and indeed were originally isolated in this host from suspected poliomyelitis patients from the town of Coxsackie, New York. Echoviruses (enteric cytopathogenic human orphan viruses), isolated in cell culture from the feces of asymptomatic people (hence "orphan" viruses, with no apparent parent disease) are not pathogenic for suckling mice.

Because the distinction between coxsackieviruses and echoviruses is neither absolute nor especially important, it was abandoned some years ago and all enteroviruses identified since 1970 have simply been allocated an enterovirus number, beginning with enterovirus 68. As is the wont of committees, however, the International Committee on Taxonomy of Viruses bowed to the preference of workers in the field to retain the existing nomenclature for all viruses named prior to that date. Thus we have to live for the moment with the compromise shown in Table 23-2. As if this were not enough, note also that human coxsackievirus A23 no longer exists, nor do echoviruses 10 and 28, all three having been reclassified; thus there are currently 68 known enteroviruses. A case has recently been made for the reclassification of echoviruses 22 and 23 within a new genus of the family.

Viral Replication

Quite early in the study of viral replication, poliovirus became the preferred model for the analysis of RNA viral multiplication. The work of Baltimore in particular has provided a detailed description of the poliovirus genome and of the mechanism of RNA replication, posttranslational proteolytic processing of polyproteins, and morphogenesis of a simple icosahedral virion, which

Table 23-2
Classification of Human Picornaviruses

Genus	Species
Enterovirus	Polioviruses 1–3
	Coxsackieviruses A1–A24 (no A23), B1–B6
	Echoviruses 1–34 (no 10 or 28)
	Enteroviruses 68–71
Rhinovirus	Rhinoviruses 1–100
Hepatovirus	Hepatitis A virus

serves as a model for picornaviruses in general. These processes, described in general terms in Chapter 3, are summarized below and in Fig. 23-2.

The cell receptor for poliovirus is a novel member of the immunoglobulin superfamily, as are those for other picornaviruses (e.g., the adhesion protein ICAM-1 for most rhinoviruses and some coxsackieviruses), in contrast to the integrin VLA-2 for certain echoviruses. Following adsorption, penetration, and intracellular uncoating, VPg is removed from the virion RNA by cellular enzymes. The virion RNA, acting as mRNA, is then translated without interruption into a single polyprotein, which is cleaved autocatalytically into the intermediates P1, P2, and P3. P1 is then further cleaved to yield first VPO, VP1, and VP3 and finally the four structural proteins VP1, VP2, VP3, and VP4 (see Fig. 3-8). The P2 region codes for three nonstructural proteins including one with protease activity, and the P3 region codes for four proteins including the RNA-dependent RNA polymerase required for RNA replication.

Viral RNA synthesis takes place in a "replication complex" which comprises RNA templates and the virus-coded RNA polymerase and several other viral and cellular proteins, tightly associated with a newly assembled smooth cytoplasmic membrane structure. Synthesis of the complementary strand is initiated at the 3′ terminus of the virion RNA and uses the protein VPg as a primer. The completed complementary strand in turn serves as a template for the synthesis of virion RNA, although the details of the process may differ. Most of the replicative intermediates found within the replication complex consist of a full-length complementary (minus sense) RNA molecule from which several nascent plus sense strands are being transcribed simul-

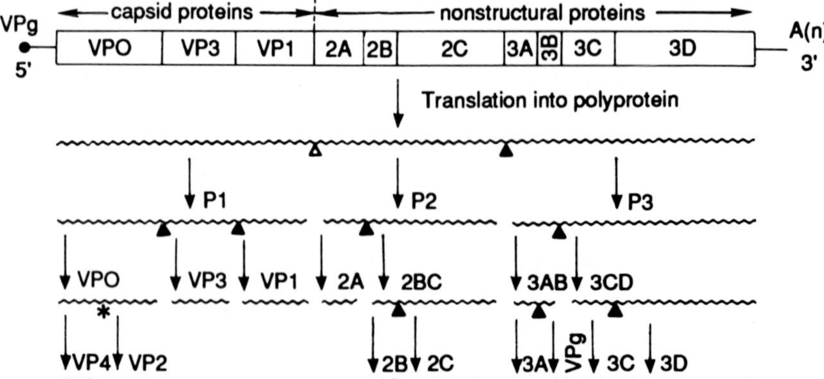

Fig. 23.2 Poliovirus RNA and posttranslational processing of the poliovirus polyprotein. (*Top*) Poliovirus RNA and its genetic organization. VPg, at the 5′ terminus, is essential for RNA replication, but the RNA is infectious if VPg is removed because it can be synthesized from region 3B. The 3′ terminus is polyadenylated, and there are short nontranslated sequences at each end (single lines). (*Bottom*) on entry into the cell, the virion RNA acting as messenger is translated into a polyprotein that is rapidly cleaved into polypeptides P1, P2, and P3 by the viral proteases 2A (open triangle) and 3C (closed triangles). P1, P2, and P3 are subsequently cleaved by protease 3C. VPO is cleaved into VP4 and VP2 by a third protease during capsid formation, so that VP1, VP2, VP3, and VP4 comprise the capsid. The organization of the genome and cleavage patterns are slightly different in different genera of picornaviruses.

taneously by viral RNA polymerase. The detailed sequence of steps involved in the morphogenesis of the virion was depicted in Fig. 3-8.

Because of the atypical absence of a 7-methylguanosine (m^7Gppp) cap on picornavirus mRNA, these viruses have been able to evolve an unusual mechanism for shutting down the translation of cellular mRNAs. Picornavirus protease 2A inactivates the cellular cap-binding complex eIF–4F, which is needed for binding of cellular mRNAs to ribosomes. Thus picornaviral replication is not only cytocidal but also very efficient, producing new virions after an eclipse period of less than 3 hours and yielding up to 10^5 virions per cell.

Polioviruses

Poliomyelitis was once a greatly feared disease; its tragic legacy of paralysis and deformity was a familiar sight in the 1950s. Today, by contrast, few medical students have seen a case, such has been the impact of the Salk and Sabin vaccines. The foundation for these great developments was laid by Enders, Weller, and Robbins in 1949, when they demonstrated the growth of poliovirus in cultures of nonneural cells. From this fundamental discovery flowed all subsequent work dependent on viral multiplication in cultured cell monolayers. Enders and colleagues were rewarded with the Nobel Prize. A new era in virology had begun.

Pathogenesis and Immunity

Following ingestion, poliovirus multiplies first in the pharynx and small intestine. It is not clear whether the mucosa itself is involved, but the lymphoid tissue (tonsils and Peyer's patches) certainly is. Spread to the draining lymph nodes leads to a viremia, enabling the virus to become disseminated throughout the body. It is only in the occasional case that the central nervous system becomes involved. Virus is carried via the bloodstream to the anterior horn cells of the spinal cord, in which the virus replicates. The resulting lesions are widely distributed throughout the spinal cord and parts of the brain, but variation in severity gives rise to a spectrum of clinical presentations, with spinal poliomyelitis the most common and the bulbar form less so. The incubation period of paralytic poliomyelitis averages 1–2 weeks, with outer limits of 3 days to 1 month. Acquired immunity is permanent but monotypic.

Pregnancy increases the incidence of paralysis, tonsillectomy increases the risk of bulbar paralysis, and inflammatory injections such as diphtheria–pertussis–tetanus (DPT) vaccine increase the risk of paralysis in the injected limb, after the usual incubation period ("provocation"). More serious in many developing countries where poliomyelitis is still common are the effects of intramuscular injections given deliberately when a child is incubating poliomyelitis ("aggravation"), a common practice in countries such as India, where injections are regarded as the best kind of therapy for all manner of illnesses.

Clinical Features of Poliomyelitis

It is important to realize that paralysis is a relatively infrequent complication of an otherwise trivial infection. Of those infections that become clinically manifest at all, most take the form of a minor illness ("abortive poliomyelitis"), characterized by fever, malaise, and sore throat, with or without headache and vomiting that may indicate some degree of aseptic meningitis. However, in about 1% of cases muscle pain and stiffness herald the rapid development of flaccid paralysis. In bulbar poliomyelitis death may result from respiratory or cardiac failure (Fig. 23-3). Otherwise some degree of recovery of motor function may occur over the next few months, but paralysis remaining at the end of that time is permanent. In some cases further muscle atrophy may be observed many years after apparent recovery ("late postpolio muscle atrophy" or "postpolio syndrome").

Laboratory Diagnosis

Virus is readily isolated from feces, and sometimes from the throat, but not from cerebrospinal fluid (CSF). Any type of human or simian cell culture is satisfactory; the virus grows so rapidly that cell destruction is usually complete within a few days. Early changes include cell retraction, increased refractivity, cytoplasmic granularity, and nuclear pyknosis (Fig. 23-4). The serotype of the isolate is identified by neutralization tests.

The ubiquity of attenuated vaccine strains poses a difficult problem for diagnostic laboratories today. Because the nucleotide sequences of all vaccine strains and prevalent wild strains are now known, the two can be readily distinguished by nucleic acid hybridization. RNA is extracted from virus isolated in cultured cells or, alternatively, may be amplified by PCR directly from a fecal specimen. The U.S. Centers for Disease Control (CDC) have-

Fig. 23-3 Totally paralyzed poliomyelitis patients in mechanical respirators during the last epidemic in the United States before the advent of universal immunization (1955). [From L. Weinstein, *J. Infect. Dis.* **129**, 480 (1974).]

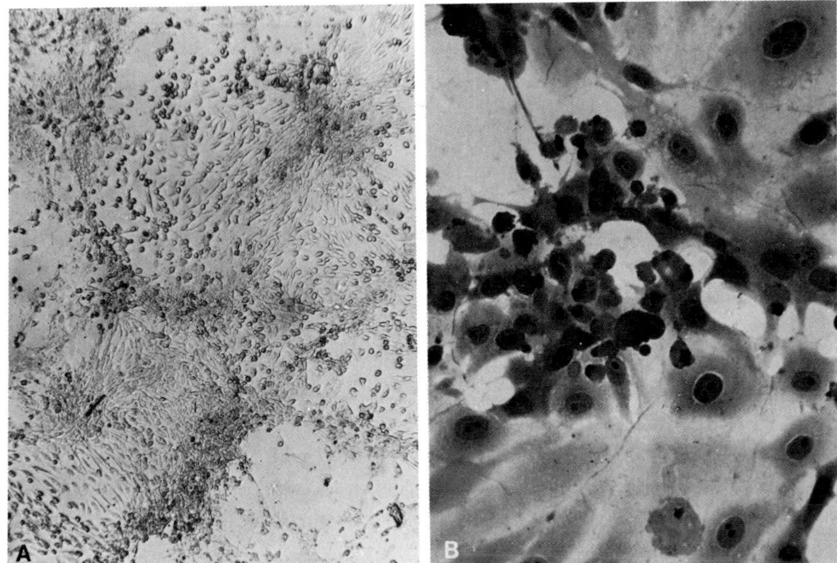

Fig. 23.4 Cytopathic effects induced by enterovirus in primary monkey kidney cell culture. (A) Unstained, low power. Note rapidly developing generalized cell destruction. (B) Hematoxylin and eosin stain, high power. Note disintegrating pyknotic cell. (Courtesy I. Jack.)

produced a panel of short labeled cDNA probes. A probe representing a conserved region of the genome is used initially for screening purposes, to identify the RNA as poliovirus; then probes specific for the three vaccine strains and for prevalent wild strains of the three serotypes identify the isolate in a dot-blot hybridization assay.

Epidemiology

Being enteric in their habitat and excreted for up to several weeks in feces, polioviruses spread mainly via the fecal–oral route. Direct fecal contamination of hands, thence food or eating utensils, is probably responsible for most case-to-case spread, especially under crowded conditions of poor hygiene and sanitation. Uncommonly, explosive epidemics have resulted from contamination of water supplies by sewage. Respiratory spread from the pharynx may also occur. In the tropics the disease is endemic throughout the year; in temperate countries before the introduction of vaccination it classically occurred in summer/autumn epidemics. The chain of infection is rarely obvious, because the vast majority of infections are inapparent.

Two major developments have greatly changed the epidemiology of poliomyelitis over the years. The first of these was the introduction of modern standards of hygiene and sanitation to the more advanced countries of the world which, paradoxically, had the effect of increasing the incidence of paralytic poliomyelitis in older children and adults. The reduction in the spread of viruses by the fecal–oral route limited the circulation of polioviruses and the incidence of infection in the community a whole. As a result, most people no longer had acquired immunity by the time they reached adolescence. The consequence was a shift in the age incidence of paralytic poliomyelitis to include young adults, making the old term "infantile paralysis" a misnomer in Western nations. Primary infection of adults is, for reasons still unknown, much more likely to result in severe paralytic disease than is primary infection of young children. This was exemplified most strikingly in virgin soil epidemics occurring, long before the vaccination era, in isolated communities with no prior experience of the virus; most of the deaths occurred in adults.

The second major influence on the epidemiology of poliomyelitis has been the development and widespread usage of a highly successful vaccine. In those countries where a vigorous policy of immunization with oral poliovaccine (OPV) has been successfully pursued, not only has the disease been virtually abolished, but wild polioviruses are no longer endemic. In North America, Australasia, and much of Europe, the incidence of paralytic poliomyelitis has been reduced 1000-fold since 1955, and wild polioviruses are found only following importations from other parts of the world. However, in the developing countries of Africa and Asia immunization has generally not met with the success anticipated. Over 250,000 new cases of paralytic poliomyelitis continue to occur annually in the Third World. The city of Bombay, for instance, regularly has more cases than the United States, Canada, Western Europe, and Australia combined. Most paralytic poliomyelitis is caused by various genotypes within serotype 1, but serotype 3 has been increasing relatively since the introduction of vaccines, presumably because of the lower seroconversion rates and titers to the type 3 component of some OPV formulations; serotype 2 is the least common, the largest reservoir being in India. The problems, and current attempts to overcome them, are discussed below.

Control by Vaccination

In the late 1940s, the U.S. National Foundation for Infantile Paralysis organized a nationwide doorknock appeal, "The March of Dimes." The response was overwhelming, and the Foundation set about sponsoring a massive research drive on several fronts, in an attempt to turn Enders' recent discovery to advantage by developing a poliomyelitis vaccine. Salk was commissioned to work toward an inactivated vaccine, Sabin toward a live one. The formalin-inactivated (Salk) poliovaccine (IPV) was the first to be licensed and was enthusiastically embraced in North America, Europe, and Australia during the mid-1950s. Sweden, the country in which paralytic poliomyelitis first became apparent in epidemic form and which for many years continued to have the highest rate of poliomyelitis in the world, eliminated the disease by the use of Salk vaccine. Meanwhile, however, the live attenuated oral poliovaccine (OPV) of Sabin was adopted in the Soviet Union and Eastern Europe, where it was demonstrated to be so successful that the whole world except Sweden, Iceland, and Holland now uses OPV because of its greater convenience and lower cost. The several advantages of such living vaccines over their inactivated counterparts were set out in detail in Chapter 13. The near abolition of poliomyelitis in much of the Western World since the introduction of poliovaccine (see Fig. 15-1) represents one of the truly great achievements of medical science.

OPV is a trivalent vaccine, consisting of attenuated strains of all three serotypes, derived empirically by serial passage in primary cultures of monkey kidney. Because of the diminishing availability of monkeys and the difficulty of ensuring that even laboratory-bred animals are free of simian viruses, manufacturers have moved to human diploid fibroblasts as a substrate for the production of poliovaccine. The vaccine strains acquired dozens of mutations during serial passage in cultured cells, leading to attenuation which was confirmed by absence of neurovirulence for monkeys. The three serotypes are pooled in carefully adjusted proportions to balance numbers against growth rate and hence to minimize the possibility of mutual interference. In the presence of molar magnesium chloride to protect the virus against heat inactivation the vaccine is stable for about a year under refrigeration.

To minimize the total number of visits OPV is most conveniently administered at the same time as other inoculations, that is, with DPT (diphtheria–pertussis–tetanus) at 2 months and 4 months of age, then again with MMR (measles–mumps–rubella) at 15 months. An additional dose should be given on entry into elementary school. No further boosters are required, except in the case of adults traveling to developing countries. The reason for the initial course of three doses is that, although one successful "take" would suffice, concurrent infection with another enterovirus may interfere with the replication of the vaccine viruses, as commonly occurs in the developing countries of the tropics. Earlier concern that breast-feeding may represent a contraindication has not been substantiated; though colostrum contains moderate titers of maternal IgA, milk itself does not contain enough antibody to neutralize the vaccine virus.

Two significant advantages of OPV over IPV follow from the fact that it multiplies in the alimentary tract. First, the subclinical infection elicits prolonged synthesis of IgA as well as the IgG antibodies which protect the individual against paralytic poliomyelitis by intercepting wild polioviruses during the viremic phase. The mucosal antibody prevents primary implantation of wild virus in the gut and hence diminishes the circulation of virulent viruses in the community. Indeed, wild polioviruses have now virtually disappeared from countries such as the United States; in sewage, vaccine strains exceed wild strains by over a millionfold. This striking replacement of wild virus by vaccine strains is facilitated by the fact that the latter are excreted in the feces and may spread to nonimmune contacts.

A consequence of the spread of vaccine virus to contacts is that it provides greater opportunities for selection of mutants displaying varying degrees of reversion toward virulence (see Chapter 7). Within 2–3 days of administration, the OPV type 3 strain recovered from the feces of the vaccinee generally displays a single nucleotide change at one particular position in the 5′ noncoding region which represents a reversion to that of the wild type. Although these partial revertants can be shown to have partially regained neurovirulence for monkeys, they only rarely produce paralysis in the vaccinee or in contacts. Acquisition of greater neurovirulence requires reversion to the wild-type nucleotide at one or more positions within the open reading frame for the capsid proteins. Very rarely (once per million vaccinees) the vaccinee, or an unvaccinated family contact, develops poliomyelitis. Because OPV-associated polio is 10,000 times more common in those who are immunocompromised in some way, only IPV should be used in such children.

Furthermore, because there are still significant numbers of parents who have never received poliovaccine, there are good arguments for immunizing unvaccinated family contacts prior to or simultaneously with their infants.

Research continues to improve the genetic stability of the type 3 component of OPV, which contains only 10 nucleotide substitutions (only 3 of which result in amino acid changes) compared with 57 (21 amino acids) in the more stably attenuated type 1 component. Genetic engineering has been employed to construct a chimera comprising a fully attenuated type 1 virus that incorporates the immunogenic region of type 3 capsid protein. Theoretically, when more becomes known about the determinants of virulence in polioviruses, it should be possible to synthesize and clone poliovirus cDNA containing appropriate nucleotide substitutions in predetermined locations. Meanwhile, it must be stressed that the OPV on which we rely today has proved itself as an outstandingly successful and safe vaccine which if delivered properly to all children throughout the world is perfectly capable of eradicating poliomyelitis forever.

In the more developed nations poliomyelitis has been so effectively conquered by OPV that it has become "the forgotten disease." Nevertheless, importation of virus remains an ever present threat, especially to pockets of unimmunized people, such as immigrants and their preschool children, who are often concentrated in urban ghettos. Thus it remains imperative that very high levels of immunization coverage are maintained to prevent such epidemics which could in turn be followed by reestablishment of endemicity.

Inactivated poliovaccine (IPV) became "the forgotten vaccine" following the development of OPV. Nevertheless, it retains a loyal band of advocates who believe it is as good or better, particularly now that IPV of improved potency is being produced as a result of a number of technologic improvements. Although human diploid cell strains do not grow sufficiently well to make them a commercial proposition for production of the very large amounts of virus required for IPV, nonmalignant, aneuploid monkey kidney cell lines such as Vero have now been approved for this purpose. The cells are grown on the surface of myriads of DEAE-Sephadex beads known as "microcarriers" suspended in large fermentation tanks. Poliovirus grown in such cells is purified by gel filtration and ion-exchange chromatography, affinity chromatography on Sepharose-immobilized antibodies, and/or zonal ultracentrifugation. The importance of removing aggregates of virions before inactivation with formaldehyde was established many years ago following the 1955 disaster in which clumped virus escaped inactivation and paralyzed numerous children in the United States. Two or three doses of today's IPV confer protection; this has encouraged a proposal that IPV replace OPV, particularly in the Third World where unsatisfactory seroconversion rates are sometimes reported even after the full course of three doses of OPV. As DPT is standard in most countries, it would not be too difficult to add IPV to the cocktail to make a quadruple vaccine. However, partly because of the expense, the current recommendation of the Expanded Programme of Immunization (EPI) is that we stick with OPV but improve its delivery.

In stark contrast to the dramatic success of the polio vaccination programs of the developed parts of the world, much less impact has been made on the disease throughout most of Asia and Africa. The reasons for this are complex, ranging from chronically poor standards of sanitation that favor continued endemicity of polioviruses to lack of a health service infrastructure adequate to ensure vaccination coverage of the whole population, so that OPV is not delivered satisfactorily to the children, particularly in the most inaccessible rural villages. Many children never receive a full course of OPV. Even when they do, there may be doubts about whether the degree of refrigeration during storage and transport in the hot tropical climate has been adequate to retain the viability of the virus; greater attention is now being paid to maintenance of the cold chain. Suboptimal rates of seroconversion also raise the question of whether the current vaccine formulation or schedule needs to be adjusted.That poliomyelitis can indeed be eradicated in developing countries by comprehensive coverage with OPV has been demonstrated in South and Central America, where no cases of poliomyelitis caused by wild polioviruses have been seen since August 1991.

Wild polioviruses were eliminated from the United States and Canada during the 1970s, and in 1985 the Pan American Health Organization resolved to eradicate poliomyelitis from the western hemisphere by 1990. Regular EPI vaccination with OPV was supplemented by annual "national vaccination days," when all children under 14 years of age were given OPV, and by "mopping-up" vaccination in localities where cases of poliomyelitis had occurred. Both of these operations required much additional vaccine, the cost of which was largely covered by a special effort by Rotary International.The progress of eradication was monitored by greatly strengthened surveillance of flaccid paralysis and the careful laboratory investigation of every such case using modern methods of molecular epidemiology described in Chapter 14.

In 1988 the World Health Assembly adopted the ambitious target of worldwide eradication by the year 2000. Elimination of poliomyelitis from many or most countries in three of the six WHO regions [Europe, the Americas, and the Western Pacific (which includes China)] is likely to succeed, but there will be massive, perhaps insurmountable problems in achieving eradication from Africa and the Indian subcontinent.

Other Enteroviruses

Pathogenesis and Immunity

Most enteroviruses are presumed to enter the body via ingestion and to grow well in both the throat and the intestinal tract, but they are shed in the feces for much longer than in respiratory secretions. Dissemination via the bloodstream is doubtless the route of spread to the wide range of target organs susceptible to attack. Little is known of the factors that determine the tropism of different enteroviruses, for example, the predilection of coxsackie B viruses for muscle or of enterovirus 70 for the conjunctiva. The incubation period is 1–2 weeks in the case of systemic enterovirus infections but may be as short as a day or two for conjunctival or respiratory disease.

Immunity is type-specific and long-lasting. Antibody appears to be relatively more important in protection and recovery from picornavirus infections than for most other virus families. Perinatal infections tend to produce no disease or mild respiratory or gastrointestinal symptoms in neonates with maternal antibody, but overwhelming disseminated disease may occasionally occur in those without antibody. In particular, children with congenital B-cell deficiencies may contract chronic disseminated enterovirus infections.

Clinical Syndromes

Most enterovirus infections are subclinical, particularly in young children. Nevertheless, they can cause a wide range of clinical syndromes involving many of the body systems (Table 23-3). It should be noted that each syndrome can be caused by several viruses, and that each virus can cause several syndromes even during the same epidemic. Rashes, upper respiratory tract infection (URTI), and "undifferentiated summer febrile illnesses" are common. Moreover, enteroviruses are the commonest cause of meningitis, albeit a relatively mild form. In general, coxsackieviruses tend to be more pathogenic than echoviruses; they cause a number of diseases rarely seen with echoviruses, for example, carditis, pleurodynia, herpangina, hand–foot-and-mouth disease, and occasionally paralysis.

Neurologic Disease

Meningitis is more commonly caused by enteroviruses than by all bacteria combined, but fortunately this "aseptic" meningitis is not nearly so serious (see Chapter 36). Numerous enteroviruses have been implicated from time to time. The most commonly involved are echoviruses 3, 4, 6, 7, 9, 11, 16, 30,

Table 23-3
Diseases Caused by Enteroviruses

Syndrome	Viruses[a]
Neurologic	
Meningitis	**Many enteroviruses**
Paralysis	**Polioviruses 1, 2, 3; enteroviruses 70, 71;** coxsackievirus A7
Chronic meningoencephalitis/dermatomyositis	Echoviruses, others
Cardiac and muscular	
Myocarditis	**Coxsackievirus B;** some coxsackievirus A and echoviruses
Pleurodynia	**Coxsackievirus B**
Skin and mucosae	
Herpangina	**Coxsackievirus A**
Hand–foot-and-mouth disease	**Coxsackieviruses A9, A16;** enterovirus 71; others
Maculopapular exanthema	**Echoviruses 9,** 16; many other enteroviruses
Respiratory	
Colds	**Coxsackieviruses A21, A24; echovirus 11, 20;** coxsackievirus B; others
Ocular	
Acute hemorrhagic conjunctivitis	**Enterovirus 70; coxsackievirus A24**
Neonatal	
Carditis, encephalitis, hepatitis	**Coxsackievirus B; echovirus 11** and others

[a] The commonest causal agents are in bold type.

coxsackieviruses A7, A9, B1–6, enterovirus 71, and (in developing countries where they are still common) polioviruses 1–3.

Paralysis is usually caused by polioviruses (in the developing world at least), but it has very rarely been associated with coxsackievirus A7. Some outbreaks of enterovirus 71 infection have been marked by meningitis, some encephalitis, and many cases of paralysis with quite a number of deaths. During the 1969–1971 enterovirus 70 pandemic, acute hemorrhagic conjunctivitis was complicated in a small minority of cases by radiculomyelitis, which manifested itself as an acute flaccid lower motor neuron paralysis, resembling poliomyelitis but often reversible.

Chronic meningoencephalitis with juvenile dermatomyositis is a fatal condition seen in children with the B-cell deficiency, X-linked agammaglobulinemia, or severe combined immunodeficiency. Meningoencephalitis is the principal presentation, and an echovirus or other enterovirus can be recovered in high titer from CSF for months or years. Chronic hepatitis may also be present. More than half of the patients have a condition known as juvenile dermatomyositis; the possible etiologic role of known and previously unknown enteroviruses in this disease is currently the subject of intensive investigation.

Cardiac and Muscular Disease

Although it has long been known that enteroviruses are the major cause of carditis in the newborn (see below), in recent years it has become clear that enteroviruses are an important cause of *carditis* at all ages, particularly in adolescents and physically active young adult males. The coxsackie B viruses and, to a lesser extent, other enteroviruses such as coxsackieviruses A4 and A16 and echoviruses 9 and 22 have been directly incriminated by demonstration of virus, viral antigen, and/or viral RNA in the myocardium itself or in pericardial fluid. The disease may present predominantly as myocarditis, pericarditis, or cardiomyopathy characterized by a greatly dilated heart. Death is uncommon; however, recrudescences occur in about 20% of cases, and permanent myocardial damage, as evidenced by persistent electrocardiographic (ECG) abnormalities, cardiomegaly, or congestive heart failure, may occur.

Pleurodynia, also known as Bornholm disease, is basically a myositis, characterized by paroxysms of stabbing pain in the muscles of the chest and abdomen. Seen mainly in older children and young adults, sometimes in epidemic form, it is caused principally by coxsackie B viruses. Despite the severity of the pain, recovery is invariable.

Exanthems and Enanthems

Rashes are a common manifestation of enteroviral infections. Generally transient, always inconsequential, their principal significance is as an index of an epidemic which may have more serious consequences for other patients. Differential diagnosis from other rashes, viral or otherwise, is quite difficult. For example, the fine maculopapular ("rubelliform") rash of the very common echovirus 9 may easily be mistaken for rubella. The "roseoliform" rash of echovirus 16 ("Boston exanthem") appears only as the fever is declining. Many other enteroviruses can also produce rubella-like or measles-like rashes with fever and sometimes pharyngitis, especially in children.

Vesicular ("herpetiform") rashes are less common but very striking. The quaint name of *hand–foot-and-mouth disease* is given to a syndrome characterized by ulcerating vesicles on these three sites. Coxsackieviruses A9 and A16 and enterovirus 71, which resembles coxsackie A viruses, are most frequently responsible. The notorious foot-and-mouth disease of cattle is caused by an unrelated picornavirus which is not transmissible to humans.

Herpangina is a severe febrile pharyngitis characterized by vesicles (vesicular pharyngitis) or nodules (lymphonodular pharyngitis), principally on the soft palate. Despite the confusing name, it has nothing to do with herpesviruses but is one of the common presentations of coxsackie A virus infections in children.

Respiratory Disease

Enteroviruses are not major causes of respiratory disease but can produce a range of febrile colds and sore throats during summer epidemics. These come under the umbrella label of URTI or "undifferentiated febrile illness." Coxsackieviruses A21 and A24 and echoviruses 11 and 20 are prevalent respiratory pathogens; coxsackie B viruses and certain echoviruses are less common.

Ocular Disease

Acute hemorrhagic conjunctivitis burst on the world as a new disease in 1969. Extremely contagious, with an incubation period of only 24 hours, the disease began in West Africa and swept across Asia to Japan, infecting 50 million people within 2 years. As the name implies, this conjunctivitis was often accompanied by subconjunctival hemorrhages and sometimes transient involvement of the cornea (keratitis). Mainly adults were affected. Resolution usually occurred within a week or two, uncomplicated by anything other than secondary bacterial infection. However, a small minority of cases, particularly in India, were complicated by neurologic sequelae, notably radiculomyelitis, discussed above. In 1981 another major epidemic of acute hemorrhagic conjunctivitis arose in Brazil and swept north through Central America and the Caribbean to the southeastern United States and the Pacific islands. The cause of the original pandemic was a previously unknown agent now called enterovirus 70, but recent epidemics have been caused by a variant of coxsackievirus A24, which is not nearly so often associated with conjunctival hemorrhage.

Neonatal Disease

Enteroviruses may infect newborn babies prenatally (transplacental), natally (fecal contamination of the birth canal) or, most commonly, postnatally (from the mother or nosocomially in hospital nurseries). Coxsackie B viruses and certain echoviruses, notable type 11, can produce fulminant infections which are often fatal. Two major presentations may be distinguished, although they tend to overlap. The first, caused by coxsackie B viruses or echovirus 11, is the *encephalomyocarditis* syndrome. Classically, a neonate suddenly becomes dyspneic and cyanosed within the first week of life; examination reveals tachycardia and ECG abnormalities, often accompanied by other signs of overwhelming systemic infection, including manifestations of meningoencephalitis. The second presentation, sometimes called the *hemorrhage–hepatitis* syndrome, is associated mainly with echovirus 11 and less commonly with other echoviruses. The baby is lethargic and feeds poorly, then develops jaundice, followed by profuse hemorrhages, hepatic and renal failure, and death within days.

Other Possible Associations

Hemolytic–uremic syndrome (hemolytic anemia, impaired renal function) is an uncommon disease of children, from which enteroviruses have frequently been isolated. Coxsackievirus A4, various coxsackievirus B types, and echovirus 11 have been incriminated, and the frequency of multiple infection has encouraged the proposition that the syndrome may be due to a Shwartzman reaction.

Juvenile-onset insulin-dependent diabetes mellitus (IDDM) has been speculatively linked with the coxsackie B viruses for decades. For example, studies have shown IgM antibodies to coxsackieviruses in many cases of IDDM, and coxsackievirus B4 recovered from a fatal case reproduced a comparable diabetic condition when inoculated into mice. In light of the known propensity of coxsackie B viruses to grow in the pancreas, especially in neonatal infections, it remains to be demonstrated whether the association with IDDM is one of cause or effect. The apparent link with HLA types DR3 and DR4 suggests that immunologic mechanisms may be involved.

Chronic (postviral) fatigue syndrome, otherwise known as epidemic neuromyasthenia, myalgic encephalomyelitis, or chronic peripheral muscle disease, is a common but vaguely defined entity much in vogue in the 1990s. The well-known predilection of coxsackie B viruses for muscle has encouraged investigations of a possible causal link, and there are reports of prolonged elevation of coxsackievirus B IgM antibodies, antigen–antibody complexes, and coxsackievirus B RNA in muscle biopsies taken up to a year after the onset of chronic fatigue syndrome. At this stage, however, the etiology and even the definition of this syndrome as a single clinical entity are very much in doubt.

Laboratory Diagnosis

Most but not all enteroviruses are readily isolated in cell culture, and this remains the diagnostic method of choice, pending the development of generally applicable more rapid alternatives. Enteroviruses are of course most easily recovered from feces, but they can often also be recovered from the throat early in the illness. In certain diseases they may be found in the eye, vesicle fluid, urine, CSF, blood, or organs such as heart or brain at autopsy, and isolation from these "sterile" sites is much more indicative of a cause/effect relationship.

Classically, primary cultures of monkey kidney have been the standard substrate for enteroviruses, but diminishing availability has led to their replacement in most laboratories by continuous cell lines. Polioviruses grow well in virtually any simian or human cell type. Human diploid embryonic lung fibroblasts (HDF or HEL) support the growth of most echoviruses but few coxsackieviruses. The monkey kidney cell line BGM is particularly susceptible to the coxsackie B viruses. A human rhabdomyosarcoma line, RD, supports the growth of some coxsackie A viruses previously regarded as cultivable, in addition to many of the echoviruses. Enterovirus 70 is fastidious but can be isolated, with difficulty, in cultures of human conjunctiva/cornea or HDF.

Cytopathic effects resemble those described above for poliovirus (Fig. 23-4) but develop more slowly, commencing with foci of rounded refractile cells which then lyse and fall off the glass. The higher numbered serotypes are generally slower growers and/or produce only incomplete CPE, so blind passage may be necessary to reveal their presence, particularly with coxsackie A viruses (even in susceptible RD cells). Provisional allocation to the family *Picornaviridae* is generally made on the basis of the characteristic CPE.

Neutralization is the only reliable method of typing the isolate. This laborious procedure can be shortcut by the use of an "intersecting" series of "polyvalent" serum pools. Each pool contains horse antiserum against, say, 10 of the 68 human enterovirus serotypes; the pools are concocted to ensure that antiserum to any given serotype is present in several pools and absent from several others, hence the isolate can be positively identified by scrutinizing the pattern of pools that neutralize. Aggregated virions may escape neutralization and hence appear untypeable; this problem can be overcome by dispersing clumps with chloroform or by using a plaque reduction assay.

The use of the infant mouse is diminishing, but a few coxsackie A viruses (e.g., types 1, 19, 22) can still be grown only in this way. Newborn mice (<24 hours of age) are inoculated intraperitoneally and intracerebrally, then observed for illness, sacrificed, and subjected to histologic examination (Fig. 23-5).

IgM serology is also used to diagnose certain enterovirus infections. For example, in coxsackievirus carditis or enterovirus 70 hemorrhagic conjunctivitis, virus-specific IgM can be identified by IgM capture EIA any time up to 2 months after symptoms appear. However, non-cross-reactive antigens are not yet generally available for most enteroviruses other than coxsackie B viruses and polioviruses. Moreover, no cross-reactive capsid antigen common to all enteroviruses is available. Thus, whether intended for identification at the genus or species level, neither IgM and IgG serology is yet of much use for the diagnosis of most other enteroviral infections.

Assays for direct antigen detection are not well developed for the enteroviruses. Immunofluorescence can be used in selected situations where the viruses concerned are difficult to culture and where clean samples of infected cells are easily accessible, for example, to identify enterovirus 70 in conjunctival scrapings or other enteroviruses in leukocytes from CSF. Exploitation of EIA for direct demonstration of antigen suffers from the same limitation as enterovirus serology generally, namely, the lack of any monoclonal antibody or polyclonal serum that recognizes all, or even most, enteroviruses. Type-specific monoclonal antibodies (MAbs) are useful only when the number of possible serotypes is strictly limited, such as identification of enterovirus 70 or coxsackievirus A24 in acute hemorrhagic conjunctivitis.

Nucleic acid hybridization has been used successfully, for example, to detect coxsackievirus RNA in cardiac biopsies, using labeled coxsackievirus cDNA as a probe, but is generally too insensitive without prior amplification by PCR. Recently, there has been good progress with the development of a versatile PCR system for enteroviruses. Two of the three conserved nucle-

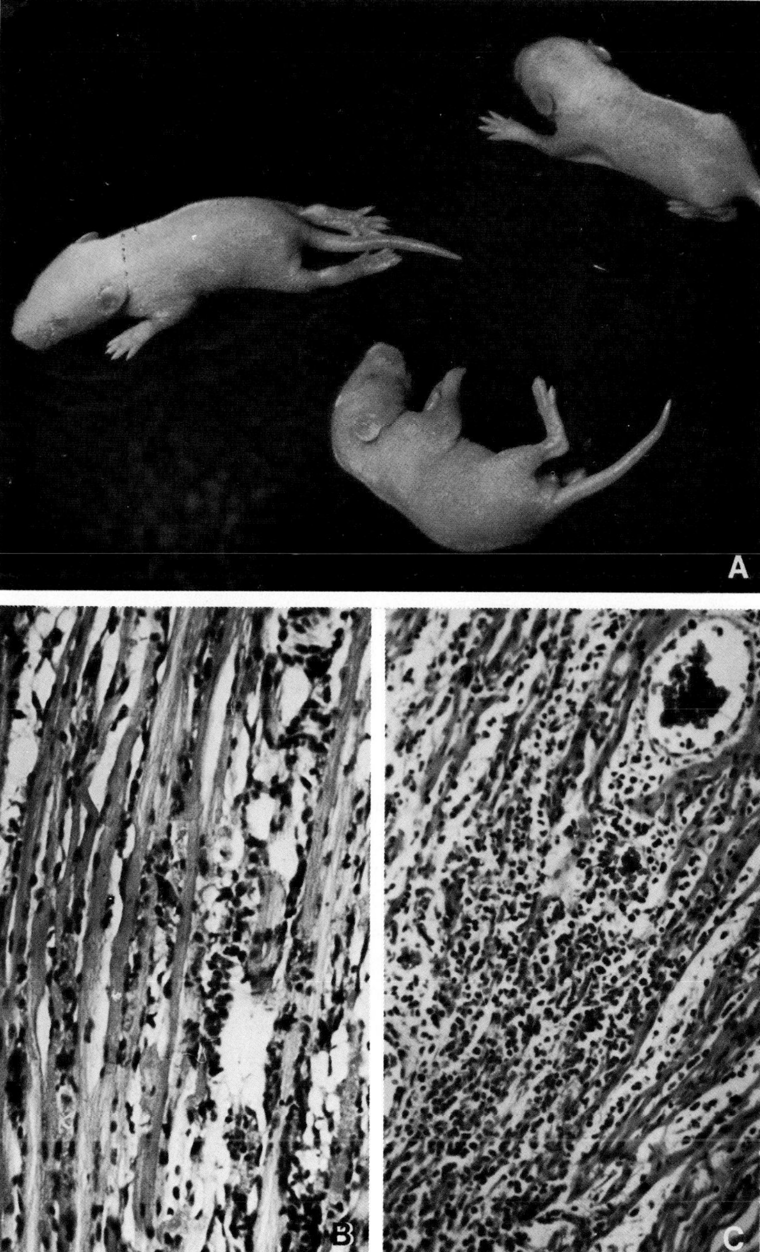

Fig. 23.5 Coxsackieviruses. (A) Infant mice 4 days after inoculation with coxsackievirus group A. The mice are moribund, with flaccid paralysis. (B) Section of muscle from an infant mouse infected with coxsackievirus group B (hemotoxylin and eosin stain; magnification: ×250). Note focal myositis. (C) Section of heart from a fatal case of coxsackievirus group B induced myocarditis in a newborn baby (hemotoxylin and eosin stain; magnification: ×100). (Courtesy I. Jack.)

otide sequences in the 5′ noncoding region, which are common to all or most enteroviruses, were chosen to construct a pair of cDNA primers for the PCR, leaving the third (nonoverlapping) conserved sequence for use as a radiolabeled probe, which successfully detected the RNA extracted from various enteroviruses in the CSF of patients with aseptic meningitis. Such sensitive and rapid diagnostic approaches should be of great value, particularly in the more serious enterovirus infections such as meningitis and carditis.

Epidemiology

Enteroviruses are transmitted mainly by close contact, via the fecal–oral route, spreading rapidly and efficiently within families. Echoviruses appear in the alimentary tract of most infants shortly after birth. In underdeveloped countries with low standards of hygiene and sanitation enteroviruses can be recovered from the feces of the majority of young children at any time—80% in one study in Karachi. In contrast, the New York "Virus Watch" program revealed an enterovirus carriage rate of only 2.4%. These differences are also reflected in enterovirus counts in sewage.

Droplet spread also occurs, more so with the coxsackieviruses, and may be more relevant to the acquisition of the upper respiratory infections for which enteroviruses are often responsible. Acute hemorrhagic conjunctivitis is highly contagious, spreading by contact, with an incubation period of only 24 hours. Over half a million cases occurred in Bombay alone during the 1971 pandemic.

Over the years the commonest enteroviruses worldwide have included echoviruses 4, 6, 9, 11, and 30, coxsackieviruses A9 and A16, and coxsackieviruses B2–B5. Outbreaks of these serotypes, and others, often reach epidemic proportions, peaking in late summer/early autumn in countries with a temperate climate. A different type tends to become prevalent the following year as immunity develops. Nevertheless, distinct serotypes do cocirculate to some degree, mainly among nonimmune infants.

Young children constitute the principal target and reservoir of enteroviruses. It is significant that the prevalence of enteroviruses tends to increase at the time children return to school after vacations. Having acquired the virus from other youngsters, schoolchildren then bring it home to the parents and siblings. In young children most infections are inapparent or involve mild undifferentiated fevers, rashes, or URTI. Disproportionately, the severe diseases—carditis, meningitis, encephalitis—are seen in older children and adults, as well as in neonates.

Rhinoviruses

In this era of organ transplantation, genetic engineering, and other dramatic demonstrations of the wonders of medical science, the man in the street perceives a certain irony in the inability of modern medicine to make the slightest impact on that most trivial of all human ailments, the common cold. An even greater irony is the fact that years of searching for the elusive common cold virus have led us finally not to one, but to over 100 separate rhinoviruses. Moreover, it is now evident that rhinoviruses cause only about 50% of all colds. A vaccine for the common cold is further away than ever.

Pathogenesis and Immunity

Consistent with their predilection for replication at 33°C, rhinoviruses usually remain localized to the upper respiratory tract. Inflammation, edema, and copious exudation begin after a very short incubation period (2–3 days) and last for just a few days. Endogenous interferons may play a role in terminating the illness and in conferring transient resistance against infection with other viruses. Acquired immunity is type-specific and correlates more with the level of locally synthesized IgA antibodies, which decline in titer within months of infection, rather than with IgG in serum, which may persist for a few years.

Clinical Features

We are all too familiar with the symptomatology of the common cold: profuse watery nasal discharge (rhinorrhea) and congestion, sneezing, and quite often a headache, mildly sore throat, and/or cough. There is little or no fever. Resolution generally occurs within a week, but sinusitis or otitis media may supervene, particularly if secondary bacterial infection occurs. Rhinoviruses may also precipitate wheezing in asthmatics or exacerbate chronic bronchitis, especially in smokers.

Laboratory Diagnosis

One would hardly ask the laboratory to assist in the diagnosis of the common cold, except perhaps in the course of epidemiologic research. Rhinoviruses are fastidious, growing only slowly, at 33°C (the temperature of the nose), in cell lines derived from embryonic human fibroblasts such as WI-38, MRC-5, or fetal tonsil, without conspicuous CPE. A simpler method is to demonstrate antigen directly in nasal washings by EIA. The PCR, using a primer based on the conserved 5′ noncoding regions in the viral RNA, can also be used to amplify the genome.

Epidemiology

Rhinovirus colds occur throughout the year, with peaks in the autumn and spring. Three or four rhinovirus serotypes circulate among the community simultaneously, one sometimes predominating; then a new set moves in after a year or so as immunity develops in the population. Preschool children, who tend to be susceptible to almost all serotypes, commonly introduce the virus to the rest of the family, with a secondary attack rate of around 50%, after an interval of 2–5 days. Virus is shed copiously in watery nasal secretions for 2–3 days or more. Sneezing and coughing generate large- and small-particle aerosols, whereas "dribbling" and contamination of hands and handkerchiefs or paper tissues mediate contact spread via fomites from hands to nose or eye.

The fact that many colds occur during winter and the "changes of seasons" has encouraged the popular myth that one is most vulnerable if exposed to cold, wet weather, but this is too simplistic a view. The influence of seasonality on transmissibility was discussed in Chapter 14.

Control

Clearly, the diversity of serotypes displaying little or no cross-protection rules out any possibility of an effective vaccine. Antiviral chemotherapy, or possibly chemoprophylaxis, seems to be the only hope.

The prospect of multibillion dollar profits has encouraged many pharmaceutical companies to persist in the pursuit of that elusive goal, "a cure for the common cold." Several viral proteins, such as the polymerase, helicase, protease, and VPg, represent potential targets of attack, but much current effort is concentrated on designing agents that block attachment or uncoating by binding directly to the virion itself. Some 91 of the 102 currently known rhinovirus serotypes use the widely distributed intercellular adhesion molecule-1 (ICAM-1) as their receptor, and recent cryoelectron microscopic analysis of complexes of rhinovirus 16 with the N-terminal domain of ICAM-1 have confirmed the precise location of the ligand, which had previously been defined by X-ray diffraction on human rhinovirus type 14 (HRV-14) to lie within a "canyon" within the capsid protein VP1. In Chapter 16 we described how X-ray crystallography has also been exploited to demonstrate how compounds that block adsorption, and particularly uncoating, of the virion occupy a hydrophobic pocket (β barrel) immediately beneath the floor of that canyon. As might be expected, optimal antiviral activity seems to be conferred by compounds that fully occupy this pocket, providing a snug fit that maximizes the number of hydrophobic interactions. It is postulated that these compounds stabilize the virion against a conformational change that occurs below pH 6 and is required for intracellular uncoating. Because serotypes may differ in their precise amino acid sequence or conformation within this pocket, it will be necessary to model a "consensus pocket" in order to design a compound that will neutralize most rhinoviruses. Even then, problems will occur with resistant mutants, which have already been demonstrated to develop in culture and *in vivo*.

Hepatitis A

The enterically transmitted disease known first as "infectious hepatitis," to distinguish it from "serum hepatitis" (hepatitis B), then as hepatitis A, was known for many years before its causal agent was identified unequivocally in 1973 by demonstration of a 27 nm icosahedral virus in patients' feces using immunoelectron microscopy (Fig. 23-6). Biophysical and biochemical studies later established hepatitis A virus as a member of the family *Picornaviridae*.

Properties of *Hepatovirus*

For a time hepatitis A virus (HAV) was classified as enterovirus type 72, but it was eventually accorded the status of a separate genus, *Hepatovirus*, on the basis of a number of differences from the enteroviruses, including stability of

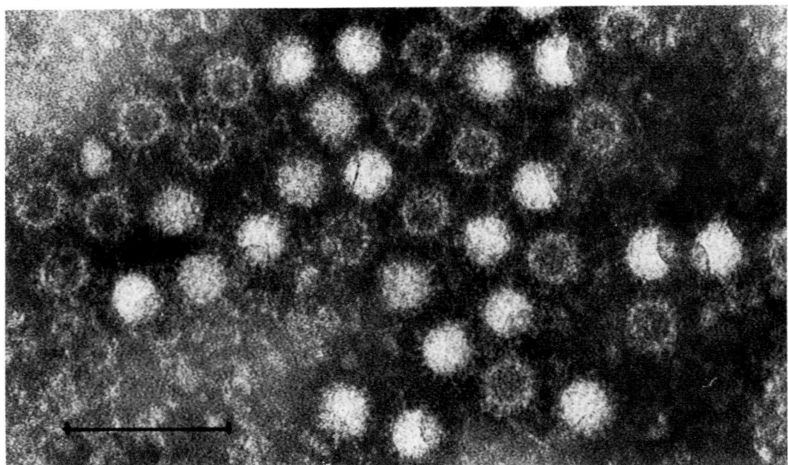

Fig. 23.6 Hepatitis A virus. Negatively stained preparation of virions clumped by antibody as seen by immunoelectron microscopy. Bar, 100 nm. (Courtesy Drs. J. Marshall and I. D. Gust.)

the virion at 60°C, lack of reactivity with an enterovirus group-specific MAb, low percentage nucleotide homology with the genome of enteroviruses (in spite of having the same gene order), and certain differences in the replication cycle (see below).

Antigenically, HAV is highly conserved, there being only a single serotype in spite of the fact that four genotypes differing by around 20% in nucleotide sequence have been described; most human strains belong to genotypes I or III. An additional three genotypes have been isolated from monkeys. Most simian HAV strains are thought to be host-restricted and differ from human HAV in immunodominant antigenic sites, but at least one, the PA21 strain from *Aotus* monkeys, is very closely related antigenically to human HAV genotype III.

Viral Replication. The replication of wild strains of HAV in cultured primate cells is slow and the yield of virus poor. Uncoating of the virion is inefficient. The 5′ nontranslated region (NTR) contains an internal ribosomal entry site facilitating cap-independent initiation of translation. The processing of the P1 polyprotein and assembly of the cleavage products into virions follows a different pathway from that of enteroviruses. Very little complementary (minus sense) RNA is detectable in infected cells; most of the newly synthesized plus strand RNA becomes rapidly encapsidated into new virions. Cellular protein synthesis is not inhibited, and there is little or no CPE; infection of cell lines is noncytocidal and persistent, with a restricted yield. After serial passage, rapidly replicating cytopathic variants emerge; they contain numerous mutations, including a 14 base reduplication in the 5′ NTR. HAV is resistant to several antiviral agents that inhibit the replication of enteroviruses.

Pathogenesis, Immunity, and Clinical Features

It is widely assumed that hepatitis A virus, known to enter the body by ingestion, multiplies first in intestinal epithelial cells before spreading via the blood stream to infect parenchymal cells in the liver, but this has yet to be formally proved. Virus is detectable (up to 10^8 virions per gram) in the feces, and in much lower titer in blood, saliva, and throat, during the week or two prior to the appearance of the cardinal sign, dark urine, and disappears soon after serum transaminase levels reach their peak. Hence the patient's feces are most likely to transmit infection before the onset of jaundice. The incubation period of hepatitis A is about 4 weeks (range 2–6 weeks).

The clinical features of hepatitis A closely resemble those already described for hepatitis B (see Chapter 22). The onset tends to be more abrupt and fever is more common, but the constitutional symptoms of malaise, anorexia, nausea, and lethargy which comprise the prodromal (preicteric) stage tend to be somewhat less debilitating and less prolonged. Hepatomegaly may produce pain in the right upper abdominal quadrant, followed by bilirubinuria, then pale feces and jaundice. Most infections worldwide occur in children, in whom they are generally subclinical (i.e., asymptomatic) or anicteric (i.e., symptomatic but without jaundice). Severity increases with age, about two-thirds of all infections in adults being icteric. The case–fatality rate is 0.5% (0.1% of all infections), resulting from liver failure (fulminant hepatitis). The illness usually lasts about 4 weeks, but a minority relapse, typically after a premature bout of drinking or heavy exercise, and symptoms may continue for up to 6 months. Also, prolonged excretion of virus in feces has been observed in neonates nosocomially infected in an intensive care unit. However, in striking contrast to hepatitis B, all nonlethal infections resolve, with complete regeneration of damaged liver parenchyma, no long-term sequelae, and no chronic carrier state.

Liver pathology resembles that for hepatitis B. It is not striking until after viral replication has peaked and the immune response is underway, suggesting that hepatocellular injury is immunopathologically mediated. Natural killer (NK) cells are mobilized and activated, as are $CD8^+$ cytotoxic T lymphocytes, known to secrete interferon γ that up-regulates expression of class I MHC protein on hepatocytes, which normally display little of this antigen. The serum antibody response to HAV is lifelong, declining significantly only in old age. As there is only one known serotype of hepatitis A virus, infection leads to lifelong immunity, and second attacks of the disease are unknown.

Laboratory Diagnosis

Markedly elevated serum alanine and aspartate aminotransferase (ALT and AST) levels distinguish viral from nonviral hepatitis but do not discriminate among viral hepatitis A, B, C, D, and E. A single serological marker, anti-HAV IgM, is diagnostic for hepatitis A. RIA or EIA are the methods of choice for detecting the IgM antibody, which is demonstrable from the time symptoms and signs appear until about 3–6 months later.

By the time the patient presents it is usually already too late to isolate virus; however, when it is necessary to do so for research purposes, virus can be recovered from feces in primary or continuous lines of primate cells, derived from monkey kidney or from human fibroblasts or hepatoma. Amplification via the PCR is more sensitive and could perhaps be applied to the detection of trace amounts of hepatitis A virus in contaminated food or water.

Epidemiology

Like poliovirus, hepatitis A virus is spread via the fecal–oral route. As might be expected, therefore, the disease is hyperendemic in the developing countries of Asia, Africa, and Central and South America where overcrowding, inadequate sanitation, and poor hygiene are rife. Where poverty and privation are extreme, infection, usually subclinical, is acquired in early childhood, so that virtually all adults have protective antibody. Most of the clinical cases are seen in children or young adults, and in visitors from the more developed countries. Direct person-to-person contact spread is most important, but contaminated food and water are also major vehicles of spread. Major common-source outbreaks may occur, particularly when wells or other communal water supplies become polluted with sewage.

In developed countries such as Sweden, in contrast, the epidemiologic picture is one of declining endemicity. The incidence of hepatitis A has been gradually declining for decades, such that only a minority of the population, notably the elderly, have antibody. Because primary infections tend to be more severe with increasing age, the peak incidence of clinical disease is in the 15–30 age group. Infection tends to be more prevalent in unsewered areas and in lower socioeconomic groups (e.g., American Indians and Hispanics in the southwestern United States where there has been a resurgence of hepatitis A in the past decade) as well as in particular high-risk occupational groups such as sewer workers and primate handlers, and in persons with high-risk behavior patterns, notably intravenous drug users and male homosexuals. In some Western nations today most sporadic cases are seen in travelers returning from Third World countries. Outbreaks, which can continue in the community for months or even years, frequently originate in communal living establishments with marginal standards of hygiene or special problems, such as children's day-care centers, homes for the mentally retarded, mental hospitals, prisons, army camps, and so forth. Infected handlers of food (especially uncooked or inadequately heated food such as salads, sandwiches, and berries) represent a particular danger in fast-food outlets, restaurants, etc. HAV can survive for months in water, hence sewage contamination of water supplies, swimming areas, or farms growing molluscs such as oysters and clams can also lead to explosive outbreaks. Special problems arise in times of war or natural disaster.

Control

As hepatitis A is transmitted exclusively via the fecal–oral route, control rests on heightened standards of public and personal hygiene. Reticulated drinking water supplies and efficient modern methods of collection, treatment, and disposal of sewage (see Chapter 15) should be the objective of every municipal government. Where this is impracticable, as in remote rural areas, particular attention needs to be given to the siting, construction, maintenance, and operation of communal drinking water supplies such as wells; for example, these should not be downhill from pit latrines because of the risk of seepage, particularly after heavy rains. Public bathing and the cultivation of shellfish for human consumption should not be permitted near sewerage outlets.

Those employed in the dispensing of food should be subject to special scrutiny and required to observe high standards of hygiene, especially hand washing after defecation. In so far as children often contract the infection at school, attention should be given to proper instruction of children in this matter. Especially difficult problems occur in day-care centers and homes for mentally retarded children. Routine precautions include separate diaper-changing areas and chemical disinfection of fecally contaminated surfaces and hands.

Passive immunization against hepatitis A is a well-established procedure which has been in use for many years. For instance, during wars or other operations involving the mass movement of personnel into endemic areas of the tropics, it has been standard to immunize visitors with normal human immunoglobulin prior to arrival and every 4–6 months thereafter. Normal immune globulin (0.02 ml/kg) is also used postexposure to protect family and institutional contacts following outbreaks in creches, schools, and other institutions.

The first hepatitis A vaccines were licensed in 1992. They are formalin-inactivated preparations of virions grown (following adaptation) in human fibroblasts or monkey kidney cell lines, adsorbed to alum as an adjuvant. Two doses injected 1 month apart, with or without a booster after 6 months, regularly elicit an excellent immune response in 99% of recipients that lasts for some years at least. Because the yield of virus from cultured cells is so low, the vaccine is expensive and hence is likely to remain a boutique vaccine targeted at small specialized markets until a live or recombinant vaccine perhaps eventually replaces it. High-risk cohorts to receive the inactivated vaccine should include travelers or long-term visitors to countries in which HAV is endemic, military personnel, sexually active homosexual men, intravenous drug users, sewage workers, primate handlers, workers in preschool day-care centers, certain staff and long-term residents of hospitals and institutions for the intellectually disabled, and workers engaged in food manufacturing and catering.

General vaccination of all infants worldwide should not be undertaken until it is clear that postvaccination immunity and/or immunologic memory are sufficient to protect for life, rather than simply to delay natural infection from childhood (when it is usually subclinical) to adulthood (when it usually causes disease). This consideration, as well as cost, argues for the development of an attenuated live virus vaccine for general use in infancy as part of the WHO Children's Vaccine Initiative. Candidate live vaccines, for parenteral not oral administration, are in the pipeline, but it has so far proved difficult to find the window of adequate attenuation without losing immunogenicity, perhaps because HAV replicates mainly or exclusively in hepatocytes.

Recombinant DNA technology offers additional approaches. Cloning of the HAV genome in insect cells, or preferably in mammalian cells containing the necessary cellular proteases for cleavage of the polyprotein, may yield HAV procapsids or at least the pentameric capsomers within which the immunodominant epitopes retain their native conformation. Alternatively, the genome can be incorporated into vaccinia virus which could then be used either as a live attenuated vaccine or as a vector for expression of properly processed HAV proteins in cultured human cells.

Further Reading

Brinton, M. A., and Heinz, F. X., eds. (1990). "New Aspects of Positive-Strand RNA Viruses." American Society for Microbiology, Plenum, New York.

Brinton M. A., and Rueckert, R. R. (1987). Positive Strand RNA Viruses. *UCLA Mol. Cell. Biol. Symp.* **54**.

Cherry, J. D. (1987). Enteroviruses: Polioviruses, coxsackieviruses, echoviruses, and enteroviruses. *In* "Textbook of Pediatric Infectious Diseases" (R. D. Feigin and J. D. Cherry, eds.), 2nd Ed., p. 1729. Saunders, Philadelphia, Pennsylvania.

Couch, R. B. (1990). Rhinoviruses. *In* "Fields Virology" (B. N. Fields, D. M. Knipe, R. M. Chanock, M. S. Hirsch, J. L. Melnick, T. P. Monath, and B. Roizman, eds.), 2nd Ed., p. 607. Raven, New York.

Grandien, M., Forsgren, M., and Ehrnst, A. (1989). Enteroviruses and reoviruses. *In* "Diagnostic Procedures for Viral, Rickettsial and Chlamydial Infections" (N. J. Schmidt and R. W. Emmons, eds.), 6th Ed., p. 513. American Public Health Association, Washington, D.C.

Grist, N. R., Bell, E. J., and Assaad, F. (1978). Enteroviruses in human disease. *Prog. Med. Virol.* **24,** 114.

Gust, I. D., and Feinstone, S. M. (1988). "Hepatitis A." CRC Press, Boca Raton, Florida.

Gwaltney, J. M. (1989). Rhinoviruses. *In* "Viral Infections of Humans: Epidemiology and Control" (A. S. Evans, ed.), 3rd Ed., p. 593. Plenum, New York.

Hendley, J. O., and Gwaltney, J. M. (1988). Mechanisms of transmission of rhinovirus infections. *Epidemiol. Rev.* **10,** 242.

Hollinger, F. B., and Ticehurst, J. (1990). Hepatitis A virus. *In* "Fields Virology" (B. N. Fields, D. M. Knipe, R. M. Chanock, M. S. Hirsch, J. L. Melnick, T. P. Monath, and B. Roizman, eds.), 2nd Ed., p. 631. Raven, New York.

Hollinger, F. B., Lemon, S. M., and Margolis, H. S., eds. (1991). "Viral Hepatitis and Liver Disease." Williams & Wilkins, Baltimore, Maryland.

Kew, O. M., Nottay, B. K., Rico-Hesse, R., and Pallansch, M. A. (1990). Molecular epidemiology of wild poliovirus transmission. *Appl. Virol. Res.* **2,** 199.

Lemon, S. M. (1992). Hepatitis A virus: Current concepts of the molecular biology, immunobiology and approaches to vaccine development. *Rev. Med. Virol.* **2,** 73.

McKinlay, M. A., Pevear, D. C., and Rossmann, M. G. (1992). Treatment of the picornavirus common cold by inhibitors of viral uncoating and attachment. *Annu. Rev. Microbiol.* **46,** 635.

Melnick, J. L. (1983). Portraits of viruses: The picornaviruses. *Intervirology* **20,** 61.

Melnick, J. L., ed. (1984). Enteric Viruses in Water. *Monogr. Virol.* **15.**

Melnick, J. L. (1989). Enteroviruses. *In* "Viral Infections of Humans: Epidemiology and Control" (A. S. Evans, ed.), 3rd Ed., p. 191. Plenum, New York.

Melnick, J. L. (1990). Enteroviruses: Polioviruses, coxsackieviruses and newer enteroviruses. *In* "Fields Virology" (B. N. Fields, D. M. Knipe, R. M. Chanock, M. S. Hirsch, J. L. Melnick, T. P. Monath, and B. Roizman, eds.), 2nd Ed., p. 549. Raven, New York.

Melnick, J. L., ed. (1990). Hepatitis A vaccine prospects. *Prog. Med. Virol.* **37,** 47.

Minor, P. D. (1992). The molecular biology of poliovaccines. *J. Gen. Virol.* **73,** 3065.

Morens, D. M., Pallansch, M. A., and Moore, M. (1991). Polioviruses and other enteroviruses. *In* "Textbook of Human Virology" (R. B. Belshe, ed.), 2nd Ed., p. 427. Mosby, St. Louis, Missouri.

Racaniello, V. R., ed. (1990) Picornaviruses. *Curr. Top. Microbiol. Immunol.* **161.**

Ross, B. C., Anderson, D. A., and Gust, I. D. (1991). Hepatitis A virus and hepatitis A infection. *Adv. Virus Res.* **39,** 209.

Rotbart, H. A. (1991). Nucleic acid detection systems for enteroviruses. *Clin. Microbiol. Rev.* **4,** 156.

Rueckert, R. R. (1990). Picornaviruses and their replication. *In* "Fields Virology" (B. N. Fields, D. M. Knipe, R. M. Chanock, M. S. Hirsch, J. L. Melnick, T. P. Monath, and B. Roizman, eds.), 2nd Ed., p. 507. Raven, New York.

Sabin, A. B. (1981). Paralytic poliomyelitis: Old dogmas and new perspectives. *Rev. Infect. Dis.* **3,** 543.

Semler, B. L., and Ehrenfeld, E., eds. (1989). "Molecular Aspects of Picornavirus Infection and Detection." American Society for Microbiology, Washington, D.C.

Semler, B. L., Kuhn, R. J., and Wimmer, E. (1988). Replication of the poliovirus genome. *In* "RNA Genetics, Volume I: RNA-Directed Virus Replication" (E. Domingo, J. J. Holland, and P. Alquist, eds.), p. 23. CRC Press, Boca Raton, Florida.

Stanway, G. (1990). Structure, function and evolution of picornaviruses. *J. Gen. Virol.* **71,** 2483.

Chapter 24

Caliciviridae and *Astroviridae*

Although it had been recognized for many years that gastroenteritis could be transmitted to human volunteers by ingestion of bacteria-free filtrates of feces from patients with diarrhea, it was not until 1972 that the first of the many types of viruses involved was identified, by immunoelectron microscopy (IEM). Norwalk virus, named for the town in Ohio that hosted the outbreak yielding the virus, was the prototype of a succession of "small round-structured viruses" whose taxonomic status remained unclear until 1991 when sequencing of the genome made it apparent that they belong to the family *Caliciviridae*. Caliciviruses had been well known for years to veterinary virologists interested in such major animal pathogens as vesicular exanthema of swine virus and feline calicivirus. In 1991 the genome of the noncultivable agent of another important enterically transmitted disease, hepatitis E, was cloned and sequenced, and the virus was provisionally classified as a member of the family *Caliciviridae*.

In addition, electron microscopists had described virions from gastroenteritis outbreaks in animals or humans which looked somewhat different in that they presented a distinct star-shaped outline (Fig. 24-1B). In 1993 these "astroviruses" were accorded the status of a separate family, *Astroviridae*.

Caliciviruses Associated with Gastroenteritis

Properties of *Caliciviridae*

The family *Caliciviridae* derives its name from the 32 cup-shaped (*calix*, cup) surface depressions that give the virion its unique appearance (Fig. 24-1A).

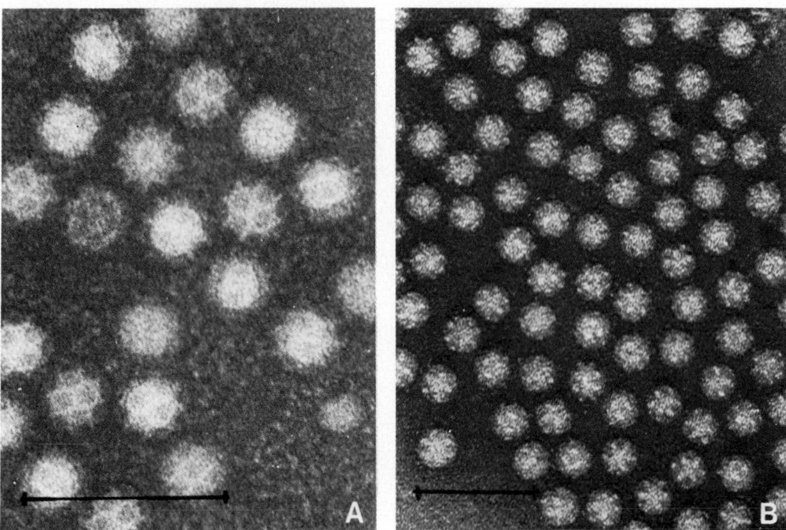

Fig. 24-1 Negatively stained electron micrographs of virions of *Caliciviridae* (A) and *Astroviridae* (B). Bars, 100 nm. (A, Courtesy Drs. M. Szymanski and P. J. Middleton; B, courtesy Dr. D. Snodgrass.)

The virion is 30–38 nm in diameter, slightly larger than the picornaviruses. The icosahedral capsid is constructed from 90 capsomers, each being a dimer of a single species of polypeptide. Although relatively resistant to inactivation by heat (60°C), stomach acid (pH 3), or minimal levels of chlorination of drinking water, the virions of several caliciviruses, including the hepatitis E virus, are particularly vulnerable to proteolysis and tend to lose their capsid definition on purification or storage.

The plus sense linear ssRNA genome of 7.5–7.7 kb resembles that of the picornaviruses in that a VPg protein is covalently attached to its 5′ terminus (except in the case of hepatitis E virus which is capped instead), the 3′ terminus is polyadenylated, and the nonstructural proteins include an RNA-dependent RNA polymerase, an ATPase/helicase, and a protease. However, the calicivirus genome has not one but three open reading frames, the long 5′ ORF encoding the nonstructural proteins, the middle ORF the capsid protein (sometimes involving a frameshift), and the 3′ ORF a protein of uncertain function (Table 24-1).

Classification

Because the human caliciviruses (and most of the other viruses causing gastroenteritis in humans) were discovered by electron microscopy (EM) and defied cultivation, the Caul and Appleton system for classifying various small, round putative gastroenteritis viruses was based on EM morphology. The system drew a distinction between (1) "typical" caliciviruses, which were 30–35 nm icosahedral virions with cup-shaped indentations, frequently presenting as a six-pointed Star of David with a dark, stain-filled central hollow (Fig. 24-1A) and (2) the "amorphous small round-structured viruses," of which the prototype was the Norwalk agent, which tended to have a wider size range (27–40 nm) and an indistinct ragged or "feathery" outline more difficult to define. In retrospect, it appears that both groups belong to the genus *Calicivirus* and that the differences observed in size and outline might reflect (a) bound coproantibody and/or (b) susceptibility of enteric caliciviruses to proteolytic degradation. Nevertheless, there may be important epidemiologic differences between the "typical" enteric caliciviruses, which tend to be associated with gastroenteritis in infants, and the Norwalk group, occurring mainly in adults and other children. The Norwalk group includes several viruses, falling into at least three "types" or "serogroups" typified by Norwalk, Snow Mountain, and Otofuke/Hawaii agents, but the antigenic relationships among them, or between them and the several strains of "typical" enteric caliciviruses, are not yet clear.

Viral Replication

Calicivirus replication is confined to the cytoplasm. Genomic RNA serves as the mRNA for the polyprotein representing the nonstructural proteins, which are derived from the precursor by proteolytic cleavage. For most caliciviruses a separate subgenomic mRNA encodes the capsid protein. Because subgenomic minus sense RNA has also been reported it is possible that the subgenomic mRNA (as well as genomic RNA) may be capable of replication via a complementary minus strand. It now appears that there may be a 3′ nested set of overlapping subgenomic mRNAs, reminiscent of coronavirus strategy. Virions assemble in the cytoplasm, sometimes forming paracrystalline arrays associated with the cytoskeleton.

Clinical Features

The disease caused by the Norwalk agent has been the most comprehensively studied, in volunteers as well as in natural outbreaks. Following a very short incubation period (generally 24–48 hours), the illness runs its course in 12–60 hours. Nausea, vomiting, diarrhea, and abdominal cramps are prominent features, with vomiting being most common in children and diarrhea in adults; headache, myalgia, and low fever are variable features.

Pathogenesis and Immunity

In Norwalk viral gastroenteritis the tips of the villi in the jejunum slough off but rapidly regenerate, there is infiltration of mononuclear cells and polymorphs into the mucosa, transient malabsorption, and delayed gastric empty-

Table 24-1
Properties of Caliciviruses

Single genus, *Calicivirus*, includes hepatitis E and human caliciviruses causing gastroenteritis
Nonenveloped spherical virion, 30–38 nm diameter, vulnerable to proteolysis
Icosahedral capsid with 32 cup-shaped depressions; 90 capsomers, one capsid protein
Linear plus sense ssRNA genome, 7.5 to 7.7 kb, polyadenylated at 3′ terminus, protein at 5′ terminus (except hepatitis E, capped), infectious
Three open reading frames: 5′, nonstructural proteins; middle, capsid protein; 3′, unknown
Cytoplasmic replication; genomic RNA ORF1 translated into polyprotein, cleaved into nonstructural proteins; subgenomic mRNA encodes capsid protein

ing. Acquired immunity is poor, lasting for a year or so at best. Indeed, when volunteers were challenged with Norwalk virus twice, at intervals of 2–3 years, those who had contracted gastroenteritis on the first occasion and had developed antibody were the very ones who became sick the second time. This was interpreted to mean (1) that prior infection confers no significant long-term immunity and (2) that unknown genetic or physiologic factors may predispose certain individuals but not others to gastroenteritis following exposure to virus. Paradoxically, it appears that antibody to the"typical" enteric caliciviruses is acquired early in life and protects adults against reinfection.

Laboratory Diagnosis

Unfortunately, human caliciviruses are extremely fastidious and can still not be grown reproducibly in cultured cells. Nevertheless, reliance on the cumbersome and insensitive diagnostic modality of IEM is now unnecessary following the recent molecular cloning of the genomes of a number of human enteric caliciviruses. For example, using baculovirus as an expression vector for the Norwalk genome, insect cells produce empty Norwalk virions which serve as antigen for immunoassays to probe for antibody, or as immunogen to make monoclonal antibodies (MAbs) to probe for antigen. EIAs based on such cloned reagents are being developed. Probes and primers have been produced for use in PCR, but special precautions must be taken first to inactivate RNases with the virion-disrupting agent guanidine isothiocyanate, and then to adsorb the viral RNA to glass beads or powder to enable soluble inhibitors of reverse transcriptase to be washed away. It must also be remembered that, since asymptomatic enteric viral infections are common and mixed infections not rare, and PCR assays for different agents may differ in sensitivity, PCR results must be interpreted with caution. The first-line diagnostic assay should be EIA for antigen detection. For epidemiologic investigation of an outbreak, EIA or RIA can be used to demonstrate a rising titer of antibody, with the specificity or cross-reactivity of the assay being determined by the choice of antigen employed; IgM capture assays circumvent the need to recall the patient for a second bleed.

Epidemiology

The Norwalk group of agents are uncommon in infants in developed countries but are seen in older children and adults, about half of all adults revealing serological evidence of past infection. Common-source outbreaks frequently occur via fecal contamination of water or food, such as salads, in restaurants, schools, camps, cruise ships, geriatric nursing homes, etc. Several major outbreaks have been traced to consumption of uncooked or partially cooked oysters, cockles, clams, or other shellfish that have been harvested from sewage-polluted estuaries. Others have been attributed to discharge of sewage into drinking water supplies, or into pools and lakes in which people swim. Secondary spread occurs from person to person within households, with transmission for up to 2 days after the resolution of symptoms. Some 40% of outbreaks of gastroenteritis in adults in the United States have been attributed to Norwalk virus. Unlike bacterial gastroenteritis or food poisoning, viral gastroenteritis occurs year-round. In developing countries these viruses are encountered much earlier in life, with the pattern of endemic exposure resembling that for the typical caliciviruses (see below).

The typical caliciviruses are endemic worldwide, infecting infants from 3 months of age until kindergarten age; most children have been infected with at least one calicivirus before leaving primary school. Outbreaks also occur in day-care centers, orphanages, maternity hospitals, schools, nursing homes, etc., with high attack rates.

Control

Control of person-to-person spread of endemic gastroenteritis requires a long-term national commitment and resources to improve the standards of sanitation, water supply, and personal hygiene (see Chapter 15), which is currently out of the reach of many impoverished countries. Common-source calicivirus outbreaks can be reduced by enforcement of legislation relating to the preparation and supply of food for public consumption. Of course viruses, unlike bacteria, cannot multiply in food or water. Careful investigation of Norwalk virus outbreaks in the United States identified an ill food-handler as the probable source in the majority of instances. Because even caliciviruses are inactivated at temperatures well below that of normal cooking, uncooked foods such as salads are the greatest risk. Shellfish concentrate viruses from fecally contaminated water and are a notorious source of caliciviruses and hepatitis viruses if served raw or inadequately cooked, accounting for fully half of the Norwalk outbreaks in the United States investigated by the Centers for Disease Control. Depuration (prolonged flushing with clean water) is only partially effective in reducing the problem. Caliciviruses are also relatively resistant to chlorination of water. The U.S. Environmental Protection Agency guidelines for municipal water systems recommend residual chlorine concentrations of at least 0.2 mg/liter, and in many localities peak levels of 5 mg/liter are used. Although these levels are adequate to destroy most bacteria and viruses, the Norwalk agent can survive at chlorine concentrations below 10 mg/liter.

Hepatitis E

The story of the original discovery of hepatitis E virus (HEV) in Soviet Central Asia is worthy of Scheherazade. An intrepid Soviet virologist, Balayan, investigating an outbreak of hepatitis in Tashkent, volunteered himself to drink a pooled filtrate of stools from the patients; sure enough he developed hepatitis. After recovering a novel 32 nm virus from his own feces, he inoculated a filtrate of that material into monkeys, which in turn developed biochemical evidence of hepatitis and excreted in their stools a virus which he identified, by immunoelectron microscopy (IEM) using convalescent human sera, to be the same virus as was present in the original patients.

Similar viruses were later recovered by others from enterically transmitted non-A, non-B hepatitis (ET-NANBH) outbreaks in India and many other countries of Asia as well as North Africa and Mexico. Bradley and colleagues then demonstrated that these several strains reacted in IEM with acute-phase sera from cases from many parts of the Third World and also with sera from monkeys in which hepatitis was induced by inoculation with geographically separated isolates.

Although the newly described agent of hepatitis E has yet to be reproducibly cultured, its genome has already been cloned and sequenced. Bile from the gallbladder of an experimentally infected cynomolgus macaque was assumed to be a relatively clean, high-titer source of the putative RNA virus. Nucleic acid was extracted, converted to cDNA by reverse transcriptase, amplified by PCR using random primers, cloned into the bacteriophage vector λgt10, and expressed in bacteria. From the resulting cDNA library several cDNA fragments corresponding to genomic sequences of the putative NANBH virus were identified using labeled cDNA probes from the same original source. The nucleotide sequence of one of the clones was found to encode the RNA-dependent RNA polymerase motif characteristic of plus sense RNA viruses, and to hybridize to a 7.5 kb polyadenylated RNA from infected macaque liver and to cDNA derived by PCR from feces of human patients. By assembling a series of such overlapping cDNA fragments it finally proved possible to sequence the whole viral genome. The resulting data, together with that obtained by electron microscopic definition of the virion itself, have led to the allocation of the hepatitis E virus within the family *Caliciviridae*, at least for the time being.

Properties of Hepatitis E Virus

The spherical nonenveloped virion of HEV resembles that of other caliciviruses except that its icosahedral capsid with the characteristic surface depressions is perhaps more readily degraded by proteolysis. The particle is extremely labile, tending to lose its outer layer even during storage for a few days at 4°C, following freeze–thawing, during concentration by ultracentrifugation, or at the high salt concentration present during purification by cesium chloride equilibrium gradient centrifugation. Thus, depending on the source of the specimen and its subsequent preparation for electron microscopy, the virion can vary in diameter from 27 to 38 (mean 32) nm, and the characteristic "fuzzy loops" projecting from the surface may or may not be evident.

The genome is a single 7.5 kb molecule of single-stranded RNA of positive sense, polyadenylated at its 3′ end and having a 5′ methylated cap (not a VPg protein cap). There are three separate but overlapping ORFs, the genes for the structural proteins being located in the middle of the genome. In these respects HEV is quite unlike the picornavirus causing hepatitis, hepatitis A virus. Five functional domains have been identified in the nonstructural polyprotein of HEV: (1) RNA-dependent RNA polymerase, (2) RNA helicase, (3) methyltransferase, (4) "X" domain of unknown function, and (5) a papain-like cysteine protease. In terms of sequence homology within the five domains as well as colinearity of genome organization (except for the position of the protease domain) the HEV genome closely resembles that of rubella virus, an enveloped virus currently classified in its own genus of the family *Togaviridae*. It is conceivable that HEV could have evolved from rubella virus (or its progenitor), either by deletion of the envelope glycoprotein genes or by recombination with a calicivirus-like genome.

Sequencing of the genome of strains from around the world reveals that the American (Mexican) isolate is genetically distinct from the Asian isolates, suggesting evolutionary divergence in the distant past, whereas numerous more minor differences are discernible between the Asian isolates. Despite this, the epitopes on the capsid protein recognized by neutralizing antibodies include one or more that are cross-reactive.

Pathogenesis and Immunity

The incubation period of hepatitis E in humans ranges from 2 to 8 weeks, with an average of 5–6 weeks, somewhat longer than for hepatitis A. Most of our limited knowledge of the pathogenesis of HEV comes from experimental studies in monkeys or chimpanzees, which are readily infectible and in which HEV can be serially passaged. Macaques and tamarins begin to excrete virus into bile and feces about 1 month postinfection. Viral antigens can be demonstrated in the cytoplasm of hepatocytes by immunofluorescence. Serum levels of liver enzymes such as alanine aminotransferase (ALT) then begin to rise, reaching their peak at about 10 weeks. The infection then resolves; there is no evidence of chronicity. Antibodies rise slowly, do not reach high titers, and appear to fall off fairly rapidly.

The most conspicuous pathologic feature of hepatitis E infection, in contrast to hepatitis A, is cholestasis. Histologically, the liver often shows intracanalicular stasis of bile and rosette formation of hepatocytes and pseudoglandular structures resembling embryonal bile ducts. It is not yet known whether HEV replicates initially in the intestine. One intriguing possibility worthy of investigation is whether the observed sensitivity of the virion to proteolysis reflects a requirement for proteolytic disruption of the capsid by trypsin or other enzymes in the intestine prior to infection of the mucosa. An important question for future research is why the disease is so severe in pregnant women (see below).

Clinical Features

Hepatitis E has so far been observed almost exclusively in the less developed parts of the world and predominantly in the 15–40 age group. Subclinical infection may be the rule in children, with icteric infection being mainly confined to young adults. Clinically, the illness closely resembles that described for hepatitis A (see Chapter 23). Bilirubin levels tend to be higher, and jaundice deeper and more prolonged. The case–fatality rate is 0.5–3%. However, the most striking feature of hepatitis E is its extraordinarily high case–fatality rate of about 10–20% in pregnant women, particularly in the final trimester. Like hepatitis A, hepatitis E does not progress to chronic hepatitis, cirrhosis, cancer, or the carrier state.

Laboratory Diagnosis

Pending the general introduction of simple diagnostic tests for hepatitis E, the first duty of the laboratory is to exclude hepatitis A (by IgM serology) and hepatitis B (HBsAg; anti-HBc IgM). Hepatitis E virus has yet to be cultured satisfactorily *in vitro*. Immunoelectron microscopy, looking for aggregated calicivirus-like particles in an acute-phase fecal specimen using polyclonal sera, was the standard approach to diagnosis and could be refined by the use of MAbs raised against clonally expressed capsid proteins or synthetic peptides, but it appears destined to be rapidly replaced by simpler and more sensitive new alternatives. Even though there seems to be relatively little antigen in feces, it may be possible to develop an enzyme immunoassay, to be followed perhaps by a confirmatory western blot. PCR assays have been developed which amplify viral RNA that has first been adsorbed from feces or from very early acute-phase serum onto glass powder (to allow inhibitors to be washed away) before reverse transcription and amplification using primers specific for the HEV RNA polymerase gene.

Antibody was originally detected and quantified by immunofluorescence, using an assay in which the binding of fluoresceinated antibody to a liver section from an infected macaque is blocked by patient's serum, but this tedious test has now outlived its usefulness. Immunoglobulin class-specific EIAs and Western blot assays have recently been developed for detection of antibody to a molecularly cloned fusion protein representing most of the HEV capsid antigen.

Epidemiology

Hepatitis E virus is now recognized to be the most important cause of epidemic hepatitis in Asia. In Central Asia hepatitis E, like hepatitis A, tends to peak in autumn, while in Southeast Asia it occurs particularly during the rainy season or following extensive flooding. Transmission is fecal–oral and is mainly water-borne. The most notorious common-source epidemic resulted from fecal contamination of a drinking-water supply in New Delhi in 1955, causing 29,000 identified cases of icteric hepatitis. Ascribed to HAV at the time, the outbreak was later reinvestigated retrospectively by testing frozen stored paired sera for HAV and HBV antibodies and was demonstrated to have been caused by a non-A, non-B hepatitis virus. Many similar water-borne outbreaks have been recorded subsequently, especially on the Indian subcontinent but also in Central Asia, China, Indonesia, North Africa, and Mexico. Particularly devastating outbreaks occurred among Ethiopian refugees encamped in Somalia and Sudan during the prolonged war in the Horn of Africa. Sporadic cases have been attributed to the consumption of shellfish from sewage-polluted waters in Italy and Spain.

Clinical attack rates during epidemics range from 1 to 10%, with most of the cases occurring in young adults and most of the mortality in pregnant women. However, the secondary attack rate among household contacts of icteric patients is low (2.4% in one outbreak in Nepal, compared with 10–20% for hepatitis A in the same locality). Perhaps this is due to relatively low numbers of infectious virions shed in feces and/or their lability. The apparent absence of hepatitis E in the developed countries of the world except in returning travelers is presumably attributable to this low secondary attack rate, as well as to piped water supplies of a standard sufficient to avoid common-source water-borne outbreaks.

Control

Reliable reticulated and chlorinated water supplies and improved standards of sanitation (see Chapter 15) throughout the developing world must underpin any long-term program to control hepatitis E, or indeed any other enterically transmitted infectious disease. During an epidemic the usual rules apply: boil the water, and for food, "cook it or peel it." Passive immunization has not been found to be effective. Research toward a vaccine should be accorded priority.

Astroviruses

Astroviruses were first described in 1975 when they were observed by electron microscopy in the feces of children with diarrhea. Other host-specific astroviruses were soon discovered in a wide range of domestic and farm animals. Human astroviruses now appear to be ubiquitous in young children but a relatively minor contributor to the overall burden of gastroenteritis.

Properties of *Astroviridae*

The 28–30 nm spherical virion, though at first glance resembling a picornavirus, is in fact highly characteristic when viewed from a certain angle; a minority of the negatively stained particles reveal a five- or six-pointed star, without any stained central hollow (Fig. 24-1B). The virion is resistant to pH 3 and to 60°C for 5 minutes. The human astrovirus genome comprises a single linear 7 kb ssRNA molecule of positive polarity, with a poly(A) tract at the 3′ terminus. In 1993 the genome of a human astrovirus was sequenced. Unlike the calciviruses, astroviruses do not encode a helicase, whereas the RNA polymerase and protease domains are brought together by a ribosomal frameshift. The capsid contains at least two or three protein species. Although there are one or more common epitopes recognized by monoclonal antibodies, at least five serotypes of human astroviruses can be distinguished with other MAbs or by neutralization (Table 24-2).

Viral Replication

Astroviruses can be isolated, in the presence of trypsin, in cultures of primary human embryonic kidney (HEK) cells and have been further adapted to growth in the LLC-MK2 continuous monkey kidney cell line. Preliminary

Table 24-2
Properties of Astroviruses

Single genus, *Astrovirus;* 5 serotypes
Nonenveloped spherical virion, 28–30 nm, resembles star; 3 capsid proteins
Virion relatively resistant to heat and acidity
Linear plus sense ssRNA genome, 7.2 kb, 3′ polyadenylated
Cytoplasmic replication; subgenomic mRNA encodes polyprotein which is cleaved into capsid proteins

studies of the viral replication cycle have revealed that replication takes place largely in the cytoplasm, and that in addition to the 7.2 kb genome, a 3′-coterminal 2.8 kb polyadenylated subgenomic mRNA is synthesized. A putative polyprotein and its cleavage products, apparently the capsid proteins, have also been described. Mature virions accumulate in the cytoplasm in crystalline arrays.

Clinical Features

Astrovirus gastroenteritis, seen principally in young children, resembles a mild form of rotavirus enteritis. An incubation period of 1–4 days is followed by watery diarrhea lasting 1–4 days or more, abdominal discomfort, vomiting, and constitutional symptoms. Immunity probably lasts for years, and most adults are resistant to experimental infection. Anamnestic responses develop following subsequent exposure to a heterologous serotype. Type 1 is by far the commonest, in the United Kingdom at least.

Laboratory Diagnosis

Although astroviruses are found in higher numbers in feces ($\sim 10^8$ per gram) than are caliciviruses ($<10^7$ per gram), IEM is still not optimally sensitive or convenient. Today, the most useful method of detection of virus in feces is an EIA using an astrovirus-cross-reactive MAb for antigen capture and a biotinylated detector antibody with avidin-labeled enzyme readout, which detects 10 ng of viral protein. Monoclonal antibodies are available to distinguish all five known serotypes. A newly developed riboprobe for dot-blot hybridization is even more sensitive and can be used to screen polluted water as well as feces, but steps must be taken to prevent degradation of RNA (see above). Probably the most suitable cell line now available for primary isolation of astroviruses, in the presence of trypsin, is the human colon carcinoma cell line CaCo-2; these cells are at their most susceptible when, after growing to form a confluent monolayer, they start to differentiate and form microvilli. The isolate can be identified by immunofluorescence or biotin–avidin EIA.

Epidemiology

Astroviruses are transmitted by the fecal–oral route, usually person to person but also probably via contaminated food or water. They are endemic throughout the world causing a relatively mild form of gastroenteritis, almost exclusively in young children, year-round but with a winter peak, like rotaviruses. A majority of children have developed protective antibodies before entering school. Many infections may be subclinical, but 4–8% of infantile gastroenteritis is attributable to astroviruses, from Great Britain to Guatemala. Epidemics also occur in the community, in day-care centers and kindergartens, nosocomially in pediatric wards, etc. Most other children and adults are immune, but outbreaks have been recorded among immunosuppressed and geriatric patients in hospitals and nursing homes.

Further Reading

Balayan, M. S. (1993). Hepatitis E virus infection in Europe: Regional situation regarding laboratory diagnosis and epidemiology. *Clin. Diag. Virol.* **1,** 1.

Bradley, D. W. (1990). Enterically-transmitted non-A, non-B hepatitis. *Br. Med. Bull.* **46,** 442.

Bradley, D. W. (1992). Hepatitis E: Epidemiology, aetiology and molecular biology. *Rev. Med. Virol.* **2,** 19.

Bradley, D. W., Krawczynski, K., Beach, M. J., and Purdy, M. A. (1991). Non-A, non-B hepatitis: Toward the discovery of hepatitis C and E viruses. *Semin. Liver Dis.* **11,** 128.

Cubitt, W. D. (1989). Diagnosis, occurrence and clinical significance of the human "candidate caliciviruses." *Prog. Med. Virol.* **36,** 103.

Greenberg, H. B., and Matsui, S. M. (1992). Astroviruses and caliciviruses: Emerging enteric pathogens. *Infect. Agents Dis.* **1,** 71.

Greenberg, H. B., Skaar, M., and Monroe, S. S. (1990). The 22 to 30 nm gastroenteritis agents of man. *In* "Viral Diarrheas of Man and Animals" (L. J. Saif and K. W. Theil, eds.), p. 137. CRC Press, Boca Raton, Florida.

Jiang, X., Graham, D. Y., Wang, K., and Estes, M. K. (1991). Norwalk virus genome cloning and characterization. *Science* **250,** 1580.

Kapikian, A. Z., and Chanock, R. M. (1990). Norwalk group of viruses. *In* "Fields Virology" (B. N. Fields, D. M. Knipe, R. M. Chanock, M. S. Hirsch, J. L. Melnick, T. P. Monath, and B. Roizman, eds.), 2nd Ed., p. 671. Raven, New York.

Koonin, E. V., Gorbalenya, A. E., Purdy, M. A., Rozanov, M. N., Reyes, G. R., and Bradley, D. W. (1992). Computer-assisted assignment of functional domains in the nonstructural polyprotein of hepatitis E virus: Delineation of an additional group of positive-strand RNA plant and animal viruses. *Proc. Natl. Acad. Sci. U.S.A.* **89,** 8259.

Kurtz, J. B., and Lee, T. W. (1987). Astroviruses: Human and animal. *Ciba Found. Symp.* **128,** 92.

Reyes, G. R., and Baroudy, B. M. (1991). Molecular biology of non-A, non-B hepatitis agents: Hepatitis C and hepatitis E viruses. *Adv. Virus Res.* **40,** 57.

Tam, A. W., Smith, M. M., Guerra, M. E., Huang, C. C., Bradley, D. W., Fry, K. E., and Reyes, G. R. (1991). Hepatitis E virus (HEV): Molecular cloning and sequencing of the full-length viral genome. *Virology* **185,** 120.

Willcocks, M. M., Carter, M. J., and Madeley, C. R. (1992). Astroviruses. *Rev. Med. Virol.* **2,** 97.

Chapter 25

Togaviridae

Early attempts to classify the large number of viruses transmitted by arthropods led to the definition of two major groups, called the Group A and Group B arboviruses. As more was learned about the member viruses, the Group A arboviruses became genus *Alphavirus* of the family *Togaviridae*, the Group B arboviruses became genus *Flavivirus* of the family *Flaviviridae* (see Chapter 26), and most of the remainder were allocated to the family *Bunyaviridae* (see Chapter 33). Later, rubella virus was placed in the family *Togaviridae* as the only member of the genus *Rubivirus*.

There are 27 member viruses in the genus *Alphavirus*, 11 of which have been shown to cause disease in humans, 8 of them producing significant epidemics. These viruses exist in particular natural habitats in which specific mosquito and vertebrate hosts play roles in virus survival, geographic extension, overwintering, and amplification (see Chapter 14). The vectors are mosquitoes; the vertebrate reservoir hosts are usually wild mammals or birds. Domestic animals and humans are usually not involved in primary transmission cycles in nature but may play a role in geographic extension and the amplification events that lead to epidemics.

Properties of *Togaviridae*

Togavirus virions are spherical, 70 nm in diameter, and consist of an icosahedral capsid surrounded by a tightly adherent lipid envelope covered with glycoprotein peplomers (Fig. 25-1). The genome is a single linear 11–12 kb

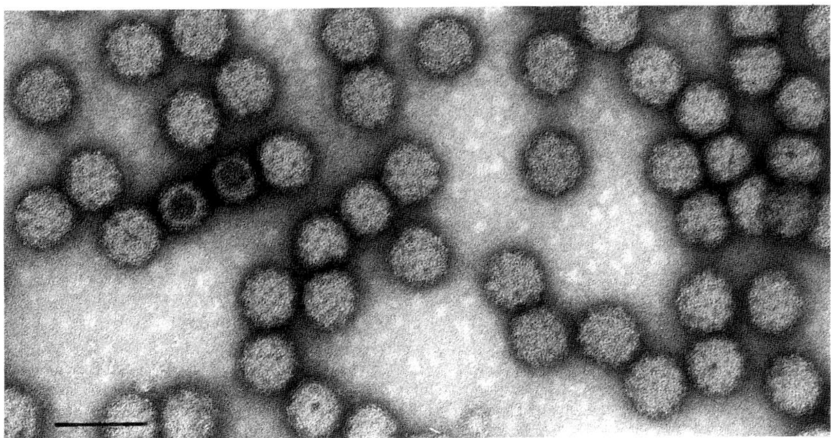

Fig. 25-1 *Togaviridae*, genus *Alphavirus*. Negatively stained virions of Semliki Forest virus. Bar, 100 nm. (Courtesy Dr. C.-H. von Bonsdorff.)

molecule of ssRNA of positive polarity, which is 5′ capped and 3′ polyadenylated, and is infectious. Togaviruses are not very stable in the environment and are easily inactivated by disinfectants. Other properties of the togaviruses are summarized in Table 25-1.

Classification

All alphaviruses share common antigenic determinants, notably on the nucleocapsid protein, C. The envelope glycoproteins carry some determinants that are common to certain species only, providing the basis for partition of the genus into six antigenic "complexes." Individual alphaviruses can be differentiated on the basis of species-specific epitopes on the envelope glycoproteins. Rubella virus, the only member of the genus *Rubivirus*, displays no antigenic cross-reactivity with any alphavirus.

Table 25-1
Properties of *Togaviridae*

Two genera: *Alphavirus* and *Rubivirus*
Spherical virion, enveloped, with peplomers, diameter 70 nm
Icosahedral capsid, diameter 40 nm
Linear, plus sense ssRNA genome, 11–12 kb (*Alphavirus*), 10 kb (*Rubivirus*); 5′ capped, 3′ polyadenylated, infectious
Genes for nonstructural proteins located at 5′ end of genome, those for structural proteins at the 3′ end
Two envelope glycoproteins, E1 and E2, containing virus-specific neutralizing epitopes and alphavirus serogroup and subgroup specificities; nucleocapsid protein, C, with broadly cross-reactive alphavirus group specificity
Full-length and subgenomic RNA transcripts; posttranslational cleavage of polyproteins
Cytoplasmic replication; budding from plasma membrane (alphaviruses) or intracytoplasmic membranes (rubella virus) of vertebrate cells, or from intracytoplasmic membranes of invertebrate cells, in which alphaviruses are noncytocidal

Viral Replication

Alphavirus virions are taken up into coated vesicles via receptor-mediated endocytosis (see Fig. 3-3); when these coated vesicles mature into phagolysosomes the low interior pH causes virions to release the genome into the cytosol, where the whole of the replication cycle takes place.

The 5′ two-thirds of the capped viral RNA genome (Fig. 25-2) is translated into a large polyprotein, which then cleaves itself into four nonstructural proteins. The viral protease, as well as a helicase involved in viral RNA replication, is contained within nonstructural protein 2 (NSP2); NSP4 is thought to represent the RNA polymerase; NSP3 converts RNA replicase to a plus strand replicase; NSP1 is involved in methylation and capping of viral RNAs and in initiation of minus strand RNA synthesis. Thus all four nonstructural proteins are required for synthesis of a full-length complementary copy of the viral genome. From this minus strand template two types of RNA are produced: full-length progeny genomic RNA and a subgenomic mRNA corresponding to the 3′ one-third of the genomic RNA which encodes all the

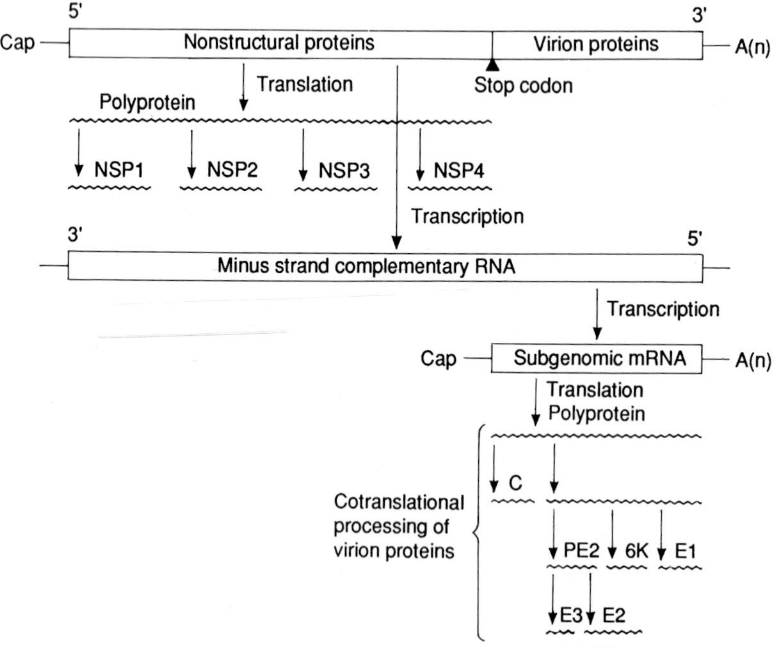

Fig. 25-2 Diagram of *Alphavirus* genome and its transcription and translation. The plus strand virion RNA is capped and polyadenylated, and there are short nontranslated sequences at each terminus (single lines). The 5′ two-thirds of the genome codes for nonstructural proteins and the 3′ one-third for the structural proteins. The portion coding for the nonstructural proteins is translated into a polyprotein which is cleaved into the four nonstructural proteins. Two of these form the RNA polymerase, which transcribes a full-length minus sense copy from the virion RNA, from which in turn two plus sense RNA species, virion RNA (not shown) and subgenomic mRNA, are transcribed. The subgenomic mRNA, which is identical to the 3′ one-third of the virion RNA, is translated into a polyprotein that is then processed into the viral structural proteins E1, E2, E3, 6K, and C.

structural proteins.This mRNA is translated into another polyprotein which also has autoprotease activity; this viral enzyme together with certain cellular proteases cleaves the polyprotein to yield the four (or five) viral structural proteins: the nucleocapsid protein (C), and the peplomer glycoproteins (E1, E2, and, in some species, E3), and a small transmembrane protein (6K). Genomic RNA and nucleoprotein are self-assembled into icosahedral nucleocapsids in the cytoplasm, and these migrate to the plasma membrane. The peplomer proteins are glycosylated in stepwise fashion as they progress from the endoplasmic reticulum through the Golgi complex to the plasma membrane. Virion assembly takes place via budding of nucleocapsids through the modified plasma membrane in vertebrate cells, but through intracytoplasmic membranes in the case of invertebrate cells, in which alphaviruses are noncytolytic and establish persistent infection.

The replication of rubella virus is basically similar to that of the alphaviruses with two significant exceptions, namely, that the rubella structural gene order is slightly different and rubella nucleocapsids acquire their envelope in mammalian cells by budding through intracytoplasmic membranes.

Pathogenesis

The infected female mosquito introduces its proboscis directly into a capillary beneath the skin and injects saliva which contains the arbovirus. Viral replication ensues in vascular endothelium and in both blood monocytes and macrophages in lymph nodes, bone marrow, spleen, and liver. Virions liberated from these cells augment the viremia and precipitate the prodromal symptoms (fever, chills, headache, muscular aches) which mark the end of the incubation period (usually 3–7 days). Particular target organs may now be infected, commonly muscles (myositis), joints (arthritis), skin (rash), or brain (encephalitis).

Virus may enter the brain via infection of capillary endothelial cells. Once in the parenchyma of the brain there are no anatomic or physiologic impediments to viral spread throughout the central nervous system, infecting and damaging neurons. Typical pathologic features include neuronal necrosis with neuronophagia and intense perivascular and interstitial mononuclear inflammatory infiltration. Virus produced in the central nervous system does not reenter the circulation; hence, it is not involved in transmission.

The immunity that follows clinical or subclinical infection with an alphavirus probably lasts for life. Partial protection may also be conferred against antigenically related viruses.

Clinical Syndromes

The majority of human alphavirus infections are asymptomatic, the only evidence of infection being seroconversion. Clinical manifestations, when they occur, often take the form of a nondescript fever. However, three alphaviruses sometimes cause potentially lethal encephalitis, and five others cause painful and sometimes persistent arthritis (Table 25-2).

Table 25-2
Human Diseases Caused by Alphaviruses[a]

Virus	Geographic distribution	Vertebrate reservoir	Vector mosquitoes	Disease	Epidemiologic features
Eastern equine encephalitis	North and South America, Caribbean	Wild birds	*Culiseta melanura*, *Coquillettidea perturbans*, *Aedes* spp.	Encephalitis	Periodic small equine epizootics and human epidemics
Western equine encephalitis	North and South America	Wild birds	*Culex tarsalis*	Encephalitis	Periodic equine epizootics and human epidemics
Venezuelan equine encephalitis	South and central America	Horses (epizootic)	*Aedes*, *Mansonia*, *Culex* spp. (epizootic subtypes)	Encephalitis	Periodic equine epizootics and human epidemics
	Enzootic subtype in Florida	Rodents (enzootic)	*Culex* (*Melanoconion*) (enzootic subtypes)		Subclinical in horses and humans
Chikungunya	Africa, south Asia, Philippines	Monkeys, humans	*Aedes aegypti*, *Aedes furcifer-taylori*	Fever, arthralgia, myalgia, rash	Urban epidemics like dengue, enzootic cycle like yellow fever
O'nyong-nyong	East Africa	Humans	*Anopheles funestus*, *Anopheles gambiae*	Fever, arthralgia, myalgia, rash	Single large outbreak in 1959–1962, not seen since
Mayaro	Tropical South America	Monkeys	*Haemagogus* sp.	Fever, arthralgia, myalgia, rash	Sporadic cases, limited outbreaks associated with deforestation
Ross River	Australia	Marsupials, rodents	*Culex annulirostris*, *Aedes vigilax*	Arthralgia, myalgia, rash	Annual outbreaks in humans and horses
	South Pacific islands	Humans	*Aedes polynesiensis*		Occasional introductions and epidemics
Sindbis	Africa, Asia, Europe, Australia	Wild birds	*Culex* sp., *Culiseta* sp., *Aedes* sp.	Fever, arthralgia, myalgia, rash	Sporadic cases, sometimes epidemics

[a] Data largely from C. H. Calisher and T. P. Monath, *Togaviridae* and *Flaviviridae*: The alphaviruses and flaviviruses. *In* "Laboratory Diagnosis of Infectious Diseases. Principles and Practice. Volume II. Viral, Rickettsial and Chlamydial Diseases" (E. H. Lennette, P. Halonen, and F. A. Murphy, eds.), p. 414. Springer-Verlag, New York, 1988.

Encephalitis

In a small proportion of persons infected with eastern, western, or Venezuelan encephalitis viruses, a few days after the onset of fever the patient develops drowsiness, often accompanied by neck rigidity, and may progress to confusion, paralysis, convulsions, and coma. Case–fatality rates average 10–20% but may be considerably higher with eastern equine encephalitis. Survivors from encephalitis caused by any of these viruses are often left with permanent neurologic sequelae such as mental retardation, epilepsy, paralysis, deafness, and blindness.

Fever/Rash/Arthritis

The fever/rash/arthritis triad of clinical features defines the disease caused by chikungunya, o'nyong-nyong, Ross River, Mayaro, and Sindbis viruses. Indeed, chikungunya and o'nyong nyong are colorful African terms describing the agony of the affected joints! Typically, after a short incubation period of 2–3 days, there is an abrupt onset of fever, chills, myalgia, and severe polyarthralgia affecting mainly the small joints; a rash, generally maculopapular, then appears. Other constitutional symptoms such as nausea, headache, backache,photophobia, and retroorbital pain may also be present but are not so diagnostic. The fever is characteristically high and sometimes of the "saddle-back" (biphasic) variety in chikungunya, but it is not at all prominent in Ross River virus infection. The arthritis usually resolves in a few weeks but may persist for months or even years in chikungunya and Ross River virus infections.

Laboratory Diagnosis

During the halcyon days of the Rockefeller Foundation's Virus Research Program through the 1950s and 1960s hundreds of previously unknown arboviruses were isolated from arthropods, wild birds and mammals, domesticated animals, and humans by intracerebral inoculation of infant mice. The suckling mouse is a particularly sensitive host, still used by some laboratories for diagnostic purposes today, but is progressively being replaced by cultured cells. Vero or BHK-21 cells are the most commonly used vertebrate cell lines, but invertebrate cell lines such as C6/36 from the mosquito *Aedes albopictus* tend to be more sensitive. Cytopathic effects do not regularly occur. The isolate is identified either by immunofluorescence (using the fixed monolayer), or hemagglutination inhibition, complement fixation, or EIA (on the culture supernatant), to place the virus in the correct serogroup, followed by neutralization using appropriate monoclonal antibodies to characterize the species. Apart from brain or other tissue sampled at autopsy, the only appropriate specimen for isolation of virus is blood taken while the patient is still viremic. Since the probability of isolating virus drops markedly after the first couple of days of illness, serology should also be used to avoid missing the diagnosis.

During an epidemic, an IgM capture EIA is routinely employed to detect antiviral IgM in a single specimen of serum; such antibodies are almost always detectable during or before the second week of the illness. However, because of serological cross-reactions between alphaviruses it is important to confirm the diagnosis by demonstrating a rise in neutralizing antibody in paired sera.

Epidemiology

Alphavirus Encephalitides

Western equine encephalitis virus is found throughout the Americas. It is spread by the mosquito *Culex tarsalis* between small birds, but this sylvatic cycle can be amplified by extension to other hosts when seasonal conditions produce large mosquito populations. Horses and humans may become infected and occasionally develop encephalitis (Fig. 25-3).

Eastern equine encephalitis virus occurs in North, Central, and South America and the Caribbean. In the United States it is maintained in freshwater swamps by *Culiseta melanura*, and this mosquito, which feeds almost exclusively on birds, is also responsible for amplification of the virus during the spring and summer (Fig. 25-4). It is not clear which of several mosquito species transmits the virus to the clinically vulnerable hosts, horses and humans. In coastal New Jersey the salt-marsh mosquito *Aedes sollicitans* has been implicated because it feeds on horses as well as birds, but this virus–vector relationship does not operate in other areas. In eastern North America, many outbreaks of eastern equine encephalitis virus infection have occurred in caged pheasants (as well as in an endangered species, the whooping crane), resulting in a substantial mortality. The virus is introduced into flocks by mosquitoes and is spread when healthy birds peck on sick, viremic birds.

Venezuelan equine encephalitis virus occurs in the forests and marshes of tropical America, including the Florida Everglades, as "enzootic" strains which circulate among rodents via *Culex* mosquitoes and are not virulent for either horses or humans. All the virulent "epizootic" strains have been isolated during equine epizootics, and horses play a major role in their amplification; nothing is known of the natural history of the virus between such outbreaks. Epizootics have occurred in drier regions of several South American countries at intervals of about 10 years and in 1969–1972 extended north as far as Texas. Although there are high death rates in horses, most infected humans suffer from fever and myalgia only; the incidence of encephalitis is low.

Alphaviruses Causing Fever and Arthritis

Five alphaviruses cause sporadic and sometimes epidemic outbreaks of fever with rash, myalgia, and arthralgia (see Table 25-2).

Chikungunya virus is the most important virus in terms of the scale of human illness, being widespread throughout Africa, India, and Southeast Asia, including the Philippines. In Africa, it is thought to be maintained in a sylvan cycle with forest species of the *Aedes furcifer-taylori* group of mosquitoes as vectors and primates as the primary vertebrate host. Devastating epidem-

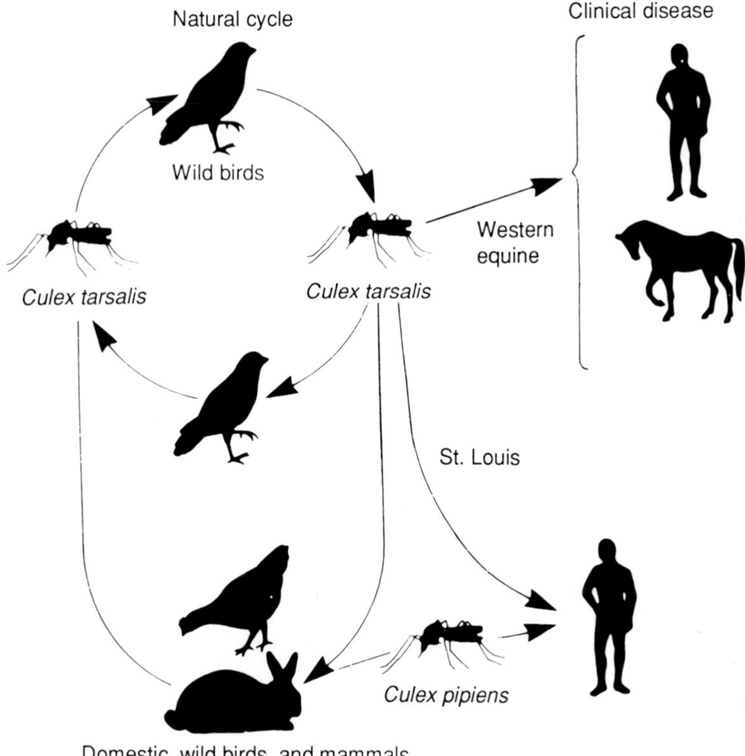

Fig. 25-3 Sylvatic cycles of western equine encephalitis virus (an alphavirus) and St. Louis encephalitis virus (a flavivirus). The natural inapparent cycle of both viruses is between *Culex tarsalis* and nestling and juvenile birds, but this cycle may be amplified by infection of domestic birds and wild and domestic mammals. Western equine encephalitis virus can multiply in the mosquito at cooler temperatures, so epidemic disease in horses and humans may occur earlier in the summer and farther north into Canada. In the western United States, St. Louis encephalitis is a rural disease transmitted primarily by *Culex tarsalis;* in the south-central United States and the Southwest it is an urban–suburban disease transmitted by *Culex pipiens.* The St. Louis encephalitis virus does not cause clinical encephalitis in horses. (Modified from R. T. Johnson, "Viral Infections of the Nervous System." Raven, New York, 1982.)

ics, sometimes lasting for years, have occurred repeatedly in Africa and Asia. Explosive urban outbreaks, observed particularly in Asia, may infect the majority of the population of the city within a few months. Under such circumstances the peridomestic mosquito *Aedes aegypti* can maintain the virus in a human–mosquito–human cycle.

O'nyong-nyong virus is spread by *Anopheles funestus* and *A. gambiae,* making it the only known human arboviral pathogen with anopheline mosquitoes as vector. The virus was responsible for an epidemic in East Africa extending from 1959 to 1962 affecting no less than 2 million people, but it has not been recognized as a cause of human disease since then. Its vertebrate reservoir is unknown.

Mayaro virus occurs in forested regions of Central and South America

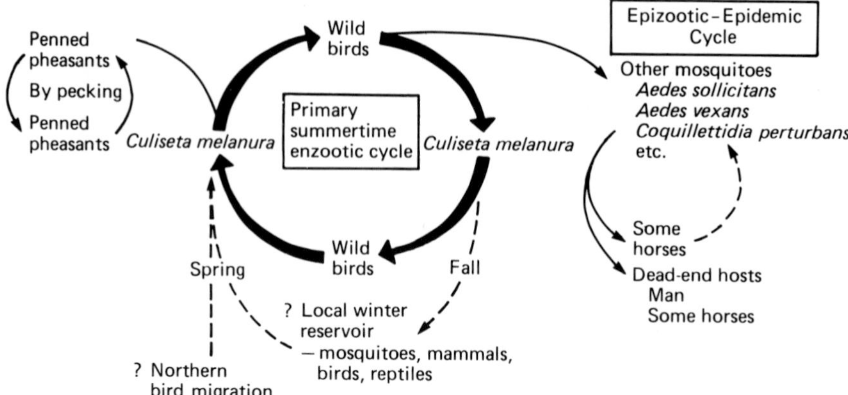

Fig. 25-4 Transmission cycle of eastern equine encephalitis virus in the United States. Known parts of the cycle are drawn in solid lines, speculative parts in broken lines. The mode of overwintering is unknown. (Courtesy Dr. T. P. Monath.)

and appears to be spread by *Haemagogus* mosquitoes, probably from a primate reservoir. Widespread human infection is demonstrable by seroconversion, and several epidemics have occurred in small towns in the Amazon basin.

Ross River virus produces a disease known as epidemic polyarthritis during late summer/early autumn in rural Australia. Several species of mosquitoes can act as vectors, and marsupials as reservoir hosts; domestic animals, rodents, and, at times of epidemics, humans, may act as amplifiers. In 1979 and 1980, following importation from Australia, probably by air travel of persons incubating the infection, explosive epidemics occurred in Fiji and spread to other islands of the South Pacific, affecting up to half the indigenous population as well as tourists.

Sindbis virus has long served as the prototype alphavirus. Human cases of fever with rash and arthralgia have been recognized in many parts of Africa, northwestern Europe, India, and Malaysia; epidemics have been described in Egypt and South Africa. Various species of wild birds serve as reservoir hosts,with *Culex, Culiseta,* and *Aedes* mosquitoes as vectors.

Control

Control of mosquito-borne diseases rests largely on avoidance of exposure, elimination of breeding sites, and destruction of mosquitoes or their larvae by methods discussed in Chapter 15. Control of alphavirus infections of humans is generally undertaken only in the face of threats of outbreaks, especially of encephalitis. Immunization of humans is currently impractical, but because encephalitis in horses is a serious disease, and because amplification in this host may increase the risk of human cases, immunization of horses each spring is an important control measure in areas subject to outbreaks. Inactivated cell culture vaccines are used for eastern, western, and Venezuelan equine encephalitides, and also an attenuated live-virus vaccine for Venezuelan equine encephalitis. Mosquito larviciding programs are carried out routinely in many areas; in short-term emergency situations, such as during an outbreak or when sentinel surveillance indicates the likelihood of an outbreak, such programs are supplemented by aerial spraying with ultra-low-volume insecticides, such as malathion or synthetic pyrethrins. Also important is public education regarding (1) *source reduction,* by destroying potential mosquito breeding sites, and (2) *contact reduction,* by avoidance of mosquito habitats, especially at dawn and dusk, and use of protective clothing, insect repellents, screening of doors and windows, bednets (particularly for infants), air-conditioning, and so on.

Currently, the only vaccines available are those described above for eastern, western, and Venezuelan equine encephalitis viruses, which are used only in horses, at-risk laboratory workers, penned pheasants, and whooping cranes! The wide prevalence of chikungunya virus in Asia and Africa, and the presence of competent *Aedes aegypti* and *Aedes albopictus* vectors in the Americas, argues for a vaccine against this agent; an experimental chikungunya vaccine is currently undergoing clinical trial in humans.

Rubella

Rubella is a trivial exanthema of childhood. However, in 1941 an Australian ophthalmologist, Gregg, noticed an unusual concentration of cases of congenital cataract among newborn babies in his practice—an epidemic of blindness. A diligent search of his records revealed that most of the mothers had contracted rubella in the first trimester of pregnancy. Further investigations revealed that these unfortunate children had also suffered a range of other congenital defects including deafness, mental retardation, and cardiac abnormalities (see Chapter 14).

Clinical Features

In childhood rubella is such a mild disease that most adults are subsequently uncertain whether they have ever contracted it. The fine, pink, discrete macules of the erythematous rash appear first on the face, then spread to the trunk and limbs, and fade after 48 hours or less. In nearly half of all infections there is no rash at all. Fever is usually inconspicuous, but a characteristic feature is that postauricular, suboccipital, and posterior cervical lymph nodes are enlarged and tender from very early in the illness. Mild polyarthritis, usually involving the hands, is a fairly frequent feature of the disease in adult females; it is usually fleeting but rarely may persist for up to several years. Thrombocytopenic purpura and postinfectious encephalopathy are rare complications. Progressive rubella panencephalitis is an even rarer and inevitably fatal complication, developing insidiously in the second decade of life, usually in children with congenital rubella.

Congenital Rubella

At least 20% of all infants infected *in utero* during the first trimester of pregnancy are born with severe congenital abnormalities, usually multiple (Fig. 25-5); most of the remainder have milder defects. The commonest con-

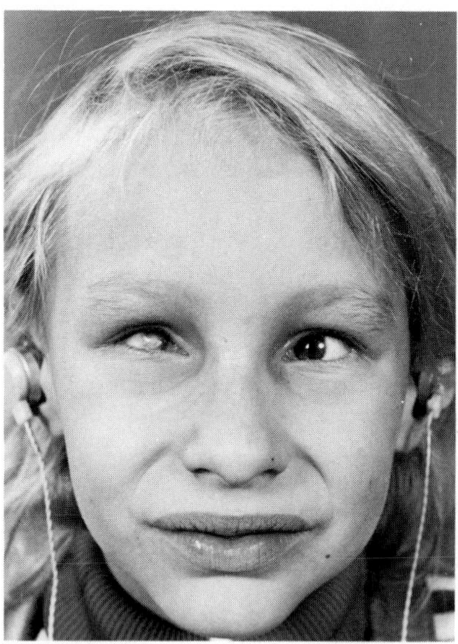

Fig. 25-5 Congenital rubella syndrome, with severe bilateral deafness and severe bilateral visual defects (microphthalmia, cataract, corneal opacity, and strabismus). (Courtesy Dr. K. Hayes.)

genital abnormalities are neurosensory deafness (total or partial, due to cochlear degeneration, becoming progressively apparent in the early years after birth), blindness (total or partial, especially cataracts, but sometimes glaucoma, microphthalmia, or retinopathy), congenital heart disease (especially patent ductus arteriosus, sometimes accompanied by pulmonary artery stenosis or septal defects), and microcephaly with mental retardation. Other common manifestations of the so-called *congenital rubella syndrome (CRS)* include bone translucency and retardation of growth, hepatosplenomegaly, and thrombocytopenic purpura. Despite the diversity and severity of this pathology, the congenital rubella syndrome is sometimes missed at birth. About 10–20% of babies with CRS die during the first year, and up to 80% develop some evidence of disease within the early years of life. Up to 20% of children with CRS develop insulin-dependent diabetes mellitus as young adults.

Pathogenesis and Immunity

The virus enters the body by inhalation to multiply asymptomically in the upper respiratory tract and then spread via local lymph nodes to the bloodstream. Not a great deal is known about the events that occupy the 18-day (range 2–3 weeks) incubation period which precedes the appearance of the rash. Virus has been found in lymphocytes and in synovial cells of joints of adults with rubella arthritis. Naturally acquired immunity to rubella lasts for many years; second infections occur occasionally but are usually subclinical and boost immunity further.

When rubella virus infects a woman during the first trimester of pregnancy there is a high probability that the baby will suffer congenital abnormalities. Severe damage (deafness, blindness, heart or brain defects) occurs in 15–30% of all infections during the first trimester, and in about 5% of those in the fourth month (usually deafness), but rarely thereafter. Minor abnormalities are even more frequent, and following spontaneous abortion or stillbirth virus can be found in practically every organ.

What makes rubella virus teratogenic, when numerous nonteratogenic viruses are so much more pathogenic postnatally and so much more cytocidal for cultured cells? Paradoxically, this relative lack of pathogenicity may hold the clue to its teratogenicity. More cytocidal viruses may destroy cells and kill the fetus, leading to spontaneous abortion (as does rubella occasionally). Rubella virus may merely slow down the rate of cell division, as has been demonstrated in cultured human fetal cells, leading to a decrease in overall cell numbers and accounting for the small size of rubella babies. Moreover, death of a small number of cells or slowing of their mitotic rate at critical stages in ontogeny might interfere with the development of key organs which are being formed during the first trimester.

Neither the mother's nor the baby's immune response is able to clear the virus from the fetus. Although maternal IgG crosses the placenta and the infected fetus manufactures its own IgM antibodies (Fig. 25-6), cell-mediated immune responses are defective and remain so postnatally. Clones of infected cells may escape immune cytolysis even though maternal IgG might restrict systemic spread of virus. Whatever the explanation, the rubella syndrome in the fetus is a true persistent infection of the chronic type. Throughout the pregnancy and for several months after birth the baby continues to shed virus in any or all of its secretions.

Laboratory Diagnosis

Rubella virus can be cultured in certain cell lines, such as RK-13 and Vero cells, but grows slowly without conspicuous cytopathic effects. Because of its lack of sensitivity, isolation in cell culture is used only as a supplementary method of diagnosing rubella from adult throat swabs, but it does have a

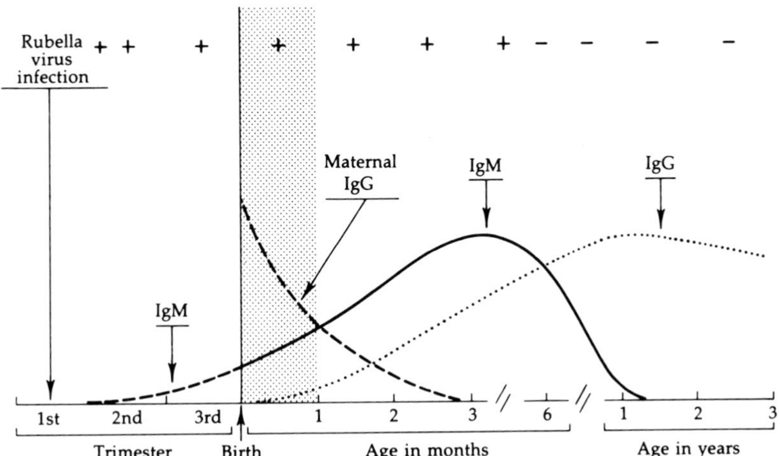

Fig. 25-6 Pattern of viral excretion (+) and the infant's antibody response in congenital rubella. (From S. Krugman and R. Ward, "Infectious Diseases of Children and Adults," 5th Ed. Mosby, St. Louis, 1973.)

place in the diagnosis of CRS; virus can be isolated from the throat, urine, cerebrospinal fluid, or leukocytes of a newborn infant with CRS.

Serology constitutes the standard approach to the laboratory diagnosis of rubella. There are three common situations in which the clinician requires help from the laboratory: (1) A woman considering vaccination wishes to know whether she has ever had rubella and is now immune; (2) An unimmunized woman develops a rash in the first trimester of pregnancy, or comes into contact with someone with rubella, and wishes to know whether she has contracted the disease and whether she should have an abortion; (3) A baby is born with signs suggestive of the rubella syndrome, or its mother is now believed to have possibly contracted rubella during the first trimester.

As there is only one serotype of rubella virus, and specific IgG continues to be demonstrable in serum for many years after clinical or subclinical infection, detection of antibody is evidence of immunity. Diagnosis of rubella in a pregnant woman requires the demonstration of either a rising titer of rubella antibody in paired sera, or rubella IgM antibody in a single specimen. If paired sera are used, the first bleed must to taken during the first week after the onset of the rash; otherwise, a fourfold rise in antibody titer may not be detected in the convalescent specimen taken 10 or more days later. On the other hand, if a pregnant women is concerned about recent exposure to a known case of rubella, the first sample should be taken as soon as possible, but the second should be delayed for at least a month to allow for the 2 to 3-week incubation period of her putative infection. Rubella IgM antibody is generally demonstrable by 1 week after the appearance of the rash and persists for at least 1 month (occasionally 3 months). IgM serology may be the only method of diagnosis in the case of a woman who first consults her doctor weeks after the rash has gone.

Diagnosis of the rubella syndrome in a newborn baby is also made by demonstrating rubella IgM in a single specimen of serum, this time the baby's. Many babies will have rubella IgG antibodies, acquired transplacentally from the mother who may have been vaccinated or infected with the virus years earlier. As IgM antibodies do not cross the placenta, rubella IgM detected in umbilical cord blood must have been synthesized by the baby itself *in utero* and is diagnostic of congenital rubella. It continues to be made in detectable amounts for 3–6 months after birth.

Traditionally, hemagglutination inhibition (HI) has been the preferred procedure for identifying and quantifying rubella antibodies in serum, but the need to first remove nonspecific inhibitors from the serum makes it technically tricky. Accordingly, HI is being supplanted by a range of simpler techniques, notably EIA and latex particle agglutination, both of which are available as disposable kits. EIA is also the procedure of choice for IgM capture assays.

Epidemiology

Rubella virus is shed in oropharyngeal secretions and is presumed to be spread by the respiratory route. It is highly transmissible, usually being acquired by school-age children, and spreads readily within the family. The disease is endemic worldwide.

Prior to the introduction of vaccines, spring epidemics occurred every few years in temperate climates, and over 80% of women had acquired immunity by the time they reached the child-bearing years. Widespread vaccination of infants in the United States commencing in 1969 reduced the incidence of rubella in that country in 1988 to less than 1% of the level 20 years earlier. Since 1988 there have been a number of outbreaks in certain unvaccinated religious communities, as well as among young adults in institutions such as prisons and colleges. Analysis of the reasons for this slight resurgence points to failure to vaccinate rather than vaccine failure, and underlines the importance of intensifying rather than relaxing the drive to eliminate CRS when the battle seems to be almost won.

Control

Live attenuated rubella vaccines, now generally based on the RA27/3 strain grown in a human fibroblast line, have been in use since 1969. A combined measles–mumps–rubella (MMR) vaccine was introduced in 1972. Either vaccine induces durable immunity in at least 95% of recipients, and over 90% have been shown to be protected for at least 15 years, not only against disease but also against the establishment of viremia, even though antibody titers are significantly lower than following natural infection. Although substantially attenuated, rubella vaccines quite commonly induce lymphadenopathy, a fleeting rash, or low-grade fever. Arthralgia is rare in children but occurs in 25% of nonimmune postpubertal females; it begins 1–3 weeks after vaccination and is usually confined to the small peripheral joints, rarely persisting for more than a few weeks. After vaccination, attenuated rubella virus is shed in small amounts from the throat but is incapable of spreading to contacts, so immunization of children is not contraindicated if the mother is pregnant.

There is no evidence that the vaccine virus is teratogenic, and hence accidental immunization during pregnancy is not an indication for termination. Nevertheless, prudence suggests that women not be deliberately immunized during the first trimester and that nonpregnant vaccinees should be advised to avoid conception for the next 2–3 months, and/or be immunized immediately postpartum.

Since 1972 the United States has practiced universal vaccination of all infants with MMR at 15 months of age. From 1991 it has been recommended also that a booster dose of MMR vaccine be routinely administered to all children, either on entering kindergarten or on entering junior high school. In addition, every effort must be made to identify and immunize those women of child-bearing age who may have missed out on the vaccine altogether; about 10% of this age group in the United States are currently seronegative. Suitable opportunities are at premarital screening (but pregnancy should be avoided for the next 3 months) or immediately postpartum (when further pregnancy is unlikely). Although initially a number of European countries selectively immunized schoolgirls at 12–13 years of age (and nonimmune older women of child-bearing age), most have now changed to the U.S. regime.

There remains the difficult problem of what to do with the woman who does in fact contract a laboratory-confirmed rubella infection during the first

3–4 months of pregnancy. There is general agreement that the risk of serious permanent damage to the baby is so substantial that the attending physician should recommend an abortion.

Babies with congenital rubella shed substantial amounts of virus from their throats for up to several months after birth and constitute a considerable risk to pregnant women, directly or via infected nursing staff in maternity hospitals or postnatal clinics. Such babies should be nursed in isolation, preferably by vaccinated or naturally immune personnel.

Further Reading

Calisher, C. H., and Monath, T. P. (1988). *Togaviridae* and *Flaviviridae:* The alphaviruses and flaviviruses. *In* "Laboratory Diagnosis of Infectious Diseases. Principles and Practice. Volume II. Viral, Rickettsial and Chlamydial Diseases" (E. H. Lennette, P. Halonen, and F. A. Murphy, eds.), p. 414. Springer-Verlag, New York.

Centers for Disease Control. (1990). Rubella prevention: Recommendations of the Immunization Practices Advisory Committee (ACIP) *Morb. Mortal. Wkly. Rep.* **39,** RR-15.

Isaacs, D., and Menser, M. (1990). Modern vaccines. Measles, mumps, rubella, and varicella. *Lancet* **335,** 1384.

Karabatsos, N., ed. (1985). "International Catalogue of Arboviruses," 3rd Ed. American Society of Tropical Medicine and Hygiene, San Antonio, Texas.

Koblet, H. (1990). The "merry-go-round": Alphaviruses between vertebrate and invertebrate cells. *Adv. Virus Res.* **38,** 343.

Monath, T. P., ed. (1988). "The Arboviruses: Epidemiology and Ecology," Vols. 1–5. CRC Press, Boca Raton, Florida.

Peters, C. J., and Dalrymple, J. M. (1990). Alphaviruses. *In* "Fields Virology" (B. N. Fields, D. M. Knipe, R. M. Chanock, M. S. Hirsch, J. L. Melnick, T. P. Monath, and B. Roizman, eds.), 2nd Ed., p. 713. Raven, New York.

Schlesinger S., and Schlesinger, M. J., eds. (1986). "The Togaviridae and Flaviviridae," Plenum, New York.

Scott, T. W., and Weaver, S. C. (1989). Eastern equine encephalitis virus: Epidemiology and evolution of mosquito transmission. *Adv. Virus Res.* **37,** 277.

Strauss, J. H., ed. (1990). Viral proteinases. *Semin. Virol.* **1,** 307.

Strauss, J. H., and Strauss, E. G. (1988). Evolution of RNA viruses. *Annu. Rev. Microbiol.* **42,** 657.

Warren, K. S., and Mahmoud, A. A. F., eds. (1990). "Tropical and Geographic Medicine," 2nd Ed. McGraw-Hill, New York.

Wolinsky, J. S. (1990). Rubella. *In* "Fields Virology" (B. N. Fields, D. M. Knipe, R. M. Chanock, M. S. Hirsch, J. L. Melnick, T. P. Monath, and B. Roizman, eds.), 2nd Ed., p. 815. Raven, New York.

Chapter 26

Flaviviridae

As noted in the preceding chapter, one of the two original major groups of arthropod-borne viruses, the Group B arboviruses, has become the genus *Flavivirus* of the family *Flaviviridae* (the other, the Group A arboviruses, became the genus *Alphavirus* of the family *Togaviridae*). Non-arthropod-borne viruses with physicochemical characteristics, gene order, and replication strategies similar to those of the arthropod-borne flaviviruses have also been placed in the family *Flaviviridae,* in the genus *Pestivirus* (which includes viruses of veterinary importance but no human pathogens) and another genus for human hepatitis C virus.

There are about 70 recognized member viruses in the genus *Flavivirus;* of these 13 cause disease in humans, varying from febrile illnesses, sometimes with rashes, to life-threatening hemorrhagic fevers, encephalitis, and hepatitis (see Table 26-2). Three of these, dengue (because of dengue hemorrhagic fever), yellow fever, and Japanese encephalitis viruses, rank among the most important viral pathogens of the developing world.

Hepatitic C virus was discovered in 1989 by a tour de force of molecular biology; even though its virion has neither been visualized nor cultivated, its complete nucleotide sequence has been determined, and it was allocated to the family *Flaviviridae* on the basis of the similarity of its genome organization to that of other members. Hepatitis C virus is now recognized to be quite as important as hepatitis A and B viruses and is today the most common cause of posttransfusion hepatitis.

Properties of *Flaviviridae*

Flavivirus virions are spherical, 40–50 nm in diameter, and consist of an inner viral core within a tightly adherent lipid envelope covered with glycoprotein peplomers (Fig. 26-1). The genome is a single linear 11 kb molecule of ssRNA of positive polarity which is 5′ capped but not 3′ polyadenylated, and is infectious. Other properties are summarized in Table 26-1.

Flaviviruses are not very stable in the environment and are easily inactivated by heat and by disinfectants containing detergents or lipid solvents.

Classification

There are no antigens shared between genera. Members of the genus *Flavivirus* have been grouped by neutralization tests into nine serogroups, each containing between three and ten members, with an additional 17 viruses currently unassigned to a group.

Viral Replication

Flaviviruses enter cells via receptor-mediated endocytosis, and replication takes place in the cytoplasm and is accompanied by proliferation of rough and smooth endoplasmic reticulum. The incoming genomic RNA serves directly as messenger; it contains one large open reading frame of over 10 kb and is translated completely from its 5′ end to produce one large precursor polyprotein which is then cleaved to produce individual viral proteins (Fig. 26-2). About one-quarter of the length of the genomic RNA from the 5′ end encodes the three structural proteins: C, the core protein; prM, a precursor which is cleaved during virus maturation to yield M, the small transmembrane protein; and E, the major peplomer glycoprotein. The rest of the genome encodes seven nonstructural proteins, the functions of most of which are not fully understood; NS3 and NS5 are believed to form the RNA-dependent RNA

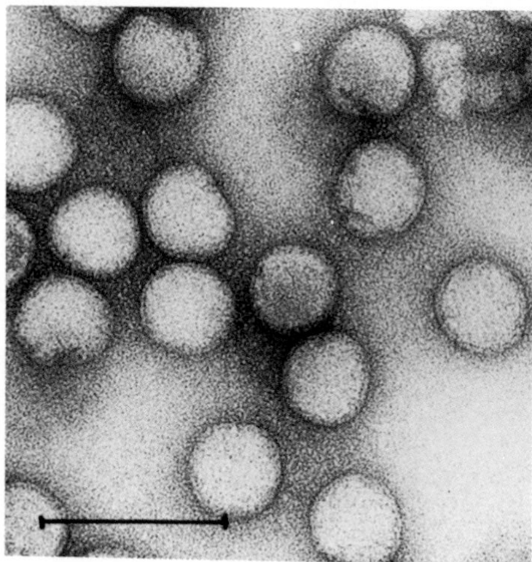

Fig. 26-1 *Flaviviridae,* genus *Flavivirus.* Negatively stained virions of tick-borne encephalitis virus. Bar, 100 nm. (Courtesy Drs. W. Tuma, F. X. Heinz, and C. Kunz.)

Table 26-1
Properties of Flaviviruses

Genera: *Flavivirus*, mostly arthropod-borne viruses; *Hepatitis C*, human hepatitis C virus
Spherical virion, enveloped, diameter 40–50 nm; envelope containing peplomers (glycoprotein E) and small membrane protein (M)
Inner core, 30 nm, composed of core protein (C)
Linear plus sense ssRNA genome, 10.5–11 kb (9.5 kb for hepatitis C virus), 5′ capped, 3′ usually not polyadenylated but looped, infectious
Genes for structural proteins located at 5′ end of genome, those for nonstructural proteins at 3′ end
Cytoplasmic replication; polyprotein translated from genomic RNA cotranslationally cleaved to yield several nonstructural proteins and three structural proteins; maturation into cytoplasmic vesicles

polymerase complex, with NS5 carrying replicase activity and NS3 possessing both helicase and protease activities. NS3 is responsible for most of the co-translational cleavage of the portion of the nascent polyprotein that yields the nonstructural proteins, whereas cellular signal peptidase effects the other primary cleavages.

Replication of the genome occurs in perinuclear foci and involves the synthesis of a complementary minus strand, which in turn serves as template for the synthesis of more plus strand molecules. During infection, plus strand synthesis is favored, suggesting complex regulatory mechanisms which appear to involve host cell constituents. Virion assembly occurs in vertebrate cells on membranes of the endoplasmic reticulum and in mosquito cells on the plasma membrane also, but preformed capsids and budding are not seen. Instead, fully formed virions appear within cisternae of the endoplasmic reticulum and are released via cell lysis. The latent period is long (12 hours or more), and virus production from any given cell continues for days without shutdown of cellular nucleic acid or protein synthesis. Some flaviviruses induce syncytia or plaques in monolayer cultures of mammalian cells, but noncytocidal persistent infections commonly occur in invertebrate cells.

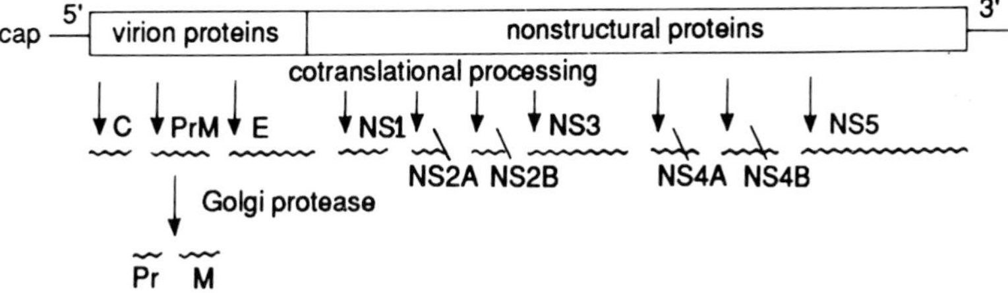

Fig. 26-2 Structure and translation of the flavivirus genome. The regions of the genome encoding the structural and nonstructural proteins are shown in the box at top; the RNA is capped at the 5′ end but is not polyadenylated. There are short nontranslated sequences at each terminus (single lines). The genome is the only mRNA and is translated into a single polyprotein which is cotranslationally cleaved by viral and cellular proteases to form the structural proteins C, M, and E and seven nonstructural proteins.

Pathogenesis and Clinical Features

Like the alphaviruses, flaviviruses may cause encephalitis or a fever/arthritis/rash syndrome, the pathogenesis and clinical features of which were described in Chapter 25, but yellow fever and dengue viruses as well as a number of tick-borne flaviviruses cause hemorrhagic fever. The striking feature of this potentially lethal syndrome is hemorrhage, which manifests as petechiae and ecchymoses on the skin and mucous membranes (see Fig. 33-4), with bleeding from any or all of the body orifices. Examination of the blood reveals thrombocytopenia and usually leukopenia. In fatal cases the patient collapses abruptly from hypotensive shock. Further details are given in Chapters 30, 32, 33, and 36.

Yellow fever is a hemorrhagic fever with a difference; the liver is the major target, with virus replicating in Kupffer cells and massive necrosis of hepatocytes leading to a decrease in the rate of formation of prothrombin as well as to jaundice. Although most cases are mild, presenting with fever, chills, headache, backache, myalgia, and vomiting, a minority progress (sometimes after a brief remission) to severe jaundice, massive gastrointestinal hemorrhages (hematemesis and melena), hypotension, dehydration, proteinuria, and oliguria signaling kidney failure. Mortality from this severe form of the disease is of the order of 20–50%.

Dengue fever is typical of the painful but nonlethal fever/arthritis/rash syndrome which is the common presentation of so many arbovirus diseases: sudden onset of fever (characteristically biphasic, or "saddle back"), chills, retroorbital headache, conjunctivitis, and severe pains in the back, muscles, and joints, followed by a rash and rapid resolution. Sometimes, however, the presentation is that of a hemorrhagic fever, in which a child can die of shock within hours—hence the name dengue hemorrhagic fever/dengue shock syndrome (DHF/DSS).

The disease is generally observed during epidemics of dengue (often type 2) in a population with a previous history of infection with another dengue serotype (e.g., type 1). The majority of the cases occur in children who can be shown to have previously acquired antibody against another type, or in young infants with maternally acquired antibodies against another type. This observation gives support to the "immune enhancement" (or "antibody-dependent enhancement") hypothesis, namely, that dengue virus, the principal target of which is cells of the monocyte/macrophage series, is opsonized by antibodies against a heterologous dengue virus serotype (which bind to the virion but do not neutralize its infectivity), hence is more avidly taken up, via Fc receptors, into the very cell in which it replicates best. Experimentally, it can readily be demonstrated that such virus–antibody complexes replicate to higher titer in macrophage cultures. On the other hand, only a small minority of people with sequential infections develop DHF/DSS, and some cases of DHF/DSS have been reported in people with no prior experience of any type of dengue; thus the possibility remains that individuals differ markedly in susceptibility and/or that strains differ substantially in virulence. Oligonucleotide fingerprinting and partial gene sequencing of numerous strains of all four dengue serotypes has demonstrated the existence of a number of discernable geographic variants (*topotypes*), especially within type 2, but no major differences in virulence have been discerned.

Laboratory Diagnosis

Virus Isolation

Dengue and yellow fever are best diagnosed by virus isolation from blood taken as early as possible during the course of the disease. This is particularly relevant to a suspected indicator case in a region that experiences occasional epidemics of dengue, since it is important to characterize the serotype causing the outbreak. However, other flaviviruses are not readily isolated from human patients, as viremia is frequently brief or undetectable. Postmortem, virus can sometimes be recovered from appropriate organs, such as the brain in cases of encephalitis and the liver in yellow fever. Most arboviruses have been isolated from collections of mosquitoes or other vectors, traditionally by intracerebral injection of newborn mice but more recently by inoculation of cultured mosquito or vertebrate cells. The cytopathic effect varies with the virus and cell type: some combinations produce syncytia, but many give undramatic cytopathology or none at all. Viral antigen can be detected by immunofluorescence.

Characterization of isolates relies on serology, combined with a knowledge of what viruses are likely to occur in particular geographic regions and ecosystems. Whereas hemagglutination inhibition is useful for assigning an arbovirus to a particular genus or group, individual flaviviruses display extensive cross-reactivity. Neutralization is the appropriate test for characterization at the species level. Even those viruses that induce no cytopathic effect in cultured cells usually produce plaques; hence, plaque reduction is the most convenient form of neutralization test.

Because of the pronounced cross-reactions between species, monoclonal antibodies have proved to be of great value in research and may become standard diagnostic reagents, particularly for viruses or geographic regions where serological cross-reactions using polyclonal antisera are a problem. For example, one monoclonal antibody might recognize an antigenic determinant common to all flaviviruses, while another recognizes a determinant common to all members of the dengue complex (dengue types 1, 2, 3, and 4), yet another recognizes only a type-specific determinant on dengue type 3, and another distinguishes subtypes of dengue 3. Oligonucleotide maps of the viral RNA discern even finer differences between isolates from different parts of the world and can be very informative in tracing their origins. For instance, such RNA fingerprints distinguish topotypes of dengue type 2 virus from various parts of the world.

Serology

Diagnosis is usually made by demonstrating a rising titer of antibody in the patient's serum. IgM serology, using EIA with anti-human IgM as capture antibody, is ideally suited to rapid diagnosis of individual patients during an epidemic, but it is not sufficiently definitive to rely on for identifying an "indicator" case prior to such an epidemic. However, the reagents used in such tests must be absolutely reliable, and IgM antibodies can persist for many months and/or be elicited by infection with heterologous viruses. Indicator cases should always be diagnosed by rising antibody titer or, where possible, by virus isolation.

Epidemiology and Control

All members of the genus *Flavivirus* are classic arboviruses (Table 26-2). For a general description of their epidemiology the reader is referred to Chapters 14 and 25; methods of control are discussed in Chapters 15 and 25. In the sections below we deal with specific features of the epidemiology and control of the major human flaviviruses, beginning with the most notorious, yellow fever, after which the family and genus were named, and continuing to other hemorrhagic fevers including dengue, today the most widespread of all arbovirus diseases.

Yellow Fever

One of the great plagues throughout history, yellow fever decimated the crews of English sailing ships visiting West Africa and hundreds of years ago was transported to the New World on slave ships, together with its vector, *Aedes aegypti*. It rapidly became entrenched in the tropical parts of South, Central, and North America, including the United States. Thousands died of yellow fever during the construction of the Panama Canal. The name of Walter Reed, the U.S. Army doctor who unraveled the epidemiology of this disease in 1900, is now legendary. One of the great epidemiologic puzzles is why yellow fever has never occurred in Asia despite the presence there of *Aedes aegypti*.

In its jungle habitat the virus is maintained in a monkey–mosquito cycle (Figs. 26-3 and 26-4). Old World monkeys develop only subclinical infections, but New World monkeys often die, reflecting the more recent introduction of the virus to the Americas. Various species of jungle canopy-feeding, treehole-breeding mosquitoes serve as vectors, generally *Aedes* spp. in Africa and *Haemagogus* spp. in the Americas. Transovarial transmission may enable the virus to be maintained in mosquitoes, and perhaps even in ticks (genus *Amblyomma*), from which it has been recovered in Africa. Some sylvatic mosquitoes can also transmit the virus to humans. However, *Aedes aegypti* is the vector responsible for most of the urban epidemics, in which the virus is maintained in a human–mosquito–human cycle.

A concerted campaign to eradicate *Aedes aegypti* from Brazil and the countries surrounding the Amazon basin has dramatically reduced the incidence of yellow fever in Latin America; some 100–300 cases are now reported annually in tropical America, mainly in adult male forest workers, but this doubtless greatly understates the actual incidence. However, in recent years the

Table 26-2
Principal Human Diseases Caused by Flaviviruses[a]

Virus	Geographic distribution	Vertebrate reservoir	Principal vectors	Disease	Epidemiologic features
Dengue 1, 2, 3, 4	Tropics worldwide	Humans, monkeys	*Aedes aegypti*, other *Aedes* spp.	Fever, arthralgia, myalgia, rash	Urban epidemics
West Nile	Africa, tropical Asia, Mediterranean	Birds	*Culex* spp.	Fever, arthralgia, myalgia, rash	Endemic in tropics with sporadic cases, summer epidemics (Mediterranean, South Africa)
St. Louis encephalitis	Americas	Birds	*Culex tarsalis*, *Culex pipiens*, *Culex* spp. (tropics)	Encephalitis	Sporadic cases (tropical South America), periodic outbreaks (North America)
Japanese encephalitis	Asia	Swine, birds	*Culex tritaeniorhynchus*	Encephalitis	Endemic (Southeast Asia), summer epidemics (north and central Asia)
Murray Valley encephalitis; Kunjin	Australia, New Guinea	Birds	*Culex annulirostris*	Encephalitis	Periodic summer epidemics
Rocio	Southeastern Brazil	Birds	*Psorophora*, *Aedes* spp.	Encephalitis	Sporadic cases, occasional epidemics
Yellow fever	Tropical Africa and Americas	Humans, monkeys	*Aedes aegypti*, *Aedes* spp., *Haemagogus* sp.	Fever, hemorrhage, jaundice	Sporadic cases related to jungle exposure, periodic epidemics (Africa)
Kyasanur Forest disease	Southwest India	Rodents	*Haemaphysalis* ticks	Fever, hemorrhage, encephalitis	Sporadic cases, occasional outbreaks related to deforestation
Omsk hemorrhagic fever	Central Russia	Rodents	*Dermacentor* ticks	Fever, hemorrhage	Sporadic cases, winter outbreaks associated with muskrat trapping
Tick-borne encephalitis	Russia, eastern Europe, Scandinavia	Rodents, birds, domestic animals	*Ixodes* ticks	Encephalitis	Sporadic cases with periodic high incidence, outbreaks from ingestion of raw milk
Louping ill	British Isles	Sheep, birds	*Ixodes* ticks	Encephalitis	Sporadic cases, by direct contact with infected sheep or tick bite
Powassan	Canada, United States, Russia	Small mammals	*Ixodes* ticks	Encephalitis	Sporadic cases

[a] Based on data from C. H. Calisher and T. P. Monath, *Togaviridae* and *Flaviviridae*: The alphaviruses and flaviviruses. *In* "Laboratory Diagnosis of Infectious Diseases. Principal and Practice. Volume II, Viral, Rickettsial and Chlamydial Diseases" (E. H. Lennette, P. Halonen, and F. A. Murphy, eds.), p. 414. Springer-Verlag, New York, 1988.

Fig. 26-3 Epidemiological investigation of yellow fever. (A) Serological survey being conducted among Indians of the Amazon forest. (B) Monkey being bled in Trinidad. (Courtesy of the World Health Organization and Cambridge University Press. From F. M. Burnet and D. O. White, "Natural History of Infectious Disease." Cambridge University Press, London, 1972.)

numbers of *Aedes aegypti* have greatly increased again throughout much of Central and South America and southern United States, and urban epidemics could readily recur.

In endemic regions of West Africa infection of children is very common; although disease is usually subclinical or mild, most of the deaths nevertheless occur in children. Major epidemics continue to occur every few years in Africa—Sudan 1940, Ethiopia 1960–1962, Nigeria 1969, Gambia 1978–1979, Ghana/Burkina Faso 1983—an estimated 100,000 human cases with 30,000 deaths having occurred during the Ethiopian epidemic alone. Since 1985 there has been a resurgence of yellow fever in Africa, where 33 countries are currently at risk of outbreaks, and thousands of cases have occurred in Nigeria.

Urban yellow fever can be prevented by eliminating or drastically reducing the population of *Aedes aegypti* mosquitoes in the vicinity of towns; the means of achieving this are discussed in Chapter 15. Vector control is not applicable to jungle yellow fever, but infection of the scattered human populations can be prevented by vaccination.

A highly effective yellow fever vaccine was developed by Theiler in the 1940s, long before the birth of modern virology. Intuitively, he derived an avirulent strain by isolating yellow fever virus from a monkey and passaging it successively through mice, various primary cell cultures, and finally chick embryos. This 17D strain is still the seed virus for today's chick embryo grown vaccine, which is supplied freeze-dried, stored refrigerated or frozen, and reconstituted for subcutaneous injection. Only 5% of vaccinees report mild constitutional side effects, and the only contraindications are known allergy

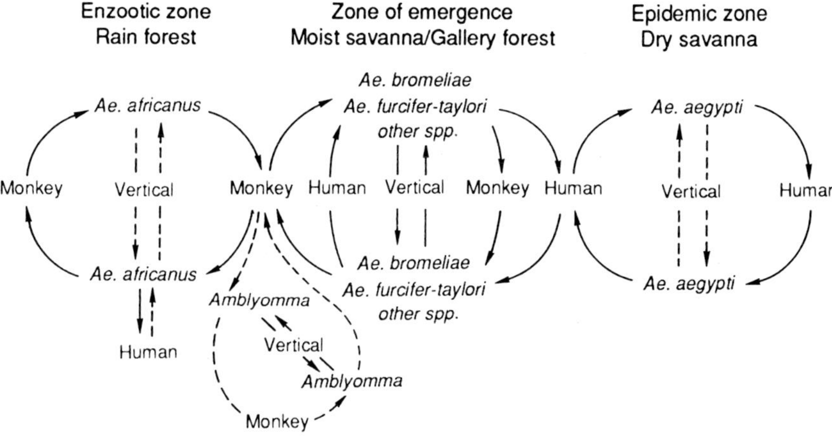

Fig. 26-4 Ecologic relationships of yellow fever virus in tropical Africa. Established parts of the cycle are drawn in solid lines, unproved or rare cycles in broken lines. In the rain forest zone of central and western Africa, the virus is maintained in a continuous cycle of low-level transmission, involving monkeys and *Aedes africanus* mosquitoes. In savanna areas with gallery forest along the watercourses, high rates of transmission to both monkeys and humans may occur during the rainy season, associated with high densities of treehole-breeding *Aedes* vectors. Humans may be infected in both of these cycles. In the very dry savanna, and in urban areas, where domestic water storage is practiced, domestic *Aedes aegypti* may be responsible for interhuman spread when the virus is introduced from endemic zones. Vertical transmission in mosquitoes probably occurs in each of these cycles. [From T. P. Monath, *in* "Fields Virology" (B. N. Fields, D. M. Knipe, R. M. Chanock, M. S. Hirsch, J. L. Melnick, T. P. Monath, and B. Roizman, eds.), 2nd Ed., p. 802. Raven, New York, 1990.]

to eggs, age younger than 6 months, and severe immunosuppression (e.g., AIDS). Over 95% of recipients develop solid immunity, which generally lasts for decades. Nevertheless, most yellow fever-free countries require that travelers through endemic areas have been vaccinated within the past 10 years. Vaccination policy within countries in which yellow fever is endemic is still not uniform. Some South American nations have undertaken routine vaccination for many years. Following the recent strengthening of the Expanded Programme of Immunization and the upsurge of yellow fever in tropical Africa, many African nations have incorporated yellow fever vaccine into routine childhood immunization programs; it is usually given at the same time as measles vaccine.

Dengue

There are four serotypes of dengue viruses. Infection with any one serotype gives lifelong homologous immunity, but there is no cross-protection. The original natural reservoir of the dengue viruses appears to have been tropical African primates, although the viruses also appear to have established reservoir status in monkeys in southeast Asia. However, like yellow fever virus, all four serotypes of dengue virus have become endemic in human populations. The principal mosquito vector for the human disease is the urban mosquito

Aedes aegypti, although both *A. albopictus* and *A. scutellaris* are efficient vectors. Dengue is now the most important human arboviral disease, with over 2 billion people at risk and millions of cases annually.

Although dengue fever has been known for over 200 years, prior to the 1950s outbreaks of dengue were rare, and because of the slow transport of viremic persons between tropical countries epidemics in any particular locality occurred at intervals of decades. During and after the Second World War millions of people in all countries of the developing world moved to the cities, resulting in rapid and unplanned urbanization and the expansion of breeding places for *Aedes aegypti.* Major epidemics involving hundreds of thousands of people have occurred in the Caribbean (1977–1981), South America (since the early 1980s), the Pacific (1979), and China (1978–1980), as well as in Southeast Asia and Africa. The simultaneous circulation of multiple serotypes led to the appearance of an essentially new disease, dengue hemorrhagic fever. Although in retrospect the disease probably occurred in northern Australia in 1898 and in Greece in the 1920s, the first outbreak of dengue hemorrhagic fever and dengue shock syndrome occurred in Manila in 1953–1954; by 1975 it was occurring at regular intervals in most countries of Southeast Asia and is today one of the leading causes of hospitalization and death among children in Southeast Asia. Dengue hemorrhagic fever is still spreading in the Asia–Pacific region, with outbreaks in French Polynesia and New Caledonia in 1989–1990 and in Sri Lanka in 1990.

Dengue had been known in the Americas for hundreds of years, but it did not become a major health problem there until about 1975. Initially this was due to the same constraints that operated in Southeast Asia, notably the slow and infrequent movement of viremic persons. The explanation for the delay after the end of the Second World War, compared with the immediate postwar occurrence of the disease in Southeast Asia, lay in the success of the campaign to eradicate *Aedes aegypti* from the Americas, primarily to control urban yellow fever. Although good results were achieved in many countries, the aim of total eradication from the region failed, and during the 1970s many cities in South American and Caribbean countries were reinvaded by *Aedes aegypti.* Coincidentally, the increased movement of people into the slums of the urban fringes led to greatly increased breeding places for the mosquito, and the increased air travel throughout the region led to increased movement of the dengue viruses, so that during the early 1980s there was a great increase in dengue transmission caused by multiple serotypes. This was exactly the situation that had led to the emergence of dengue hemorrhagic fever in Southeast Asia in the 1950s, and the result was the same.

The first major outbreak of dengue hemorrhagic fever in the Americas occurred in Cuba in 1981, and since then there have been epidemics in Mexico (1984), Nicaragua (1985), Puerto Rico (1986), El Salvador (1987), Venezuela (1989), and Colombia (1990), as well as sporadic cases in other countries. Dengue hemorrhagic fever is now established as part of the health problem of South and Central America and the Caribbean. Because of the widespread occurrence of *Aedes aegypti* and the recent extensive spread of *Aedes albopictus,* it is also a threat to the southern states of the United States.

The prospects for control are poor, at least in the immediate future. Although an experimental live attenuated vaccine containing all four dengue serotypes (because of the risk of immune enhancement a multiple serotype vaccine is essential) is undergoing clinical trials, and proteins E and NS1 produced by genetically engineered baculovirus in insect cells have been shown to protect mice, safe and effective dengue vaccines for general use are probably still years away. Mosquito control is the only tool now available, but the common methods now used (spraying of insecticide aerosols from airplanes or trucks) fail to kill all *Aedes aegypti,* because the mosquitoes often rest in places not reached by the aerosol. Further, over the last several decades the extensive use of pesticides has led to pesticide resistance. Control of breeding is theoretically feasible, and governments should legislate to make it the responsibility of citizens to remove abandoned containers, tires, etc., and to drain stagnant pools of water in the vicinity of dwellings, but these measures are increasingly difficult in the slums of the urban fringes in the expanding cities. Dengue and dengue hemorrhagic fever are here to stay, and it may well become even more widely distributed before effective control measures are introduced; the disease is "a ticking time bomb of huge epidemic potential."

Flavivirus Encephalitides

Japanese Encephalitis

Japanese encephalitis virus is a member of a serogroup containing nine other viruses, including St. Louis encephalitis virus, Murray Valley encephalitis virus, and West Nile and Kunjin viruses. It is the world's most important encephalitic arbovirus, is widely distributed across China, Korea, Japan, the Philippines, and the countries of southeastern Asia, and has recently extended its range westward into India, Nepal, and Sri Lanka and eastward into the Pacific islands of Saipan and the northern Marianas. In tropical areas where the virus is endemic, sporadic cases occur throughout the year, and many young children are infected subclinically; outbreaks are occasionally seen at the end of the wet season. In temperate zones outbreaks tend to occur in late summer/early autumn.

The mosquito *Culex tritaeniorhynchus,* which breeds in irrigated rice fields and feeds on birds, swine, and humans, is the most common vector. Swine are the most abundant species of domestic animal in many parts of Asia; they have a short life span and continuously provide generations of susceptible hosts. The mosquito–swine–mosquito transmission cycle serves as an efficient mode of virus amplification.

It is instructive to observe how successful the Japanese have been in controlling Japanese encephalitis by a well-conceived multipronged attack on the problem: draining of rice paddies at the time when *C. tritaeniorhynchus* normally breeds, removal of the principal amplifier host, pigs, from the vicinity of human habitation, and widespread vaccination of swine, horses, and children with formalin-inactivated, mouse brain-derived vaccine. Inactivated, cell culture grown vaccine is being used effectively in China, while attenuated live-virus vaccines for human and animal use are under development; these vaccines offer the possibility of lower costs and may be suitable for use in large areas of southeastern Asia.

St. Louis Encephalitis

Epidemic St. Louis encephalitis is the most devastating encephalitic disease in the United States. The natural cycle of the virus occurs between *Culex tarsalis* and nesting and juvenile birds, but when mosquitoes are numerous the cycle may be amplified by infection of domestic birds and wild and domestic mammals, which may then lead to encephalitis in a minority of infected humans (see Fig. 25-3). In the eastern United States, *Culex pipiens,* which breeds in stagnant, polluted water, may set up epidemic St. Louis encephalitis in urban areas. A less virulent subtype transmitted by *Culex tarsalis* is endemic in rural agricultural areas of western United States.

Murray Valley (Australian) Encephalitis

The third encephalogenic member of the complex, Murray Valley encephalitis, is enzootic in Papua New Guinea and northern Australia, where sporadic cases of encephalitis occur. Epidemics involving humans occur in the Murray River Valley of southeastern Australia only in occasional summers, following heavy rainfall with extensive flooding. These conditions encourage explosive increases in the populations of waterbirds, which appear to be the principal reservoir, and the mosquito vectors, notably *Culex annulirostris.*

Rocio Encephalitis

The flavivirus responsible for Rocio encephalitis, first isolated in 1975, has been associated with several outbreaks of encephalitis in Brazil. Wild birds are the probable vertebrate reservoir, and potential vectors include *Psorophora ferox* and *Aedes scapularis.*

West Nile Fever and Kunjin Virus Infection

West Nile virus is maintained in a *Culex*–bird cycle; it also infects ticks. It is endemic in rural Africa, southern Europe and Central Asia, India, and the Far East. Usually it causes a dengue-like illness, very rarely encephalitis. Unlike the case with many other flaviviruses, virus can often be isolated from the blood early in the illness. The related Kunjin virus is quite common in Australia and occasionally causes encephalitis in humans; like West Nile virus, it can be lethal in horses.

Tick-Borne Flavivirus Encephalitis

The tick-borne encephalitis viruses are members of a serocomplex of 15 related tick-borne viruses that includes Russian spring–summer encephalitis virus and Central European encephalitis virus, as well as Omsk hemorrhagic fever virus and Powassan virus. Unlike mosquitoes, ticks are present throughout the year even in temperate climates and often live through more than a single breeding cycle of their host. The eggs of ticks develop successively through stages (larva to nymph to adult), and a blood meal is generally required at each stage. Tick-borne flaviviruses are passed from one developmental stage (instar) to another (*transstadial transmission*) as well as from one generation of tick to the next (*transovarial transmission*) (see Fig. 14-3). Some species spend their whole lives on one vertebrate host, whereas others fall off, molt, then find a different host after each meal. The larvae and nymphs generally parasitize birds or small mammals such as rodents, whereas adult ticks prefer larger animals. Domestic and farm animals such as cattle, sheep, and goats are important in the spread of these viruses to humans, who can be infected by tick bite or by the ingestion of raw milk. Formalin-inactivated vaccines are available for use against tick-borne encephalitis in Eastern Europe.

The closely related flavivirus of Britain, louping ill, is maintained in a tick–grouse cycle and is transmissible to sheep, in which it causes a rapidly lethal disease. Rarely, louping ill virus is transmitted to humans by ticks or occupationally by contact with infected sheep tissues, producing a relatively mild meningoencephalitis.

Tick-Borne Flavivirus Hemorrhagic Fevers

The Omsk hemorrhagic fever virus of Siberia, the Kyasanur Forest disease virus of India, and the Powassan virus of North America and Russia are closely related to the tick-borne encephalitis viruses but cause hemorrhagic fever instead. Water voles are the reservoir hosts of Omsk hemorrhagic fever virus, but humans are usually infected by direct contact with the blood or excreta of the muskrat, an animal imported from America in the early 1900s. Rodents are the principal vertebrate reservoir of Kyasanur Forest virus, but monkeys may serve as amplifiers; moreover, infected ticks are also carried to domestic cattle and goats and the deer and antelopes in nature reserves, hence the concentration of human cases acquired along cattle tracks in Mysore State forests. Powassan virus infects small mammals in the northern United States, Canada, and Russia and is rarely transmitted to humans.

Hepatitis C

With the introduction of sensitive assays for screening blood for hepatitis B virus in the late 1970s, it was anticipated that posttransfusion hepatitis would be virtually eliminated, but this was not to be. There remained a substantial residue of cases which were called non-A, non-B hepatitis (NANBH). The causative agent remained frustratingly elusive for over a decade and has still not been convincingly cultured *in vitro* nor visualized by electron miscroscopy. Nevertheless, in 1989, a team of molecular biologists in the United States succeeded in an ambitious assignment which seemed to many to be unachievable. The ingenious protocol they devised serves as a prototype which could be applied in the future to the discovery not only of additional NANBH viruses but also of many other currently unknown noncultivable infectious agents.

Bradley and colleagues had previously demonstrated that hepatitis could be transmitted to chimpanzees by inoculation of factor VIII known to have been contaminated with an agent that had caused non-A, non-B hepatitis in hemophiliacs. Filtration studies had indicated the size of the causative agent to be of the order of 40–50 nm, while a buoyant density of 1.1 g/cm^3 and sensitivity to chloroform suggested it to be an enveloped virus. The reasonable assumption was made that plasma from chimpanzees that had developed chronic hepatitis following inoculation with factor VIII would constitute a good source of the putative virus, and that virions could be concentrated from the plasma by ultracentrifugation. Total DNA and RNA was extracted from the pellet, denatured to single strands, and reverse-transcribed using random primers that prime the transcription of any single-stranded nucleic acid. The resulting complementary DNA was cloned into the bacteriophage λgt11 expression vector, which was then used to infect bacteria. Bacterial colonies were then screened for production of any antigenic polypeptide sequence recognizable in an immunoassay by serum taken from chronically infected patients, assumed to contain antibody against the putative virus. After screening about a million such random cDNA clones, a single clone capable of binding antibodies from several infected individuals was found! This DNA was then used as a hybridization probe to derive a larger overlapping clone from the cDNA library, which in turn was used to identify the full-length (9.5 kb) plus sense ssRNA hepatitis C viral genome. Eventually the complete nucleotide sequence of the genome was determined by isolating overlapping cDNA clones using hybridization probes based on the sequences of previous clones.

Properties of Hepatitis C Virus

The genome of hepatitis C virus (HCV) is a single 9.5 kb molecule of ssRNA of positive polarity, with a gene order characteristic of the family *Flaviviridae.* A single long ORF encoding a polyprotein of about 3000 amino acids is flanked by untranslated 5′ and 3′ sequences, each containing short direct repeats and taking the form of a hairpin. The 3′ terminus is not polyadenylated. The structural proteins occupy the 5′ quarter of the ORF, and the nonstructural proteins the remainder. The gene order, namely, 5′-C, E1, E2/NS1, NS2, NS3, NS4, NS5-3′, closely resembles that of other members of the *Flaviviridae.* Structural protein C is highly basic and presumably represents the core (capsid) of the virion; E1 and E2 are glycoproteins, presumably both membrane proteins as in the genus *Pestivirus,* although it is possible that E2 may be the equivalent of the nonstructural protein NS1 of the genus *Flavivirus,* which is otherwise missing from hepatitis C virus. Of the four other nonstructural proteins, it can be deduced from characteristic motifs that, as for other members of the *Flaviviridae,* NS3 carries serine protease activity in its amino-terminal half and helicase activity in its carboxy-terminal half, whereas NS5 has RNA-dependent RNA polymerase activity, and NS2 and NS4 may be comparable with the membrane-binding proteins postulated to be required by other flaviviruses during membrane-associated replication.

Classification

Analysis of PCR products after reverse transcription of the genome of hepatitis C viruses from around the world reveals significant heterogeneity in nucleotide sequence, indicating the existence of at least a dozen genotypes. Whereas differences occur in all genes, there are hypervariable regions within the glycoproteins E1 and E2, suggesting that these envelope proteins are subject to immune selection *in vivo.* However, the relationship between genotypes and serotypes has yet to be sorted out.

Viral Replication

Little is currently known about the replication cycle, as no simple cell culture system has yet been discovered. Replication probably occurs exclusively in the cytoplasm. Dense reticular cytoplasmic inclusions and convoluted membranes are conspicuous by electron microscopy in hepatocytes from infected chimpanzees. Immunofluorescence reveals viral proteins, mainly NS3 and NS4, confined to the cytoplasm. Full-length plus and minus RNA strands as well as a full-length double-stranded RNA (replicative form) have been found by *in situ* hybridization in the cytoplasm of infected liver, and apparently also in T lymphocytes. Consistent with the transcription strategy of other flaviviruses, no subgenomic RNAs have been detected by northern blot analysis of hepatocytes; it is reasonable to assume that a single polyprotein translated from the single long ORF is systematically cleaved at the appropriate motifs by viral and cellular proteases.

Pathogenesis and Clinical Progression

In many developed countries today hepatitis A, B, and C are about equally common. Acute hepatitis C is clinically similar to hepatitis A and B, and the reader is referred back to Chapters 22 and 23 for descriptions. The major differences are as follows. The incubation period of hepatitis C, though ranging up to several months, averages 6–8 weeks. About 75% of infections are subclinical. Clinical infections are generally less severe than hepatitis B, having a shorter preicteric period, milder symptoms, absent or less marked jaundice, and somewhat lower serum alanine aminotransferase (ALT) levels, which often fluctuate widely. The case–fatality rate from fulminant hepatitis is 1% or less. However, HCV leads much more commonly to chronic liver disease than does HBV. At least 50% of all patients with hepatitis C remain continuously or erratically viremic with moderate elevation of ALT levels for at least a year or two, and often much longer. Most of these are asymptomatic carriers or mild cases of chronic persistent or chronic active hepatitis which spontaneously resolve, but up to 20% progress to cirrhosis. Indeed, one U.S. study suggests that hepatitis C may be as important a cause of cirrhosis in that country as alcoholism, and more common than hepatitis B.

There is also a clear correlation between chronic HCV infection and the development of hepatocellular carcinoma (HCC), more than 90% of HBV-negative cases in some countries being HCV antibody positive, but proof of etiological association may have to await controlled prospective studies. Because the HCV genome is RNA, and reverse transcriptase is not involved in its replication, it is unlikely that integration of cDNA is involved as with HBV-associated HCC.

Little is known of the pathogenesis of HCV infection. Using the limited range of assays currently available to study the progress of infection in chimpanzees and humans, certain facts have emerged. The major target cell seems to be the hepatocyte, but there is also evidence suggesting viral replication in leukocytes and perhaps other cells. Viremia is detectable by PCR within days of infection and lasts for weeks or months before resolution in most cases, but it may persist for years in chronic carriers, often fluctuating erratically. Whether these swings reflect reactivation of virus, with or without accompanying relapses in clinical hepatitis, and if so what triggers them have yet to be resolved. Although antibodies continue to be made, and most of the virus is bound in virus–antibody complexes, the infection is not eliminated. Indeed, chimpanzees that have recovered from HCV infection can be reinfected with the homologous (or heterologous) strain of virus. Humans also frequently experience multiple episodes of acute hepatitis C, but it is currently unclear whether these are exogenous reinfections with the same or another strain or reactivation of the original infection.

Laboratory Diagnosis

In 1990 the first specific diagnostic test for hepatitis C virus was licensed for the screening of blood donors in the United States. A cDNA clone representing part of the HCV genome was expressed in yeast to produce a recombinant antigen (fusion protein) corresponding to a large portion of the nonstructural protein, NS4, and this antigen was employed in an enzyme immunoassay to detect antibodies to that particular protein in the serum of potential blood or organ donors. However, this first-generation EIA was not sufficiently sensitive or specific, and it was replaced by a second-generation version based on a recombinant yeast chimeric protein comprising the three most conserved HCV proteins, C, NS3, and NS4. Although this cloned antigen, which lacks the hypervariable E1 and E2 proteins, fails to distinguish between putative serotypes or strains of the virus, it does identify hepatitis C antibody in up to 95% of posttransfusion non-A, non-B hepatitis cases and has replaced the first-generation EIA in blood banks and elsewhere. Because of the slow and variable development of antibodies postinfection, there is a window period of about 3 months before the test registers positive.

Of course, assays for antibody do not distinguish between acute, chronic, and past infection. Because about 50–60% of HCV infections progress to the chronic carrier state it is currently prudent for blood banks to discard all HCV antibody-positive blood, but for the purposes of clinical management it is important to develop additional tests. Assays for IgM antibody have not proved to be particularly useful. Immunoassays probing for antigen are not sufficiently sensitive to pick up the low titer of virus present in serum. The PCR is much more sensitive and successfully detects the viral genome from about 2 weeks postinfection but may miss some chronic carriers because of the fluctuating levels of viremia in any given patient over the years. Immunofluorescence and *in situ* nucleic acid hybridization are invaluable tools for studying pathogenesis *in vivo* but are of diagnostic value only on biopsy and autopsy specimens. There is an obvious need for a reproducible system for isolation of HCV in cultured cells.

Epidemiology

In most Western countries the prevalence of antibodies to hepatitis C virus is around 1% in blood donors, but this may underestimate the prevalence in the community as a whole. While blood transfusion and factor VIII have been major sources of infection in the past, they have already declined as a result of more rigorous selection of donors and can be expected to fall off much more rapidly with the introduction of second-generation (and later) EIAs for screening in blood banks. Today the most clearly identifiable infected cohort are intravenous drug users, the majority of whom are infected. Whereas these three parenteral risk groups account for about half of all HCV infections, the route of infection of the remainder is a mystery. There is some evidence that, like HBV, HCV may be shed in genital secretions and saliva, as well as in blood. Heterosexual promiscuity seems to correlate with HCV seropositivity; however, presumably because of the lower circulating virus titer, sexual transmission does not appear to be as important as with HBV. The same applies to perinatal transmission: it appears to be relatively uncommon except when the mother is infected with HIV as well as HCV.

Control

The surprisingly poor homologous immunity that has been reported to follow infection of chimpanzees does not augur well for prevention of hepatitis C by vaccination. Nevertheless, experimental vaccines based on molecularly cloned envelope proteins are in the pipeline. Provided the preparation is highly immunogenic and attention is given to any antigenic variation evident in natural strains, there is some hope of success.

Interferon α injected subcutaneously at a dosage of 2–3 million units thrice weekly for 6 months reduces ALT levels to normal in about half of chronic hepatitis C patients, but all except about 20–25% of these relapse after withdrawal of the drug. Higher doses are no advantage, but patients with less severe disease or those treated before cirrhosis sets in appear to have a better chance of long-term response. Ribavirin may have a comparable effect. HCV-induced cirrhosis is one of the commonest indications for liver transplantation, but endogenous recurrence of infection is a major problem.

Further Reading

Bradley, D. W. (1990). Hepatitis C virus: The major causative agent of viral non-A, non-B hepatitis. *Br. Med. Bull.* **46,** 423.

Bradley, D. W. (1990). Hepatitis non-A, non-B viruses become identified as hepatitis C and E viruses. *Prog. Med. Virol.* **37,** 101.

Bradley, D. W., ed. (1993). Hepatitis Viruses. *Semin. Virol.* **4(5),** 269.

Bradley, D. W., Krawczynski, K., Beach, M. J., and Purdy, M. A. (1991). Non-A, non-B hepatitis: Toward the discovery of hepatitis C and E viruses. *Semin. Liver Dis.* **11,** 128.

Calisher, C. H., and Monath, T. P. (1988). *Togaviridae* and *Flaviviridae:* The alphaviruses and flaviviruses. *In* "Laboratory Diagnosis of Infectious Diseases. Principles and Practice. Volume II, Viral, Rickettsial and Chlamydial Diseases" (E. H. Lennette, P. Halonen, and F. A. Murphy, eds.), p. 414. Springer-Verlag, New York.

Chambers, T. J., Hahn, C. S., Galler, R., and Rice, C. M. (1990). Flavivirus genome organization, expression and replication. *Annu. Rev. Microbiol.* **44,** 649.

Chien, D. Y., Choo, Q.-L., Tabrizi, A., Kuo, C., McFarland, J., Berger, K., Lee, C., Shuster, J. R., Nguyen, T., Moyer, D. L., Tong, M., Furuta, S., Omata, M., Tegtmeier, G., Alter, H., Schiff, E., Jeffers, L., Houghton, M., and Kuo, G. (1992). Diagnosis of hepatitis C virus (HCV) infection using an immunodominant chimeric polyprotein to capture circulating antibodies: Re-evaluation of the role of HCV in liver disease. *Proc. Natl. Acad. Sci. U.S.A.* **89,** 10011.

Choo, Q.-L., Weiner, A. J., Overby, L. R., Kuo, G., Houghton, M., and Bradley, D. W. (1990). Hepatitis C virus: The major causative agent of viral non-A, non-B hepatitis. *Brit. Med. Bull.* **46,** 423.

Halstead, S. B. (1988). Pathogenesis of dengue: Challenges to molecular biology. *Science* **239,** 476.

Hollinger, F. B., Lemon, S. M., and Margolis, H. S., eds. (1991). "Viral Hepatitis and Liver Disease." Williams & Wilkins, Baltimore, Maryland.

Karabatsos, N., ed. (1985). "International Catalogue of Arboviruses," 3rd Ed. American Society of Tropical Medicine and Hygiene, San Antonio, Texas.

Monath, T. P., ed. (1988). "The Arboviruses: Epidemiology and Ecology," Vols. 1–5. CRC Press, Boca Raton, Florida.

Monath, T. P. (1990). Flaviviruses. *In* "Fields Virology" (B. N. Fields, D. M. Knipe, R. M. Chanock, M. S. Hirsch, J. L. Melnick, T. P. Monath, and B. Roizman, eds.), 2nd Ed., p. 763. Raven, New York.

Monath, T. P. (1991). Yellow fever: *Victor, Victoria?* Conqueror, conquest? Epidemics and research in the last forty years and prospects for the future. *Am. J. Trop. Med. Hyg.* **45,** 1.

Reed, W. (1902). Recent researches concerning the etiology, propagation, and prevention of yellow fever, by the United States Army Commission. *J. Hyg.* **2,** 101.

Reyes, G. R., and Baroudy, B. M. (1991). Molecular biology of non-A, non-B hepatitis agents: Hepatitis C and hepatitis E viruses. *Adv. Virus Res.* **40,** 57.

Rice, C. M., Lenches, E. M., Eddy, S. R., Shin, S. J., Sheets, R. L., and Strauss, J. H. (1985). Nucleotide sequence of yellow fever virus; implications for flavivirus gene expression and evolution. *Science* **229,** 726.

Schlesinger, S., and Schlesinger, M. J. (1986). "The Togaviridae and Flaviviridae." Plenum, New York.

Shikata, T., Purcell, R. H., and Uchida, T., eds. (1991). "Viral Hepatitis C, D and E." Elsevier, Amsterdam.

Technical Advisory Group on Haemorrhagic Fever/Dengue Shock Syndrome. (1986). "Dengue Haemorrhagic Fever: Diagnosis, Treatment and Control." World Health Organization, Geneva.

Trent, D. W., Manske, C. L., Fox, G. E., Chu, M. C., Kliks, S. C., and Monath, T. P. (1990). The molecular epidemiology of dengue viruses: Genetic variation and microevolution. *Appl. Virol. Res.* **2,** 293.

Vaughn, D. W., and Hoke, C. H., Jr. (1992). The epidemiology of Japanese encephalitis: Prospects for prevention. *Epidemiol. Rev.* **14,** 197.

World Health Organization. (1986). "Dengue haemorrhagic fever: Diagnosis, treatment and control." World Health Organization, Geneva.

Chapter 27

Coronaviridae

The family *Coronaviridae* embraces over a dozen major host-specific pathogens of mammals and birds, displaying tropism for the respiratory tract (e.g., avian infectious bronchitis virus), the enteric tract (transmissible gastroenteritis virus of swine), or the liver and brain (murine hepatitis virus). No convincing evidence has yet been obtained to link human coronaviruses with serious disease affecting any of these systems, but they are an important cause of that trivial but annoying disease, the common cold. In addition, particles morphologically resembling coronaviruses are often seen by electron microscopy in feces, but it has yet to be established whether they cause gastroenteritis in humans. Coronaviruses have the largest genome of all the RNA viruses and exhibits a unique transcription strategy of considerable interest to molecular biologists.

Properties of *Coronaviridae*

The coronaviruses were so named because the unusually large club-shaped peplomers projecting from the envelope give the particle the appearance of a solar corona (Fig. 27-1). The virion is pleomorphic, being roughly spherical in the case of the genus *Coronavirus* but often disk-, kidney-, or rod-shaped in the case of the genus *Torovirus*, and can range in size from 60 to 220 nm. The tubular nucleocapsid, difficult to discern in electron micrographs, is composed of a phosphorylated nucleoprotein (N) and seems to be connected directly to an unusual transmembrane protein, M, which spans the viral envelope three times and performs the role normally filled by matrix protein in other enveloped viruses. A very large (200K), heavily glycosylated envelope glycoprotein S (for spike) forms the bulky peplomers and carries cell

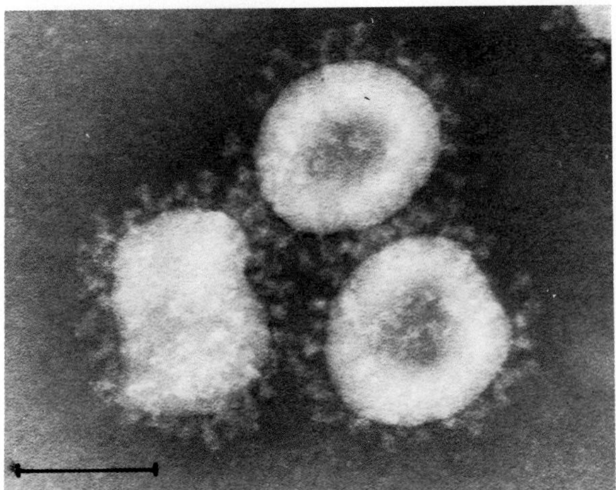

Fig. 27-1 *Coronaviridae.* Negatively stained preparation of virions. Bar, 100 nm. (Courtesy Dr. F. A. Murphy.)

binding, membrane fusion, and hemagglutinating activities. The human coronavirus OC43 also possesses a third glycoprotein, HE (for hemagglutinin–esterase) which closely resembles a similar cell-binding/receptor-destroying enzyme found in influenza C virus.

The genome consists of a single linear molecule of ssRNA of positive polarity, about 30 kb in size, which is 5′ capped and 3′ polyadenylated, and is infectious (Table 27-1).

Classification

Two genera, *Coronavirus* and *Torovirus,* contain viruses infecting humans. The genus *Coronavirus* includes two human serotypes causing respiratory disease, the prototype strains being HCV-229E and HCV-OC43, which are clearly distinguishable from one another by neutralization or hemagglutination inhibition. The status of the "human enteric coronaviruses" and human toroviruses has yet to be determined.

Table 27-1
Properties of *Coronaviridae*

Pleomorphic spherical virion, *Coronavirus* 60–220 nm (average 100 nm), *Torovirus* 120–140 nm
Envelope with large widely spaced, club-shaped peplomers
Tubular nucleocapsid with helical symmetry
Linear plus sense ssRNA genome, 30 kb (range 27–33 kb), capped, polyadenylated, infectious
Three or four structural proteins: nucleoprotein (N), peplomer glycoprotein (S), transmembrane glycoprotein (M), sometimes hemagglutinin–esterase (HE)
Replicates in cytoplasm; genome transcribed to full-length minus sense RNA, from which is transcribed a 3′-coterminal nested set of mRNAs, only the unique 5′ sequences of which are translated; budding into endoplasmic reticulum and Golgi cisternae; virions released by exocytosis

Viral Replication

The strategy of expression of the coronavirus genome is unique (Fig. 27-2). The infectious, plus sense viral RNA is translated directly, the product of the 5′ two-thirds of the genome being two polyproteins, the larger of which is produced by ribosomal frame-shifting; each is posttranslationally self-cleaved and the resulting polypeptides assembled to form an RNA polymerase. This enzyme is then employed to transcribe a full-length complementary (minus sense) RNA, from which in turn are transcribed not only full-length new plus strands, but also a 3′-coterminal *nested set of subgenomic mRNAs*. The nested set comprises 5–7 overlapping species of plus sense mRNAs which extend for different lengths from their common 3′ ends and share a common 70 nucleotide 5′ leader sequence. They are generated by a leader-primed mechanism of discontinuous transcription, whereby the polymerase first transcribes the noncoding leader sequence from the 3′ end of the minus sense antigenome, then the capped leader RNA dissociates from the template and reassociates with a complementary sequence at the start of any one of the genes to continue copying the template right through to its 5′ end. Only the unique sequence toward the 5′ end which is not shared with the next smallest mRNA in the nested set is translated, the product therefore being a unique protein in most cases. A puzzling finding is that subgenomic minus sense RNA species complementary to the nested set of plus sense mRNAs are also present in infected cells, giving rise to the intriguing possibility that coronavirus mRNAs may be self-replicating.

The synthesis, processing, oligomerization, and transport of the several envelope glycoproteins have been studied in depth and display some unusual features. For example, the envelope protein M, which in some coronavirus

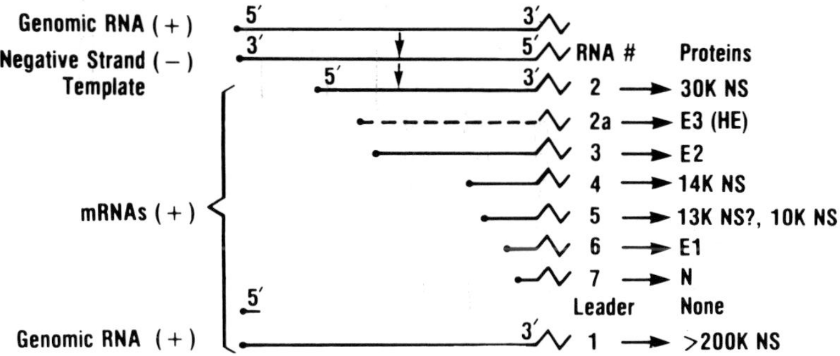

Fig. 27-2 Coronavirus transcription and translation. After release of the plus strand genomic RNA in the cytoplasm, an RNA-dependent RNA polymerase is synthesized which transcribes a full-length minus strand RNA, from which are synthesized (a) new genomic RNA, (b) an overlapping series of subgenomic mRNAs, and (c) leader RNA. The genomic RNA and mRNAs are capped and polyadenylated (zigzag line) and form a "nested set" with common 3′ ends and a common leader sequence on the 5′ end. Only the unique sequence of the mRNAs toward the 5′ end is translated, to produce several nonstructural proteins (NS) and four structural proteins: M (E1), transmembrane glycoprotein; S (E2), peplomer glycoprotein; N, nucleoprotein; and in some coronaviruses HE (E3), hemagglutinin–esterase glycoprotein. [Modified from K. V. Holmes, *in* "Fields Virology" (B. N. Fields, D. M. Knipe, R. M. Chanock, M. S. Hirsch, J. L. Melnick, T. P. Monath, and B. Roizman, eds.), 2nd Ed., p. 847. Raven, New York, 1990.]

species contains O-linked rather than N-linked glycans, is directed exclusively to certain internal membranes, namely, cisternae of the endoplasmic reticulum and the Golgi complex, as a result of which virions bud only from these regions and not from the plasma membrane. The virions are then transported in vesicles to the plasma membrane for exit from the cell by exocytosis (see Fig. 3-9B). Following their release many of the mature enveloped virions remain adherent to the outside of the cell. The whole of the replication cycle is confined to the cytoplasm, and indeed can occur in enucleate cells.

Genetic recombination occurs at high frequency between the genomes of different but related coronaviruses. This may be an important mechanism for generation of genetic diversity in nature.

Pathogenesis and Immunity

The virus remains localized to the epithelium of the upper respiratory tract and elicits a poor immune response. No doubt local IgA is important, but this has not yet been formally proved. Immunity to the homologous strain lasts for a few years at most, hence there is a high rate of reinfection. There is no cross-immunity between the HCV-229E and HCV-OC43 serotypes. Moreover, it is likely that new strains are continually arising by antigenic drift (mutation) and shift (recombination).

Although it is reasonable to assume that virus particles morphologically resembling coronaviruses and/or toroviruses which have repeatedly been visualized by electron microscopy in feces of patients with gastroenteritis replicate in the gut, it has yet to be demonstrated unequivocally that they cause disease in the human enteric tract.

Clinical Features

The typical coronavirus cold is marked by nasal discharge and malaise; cough and sore throat are generally not as prominent as in rhinovirus colds, and there is little or no fever. The illness lasts about a week and is of no real consequence. The lower respiratory tract is not involved, but coronavirus infection may occasionally precipitate attacks of wheezing in asthmatic children, or exacerbate chronic bronchitis in adults.

Laboratory Diagnosis

Coronaviruses are difficult to grow in cultured cells and hence are rarely recovered from humans. HCV-OC43 and related strains were originally isolated in organ cultures of human embryonic trachea or nasal epithelium. Organ culture is too intricate a technique for a diagnostic laboratory, but some strains can be isolated directly in diploid fibroblast lines from human embryonic lung or intestine. Foci of "granular" cells become evident after a week and may progress to vacuolation before disintegrating; syncytia may form in some cell types. Hemadsorption and hemagglutination are demonstrable with strain OC43 only.

Though rarely called for, the diagnostic method of choice is direct demonstration of coronavirus antigens in nasopharyngeal aspirates or nasal swabs by EIA using appropriate monoclonal antibodies to differentiate strain 229E from OC43. EIA is also the most convenient method of screening for antibodies in serological surveys. Identification and characterization of enteric coronaviruses or toroviruses will not become practicable until a more reproducible method of culturing them is developed.

Epidemiology

Colds are readily transmissible to human volunteers by intranasal administration of human coronaviruses. After an incubation period of 2–5 days, symptoms develop in about half of those inoculated, and virus is shed for about a week.

Coronavirus colds occur mainly in the winter and early spring. A longitudinal study in the United States indicated that large outbreaks of either 229E or OC43 infection tend to occur with a periodicity of 2–4 years. Adults are affected as well as children because acquired immunity is so ephemeral. Overall, it is estimated that coronaviruses are responsible for about 15% of all colds.

The role, if any, of enteric coronaviruses and/or toroviruses in human gastroenteritis remains uncertain. Transmission is presumably by the fecal–oral route.

Further Reading

Brinton, M. A., and Heinz, F. X., eds. (1990). "New Aspects of Positive-Strand RNA Viruses." Plenum, New York.

Cavanagh, D., and Brown, T. D. K., eds. (1990). "Coronaviruses and Their Diseases." Plenum, New York.

Holmes, K. V. (1990). Coronaviridae and their replication. *In* "Fields Virology" (B. N. Fields, D. M. Knipe, R. M. Chanock, M. S. Hirsch, J. L. Melnick, T. P. Monath, and B. Roizman, eds.), 2nd Ed., p. 841. Raven, New York.

Koopmans, M., and Horzinek, M. C., (1993). Toroviruses of animals and humans: A review. *Adv. Virus Res.* **43,** 233.

Lai, M. M. C. (1990). Coronaviruses: Organization, replication and expression of the genome. *Annu. Rev. Microbiol.* **44,** 303.

Snijder, E. J., and Horzinek, M. C. (1993). Toroviruses: Replication, evolution and comparison with other members of the coronavirus-like superfamily. *J. Gen. Virol.* **74,** 2305.

Spaan, W., Cavanagh, D., and Horzinek, M. C. (1988). Coronaviruses: Structure and genome expression. *J. Gen. Virol.* **69,** 2939.

Wege, H., Siddell, S., and ter Meulen, V. (1982). The biology and pathogenesis of coronaviruses. *Curr. Top. Microbiol. Immunol.* **99,** 165.

Chapter 28

Paramyxoviridae

As the name implies, the paramyxoviruses were originally classified with the orthomyxoviruses under the name "myxoviruses," because of the shared properties of hemagglutination and neuraminidase activity exhibited by some members of the family. However, these two groups differ in a number of vital respects including the nature of the genome and the strategy of replication, in which the *Paramyxoviridae* resemble more closely the other families in the order *Mononegavirales*, namely, *Rhabdoviridae* and *Filoviridae*. The family *Paramyxoviridae* is in turn subdivided into two subfamilies, containing a total of four genera (Table 28-1). Though few in number, almost all the human paramyxoviruses are important causes of respiratory disease in children. The parainfluenza and respiratory syncytial viruses are responsible for nearly half of the croup, bronchiolitis, and pneumonia in infants. Measles and mumps used to be familiar to every mother but have now receded significantly in developed countries as a result of successful immunization campaigns.

Properties of *Paramyxoviridae*

The paramyxoviruses range from 150 to 300 nm or more in diameter and are enclosed by a lipid envelope which is extremely fragile, rendering the virion vulnerable to destruction by storage, freeze–thawing, or even preparation for electron microscopy. Accordingly, particles often appear distorted in electron micrographs and may rupture to reveal the internal nucleocapsid which occurs as a single helix up to 1 μm long (Fig. 28-1; Table 28-2).

In contrast to the genome of orthomyxoviruses, that of paramyxoviruses is not segmented. It consists of a single 15–16 kb molecule of single-stranded RNA of negative polarity. The genome comprises 6–10 genes separated by

Table 28-1
Taxonomy of Family *Paramyxoviridae*

Subfamily	Genus	Human species
Paramyxovirinae	*Paramyxovirus*	Human parainfluenza viruses types 1 and 3
	Rubulavirus	Human parainfluenza viruses types 2, 4a, and 4b, mumps virus
	Morbillivirus	Measles virus
Pneumovirinae	*Pneumovirus*	Human respiratory syncytial virus

conserved noncoding sequences that contain termination, polyadenylation, and initiation signals. The gene order is generally conserved within the family though viruses of the genera *Paramyxovirus* and *Morbillivirus* contain 6 genes, *Rubulavirus* 7, and *Pneumovirus* 10. Whereas the 10 *Pneumovirus* genes encode 10 proteins, the 6–7 genes of the other genera encode 10–12 proteins (see

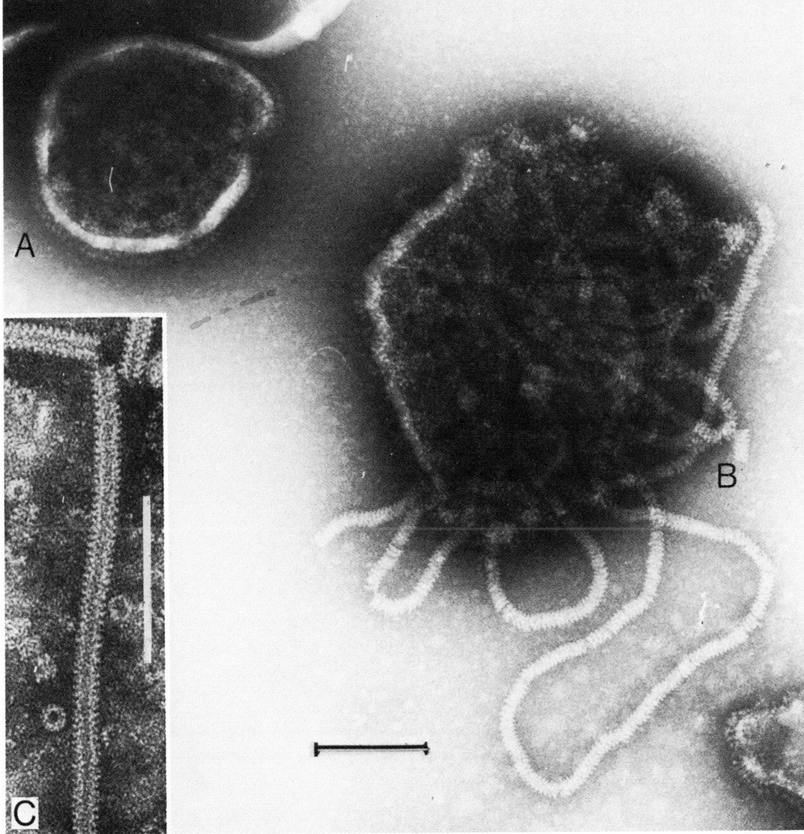

Fig. 28-1 *Paramyxoviridae.* Negatively stained virions of a paramyxovirus. (A) Intact virion with peplomers visible at lower edge. (B) Partially disrupted virion, showing nucleocapsid. (C) Enlargement of a portion of the nucleocapsid, in longitudinal and cross section. Bars, 100 nm. (Courtesy Dr. A. J. Gibbs.)

Table 28-2
Properties of *Paramyxoviridae*

Pleomorphic spherical virion, 150–300 nm, sometimes filamentous
Envelope containing two glycoproteins: F (fusion protein) and attachment protein HN (hemagglutinin–neuraminidase) or H or G; also nonglycosylated membrane protein, M
Helical nucleocapsid, 18 nm (*Paramyxovirinae*) or 13 nm diameter (*Pneumovirinae*), NP (or N) protein with associated transcriptase (L) and phosphoprotein (P)
Linear minus sense ssRNA genome, 15–16 kb, with 6–7 genes encoding 10–12 proteins (*Paramyxovirinae*) or 10 genes encoding 10 proteins (*Pneumovirinae*)
Cytoplasmic replication, with budding from plasma membrane
Syncytium formation, cytoplasmic inclusions (also nuclear with *Morbillivirus*)

Fig. 28-2 and below). Most of the gene products are structural proteins found in the virion itself; their functions are set out in Table 28-3.

Special mention should be made of the two major membrane glycoproteins because they play key roles in the pathogenesis of all paramyxovirus infections. Cell attachment is mediated via the glycoprotein variously known, for different genera, as H (for hemagglutinin), HN (carrying neuraminidase as well as hemagglutinating activity) or G (carrying neither activity, as a result of which hemadsorption cannot be used as a laboratory diagnostic tool for respiratory syncytial virus). This glycoprotein elicits neutralizing antibodies which inhibit adsorption of virus to cell receptors. The other major envelope glycoprotein is known as the fusion protein (F) because it enables the virus to fuse cells together to form the syncytia so characteristic of this family (see Chapter 5). To acquire biological activity the F protein precursor (F_0) must be cleaved by a cellular protease into two disulfide-linked polypeptides, F_1 and

Table 28-3
Functions and Terminology of Virion Proteins in Paramyxoviruses

Protein Function	Genus: *Paramyxovirus* *Rubulavirus*	Genus: *Morbillivirus*	Genus: *Pneumovirus*
Attachment protein: hemagglutinin, induction of immunity	HN[a]	H	G[b]
Fusion protein: virus penetration, cell–cell spread, induction of immunity	F	F	F
Nucleoprotein: protection of genome RNA	NP	N	N
Transcriptase: genome transcription[c]	L and P[d]	L and P	L and P
Matrix protein: virion core stability	M	M	M
Small membrane proteins	SH[e]	—	SH[f], M2[g]

[a] Also carries neuraminidase activity.
[b] No hemagglutinating activity.
[c] Associated with RNA polymerase, adenylyl transferase, and mRNA guanylyl- and methyltransferase activities.
[d] Phosphoprotein.
[e] *Rubulavirus* only, certain species only.
[f] Also known as 1A.
[g] Also known as 22K.

F_2; in certain types of nonpermissive host cells this fails to occur and the virions are noninfectious. The F protein is essential for viral penetration of the host cell by fusion of the viral envelope with the plasma membrane, and for direct intercellular spread by cell-to-cell fusion, as well as perhaps contributing to the maintenance of persistent infections (see Chapter 10). Paramyxovirus vaccines, to be effective, must elicit antibodies against the F protein as well as against the other glycoprotein.

All the paramyxoviruses are labile to the effects of heat or desiccation; hence, clinical specimens must be maintained at 4°C and inoculated into cell cultures without delay. They do not cause extensive cell destruction, but syncytium formation is a regular feature (see Fig. 5-2C), with acidophilic inclusions in the cytoplasm (and also in the nucleus in the case of *Morbillivirus*) (see Fig. 5-3C).

Viral Replication

Paramyxoviruses multiply entirely within the cytoplasm; indeed, some have been shown to replicate in enucleated cells. The infecting virion attaches to a sialoglycoprotein or glycolipid receptor via the envelope glycoprotein variously known as G, H, or HN. The F protein then mediates fusion of the viral envelope with the plasma membrane, at physiologic pH. The liberated nucleo-

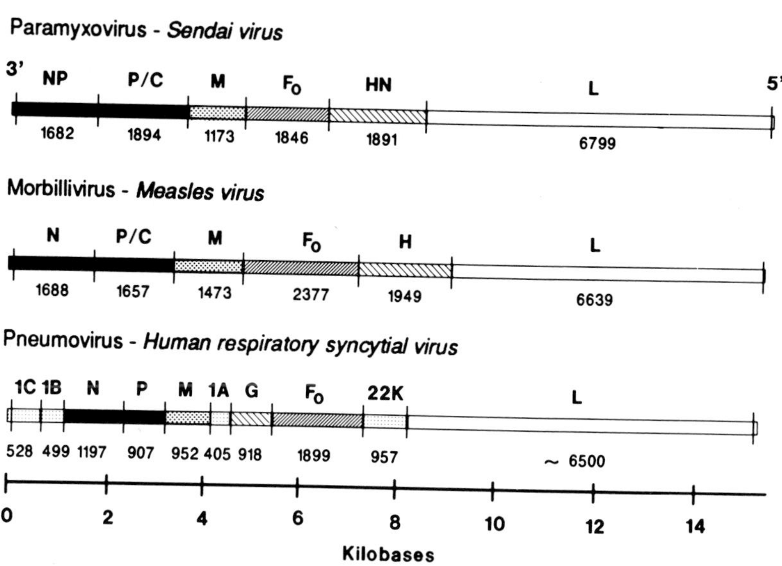

Fig. 28-2 Genetic map of a typical member of each of three genera of *Paramyxoviridae*. Since this diagram was drawn it has been proposed that the genus *Paramyxovirus* be divided into two genera, *Paramyxovirus* and *Rubulavirus* (not shown, but whose genome resembles that of *Paramyxovirus* except that it contains a seventh gene, SH, between the F and HN loci). Homologous genes are indicated by the style of shading; the number beneath each gene indicates the number of bases it contains. [From D. W. Kingsbury, *in* "Fields Virology" (B. N. Fields, D. M. Knipe, R. M. Chanock, M. S. Hirsch, J. L. Melnick, T. P. Monath, and B. Roizman, eds.), 2nd Ed., p. 948. Raven, New York, 1990.]

capsid remains intact, all three of the associated proteins (N, P, and L) being required for transcription by the viral polymerase. The minus sense genome is transcribed progressively into 6–10 subgenomic, generally monocistronic mRNAs.

Whereas most of the genes encode a single protein, the P gene encodes 2–5 distinct proteins in the case of the various genera of the subfamily *Paramyxovirinae.* It is extraordinary that this applies only to one particular gene, yet that different strategies for maximizing the coding potential of this gene have evolved in different genera. For example, the *Morbillivirus* and *Respirovirus* P/C/V gene encodes four proteins, the production of which entails two distinct mechanisms: internal initiation of translation, and insertion of nontemplated G residues into mRNA to shift the reading frame to that of an otherwise inaccessible open reading frame. Whereas the P protein itself is translated from a faithful mRNA copy of the complete gene, the smaller C protein is read in a different reading frame following initiation of translation from an internal initiation codon. On the other hand, the V (NS1) protein displays N-terminal homology with P but thereafter has a distinct amino acid sequence because an extra G residue is inserted into its mRNA during transcription by polymerase site-specific stuttering ("editing," see Chapter 3). Since this frame is distinct also from that used to translate protein C, all three possible reading frames are utilized to read that section of the P/C/V gene! In the case of parainfluenza virus type 3, a fourth protein, D, is translated by insertion of two nontemplated G residues.

The genus *Rubulavirus* was recently separated from the genus *Paramyxovirus*, largely because of a different P gene strategy. Faithful transcription of this gene yields mRNA that is translated into the V (NS1) protein, whereas insertion of two nontemplated G residues shifts the reading frame to produce protein P; with mumps virus insertion of four G residues produces yet another protein, I. Rubulaviruses also lack a C protein ORF, but some species have a seventh gene which encodes a small integral membrane protein, SH. Pneumoviruses have 10 genes, all of which encode a single protein, except for some overlap between M2 and L; the four additional genes not found in the paramyxoviruses and morbilliviruses are two small integral membrane proteins, SH and M2, plus two nonstructural proteins, NS1 and NS2.

As the concentration of N protein builds up in the cell, these molecules are believed to associate either with the transcriptase or with the nascent plus strands, preventing the polyadenylation, termination, and reinitiation events that characterize the "transcription mode" and enabling a switch to the "replication mode" in which full-length plus strands ("antigenomes") are made. These, in turn, serve as templates for RNA replication. Newly synthesized minus strands associate with N protein and transcriptase to produce nucleocapsids, which eventually associate with the M protein attached to those areas on the apical surface of the plasma membrane into which the HN (or H or G) and F glycoproteins have migrated. Enveloped virions are released by budding. The F protein, which must be proteolytically cleaved to render the virion infectious, causes infected cells to fuse with adjacent uninfected cells, thus enabling transfer of infectious nucleocapsids and evading circulating antibodies.

Measles

Until recently measles was perhaps the best known of all the common childhood diseases. The characteristic maculopapular rash, with coryza and conjunctivitis (Fig. 28-3), was familiar to every mother. However, as a consequence of the development of an effective live vaccine and vigorous implementation of a policy of immunization of every child, the United States and a few other countries have reduced the incidence of this disease so dramatically that many young doctors have never seen a case. The objective now must be to match this performance in the rest of the world, including the developing countries, where the mortality from measles in malnourished infants makes it one of the leading causes of death in children.

Pathogenesis and Immunity

Infection occurs via the respiratory tract. Virions enter the local lymphatics, either free or associated with macrophages, and are transported to the local lymph nodes (Fig. 28-4). Here the virus multiplies without producing much cell damage, and there is early spread to other lymph nodes and to the spleen. Infected mononuclear cells give rise to multinucleated giant cells, and T lymphocytes are susceptible to infection when they are mitotically active. About 6 days after the initial infection viremia occurs and virions are seeded to all the epithelial surfaces of the body: oropharynx, conjunctiva, skin, respiratory tract, bladder, and alimentary canal. Virus deposited in these sites from the local blood vessels produces foci which spread toward the epithelial surface. Because the epithelia of the conjunctiva and respiratory tract are only one or two cells deep, they undergo necrosis first, some 9–10 days after infection. There is an abrupt onset of illness with a cough, running nose, and

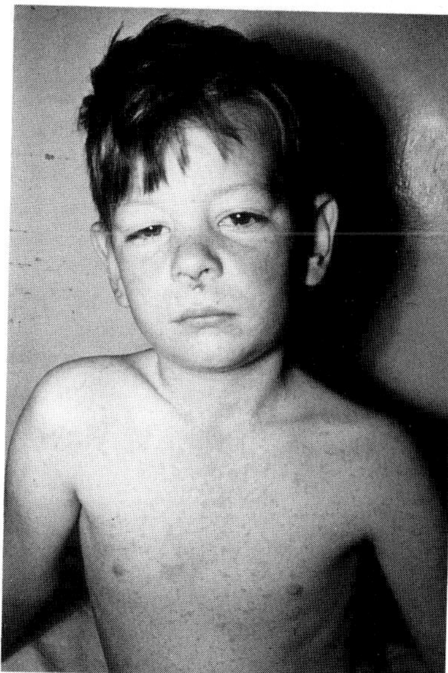

Fig. 28-3 Measles. Note maculopapular rash and conjunctivitis in young child with upper respiratory infection. (Courtesy Dr. J. Forbes.)

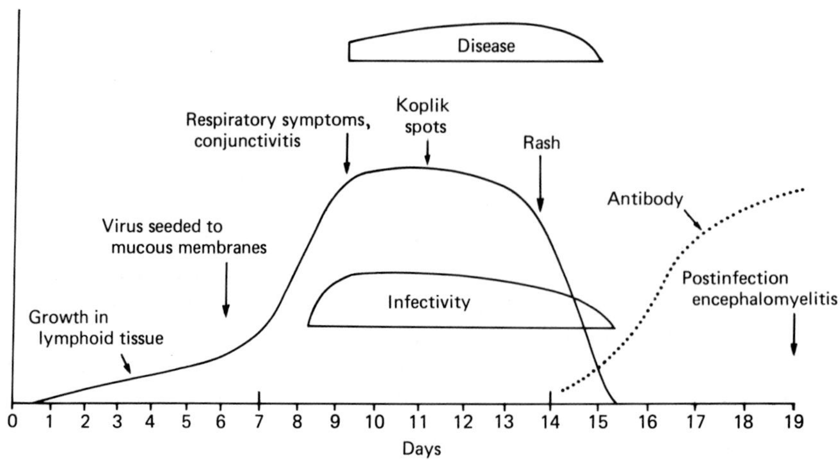

Fig. 28-4 Pathogenesis of measles. (From C. A. Mims and D. O. White, "Viral Pathogenesis and Immunology." Blackwell, Oxford, 1984.)

inflamed conjunctivae. It seems likely that immune responses contribute to the respiratory damage, malaise, and fever, which appear at this stage and get steadily worse until the rash appears. Mucosal foci ulcerate on about the eleventh day, to produce the characteristic Koplik's spots in the mouth. By the fourteenth day, just as circulating antibodies become detectable, the characteristic maculopapular rash appears and the fever falls. This skin rash is due in large part to cell-mediated immune responses to viral antigens (type IV hypersensitivity; see Chapter 9). Measles decreases the resistance of the respiratory epithelium to secondary bacterial infection, hence pneumonia, sinusitis, or otitis media may supervene. In immunocompromised patients, especially those with impaired cell-mediated immunity, giant cell pneumonia may occur, sometimes several months after the acute infection and often with fatal consequences. Measles also provides the classic example of increased severity of a disease due to the effects of malnutrition, as discussed in Chapter 7.

Sometimes central nervous system disease occurs, usually after infection in visceral organs has come to an end. Three syndromes have been recognized. Acute postinfectious measles encephalitis is rare in children less than 2 years of age but occurs in about 1 in 1000 cases of measles in older children, with a case–fatality rate of 15%, and is the principal reason for vaccination against measles in the First World. It develops during the first week after the onset of the rash, with a sudden onset and recurrence of fever. There is little or no production of virus in the brain; however, myelin basic protein is found in the cerebrospinal fluid (CSF), and patients' T lymphocytes are often reactive to myelin. The pathogenesis appears to involve autoimmune demyelination (see Chapter 9). In contrast, subacute measles encephalitis occurs only in immunocompromised children, usually within 6 months of the rash; it may be rapidly progressive and is attributable to failure to eliminate virus-infected cells because of the lack of cytotoxic T cells. Finally, subacute sclerosing panencephalitis (SSPE) occurs years after the acute disease; it too is always fatal, and is characterized by very slow replication and spread of measles virus in the brain. Although it is difficult to isolate measles virus from the brain of a patient with SSPE, some neurons contain very large accumulations of measles virus nucleocapsids; sequencing of the viral genome reveals numerous mutations in the M gene and to a lesser extent in other genes. The pathogenesis of SSPE was discussed in Chapter 10.

In the less developed countries of Africa and South America the case–fatality rate of measles is of the order of 3–6%, sometimes higher; this is several hundred times that in developed nations. Factors associated with the increased severity of measles in developing countries include young age at the time of infection, lower socioeconomic status, crowding, concomitant diarrhea, malnutrition (including vitamin A deficiency), lack of access to health care, and underlying immunodeficiency from a variety of causes. In addition to the direct mortality caused by measles, increased overall mortality rates are found in children who have had measles during the previous 9 months compared with children who have not.

Essentially all primary measles infections give rise to clinically manifest disease. The resulting immunity is effectively lifelong; second attacks of measles are virtually unknown, even in totally isolated island communities where subclinical boosts to immunity cannot have occurred for decades (see Chapter 14). As these data suggest, there is only one serotype of measles virus; although a small number of amino acid substitutions are demonstrable in the H protein over the years they do not appear to be sufficient to provide variants with significant selective advantage in a naturally infected population. Although adoptive transfer of neutralizing antibodies against either the H or F glycoprotein confers excellent short-term protection against challenge, T cells raised to the internal nucleoprotein (N) are also protective, presumably by expediting recovery.

Clinical Features

Following a prodrome marked by fever, cough, coryza, and conjunctivitis, an exanthem appears on the head and spreads progressively over the chest, trunk, then limbs. The rash consists of flat macules that fuse to form blotches rather larger than those of other viral exanthems. They are slow to fade and often leave the skin temporarily stained.

Common complications include otitis media, croup, bronchitis, and bronchopneumonia. Bacterial pneumonia is the usual cause of death when measles kills malnourished children. Immunologically deficient children can die from measles virus-induced giant-cell pneumonia or from acute progressive infectious encephalitis (measles inclusion body encephalitis) with no sign of a rash. However, the most dangerous complication of measles is acute postinfectious encephalitis, which occurs in about 1 in every 1000 cases and inflicts a mortality of about 15%, with permanent neurologic sequelae in many of the survivors. Subacute sclerosing panencephalitis (SSPE) is a very much rarer complication, developing in only about 1 in every 300,000 cases, some years after apparent recovery from the original infection.

Laboratory Diagnosis

The clinical diagnosis of measles is so straightforward that the laboratory is rarely called on for help. Cultivation of the virus is difficult and slow, hence is not usually attempted; if cultivation is successfully accomplished, multinucleated giant cells containing numerous acidophilic inclusions in cytoplasm and nuclei are diagnostic (see Fig. 5-3C). Measles antigen can be identified by immunofluorescence in cultured cells or, more simply and quickly, in cells aspirated directly from the nasopharynx. However, the best approach nowadays is IgM capture EIA using as antigen either sonicated virus or recombinant DNA cloned measles N protein. Serology is also employed to screen populations for their immune status. Hemagglutination inhibition (HI) was traditionally used for this purpose, but EIA is more convenient and more sensitive.

Epidemiology and Control

The epidemiology of measles prior to the introduction of vaccination in 1963 was discussed fully in Chapter 14, while in Chapter 7 we said something of the tragic situation in the developing world where over a million malnourished infants die annually from this readily preventable disease. In Chapter 15 we discussed prospects of eradication of measles by vaccination.

A live measles vaccine was developed originally by Enders then further attenuated to produce the Schwarz vaccine used today. In the developed world this vaccine should be administered subcutaneously to every child at about 15 months, after maternal antibody has completely disappeared. Seroconversion occurs in 95–98% of recipients at this age, compared with as few as 50% if vaccinated at 6 months. The antibody titers that result from vaccination, even at the optimal age, are anything up to 10-fold lower than following natural infection, but they do generally persist for many years at protective levels. Trivial side effects are not uncommon, particularly mild fever (in about 10%) and/or transient rash (5%).

By 1983 indigenous measles had all but disappeared from the United States, where immunization of children prior to entry into school or day-care centers is required by law, and the declared goal of measles eradication from that country seemed within reach. Since 1989, however, there has been a marked increase in the annual number of cases, particularly among unimmunized preschool infants of racial and ethnic minorities in inner city areas. Thus, it is important to redouble the effort to vaccinate all children, concentrating particularly on unimmunized immigrants, the poor in inner city ghettos, and if possible the conscientious objectors. In addition, however, several outbreaks have occurred in college students, most of whom were vaccinated in infancy. Serological surveys indicate that the percentage of people with protective levels of immunity falls progressively from 95–98% over the years following vaccination. Moreover, with relatively little virus circulating in the community to boost such waning immunity, there is a danger that nonimmune vaccinated or unvaccinated adults might first encounter the virus at an age when complications are more severe. Accordingly, it is now recommended that all children receive two doses of measles vaccine: the first at (12 to) 15 months (as combined measles–mumps–rubella vaccine, MMR) plus a booster just prior to entering either kindergarten/first grade (4–6 years of age) or junior high school (11–12 years).

In the developing world, where there is high mortality from measles in infants in the first year of life, immunization is a top priority of the World Health Organization (WHO), as part of the Expanded Immunization Programme. In these areas maternal immunity declines more rapidly than in developed countries, and infants become susceptible to measles, and to measles vaccination, by 6–9 months of age. The WHO recommends measles vaccination at 9 months or as soon as possible thereafter. Ideally, the optimum age for vaccination should be determined for each individual country, and a second dose should be given later in case of primary vaccination failure. Maintenance of the "cold chain" is also vital in the tropics. Today's vaccines are less heat-labile and are supplied freeze-dried, permitting storage at 4°C until reconstitution immediately before use.

In 1990 the WHO recommended the use in such high-risk areas as West Africa of very high dosages of a less attenuated live virus strain, the Edmonston–Zagreb vaccine, which induces immunity at 6 months of age even in the presence of residual maternal antibody. However, it was soon demonstrated statistically that vaccinees were more likely to die of infectious diseases other than measles than were those immunized at 9 months with the standard Schwarz vaccine. This paradox suggests that this particular strain, at such high dosage, suppresses T-cell immunity to a dangerous degree. Even more puzzling is the finding that girls are affected significantly more often than boys. In 1992 the WHO acted promptly to suspend the use of the high-titer Edmonston–Zagreb vaccine.

Passive immunization still has a place in protecting unvaccinated children, particularly immunocompromised children, following exposure to measles. If administered promptly, pooled normal human immunoglobulin will abort the disease; if given a few days later, the disease may still be modified. No antiviral agent is effective against measles, but bacterial superinfections, such as pneumonia or otitis media, require vigorous chemotherapy.

Mumps

Clinical Features

The comical spectacle of the unhappy young man with face distorted by painful edematous enlargement of parotid and other salivary glands, unable to eat or talk without discomfort, is familiar music-hall fare, often spiced with the innuendo that the case is complicated by a well-deserved orchitis! Epididymoorchitis is indeed a painful development which occurs in 25% of all mumps cases in postpubertal males and may lead to atrophy of the affected testicle. A wide variety of other glands may be involved, including the pancreas (quite commonly), ovary, thyroid, and breast (more rarely). Benign meningeal signs are detectable in at least 10% of all cases of mumps, and clinical meningitis, sometimes presenting without parotid involvement, occurs often enough to make mumps the most common single cause of this disease. Fortunately, the prognosis is very much better than with bacterial meningitis, and sequelae are rare. Mumps encephalitis, on the other hand, though much less frequent, is a more serious development. Unilateral nerve deafness is an uncommon but important long-term consequence. In infants, mumps infections are often symptomless or present as respiratory infections.

Laboratory Diagnosis

The classic case of mumps (Fig. 28-5) can be identified without help from the laboratory, but atypical cases and meningoencephalitis present a diagnostic problem. Virus can be isolated from saliva (or from swabs taken from the orifice of Stensen's duct), from urine (mumps virus being one of the few viruses readily isolated from the latter source), or from CSF in patients with meningitis. Primary cultures of primate kidney cells have been largely replaced by continuous human cell lines such as H292, derived originally from a human lung mucoepidermoid carcinoma. The readout is hemadsorption, and immunofluorescence confirms the diagnosis. However, mumps is more commonly diagnosed by serology. IgM capture EIA allows rapid diagnosis of mumps meningoencephalitis, while EIA can also be used to screen for IgG to monitor immune status in vaccination studies.

Epidemiology

Mumps is transmissible by direct contact with saliva or by droplet spread, from a few days before the onset of symptoms until about a week after. Mumps does not show the dramatic periodicity of the other paramyxoviruses but tends to cause sporadic cases throughout all seasons, with winter–spring epidemics every few years. A dramatic decline in the incidence of mumps has been recorded in the United States since the introduction of the vaccine (see Fig. 15-1).

Control

The live attenuated vaccine, derived by passage in chick fibroblasts, may be used alone in adolescent males, for example, military recruits, but in many countries including the United States is administered to 15 month old infants in the form of a combined measles/mumps/rubella (MMR) vaccine. Protective levels of antibody against mumps are conferred by a single subcutaneous injection in at least 90% of recipients and persist for at least 20 years. The vaccine is relatively free of side effects, although occasional mild allergic reactions occur.

In the United States, there have been indications recently that the 98% reduction in the incidence of mumps since the introduction of the vaccine is not being consistently maintained and that cases are now occurring in unvaccinated adolescents and young adults, in whom the risk of complications is higher. As with measles, this points up the importance of attaining universal coverage whenever a policy of widespread immunization of infants is adopted, and reinforces the desirability of a booster dose of MMR vaccine on entering high school or college.

Fig. 28-5 Mumps. Note swelling of parotid and other salivary glands. (Courtesy Dr. J. Forbes.)

Parainfluenza Viruses

Parainfluenza viruses are common human respiratory pathogens. In the main they produce relatively harmless upper respiratory tract infections (URTI), but they are also the commonest cause of a more serious condition in young children known as "croup" and occasionally cause pneumonia. Human parainfluenza virus types 1 and 3 belong to the genus *Paramyxovirus,* whereas types 2, 4a, and 4b are now classified with mumps virus in the genus *Rubulavirus.*

Clinical Features

Primary infection, typically in a young child, generally manifests itself as coryza and pharyngitis, often with some degree of bronchitis and low fever. However, there are two more serious presentations which are seen in 2–3% of infections (Table 28-4). In an infant, especially under the age of 6 months, parainfluenzavirus type 3 may cause bronchiolitis and/or pneumonia clinically indistinguishable from that more commonly caused by the respiratory syncytial virus (see below). In somewhat older children (6 months to 5 years) parainfluenzavirus type 1 and, to a lesser extent, type 2 are the major cause of croup (laryngotracheobronchitis). The child presents with fever, cough, stridor, and respiratory distress which may occasionally progress to laryngeal obstruction requiring intubation or tracheotomy.

Laboratory Diagnosis

Traditionally, parainfluenza viruses have been isolated in primary human or monkey kidney cells; since these cells have become no longer generally ob-

Table 28-4
Clinical and Epidemiologic Features of Parainfluenza and Respiratory Syncytial Virus Infections

Virus	Major syndrome	Age	Epidemiology
Parainfluenza 1 and 2	Croup	First 5 years	Autumn epidemic
Parainfluenza 3	Bronchiolitis, pneumonia	First year	Endemic
Parainfluenza 4	Upper respiratory infection	Children	Endemic
Respiratory syncytial virus	Bronchiolitis, pneumonia	First year	Winter epidemic

tainable, they have been replaced by continuous cell lines such as H292. Trypsin is added to the maintenance medium to ensure cleavage of the viral F protein. Parainfluenza viruses multiply rather slowly and cause little cytopathic effect (CPE), except in the case of type 2 which induces syncytia. Viral growth is detected by hemadsorption of guinea pig red cells (see Fig. 5-2D). Differentiation from other hemadsorbing respiratory viruses may then be made by fluorescent antibody staining of the infected monolayer or by hemagglutination inhibition using virus from the cell culture supernatant.

Today, however, the pendulum has swung toward more rapid diagnostic techniques, notably immunofluorescence (IF) and EIA on nasopharyngeal aspirates. Immunofluorescence is used to demonstrate antigen in exfoliated cells, whereas EIA or RIA is sufficiently sensitive to detect free antigen in mucus suitably solubilized to liberate intracellular as well as extracellular protein.

Epidemiology

Like other respiratory agents, parainfluenza viruses are spread by droplets and by contact with respiratory secretions. The incubation period ranges from 2 to 6 days, and shedding continues for about a week. Parainfluenza viruses are highly transmissible, infecting most children by the age of 5 years. Indeed, type 3 infects the majority of infants within the first year or two of life and can spread within hospitals and babies' homes, causing cases of pneumonia and bronchiolitis. Types 1 and 2 tend to cause croup, whereas type 4 produces only trivial illness. Reinfections with any given type of parainfluenza virus commonly occur, although clinical disease is generally mild and confined to the upper respiratory tract the second time round.

Control

Several types of experimental parainfluenza type 3 vaccines have shown some degree of protection in rodent or primate models. However, formidable problems militate against parainfluenza vaccines achieving wide acceptability. First, the immunity that follows natural infection with parainfluenza viruses is poor, and it is unrealistic to expect any vaccine to do more than limit the challenge virus to the upper respiratory tract. Second, in order to protect infants during the first months of life when parainfluenza 3 can be a threat, any vaccine would need to be administered shortly after birth, when immune responses are weakest and maternal antibodies are present. Third, in light of the worrying experience with earlier inactivated respiratory syncytial virus and measles vaccines, which actually potentiated the disease occurring on subsequent challenge, subunit vaccines would also need to undergo very careful clinical trials before licensing.

Respiratory Syncytial Virus

Respiratory syncytial virus (RSV) is the most important respiratory pathogen of childhood, being responsible for about half of all cases of bronchiolitis and a quarter of all cases of pneumonia during the first few months of life.

Pathogenesis and Immunity

The virus multiplies in the mucous membranes of the nose and throat; in the very young and very old it may involve the trachea, bronchi, bronchioles, and alveoli. The incubation period is 4–5 days. Fatal cases usually show extensive bronchiolitis and pneumonitis with scattered areas of atelectasis and emphysema resulting from bronchiolar obstruction (Figs. 28-6, 36-1, and 36-2).

A challenging unanswered question is why severe lower respiratory disease develops only in certain very young infants. Undoubtedly the airways of such young babies, being narrower than those of older children, are much more readily obstructed by inflammation, edema, and shedding of necrotic cells into copious mucus. However, this does not explain why only a minority of these small infants develop bronchiolitis. There is evidence that the condition may have an immunologic basis.

Many years ago, an experimental formalin-inactivated RSV vaccine was tested in children. When the immunized children encountered RSV during a subsequent epidemic, they actually suffered significantly more serious lower respiratory disease than did unimmunized controls. It was subsequently demonstrated that, although the formalin-inactivated virus induced a good antibody response, the antibodies did not have good neutralizing activity. This alarming occurrence led not only to the abandonment of the killed vaccine but also to speculation about the immunologic basis of the episode and its possible relationship to natural RSV bronchiolitis. The first hypothesis was that maternal IgG, present only in young infants, might react with virus multiplying in the lung to produce a hypersensitivity reaction of the Arthus type. However plausible, this idea had to be abandoned when a negative correlation was demonstrated between the titer of maternal antibody and the severity of the RSV-induced illness. Anti-RSV IgE as well as histamine and leukotrienes have been reported in respiratory secretions of infants with RSV bronchiolitis. There is also evidence in rodent models that cytotoxic T cells, while expediting recovery by eliminating infected epithelial cells, may nevertheless exacerbate the symptomatology by augmenting the inflammatory response. The last word has yet to be spoken on this important but complex question.

Immunity acquired as a result of RSV infection is notoriously poor. For example, during the annual winter epidemics that plague babies' homes, the majority of children become reinfected; the severity of the illness is generally,

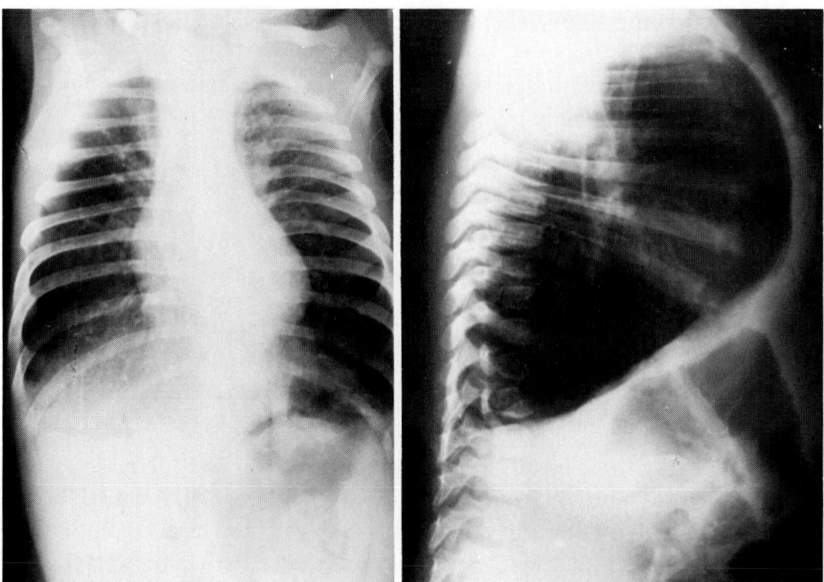

Fig. 28-6 Radiographs of lungs of a baby with respiratory syncytial viral bronchiolitis and pneumonitis. Note grossly overinflated lung fields with depression of the diaphragm and bulging of the anterior chest wall in the lateral view. (Courtesy Dr. H. Williams and Dr. P. Phelan.)

but not always, diminished the second time. This poor immunity is partially explicable by the fact that the immune response of infants to the protective F and G glycoproteins is very weak. However, studies in adult volunteers showed that they, too, became reinfectable experimentally within a year or so following natural infection with a similar strain of virus. Cotton rats can be protected against RSV pneumonia by immunization with F or G protein, or with high-titer neutralizing monoclonal antibodies raised against either. The F protein generally induces the higher antibody titers and confers a degree of heterotypic as well as homotypic immunity; the less immunogenic G protein confers protection only against strains within the same group or subgroup.

None of this answers the question of whether naturally acquired immunity, such as it is, is principally attributable to antibody and, if so, whether serum IgG or secretory IgA is paramount. It is possible that IgA protects (poorly) against infection of the upper respiratory tract but that IgG protects the lungs rather more effectively. Clearly, maternal antibody fails to protect infants beyond the first 6 weeks or so of life, even though it is generally transferred at quite high levels. A role for cell-mediated immunity, presumably Tc cells, is indicated by the finding that immunization with a vaccinia recombinant bearing the RSV internal N nucleoprotein is also protective, not by preventing infection, but by expediting recovery. Children with congenital T-cell immunodeficiencies have been shown to excrete RSV from their lungs for months following infection. Furthermore, RSV is a much less effective inducer of interferon synthesis in normal infants than are influenza and parainfluenza viruses.

Clinical Features

The commonest manifestation of RSV infection in all age groups is a febrile rhinitis and/or pharyngitis with limited involvement of bronchi. However, the consequences may be much more serious in certain babies between the second and sixth months of life. Almost 1% of all babies develop an RSV infection severe enough to require admission to hospital, and of these about 1% die, particularly those with congenital heart defects, bronchopulmonary dysplasia, very low birth weight, or immunodeficiency. Characteristically, an infant with rhinorrhea develops a pronounced cough and wheezing, progressing to dyspnea with a markedly elevated respiratory rate and hypoxemia (Fig. 28-7). Death may occur very rapidly and may account for a proportion of cases of the so-called sudden infant death syndrome (SIDS). RSV infections in children also frequently involve the middle ear, making this virus the most important causal agent of otitis media. Of the children who recover from a severe pulmonary infection with RSV, some retain evidence of impaired lung function for years and in particular are predisposed to recurrent bouts of wheezing (asthma).

In older children and adults, RSV infections are reinfections against a background of partial immunity. The disease resembles a cold, with or without cough and fever. However, in the elderly, during winter epidemics of RSV, and in immunosuppressed transplant patients, severe pneumonia can occur.

Although all strains of RSV are considered to belong to a single species, they may be divided into two major groups by genomic and antigenic analysis. The F protein is serologically cross-reactive, but the G protein is group-specific; strains within each group can be differentiated using selected anti-G monoclonal antibodies. Group A strains are somewhat commoner than B and may be more often associated with severe disease.

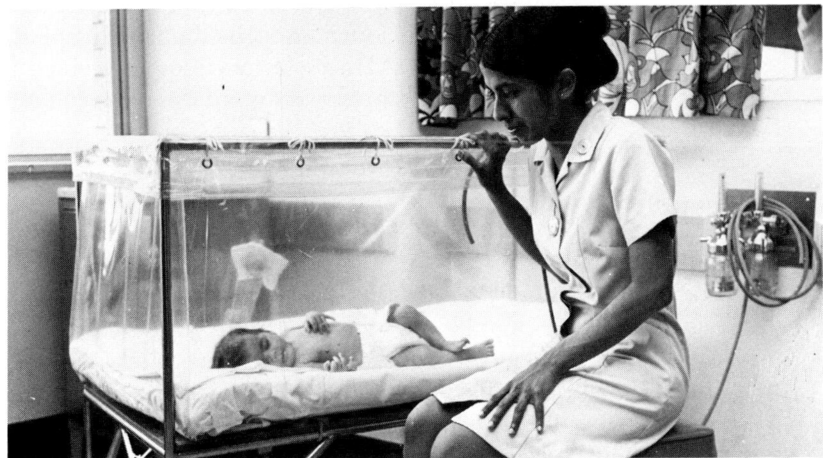

Fig. 28-7 Respiratory syncytial viral bronchiolitis. The baby is maintained in an oxygen tent. (Courtesy Dr. H. Williams and Dr. P. Phelan.)

Laboratory Diagnosis

Three diagnostic methods of approximately equal sensitivity are currently favored by different laboratories: (1) isolation of the virus in cell culture, (2) immunofluorescence on exfoliated cells, and (3) enzyme immunoassay on antigen from nasopharyngeal mucus.

Virus may be recovered from a nasopharyngeal aspirate (see Fig. 12-1E) or nasal wash by inoculation of cultured cells. The extreme lability of RSV makes it mandatory that the specimen be taken early in the illness and that it be added to cultured cells without delay and without preliminary freezing. Human heteroploid cell lines such as HeLa or HEp-2 are the most sensitive, though human embryonic lung fibroblasts may also be used. Up to 10 days may elapse before the characteristic syncytia become obvious (see Fig. 5-2C), although a trained eye can usually detect early CPE by about the third to fifth day. Fixation and staining generally reveal extensive syncytia containing acidophilic cytoplasmic inclusions, but some strains produce only rounded cells. Absence of hemadsorption distinguishes RSV from all the other paramyxoviruses. Definitive identification can be established by IF as soon as early CPE first becomes apparent.

Although only one species of human RSV is currently recognized, strains can be allocated into groups A and B, then into various subgroups by EIA using monoclonal antibodies to the G protein. Monoclonal antibody can also be used for IF on exfoliated cells aspirated from the nose and/or oropharynx. This method has the advantage of speed and, in addition, can produce a positive result even if the specimen is taken too late to expect viable virus still to be available for culture, or if the virus is neutralized by bound IgA. However, the preferred option today is detection of detergent-solubilized antigen by EIA, which is as sensitive and as specific as either of the foregoing methods. Using appropriate monoclonal antibodies as capture and detector reagents, EIA can detect as little as 10–50 ng of antigen and lends itself to automation in large diagnostic or reference laboratories.

For epidemiologic surveys, serum antibodies can be assayed most readily by EIA using recombinant DNA-cloned antigens, but serology is not customarily used for diagnosis because (1) satisfactory antibody responses do not regularly occur and (2) venipuncture of a tiny infant is not to be undertaken lightly.

Epidemiology

Respiratory syncytial virus is highly contagious, being shed in respiratory secretions for several days, sometimes weeks, and conveyed by contact to a generally susceptible population, including those with prior experience of the virus but with a negligible degree of acquired immunity. Not surprisingly, therefore, RSV causes a sharply defined epidemic every winter (see Fig. 14-1). Most children become infected in their first year or two, then reinfections occur repeatedly throughout life.

During a given epidemic, one or more strains from both group A and group B may cocirculate, even in the same city. This resembles the pattern observed with influenza type B but differs from that of influenza A, in which one variant (or a small number of variants) with the particular constellation of mutations best equipping it to escape herd immunity dominates the scene to the exclusion of all others. The epidemiologic, immunologic, and clinical significance of these RSV variants remains to be established.

Nosocomial infections are frequent. Outbreaks occur in neonatal wards of maternity hospitals, sometimes inflicting high mortality. Moreover, hospital staff and parents of babies with RSV bronchiolitis often develop febrile colds and/or pharyngitis, and staff are largely responsible for spreading the virus within the ward via aerosol, fomites, and direct contact.

Control

Improvements in intensive care facilities have led to a marked reduction in the mortality from RSV pneumonia. Ribavirin (see Chapter 16) reduces the severity and duration of the illness as well as the titer of virus in the lungs. A nebulizer is used to generate a small-particle aerosol into the oxygen tent, hood, ventilator, or mask for 12–18 hours per day for 3–6 days. Since the magnitude of the benefit is still in dispute and the cost of the treatment is very high, it is currently advocated only for infants who are severely ill, very young, or otherwise considered to be at high risk.

The success of neutralizing anti-G or anti-F antibodies in preventing or even treating experimental RSV infections in cotton rats encouraged clinical trials of purified, high-titer human IgG administered intravenously for the treatment of RSV bronchiolitis/pneumonia in infants. Some benefit was demonstrated and the trials have been extended to prophylaxis of RSV in high-risk infants during winter epidemics. Recently, human monoclonal Fab fragments against the F protein, isolated from a combinatorial antibody library expressed on the surface of bacteriophage, have been shown to neutralize RSV. Whereas Fab fragments may suffice for delivery by aerosol, intact human monoclonal antibodies will be required for parenteral administration to humans.

Nosocomial transmission via medical and nursing staff must be minimized by giving proper attention to such matters as hand washing, wearing gowns, gloves, and perhaps masks when undertaking high-risk duties, as well as segregation of RSV-infected patients, and temporary redeployment of staff with respiratory infections to avoid transmission of RSV to high-risk infants.

Following the disappointment with the inactivated vaccine in the late 1960s, workers at the U.S. National Institutes of Health have been struggling valiantly for over 20 years to develop a genetically stable, cold-adapted, and temperature-sensitive RSV strain for use as a live vaccine for intranasal administration. Others are working on vaccines comprising just the F protein, either purified from lysed virions by affinity chromatography or genetically engineered using a baculovirus expression system. In light of the paradoxical experience with formalin-inactivated vaccine, however, the testing of any RSV vaccine will have to be approached with special caution. Moreover, the principal target group for a safe RSV vaccine would be very young, especially sickly infants in babies' homes and hospitals. A satisfactory immune response to a vaccine at this very early age, in the presence of maternal antibody, may be exceedingly difficult to achieve.

Further Reading

Anderson, L. J., Hendry, R. M., Pierik, L. T., Tsou, C., and McIntosh, K. (1991). Multicenter study of strains of respiratory syncytial virus. *J. Infect. Dis.* **163,** 687.

Atkinson, W. L., Orenstein, W. A., and Krugman, S. (1992). The resurgence of measles in the United Stats, 1989–1990. *Annu. Rev. Med.* **43,** 451.

Bishop, D. H. L., and Compans, R. W., eds. (1984). "Non-Segmented Negative Strand Viruses: Paramyxoviruses and Rhabdoviruses." Academic Press, New York.

Black, F. L. (1989). Measles active and passive immunity in a worldwide perspective. *Prog. Med. Virol.* **36,** 1.

Centers for Disease Control. (1989). Mumps prevention. *Morb. Mortal. Wkly. Rep.* **38,** 388.

Chanock, R. M., and McIntosh, K. M. (1990). Parainfluenza viruses. *In* "Fields Virology" (B. N. Fields, D. M. Knipe, R. M. Chanock, M. S. Hirsch, J. L. Melnick, T. P. Monath, and B. Roizman, eds.), 2nd Ed., p. 963. Raven, New York.

Galinski, M. S. (1991). Paramyxoviridae: Transcription and replication. *Adv. Virus Res.* **39,** 129.

Gindler, J. S., Atkinson, W. L., and Markowitz, L. E. (1992). Update—The United States measles epidemic, 1989–1990. *Epidemiol. Rev.* **14,** 270.

Kingsbury, D. W. (1990). Paramyxoviridae and their replication. *In* "Fields Virology" (B. N. Fields, D. M. Knipe, R. M. Chanock, M. S. Hirsch, J. L. Melnick, T. P. Monath, and B. Roizman, eds.), 2nd Ed., p. 945. Raven, New York.

Kingsbury, D. W., ed. (1991). "The Paramyxoviruses." Plenum, New York.

Kolakofsky, D., and Mahy, B. W. J., eds. (1989). "Genetics and Pathogenicity of Negative Strand Viruses." Elsevier, Amsterdam.

McIntosh, K. M., and Chanock, R. M. (1990). Respiratory syncytial viruses. *In* "Fields Virology" (B. N. Fields, D. M. Knipe, R. M. Chanock, M. S. Hirsch, J. L. Melnick, T. P. Monath, and B. Roizman, eds.), 2nd Ed., p. 1045. Raven, New York.

Mahy, B. W. J., and Kolakofsky, D., eds. (1987). "The Biology of Negative Strand Viruses." Elsevier, Amsterdam.

Murphy, B. R., Prince, G. A., Collins, P. L., van Wyke Coelingh, K., Olmsted, R. A., Spriggs, M. K., Parrott, R. H., Kim, H.-W., Brandt, C. D., and Chanock, R. M. (1988). Current approaches to the development of vaccines effective against parainfluenza and respiratory syncytial viruses. *Virus Res.* **11,** 1.

Norrby, E., and Oxman, M. N. (1990). Measles virus. *In* "Fields Virology" (B. N. Fields, D. M. Knipe, R. M. Chanock, M. S. Hirsch, J. L. Melnick, T. P. Monath, and B. Roizman, eds.), 2nd Ed., p. 1013. Raven, New York.

Pringle, C. R. (1987). Paramyxoviruses and disease. *Soc. Gen. Microbiol. Symp.* **40,** 51.

Stott, E. J., and Taylor, G. (1985). Respiratory syncytial virus: Brief review. *Arch. Virol.* **84,** 1.

Wolinsky, J. S., and Washam, M. N. (1990). Mumps virus. *In* "Fields Virology" (B. N. Fields, D. M. Knipe, R. M. Chanock, M. S. Hirsch, J. L. Melnick, T. P. Monath, and B. Roizman, eds.) 2nd Ed., p. 989. Raven, New York.

Chapter 29

Rhabdoviridae

The family *Rhabdoviridae* encompasses more than 150 viruses of vertebrates, invertebrates (mostly arthropods), and plants, the virions of each having a distinctive bullet-shaped morphology. The only important human pathogen is rabies virus; vesicular stomatitis, which sometimes causes severe epizootics in horses, is an occasional, mild zoonotic disease.

Rabies virus is the cause of one of the oldest and most feared diseases of man and animals and was recognized in Egypt before 2300 B.C. and in ancient Greece, where it was well described by Aristotle. The most lethal of all infectious diseases, rabies also has the distinction of having stimulated one of the great early discoveries in biomedical research. In 1885, before the nature of viruses was comprehended, Louis Pasteur developed, tested, and applied a rabies vaccine, thereby opening the modern era of infectious disease prevention by vaccination.

Properties of *Rhabdoviridae*

Rhabdoviruses are approximately 70 nm wide and 170 nm long, and consist of a lipid envelope with glycoprotein peplomers surrounding a helically wound nucleocapsid, which gives the viruses their distinctive bullet-shaped or conical morphology (Fig. 29-1). The viruses contain a single linear molecule of minus sense ssRNA, 11–12 kb in size. Rhabdovirus virions contain five major proteins which, for the genus *Lyssavirus*, are designated as follows: L [transcriptase, with 5′-cap methylase, 3′-poly(A) polymerase, and protein kinase activities]; G (glycoprotein peplomer with hemagglutinin activity, target of neutralizing antibodies); N (nucleoprotein); NS, or P, or M1 (phosphoprotein, binds to L and promoter); and M, or M2 (matrix protein).

Table 29-1
Properties of Rhabdoviruses

Two genera include human pathogens: *Lyssavirus* (rabies virus) and *Vesiculovirus* (vesicular stomatitis virus)
Bullet-shaped enveloped virion, 170 × 70 nm, with glycoprotein peplomers, matrix protein under lipoprotein envelope
Nucleocapsid with helical symmetry
Linear minus sense ssRNA genome, 11–12 kb
Cytoplasmic replication; viral transcriptase transcribes five monocistronic mRNAs which are translated into five proteins: transcriptase (L+P), nucleoprotein (N), matrix protein (M), glycoprotein peplomer (G), and phosphoprotein (P or NS)
Maturation by budding through plasma membrane
Vesicular stomatitis causes rapid cytopathology; rabies virus is noncytopathogenic

Classification

Two genera, *Vesiculovirus* and *Lyssavirus,* have been defined among the rhabdoviruses of animals; within each genus, species are distinguished by neutralization tests, which recognize epitopes on the G glycoprotein. The genus *Vesiculovirus* includes some 35 serologically distinct viruses, only one of which causes human infection. The genus *Lyssavirus* comprises rabies virus and three rabies-like viruses from Africa: Mokola, Lagos bat, and Duvenhage viruses. Each of these viruses is capable of causing rabies-like disease in animals and humans.

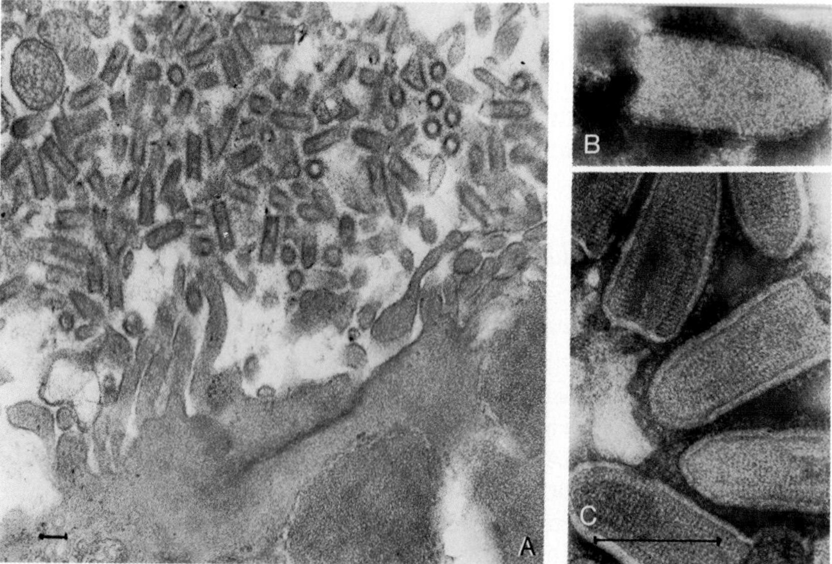

Fig. 29-1 *Rhabdoviridae.* (A) Section of fox salivary gland infected with rabies virus. (B, C) Negatively stained preparations of (B) vesicular stomatitis virus and (C) rabies virus. Bars, 100 nm. (Courtesy Dr. F. A. Murphy.)

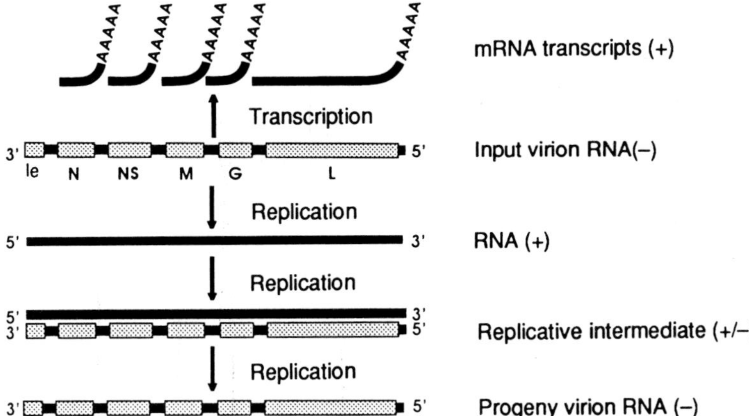

Fig. 29-2 Genome structure of vesicular stomatitis virus and its mode of replication. Wide bars indicate genes and their relative sizes; narrow bars indicate noncoding nucleotide sequences; the long narrow bar indicates (+) strand complementary RNA. The N protein–RNA core plus the NS and L proteins comprise the transcription complex. Polyadenylation, signaled by the conserved sequence AUACUUUUUUU at the end of each gene, occurs by a "stuttering" mechanism. RNA replication occurs through a replicative intermediate (+ / −) dsRNA. le, leader; N, nucleoprotein; P (NS), nonstructural protein; M, matrix protein; G, peplomer glycoprotein; L, with P, comprises RNA polymerase.

Viral Replication

Laboratory-adapted ("fixed") and wild-type ("street") rabies virus, and vesicular stomatitis virus, replicate to high titer in the brains of suckling mice and in many kinds of cell cultures. Most of our knowledge of rhabdovirus replication comes from studies of vesicular stomatitis virus (Fig. 29-2).

Virions bind to receptors via glycoprotein G, enter the cell by endocytosis, and are uncoated, releasing the nucleocapsid into the cytoplasm where all subsequent events occur. Primary transcription by the viral transcriptase complex (L + P) produces five monocistronic mRNA species in the order N, P (NS), M, G, and L. Each of these is 5′ capped and 3′ polyadenylated by the multifunctional enzyme encoded by the L protein. Because there is only a single promoter it is postulated that polyadenylation occurs via polymerase slippage at each intergenic stretch of seven U residues and that, with an efficiency of less than 100%, the transcriptase complex then moves on to the next open reading frame. This is consistent with the observation that the five genes are transcribed in decreasing molar abundance from N at the 3′ end through to L at the 5′ end of the genome. Transition to the replication mode requires synthesis of a protein, thought to be N protein, which somehow enables the polymerase to read through all the intergenic transcription termination, polyadenylation, and capping signals to produce a full-length authentic complementary copy of genomic RNA. Both plus and minus strands associate with nucleoprotein, but only minus strands associate with M and trigger budding from those areas of plasma membrane that contain glycoprotein G.

Replication of vesicular stomatitis virus usually causes rapid cytopathology (thought to be attributable to inhibition of cellular mRNA transcription by protein M), but the replication of rabies virus is usually noncytopathic. Defective interfering (DI) virus particles are commonly formed during rhabdovirus replication. These are shorter and have a smaller RNA molecule than normal infectious particles, with complex deletion mutations in their genome (see Chapter 4).

Rabies

Rabies virus can infect all warm-blooded animals, and in nearly all instances the infection ends in death. Rabies occurs throughout the world, with the exception of Australia, Japan, Great Britain, and many smaller islands such as Hawaii and most of the islands of the Caribbean basin. Dog rabies is still important in many parts of the world; virus in the saliva of infected dogs causes most of the estimated 75,000 human rabies cases that occur each year worldwide. In many countries of Europe, and in the United States and Canada, wildlife rabies has become of increasing importance as a threat to humans.

Pathogenesis and Immunity

Infection by the bite of a rabid animal usually results in deposition of rabies-infected saliva deep in the striated muscles, but rabies can also occur, albeit with less certainty, after superficial abrasion of the skin. Initially, virus replicates in the muscle cells or cells of the subepithelial tissues until it has reached a sufficient concentration to infect motor and sensory nerves in the muscle or skin by binding specifically to the acetylcholine receptor or other receptors and entering nerve endings. Neuronal infection and centripetal passive movement of the viral genome within axons delivers virus to the central nervous system, usually via the spinal cord initially. An ascending wave of neuronal infection and neuronal dysfunction then occurs. Virus reaches the limbic system, where it replicates extensively, and the release of cortical control of behavior leads to "furious" rabies. Spread within the central nervous system continues, and when replication occurs in the neocortex the clinical picture changes to "dumb" rabies. Depression, coma, and death from respiratory arrest follow.

In the many species of animals that transmit rabies by biting, virus moves centrifugally from the central nervous system down peripheral nerves to a variety of organs: adrenal cortex, pancreas, and most importantly the salivary glands. In the nervous system most virus is assembled on intracytoplasmic membranes; the cells are not lysed, so that little viral antigen is released to stimulate host defense mechanisms. In the salivary gland, however, virions bud apically from plasma membranes at the luminal surface of mucous cells and are released in high concentrations into the saliva. Thus, at the time when viral replication within the central nervous system causes the infected animal to become furious and to bite indiscriminately, the saliva is highly infectious.

On histopathologic examination there is little evidence of brain damage, yet electron microscopic or fluorescent antibody studies show that almost all neurons are infected. There is minimal cellular destruction to match the extensive neurologic dysfunction seen in the disease.

Although rabies proteins are highly immunogenic, neither humoral nor cell-mediated responses can be detected during the stage of movement of virus from the site of the bite to the central nervous system, probably because very little antigen is delivered to the immune system (most is sequestered in muscle cells or within nerve axons). However, this early stage of infection is accessible to antibody, hence the efficacy of the classic Pasteurian postinfection vaccination, especially if combined with the administration of hyperimmune immunoglobulin. Immunologic intervention is effective during the long incubation period because of the delay between the initial viral replication in muscle cells and the entry of virus into the protected environment of the nervous system.

Clinical Features

Following the bite of a rabid animal the incubation period is usually between 14 and 90 days, but may be considerably longer. Cases have been observed in rabies-free countries, like Australia, in which the last opportunity of infection occurred up to 6 years earlier. After a prodromal phase of fever, malaise, and often paresthesia around the site of the bite, muscles become hypertonic and the patient becomes anxious, with episodes of hyperactivity, aggression, and convulsions. Paralysis is often a major feature. Delirium, coma, and death follow.

Laboratory Diagnosis

It is important that the laboratory diagnosis of rabies in animals be undertaken in approved laboratories by qualified, experienced personnel. The most common request is to determine whether an animal known to have bitten a human is rabid. If clinical observation by a veterinarian suggests rabies, the suspect animal must be killed and brain tissue collected for testing by direct immunofluorescence to demonstrate rabies antigen in touch impressions of brain tissue (medulla, cerebellum, and hippocampus; see Fig. 12-6). If necessary, postmortem diagnosis can also be performed using the polymerase chain reaction (PCR) with primers that amplify both genomic RNA and viral mRNA sequences from the brain. For antemortem diagnosis in humans and animals, immunofluorescence or PCR assays on skin biopsy, corneal impression, or saliva specimens can be used. Paraesthesia at the site of a known animal bite is a strong indication for testing humans for rabies infection, but by the time signs of encephalitis appear it is too late to expect treatment to be effective. Only positive results are of diagnostic value, since the lack of sensitivity of these procedures does not exclude an infection. Virus can be isolated in a high-security laboratory by intracerebral inoculation of suckling mice, or in a neuroblastoma cell line, with confirmation by immunofluorescence.

Epidemiology, Prevention, and Control

Rabies virus is not stable in the environment and in usual circumstances is only a risk when transmitted by the bite or scratch of a rabid animal, although in bat caves, where the amounts of virus may be very high, it can be transmitted via aerosol. Human to human transmission (via saliva?) has been reported only very rarely and has never been proved, except iatrogenically, for example, via corneal transplantation from donors dying of undiagnosed rabies.

The control of rabies in different countries of the world poses very different problems, depending on whether they are free of the disease, whether they are industrialized or developing countries, and whether vampire bat rabies is a problem.

Rabies-Free Countries

Rigidly enforced quarantine of all dogs and cats for 6 months before importation has been effectively used to exclude rabies from Australia, Japan, New Zealand, Hawaii, and several other islands. Rabies did not become endemic in wildlife in the United Kingdom and was eradicated from dogs in that country in 1902 and again in 1922, after its reestablishment in the dog population in 1918.

Developing Countries

In most countries of Asia, Latin America, and Africa, enzootic dog rabies is a serious problem, marked by significant domestic animal and human mortality. In these countries, large numbers of doses of human vaccines are used, and there is a need for comprehensive, professionally organized, and publicly supported agencies active in the following areas: (1) stray dog and cat elimination, and control of the movement of pets (quarantine may be called for in emergencies); (2) immunization of dogs and cats, so as to break the chains of virus transmission; (3) laboratory diagnosis, to confirm clinical observations and obtain accurate incidence data; (4) surveillance, to measure the effectiveness of all control measures; and (5) public education programs to ensure cooperation.

Industrialized Countries

Fox rabies is enzootic in several countries of Western Europe, in the Appalachian Mountain regions of the United States, in Ontario, and in polar areas inhabited by the arctic fox. Skunk rabies is common in central North America, from Texas to Saskatchewan, where it is the principal cause of rabies in cattle. Racoon rabies in the United States began a gradual northern movement from Florida in the 1950s, following the importation of raccoons for sporting purposes, causing an explosive epidemic in Virginia, Maryland, Pennsylvania, and the District of Columbia in the 1980s, and in New Jersey, New York, and Connecticut in the 1990s. Historically, rabies control in wildlife has been based on animal population reduction by trapping and poisoning, but in the past few years, fox immunization, by the distribution of baits containing an attenuated live-virus rabies vaccine, appears to have been highly successful in reducing transmission in Switzerland and Germany. The question of whether immunization of other wildlife species will be useful, especially in more complex ecosystems, will depend on (1) the population density of the target species, (2) further research on the safety and efficacy of orally ingested wildlife vaccines, (3) delivery systems appropriate for each reservoir host species, and (4) solution of legal and jurisdictional problems. Many of these problems are now being solved and comprehensive field studies are in progress, including trials with a vaccinia virus–rabies glycoprotein recombinant.

Latin America

In several countries of Latin America vampire bat rabies is a problem to both humans and livestock industries. Here control efforts have depended on the use of bovine vaccines and more recently on the use of anticoagulants such as diphenadione and warfarin. When vampire bats feed on the blood of treated cattle, they suffer fatal hemorrhages in their wing capillaries.

Vaccination

Rabies is the only human disease that can be prevented by active immunization after infection ("postexposure" vaccination), because the infecting event

Table 29-2
Rabies: Guide for Human Postexposure Prophylaxis[a]

Animal species	Condition of animal at time of attack	Treatment of exposed person[b]
Domestic animals		
Dog, cat	Healthy and available for 10 days of observation	None, unless animal develops signs of rabies[c]
	Rabid or suspected rabid	Immediate rabies immune globulin[d] and vaccine[e]
	Unknown (escaped)	Consult public health official; if treatment is indicated, give rabies immune globulin and vaccine
Wild animals		
Skunk, bat, fox, coyote, raccoon, bobcat, woodchuck, other carnivores	Regard as rabid unless proved negative by laboratory tests[f] or from geographic area known to be rabies-free	Rabies immune globulin[d] and vaccine[e]
Other		
Livestock, rodents, rabbits, hares	Consider individually; public health officials should be consulted about the need for rabies prophylaxis; bites of squirrels, hamsters, guinea pigs, gerbils, chipmunks, rats, mice, other rodents, rabbits, and hares almost never call for antirabies prophylaxis	

[a] In applying these recommendations, take into account the animal species involved, the circumstances of the bite or exposure, the vaccination status of the animal, and presence of rabies in the region. Public health officials should be consulted if questions arise about the need for rabies prophylaxis.
[b] All bites and wounds should immediately be thoroughly cleansed with soap and water. If antirabies treatment is indicated, both rabies immune globulin and vaccine should be given as soon as possible, regardless of the interval from exposure.
[c] If during the 10-day observation period a dog or cat should exhibit clinical signs of rabies, it should be immediately killed and tested, and treatment of the exposed individual with serum and vaccine should be started.
[d] If rabies immune globulin is not available, use equine antirabies serum. Do not use more than the recommended dosage. Anticipate possible need to treat for serum sickness.
[e] Five 1-ml intramuscular doses to be given on days 0, 3, 7, 14, and 28. The WHO recommends an optimal sixth dose at 90 days. Local reactions to vaccines are common and do not contraindicate continuing treatment. Discontinue vaccine if fluorescent antibody tests of the animal are negative.
[f] The animal should be killed and tested as soon as possible; holding for observation is not recommended.

is recognizable and the incubation period is long. In addition, "preexposure" vaccination of occupationally at-risk humans is regularly and successfully practiced.

Rabies vaccines have come a long way since the days of Pasteur. Early vaccines were made from infected brain material and were associated with serious side effects, but excellent cell culture vaccines are now available. Today's vaccines are grown from attenuated virus in human diploid fibroblasts or Vero cells, then inactivated with β-propiolactone or split into subunits with tri-*n*-butyl phosphate. Moreover, poxvirus–rabies glycoprotein recombinant vaccines are undergoing clinical trials in humans. In addition to active immunization, persons at risk should be given rabies immune globulin intramuscularly and around the bite. Remarkably, prompt administration of rabies immune globulin (after thorough cleansing of the wound) and commencement of a full course of vaccine reduce the mortality from this frightening disease from virtually 100% to zero. A comprehensive guide on human postexposure treatment is provided by the U.S. Centers for Disease Control (Table 29-2).

Individuals occupationally or otherwise at risk of rabies should be prophylactically immunized. These include laboratory personnel working with rabies virus, veterinarians, animal control and wildlife workers in rabies-enzootic areas, and certain travelers visiting such areas. Preexposure immunization consists of three doses of modern cell-culture vaccine, 1.0 ml intramuscularly, one each on days 0, 7, and 28, with a booster (or serological confirmation of adequate antibody level) every 2 years.

Vesicular Stomatitis

Vesicular stomatitis virus is zoonotic, being transmissible to humans (typically, farmers and veterinarians) from vesicular fluids and tissues of infected animals, but there are no practical measures of preventing occupational exposure. The disease in humans resembles influenza, presenting with an acute onset of fever, chills, and muscle pain. It resolves without complications within 7–10 days. Human cases are not uncommon during epizootics in cattle and horses, but because of lack of awareness few cases are reported. Human cases can be diagnosed retrospectively by serological methods.

Further Reading

Advisory Committee on Immunization Practices, Centers for Disease Control. (1991). Rabies prevention—United States, 1991. *Morb. Mortal. Wkly. Rep.* **40**(RR-3), 1.

Baer, G. M., ed. (1991). "The Natural History of Rabies," 2nd Ed., Vols. 1 and 2. CRC Press, Boca Raton, Florida.

Baer, G. M., Birdbord, K., Hui, F. W., Shope, R. E., and Wunner, W. H., eds. (1988). Research toward Rabies Prevention. *Rev. Infect. Dis.* **10**(Suppl. 4).

Baer, G. M., Bellini, W. J., and Fishbein, D. B. (1990). Rhabdoviruses. *In* "Fields Virology" (B. N. Fields, D. M. Knipe, R. M. Chanock, M. S. Hirsch, J. L. Melnick, T. P. Monath, and B. Roizman, eds.), 2nd Ed., p. 883. Raven, New York.

Banerjee, A. K., and Barik, S. (1992). Gene expression of vesicular stomatitis virus genome. *Virology* **188,** 417.

Bishop, D. H. L., ed. (1980). "Rhabdoviruses," Vols. 1–3. CRC Press, Boca Raton, Florida.

Calisher, C. H., Karabatsos, N., Zeller, H., Digoutte, J.-P., Tesh, R. B., Shope, R. E., Travassos da Rosa, A. P. A., and St. George, T. D. (1989). Antigenic relationships among rhabdoviruses from vertebrates and hematophagous arthropods. *Intervirology* **30,** 241.

Reif, J. S. (1994). Vesicular stomatitis. *In* "Handbook of Zoonoses, 2nd Ed., Section B, Viral" (G. W. Beran, ed.), p. 179. CRC Press, Boca Raton, Florida.

Smith, J. S. (1989). Rabies virus epitopic variation: Use in ecologic studies. *Adv. Virus Res.* **36,** 215.

Smith, J. S., Orciari, L. A., Yager, P. A., Seidel, H. D., and Warner, C. K. (1992). Epidemiologic and historical relationships among 88 rabies virus isolates as determined by limited sequence analysis. *J. Infect. Dis.* **166,** 296.

Tsiang, H. (1993). Pathophysiology of rabies virus infection of the nervous system. *Adv. Virus Res.* **42,** 375.

Wagner, R. R. (1990). Rhabdoviridae and their replication. *In* "Fields Virology" (B. N. Fields, D. M. Knipe, R. M. Chanock, M. S. Hirsch, J. L. Melnick, T. P. Monath, and B. Roizman, eds.), 2nd Ed., p. 867. Raven, New York.

World Health Organization. (1984). WHO Expert Committee on Rabies. Seventh report. *WHO Tech. Rep. Ser. 709.* World Health Organization, Geneva.

Chapter 30

Filoviridae

In 1967 31 cases of hemorrhagic fever, with 7 deaths, occurred among laboratory workers in Marburg, Germany, and Belgrade, Yugoslavia, who were processing kidneys from African green monkeys (*Cercopithecus aethiops*) that had been imported from Uganda. A bizarre new virus with very long, filamentous virions was isolated from both patients and monkeys; it was named Marburg virus.

Nine years later two further extraordinary epidemics of hemorrhagic fever occurred, one in villages in the rainforests of Zaire and then in the local hospital, and the other in a cotton factory and then in the local hospital in southern Sudan, 600–700 km away. Altogether there were more than 550 cases and 430 deaths. A virus morphologically identical to but antigenically distinct from Marburg virus was isolated from patients in each location; it was named Ebola virus. The viruses from Zaire and Sudan were slightly different and are designated Ebola-Z and Ebola-S, respectively. Since then, sporadic human cases of hemorrhagic fever due to Marburg virus have been recognized in southern Africa and sporadic infections with Ebola virus in Sudan, Zaire, and Kenya.

In 1989 and 1990 several shipments of monkeys imported from the Philippines into the United States were found to have been infected with a filovirus morphologically identical and serologically related to Ebola virus. Infected monkeys at a holding facility at Reston, Virginia (hence the name Ebola–Reston virus), became ill and some died. There were no human cases of disease, but 14% of persons having close occupational contact with the monkeys tested positive for filovirus antibodies. Serological examination of several thousand cynomolgus monkeys imported into the United States in 1989, from the Philippines and Indonesia, showed that 11.7% had antibody to the Reston virus.

Properties of *Filoviridae*

Virions of members of the family *Filoviridae* (*filo*, threadlike) are very long filamentous rods or more compact convoluted forms, each composed of a lipid bilayer envelope covered with peplomers surrounding a helically wound nucleocapsid (Fig. 30-1). The virions are 80 nm in diameter and have a unit nucleocapsid length of 800–1000 nm, but particles as long as 14,000 nm have been seen. The genome is a single molecule of minus sense ssRNA, 19 kb in size, with complementary end sequences. Virions contain seven proteins; the nucleocapsid contains proteins L, NP, VP35, and VP30 (see Fig. 30-2).

Filoviruses replicate well in Vero (African green monkey) cells, as well as infecting guinea pigs, hamsters, and monkeys. Viral replication in the cytoplasm of host cells is marked by the formation of large inclusion bodies, and maturation occurs via budding from the plasma membrane. More detailed studies of the relationships between different filoviruses have been handicapped by the difficulty in developing a virus neutralization test. However, Marburg and Ebola viruses are distinguishable by small differences in genome size and protein profiles, as well as by the absence of antigenic cross-reactivity as determined by antibody-binding assays. There are also consistent differences in tryptic peptide and oligonucleotide fingerprints of the Sudan and Zaire strains of Ebola virus, but strains isolated in different places and in different years in each country are similar. The Ebola–Reston virus, which was initially identified with polyclonal Ebola antisera, reacts with most but not all Ebola virus monoclonal antibodies, but it differs in its pathogenicity for humans.

The sensitivity of filoviruses to lipid solvents and inactivating agents re-

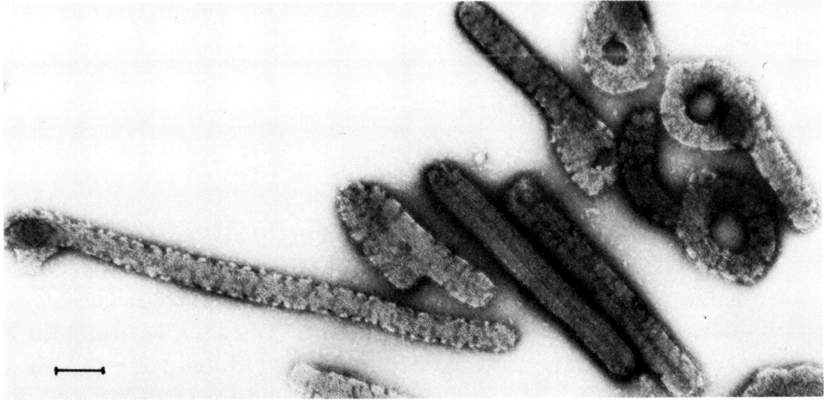

Fig. 30-1 *Filoviridae*. Negatively stained preparation of virions of Ebola virus. Bar, 100 nm. (Courtesy Dr. F. A. Murphy.)

Fig. 30-2 Gene organization of the single-stranded negative sense Marburg virus RNA (19 kb). Genes are indicated as stippled boxes, noncoding regions as open areas, and conserved intergenic sequences as heavy vertical lines. The arrow indicates the position of an mRNA overlap between VP30 and VP24. [Modified from H. Feldmann, E. Muhlberger, A. Randolph, C. Will, M. P. Kiley, A. Sanchez, and H.-D. Klenk, *Virus Res.* **24,** 1 (1992).]

sembles that of other enveloped RNA viruses. They retain infectivity at room temperature for several days.

Viral Replication

Virions appear to enter cells by endocytosis, and replication occurs in the cytoplasm, in which nucleocapsids accumulate to form prominent inclusion bodies. Maturation occurs by budding through plasma membranes. The minus sense genome contains seven open reading frames (Fig. 30-2), which code for the seven known structural proteins. At the gene boundaries there are conserved transcriptional stop and start signals and a highly conserved intergenic sequence of five nucleotides.

Pathogenesis

Of all the hemorrhagic fevers, Marburg, Ebola-Z, and Ebola-S have the highest case–fatality rates, the most severe hemorrhagic manifestations, and the most pronounced liver necrosis. The pathophysiologic changes are still obscure. There is an early and profound leukopenia, followed by a dramatic neutrophilia with a shift to the left, and very little monocyte infiltration in sites of parenchymal necrosis in the liver, but no unequivocal evidence of disseminated intravascular clotting. Antigen is localized in the liver, spleen, kidney, and adrenal glands, where virus particles can also be seen by electron microscopy.

Clinical Features

Infections of humans with Marburg and Ebola viruses cause similar syndromes, with severe hemorrhages, vomiting, abdominal pain, myalgia, pharyngitis, conjunctivitis, and proteinuria. The onset is sudden, and progress to prostration, profound hypotension, and death is rapid. The mortality rate in the known outbreaks has been very high; 25% with Marburg virus, and 60% in the Sudan outbreak and 90% in the Zaire outbreak of Ebola virus infection. However, although several infections of laboratory workers and animal handlers with the Reston virus were recognized serologically, they were not associated with human disease.

Laboratory Diagnosis

The filoviruses are classified as Biosafety Level 4 pathogens and must be handled only in maximum security laboratories. Filoviruses can be isolated from the blood during the febrile phase, and most strains grow well in Vero cells. The key to identification is the distinctive morphology as seen by electron microscopy. Direct immunofluorescence can be used to identify viral antigen in cultured cells, and indirect immunofluorescence is used for serological surveys.

Epidemiology

Marburg, Ebola, and Reston viruses are transmissible to humans from primates. In all outbreaks, secondary spread between humans appears to have been principally due to contact with body fluids from an acute case, although respiratory infection may also occur. In the major African outbreaks spread was largely within hospitals, owing to the reuse of blood-contaminated syringes and/or needles. Serological surveys of people in several parts of tropical Africa have yielded a few positive results in several areas where no cases of illness have been recognized, suggesting that in such environments milder human cases may occur.

Serological surveys suggest that the Reston virus is widespread among wild monkeys of several species in the Philippines, Thailand, and Indonesia; Ebola-Z, Ebola-S, and Marburg viruses occur among several species of African monkeys. Experience in the United States showed that Reston virus caused severe disease in cynomolgus monkeys and spread readily between them and to humans, apparently by the aerosol route. It is not known whether monkeys constitute the principal reservoir hosts for any of the filoviruses or whether they are merely serving as amplifying hosts.

Prevention

Wild-caught monkeys are no longer widely used for vaccine production, partly because of export prohibitions established by source countries for conservation purposes, and most importing countries operate import quarantine procedures. In 1990, following the discovery of the Reston virus in monkeys imported from Asia, additional local, national, and international primate transport and import restrictions were imposed, and protocols to prevent filovirus infection in workers in primate facilities were improved.

If filoviruses are suspected as the cause of disease in humans or in monkeys, consultation and laboratory diagnosis are available from the maximum containment laboratories that operate in several countries, such as the Centers for Disease Control in the United States. Specimen transport must be arranged in conformance with national and international regulations. If human cases are diagnosed, they should be treated in isolation wards, with barrier nursing and careful attention to prevent nosocomial or respiratory spread.

Further Reading

Jahrling, P. B., Kiley, M. P., Klenk, H.-D., Peters, C. J., Sanchez, A., and Swanepoel, R. (1994). Family *Filoviridae*. *In* "Virus Taxonomy. The Classification and Nomenclature of Viruses. Sixth Report of the International Committee on Taxonomy of Viruses." (F. A. Murphy, C. M. Fauquet, D. H. L. Bishop, S. A. Ghabrial, A. W. Jarvis, G. P. Martelli, M. A. Mayo, and M. D. Summers, eds.) Springer-Verlag, Wien, New York, in press.

Johnson, K. M. (1990). Marburg and Ebola viruses. *In* "Principles and Practice of Infectious Diseases" (G. L. Mandell, ed.), 3rd Ed., p. 1303. Churchill-Livingstone, New York.

McCormick, J. B., and Fisher-Hoch, S. (1990). Viral hemorrhagic fevers. *In* "Tropical and Geographic Medicine" (K. S. Warren and A. A. F. Mahmoud, eds.), 2nd Ed., p. 700. McGraw-Hill, New York.

Murphy, F. A., Kiley, M. P., and Fisher-Hoch, S. P. (1990). Filoviridae. Marburg and Ebola viruses. *In* "Fields Virology" (B. N. Fields, D. M. Knipe, R. M. Chanock, M. S. Hirsch, J. L. Melnick, T. P. Monath, and B. Roizman, eds.), 2nd Ed., p. 933. Raven, New York.

Peters, C. J., Johnson, E. D., and McKee, K. T., Jr. (1991). Filoviruses and management of viral hemorrhagic fevers. *In* "Textbook of Human Virology" (R. B. Belshe, ed.), 2nd Ed., p. 699. Mosby, St. Louis, Missouri.

Peters, C. J., Johnson, E. D., Jahrling, P. B., Ksiazek, T. G., Rollin, P. E., White, J., Hall, W., Trotter, R., and Jaax, N. (1993). Filoviruses. *In* "Emerging Viruses" (S. S. Morse, ed.), p. 159. Oxford University Press, New York.

Chapter 31

Orthomyxoviridae

Few viruses have played a more central role in the historical development of virology than that of influenza. The pandemic that swept the world in 1918, just as the First World War ended, killed 20 million people—more than the war itself. The eventual isolation of the virus in ferrets in 1933 was a milestone in the development of virology as a laboratory science. During the ensuing two decades Burnet pioneered technological and conceptual approaches to the study of the virus in embryonated eggs. His system became the accepted laboratory model for the investigation of viral multiplication and genetic interactions until the early 1950s, when newly discovered cell culture techniques transferred the advantage to poliovirus. Hemagglutination, discovered accidentally by Hirst when he tore a blood vessel while harvesting influenza-infected chick allantoic fluid, provided a simple assay method, subsequently extended to many other viruses. The imaginative investigations of Webster and Laver into the continuing evolution of influenza virus by antigenic drift and shift established the discipline of molecular epidemiology. The pièce de résistance was the description by Wiley, Wilson, and Skehel of the location of the antigenic sites on a three-dimensional model of the influenza HA molecule derived by X-ray crystallography (see Fig. 4-3).

Properties of *Orthomyxoviridae*

Typical virions of influenza virus are spherical and about 100 nm in diameter (Fig. 31-1), but larger pleomorphic and filamentous forms are commonly seen, especially on initial isolation. The nucleocapsid of helical symmetry occurs in eight segments, each forming a loop at one end, composed of nucleoprotein

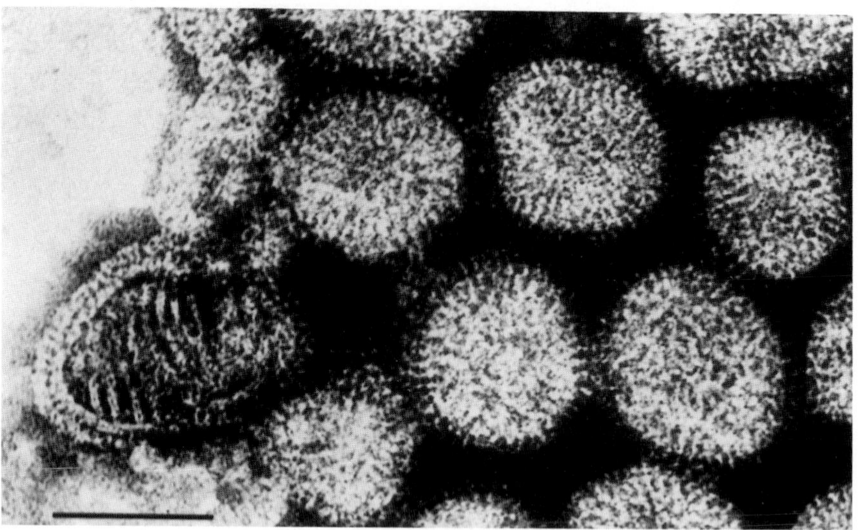

Fig. 31-1 *Orthomyxoviridae*. Negatively stained preparation of virions of influenza A virus (bar, 100 nm). (Courtesy Dr. R. Rott.)

(NP) enclosing a segmented RNA genome associated with an RNA polymerase (P) complex (PA, PB1, PB2). The envelope is lined on the inside by matrix protein (M1) and is spanned by a small number of ion channels composed of tetramers of protein M2. There are two kinds of peplomers: rod-shaped hemagglutinin (HA or H) molecules, which are homotrimers of a class I membrane glycoprotein (amino-terminal signal sequence, carboxyl-terminal hydrophobic transmembrane anchor), and mushroom-shaped neuraminidase (NA or N) molecules, which are tetramers of a class II membrane protein (N-terminal membrane anchor) (see Fig. 1-2C,D).

The minus sense ssRNA genome of influenza A and B viruses occurs as eight separate molecules (Table 31-1). Genetic reassortment can occur in cells infected with two different strains of influenza A virus, acquisition of the gene for a novel hemagglutinin producing a new human subtype (antigenic shift). Mutations in these genes cause antigenic drift. Both of these processes occur in nature and generate the diversity which is responsible for the occurrence of epidemics (see Chapter 4).

Influenza viruses are sensitive to heat (56°C, 30 minutes), acid (pH 3), and lipid solvents, and are thus very labile under ordinary environmental conditions.

Classification

Influenza A and Influenza B Viruses

Influenza A virus (usually known as type A, though more properly a species, since it shares no antigens with influenza B) is divided into subtypes, all of which share related nucleoprotein and matrix proteins but differ in their hemagglutinin and/or neuraminidase. So far, 14 subtypes of HA (H1–H14) and 9 of NA (N1–N9) have been described in birds, some of which have been found in various combinations in mammals including humans. Strains arising naturally by antigenic drift are codified as follows: type (A or B), animal species (omitted if human), place of isolation, number of the isolate, year of first isolation, followed in parentheses by the H and N subtypes. For example, the three strains chosen for incorporation in the Australian vaccine for the 1993 winter were A/Texas/36/91 (H1N1), A/Shanghai/24/90 (H3N2), and B/Panama/45/90.

Influenza C Virus

Influenza C virus comprises a distinct genus. Its genome contains only seven RNA segments. A single envelope glycoprotein, known as HEF, functions as hemagglutinin, fusion protein, and an esterase which has been postulated to cleave the cellular receptor (9-*O*-acetyl-*N*-acetylneuraminic acid) to facilitate entry/uncoating. Influenza C virus shares no antigens with influenza A or B. It infects children quite commonly but does not cause significant disease. A similar virus has been isolated from swine in China.

Tick-borne Orthomyxoviruses

A surprising recent discovery is that a little-known group of tick-borne arboviruses infecting humans and livestock in Africa, Europe, and Asia is genetically related to influenza virus. The group has been allocated to an unnamed genus within the family *Orthomyxoviridae.* The minus sense RNA genome comprises only six or seven segments (for Thogoto and Dhori viruses, respectively), total 10 kb. The acariviruses display no antigenic relationship to influenza virus and are not known to cause disease in humans.

Viral Replication

Influenza virus HA attaches to sialic acid on glycoprotein or glycolipid receptors, and the virion is taken into the cell by endocytosis. At pH 5.5 within endosomes, a conformational change in the HA exposes the hydrophobic amino terminus of HA_2, precipitating fusion of the viral envelope with endosomal membrane. At the same time, protons pass into the virion through the ion channels formed by the M2 protein, dissociating M1 from the ribonucleo-

Table 31-1
Properties of Orthomyxoviruses

Three genera: *Influenzavirus A,B; Influenzavirus C;* unnamed Thogoto-like viruses
Pleomorphic spherical or filamentous virion, diameter 80–120 nm
Envelope containing hemagglutinin (H) and neuraminidase (N) peplomers, a matrix protein (M1) on the inner surface, and a small number of pores composed of protein M2
Nucleocapsid of helical symmetry in 8 segments, composed of nucleoprotein (NP) and RNA polymerase complex (PA, PB1, PB2) in association with genome
Genome linear minus sense ssRNA in 8 segments, total size 13.6 kb (for influenza A and B); 7 segments (influenza C); 6 or 7 segments, 10 kb (tick-borne orthomyxoviruses)
Transcription and RNA replication in the nucleus; capped 5′ termini of cellular RNAs cannibalized as primers for mRNA transcription; budding from plasma membrane
Defective interfering particles and genetic reassortment frequently occur

protein and releasing the nucleocapsid into the cytosol. The nucleocapsid–polymerase complex is transported into the nucleus where transcription and replication of RNA take place.

Primary transcription involves a unique phenomenon known as cap snatching: the viral endonuclease (PB2) cleaves the 5′ methylguanosine cap plus about 10–13 nucleotides from heterogeneous nuclear RNA, and this is used as a primer for transcription by the viral transcriptase (PB1). Of the primary RNA transcripts so produced from the eight minus sense gene segments, six are monocistronic mRNAs which are translated directly into the proteins HA, NA, NP, PA, PB1, and PB2. The other two primary RNA transcripts undergo splicing, each yielding two mRNAs which are translated mainly in different reaching frames to produce pairs of proteins (M1 plus M2, and NS1 plus NS2) sharing only a few amino-terminal amino acids in common.

Influenza B virus adopts another strategy to increase the coding capacity of a short genome, namely, alternative translation start sites: the bicistronic mRNA transcribed from genome segment 6 contains two nearby AUG initiation codons from which two completely different proteins, NA and NB, with distinct amino acid sequences are translated in different reading frames. Segment 7 encodes two proteins, M1 and BM2, the translation of which seems to be coupled because the BM2 initiation codon overlaps the M1 termination codon.

Replication of the eight viral RNA segments requires the synthesis of eight full-length complementary plus strands which, unlike the corresponding mRNA transcripts, must lack the 5′ capped primer and the 3′ poly(A) tract. Newly synthesized nucleoprotein (NP) binds to this cRNA, serving as the scaffold on which the template is copied. Later, the matrix protein M1, which contains the zinc-finger motif characterizing proteins that bind nucleic acid, enters the nucleus and binds to the nucleoprotein–transcriptase–minus sense RNA complex, down-regulating transcription and permitting its export from the nucleus prior to assembly into virions.

The HA and NA proteins undergo glycosylation (see Fig. 3-7), polymerization, and acylation. In permissive cells HA is also subjected to cleavage into polypeptides HA_1 and HA_2 by cellular protease(s) such as furin which recognizes a polybasic motif of arginine residues (see Chapter 7). M2 is believed to be instrumental in protecting the native HA against conformational change in the trans-Golgi vesicles by countering acidification by cellular proton pumps. HA, NA, and M2 migrate preferentially to plasma membrane on the apical surface of the cell. M1 recognizes and binds to the cytoplasmic moiety of the HA trimer and precipitates budding (see Fig. 3-9). It is not known by what mechanism one copy of each the RNA segments is selected for incorporation into a virion. One possibility is that the eight nucleocapsid fragments are in fact loosely linked in a fixed sequence by some complex system of RNA/protein recognition signals. Alternatively, the segments may be packaged at random, with the average virion enveloping substantially more than eight segments to ensure that a reasonable proportion of particles happen to contain at least one copy of each gene. Following budding, the envelope enzyme neuraminidase facilitates the release of virions by destroying sialic acid receptors on the plasma membrane which would otherwise recapture the virions and hold them at the cell surface.

Defective interfering (DI) particles, originally known as "incomplete virus," are quite often produced, especially following infection at high multiplicity (see Chapter 4).

Pathogenesis and Immunity

The reader is referred back to Chapter 7 where we discussed factors that determine the virulence of influenza viruses. Best understood are the mutations that alter host and/or tissue tropism by changing the affinity of the HA ligand for receptors in which sialic acid is linked to galactose in different configurations, and second the mutations that affect uncoating of the endocytosed virion by changing the cleavability of HA and thus its capacity to mediate fusion. However, it is clear that most of the viral genes, especially NP, P, and NA, as well as HA, influence virulence in one way or another.

In Chapter 9 we discussed the pathology of influenza, mentioning some host factors affecting the outcome. In brief, the virus replicates principally in ciliated columnar epithelium of the respiratory tract, producing tracheobronchitis; features include enhanced airway reactivity (bronchospasm) and impaired mucociliary clearance. Primary viral pneumonia is uncommon, but secondary bacterial pneumonia is an important cause of death in the elderly, the very young, the chronically ill, or the immunocompromised.

Factors contributing to innate resistance include the following: (1) the mucus blanket that protects the underlying epithelium, and the continuous beating of cilia that clears invaders from a healthy respiratory tract, (2) soluble mannose-binding lectins, lung surfactants, and sialylglycoproteins present in mucus and transudates (see Chapter 7), and (3) alveolar macrophages. These and/or other natural defense mechanisms may be suboptimal in the aged, the premature, the pregnant, the pulmonary invalid, the immunocompromised, or the smoker. In Third World countries vulnerability to influenza is increased by inhaled pollutants (tobacco or wood smoke), malnutrition, prior or coincident infection, or immunosuppression (see Chapter 7).

If the individual has been infected within the past few years by a closely related strain of the same influenza HA subtype, anti-HA antibodies may intercept and neutralize the infecting virions. Argument continues on the relative importance of IgA and IgG antibodies and on their mechanisms of neutralization of influenza virus. Secretory IgA is generally believed to be the most relevant antibody in the upper respiratory tract at least, but serum IgG may provide protection in the lung. At high concentration, polymeric IgA may block attachment of virions to their receptors more effectively than IgG. At low concentration, anti-HA antibodies of both classes have been shown to inhibit uncoating; IgA has been demonstrated to do this by inhibiting both the fusion of HA to endosomal membrane and the separation of M1 from the ribonucleoprotein complex, thus preventing the latter from entering the nucleus.

If preexisting antibody proves inadequate to block the establishment of infection, recovery is dependent on two nonspecific cell types, activated macrophages and natural killer cells, plus two important cytokines, interferon γ and interleukin-2, and immunologically specific T lymphocytes of two subclasses, $CD4^+$ and $CD8^+$. Class I restricted $CD8^+$ cytotoxic T cells, recognizing determinants on any of the virus-coded proteins, are the most effective in clearing virus from the lower respiratory tract. Most of these Tc cells are elicited by conserved determinants on the internal proteins NP, M1, and P, or the nonstructural protein NS1, ensuring a useful degree of cross-protection when long-lived memory T cells are activated by a different strain of influenza virus (of the same species, A or B) some years later. Class II restricted $CD4^+$ T cells, particularly if they have cytotoxic potential, can clear low-level infection or less virulent strains, but they are more important as Th cells, and as Td cells secreting cytokines that attract and activate macrophages, NK cells, and T cells. Paradoxically, however, the inflammatory response produced by $CD4^+$ T cells also contributes to the lung consolidation (pneumonia) which may kill the patient.

Clinical Features of Influenza

There is a tendency for patients, and regrettably even some doctors, to label all respiratory ailments as "flu," being reluctant to confess to taking a few days off work for anything less! In reality, of course, influenza is a distinct clinical entity characterized by abrupt onset of fever, sore throat, nonproductive cough, myalgia, headache, and malaise. The uncomplicated syndrome is over in 3–7 days, but the cough and weakness may sometimes persist for another week or more.

Complications depend on the age of the patient. Young children may develop croup, pneumonia, or middle ear infection. However, most deaths occur in the elderly and are most frequently attributable to secondary bacterial pneumonia (usually due to *Staphylococcus aureus, Streptococcus pneumoniae,* or *Haemophilus influenzae*) and/or to exacerbation of a preexisting chronic condition such as obstructive pulmonary disease or congestive cardiac failure. Some of the elderly seem just to fade away.

Whereas 80–90% of influenza-related deaths occur in people over the age of 65 (especially over 75), the risk is at least as great for invalids of any age who suffer from chronic conditions affecting the pulmonary, cardiac, renal, hepatic, or endocrine systems. Further, the success of modern medicine in keeping alive so many children with congenital diseases such as cystic fibrosis and immunodeficiencies, and patients of any age with organ transplants or AIDS, has increased the number of younger people at risk of death during an influenza epidemic.

Laboratory Diagnosis

The most convenient specimen is a throat swab, but a higher isolation rate is obtained from a gargle (throat washing), nasal wash, or nasopharyngeal aspirate. Traditionally, the chick embryo was the standard host for cultivation of influenza viruses and is still used in addition to cell culture by some reference laboratories. Routinely, however, virus is isolated in the MDCK cell line at 33°–34°C, in the presence of trypsin to cleave the HA of progeny virions and enable spread to other cells. Cytopathic effect is not usually conspicuous, but growth of virus can be recognized after 1–10 (typically 3–7) days by hemadsorption (Fig. 5-2D), and the isolate identified by immunofluorescence on the fixed monolayer. Probably the optimal combination for rapid diagnosis is 24–48 hours of cultivation in MDCK cells, followed by EIA to identify antigen in detergent-disrupted cells using enzyme-labeled anti-NP monoclonal antibody specific for influenza type A or B, respectively. If positive, the isolate is then identified more precisely by hemagglutination inhibition (HI) on the culture supernatant, using ferret or chicken antisera against the prevalent strains of influenza A (H1N1), A(H3N2), and B which have been treated appropriately to remove nonspecific inhibitors of hemagglutination (see Fig. 12-11).

Direct detection of antigen by immunofluorescence or EIA is practicable if the specimen is satisfactory, with EIA being somewhat more sensitive, quicker, and easier. Monoclonal antibodies enhance both sensitivity and specificity.

Serological diagnosis, by demonstrating a rising HI antibody titer, is not satisfactory. All but very young patients have previous experience of, and therefore antibody against, one or more strains of influenza virus; on subsequent infection with a later strain of the same subtype they develop a memory antibody response directed against earlier strains to which their B cells are primed, which complicates the interpretation of HI results.

Epidemiology

The reader is referred back to Chapter 4 for a detailed description of the ongoing evolution of influenza virus by antigenic shift and drift. Since the reemergence of the H1N1 subtype in 1977, strains of two subtypes of influenza A, H1N1 and H3N2, have cocirculated with two separate lineages of influenza B in humans. One or sometimes two strains tend to dominate at any given time in a particular region of the world. In the tropics influenza can occur throughout the year, but in the temperate and colder regions of both hemispheres the disease occurs as winter epidemics (Fig. 31-2). These epidemics escalate with alarming speed because (1) the incubation period is exceptionally short (1–4 days), (2) copious numbers of virions are shed in droplets discharged by sneezing and coughing for a week (or longer in young children), (3) many symptomatic people (and all those who are subclinically infected) remain in circulation, and (4) the population lacks immunity to any novel subtype arising by antigenic shift, or displays inadequate immunity to strains arising by antigenic drift. The school is the principal marketplace for trading parasites, and the children bring the virus home to the family; parents then play their part in disseminating the infection in the workplace.

The most reliable indicator of the scale of an epidemic is a sudden leap in the incidence of absenteeism from schools and large industries; this is followed by an increase in hospital admissions and deaths, particularly among the elderly. Epidemiologists use an indicator known as "excess deaths," which measures the increase in mortality during an influenza epidemic in comparison with the average number during comparable winters without an influenza epidemic. In each of the 20 influenza epidemics recorded in the United States during the period 1957–1987, at least 10,000 (and occasionally up to 50,000) people have died as a direct or indirect consequence of the

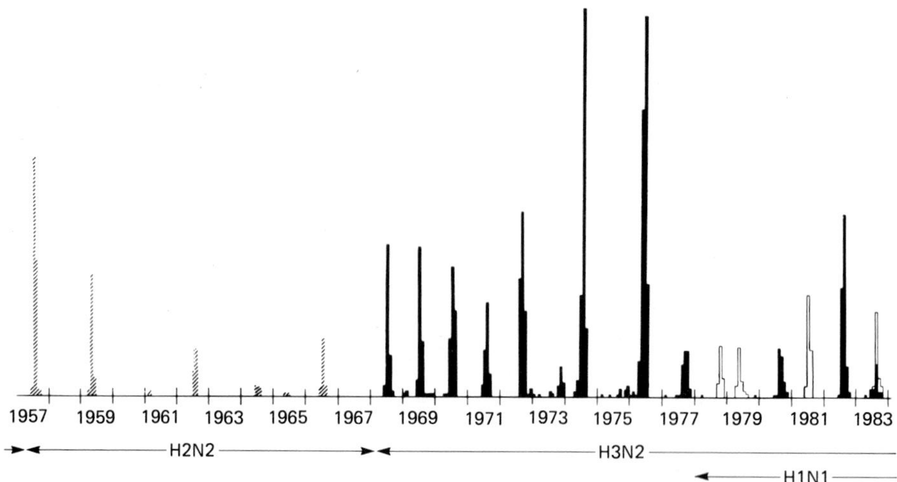

Fig. 31-2 Epidemic occurrence of influenza A. The histograms show the monthly isolations of influenza A viruses from patients admitted to Fairfield Hospital for Infectious Diseases, Melbourne, from 1957 to 1983. Note annual or biennial epidemics in the Australian winter (June to August). The major peak in 1957 represents the country's first experience of the H2N2 subtype which caused the 1957 pandemic of "Asian flu." The next major peak corresponds with the 1968/1969 pandemic of "Hong Kong flu," a new subtype (H3N2). The sporadic winter outbreaks during the 1970s, sometimes quite extensive as in 1974 and 1976, were due to a succession of H3N2 variants arising by antigenic drift. However, almost all of the cases in 1978, 1979, and 1981 were of the H1N1 subtype, which reemerged after being totally absent since 1950. Most of the H1N1 cases were in young people, born after 1950. The 1983 outbreak involved a mixture of H3N2 and H1N1. Since then, mixed epidemics have occurred quite often through the 1980s and early 1990s, and strains of the less virulent type B have been prominent. (Data courtesy A. A. Ferris, F. Lewis, N. Lehmann, and I. D. Gust.)

infection. Indeed, influenza consistently ranks among the top 10 causes of death in the United States.

Only influenza type A undergoes antigenic shift, unpredictably every 10–40 years. Influenza A and B undergo antigenic drift. Considerable effort has been expended in trying to devise ways of anticipating influenza epidemics, but this remains a very inexact science. We cannot predict the next strain of virus, when it will strike, how widely it will spread, nor how virulent it will be. In general, however, a novel subtype of influenza A arising by antigenic shift causes a pandemic with high morbidity and significant mortality. Between pandemics a succession of strains arise by antigenic drift and, as time goes on, infect people with partial immunity because of prior infection with another strain of that subtype. Since the reappearance of the H1 subtype of influenza A in 1977 most people old enough to have experienced that subtype before it disappeared in the 1950s have been found to display a degree of immunity to current H1 strains. Type B is less pathogenic, infects mainly children, and produces fewer dangerous complications than type A.

Control

No disease illustrates better than influenza the difficulties of control by immunization and the ingenuity required to overcome them. Existing vaccines are continually being rendered obsolete by antigenic shift and drift. The World Health Organization, through its extensive network of laboratories around the globe, is constantly on the alert for such changes. A WHO Reference Laboratory supplies the vaccine manufacturers with seed stocks of the latest strain(s) of influenza so that their product may be updated if necessary prior to each influenza season.

Vaccines

The current strains of influenza A(H1N1), A(H3N2), and B are grown separately in the allantois of chick embryos, inactivated with an appropriate chemical such as β-propiolactone, then purified by zonal ultracentrifugation, disrupted with detergent, and pooled. The resulting polyvalent inactivated vaccine is inoculated each autumn. Although some countries direct their program at schoolchildren with a view to limiting the circulation of virus in the community, thereby protecting the whole population, most do not regard this as a realistic proposition. It is generally not considered cost-effective or necessary to immunize the whole community, but only the most vulnerable cohorts, namely, (1) the elderly (>65 years), (2) residents of nursing homes and other chronic care facilities, (3) those (of any age) with chronic debilitating disease of the pulmonary, cardiovascular, renal, or endocrine systems (asthma, emphysema, chronic bronchitis, cystic fibrosis, diabetes, etc.), and (4) those with compromised immune function. In bad epidemic years there may also be a case for immunizing medical personnel and others providing vital community services, as well as close relatives and home-care personnel attending high-risk invalids.

The commonest side effect is a mild local reaction: some tenderness, redness, and swelling occur around the injection site in about 15% of recipients. Less frequently fever, malaise, and myalgia may develop within hours and disappear a day later. The only major contraindication for influenza vaccine use is known allergy to eggs; though the allergy is rare, inoculation of such persons can produce an immediate allergic reaction. A particularly serious problem, the Guillain-Barré syndrome, was encountered in 1 in every 100,000 Americans vaccinated against influenza A/New Jersey/76 (H1N1), during a mass campaign in 1976–77 to protect the population against an outbreak of "swine flu" which did not spread widely anyway, but no such association has been reported with any previous or subsequent influenza vaccine (see Chapter 9).

Efficacy is highest in the young and lowest in the old. There are probably two reasons for this difference. First, immune responsiveness declines with age; second, "original antigenic sin" tends to divert the response to influenza vaccines in the elderly, which is a great pity because that cohort constitutes the principal target group of annual vaccination campaigns. However, even though antibody rises following vaccination in the elderly are often disappointing and there is generally only a 30–70% reduction in the incidence of influenza, there is a 60–90% reduction in pneumonia, hospitalization, and mortality, providing that there is a good match between the vaccine strain and the challenge strain of virus. In so far as severe illness and death are what we are really aiming to prevent by immunization of the elderly, it can be asserted that current influenza vaccines, while less than perfect, offer a worthwhile degree of protection.

There has been much debate about whether the vaccine virus should continue to be grown in embryonated hen's eggs rather than in cultured mammalian cells such as MDCK, in light of the observation that replication in chick embryos tends to select for variants with preferential affinity for the avian cell receptor; the favored mutations affecting the HA ligand sometimes also alter the antigenic characteristics of the virus. The current view is that the greater convenience of and experience with the chick embryo argue for its continuation, provided that the vaccine seed stock is a clone demonstrated to be antigenically identical to the original human isolate.

Research is progressing on a number of fronts to find a better approach to vaccination against influenza. Some approaches are directed simply at increasing the immunogenicity of inactivated virions, or of solubilized or genetically cloned HA (± NA), for example, by addition of adjuvants, coupling to carriers, or incorporation into liposomes, virosomes, or iscoms (see Chapter 13). However, none of these approaches addresses the most critical problems of all, namely, (1) antigenic shift and drift and (2) the inability of inactivated vaccines to generate either a local IgA or a cell-mediated immune response. A topically administered live vaccine producing a mucosal IgA, IgG, and T-cell memory response of broad cross-reactivity within a type or subtype may be the answer.

Cold-adapted (*ca*) variants of influenza virus with mutations in every gene have been developed and used as master stains for genetic reassortment (see Chapter 4) with contemporary strains of influenza to produce an attenuated vaccine virus containing 6 *ca* genes plus wild-type HA and NA genes (see Chapter 13). When the *ca* vaccine is administered by aerosol spray or intranasally, the replication of virions is restricted to the nasopharynx. During a long history of successful use in Russian schoolchildren and clinical trials in United States, these *ca* vaccines have been demonstrated to be stably attenuated for humans, with no reversion to virulence and no evidence of secondary transmission. They display good immunogenicity and efficacy in influenza-naive infants, but replicate poorly and are much less effective in adults, and are of no value in the elderly. A case can be made for their use in high-risk infants, provided evident problems of interference can be overcome by careful balancing of the titers of the A/H1, A/H3, and B strains in trivalent live vaccines.

Attention is also turning to the possibility of oral delivery of influenza vaccine, taking advantage of the common mucosal system. Inactivated or live virions have been mixed with potent adjuvants such as the heat-labile *Escherichia coli* toxin (LT), or enclosed in biodegradable microspheres, and shown by oral administration to mice to elicit a protective (and sometimes cross-reactive) secretory IgA antibody response in the lungs.

Chemoprophylaxis

Though no substitute for vaccination, chemoprophylaxis with amantadine, or preferably rimantadine, by mouth has a place in protecting unimmunized high-risk people during a major epidemic of influenza A. The use of these agents was discussed fully in Chapter 16, as was the exciting new X-ray crystallography/computer modeling approach to the design of compounds that may inhibit attachment of virus to host receptors (by occupying the corresponding ligand on influenza HA) or inhibit release of virus from the plasma membrane following budding (by occupying the active site on the viral neuraminidase).

Further Reading

Air, G. M., and Laver, W. G. (1986). The molecular basis of antigenic variation in influenza virus. *Adv. Virus Res.* **31,** 53.

Centers for Disease Control. (1993). Prevention and control of influenza: Recommendations of the Immunization Practices Advisory Committee (ACIP). *Morb. Mortal. Wkly. Rep.* **42** (RR-6).

Douglas, R. G., ed. (1987). Prevention, Management and Control of Influenza: A mandate for the 1980s. *Am. J. Med.* **82**(Suppl. 6A).

Hannoun, C., Kendal, A. P., Klenk, H.-D., and Ruben, F. L., eds., (1993). "Options for the Control of Influenza II." Excerpta Medica, Elsevier, Amsterdam.

Kendal, A. P., and Harmon, M. W. (1988). *Orthomyxoviridae:* The influenza viruses. *In* "Laboratory Diagnosis of Infectious Diseases. Principles and Practice. Volume II. Viral, Rickettsial and Chlamydial Diseases" (E. H. Lennette, P. Halonen, and F. A. Murphy, eds.), p. 602. Springer-Verlag, New York.

Kilbourne, E. D., ed. (1987). "Influenza." Plenum, New York.

Kingsbury, D. W. (1990). Orthomyxoviridae and their replication. *In* "Fields Virology" (B. N. Fields, D. M. Knipe, R. M. Chanock, J. L. Melnick, T. P. Monath, and B. Roizman, eds.), 2nd Ed., p. 1075. Raven, New York.

Klenk, H. D., and Rott, R. (1988). The molecular biology of influenza virus pathogenicity. *Adv. Virus Res.* **34,** 247.

Krug, R. M., ed. (1989). "The Influenza Viruses." Plenum, New York.

Lamb, R. A., and Choppin, P. W. (1983). The gene structure and replication of influenza virus. *Annu. Rev. Biochem.* **52,** 467.

Murphy, B. R., and Webster, R. G. (1990). Orthomyxoviruses. *In* "Fields Virology" (B. N. Fields, D. M. Knipe, R. M. Chanock, J. L. Melnick, T. P. Monath, and B. Roizman, eds.), 2nd Ed., p. 1091. Raven, New York.

Webster, R. G., Laver, W. G., Air, G. M., and Schild, G. C. (1982). Molecular mechanisms of variation in influenza viruses. *Nature (London)* **296,** 115.

Wiley, D. C., and Skehel, J. J. (1987). The structure and function of the hemagglutinin membrane glycoprotein of influenza virus. *Annu. Rev. Biochem.* **56,** 365.

Wilson, I. A., and Cox, N. J. (1990). Structural basis of immune recognition of influenza virus hemagglutinin. *Annu. Rev. Immunol.* **8,** 737.

Wright, P. F., and Karzon, D. T. (1987). Live attenuated influenza vaccines. *Prog. Med. Virol.* **34,** 70.

Chapter 32

Arenaviridae

The prototype arenavirus is lymphocytic choriomeningitis virus, which produces a clinically inapparent lifelong infection in mice and is occasionally transmitted to humans, in whom it causes disease ranging in severity from mild fever to meningitis. First isolated over 60 years ago, the virus has provided an important model for studies of persistent infections, immunological tolerance, virus-induced immune complex disease, and the role of the MHC complex (see Chapters 9 and 10). In 1969 another arenavirus made newspaper headlines. After a nurse from a mission in Lassa, Nigeria, had died in the hospital, a nurse who had attended her also died, and another who had assisted at her autopsy became desperately ill but recovered after intensive care following evacuation to the United States. A virus was isolated from her blood by virologists at Yale University, one of whom became ill but survived following transfusion with immune plasma from the previous patient; however, one of the Yale laboratory technicians later died. Lassa virus, like a number of other arenaviruses isolated from humans during outbreaks of hemorrhagic fever in South America, occurs as a lifelong, persistent, inapparent infection of its natural rodent host (Table 32-1).

Properties of *Arenaviridae*

The family *Arenaviridae* derives its name from the presence within virions of cellular ribosomes, which in electron micrographs resemble grains of sand (*arena,* sand), and are incorporated into virions coincidentally during budding (Table 32-2). Arenaviruses are pleomorphic, 110–130 nm (rarely up to 300 nm) in diameter (Fig. 32-1), and are composed of a lipoprotein envelope covered with glycoprotein peplomers surrounding two circular nucleocapsid seg-

Table 32-1
Distribution and Rodent Hosts of Arenaviruses Pathogenic for Humans

Virus	Disease	Geographic distribution	Natural host
Lymphocytic choriomeningitis	Meningitis	Europe, Americas	*Mus musculus*
Junin	Argentine hemorrhagic fever	Argentina	*Calomys* spp.
Machupo	Bolivian hemorrhagic fever	Bolivia	*Calomys callosus*
Guanarito	Venezuelan hemorrhagic fever	Venezuela	*Sigmodon* sp.
Lassa	Hemorrhagic fever	West Africa	*Mastomys natalensis*

ments, each looking like a string of beads. The genome comprises two linear segments of ssRNA, L and S, 7.2 and 3.4 kb, respectively; each segment forms a circle by hydrogen bonding of its ends. Conserved nucleotide sequences at the 3′ end of each RNA, which are shared by most arenaviruses, are complementary to sequences at the 5′ end. Most of the genome is of minus sense, but the 5′ half of the S segment and a short sequence at the 5′ end of the L segment are of plus sense; the term *ambisense* has been coined to describe this unusual genome arrangement, which is also found in some members of the family *Bunyaviridae* (see Chapter 33).

The family *Arenaviridae* comprises a single genus, *Arenavirus*, which is divided into two serogroups (or complexes) corresponding in general with their geographic distribution: the Old World arenaviruses and the New World arenaviruses. The Old World serogroup (LCM–LAS complex) contains lymphocytic choriomeningitis (LCM) virus and various African arenaviruses of which Lassa is the most important; the New World serogroup (Tacaribe complex) comprises the South American arenaviruses Tacaribe, Junin, Machupo, Guaranito, and others.

Viral Replication

Arenaviruses grow to high titer in cell cultures, replicating in the cytoplasm and maturing by budding from the plasma membrane. They have limited lytic

Table 32-2
Properties of Arenaviruses

Two serogroups: Old World (LCM–LAS complex) and New World (Tacaribe complex)
Spherical enveloped virion, 110–300 (generally 110–130) nm
Virion contains nonfunctional ribosomes
Two closed-circular, loosely helical nucleocapsids with associated transcriptase
Genome comprises large (L, 7.2 kb) and small (S, 3.4 kb) segment of ssRNA, both ambisense
Viral proteins: nucleoprotein (N), RNA polymerase (L), glycoproteins (G1, G2), zinc-binding protein (Z), plus poly(U) and poly(A) polymerases, and protein kinase
Replication occurs in cytoplasm, generally noncytocidal, causes persistent infection
Genetic reassortment occurs during replication
Maturation by budding from plasma membrane

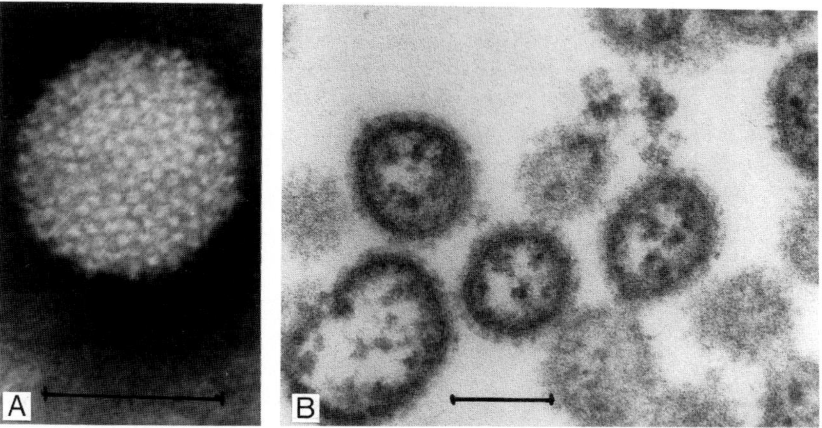

Fig. 32-1 *Arenaviridae.* (A) Negative stain, Tacaribe virus. (B) Lassa virus after budding from an infected cell (thin section). Bars, 100 nm. (Courtesy Dr. F. A. Murphy.)

capacity, usually leading to carrier cultures in which defective interfering (DI) particles are produced. After entry and uncoating of the virion, subgenomic mRNA encoding the nucleoprotein (N) is transcribed by the virion transcriptase from the minus sense (3′) half of the ambisense S segment of the genome (Fig. 32-2). A short hairpin configuration in the intergenic region in the middle of the S gene segment is thought to serve as the transcription termination signal. The mRNA appears to derive its 5′ cap by cap snatching from heterogeneous cellular RNA, as do orthomyxoviruses. Similarly, subgenomic mRNA encoding the transcriptase (L) is transcribed from the minus sense (3′) portion of the ambisense L segment of the genome. Translation of both N and L proteins is required prior to replication of the viral genome, which requires the synthesis of full-length complementary copies of both ambisense genome

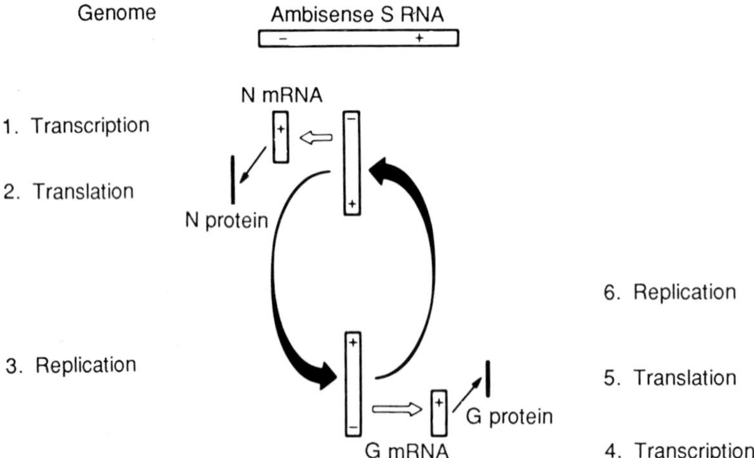

Fig. 32-2 Coding, transcription, translation, and replication strategies of arenavirus S RNA species, indicating the sequence of events necessary to obtain the S-coded gene products N and G. There are similar strategies for the transcription of the ambisense L RNA species. [From D. H. L. Bishop, *in* "Fields Virology" (B. N. Fields, D. M. Knipe, R. M. Chanock, M. S. Hirsch, J. L. Melnick, T. P. Monath, and B. Roizman, eds.), 2nd Ed., p. 1234. Raven, New York, 1990.]

segments, a process that must involve read-through of the transcription termination signals. Only then can mRNA for the glycoprotein (G) and the putative zinc-binding protein (Z) be transcribed from the other end of the complementary copy of S and L genome segments, respectively. Following glycosylation and proteolytic cleavage of G (into G1 and G2), budding of virions occurs from the plasma membrane. There is a lack of fidelity in this process, as indicated by accidental envelopment of nearby ribosomes, and packaging of multiple copies of one or both genome segments may occur such that the resulting virions are often diploid or even multiploid. When a cell is infected concurrently with two different arenavirus species, genetic reassortants regularly emerge.

Laboratory Diagnosis

Biosafety Level 4 containment (Fig. 32-3) is required for laboratory diagnosis of all arenaviruses (except LCM virus, where level 3 suffices). Specimen trans-

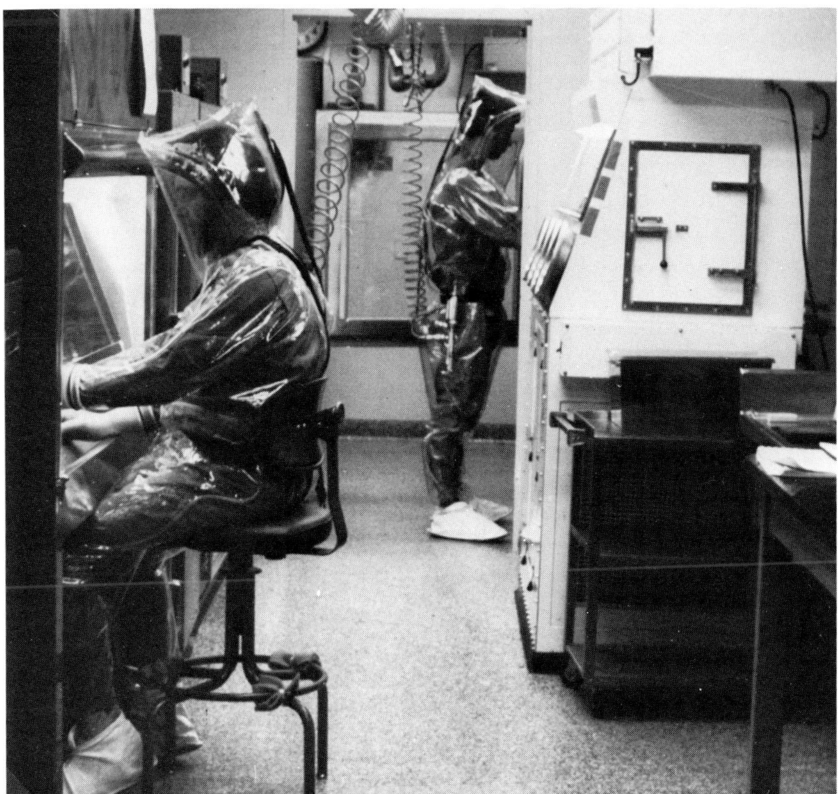

Fig. 32-3 Maximum containment laboratory, Centers for Disease Control, Atlanta, Georgia. Workers are protected by positive-pressure suits with an independent remote source of breathing air. Primary containment of aerosols is achieved by use of filtered vertical laminar-flow work stations. The ultracentrifuge is contained in an explosion-proof laminar-flow hood. A full range of equipment is available, and animals as large as monkeys can be used for experiments. [From J. S. Mackenzie (ed.), "Viral Diseases in South-East Asia and the Western Pacific." Academic Press, New York, 1982. Courtesy Dr. K. M. Johnson.]

port must be arranged in keeping with national and international regulations. Diagnosis is usually based on serology to demonstrate antibodies of the IgM class and/or a rising titer of antibody in paired sera (although any arenavirus antibody is highly suggestive in a traveler returning from Africa or South America). Traditionally, the method used has been indirect immunofluorescence (IFA), using previously prepared "spotslides" of virus-infected, acetone-fixed, γ-irradiated (for safety) Vero E6 cell monolayers. However, the production of safe, genetically cloned antigens may favor a shift to enzyme immunoassay (EIA).

Arenaviruses can be readily isolated in rodents, such as suckling mice, hamsters, or guinea pigs, from blood, throat, urine, or sometimes cerebrospinal fluid (CSF) or autopsy specimens. More convenient today are cultured Vero E6 cells or BHK-21 cells; the growth of arenaviruses, several of which are noncytopathogenic, can be confirmed within 2–3 days by IFA or EIA, using broadly reactive antibodies for initial detection, followed by more specific monoclonal antibody for precise characterization. Detection of arenavirus antigen in specimens taken directly from the body has not so far proved to be sufficiently reliable. Polymerase chain reaction assays are being developed.

Epidemiology

Each arenavirus is maintained in nature by a rodent species in which persistent infection occurs, with chronic viremia and virus shedding in urine and saliva. In *Mus musculus* (lymphocytic choriomeningitis virus) and *Mastomys natalensis* (Lassa virus) infection is completely subclinical. However, Machupo virus renders *Calomys callosus* sterile and induces hemolytic anemia and splenomegaly, and Junin virus induces fetal death in its rodent hosts. Vertical transmission occurs in infected rodents, by transovarial, transuterine, and various postpartum routes, including milk, saliva, and urine. The principal reservoir hosts of the zoonotic arenaviruses are shown in Table 32-1.

Chronically infected animals shed virus in their urine, which contaminates their environment and then spreads to humans by contact (perhaps facilitated by skin abrasions or cuts) or by aerosol. The natural history of the human diseases is determined by the pathogenicity of the virus, the geographic distribution, habitat and habits of the rodent reservoir host, and the nature of the human–rodent interaction.

Diseases Caused by Arenaviruses

Human infections with arenaviruses range in severity from an influenza-like syndrome, sometimes with meningitis, to severe hemorrhagic fever. Aspects of the pathogenesis of hemorrhagic fever are described in Chapter 9 and clinical comparisons in Chapter 36. In brief, arenaviruses are thought to enter the body via skin abrasions or inhalation. Viremia follows, and these pantropic viruses replicate in a variety of organs during the 1 to 2-week incubation period; however, there is little evidence of direct virus-induced cytopathology. For example, Lassa virus produces some focal necrosis of the liver, interstitial pneumonitis, facial edema, and encephalopathy, but death is attributable to the sudden onset of hypovolemic shock in the second week of the illness. The pathophysiology of the hemorrhagic fever/shock syndrome is not fully understood. One major point of difference from hemorrhagic fevers caused by viruses of other families is that disseminated intravascular coagulation does not appear to play a significant role until perhaps the terminal phase of the illness. In Lassa fever it appears that an inhibitor of platelet function may be responsible for the bleeding and perhaps also for endothelial dysfunction. Bleeding with severe thrombocytopenia is more prominent in Junin and Machupo viral hemorrhagic fevers.

Lymphocytic Choriomeningitis Virus Infections

The LCM virus is focally distributed throughout the world, especially in Europe and the Americas, in the common house mouse (*Mus musculus*). The distribution of human cases is correspondingly focal, and also seasonal, probably because mice move into houses and barns in winter. Feral mice may also introduce the virus into laboratory and commercial mouse, rat, hamster, guinea pig, and rhesus monkey colonies. In the United States, the virus has been a particular problem in colonies of hamsters and immunocompromised (nude, SCID, etc.) mice, resulting in contaminated diagnostic reagents, failed research protocols, and clinical disease in laboratory and animal care personnel. The increasing popularity of hamsters as pets has resulted in many human disease episodes, some involving hundreds of cases.

Human infection with LCM virus is commonly asymptomatic but may present as one of three syndromes: (1) most commonly as an influenza-like illness with fever, headache, myalgia, and malaise; (2) less often as an aseptic meningitis; (3) rarely as a severe encephalomyelitis.

Lassa Fever

Lassa virus is enzootic in the West African multimammate mouse, *Mastomys natalensis,* which is a peridomestic rodent that lives in or near human dwellings, breeds year-round, and transmits the virus vertically to its offspring and horizontally to humans by contaminating the house with urine. Uniquely among the arenaviruses, person-to-person spread of Lassa virus is also common, although not as frequent as infection from a rodent source; the fact that Lassa virus does not appear to be spreading outside of West Africa suggests that the virus is not capable of sustained person-to-person transmission.

In humans, Lassa fever is now recognized to be endemic in rural West Africa, with sporadic cases occurring from northern Nigeria to Guinea. Serological surveys show that millions of West Africans have antibody; there are over 100,000 new human infections a year, and at least 1000–3000 deaths, with a disease to infection ratio of about 20% and a case–fatality rate of 5–15%. Changing social circumstances have led to several major outbreaks. For example, in eastern Sierra Leone, where surface diamond mines brought together nearly 100,000 people from all over West Africa, "instant" villages and a currency-based economy resulted in great increases in *Mastomys* populations and their contact with humans. Nearly half the febrile patients admitted to two hospitals in the region had Lassa fever, and the 16% case–fatality rate accounted for 30% of the deaths in hospital wards. Since the late 1980s urban outbreaks have occurred in Nigerian cities, displaying similarly alarming mortality, with nosocomial spread in hospitals reminiscent of the original 1969 occurrence.

Lassa fever is very variable in its presentation, making it difficult to diagnose, whether in endemic areas or in returning travelers. It may present with insidious development of fever, headache, and malaise, progressing to a very sore throat, pains in the back, chest, and joints, vomiting, and proteinuria. In severe cases, conjunctivitis, pneumonitis, carditis, hepatitis, encephalopathy, nerve deafness, and/or hemorrhages are seen, death occurring in about 20% of hospitalized cases, usually following cardiovascular collapse. Mortality is higher during the third trimester of pregnancy, and fetal loss is almost invariable.

South American Arenavirus Hemorrhagic Fevers

There are five hemorrhagic fevers in South America: yellow fever, dengue hemorrhagic fever, and three caused by arenaviruses, of three different viral species, in three different countries, and with three different principal reservoir hosts (Table 32-1). All have a similar natural history. They cause lifelong infections of rodents, with lifelong excretion of large amounts of virus in the urine. To become infected with these viruses, humans must come into close contact with the reservoir host rodents.

Argentine Hemorrhagic Fever

In Argentina an arenavirus called Junin virus is enzootic in the voles *Calomys musculinus* and *C. laucha,* and can also infect other wild rodents. Fifty years ago human infections with this virus were unknown, because humans did not come into contact with these rodents. However, the introduction of widespread planting of maize in the pampas in the late 1940s favored this rodent over other indigenous rodents and increased its contact with humans. When the virus infects humans it causes a severe hemorrhagic fever, and Junin virus was first isolated from such cases in 1958. Since then the virus has spread to over 100,000 square kilometres of what is now rich farmland of the humid pampas. Argentine hemorrhagic fever has always been an occupational disease, affecting adult males harvesting grain crops. Prior to mechanization, workers harvesting by hand were the victims; today it is the combine harvester and grain truck drivers who bear the brunt of the infection. There is a 3 to 5-year cyclic trend in the incidence of human cases which exactly parallels changes in the density of *Calomys.*

Argentine hemorrhagic fever is a more typical hemorrhagic fever than Lassa fever. The pathology is largely confined to the circulatory system. There is relatively little inflammation or necrosis; prominent features are hemorrhage, thrombocytopenia, leukopenia, hemoconcentration, proteinuria, and hypotension, sometimes culminating in death from hypovolemic shock.

Bolivian Hemorrhagic Fever

Machupo virus emerged in Bolivia in 1952 when a revolution forced the people living in the plains to attempt subsistence agriculture at the forest edge. *Calomys callosus,* a forest rodent and the reservoir host of Machupo virus, is well adapted to human contact, and some of the later outbreaks resulted from rodent invasion of villages, cases tending to cluster in particular houses in which numbers of infected rodents were subsequently trapped. There were over 1000 cases in 1962 and 1964 with a 20% mortality, but control of *Calomys* in dwellings in the endemic area by trapping has resulted in the disappearance of the disease, no cases having been recognized since 1974.

Venezuelan Hemorrhagic Fever

The most recent emerging virus disease in South America is Venezuelan hemorrhagic fever, occurring in rural areas of central Venezuela. The first cases, recognized in 1989, were at first thought to be cases of dengue hemorrhagic fever. In 1990–1991 a total of 104 cases were seen, with 26 deaths. An arenavirus, which has been called Guanarito virus, was isolated from cases. The epidemiology has still to be worked out, but the sex and age distribution of cases suggests that, like Bolivian hemorrhagic fever, transmission occurs in and around houses. Virus has been isolated from a cotton rat (*Sigmodon hispidus*)and antibodies found in sera from a rice rat.

Prevention and Treatment

Rodent control, by trapping, poisoning, and/or cats, on a village-wide basis is applicable where the reservoir host is a commensal animal, as in lymphocytic choriomeningitis (*Mus musculus*), Lassa fever (*Mastomys natalensis*), or Bolivian hemorrhagic fever (*Calomys callosus*). However, it is difficult in rural settings such as those characteristic of Argentine hemorrhagic fever. A live attenuated Junin virus vaccine is undergoing phase 2 clinical trials in Argentina, and a vaccinia recombinant carrying the Lassa virus glycoprotein gene has been shown to protect monkeys against challenge.

Isolation and barrier nursing are required to prevent nosocomial spread of arenaviruses to other patients and nursing staff. Fluid, electrolyte, and osmotic imbalances need correction. The main challenge is timely anticipation and management of the profound hypotension and shock, sometimes associated with acute pulmonary edema which may suddenly develop during the second week of the illness.

Ribavirin, administered intravenously in high dosage during the first 6 days of Lassa fever, has been shown to reduce viremia and mortality significantly. Oral ribavirin also has some effect and has been recommended for prophylaxis in close contacts of known cases.

The course of Argentine hemorrhagic fever has been modified by administration of high-titer convalescent-phase plasma within the first week of the illness, although late, generally reversible, neurologic sequelae occurred in about 10% of the treated patients. This unusual approach to treatment of acute infection (rather than prophylaxis) is being explored for other arenaviruses, but the high incidence of HIV infection in Africa would discourage its use on that continent at least.

Further Reading

Bishop, D. H. L. (1990). Arenaviridae and their replication. *In* "Fields Virology" (B. N. Fields, D. M. Knipe, R. M. Chanock, M. S. Hirsch, J. L. Melnick, T. P. Monath, and B. Roizman, eds.), 2nd Ed., p. 1231. Raven, New York.

Lopez, C., ed. (1988). "Immunology and Pathogenesis of Persistent Virus Infections," Chapters 6–12, pp. 79–170. American Society for Microbiology, Washington, D.C.

McCormick, J. B. (1990). Arenaviruses. *In* "Fields Virology" (B. N. Fields, D. M. Knipe, R. M. Chanock, M. S. Hirsch, J. L. Melnick, T. P. Monath, and B. Roizman, eds.), 2nd Ed., p. 1245. Raven, New York.

McCormick, J. B., and Fisher-Hoch, S. P. (1990). Viral hemorrhagic fevers. *In* "Tropical and Geographic Medicine" (K. S. Warren and A. A. F. Mahmoud, eds.), p. 700. McGraw-Hill, New York.

Oldstone, M. B. A., ed. (1987). Arenaviruses. *Curr. Top. Microbiol. Immunol.* **133, 134.**

Peters, C. J. (1991). Arenaviruses. *In* "Textbook of Human Virology" (R. B. Belshe, ed.), 2nd Ed., p. 541. Mosby, St. Louis, Missouri.

Salvato, M. S., ed. (1993). "The Arenaviridae." Plenum, New York.

Chapter 33

Bunyaviridae

The largest family of viruses affecting mammals, *Bunyaviridae,* was one of the last to be recognized. Most of the over 300 members are arthropod-borne, with a variety of vectors and life cycles involving mammalian or avian vertebrate hosts. Many persist in their arthropod vectors via transovarial transmission; in an infected female arthropod the virus infects the eggs so that larvae, nymphs, and adults of succeeding generations are infected and are thus capable of transmitting the virus to vertebrate hosts. This is an important mechanism of overwintering. Humans are infected when bitten by the arthropod. However, viruses of the genus *Hantavirus* are transmitted by urine and saliva between reservoir rodent hosts; humans are infected when they come into contact with these rodents. More than a dozen bunyaviruses, from four genera, are pathogenic for humans, causing diseases ranging from undifferentiated fever, sometimes with a rash, to potentially lethal encephalitis and hemorrhagic fever.

Properties of *Bunyaviridae*

The virions are spherical, 90–100 nm in diameter (Fig. 33-1), and are composed of a lipid envelope with glycoprotein peplomers enclosing three circular nucleocapsids of helical symmetry (Table 33-1). The genome consists of three linear segments of ssRNA, designated large (L), medium (M), and small

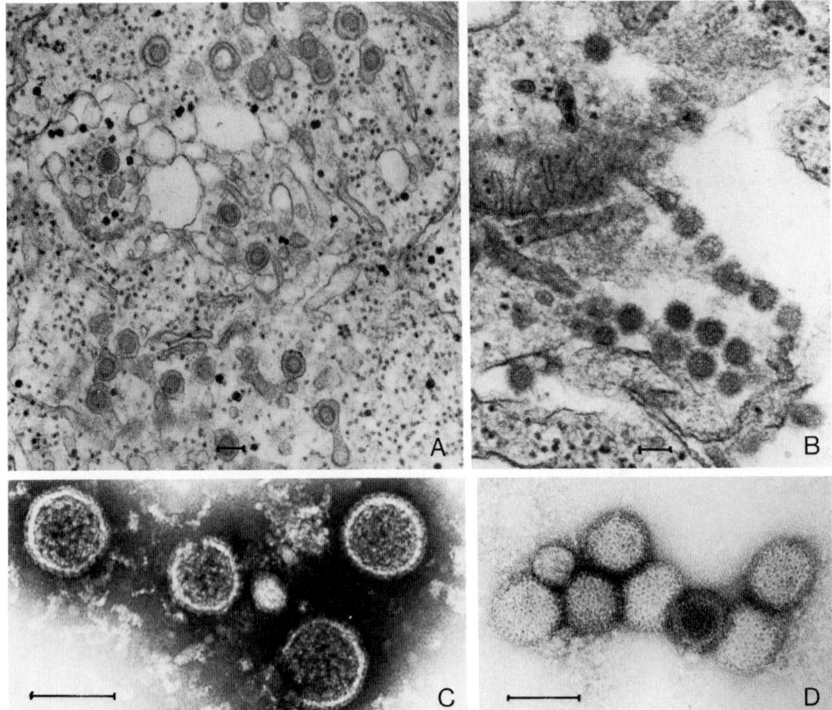

Fig. 33-1 *Bunyaviridae.* (A, B) Sections of cultured cells. (A) Virions in Golgi vesicles. (B) Extracellular virions. (C, D) Negatively stained preparations. (C) Hantaan virus. (D) Rift Valley fever virus. Bars, 100 nm. (A, B, C, courtesy Dr. F. A. Murphy;C, courtesy Dr. J. McCormick and Dr. E. L. Palmer; D, courtesy Dr. E. L. Palmer.)

(S), of approximately 7, 4, and 1–2 kb, respectively, each formed into a circle by hydrogen bonding of the ends; as with the arenaviruses, conserved 3′-terminal sequences common to all segments of the genome of a given genus are complementary to conserved 5′ sequences. The genome is generally of minus sense, but the S segment of some genera is ambisense. Virions contain four major proteins: a transcriptase (L), a nucleocapsid protein (N), and two glycoproteins (G1 and G2), which form the surface peplomers.

Table 33-1
Properties of *Bunyaviridae*

Four genera infect vertebrates: *Bunyavirus, Phlebovirus,* and *Nairovirus,* all arthropod-borne; *Hantavirus,* non-arthropod-borne
Spherical, enveloped virion, 90–100 nm
Glycoprotein peplomers but no matrix protein in envelope
Three circular nucleocapsids with helical symmetry
Segmented minus sense ssRNA genome; three segments of 7, 4, and 1–2 kb; complementary 3′ and 5′ termini; segment S of *Phlebovirus* genome is ambisense
Capped 5′ termini of cellular RNAs cannibalized as primers for mRNA transcription
Cytoplasmic replication; budding into Golgi vesicles
Generally cytocidal for vertebrate cells but noncytocidal persistent infection in invertebrate cells
Genetic reassortment occurs between closely related viruses

Classification

The family *Bunyaviridae* (Bunyamwera, locality in Uganda) contains four genera: *Bunyavirus,* with more than 160 viruses; *Phlebovirus,* with more than 50 viruses; *Nairovirus,* with at least 30 viruses; and *Hantavirus,* with 6 viruses; many other members of the family remain to be characterized (Table 33-2). Assignment to genera is based on the characteristics and coding strategy of the genome, common nucleocapsid antigens (conventionally determined by complement fixation), and mode of transmission. Members of the genus *Bunyavirus* are transmitted predominantly by mosquitoes, nairoviruses by ticks, and phleboviruses by sandflies, gnats, or ticks; hantaviruses have no known arthropod vectors. Within genera, species are arranged into serogroups based on partial cross-reactivity (some shared epitopes) between the envelope glycoproteins, as measured by cross-neutralization and hemagglutination inhibition.

Genetic reassortment is readily demonstrable when cultured cells or mosquitoes are coinfected with closely related bunyaviruses and has probably played a part in the evolution of the family. Within its particular ecologic niche each bunyavirus continues to evolve by genetic drift; for example, oligonucleotide fingerprinting or sequencing of the RNA genome from different isolates of La Crosse virus frequently reveal point mutations, deletions, or duplications.

Table 33-2
Members of Family *Bunyaviridae* Causing Disease in Humans

Virus	Genus	Geographic distribution	Arthropod vector	Amplifier or reservoir host	Human disease
Rift Valley fever	*Phlebovirus*	Africa	Mosquito	Sheep, cattle, buffalo, goats	Fever, myalgia, retinitis, hemorrhagic fever
Sandfly fever	*Phlebovirus*	Mediterranean, South America	Sandfly	?Gerbil, forest rodents	Fever, myalgia, conjunctivitis
La Crosse	*Bunyavirus*	Western United States	Mosquito	Chipmunk, squirrel	Encephalitis
Oropouche	*Bunyavirus*	Brazil	Mosquito, midge	Sloths, monkeys	Fever, arthralgia, myalgia
Crimean–Congo hemorrhagic fever	*Nairovirus*	Asia, Eastern Europe, Africa	Tick	Sheep, cattle, goats	Hemorrhagic fever
Hantaan	*Hantavirus*	Asia, Europe	Nil	Rodents (bats, birds)	Hemorrhagic fever, nephropathy
Puumala	*Hantavirus*	Scandinavia	Nil	Rodents, humans	Nephropathy
Belgrade	*Hantavirus*	Balkans	Nil	Rodents	Hemorrhagic fever, nephropathy
Seoul	*Hantavirus*	Worldwide	Nil	Rats	Nephropathy, sometimes hemorrhagic fever
Muerto Canyon	*Hantavirus*	United States	Nil	Rodents	Fever, myalgia, respiratory failure

Viral Replication

Viral entry into its host cell is thought to occur by receptor-mediated endocytosis. All subsequent steps take place exclusively in the cytoplasm. After penetration of the host cell, the virion transcriptase is activated and transcribes subgenomic mRNAs from each of the three virion RNA segments while still associated with their nucleocapsids. The enzyme carries not only RNA polymerase activity but also an endonuclease which is thought to effect 5′-methylated capping of viral RNA transcripts by cleaving 12–15 nucleotides from the 5′ end of cytoplasmic cellular mRNA molecules. After translation of these mRNAs, replication of the virion RNA can occur and a second round of transcription begins. The L RNA segment codes for the transcriptase; the M segment codes for the G1 and G2 glycoproteins, and in some genera for a nonstructural protein, NS_M; the S RNA segment codes for the nucleocapsid protein and, in some genera, a nonstructural protein NS_S, which is postulated to be involved in RNA replication/transcription. In the genus *Bunyavirus*, NS_S is read from the 3′ end of the N gene but in an overlapping reading frame. In the genus *Phlebovirus*, the S RNA segment also codes for these two proteins, but it employs a remarkable ambisense transcription strategy (Fig. 33-2).

The viral glycoproteins G1 and G2 are generally derived by cotranslational cleavage in the rough endoplasmic reticulum and accumulate in membranes of the Golgi complex, where they are terminally glycosylated and then associate with nucleocapsids. Virions bud into Golgi cisternae before transport in smooth-walled vesicles to fuse with the basolateral surface of the plasma membrane and release by exocytosis (Figs. 33-1A and 3-9B).

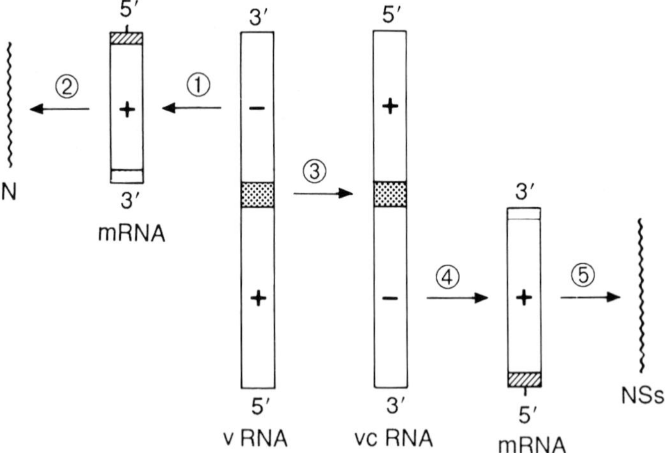

Fig. 33-2 Expression of the ambisense S segment of the phlebovirus genome. Genomic RNA (vRNA) comprises a 3′ half of minus sense and a 5′ half of plus sense separated by a noncoding intergenic region (stippled). (1) First, the open reading frame at the 3′ end of the genome is transcribed (from its 3′ end) by the virion transcriptase to produce a subgenomic mRNA of complementary sense, which is capped at its 5′ terminus by a primer comprising 12–15 nucleotides (hatched) cannibalized from cellular mRNA by the viral endonuclease. (2) This mRNA is translated into nucleoprotein N. (3) Genomic replication produces a full-length complementary copy of the viral genome, vcRNA. (4) The open reading frame in the minus sense of this antigenome now serves as the template for transcription, from its 3′ end, to yield mRNA which is translated (5) to give the nonstructural protein NS_S.

Pathogenesis

The segmented genome of bunyaviruses facilitates assignment of biological functions to individual genome segments by experimentation based on genetic reassortment. For example, studies in mice with members of the California serogroup of the genus *Bunyavirus* indicated that neurovirulence maps to segment L (encoding the transcriptase) and to a lesser extent to segment S (nucleocapsid), whereas neuroinvasiveness maps to segment M. Many of the properties affecting pathogenesis (e.g., host range, tissue tropism, virulence, and transmissibility by the invertebrate vector) segregate to segment M, which encodes the envelope glycoproteins G1 and G2.

Laboratory Diagnosis

Classically, bunyaviruses have been isolated by intracerebral inoculation of suckling mice, and this remains an important method of discovery of new arboviruses from mosquito pools as well as from putative amplifier hosts or humans. However, most bunyaviruses also grow well in more convenient cultured vertebrate cells (e.g., Vero-E6 or BHK-21) or invertebrate cells (e.g., *Aedes* mosquito cell line). They are cytolytic for mammalian cells (except for hantaviruses and some nairoviruses), but noncytolytic for invertebrate cells.

For diagnosis in suspected cases of human disease, appropriate specimens are blood taken very early in the illness, while the patient is still viremic, or relevant tissue taken at autopsy. Growth of virus in cultured cells can be detected by immunofluorescence, hemagglutination (for some bunyaviruses), or EIA. The isolated virus is assigned to a particular serogroup by a complement fixation test using serogroup-specific antisera directed mainly toward the nucleocapsid protein. Thereafter, the species or serotype is identified by hemagglutination inhibition (goose erythrocytes, pH approximately 6) or neutralization (plaque reduction in cell culture), both of which measure antibodies directed at the membrane glycoproteins. Finer distinctions can be made by oligonucleotide fingerprinting of the RNA genome.

Serology is often used to screen populations for antibody, or to diagnose illness using paired sera or specific IgM assays. Complement fixation, EIA, neutralization, and immunofluorescence can all be used for these purposes. While this approach is useful for screening for antibody to an exotic virus in a traveler, it can yield ambiguous information in people who have lived for years in a part of the world where other members of the same genus or serogroup are endemic.

Rift Valley Fever

Rift Valley fever has been known for many years as a devastating disease of ruminants which breaks out every decade or so in East or South Africa, killing lambs and calves and inducing abortion in pregnant ewes and cows. At the time of such epizootics occasional cases of a nonlethal dengue-like illness were observed in people who came into contact with sick animals or handled their carcasses. Suddenly in 1977 an epizootic of unprecedented scale occurred in the delta and valley of the Nile, with many hundreds of thousands of cases in sheep and cattle, and for the first time large numbers of humans were affected. The disease broke out again in 1978. Over 200,000 Egyptians contracted the disease, and more than 600 died. The extent and severity of this epizootic/epidemic may have been due to the high population densities of fully susceptible animals and humans. In 1987–1988, further virus activity was detected in eastern Africa and in western Africa, with hundreds of human deaths in Senegal and Mauritania.

Clinical Features

The disease begins after a very short incubation period (2–6 days) with fever, severe headache, retroorbital pain, photophobia, and generalized myalgia. In the Egyptian epidemic a proportion of the patients progressed to one of three complications: encephalitis (relatively mild and usually without sequelae), retinitis (with diminution of visual acuity sometimes leading to permanent loss of central vision), or hemorrhagic fever (marked by jaundice and widespread hemorrhage, with a mortality of 5–10%).

Epidemiology

The native vertebrate reservoir host of Rift Valley fever virus in sub-Saharan Africa has not been precisely identified but is presumably one or several species of wild ungulates. The virus survives in a silent enzootic cycle for many years and then, when there is exceptionally heavy rainfall, explodes in epizootics of great magnitude among domestic sheep and cattle. In such epizootics, Rift Valley fever virus is transmitted by many species of *Culex* and *Aedes* mosquitoes, which are very numerous after heavy rains or when improper irrigation techniques are used. The mosquitoes are infected when feeding on viremic sheep or cattle, which maintain a very high level of viremia for 3–5 days. This amplification of the transmission cycle, together with mechanical transmission by biting flies, results in infection and disease in a very high proportion of animals and humans at risk.

These epizootic cycles are started by infected mosquitoes occupying an unusual ecologic niche. Throughout the grassy plateau regions of sub-Saharan Africa there are dry depressions in which floodwater *Aedes* species live, surviving long periods of drying as eggs and emerging only when the depressions are filled by exceptional rainfall. These mosquitoes are transovarially infected with Rift Valley fever virus and are capable of transmitting it to a few sheep, cattle, and wild ruminants, thus starting epizootics which are maintained and amplified by other mosquito species.

In its epizootic cycles Rift Valley fever virus may also be spread directly by fomites, by direct contact, and mechanically by arthropods such as tabanid flies. Infected sheep have a very high level of viremia, and pregnant ewes and cows almost invariably abort; transmission to humans at the time of abortion via contaminated placenta and fetal and maternal blood is a particular problem. Abattoir workers, farmers, and veterinarians are often infected directly.

Control

The 1977–1978 epidemic in Egypt raised the specter of future spread of Rift Valley fever to the Middle East and perhaps beyond. To minimize the chance of this happening it may become necessary to police the international movement of livestock as well as to screen animals serologically from time to time. Spread to humans during outbreaks might be reduced by vigorous mosquito control and by implementing safer procedures for the killing and disposal of animals. Live attenuated and formalin-inactivated Rift Valley fever vaccines are available for immunization of susceptible livestock in enzootic areas, but they would need to be used systematically on a very large scale to prevent outbreaks. Persons who are particularly at risk, namely, laboratory personnel, veterinarians, and slaughtermen, in East and South Africa, should be vaccinated when a suitable vaccine is licensed for general use in humans.

Sandfly Fever

Sandfly (phlebotomus) fever is a common but nonlethal disease caused by a *Phlebovirus* and transmitted to humans by peridomestic sandflies (*Phlebotomus papatasii*) in countries around the Mediterranean Sea and eastward to central Asia and India. The indigenous people are usually immune as a result of childhood infection; however, travelers are at risk, and epidemics have occurred in armies throughout history. A second focus occurs in Central and South America where forest-dwelling phlebotomines of the genus *Lutzomyia* are the vectors. No vertebrate host other than man has been definitely incriminated, but gerbils are the principal suspects in Europe and Asia and forest rodents in South America. The virus also persists in phlebotomines by transovarial and transstadial transmission. The human disease is a self-limiting dengue-like syndrome marked by fever, headache, myalgia, retroorbital pain, conjunctivitis, and leukopenia. Genetic reassortment between closely related viruses has been shown to occur in nature.

California Encephalitis

The California serogroup within the genus *Bunyavirus* comprises more than a dozen viruses isolated from mosquitoes or vertebrates in various parts of the world. The best studied and most important human pathogen in the group is La Crosse virus, which causes encephalitis in children and forestry workers in wooded areas of the United States. The virus is endemic, with about 100 cases of encephalitis reported annually, but serological surveys indicate that there are about 300,000 human infections annually, mostly in summer, throughout the Midwest; fortunately fewer than 1 case occurs for every 1000 infections in children. The ecology of the virus is described in the legend to Fig. 33-3. It has been said that "no two isolates are the same," probably because of the high frequency of genetic drift and perhaps occasional genetic reassortment.

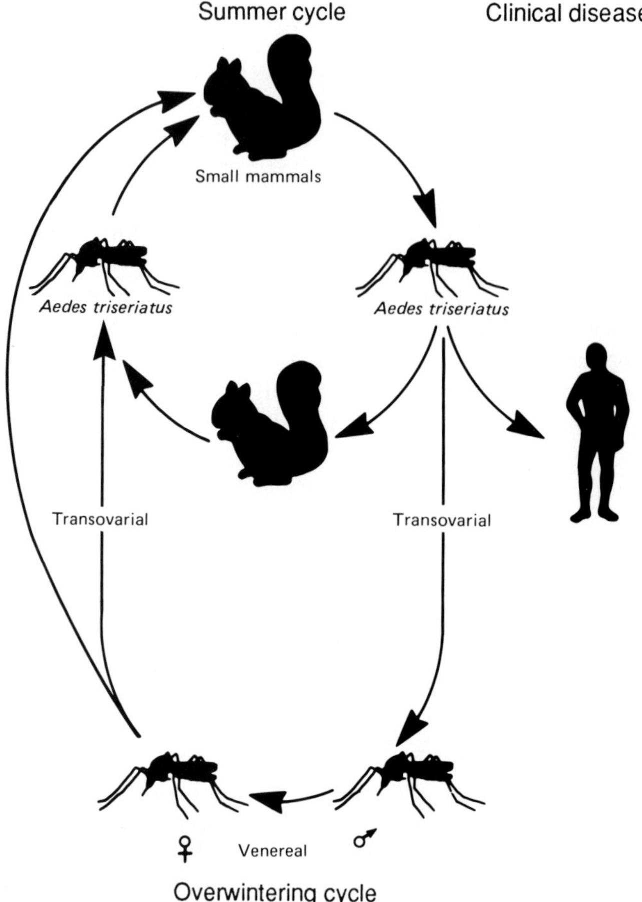

Fig. 33-3 Cycle of California (La Crosse) encephalitis virus. During the warmer months there is a cycle of the La Crosse virus between *Aedes triseriatus,* a woodland mosquito, and chipmunks and tree squirrels. The virus is also maintained indefinitely in mosquitoes by transovarial transmission and is amplified by venereal transmission between male and uninfected female mosquitoes, which can in turn transmit either to a vertebrate by biting or to the next generation of mosquito by transovarial transmission. Humans, who are dead-end hosts, are the only host known to develop clinical disease. (Modified from R. T. Johnson, "Viral Infections of the Nervous System." Raven, New York, 1982.)

Oropouche Fever

Oropouche virus, a member of the genus *Bunyavirus,* has caused repeated epidemics with thousands of cases in northern Brazil, usually in the rainy season, and may occur more widely in northern South America. The human disease is characterized by fever, headache, myalgia, arthralgia, and prostration, but no mortality. The major urban vector is the midge *Culicoides paraensis,* and the virus may be maintained in a midge–human cycle during epidemics. The sylvatic cycle involves sloths, monkeys, and probably jungle mosquitoes.

Crimean–Congo Hemorrhagic Fever

Crimean hemorrhagic fever has been recognized for many years in central Asia and eastern Europe as a severe zoonotic disease affecting people coming in contact with livestock, as well as woodcutters and other people coming in contact with ticks. The causative virus, a member of the genus *Nairovirus*, is identical to a virus, originally named Congo virus, which causes a nonfatal febrile disease in humans in central Africa. The distribution of Crimean–Congo hemorrhagic fever virus is now known to extend from China through central Asia to India, Pakistan, Afghanistan, Iran, Iraq, other Persian Gulf countries, the Middle East, eastern Europe, to most of Saharan and sub-Saharan Africa.

Clinical Features

Crimean–Congo hemorrhagic fever commences abruptly with fever, headache, and severe back and abdominal pain, and progresses to extensive hemorrhages from almost any site, with melena, hematemesis, hematuria, and a hemorrhagic skin rash (Fig. 33-4). Leukopenia, thrombocytopenia, proteinuria, and hepatitis are key findings. Blood loss from internal bleeding leads to shock, pulmonary edema, and death. Case–fatality rates range from 5 to 50%, depending on the availability of modern medical care.

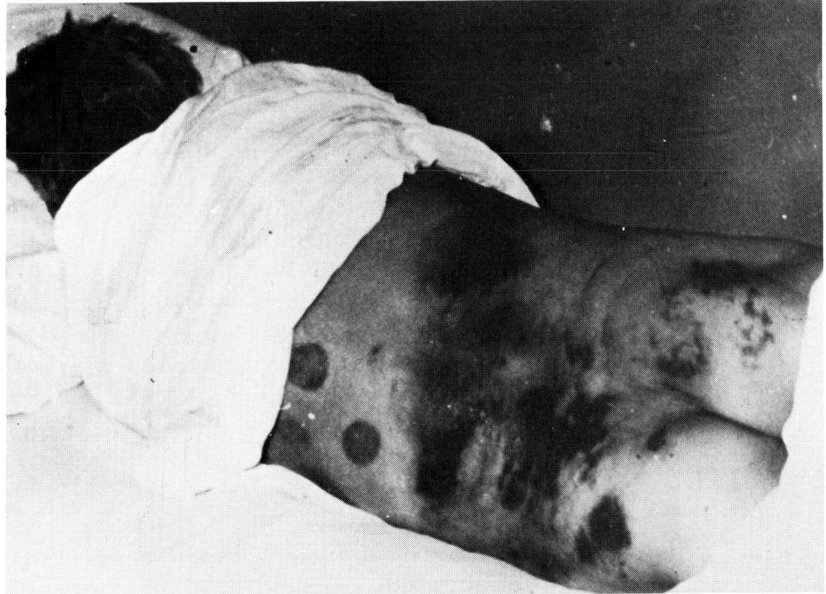

Fig. 33-4 Crimean–Congo hemorrhagic fever, showing hemorrhagic skin rash. Petechiae progress to extensive ecchymoses. The circular sites on the left were caused by capillary fragility tests. [From J. Casals, B. E. Henderson, H. Hoogstraal, K. M. Johnson, and A. Shelekov, *J. Infect. Dis.* **122,** 437 (1970). Courtesy Dr. A. Shelekov.]

Epidemiology

The virus is maintained by a cycle involving transovarial/transstadial transmission in ixodid ticks, its distribution worldwide coinciding with that of ticks of the genus *Hyalomma,* which is the most important of the seven genera from which the virus has been isolated. Vertebrates are important amplifier hosts. The tick larvae and nymphs generally parasitize birds and small animals such as hares, whereas adult ticks are mainly responsible for transmitting the virus to larger animals like cattle, sheep, goats, camels, and humans.

The virus is also transmitted to humans by direct contact with subclinically infected viremic animals, for example, during sheep shearing or veterinary procedures, and it is also transmitted from person to person, especially nosocomially, for example, to surgeons stemming internal hemorrhage. Crimean–Congo hemorrhagic fever is an increasing problem, with more and more cases being reported each year from many parts of the world, and an increasing percentage of animals being found to be seropositive.

Control

Tick control measures in domestic animals, spraying camp sites and limited areas with acaricides, and impregnation of clothing with repellent, especially during spring and summer, can reduce the incidence of Crimean–Congo hemorrhagic fever. Strict isolation of patients and barrier nursing with particular attention to blood and vomitus are required to prevent nosocomial spread. There is a need for a safer vaccine than the mouse brain-derived version currently available.

Hemorrhagic Fever with Renal Syndrome

During the Korean war of 1950–1952, thousands of United Nations troops developed a disease marked by fever, hemorrhagic manifestations, and acute renal failure with shock; the case–fatality rate was 5–10%. The etiologic agent of this disease remained a mystery until 1978 when a virus, named Hantaan virus, was isolated in Korea from the field rodent *Apodemus agrarius* and identified as a unique bunyavirus. Since then, several related viruses have been found in other parts of the world in association with other rodents. These viruses comprise the genus *Hantavirus.* Five hantaviruses, Hantaan, Puumala, Belgrade, Seoul, and Muerto Canyon viruses, are associated with human diseases with different epidemiologic patterns, varying clinical manifestations, and a variety of local names (see Table 33-2).

Clinical Features

As the descriptive if cumbersome name hemorrhagic fever with renal syndrome implies, infections with Hantaan virus produce hemorrhagic fever with profound renal tubular involvement. Lumbar abdominal pain and proteinuria are prominent during the febrile phase, which gives way to a hypotensive hemorrhagic phase, then an oliguric phase with abnormal renal function, and finally a diuretic phase heralding recovery. Hemorrhage can occur from different sites in different patients, for example, a petechial skin rash, massive gastrointestinal bleeding (sometimes presenting as an acute abdomen), or hemorrhagic pneumonia. Belgrade virus produces a similar syndrome in the Balkans. The urban and laboratory-associated disease caused by Seoul virus is usually milder and associated with hepatic rather than renal dysfunction. Nephropathia epidemica caused by Puumala virus is a nonlethal form of hantavirus infection encountered in Europe, especially Scandinavia, in which there is little hemorrhage and no shock. In 1993 a new hantavirus was identified in the southwestern United States as the cause of a pulmonary syndrome with a case-fatality rate of over 50%; fever and myalgia progress rapidly to dyspnea, respiratory insufficiency, and hemodynamic collapse.

Epidemiology

Unlike other members of the family *Bunyaviridae,* hantaviruses are not arboviruses but are transmitted to humans from their rodent hosts by inhalation of aerosolized rodent urine or by direct contact with rodent excreta or contaminated fomites. The hantaviruses produce lifelong inapparent infection in rodents, with persistent shedding in urine and saliva, although unlike arenaviruses they do not appear to be passed congenitally.

Epidemiologically, there are three disease patterns: rural, urban, and laboratory-acquired, each with a different rodent–virus combination. The rural type, which is much the commonest, accounting for over 100,000 cases annually in eastern China, is caused by Hantaan virus in the field mice *Apodemus agrarius* in Korea, China, and eastern Russia and *Apodemus flavicollis* in the Balkans, but by Puumala virus in the bank vole *Clethrionomys glareolus* in Scandinavia and eastern Europe. The principal host of Muerto Canyon virus in southwestern United States is *Peromyscus maniculatus.* All these species are field rodents, hence human infections occur mainly in rural workers, sometimes as sporadic cases, sometimes as small outbreaks when a number of persons are exposed to a contaminated focus. Outbreaks in military personnel are associated with their exposure to such foci. In China and Korea there are two seasonal peaks, associated with the harvesting of wheat (summer) and rice (late fall) when the populations of *Apodemus* peak and much dust is generated. In Scandinavia maximum transmission occurs during the winter, when the reservoir host *Clethrionomys* invades houses.

Seoul virus is widespread among urban rats (*Rattus norvegicus*) throughout the world, especially in seaports, and has recently been recognized as the cause of an occupational disease transmitted to animal caretakers and research personnel from laboratory rats and wild reservoir host rodents brought into laboratories.

In May 1993 a mysterious new disease broke out near the confluence of New Mexico, Arizona, Utah and Colorado. The victims, mainly healthy young adults, developed influenza-like symptoms and then rapidly became very sick, and many died of respiratory failure. The Centers for Disease Control and Prevention (CDC) screened patients for antibodies to a wide range of possible infectious agents, and they were found to have antibodies cross-reacting with hantaviruses. Oligonucleotide primers corresponding to the ends of a conserved sequence of the hantavirus genome were synthesized and used in a reverse transcriptase-PCR assay to amplify the genome of the putative new hantavirus obtained from appropriate specimens, including autopsy material. All patients tested were positive for hantavirus RNA by PCR, and hantaviral antigen was demonstrable by immunofluorescence in capillary endothelial cells throughout many organs. The region had recently suffered a plague of the deer mouse, *Peromyscus maniculatus,* and many mice trapped in the area were hantavirus-positive by serology and/or PCR. Subsequently, sporadic human cases have been reported in many states and various other rodent species have also been found to be carriers of the virus.

Control

Rodent control by trapping, poisoning, or cats is recommended in situations where human infections are acquired from rodents in houses, as with the Seoul virus and Puumala virus. Because rodent control in an agrarian environment is impracticable, there is a clear need for a vaccine against Hantaan virus. An inactivated vaccine derived from suckling mouse brain was licensed in Korea in 1990, but a cell culture grown inactivated vaccine or a genetically engineered vaccine would be preferable in the long term. Special precautions are recommended for laboratory personnel who handle rats.

Further Reading

Bishop, D. H. L. (1990). Bunyaviridae and their replication, Part I. *In* "Fields Virology" (B. N. Fields, D. M. Knipe, R. M. Chanock, M. S. Hirsch, J. L. Melnick, T. P. Monath, and B. Roizman, eds.), 2nd Ed., p. 1155. Raven, New York.

Bouloy, M. (1991). Bunyaviridae: Genome organization and replication strategies. *Adv. Virus Res.* **40,** 235.

Calisher, C. H. (1993). Family *Bunyaviridae,* genus *Bunyavirus* (the bunyaviruses). *In* "Handbook of Infectious Diseases" (E. H. Kaas, T. H. Weller, S. M. Wolfe, and D. A. J. Tyrrell, eds.), in press. Dekker, Burlington, Ontario.

Elliott, R. M. (1990). Molecular biology of the Bunyaviridae. *J. Gen. Virol.* **71,** 501.

Gonzalez-Scarano, F., and Nathanson, N. (1990). Bunyaviruses. *In* "Fields Virology" (B. N. Fields, D. M. Knipe, R. M. Chanock, M. S. Hirsch, J. L. Melnick, T. P. Monath, and B. Roizman, eds.), 2nd Ed., p. 1195. Raven, New York.

Johnson, K. M. (1989). Hantaviruses. *In* "Viral Infections of Humans: Epidemiology and Control" (A. S. Evans, ed.), 3rd Ed., p. 341. Plenum, New York.

Karabatsos, N., ed. (1985). "International Catalogue of Arboviruses including Certain Other Viruses of Vertebrates," 3rd Ed. American Society of Tropical Medicine and Hygiene, San Antonio, Texas.

Kolakofsky, D., ed. (1991). Bunyaviridae. *Curr. Top. Microbiol. Immunol.* **169,** 1.

Le Duc, J. (1987). Epidemiology of Hantaan and related viruses. *Lab. Anim. Sci.* **37,** 413.

Lee, H. W., and van der Groen, G. (1989). Hemorrhagic fever with renal syndrome. *Prog. Med. Virol.* **36,** 62.

McCormick, J. B., and Fisher-Hoch, S. (1990). Viral hemorrhagic fevers. *In* "Tropical and Geographic Medicine" (K. S. Warren and A. A. F. Mahmoud, eds.), 2nd Ed., p. 700. McGraw-Hill, New York.

McKee, K. T., Jr., LeDuc, J. W., and Peters, C. J. (1991). Hantaviruses. *In* "Textbook of Human Virology" (R. B. Belshe, ed.), 2nd Ed., p. 615. Mosby, St. Louis, Missouri.

Meegan, J. M. (1979). The Rift Valley fever epizootic in Egypt 1977–1978. I. Description of the epizootic and virological studies. *Trans. R. Soc. Trop. Med. Hyg.* **73,** 618.

Monath, T. P., ed. (1988). "The Arboviruses: Epidemiology and Ecology," Vols. 1–5. CRC Press, Boca Raton, Florida.

Nichol, S. T., Spiropoulou, C. F., Morzunov, S., Rollin, P. E., Ksiazek, T. G., Feldmann, H., Sanchez, A., Childs, J., Zaki, S., and Peters, C. J. (1993). Genetic identification of a hantavirus associated with an outbreak of acute respiratory illness. *Science* **262,** 914.

Peters, C. J., and LeDuc, J. W. (1991). *Bunyaviridae:* Bunyaviruses, phleboviruses, and related viruses. *In* "Textbook of Human Virology" (R. R. B. Belshe, ed.), 2nd Ed., p. 571. Mosby, St. Louis, Missouri.

Schmaljohn, C. S., and Patterson, J. L. (1990). Bunyaviridae and their replication, Part II. *In* "Fields Virology" (B. N. Fields, D. M. Knipe, R. M. Chanock, M. S. Hirsch, J. L. Melnick, T. P. Monath, and B. Roizman, eds.), 2nd Ed., p. 1175. Raven, New York.

Chapter 34

Reoviridae

The name reovirus is an acronym, short for respiratory enteric orphan because the first members of this family to be discovered, now classified as the genus *Orthoreovirus* of the family *Reoviridae,* were found to inhabit both the respiratory and the enteric tract of humans and animals, but to be "orphans" in the sense that they are not associated with disease. The discovery of the human rotaviruses in 1973 changed all that, for members of the genus *Rotavirus* are recognized to be the most important cause of infantile gastroenteritis throughout the world. In addition, dozens of arboviruses, at least one of them causing disease in humans, have been allocated to the genera *Orbivirus* and *Coltivirus.* Yet other genera contain pathogens that infect both plants and insects, raising the question of whether these fascinating viruses that cross kingdoms so readily might have evolved in insects.

Furthermore, reoviruses have attracted much attention from molecular biologists because of the unique nature of the genome. Composed of double-stranded RNA, the genome is segmented into 10–12 separate molecules, each representing a different, generally monocistronic, gene. For several reoviruses, each gene has been cloned and sequenced and its protein product characterized. Moreover, the facility with which these viruses undergo genetic reassortment has been exploited to exchange genes from temperature-sensitive (*ts*) mutants and thus determine the role of individual genes in pathogenesis and virulence.

Properties of *Reoviridae*

Members of the family *Reoviridae* have striking icosahedral or quasi-spherical virions (Fig. 34-1), with two concentric icosahedral capsids surrounding an inner core. The virions of rotaviruses and orthoreoviruses are relatively stable to heat and to pH 3, and their infectivity is actually increased by exposure to proteolytic enzymes encountered in the gut. The genome comprises 10–12 unique molecules of dsRNA, total size 18–27 kilobase pairs, depending on the genus (Table 34-1). The positive strands of each duplex are 5′ capped and the negative strands 5′ phosphorylated; the 3′ termini are not polyadenylated. Genera share no antigens, but within genera there are genus-specific and species-specific antigens.

Viral Replication

The replication cycle has been studied in most detail with reovirus 3, a member of the genus *Orthoreovirus.* The intact virion may enter the cell by receptor-mediated endocytosis, or, alternatively, intermediate subviral particles, resulting from digestion with chymotrypsin in the intestine, may pass into the cytoplasm either via the endosomal pathway or directly. Both types of particles are then degraded to become "core particles," within which the virion-associated transcriptase and capping enzymes repetitively transcribe 5′ capped (but not 3′ polyadenylated) mRNA molecules, which are extruded through the hollow apices of the icosahedral core particles. Only certain genes are transcribed initially; the others are derepressed following the synthesis of an early viral protein. The mechanism of replication of the genome is complex and not yet fully understood. Various nonstructural and structural viral proteins associate with a complete set of mRNA molecules to form a subviral particle within which complementary minus sense RNA strands are synthesized, producing dsRNA molecules. These in turn serve as templates for the transcription of more mRNA which is translated preferentially to yield a large pool of viral structural proteins. New virions self-assemble and accumulate in cytoplasmic inclusions before being released by cell lysis.

Although the replication of rotaviruses has not yet been studied in so much detail, it is apparent that the general principles are similar. However, certain distinctive features mark the early and late stages of the rotavirus replication cycle. Rotaviruses bind to sialic acid on receptors, via a ligand on the outer capsid hemagglutinating protein VP4; cleavage of VP4 by trypsin is required to enable the virion to enter the cytoplasm, which apparently occurs

Table 34-1
Properties of *Reoviridae*

Three of nine genera contain human viruses: *Rotavirus, Coltivirus,* and *Orthoreovirus*
Nonenveloped spherical virion (*Orthoreovirus* and *Coltivirus,* 80 nm; *Rotavirus,* 70 nm)
Outer and inner icosahedral capsid surrounding inner core; core contains RNA polymerase and associated enzymes involved in mRNA synthesis (RNA elongation, capping, methylation, ?helicase)
Double-stranded RNA segmented genome; *Orthoreovirus,* 10 segments, total size 24 kbp; *Rotavirus,* 11 segments, 18 kbp; *Coltivirus,* 12 segments, 27 kbp
Cytoplasmic replication; entry may require cleavage of capsid protein; transcriptase in core transcribes early genes; genome replication by transcription from mRNA molecules to form dsRNA in subviral particles; transcription of late genes for structural proteins; virions undergo self-assembly, released by cell lysis
Genetic reassortment occurs between species within each genus

by direct penetration of the plasma membrane. Newly synthesized outer capsid glycoprotein VP7 is retained in the rough endoplasmic reticulum (RER). Assisted by a nonstructural glycoprotein, the rotavirus progenitor particle buds into cisternae of the RER, acquiring a temporary envelope, which is then removed so that VP7 can assemble to complete the outer capsid.

Rotaviruses

In 1973 a new virus was discovered in Australia, first in duodenal biopsies then in feces of Melbourne children with gastroenteritis. Electron micrographs revealed virions with a unique appearance (*rota,* wheel). Quickly it became apparent that rotaviruses are the commonest cause of gastroenteritis in infants and are responsible for about a million deaths each year in the Third World.

Properties of Rotaviruses

The rotavirus is particularly photogenic (Fig. 34-1). Smooth and round in outline, it is seen to have two concentric icosahedral shells. The 50 nm core of the virion, composed mainly of the structural protein VP2 (but also the transcriptase/replicase VP1 and the guanylyltransferase VP3) is surrounded by a 60 nm inner icosahedral capsid composed of the major group-specific protein VP6, which in turn is enclosed within a 70 nm outer icosahedral capsid composed mainly of the glycoprotein VP7, with dimers of the hemagglutinin/cell attachment protein VP4 projecting from the surface as 60 short spikes; 132 channels pass through both the outer and inner capsid to communicate with the core. The genome consists of 11 molecules of dsRNA, total size 18 kbp.

Rotaviruses are ubiquitous. Essentially every species of domestic animal

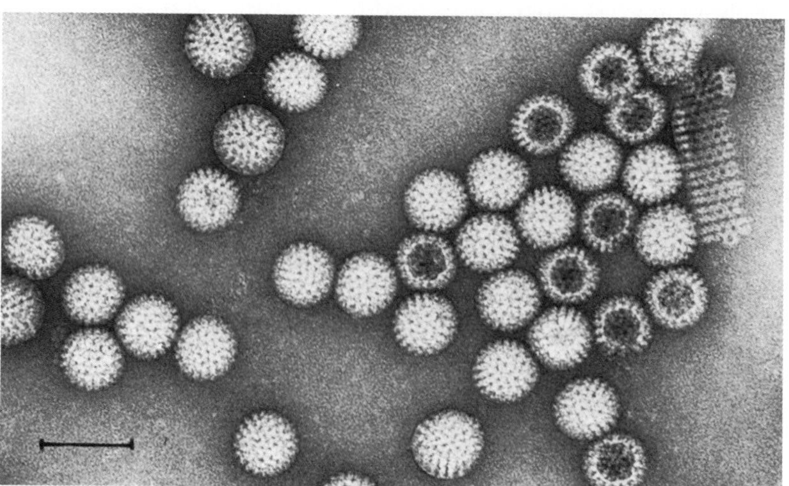

Fig. 34-1 *Reoviridae.* Negatively stained preparation of virions of human rotavirus. Bar, 100 nm. (Courtesy Dr. E. L. Palmer.)

or bird that has been thoroughly searched has at least one indigenous rotavirus, causing diarrhea ("scours") in the newborn. Rotaviruses are classified into six serogroups, A–F, on the basis of differences between the major group-specific capsid antigen VP6. Most human rotaviruses fall into group A, but others, once called "atypical rotaviruses" or "pararotaviruses," belong to recently defined groups B and C.

Differentiation into serotypes within group A is based on neutralization tests. As both of the outer capsid proteins carry type-specific epitopes recognized by neutralizing antibodies, a binary system of classification of serotypes has been developed, akin to that used for influenza viruses. So far, 14 serotypes (G1–G14) have been defined on the basis of different VP7 antigens, whereas 8 serotypes (P1–P8) have been defined on the basis of different VP4 antigens. Human group A rotaviruses fall into nine G serotypes, 1, 2, 3, 4, and rarely 6, 8, 9, 10, or 12, and into four P serotypes, 1, 2, 3, and 4. Monoclonal antibodies can be used to make finer distinctions, for example, between subtypes within the G1 serotype, which is the major cause of severe rotavirus enteritis worldwide.

Large numbers of electropherotypes can be differentiated by polyacrylamide gel electrophoresis (PAGE) of the viral RNA (Fig. 34-2). Because the patterns reflect differences in the migration of any of the 11 RNA segments, there is no direct relationship between electropherotypes and serotypes, but they are a valuable tool for tracking the epidemiology of individual strains.

Pathogenesis and Immunity

The pathogenesis of rotavirus infection was discussed in Chapter 9 as an example of the mechanism and consequences of viral damage to the enteric tract (see Fig. 9-2 and associated text).

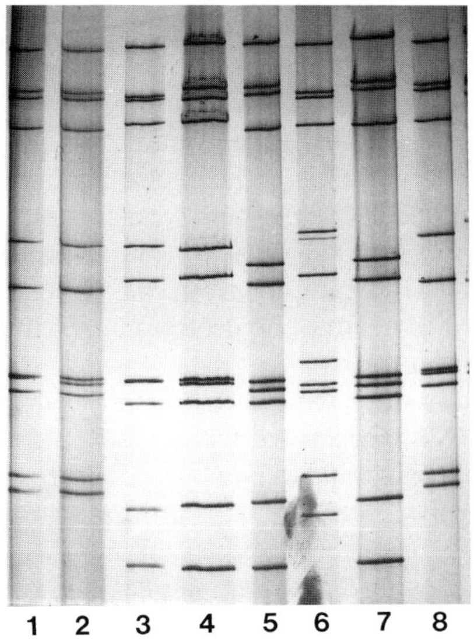

Fig. 34-2 Electropherotypes of rotavirus. Seven separate isolates of rotaviruses obtained from infants in Noumea between 1980 and 1983 were analyzed by polyacrylamide gel electrophoresis of their RNA. Tracks 5 and 7 show two samples of the same electropherotype. (Courtesy Drs. G. Panon and I. H. Holmes.)

Maternal antibodies of the IgG class transmitted across the placenta do not protect the newborn, but IgA antibodies in the colostrum do. Breast-fed infants are significantly less susceptible than bottle-fed infants during the first few days of life. Thereafter, the antibody titer in milk drops off rapidly, but infections acquired during the first few months still tend to be inapparent. Neonatal infections themselves elicit a transient neutralizing secretory IgA ("coproantibody") response in the gut, as well as a neutralizing IgG response in serum that persists for a few months, but they do not confer significant immunity against reinfection. Reinfections do, however, tend to be clinically less severe, presumably as a result of priming for an anamnestic response. Immunity following clinical disease occurring at a later age may be more durable, and there is some evidence for low levels of cross-reactive (heterotypic), as well as homotypic, immunity. Severe disease is less common on reinfection, and in older children. However, longitudinal studies are required to sort out whether increasing resistance is simply age-related or attributable to homotypic or heterotypic immunity, whether secretory IgA, serum IgG, or possibly T-cell mediated immunity is the most important, and whether neutralizing antibodies to VP4 are more relevant than those to VP7.

Clinical Features

Asymptomatic infection is the rule in neonates and is quite common in older children and adults. Clinical illness is seen principally between the ages of 6 and 24 months, with the peak around 12 months. After an incubation period of 1–3 days, vomiting generally precedes diarrhea, which lasts for 4–5 days and can lead to severe dehydration. Death is rare in well-nourished children, but large numbers die in the poorer tropical countries.

Laboratory Diagnosis

Rotaviruses were discovered by electron microscopy, which remains a satisfactory approach to rapid diagnosis (see Chapter 12); the virions are plentiful in feces and are so distinctive that they cannot be mistaken for anything else. However, enzyme immunoassay (EIA) is a more practicable and more sensitive method for the average laboratory to detect rotavirus in feces. Latex particle agglutination or reverse passive hemagglutination are also sensitive, specific, and simple (see Chapter 12 for details). The specificity of any of these tests can be manipulated at will by selecting either type-specific or broadly cross-reactive monoclonal antibodies (MAbs) as capture and/or indicator antibodies in the antigen-capture assay. For example, group B or C rotaviruses can be distinguished by using antibodies against the group-specific internal antigen VP6.

Recently, attention has turned to improving the sensitivity of diagnosis by identifying the viral genome in RNA extracted directly from feces. Reference has already been made to the use of polyacrylamide gel electrophoresis to resolve the 11 gene segments; for instance, rotavirus groups A, B, and C can be clearly distinguished by RNA electrophoretic pattern alone. Another possible approach is dot hybridization, using a radioactively labeled cDNA probe to anneal with viral RNA that has been extracted from feces, denatured, and immobilized on nitrocellulose filters; again, a VP6 gene sequence could be selected as a probe to place the virus in the correct group, or a VP7 gene sequence to identify the G type. Finally, the polymerase chain reaction can be used to amplify viral dsRNA extracted from feces; the RNA is purified and then used as a template for reverse transcription and polymerase amplification using primer pairs appropriate for the degree of specificity desired.

Rotaviruses are difficult to culture *in vitro*. The secret of success is that trypsin must be incorporated in the (serum-free) medium, to cleave the relevant outer capsid protein VP4 and thus facilitate entry and uncoating of the virus. The monkey kidney cell line MA104 and the human colon carcinoma line CaCo-2 are the most commonly used for virus isolation. The cytopathic effect is not striking; immunofluorescence is used to identify rotaviral antigen in infected cells. Neutralization tests using appropriate polyclonal antisera or MAbs (plaque reduction or fluorescent focus reduction) can be used to determine the serotype of the isolate. Serum antibodies can be measured by EIA or neutralization tests.

Epidemiology

Rotaviruses are shed in large numbers, up to 10^{10} particles per gram of feces, for 3–7 days, or longer in the case of immunocompromised children. Spread presumably occurs mainly by contact with feces and can be decreased by rigorous attention to hygiene, including hand washing, disinfection, and proper disposal of contaminated diapers. Nevertheless, nosocomial outbreaks in hospitals, nurseries, and day-care centers are common. Waterborne epidemics involving adults also occur, the virus being resistant to chlorination. Descriptions of respiratory symptoms in some children and the winter prevalence of rotavirus in temperate climates raise the question of whether transmission can also occur via the respiratory route. The antigenic similarity between certain serotypes of animal and human rotavirus and the facility with which human rotaviruses can be grown in piglets, calves, and dogs raise the possibility of zoonotic infections.

Most human rotavirus infections are caused by group A serotypes, principally in infants 6–24 months of age. The viruses are endemic year-round in the tropics, but display an autumn, winter, or spring peak in temperate countries. Group B human rotaviruses are much less common but have been responsible for some extensive waterborne outbreaks in China, involving adults as well as children. Group C rotaviruses occur mainly in pigs and affect humans occasionally.

Of the group A rotaviruses, G types 1–4 are ubiquitous, with G1 being responsible for most of the severe disease worldwide. The "nursery" strains of group A isolated from nosocomial infections in newborn babies may be of G type 1, 2, 3, or 4, and are generally of P type 3. They appear to be relatively avirulent, although secretory IgA and trypsin inhibitors in breast milk, and the delayed appearance of proteolytic enzymes in the neonatal gut, may explain the mild clinical outcome.

Long-term prospective studies indicate that in some hospital nurseries and pediatric wards a high proportion of all newborn infants may become infected. Usually no disease occurs at this early age, but a degree of immunity develops. Disease is commoner in the 6 month to 2 year group. A 7-year prospective investigation of the molecular epidemiology of rotaviruses in children admitted to a Melbourne hospital revealed that, while one particular strain tends to predominate for months or years before being replaced by another, several electropherotypes are often found to be circulating simultaneously. Under these circumstances, opportunities abound for genetic reassortment. A logical extension of these studies will be to determine whether these genetic changes are accompanied by evidence of antigenic shift or drift analogous to that seen with influenza.

Control

Raising the standard of nutrition and hygiene in the developing world is the long-term answer to the horrific mortality from infantile gastroenteritis. In the meantime many lives can be saved by fluid and electrolyte replacement. A suitable mixture of glucose and electrolytes for administration by mouth is approved by the World Health Organization for universal use. Indeed, such has been the success of oral therapy that intravenous therapy may be necessary only for those infants with shock or unusually severe vomiting.

Patently there is a need for a vaccine, and several candidates have already undergone clinical trials. Exploiting the so-called Jennerian approach, U.S. workers have tested various animal rotavirus strains as attenuated oral vaccines in humans, with no consistent success so far. Human "nursery" strains are also about to be tested. A third approach is to use recombinant live viral or bacterial vectors (e.g., vaccinia virus or an attenuated *Salmonella*) for oral delivery of genes encoding VP4 and/or VP7. Finally, protection of mice has been obtained by injection of viral capsid proteins produced in insect cells infected with a recombinant baculovirus vector. Any vaccine would need to be administered early in life to protect the infant during the all-important first 2 years. It remains to be seen whether the growth of a live vaccine given at that time will be seriously impeded by maternal antibody, how long the resultant immunity will last, and whether it is cross-protective against serotypes not included in the vaccine. Oral delivery offers the obvious advantages of eliciting mucosal immunity and being capable of delivery to all infants in combination with oral poliovaccine.

Thought is also being given to passive immunization, using rotavirus antibody. Cows have been immunized with human or bovine rotaviruses to produce colostrum and milk containing high-titer rotavirus antibodies for oral administration to babies. It is envisaged that such preparations could be used either prophylactically, for control of cross-infection in hospitals, or therapeutically, to treat immunocompromised infants who are chronic rotavirus excretors.

Coltiviruses and Orbiviruses

The genus *Orbivirus* gets its name from the large doughnut-shaped or ring-shaped "capsomers" (*orbis*, ring or circle) that make up the inner capsid. The diameter of the virion is 80 nm, and the genome comprises 10 molecules of dsRNA. There are dozens of species, all being arboviruses. The genus includes important animal pathogens causing bluetongue of sheep and African horse sickness. The only orbiviruses believed to infect humans are the tick-borne Kemerovo viruses of Siberia. A much better known tick-borne agent, causing Colorado tick fever in North America, has recently been reassigned to a separate genus, *Coltivirus*, because its genome contains not 10 but 12 separate segments.

Colorado Tick Fever

Colorado tick fever is contracted from ticks by campers, hikers, hunters, and forest workers in the Rocky Mountains of North America. The virus is maintained in the wood tick, *Dermacentor andersoni*, being transmitted transstadially and overwintering in hibernating nymphs and adults. Nymphs feed on small mammals such as squirrels and other rodents, which serve as a reservoir for the virus. Adult ticks feed on larger mammals including humans during the spring and early summer.

After an incubation period of 3–6 days, the onset of illness is sudden. A characteristic "saddle-back" fever, headache, retroorbital pain, severe myalgia in the back and legs, and leukopenia are the cardinal features; convalescence can be protracted, particularly in adults. More serious forms of the disease, notably meningoencephalitis and hemorrhagic fever, occur in perhaps 5% of cases, mainly in children. Virus can be isolated from red blood cells or detected inside them by immunofluorescence, even several weeks after symptoms have disappeared. This is a remarkable situation, as erythrocytes have no ribosomes and cannot support viral replication. It seems that the virus multiplies in erythrocyte precursors in bone marrow, then persists in the mature red cell throughout its life span, protected from antibody during a prolonged viremia.

Eyach and Kemerovo Viruses

Eyach virus is a coltivirus present in ticks in Europe; antibodies have been reported in some patients with meningoencephalitis and polyneuritis, but no causal relationship has been established. Kemerovo virus, a member of the genus *Orbivirus*, is an arbovirus carried by ticks in Siberia; antibodies to it have been recorded in patients suffering from febrile illnesses, and the virus has been isolated from the blood and cerebrospinal fluid of patients with meningoencephalitis from the Kemerovo region of Siberia.

Orthoreoviruses

The first reovirus was isolated in 1951 from the feces of an Australian aboriginal child, by inoculation of infant mice. We now recognize this to be one of three serotypes of mammalian reoviruses belonging to the genus *Orthoreovirus*, which are not only ubiquitous in humans but also widespread in virtually every species of mammal that has been carefully studied. The orthoreoviruses replicate well in many types of cell culture. CPE is slow to develop and not particularly striking; however, on staining, characteristic crescentic perinuclear acidophilic inclusions are seen in the cytoplasm (see Fig. 5-3B). The three serotypes share common antigens but can be distinguished by hemagglutination inhibition.

Orthoreoviruses have been intensively studied by molecular biologists because the segmented genome lends itself to detailed biochemical and genetic analysis. The position and role of each type of viral protein in the virion, the function of individual gene products in replication, and the molecular pathogenesis of infection in mice were described in Chapter 7. We have relegated the description of their clinical significance to this postscript because they have never been positively demonstrated to cause significant disease in humans, other than some early studies that suggested they can produce mild upper respiratory symptoms. Yet they must commonly cause inapparent infection because the majority of people acquire antibodies to all three serotypes by adulthood. The virus is shed in feces for up to several weeks, is regularly found in sewage and polluted water, and is assumed to spread mainly by the fecal–oral route but also by the respiratory route.

Further Reading

Bellamy, A. R., and Both, G. W. (1990). Molecular biology of rotaviruses. *Adv. Virus Res.* **38,** 1.

Bishop, R. F. (1993). Development of candidate rotavirus vaccines. *Vaccine* **11,** 247.

Centers for Disease Control. (1990). Viral agents of gastroenteritis: Public health importance and outbreak management. *Morb. Mortal. Wkly. Rep.* **39,** RR-5.

Estes, M. K., and Cohen, J. (1989). Rotavirus gene structure and function. *Microbiol. Rev.* **53,** 410.

Holmes, I. H. (1988). *Reoviridae:* The rotaviruses. *In* "Laboratory Diagnosis of Infectious Diseases. Principles and Practice. Volume II, Viral, Rickettsial and Chlamydial Diseases" (E. H. Lennette, P. Halonen, and F. A. Murphy, eds.), p. 384. Springer-Verlag, New York.

Joklik, W. K., ed. (1983). "The Reoviridae." Plenum, New York.

Kapikian, A. Z., and Chanock, R. M. (1990). Rotaviruses. *In* "Fields Virology" (B. N. Fields, D. M. Knipe, R. M. Chanock, M. S. Hirsch, J. L. Melnick, T. P. Monath, and B. Roizman, eds.), 2nd Ed., p. 1353. Raven, New York.

Knudson, D. L., and Monath, T. P. (1990). Orbiviruses. *In* "Fields Virology" (B. N. Fields, D. M. Knipe, R. M. Chanock, M. S. Hirsch, J. L. Melnick, T. P. Monath, and B. Roizman, eds.), 2nd Ed., p. 1405. Raven, New York.

Matsui, S. M., Mackow, E. R., and Greenberg, H. B. (1989). Molecular determinants of rotavirus neutralization and protection. *Adv. Virus Res.* **36,** 181.

Ramig, R. F., and Ward, R. L. (1991). Genomic segment reassortment in rotaviruses and other *Reoviridae*. *Adv. Virus Res.* **39,** 163.

Schiff, L. A., and Fields, B. N. (1990). Reoviruses and their replication. *In* "Fields Virology" (B. N. Fields, D. M. Knipe, R. M. Chanock, M. S. Hirsch, J. L. Melnick, T. P. Monath, and B. Roizman, eds.), 2nd Ed., p. 1275. Raven, New York.

Chapter 35

Retroviridae

The novelty and intrinsic importance of retroviruses have been underlined by the award of Nobel Prizes on no less than three separate occasions. The immense significance of the discovery by Peyton Rous in 1911 that a malignant sarcoma of chickens could be transmitted by a cell-free filtrate containing an infectious virus was belatedly recognized by the award of a Nobel Prize in 1966, more than half a century later. Equally controversial initially was the iconoclastic proposal by Howard Temin that genetic information could flow "against the tide," from RNA to DNA, which was confirmed unequivocally by the subsequent discovery in 1970, independently by Temin and David Baltimore, of the enzyme reverse transcriptase in retroviruses, for which they received the Nobel Prize in 1975. Among other things, this enzyme has underpinned many of the subsequent spectacular advances in recombinant DNA technology and genetic engineering. Michael Bishop and Harold Varmus received the Nobel Prize in 1989 for the discovery of oncogenes and their role in oncogenic viruses and in cancer generally. In 1980 Bob Gallo and colleagues described the first human retrovirus, the cause of a form of adult T-cell leukemia, and in 1984 Gallo was also instrumental in confirming and extending the discovery by Luc Montagnier of the causal agent of the devastating pandemic of the acquired immunodeficiency syndrome (AIDS).

The threat posed by AIDS has triggered an unprecedented effort, by research scientists and governments alike, to understand and conquer this disease. We already know more about the human immunodeficiency virus (HIV) than about any of the viruses of longer standing. Indeed, HIV now sets the pace in virus research. Many outstanding virologists have moved across to HIV, and new concepts and techniques pioneered by HIV virologists in every area of the discipline—from regulation of viral replication, through molecular pathogenesis, to laboratory diagnostic methods and novel approaches to antiviral therapy and vaccinology—now represent the gold standard to which others aspire.

Properties of *Retroviridae*

Retrovirus virions are spherical, 80–100 nm in diameter, and have a unique three-layered structure. Innermost is the genome–nucleoprotein complex, closely associated with several molecules of the viral enzymes reverse transcriptase, integrase, and protease. This structure is enclosed within a capsid, which appears icosahedral and is centrally located in human T-cell lymphotropic virus (HTLV), spumaviruses, and the so-called C-type retroviruses of animals and birds (Fig. 35-1) but appears as a cone or rod in the case of the lentiviruses. This in turn is surrounded by a matrix protein, then a lipid envelope from which project glycoprotein peplomers.

The retroviral genome is unique among viral genomes in several respects: (1) it is the only diploid genome; (2) it is the only viral RNA that is synthesized and processed by the mRNA-processing machinery of the cell; (3) it is the only genome associated with a specific tRNA whose function is to prime replication; and (4) it is the only plus sense ssRNA genome that does not serve as mRNA soon after infection. Each haploid segment of the genome is a linear, single-stranded, plus sense molecule of 7–10 kb, with a 3′ polyadenylated tail and a 5′ cap (see Fig. 11-2). A molecule of a particular cellular tRNA, which serves as the primer for transcription, is base-paired to a site near the 5′ end of each viral RNA monomer.

The genome of all nondefective retroviruses contains three major genes, each coding for two or more polypeptides. Reading from the 5′ end, the *gag* gene (standing for group-specific antigen) encodes the virion core (capsid) proteins, the *pol* gene encodes the reverse transcriptase (polymerase), and the *env* gene encodes the peplomer proteins (envelope). Unlike most exogenous oncogenic retroviruses of animals and birds (see Chapter 11), the only known human oncogenic retrovirus, HTLV-1, contains no oncogene; it does, however, contain two regulatory genes one of which, *tax*, transactivates expression of all of its own genes as well as certain cellular genes. The genome of the lentiviruses, which are nononcogenic, contains half a dozen regulatory genes. The genome termini have several distinctive components (see Fig. 11-2) which are functionally important in transcription, integration, and regulation of expression of the integrated cDNA provirus. Other properties of retroviruses are summarized in Table 35-1.

Retroviruses are widely distributed among vertebrates as exogenous or endogenous agents (see Chapter 11) and are currently classified into seven genera. Only three genera contain human retroviruses (Table 35-2). Genera differ in virion morphology, regulatory genes, and replication strategy; they share no antigens. Within genera, however, there are some shared "group-specific" antigenic determinants. Antibodies to type-specific determinants on envelope glycoproteins neutralize viral infectivity.

Viral Replication

Many retroviruses replicate only in dividing cells. Most mammalian and avian retroviruses are not cytocidal and do not dramatically alter the metabolism of the cells that they infect, but this is not the case with the lentiviruses.

Virions adsorb to specific cell receptors via one of the two envelope gly-

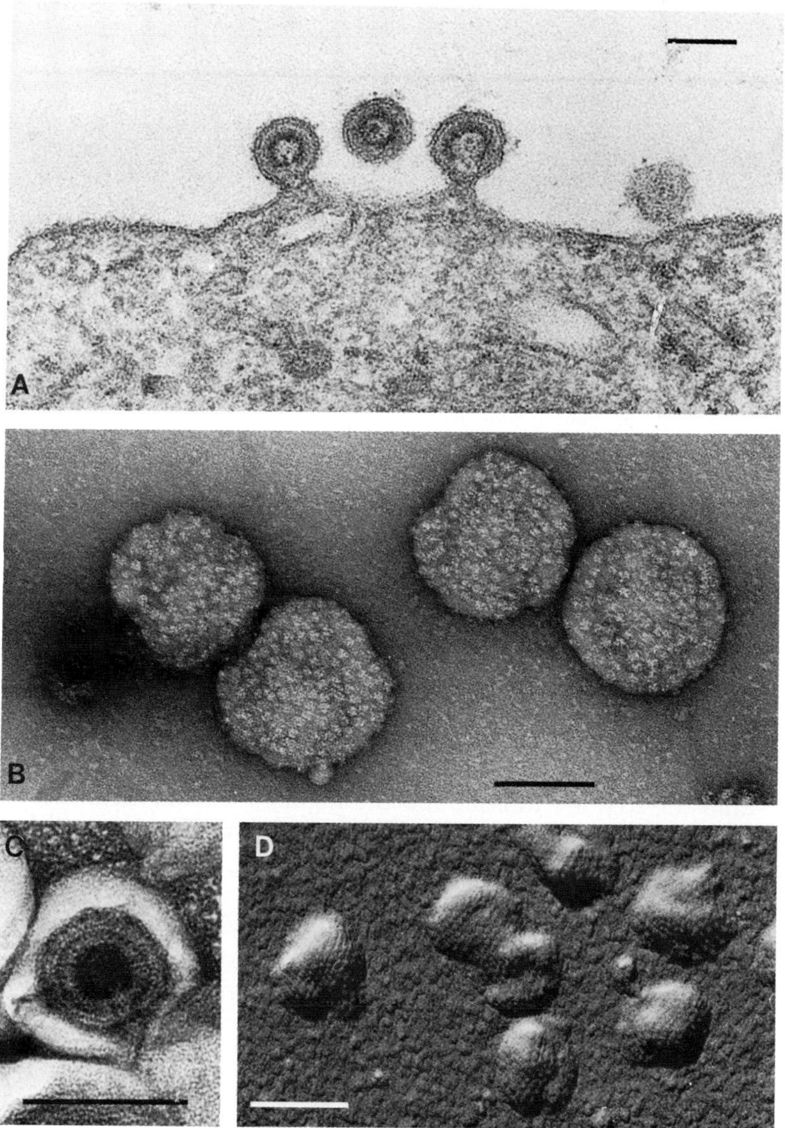

Fig. 35-1 *Retroviridae.* Murine leukemia virus, a typical type C retrovirus. (A) Budding of virions from a cultured mouse embryo cell. (B) Virions negatively stained with uranyl acetate, showing peplomers on the surface. (C) Virion somewhat damaged and penetrated by uranyl acetate, so that the concentric arrangement of core, shell, and nucleoid becomes visible. (D) Cores isolated by ether treatment of virions, freeze-dried and shadow-cast. The hexagonal arrangement of the subunits of the shell around the core is recognizable. Bars, 100 nm. (Courtesy Drs. H. Frank and W. Schafer.)

coproteins. Entry occurs either by receptor-mediated endocytosis or by direct fusion with plasma membrane, for different retroviruses. Uncoating releases the nucleoprotein complex into the cytoplasm. Using the genome-associated tRNA as a primer, the reverse transcriptase transcribes a minus sense cDNA from the viral genome (see Fig. 11-2B). Concomitantly, the viral RNA template

Table 35-1
Properties of Retroviruses

Spherical enveloped virion, 80–100 nm diameter
Ribonucleoprotein in central nucleoid (concentric in type C viruses; truncated cone in lentiviruses) within icosahedral capsid; surrounded by envelope with glycoprotein peplomers
Two copies of linear plus sense ssRNA genome, each 7–10 kb, covalently linked; *gag, pol, env* genes; some also contain regulatory genes; some carry an oncogene
Reverse transcriptase transcribes DNA from virion RNA; following formation of long terminal repeats, dsDNA is integrated into cellular chromosomes as provirus
Full-length RNA transcripts encode core and capsid proteins, protease, reverse transcriptase/RNase H, integrase; shorter spliced transcripts encode envelope glycoproteins and regulatory proteins
In productive infections, virions assemble at and bud from plasma membrane
Infection may be cytocidal, noncytocidal, or transforming; oncogenic retroviruses may be replication-competent or defective, and induce malignancy by transduction, *cis*-activation, or *trans*-activation

is digested by a second domain of the reverse transcriptase molecule, which carries RNase H enzymatic activity. Oligonucleotides resulting from this hydrolysis then serve as primers for the synthesis of plus sense cDNA using the newly made minus sense cDNA as template. The resulting linear double-stranded DNA contains long terminal repeats (LTRs) composed of sequences duplicated from the 3′ (U3) and 5′ (U5) ends of the viral RNA, so that each end of the provirus contains an LTR consisting of a U3, R, and U5 region; within the LTR are to be found the regulatory sequences, including an enhancer/promoter region with binding sites for viral and/or cellular regulatory proteins. The dsDNA moves to the nucleus, and several such molecules become integrated as provirus at random sites in the cell DNA. Integration requires the viral integrase and involves removal of two nucleotides from the ends of the viral DNA and generation of a short duplication of cell sequences at the integration site, enabling joining of the ends of the viral DNA to the cell DNA.

Integration is a prerequisite for virus replication. The integrated provirus is transcribed by cellular RNA polymerase II. The complete RNA transcript is identical to the original genomic RNA and serves as mRNA for translation into the Gag (core) polyprotein and, after frameshifts, the protease and the Pol polyprotein. A second, shorter mRNA, spliced from the full-length RNA, is translated, again in a different reading frame, to yield the Env precursor. In

Table 35-2
Human Retroviruses

Genus	Virus	Disease
BLV-HTLV retrovirus	HTLV-1	Adult T-cell leukemia/lymphoma, tropical spastic paraparesis
	HTLV-2	?Nil
Lentivirus	HIV-1, HIV-2	AIDS
Spumavirus	Human foamy virus	?Nil

the case of HTLV and HIV, a range of differently spliced mRNAs encode various regulatory proteins. All the mRNAs share a common sequence at their 5′ end. The viral protease is responsible for posttranslational cleavage of the Gag polyprotein to yield the matrix, capsid, and nucleocapsid proteins, and also for cleavage of the Pol polyprotein to yield the reverse transcriptase/RNase H and the integrase. On the other hand, the Env polyprotein is cleaved by a cellular protease to yield two envelope glycoproteins: a transmembrane protein and a receptor-binding protein.

The details of assembly of the virion are not fully understood and differ from genus to genus. Two covalently linked molecules of the full-length plus sense genomic RNA assemble into a core structure by interaction of a packaging signal in the leader sequence of the RNA with a zinc-finger motif in the nucleocapsid protein. Capsids generally assemble at the cell surface. Myristylated and glycosylated envelope proteins enter the plasma membrane and form the peplomers of the envelope which the virion gains by exocytosis. The final proteolytic cleavage steps occur during and even after budding.

Human T-Cell Leukemia (Lymphotropic) Viruses

HTLV-1 was the first human retrovirus to be discovered. Like most of the previously known retroviruses of other animals and birds, HTLV-1 was found to be oncogenic—in this case responsible for an unusual form of T-cell leukemia in adults—but the mechanism of oncogenesis turned out to be unique. A second serotype, HTLV-2, is prevalent in intravenous drug users in the United States, but has not yet been demonstrated to cause disease.

Viral Replication

HTLV-1 is a replication-competent retrovirus which can induce cancer in spite of the fact that its genome (1) carries no viral oncogene and (2) does not, following integration of proviral DNA, *cis*-activate any nearby cellular oncogene. It contains a regulatory gene, *tax* (Fig. 35-2), the protein product of which, Tax, acts in *trans* both to up-regulate transcription of all the viral genes in the integrated HTLV proviral DNA and also to initiate the leukemogenic process (see Chapter 11). A second HTLV regulatory gene, *rex*, encodes a protein Rex that regulates the splicing of RNA transcripts by promoting the production and transport to the cytoplasm of unspliced viral RNA and the incompletely spliced *gag/pol* and *env* mRNAs, so allowing the production of new virions. Rex has a negative effect on its own production and that of Tax, because multiple splicing is required to produce the mRNAs for these two regulatory proteins (Fig. 35-2). This negative regulation of Tax and Rex expression by Rex would then lead to decreased expression of all the viral genes, and might reestablish latency. Thus it has been postulated that Rex may orchestrate a state of latency alternating with periods of virus production.

Pathogenesis

HTLV probably enters the body principally inside infected $CD4^+$ lymphocytes in semen or blood, as well as vertically from mother to infant via breast milk

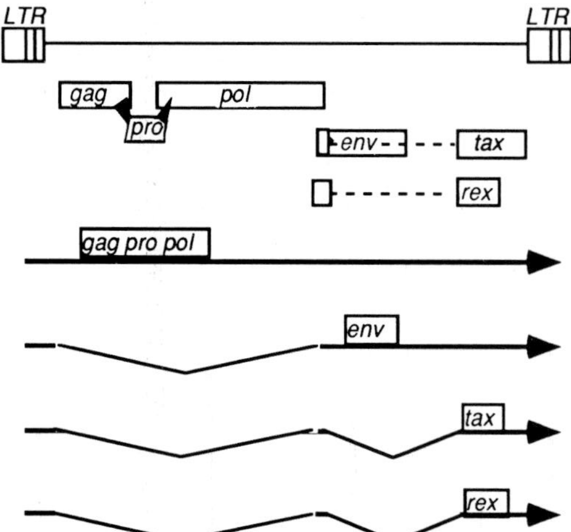

Fig. 35-2 HTLV provirus genome, showing the position of the LTRs, encoded structural genes (*gag, pro, pol, env*), and nonstructural genes (*tax, rex*), which are divided genes as indicated by the dashed lines. The short arrowheads indicate ribosomal frameshift sites. Beneath are shown the RNA transcripts that serve as mRNAs for translation into particular proteins (long arrows); the thinner V-shaped lines represent introns that are removed from the original RNA transcript to yield the functional spliced mRNA molecules. (Courtesy Dr. J. M. Coffin.)

and possibly transplacentally. It establishes a lifelong persistent infection which generally remains subclinical but occasionally induces disease after an incubation period of 10–40 years. The pathogenesis of one of the two major clinical manifestations, T-cell leukemia/lymphoma, was discussed in detail in Chapter 11. Briefly, the oncogenic potential of HTLV is ascribed to its regulatory gene *tax,* the product of which transactivates transcription not only from the proviral LTR but also from certain cellular oncogenes as well as from the cellular gene encoding the interleukin-2 (IL-2) receptor, thereby inducing autocrine stimulation of T-cell proliferation and setting in motion a chain of events that may eventually lead to cancer. As will be seen later in this chapter, expression of the HTLV *tax* gene also induces the production of the pleiotropic cellular DNA-binding transcription factor NF-κB and may thereby serve as a cofactor in the progression of AIDS by inducing transcription of the integrated HIV provirus.

Clinical Features

Although most HTLV-1 infections remain asymptomatic, there is a 1–4% risk of development of disease after an incubation period of 10–40 years, usually between the ages of 30 and 50. There are two distinct clinical manifestations, only rarely encountered in the same patient.

Adult T-cell leukemia/lymphoma (ATLL or ATL) generally takes the form of an acute aggressive leukemia of mature $CD4^+$ T lymphocytes in middle-aged adults. The patient presents with lymphadenopathy, hepatosplenomegaly, hypercalcemia, lytic bone lesions, and sometimes leukemic cell infiltrates in the skin. The malignant T cells are pleomorphic with large convoluted nuclei. The consequent immunosuppression may lead to various opportunistic infections. Death generally occurs within a year. Less commonly, the disease may follow a more chronic course with fewer leukemic cells in the blood but resembling a non-Hodgkin's lymphoma, with or without skin deposits. There also appears to be an association with polymyositis.

Tropical spastic paraparesis (TSP), otherwise known as HTLV-1 associated myelopathy, is a progressive demyelination of the long motor neuron tracts in the spinal cord. Seen mainly in 20- to 50-year-old women, the affliction starts with lumbar back pain radiating down the legs and progresses to weakness and spastic paralysis of both lower limbs, with dysethesia, urinary frequency or retention, and sometimes visual changes. Unlike multiple sclerosis, there are no remissions.

Laboratory Diagnosis

ATL is characterized by pleomorphic lobular leukemic cells with mature $CD4^+$ T cell markers. HTLV-1 infection, usually subclinical, is diagnosed by demonstrating antibodies in an enzyme immunoassay (EIA) or gelatin particle agglutination. Today, genetically cloned HTLV antigen is preferred to lysates of virus-transformed cell lines as capture antigen, in order to reduce the excessive number of false positives. Because of extensive sharing of epitopes between HTLV-1 and HTLV-2, appropriate synthetic peptides are required to distinguish the respective antibodies. Western blotting is used to confirm provisionally positive results from EIA.

The polymerase chain reaction (PCR) followed by Southern blotting is used to demonstrate HTLV DNA from lymphocytes and has the additional advantage of discriminating between HTLV-2 and HTLV-1, providing suitable pairs of primers are chosen. However, as always, the exquisite sensitivity of PCR is a two-edged sword; HTLV infection can be recognized even in seronegative individuals, but false positives may occur in any but the most experienced hands.

Virus can be propagated only with difficulty, classically by *in vitro* culture of the patient's peripheral blood leukocytes, either with a mitogen or with an HTLV-transformed T-cell line. Because virus is almost exclusively cell-associated, infection occurs by fusion of infected and uninfected $CD4^+$ lymphocytes, leading to transformation and immortalization of the latter. Cells other than T lymphocytes can also be infected, albeit with difficulty, by cocultivation with infected $CD4^+$ T cells.

Epidemiology

The geographic distribution of HTLV-1 is patchy, with a tendency to cluster in certain countries, mainly in the tropics. The highest prevalence has been observed in southern Japan, the Caribbean, equatorial Africa, parts of South America, eastern Siberia, Pacific islands such as Papua New Guinea, and certain Eskimo populations. This suggests that the virus is of ancient origin and has been maintained by vertical transmission in isolated racial groups, dating back to the days before the major waves of intercontinental migration 40,000 to 100,000 years ago when the average human life span was shorter than the incubation period of the diseases we see in HTLV-positive patients today. Spread to the Caribbean and southeastern United States may have coincided with the much more recent slave trade from Africa.

The observed clustering of infection within particular families indicates that vertical transmission or close contact is required. Infants of seropositive mothers acquire infection more readily if breast-fed. However, the existence of additional currently undefined routes of transmission from mother to child is suggested by the observation that the prevalence of infection in the above-mentioned endemic areas is diminishing in parallel with improving living standards. The second important route of transmission is sexual intercourse, with the receptive partner being most at risk, so accounting for the somewhat higher incidence in females. Third, parenteral transmission is important.

Prevalence in blood donors varies from less than 0.1% in the United States to 5% or more in countries of high endemicity. Transmission by blood transfusion will fall away following more widespread introduction of screening of blood donors, now routine in many developed nations. Hemophiliacs appear not to be at risk, presumably because the virus is cell-associated and is effectively lost during the extraction of factor VIII from plasma. On the other hand, intravenous drug abuse is an increasing hazard, so much so that around 5% of IV drug users are already infected with HTLV; fortunately, most infections appear to be due to HTLV-2, with which no disease has yet been linked. Little is yet known of the natural history of HTLV-2, which has been found to be common in postpubertal Guaymi Indians in Panama, among whom it is considered to spread by sexual intercourse. The search continues for additional HTLV types and for possible additional disease associations.

Control

Screening of blood and organ donors for HTLV antibody will essentially abolish transmission by that route. Counseling of high-risk groups is about all that can be done to reduce transmission by other routes. Intravenous drug users must be discouraged from sharing needles and syringes. Known positive mothers should not breast-feed their infants. Seropositive men should be advised to use condoms with any seronegative partner and to consider the virtues of monogamy. Currently, experimental recombinant vaccines are being tested in laboratory animal models, but interruption of cell-mediated transmission will not be easy; moreover, human trials would necessarily go on for decades and would presumably not be contemplated unless a putative vaccine had been shown to prevent infection. Antiviral therapy, for example, using inhibitors of reverse transcriptase, seems unlikely to be of relevance in treating an established malignancy.

Human Immunodeficiency Viruses

Although the first case of AIDS was described as recently as 1981, the pandemic has escalated at such a rate that by 1993 the World Health Organization (WHO) estimated that over 13 million young adults had been infected with HIV and over 2 million had already developed AIDS, most of whom had died (Fig. 35-3). Approximately 5000 new infections occur every day. By the turn of

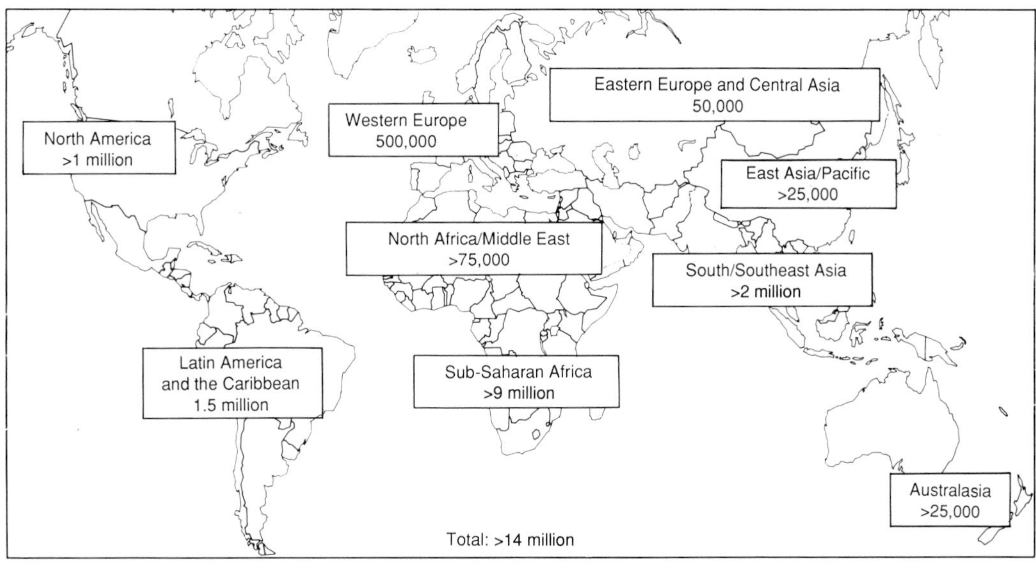

Fig. 35-3 Estimated distribution of cumulative HIV infections in adults, by continent or region, as of late 1993. [Based on *Weekly Epidemiological Record,* **69,** 7 (1994) .]

the century it is projected that 30–40 million will have been infected and that over 1 million will die every year. Moreover, because the principal modes of spread are heterosexual intercourse and perinatal transmission, millions of children will either die from AIDS or be orphaned by the year 2000. In many of the large cities of the United States AIDS is now the commonest cause of death in young adults. However, the situation is even worse in Africa, where 20–30% of the sexually active age group in several cities are already infected, and in Southeast Asia and India, where poverty breeds prostitution and HIV is endemic among this underclass. AIDS threatens the very fabric of society, increasingly targeting the poor and underprivileged. The pandemic is out of control. We have no effective antiviral agent and no immediate prospect of a vaccine. Indeed, if a vaccine with 100% efficacy were discovered tomorrow, millions of currently infected people would still die over the next decade.

Properties of Human Immunodeficiency Viruses

The HIV virion differs from that of other retroviruses principally in that the dense core appears cone-shaped and contains some additional minor proteins (Figs. 35-4 and 35-5). The 72 peplomers projecting from the envelope are oligomers of a glycoprotein, gp160, which has been cleaved into two noncovalently linked components, gp120 and gp41. The receptor-binding ligand and the most important antigenic domains, notably the V3 loop, are present in the gp120 molecule; gp120 is the most extensively glycosylated viral protein known, and it is presumed that this "sugar coating" is a protective device to impede access of neutralizing antibodies. The hydrophobic membrane anchor is provided by gp41, which is responsible for viral entry into the host cell by membrane fusion. The viral envelope also contains some cellular proteins, notably class I and II MHC antigens.

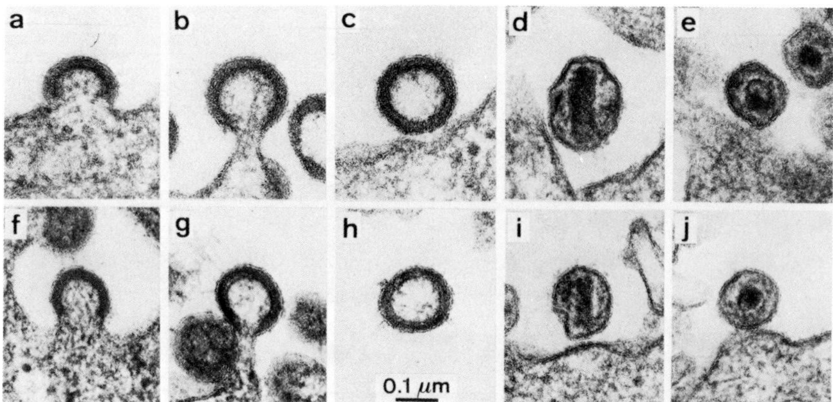

Fig. 35-4 Electron micrographs of thin sections of cells infected with HIV-1 (a to e) and visna virus, a classic animal lentivirus, showing budding (a,b), immature virion (c), and mature virions with cone-shaped or cylindrical core (d,e). [From M. A. Gonda, F. Wong-Stahl, R. C. Gallo, J. E. Clements, O. Narayan, and R. V. Gillen, *Science* **227,** 173 (1985). Copyright 1985 by the AAAS. Courtesy Dr. M. A. Gonda.]

The inner surface of the envelope is lined by a myristylated matrix protein, p17, a cleavage product of the 55 kDa *gag* gene product. The most abundant viral protein is another *gag* gene product, the phosphoprotein p24, of which the icosahedral nucleocapsid is constructed; the other two are p9 and p7, which are closely associated with the genome in the core of the virion. Also associated with the genome in the core are three viral enzymes: reverse transcriptase (also carrying RNase H activity), integrase (endonuclease), and protease.

The diploid (2 $\times$ 9.2 kb) plus sense ssRNA genome of HIV contains, in

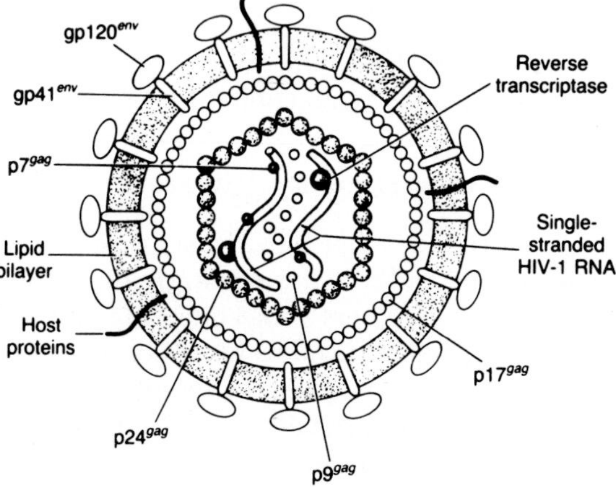

Fig. 35-5 Schematic diagram of the HIV-1 virion. The location of each of the structural proteins of the virion is shown (see text for details). The reverse transcriptase (with its associated enzyme activities) is placed in apposition to the viral RNA, which is depicted as two molecules to indicate the diploid nature of the genome, although the two are in fact linked together end to end. [From W. C. Greene, *N. Engl. J. Med.* **324,** 308 (1991).]

addition to the standard *gag, pol,* and *env* genes, six additional genes, all of which have regulatory functions, described below (Fig. 35-6).

Two types of HIV are currently recognized. HIV-2 displays only about 40% nucleotide sequence similarity (homology) with HIV-1 and contains a unique gene, *vpx,* in lieu of the *vpu* of HIV-1. More recently described, HIV-2 shows a close homology with the sooty mangabey simian immunodeficiency virus, is currently largely confined to Africa, and appears to be less virulent than HIV-1.

Five major genotypic subtypes (*clades*) of HIV-1, differing by 30–35% in their *env* and *gag* sequences, have been defined, only one of which occurs in the United States. Overall, within a clade HIV-1 strains can differ by up to 20% in nucleotide sequence. The extent of genetic drift in an individual patient is much less, usually less than 3%, hence PCR and sequencing can be used to trace the source of infection (e.g., from a particular batch of factor VIII, from mother to child, or from a single blood donor).

Replication of HIV

The replication of HIV is notable in a number of respects. (1) Since $CD4^+$ T lymphocytes support HIV replication only when growing, yet most T cells in the body are resting, many virus–cell interactions result in a long-standing latent infection which may later be converted to a productive infection following activation of the T cell. (2) Both cellular and viral proteins are involved in complex pathways regulating almost every step in the viral replication cycle. (3) Mammalian lentiviruses are unique in that patterns of transcription can be determined by binding of certain regulatory proteins (Tat and Rev) not to the DNA provirus but to the RNA transcript itself. (4) Although the HIV genome contains only nine genes, three encoding structural proteins and six encoding regulatory proteins, multiple splicing options yield over 30 distinct RNA species, the functions of many of which are not yet understood. (5) HIV spreads much more efficiently from cell to cell by cell–cell fusion to form a syncytium, yet most of our knowledge of the biochemistry of the replication cycle comes

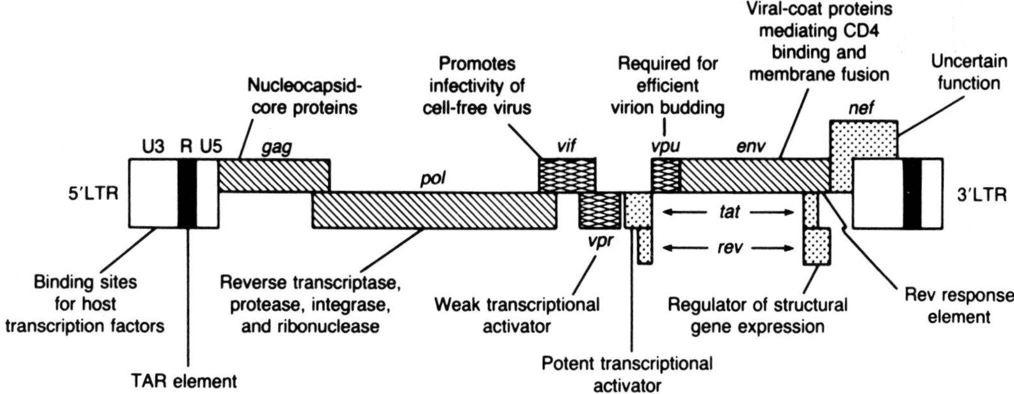

Fig. 35-6 Genomic organization of HIV-1 and overview of the known functions of each gene product (see text for details). [From W. C. Greene, *N. Engl. J. Med.* **324,** 308 (1991).]

from cell culture studies involving conventional infection by free virions; there are probably important differences between the two.

The virion attaches via its gp120 envelope protein to the CD4 differentiation antigen, which is the primary HIV receptor on helper T lymphocytes and cells of the macrophage lineage. A protease (CD26) has been postulated to serve as an accessory receptor, cleaving gp120 within its V3 loop, inducing a conformational change in the molecule and exposing the fusogenic domain of gp41, but this remains to be proved. Entry then occurs by direct (pH-independent) fusion of gp41 with the plasma membrane, liberating the core of the virion into the cytoplasm. Reverse transcription to produce dsDNA copies of the genome, followed by integration of some of these cDNA molecules into cellular chromosomes, occurs as described earlier for other retroviruses.

The integrated provirus may remain latent indefinitely. Transcription by cellular RNA polymerase II cannot occur in resting T cells but only in replicating cells. Activation of the T cell can be brought about by mitogens, cytokines, or transactivating proteins encoded by certain other viruses such as HTLV-1, herpes simplex virus (HSV), Epstein-Barr virus (EBV), cytomegalovirus (CMV), human herpesvirus 6 (HHV-6), or hepatitis B virus (Fig. 35-7). Many of these agents activate the cell by activating expression of members of the NF-κB family of cellular proteins which are usually sequestered in the cytoplasm as a complex with an inhibitor; activation promotes disassembly of the complex and translocation of NF-κB to the nucleus. The normal function of these host transcription factors is to bind to discrete enhancer elements in cell DNA which control expression of certain cellular genes, but they also recognize and bind to two κB enhancer elements in the U3 region of the HIV proviral LTR. Binding of NF-κB, together with several additional host transcriptional factors which are expressed constitutively, enables the proviral genome to be transcribed, albeit at a relatively low level.

There are three major classes of HIV mRNA, which result from different splicing protocols: unspliced full-length (9.2 kb) RNA transcript, corresponding to the viral genome, is used as mRNA for translation of the Gag and Pol polyproteins; singly spliced 4.5 kb RNAs encode the Env, Vif, Vpr, and Vpu proteins; several multiply spliced 2 kb RNA transcripts encode the Tat, Rev, and Nef proteins. In the initial round of transcription, the only RNA transcripts exported to the cytoplasm are the multiply spliced 2 kb mRNAs encoding regulatory proteins Tat, Rev, and Nef.

Tat is a potent transactivator which binds to a responsive element known as TAR which is present in the LTR of both proviral DNA and the common 5′ end of all the viral RNA transcripts. In association with cellular factors, Tat enhances the efficiency of transcription by cellular RNA polymerase from the HIV promoter 1000-fold, mainly by preventing premature termination of transcription. All the viral genes are transcribed, but transcripts encoding all proteins other than Tat, Rev, and Nef are retained in the nucleus until significant levels of Rev have accumulated.

Rev is the second essential regulatory protein encoded by HIV. It is a phosphoprotein that binds to a sequence known as RRE (Rev responsive element) in viral RNA. Because RRE is located within the *env* gene, it is present in the 9.2 kb and 4.5 kb HIV RNA transcripts but not in the 2 kb transcripts encoding Tat, Rev, or Nef. In the absence of Rev, the 9.2 and 4.5 kb transcripts are retained unspliced in the nucleus. However, as Rev accumu-

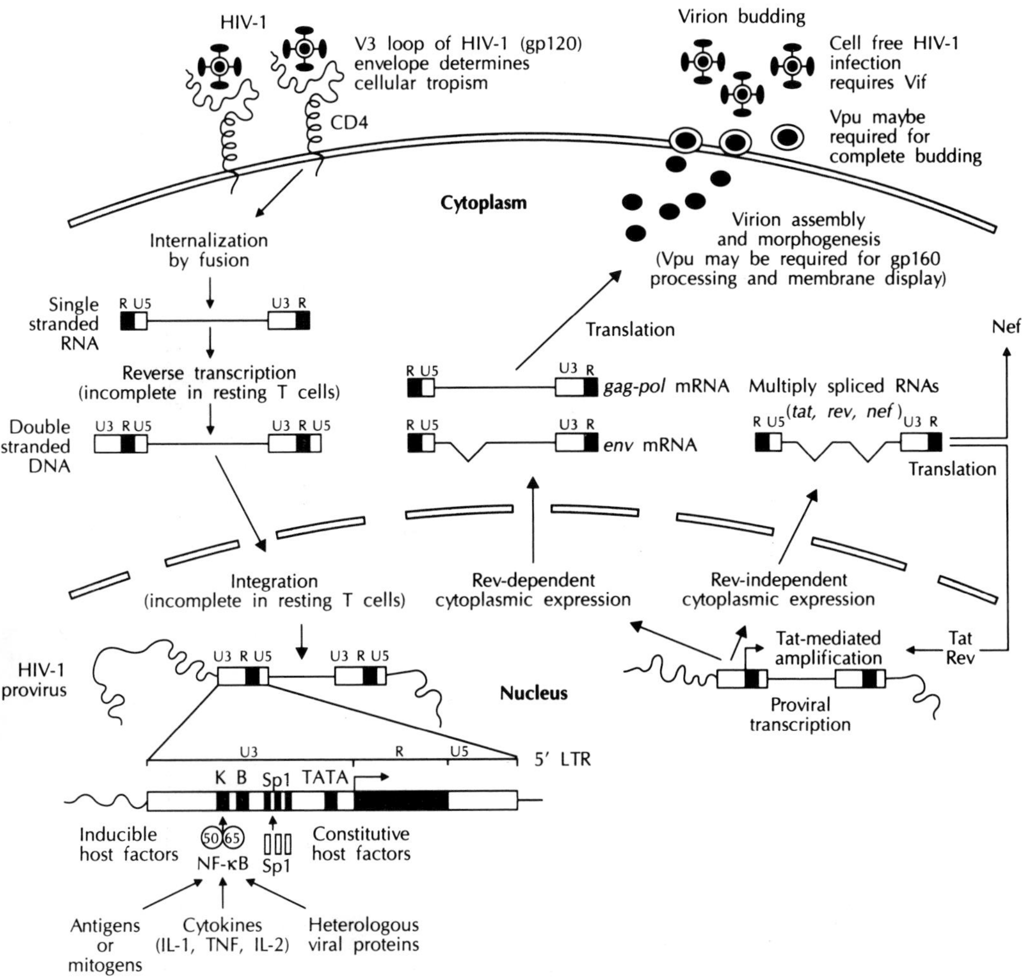

Fig. 35-7 Overview of the replication cycle of HIV-1 in a T cell. Infection begins (top left) with attachment of the virion via gp120 to its receptor, CD4. Following entry by gp41-mediated fusion, the genome becomes available for reverse transcription to produce a dsDNA with long terminal repeats (LTRs) composed of sequences duplicated from the 3′ (U3) and 5′ (U5) ends of the viral RNA. Complete reverse transcription and integration of the DNA provirus into a chromosome occur efficiently only in activated and proliferating T lymphocytes. Transcriptional activity of the HIV provirus is regulated by constitutive host cell transcription factors (Spl and the TATA-binding factors) and by activation-inducible members of the NF-κB family of host transcription factors (p50 and p65), both of which bind to specific sequences in the proviral regulatory region, LTR (an enlarged version of which is shown at bottom left). The virus-coded regulatory protein, Tat, is one of the early gene products; Tat binds to the TAR region in the LTR and greatly amplifies transcription of all the viral genes (bottom right). Following synthesis of a full-length RNA transcript, a complex array of alternatively spliced viral mRNAs can be produced. The differential expression of distinct species of viral mRNAs is controlled by a second HIV regulatory protein Rev. Early in infection, when the level of Rev is low, only the multiply spliced mRNAs for the regulatory proteins Tat, Rev, and Nef are exported to the cytoplasm for translation. Once a sufficient level of Rev accumulates, the unspliced and singly spliced mRNAs that provide new viral genomes and also encode the structural proteins (Gag and Env), the enzymes (Pol), and the remaining regulatory proteins (Vif, Vpr, and Vpu) are exported to the cytoplasm and translated. Encapsidation of the viral genome is followed by budding from the plasma membrane (top right, see text for details). [From M. B. Feinberg and W. C. Greene, *Curr. Opin. Immunol.* **4,** 466 (1992).]

lates in the nucleolus, splicing and export of these RRE-containing transcripts to the cytoplasm occur, and the full range of HIV mRNAs becomes available for translation. Evidently some characteristic of RRE-containing transcripts holds them in the nucleus, and/or perhaps inhibits spliceosome function, until binding of Rev overcomes the block.

Nef is the third important HIV regulatory protein. It is a myristylated cytoplasmic protein which, though not required for viral replication in cultured lymphocytes, is essential for replication in macrophages and also for the development of simian AIDS in the simian immunodeficiency virus (SIV)–rhesus monkey model. It down-regulates expression of CD4 and IL-2 and may be a virokine that alters the state of activation of the target cell *in vivo.*

Vif (viral infectivity factor), while not required for direct cell-to-cell transmission of HIV by fusion, is required for some step in morphogenesis that determines infectivity of the virion for the next cell to which it spreads. Vpu is an integral membrane phosphoprotein resembling the ion channel M2 protein of influenza virus. It promotes maturation of the Env glycoprotein and release of the virion by budding. Vpr, which is found in the virion itself, is a weak transcriptional activator of the HIV LTR and of various cellular promoters, perhaps stimulating cell differentiation.

The *gag–pol* transcript is translated into a Gag polyprotein and, following a ribosomal frameshift, into a Pol polyprotein, the latter being produced with much lower efficiency than the former. The viral genome assembles in the cytoplasm with the Gag and Gag–Pol precursors of the core proteins and migrates to associate with areas of the plasma membrane containing the Env glycoprotein gp160. The Pol polyprotein is cleaved to yield the four enzymes contained within the core of the virion, whereas the Gag polyprotein is cleaved to yield the four structural proteins p24, p17, p9, and p7. The viral protease is responsible for these processing events, which occur during and even after budding of the virion.

Pathogenesis

The hallmark of AIDS is depletion of $CD4^+$ T lymphocytes, and these cells are the principal target of HIV infection in lymph nodes and blood. $CD4^+$ T cells support the replication of the virus only when activated, but viral cDNA can persist in resting lymphocytes as a latent infection. In addition, cells of the monocyte lineage are susceptible, and macrophages constitute a second major reservoir of replicating virus in tissues, including the brain.

HIV is normally transmitted from person to person via genital secretions, notably semen, or less commonly via blood. The first cells to become infected may be resident tissue macrophages or submucosal lymphocytes in the genital tract or rectum, but the virus is then transported to the draining lymph nodes, where it replicates extensively. One to three weeks after infection, most patients exhibit a brief glandular fever-like illness, associated with a high titer of virus in the blood and a decline in $CD4^+$ T cells (Fig. 35-8). A vigorous cellular and humoral immune response ensues, and within a month or so the viremia declines to a near undetectable level. $CD8^+$ cytotoxic T cells, natural killer (NK) cells, and antibody-dependent cell-mediated cytotoxicity (ADCC) may all contribute to this decline by lysing infected cells, while neutralizing antibodies mop up free virions.

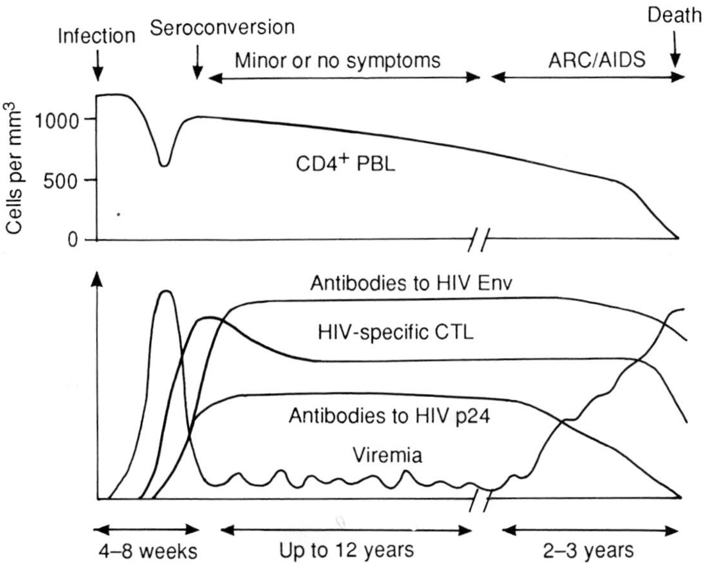

Fig. 35-8 Time course of HIV infection. (*Top*) Decline of CD4+ T lymphocytes in peripheral blood, in relation to the early illness that precedes seroconversion, followed by the long asymptomatic period and the later development of the constellation of symptoms sometimes known as the AIDS-related complex (ARC), and finally AIDS itself. (*Bottom*) Time course of development and decline of viremia, antibodies against the two key proteins (Env and p24), and HIV-specific cytotoxic T lymphocytes (CTL). [From R. A. Weiss, *Science* **260,** 1273 (1993). Copyright 1993 by the AAAS. Courtesy Dr. R. A. Weiss.]

There then follows a long asymptomatic period of "clinical latency," lasting anything from 1 to perhaps 15 years or longer (average about 10 years), before any further clinical evidence of disease becomes apparent. During this period, only very low titers of virus are demonstrable in blood by conventional culture techniques, and only a tiny minority of circulating CD4+ T cells are producing virus. However, much higher levels of virus, and at least 10-fold as many infected cells, are detectable in lymph nodes. Follicular hyperplasia is demonstrable in these and other lymphoid organs, where follicular dendritic cells entrap free virus and present it to lymphocytes that move through the organ. As time passes there is a steady decline in the number of CD4+ T cells, with involution of lymphoid follicles. When CD4+ cell counts fall below about 200–400 per μl, opportunistic infections with various microorganisms may occur, and eventually the depleted immune system is unable to cope. Activation of CD4+ T cells renders them permissive for HIV replication, and other viruses such as HHV-6, CMV, HTLV, and hepatitis B virus transactivate HIV expression also. Large numbers of virions spill over from the degenerating lymph nodes into the blood. The increasing burden of virus and the decreasing numbers of immunocompetent helper T cells collude to convert a low-level chronic infection to a more rapidly progressive infection. Death follows as a result of infections, malignancy, or a cachexia-like state.

Much remains to be learned about the pathogenesis of HIV (Table 35-3). For example, there are several unproved hypotheses about the mechanism of destruction of CD4+ T cells. Most obviously, HIV infection can be directly cytolytic for activated CD4+ T cells, or can cause them to fuse (via HIV envelope glycoprotein) with uninfected CD4+ cells to form a labile syncytium.

Table 35-3
Immunosuppressive Interactions between HIV and $CD4^+$ T Cells

Viral cytolysis of $CD4^+$ T cells, especially memory cells
Fusion of infected with uninfected cells to form syncytia
Inhibition of cytokine expression, CD4 expression, and biological functions of $CD4^+$ T cells
Lysis of infected $CD4^+$ T cells by cytotoxic T cells, NK cells, ADCC, antibody/complement
Induction of apoptosis of infected $CD4^+$ T cells
Infection of stem cells
Dysregulation of helper T-cell differentiation (Th_1 decrease, Th_2 increase)
Autoimmune destruction of infected $CD4^+$ T cells

Second, infected T cells are subject to immune cytolysis by cytotoxic T lymphocytes, ADCC, or antibody/complement-mediated lysis. Third, it has been proposed that, as a result of cross-linking of CD4 by HIV gp120, or by other mechanisms, T cells may be activated to commit suicide by apoptosis (programmed cell death). To explain the fact that dying $CD4^+$ T cells are not totally replenished, it has been postulated that HIV might also infect stem cells, or other cells that normally secrete cytokines required for such replacement. Moreover, early in the asymptomatic phase, HIV preferentially destroys memory $CD4^+$ T cells. Later, there is evidence of dysregulation of the process of differentiation of Th_0 into Th_1 and Th_2 cells such that the Th_1 [inflammatory, or delayed hypersensitivity (DTH)] cell activity needed to promote cell-mediated immune responses is progressively lost and replaced by a Th_2 (help for antibody production) response; this switch is a predictor of imminent decline in the $CD4^+$ T-cell population to be followed shortly by AIDS and death. Finally, it has been proposed that AIDS may be an autoimmune disease, sometimes but not solely attributable to molecular mimicry, but the evidence for this is not strong. The immunological circuitry is so interconnected that it is extremely difficult to assign blame to any particular mechanism, or even to dissociate cause from effect.

Macrophages are the principal target cell for certain lentiviruses of other animals which are not notably immunosuppressive but which, like HIV, attack the brain. HIV replicates slowly in differentiated nonreplicating human tissue macrophages, and less well in monocytes. Chronic infection of these phagocytes can continue for weeks, with little cytopathology or syncytium production, although functions such as cytokine production may be affected. HIV buds into cytoplasmic vacuoles rather than through the plasma membrane of macrophages, and this may insulate these cells from immune cytolysis. Infection of macrophages may predispose the patient to opportunistic infections such as tuberculosis and toxoplasmosis.

There is evidence from cell culture studies that HIV can infect other types of cells, including neurons, endothelial cells, oligodendrocytes, and astrocytes in the brain, and M cells and enterochromaffin cells in intestinal mucosa, thus accounting, respectively, for the diverse CNS manifestations of AIDS and the regular occurrence of chronic diarrhea and malabsorption. As several of these cells lack CD4 antigen, there would be a need for an alternative receptor (e.g., galactosylcerebroside on neurons). There is debate about whether the protean effects of HIV on the CNS are attributable solely to infection of the principal target cell, the resident brain macrophage known as microglia, with a neurotropic variant of the virus, and whether the damage is attributable to direct lysis of cells, to their destruction by cytotoxic T cells, or to release of neurotoxic cytokines, proteases, gp120, etc., or to more subtle effects such as interference with neurotransmitters.

Genetic Variation in HIV

Such is the infidelity of reverse transcription that the virus recoverable from any given individual is found to have diverged steadily from the original infecting strain to comprise a constellation of countless distinct mutants, jointly known as a "quasi-species." This variation is most marked in the "variable" regions of the *env* gene. Phenotypically, such mutants, (or less commonly recombinants) can be allocated to at least three classes, differing in tropism, virulence, or antigenicity respectively.

T-cell tropic variants tend to grow rapidly to high titer ("rapid–high" strains) in cultures of peripheral blood lymphocytes (PBL), as well as to be syncytium-inducing (SI), to predominate in $CD4^+$ T lymphocytes *in vivo,* and to be increasingly evident as AIDS-related complex (ARC) then AIDS develop in the patient. On the other hand, macrophage-tropic variants tend to grow slowly and to low titer ("slow–low" strains) in cultures of PBL or macrophages (thus are generally "dual-tropic"), and to be non-syncytium-inducing (NSI), to predominate *in vivo* in monocytes in blood and glial cells in the brain, and to be favored in the early stages of infection acquired by sexual transmission.

In so far as the emergence of T-cell tropic SI strains toward the end of the long asymptomatic period presages the development of AIDS, such strains are considered to be the more virulent. However, there are also differences in virulence between strains of HIV-1 displaying similar tropism. Moreover, HIV-2 appears to be less virulent than HIV-1.

Some animal lentivirus infections are notorious for the progressive emergence of antigenic variants during prolonged chronic infection of a given animal. Neutralization-escape mutants can be selected readily by growing HIV in the presence of monoclonal antibodies *in vitro,* or even in the presence of monoclonal cytotoxic T cells. Such escape mutants often display amino acid substitutions in the hypervariable regions of the envelope glycoprotein(s), such as are found in natural human isolates, but it is difficult to prove whether the latter are selected by the immune response *in vivo,* whether they are random, or whether they are associated with favored changes in tropism or virulence which may also map to the hypervariable V3 loop of gp120, for example.

Thus, as the long period of asymptomatic infection progressively gives way to advanced disease, the viral mutants that emerge tend to be those displaying the properties of extended cellular tropism, enhanced cytopathogenicity for $CD4^+$ T cells, more rapid kinetics of replication, and escape from neutralizing antibodies.

Latency

Although it is now clear that HIV infection is more properly regarded as a chronic rather than a latent infection because some virus can be detected in the patient at all times by sufficiently sensitive techniques, it is nevertheless true that, at any given time, many more $CD4^+$ T cells are latently infected than are producing virus. Nucleic acid hybridization reveals at least two types of cellular latency: (1) true latency, in which viral cDNA is integrated and may subsequently be induced to allow viral replication following activation of the T cell by an appropriate cytokine or transactivation by another virus, and (2) "blocked early-stage latency," in which the cDNA transcript remains extrachromosomal in the resting (G_0) T cell but may become integrated and go on to produce infectious virions if that cell is activated within the next few days.

HIV as the Cause of AIDS

There was some initial skepticism about whether the newly discovered virus was indeed the cause of AIDS rather than an incidental passenger like the many opportunistic invaders that superinfect the patient as the immune system eventually collapses. However, the case for declaring HIV the sole etiologic agent is now overwhelming. First, HIV can be cultured (and/or the viral genome detected by PCR) in virtually all cases of AIDS, not only during overt disease but throughout the several years of silent infection that precedes it. Second, HIV grows preferentially in and destroys $CD4^+$ T lymphocytes, macrophages, and brain cells, both *in vitro* and *in vivo,* accounting persuasively for the observed pathology of the disease. Most compelling, however, are the numerous prospective studies that have followed at-risk populations for years, commencing prior to the acquisition of HIV infection and following through to death from AIDS. Not only are all AIDS patients found to be infected with HIV, but most HIV-infected persons eventually develop AIDS. Furthermore, there are several studies in which deaths from AIDS in blood transfusion recipients or hemophiliacs have been traced back to a particular HIV-infected blood donor or batch of factor VIII, and vice versa. Moreover, when a baby dies of AIDS, serological screening of the mother invariably reveals that she is carrying HIV, even though the baby may predecease her.

Clinical Features

In 1986 the U.S. Centers for Disease Control and Prevention (CDC) introduced, mainly for epidemiologic and clinical reporting purposes, a system for classifying HIV-induced disease into four major categories: I, acute infection; II, asymptomatic infection; III, persistent generalized lymphadenopathy (PGL); IV, other disease. Only certain listed clinical conditions within group IV qualified to establish the diagnosis of AIDS; groups I, II, and III were regarded as pre-AIDS conditions which may or may not progress to AIDS. Recently, the CDC expanded its definition of AIDS to include HIV-infected persons with $CD4^+$ lymphocyte counts of less than 200 per μl. This reflects the reality that most HIV-infected individuals whose $CD4^+$ cell numbers have dropped below 200 per μl are already showing clinical abnormalities resulting from this degree of immunosuppression and will go on to develop clinical AIDS within the next 2 years. Advancing the diagnosis enables more realistic case recording and early commencement of treatment.

HIV Seroconversion Illness

Two to three weeks after infection there is often a brief illness similar to but distinguishable from mononucleosis. Features include acute-onset fever with or without night sweats, myalgia, arthralgia, lethargy, malaise, diarrhea, depression, lymphadenopathy, sore throat, skin rash and mucocutaneous ulceration, and sometimes neurologic manifestations, often presenting clinically as headache, photophobia, and retroorbital pain. Examination of the blood reveals a temporary reduction in $CD4^+$ (and $CD8^+$) T cell count, followed by a predominantly $CD8^+$ lymphocytosis. This illness is often disregarded or misdiagnosed; a high level of clinical suspicion should be triggered if there are relevant lifestyle considerations. Virus, viral nucleic acid, or viral p24 antigen may be detectable during the illness. Seroconversion (development of antibodies) coincides with resolution of the illness or follows shortly thereafter.

Asymptomatic Infection

With few exceptions, the patient recovers from the primary (seroconversion) illness within 2–3 weeks, and the majority go on to enjoy at least 5 years of relatively good health. PGL, otherwise known as lymphadenopathy syndrome (LAS), may or may not become apparent toward the end of this period but does not have any clear prognostic relevance. Reflecting the polyclonal activation of the immune system, autoimmune conditions may occur during this period. They include the Guillain-Barré syndrome, chronic demyelinating neuropathy, idiopathic thrombocytopenia, Reiter's syndrome, polymyositis, cranial nerve palsy, and Sjögren's syndrome.

Symptomatic HIV Infection

When the $CD4^+$ T-cell count falls below about 400 per μl the patient may develop a constellation of constitutional symptoms (fever, night sweats, oral candidiasis, diarrhea, and weight loss) which used to be known as AIDS-related complex (ARC). Early opportunistic infections begin to be seen. At this intermediate stage of immune depletion these infections are generally not life-threatening. They particularly include infections of the skin and mucous membranes such as tinea, seborrheic dermatitis, bacterial folliculitis, warts, molluscum contagiosum, gingivitis, oral and esophageal candidiasis, oral hairy leukoplakia (Fig. 35-10D), and chronic sinusitis. Reactivation of latent herpesviruses, particularly herpes simplex and zoster, also occurs (see Chapter 20). Gastrointestinal infections, caused by any of a wide variety of organisms, including the yeast *Candida albicans* and parasites such as cryptosporidia, are common. Mycobacterial infections are also common in these patients, and this has led to an alarming resurgence of tuberculosis in some countries such as the United States.

When $CD4^+$ T cells drop below 200 per μl their numbers generally begin to decline at an accelerated rate, the titer of virus in the blood increases markedly, and the tempo of progression of the illness increases. This stage of HIV infection is associated with more severe opportunistic infections. Particular infections tend to appear at a relatively predictable level of $CD4^+$ T cells

(Fig. 35-9). By far the commonest of these opportunistic infections (in the absence of chemoprophylaxis) is pneumonia caused by *Pneumocystis carinii*, an organism rarely encountered in immunocompetent people (Fig. 35-10A). Others include EBV-associated interstitial pneumonitis (particularly in pediatric AIDS), esophageal candidiasis, cryptosporidial and microsporidial enteritis, infections with members of the *Mycobacterium avium* complex, cytomegalovirus retinitis (Fig. 35-10C), enteritis, or encephalitis, and infections of the brain with *Toxoplasma gondii* or *Cryptococcus neoformans*.

Malignant tumors may also appear as CD4 cell counts decline. The most common of these, seen mainly in male homosexuals and therefore thought perhaps to be caused by an unknown sexually transmitted agent, is Kaposi's sarcoma (Fig. 35-10B). Previously known best as a relatively harmless collection of indolent skin tumors in old people of Mediterranean descent, Kaposi's sarcoma presents in young gay HIV-infected males in a more invasive but slowly progressive form, often quite early in the symptomatic phase of AIDS. Other types of malignancy, notably aggressive B-cell lymphoma or non-Hodgkin's lymphoma, may develop in patients with very low CD4 cell counts; genital cancers are also being increasingly recorded.

Neurologic disease is now recognized to be an extremely common and sometimes the first observed manifestation of HIV infection. The spectrum of HIV-associated neurologic disease includes dementia and its early forms, a severe encephalopathy (especially in children), myelopathy, and motor dysfunction. The patients may notice diminished concentration and memory, together with motor disturbances such as action tremor and loss of balance, as in Parkinson's disease. They often also display signs of coexistent myelopathy and peripheral neuropathy (e.g., ataxia and parasthesia). Other CNS manifestations include cerebral toxoplasmosis, cryptococcal meningitis, primary CNS lymphoma, CMV-associated encephalomyelitis, and progressive multifocal leukoencephalopathy (see Fig. 18-4).

Laboratory Diagnosis

There are several distinct situations in which laboratory diagnosis of HIV infection may be required: (1) diagnosis of infection in an individual with or without symptoms; (2) monitoring disease progression, with or without antiviral therapy; (3) screening of blood and organ donors, voluntary screening of individuals considered to be at risk, and seroepidemiologic surveys; and (4) evaluation of new antivirals or vaccines. The preferred methodology differs, depending on the particular objective.

Clinicians dealing with symptomatic HIV-infected patients draw on a number of indicators of immunologic function, the most useful of which is the $CD4^+$ T lymphocyte count (as well as the CD4:CD8 ratio). Assays of helper T-cell function, such as skin tests for DTH to common antigens or production of IL-2 *in vitro*, are also useful. Surrogate markers of T-cell activation, for example, serum β_2-microglobulin levels in excess of 4 mg/liter, are often used to forecast clinical progression. Serum or urinary neopterin elevation as a marker of macrophage activation is relatively nonspecific. While useful adjuncts, none of these assays is diagnostic of HIV infection.

Because essentially all HIV infections persist for life, detection of antibody

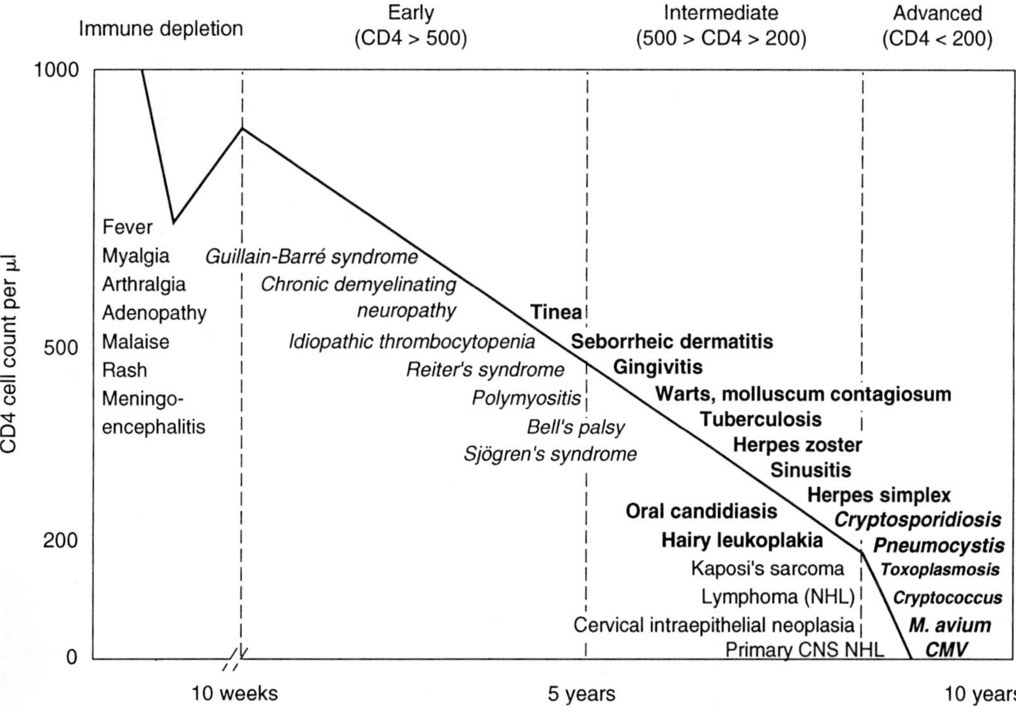

Sero-conversion illness	*Clinical latency (symptom free)* *Sporadic autoimmune phenomena/disease*	**Less severe infections** Malignancy Dementia	***Severe opportunistic infections***
Primary HIV infection with viremia	*Immune activation and control* *HIV mostly within lymph nodes*	**Increasing HIV virulence** **Decreasing immune control**	***Severe immune deficiency with viremia***
	Gradually increasing central nervous system pathology		

Fig. 35-9 Chronology of HIV-induced disease. The course of infection may be divided into four clinically relevant periods shown at bottom in relation to key features of the pathogenesis: (1) the brief seroconversion illness in the first few weeks; (2) the long asymptomatic period of clinical latency, interrupted perhaps by episodes of autoimmune disease; (3) the early clinical phase marked by non-life-threatening opportunistic infections and possibly cancer; (4) AIDS, defined by severe opportunistic infections after the $CD4^+$ T cell count falls below 200 per μl. The graph at top illustrates that the timing of particular infections is often (but by no means always) related in a fairly predictable fashion to the CD4 T cell count. NHL, non-Hodgkin's lymphoma. The typeface used in the upper graph coincides with that used for each of the four stages of the disease shown in the lower panel. [From G. J. Stewart, "Could It Be AIDS?" *Med. J. Aust.*, special issue (1993). Courtesy Dr. G. J. Stewart.]

can be taken as an indication of infection rather than of recovery. Accordingly, the standard procedure for diagnosing HIV infection is an enzyme immunoassay (EIA) probing for HIV antibody. In the first-generation assays the capture antigen was a relatively crude lysate of HIV-infected lymphocytes. This test has proved to be extremely sensitive, with less than 0.1% false

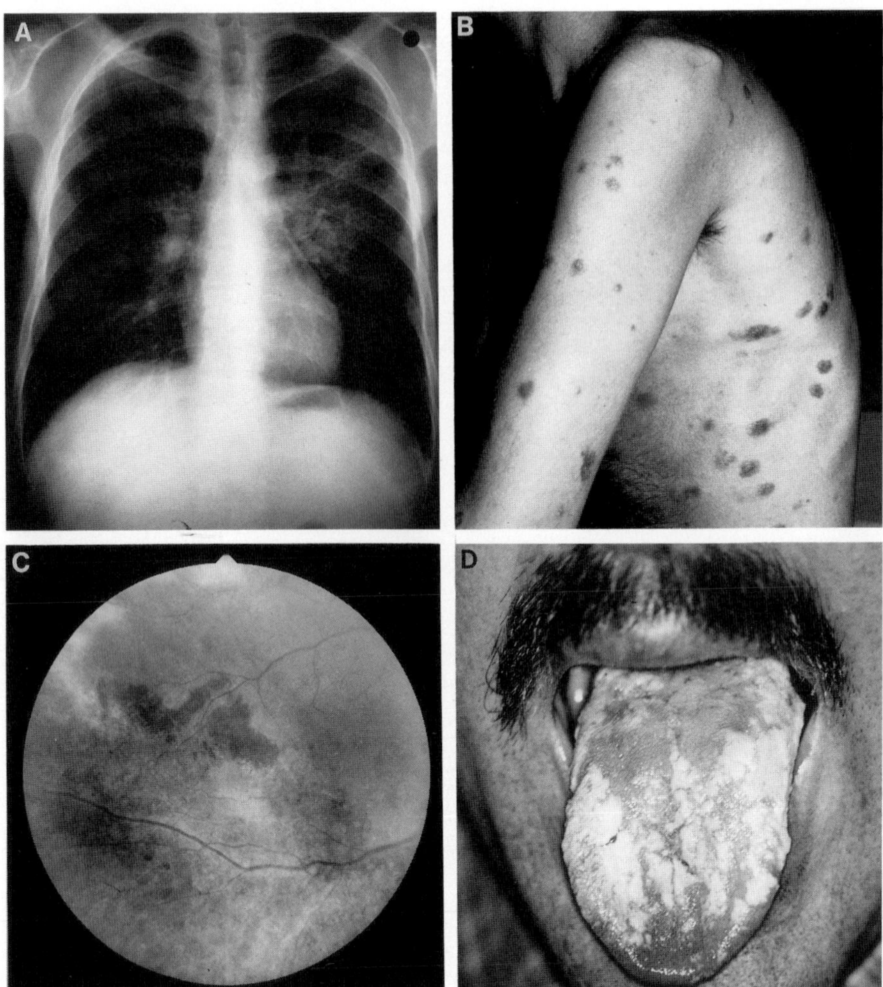

Fig. 35-10 Common opportunistic infections encountered in HIV infections. (A) Chest radiograph of *Pneumocystis carinii* pneumonia. (B) Kaposi's sarcoma. (C) Cytomegalovirus chorioretinitis. (D) Oral hairy leukoplakia. (See also Fig. 18-4.) (Courtesy Dr. R. L. Doherty.)

negatives. However, a lower level of specificity (1% false positives) meant that the vast majority of positive persons recorded in a low-risk population such as blood donors are false positives, that is, the positive predictive value of the test is poor (see Chapter 12). Second-generation EIAs use recombinant HIV antigen for capture; synthetic peptides representing immunodominant epitopes from the envelope glycoprotein(s) are also being developed as capture antigens.

To protect donated blood supplies, most countries instigated rigid HIV testing routines from about mid-1985 onward. These procedures included requirements for retesting all positive specimens, first by EIA again, then, if repeatedly positive, by Western blotting (Fig. 35-11). Immunoblotting offers greater specificity in that it is capable of detecting antibodies against any of the HIV proteins, and these are separately identifiable. In practice, any particular patient's serum at a given time may contain antibodies of sufficiently high titer only against certain antigens. The WHO now prescribes that, to be regarded as positive, a Western blot must reveal antibodies against at least two envelope proteins (gp160, gp120, or gp41), or one envelope protein and at least one Gag protein (p55, p40, p24, or p17), or one envelope protein plus one Pol protein (p66, p51, or p32). If the Western blot gives an indeterminate result (e.g., a single band corresponding to antibodies against p24, which are generally the first antibodies to become demonstrable following HIV infection, yet is the commonest nonspecific finding), serial testing of additional serum samples will usually clarify the situation. Alternative tests for HIV antibodies, such as radioimmunoprecipitation, and immunofluorescence on live infected T cells, may occasionally be of value. Simpler "desktop" assays for HIV antibody, such as a latex particle agglutination test, a membrane spot test, and an ingenious red cell agglutination test, are not considered to be sufficiently reliable for uncontrolled use in the diagnosis of a disease where life or death is at stake. Separate EIAs and confirmatory Western blot assays have been developed for HIV-2, which is currently seen mainly in Africa but can be expected to spread more widely.

The "window" period of 1–3 months (occasionally longer) between HIV infection and full seroconversion (as defined by Western blot) poses a dilemma. Prospective blood donors in some countries such as Australia are required to sign a legally binding declaration that they have not engaged in any of certain defined risky behaviors during the preceding six months. On the other hand, where HIV infection is suspected clinically, there are two general solutions. The first is simply to repeat the antibody assay after a suitable interval. The second is to select a more sensitive test capable of establishing a diagnosis earlier after infection. There are three approaches to the latter, involving detection of virus, viral antigen, or the viral genome.

Virus can be isolated from quite early in infection from peripheral blood leukocytes (PBL), or less readily from plasma, cerebrospinal fluid (CSF), genital secretions, or various organs such as the brain or bone marrow. The procedure is slow, tricky, and of course must be conducted under conditions of strict biocontainment. The patient's PBL are cocultured with mitogen-activated PBL from a seronegative donor; after several days or weeks, reverse transcriptase or p24 antigen can be detected in the culture medium. In general, virus is present in PBL in high titer only early and late in the prolonged course of HIV infection (Fig. 35-8), and virus isolation is not routinely attempted except in certain research and reference laboratories. On the other hand, there is an increasing demand for quantification of "viral load" as an indicator of disease progression or efficacy of antiviral therapy. As currently available plaque assays or syncytium assays are not regarded as sufficiently reproducible for this purpose, p24 and quantitative PCR assays are preferred.

The core antigen, p24, is one of the most abundant proteins made by HIV-infected cells and is detectable in the plasma by EIA a few weeks before seroconversion. It disappears when antibodies appear, remains undetectable for years, and rises again around the time clinical manifestations of AIDS develop. The reappearance of antigenemia is associated with a poor prognosis. Conversely, decline in the titer of p24 antigen is often used to monitor the success of antiviral therapy.

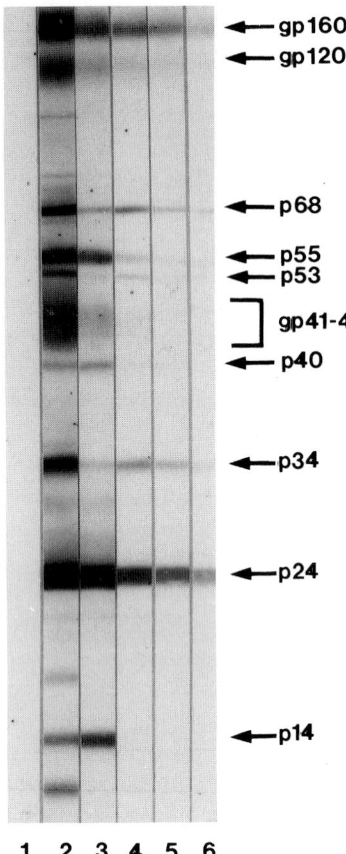

Fig. 35-11 Western blot for identification of HIV antibodies. HIV-1 was solubilized with sodium dodecyl sulfate (SDS), subjected to polyacrylamide gel electrophoresis (PAGE) to separate the constituent proteins on the basis of M_r, and the bands transferred by electrophoresis to a nitrocellulose membrane. After blocking unoccupied sites, the membrane was cut into strips and each strip incubated with a 1:100 dilution of serum from different subjects, then rinsed and treated with an enzyme-labeled anti-human immunoglobulin, rinsed again, and treated with a substrate that produces an insoluble precipitate in the presence of the enzyme. Lane 1, Negative control; lane 2, strong known positive control; lane 3, subject under test (positive); lanes 4–6, quality control, namely, twofold dilutions from 1:12,800 of a known positive standard. Note that not all infected persons produce antibody against all HIV proteins, and that the titers of antibody against certain proteins are consistently higher than others. Certain proteins (gp160, p55, p40) are precursors or intermediates in the cleavage pathways that produce the final products (gp120, gp41–45, p24, etc.). (Courtesy National HIV Reference Laboratory, Melbourne, Australia.)

More recently, the exquisite sensitivity of the PCR has been harnessed to detect viral RNA or cDNA in virions or infected cells, even in patients who are still seronegative (during the window period) or who are seropositive but asymptomatic, producing very little virus. It also provides a method for diagnosing HIV infection in neonates, whose serological status is ambiguous as a result of transplacental transmission of maternal antibodies. Properly conducted and controlled, the PCR assay is much simpler and quicker than virus isolation, and much more sensitive and informative than p24 detection. Furthermore, by careful selection of appropriate primers it can be used to distinguish HIV-2 from HIV-1, as well as some drug-resistant mutants from sensitive strains. Moreover, quantitative PCR methodologies have now been developed which permit a reasonable estimate to be made of the total viral load. For example, a new quantitative competitive PCR is internally controlled by mixing the target HIV RNA with dilutions of known concentrations of a comparable HIV RNA template differing only by an internal deletion introduced as a marker to permit later separation of the two PCR products by electrophoresis. Such quantitative PCR methods should greatly advance our capacity to distinguish various types of latent and productive virus–cell interactions in different tissues at different stages throughout the long course of

HIV infection, as well as to monitor disease progression and the efficacy of antiviral therapy.

Epidemiology of HIV Infection

There are three principal modes of transmission of HIV, as worked out commendably quickly within a couple of years of the recognition of the first cases of AIDS in Los Angeles: sexual intercourse, exchange of blood, and perinatal transmission.

Sexual intercourse now accounts for some 80% of all infections. The risk of transmission is higher to the passive (receptive) partner, generally the female, and higher following anal intercourse (estimated at 1% per episode) than vaginal intercourse (estimated about 0.1% per episode). Another concurrent sexually transmitted disease enhances the risk by an order of magnitude, especially if genital ulcers are present, as in syphilis or chancroid. Obviously the risk is also greatly enhanced the greater the number of different sexual partners.

Perinatal infection currently accounts for about 10% of all HIV infections. The risk of infection from an infected mother to her baby has been estimated as 12–40% in various studies. The precise routes are not clearly understood but probably include prenatal transmission across the placenta, intrapartum transmission via blood and/or genital secretions during parturition, and postnatal transmission via breast milk. Risk factors have been shown to include the stage and severity of the maternal infection (maternal virus titers being highest early and late in the mother's infection), low maternal titers of neutralizing antibody, prematurity at delivery, and breast-feeding; obstetric procedures may also be significant.

Blood transfusion and the administration of blood products such as factor VIII to hemophiliacs account for only 3–5% of all past and present HIV infections and declined rapidly following the introduction in 1985 of routine screening procedures in blood banks. The efficiency of transmission of HIV infection by transfusion of infected blood is greater than 90%. Another 5–10% of all infections result from sharing of needles by injecting drug users (IDUs). The risk per episode will vary with the precise circumstances but may be comparable with that of needle-stick injuries to medical or laboratory workers receiving small volumes of infected blood (about 0.5%).

Although trace amounts of virus can be detected from time to time in a range of bodily secretions including saliva, there is no evidence that infection can be transmitted from person to person by kissing or any form of casual contact other than sexual intercourse or the exchange of blood.

Worldwide, three different epidemiologic patterns were observed during the 1980s. Pattern I, seen in North America, Western Europe, and Australasia, where AIDS was first reported in 1981, implying that HIV infection had probably been present since the 1970s, was characterized by spread principally via anal intercourse among male homosexuals, and to a lesser extent by needle sharing among IDUs. In recent years HIV has moved increasingly into the female population, and heterosexual transmission is occurring, especially among teenagers. The annual incidence of AIDS is projected to peak in the mid-1990s.

Pattern II, seen in sub-Saharan Africa, where AIDS has been present at least since the 1970s, is characterized by heterosexual transmission, with infection being equally common in males and females, and a correspondingly high incidence of perinatal transmission to infants. Massive economic problems including inflation and unemployment have led to urban immigration, breakdown of traditional tribal values, and sexual promiscuity especially involving casual liaisons between female prostitutes and males separated from their families. Prevalence of HIV infection in Central and East Africa has escalated alarmingly during the past decade, and the situation appears certain to deteriorate further.

Pattern III applies to Asia, Eastern Europe, North Africa, and the Middle East. The virus was not introduced into these areas until the early to mid-1980s. However, an alarming change became apparent in 1988 when an outbreak caused by HIV-1 genotype B (resembling the United States strains) occurred in IDUs in the Golden Triangle (at the junction of China, India, Myanmar, and Thailand). Then in 1989 it became clear that female prostitutes in India and Thailand had become infected with genotype A (resembling African strains), and this genotype has since spread rapidly by heterosexual intercourse. It is predicted that by the turn of the century the number of cases in Asia will exceed the number in Africa.

These three official patterns are not static, and the distinctions are no longer clear-cut. For instance, the situation in Central and South America was originally comparable to that in the United States, but spread to the heterosexual population has occurred much more quickly. Whereas the classification was invaluable for some years, it is now clear that AIDS everywhere will sooner or later become a heterosexually transmitted disease, with intravenous drug use also making a significant contribution. The principal difference between different geopolitical regions is likely to be socioeconomic: AIDS will become increasingly a disease of the poor and underprivileged, as has been their lot with regard to other devastating infectious diseases throughout history. In addition to recognizing the reality of global health interdependence, the wealthier nations will be called on to display a high degree of altruism if the pandemic is to be controlled.

Control

Containment of the AIDS epidemic is a global problem requiring a concerted, multifaceted, international approach. Although the detailed implementation of national programs will be greatly influenced by such variables as the extent of the epidemic in the region, the general level of health, education, and affluence, social customs and superstitions, political realities, and available funding, the major objectives are universal.

The first and most difficult objective is to bring about a major change in sexual behavior. The primeval urge is so strong that neither legislation nor appeals to altruism are likely to be as persuasive as self-preservation. The whole community needs to be educated about the danger and routes of transmission of HIV, what constitutes high-risk behavior, and practical approaches to minimize those risks. The message needs to be targeted in different ways to different audiences, but particularly to the young and sexually active, through the mass media, through the schools, and through relevant, nonthreatening peer-group leaders speaking the same language. It is crucial not to alienate high-risk groups at the social margins by judgmental or discriminatory attitudes. The message is the following:

1. Reduce the number sex partners. Ideally, practice "abstinence before marriage, fidelity after marriage," to borrow a slogan from a recent advertising campaign in one African country. "Mutual monogamy" should be the objective of all heterosexual or homosexual partners.
2. Use condoms. "Unprotected" sex must be discouraged, although to propagate the view that a condom offers "safe sex" is an exaggeration; it is certainly safer sex, but not as safe as nonpenetrative forms of sex.
3. Treat all sexually transmitted diseases (STD) promptly. Not only does prompt treatment decrease the chance of contracting HIV, but patients attending an STD clinic constitute a captive audience ready to receive and implement advice about AIDS prevention.

There is a real need for an effective vaginal microbicidal agent; this would empower women to control their own destinies with a greater degree of certainty.

The second target group includes injecting drug users. This needle-sharing subculture is particularly difficult to access. Some countries have adopted the daring approach of providing all registered IDUs with new disposable syringes and needles free of charge, on request. This appears to have been particularly successful in restricting the spread of HIV (and hepatitis B, C, and D) among this vulnerable cohort in Australia.

Medical and laboratory personnel are required to follow strict standards of aseptic technique in handling potentially infectious materials in the ward or laboratory. The details are beyond the scope of this text, but some of the procedures were described in Chapter 12.

Almost all countries have implemented effective procedures for the routine screening of blood, sperm, and organ donors for HIV antibody. Furthermore, blood products such as factor VIII used for the management of hemophilia are now routinely heat-treated to destroy the labile HIV virion.

In most countries compulsory screening for HIV infection (by EIA for antibody) is confined to blood, semen, and organ donors; universal screening is considered to be neither cost-effective nor socially acceptable. However, voluntary screening of those at high risk of infection is to be strongly encouraged. High-risk categories include the following: (1) sexually active gay or bisexual men, (2) patients attending STD clinics, (3) travelers returning after unprotected sex in countries of high HIV endemicity, (4) injecting drug users, (5) prisoners, (6) anyone who received blood or a blood product between 1980 and 1985, (7) health workers accidentally exposed (e.g., via needle stick or blood spill), and (8) anyone with an "AIDS-defining" condition. Regarding the last, clinicians should be increasingly alert to the possibility of AIDS and should automatically recommend testing of anyone with a disorder virtually unique to AIDS, such as Kaposi's sarcoma, oral hairy leukoplakia, or such "AIDS-defining" opportunistic infections as *Pneumocystis* pneumonia. A sexually active patient presenting with a disorder that is uncommon but not unique to AIDS, such as oral candidiasis or Guillain-Barré syndrome, should also be tested. Those presenting with disorders more commonly unrelated to AIDS should be tested if lifestyle clues suggest any of the risk groups listed above.

Counseling before and after testing is important. Management of seropositive individuals also involves regular medical surveillance and preemptive interventions as well as reactive responses to episodes of opportunistic infection. Confidentiality must be respected. The patient must also appreciate his/her special obligation to implement lifestyle changes, particularly with regard to sexual encounters; partners should be contacted, notified, counseled, and monitored medically. A particularly sad situation is that of the HIV-positive woman who must be informed of the substantial risk of transmission of HIV to her infant should she ever become pregnant.

Vaccines

Because chemotherapy can never eliminate HIV infection, the ultimate control of AIDS can be achieved only by the use of an effective vaccine. In the developed world the spread of HIV could be restricted by administration of such a vaccine to the major risk groups, namely, sexually promiscuous male homosexuals, prostitutes, injecting drug users, and newborn children of infected mothers. However, in the Third World, where AIDS is spread heterosexually and perinatally, control would only be achieved if the effective vaccine were to be administered to every infant. Ultimately, universal vaccination might be expected to become the policy worldwide.

Numerous problems stand in the way of the development of an effective HIV vaccine (Table 35-4). HIV is notoriously mutable as a result of the lack of fidelity of reverse transcription, the absence of error-correcting mechanisms, and the additional possibility of recombination between different HIV strains. The prolonged asymptomatic period provides ample time for significant antigenic drift to occur, and this continues after transmission to sexual contacts, thus generating a daunting range of very different strains around the world. Any vaccine would need to incorporate at least the major subtypes of HIV-1 and ideally HIV-2, or to contain immunodominant cross-reactive epitopes.

HIV may enter the body not only as free virions, neutralizable by antibody, but as infected leukocytes (in semen or blood), in which the virus or provirus would be protected against antibody or T-cell-mediated cytolysis. Unless the infecting cells were rapidly destroyed by allograft rejection, infection could then spread directly from cell to cell by fusion to produce syncytia. Also, for every T cell or monocyte actively producing virus, there are many more carrying a cDNA copy of the HIV genome in the chromosomes of latently infected cells which do not express viral proteins and are therefore inaccessible to antibody or cell-mediated immunity. Moreover, infected cells may be sequestered from the immune response in such sites as the CNS.

As the normal mode of transmission of HIV is via the mucosal route, the first line of defense in a vaccinated host would preferably be mucosal IgA antibodies and submucosal lymphocytes. This would argue for mucosal delivery, preferably of a live or recombinant vaccine, perhaps vaginally, or orally to take advantage of the common mucosal pathway. However, what constitutes protective immunity against HIV has yet to be established (in humans or simian models).

Table 35-4
Problems of Vaccination against HIV Infection

Numerous HIV strains occur around the world; antigenic drift and possibly genetic recombination ensure continuing evolution
Virus normally enters by the mucosal route
HIV may be transmitted by infected cells as well as virions; infection spreads readily from cell to cell by fusion
Virus readily establishes lifelong latent infection
HIV grows selectively in cells of the immune system, notably helper T lymphocytes and cells of the monocyte/macrophage lineage

Because of the lethality of the virus, the certainty of establishment of latent infection, and the likelihood of subsequent mutation, it is improbable that any live attenuated HIV vaccine would be licensed for human use, with the possible exception of a deletion mutant analogous to the *nef*$^-$ mutant of the simian immunodeficiency virus, which is avirulent for rhesus monkeys. The options are therefore limited to conventional inactivated whole-virus vaccines, purified envelope glycoproteins, or recombinant live vectors (e.g., vaccinia virus or avipoxvirus) carrying at least the gene for the HIV envelope protein(s) (gp120 ± gp41). Experimental vaccines of all these kinds have been produced (Table 35-5) and tested in chimpanzees and/or humans. Chemically inactivated virions, or gp160, gp120, or gp41 produced by recombinant DNA technology in mammalian cells or in insect cells using baculovirus as a vector, have all now been tested, with very limited success so far. Whereas the ideal vaccine would prevent infection, we may be obliged to lower our sights somewhat in the case of HIV and settle for reduction in the virus load such that latent infection is established on only a limited scale, in the hope that the immune response might reduce the rate of spread and delay the onset of disease. Some consideration is even being given to the strategy of "therapeutic" vaccination, in which vaccine is administered to those already infected in the hope of retarding the disease.

Antiviral Therapy

The complexity of the HIV replication cycle, in particular the requirement for so many different viral enzyme functions and so many viral regulatory proteins, offers hope that truly effective antiviral therapy for AIDS may be just around the corner. However, in spite of over a decade of frenetic research

Table 35-5
Alternative Approaches to Development of HIV Vaccine

Chemically inactivated whole-virus vaccine
Purified gp160, gp120, or gp41 envelope glycoprotein
Envelope glycoprotein produced by gene cloning in mammalian or insect cells
Recombinant live virus (e.g., vaccinia) incorporating gene for HIV envelope glycoprotein and perhaps other proteins
Live attenuated deletion mutant

activity by a large number of research laboratories in industry, universities, and research institutes, the harvest of effective antiviral agents has been meager. Only the dideoxynucleoside analogs that inhibit reverse transcriptase have shown an acceptable combination of efficacy and lack of toxicity *in vivo* (see Chapter 16 for detailed discussion). Zidovudine (AZT, azidothymidine), which has been the standard treatment for several years, reduces the incidence and severity of opportunistic infections somewhat but does not consistently arrest the progress of the disease even when commenced during the asymptomatic phase when the CD4 counts are in the range of 200–400 per μl. Moreover, resistant mutants consistently emerge within months. Combination or alternation therapy with AZT plus ddI or ddC generally offers little or no significant advantage. Nonnucleoside inhibitors of reverse transcriptase such as nevirapine and the "TIBO" compounds rapidly induce resistance and probably do not have a role as single agents.

Attention has turned to other vulnerable steps in the replication cycle (see Table 16-1). Inhibitors of aspartyl protease, some of which display synergism with dideoxynucleosides *in vitro,* are currently undergoing clinical trials. Theoretically, it might be possible to find agents that bind and inactivate one of the regulatory proteins such as Tat or Rev, or its nucleic acid recognition sequence (TAR or RRE), although the first of the promising Tat inhibitors has recently been found wanting *in vivo.* The assessment of these and other compounds will require painstaking and frustratingly slow clinical trials.

The infidelity of reverse transcription expedites the emergence of resistant mutants, whatever the target of the drug in question. Thus, it is likely that the future of anti-HIV therapy will lie with combinations of drugs with totally different modes of action, preferably addressing different viral proteins. Potentially, such a strategy also presents the advantages of synergy and reduced toxicity (by lowering the required dose of each drug), as well as delayed emergence of resistance. Of course, eradication of the nonreplicating latent provirus from quiescent cells will be impossible, so anti-HIV therapy must be continued for life.

More effective chemotherapy is available for most of the opportunistic infections with bacteria, fungi, protozoa, and some of the viruses that are a feature of the later stages of HIV infection. Acyclovir and ganciclovir are effective against reactivations of herpes simplex and zoster viruses or cytomegalovirus (see Chapters 16 and 20). Control of these secondary infections can be achieved by one of three main strategies: primary prophylaxis (at critical times), therapy (treatment of infections as they arise), or secondary prophylaxis (prevention of recurrences by using a low dose of drug for life).

Simian Immunodeficiency Virus

The primate lentivirus group of the genus *Lentivirus* also contains the simian immunodeficiency virus (SIV), strains of which have been recovered from a range of African monkey species. Although they cause inapparent infections in African monkeys, SIV causes simian AIDS in rhesus (Asian) monkeys. SIV differs from HIV-2 by a deletion in the *nef* gene. That humans can occasionally become infected by SIV is indicated by a recent survey of monkey handlers and SIV researchers, a few of whom (<1%) had seroconverted to SIV: the virus was successfully isolated from one case. However, there is no indication that SIV causes any disease in humans.

Human Spumaviruses

The genus *Spumavirus* comprises a group of "foamy viruses," named for their characteristic cytopathology, namely, vacuolation in cultured cells. They are exogenous retroviruses which contain no oncogene. Species-specific spumaviruses are widespread throughout the animal kingdom, including humans. They cause inapparent persistent infections, but have not yet been proved to cause any disease, although association with certain degenerative and autoimmune diseases has been reported.

Further Reading

Blattner, W. A., ed. (1990). "Human Retrovirology: HTLV." Raven, New York.

Bolognesi, D. P. (1993). Human immunodeficiency virus vaccines. *Adv. Virus Res.* **42,** 103.

Cann, A. J., and Chen, I. S. Y. (1990). Human T-cell leukemia virus types I and II. *In* "Fields Virology" (B. N. Fields, D. M. Knipe, R. M. Chanock, M. S. Hirsch, J. L. Melnick, T. P. Monath, and B. Roizman, eds.), 2nd Ed., p. 1501. Raven, New York.

Coffin, J. M. (1990). Retroviridae and their replication. *In* "Fields Virology" (B. N. Fields, D. M. Knipe, R. M. Chanock, M. S. Hirsch, J. L. Melnick, T. P. Monath, and B. Roizman, eds.), 2nd Ed., p. 1437, Raven, New York.

Cullen, B. R. (1991). Human immunodeficiency virus as a prototypic complex retrovirus. *J. Virol.* **65,** 1053.

Cullen, B. R. (1991). Regulation of human immunodeficiency virus replication. *Annu. Rev. Microbiol.* **45,** 219.

Fauci, A. S. (1988). The human immunodeficiency virus: Infectivity and pathogenesis. *Science* **239,** 617.

Fauci, A. S., ed. (1993). The Immunology of HIV Infection. *Semin. Immunol.* **5,**(3), 147.

Gallo, R. C., and Jay, G., eds. (1991). "The Human Retroviruses." Academic Press, New York.

Greene, W. C. (1990). Regulation of HIV-1 gene expression. *Annu. Rev. Immunol.* **8,** 453.

Greene, W. C. (1991). The molecular biology of human immunodeficiency virus type 1 infection. *N. Engl. J. Med.* **324,** 308.

Haseltine, W. A. (1988). Replication and pathogenesis of the AIDS virus. *J. AIDS* **1,** 217.

Haseltine, W. A., and Wong-Staal, F., eds. (1991). "Genetic Structure and Regulation of HIV." Raven, New York.

Hirsch, M. S., and Curran, J. (1990). Human immunodeficiency viruses: Biology and medical aspects. *In* "Fields Virology" (B. N. Fields, D. M. Knipe, R. M. Chanock, M. S. Hirsch, J. L. Melnick, T. P. Monath, and B. Roizman, eds.), 2nd Ed., p. 1545. Raven, New York.

Jasny, B., ed. (1993). AIDS: the unanswered questions. *Science* **260,** 1253.

Koff, W. C., Wong-Staal, F., and Kennedy, R. C., eds. (1991 on). "AIDS Research Reviews," Vols. 1 on (annually). Dekker, New York.

Levy, J. A. (1988). *Retroviridae:* Human immunodeficiency viruses. *In* "Laboratory Diagnosis of Infectious Diseases. Principles and Practice. Volume II, Viral, Rickettsial, and Chlamydial Diseases" E. H. Lennette, P. Halonen, and F. A. Murphy, eds.), p. 677. Springer-Verlag, New York.

Levy, J. A. (1993). Pathogenesis of human immunodeficiency virus infection. *Microbiol. Rev.* **57,** 183.

Levy, J. A., ed. (1993). "The Retroviridae." Plenum, New York.

McFarlin, D. E., and Blattner, W. A. (1991). Non-AIDS retroviral infections in humans. *Annu. Rev. Med.* **42,** 97.

Myers, G., Rabson, A. B., Josephs, S. F., Smith, T. F., Berzofsky, J., and Wong-Staal, F. (1992).

"Human Retroviruses and AIDS 1992." Los Alamos National Laboratory, Los Alamos, New Mexico.

Narayan, O., ed. (1992). The Lentiviruses. *Semin. Virol.* **3,**(3), 135.

Piel, J., ed. (1988). What Science Knows about AIDS. *Sci. Am.* **259**(4).

Richman, D. D. (1991). Antiviral therapy of HIV infection. *Annu. Rev. Med.* **42,** 69.

Rosenberg, Z. F., and Fauci, A. S. (1989). Immunopathogenesis of HIV infection. *Adv. Immunol.* **47,** 377.

Schochetman, G., and George, J. R. (1991). "AIDS Testing: Methodology and Management Issues." Springer-Verlag, New York.

Schochetman, G., Epstein, J. S., and Zuck, T. F. (1989). Serodiagnosis of infection with the AIDS virus and other human retroviruses. *Annu. Rev. Microbiol.* **43,** 629.

Schüpbach, J. (1990). "Human Retrovirology." Springer-Verlag, Berlin.

Swanstrom, R., and Vogt, P. K., eds. (1990). Retroviruses: Strategies of Replication. *Curr. Top. Microbiol. Immunol.* **157,** 1.

Varmus, H. (1988). Retroviruses. *Science* **240,** 1427.

Weiss, R. A., Teich, N., Varmus, H. E., and Coffin, J. M., eds. (1985). "RNA Tumor Viruses," 2nd Ed., Vols. 1 and 2. Cold Spring Harbor Laboratory, Cold Spring Harbor, New York.

Wong-Staal, F. (1990). Human immunodeficiency viruses and their replication. *In* "Fields Virology" (B. N. Fields, D. M. Knipe, R. M. Chanock, M. S. Hirsch, J. L. Melnick, T. P. Monath, and B. Roizman, eds.), 2nd Ed., p. 1529. Raven, New York.

Yarchoan, R., Mitsuya, H., and Broder, S. (1993). Challenges in the therapy of HIV infection. *Immunol. Today* **14,** 303.

Chapter 36

Viral Syndromes

In the preceding 19 chapters we have examined individually the contribution of various members of each of the relevant families of viruses to the spectrum of human disease. In this final chapter we examine the same scene from the opposite aspect, namely, to focus in turn on each of the major clinical syndromes. Of course, this is not a textbook of medicine nor even of infectious diseases, but of virology. We cannot devote space to detailed clinical descriptions of viral diseases and certainly not to their differential diagnosis from diseases caused by nonviral infectious agents, nor to their clinical management. Appropriate reference works on infectious diseases have been listed under Further Reading. What follows is intended to provide only a bird's-eye view of the commoner syndromes so that we may assess the contribution of particular viruses to each one. In so far as the clinical features as well as the pathogenesis and epidemiology of all the major human viral infections were dealt with in the previous 19 chapters, this final chapter should be regarded as little more than an appendix which brings all these virus–disease associations together in a number of summary tables for ready reference.

Viral Diseases of the Respiratory Tract

Respiratory infections are the most common afflictions of humans, and most are caused by viruses. Children contract on average about half a dozen respiratory illnesses each year, and adults perhaps two or three. Admittedly these are mainly trivial colds and sore throats, but they account for millions of lost working hours and a significant proportion of all visits to family physicians. More serious lower respiratory tract infections tend to occur at the extremes of life, and in those with preexisting pulmonary conditions. The most important human respiratory viruses are influenza and respiratory syncytial viruses (RSV), the former killing mainly the aged and the latter the very young. Of the estimated 5 million deaths from respiratory infections in children annually worldwide, at least 1 million are viral in origin.

Altogether, there are about 200 human respiratory viruses, falling mainly within six families: orthomyxoviruses, paramyxoviruses, picornaviruses, coronaviruses, adenoviruses, and herpesviruses. Here we shall confine ourselves to those that enter the body via the respiratory route and cause disease confined largely to the respiratory tract. Many other "respiratory" viruses are disseminated via the bloodstream to produce a more generalized disease, as is the case with most of the human childhood exanthems such as measles, rubella, and varicella. Other viruses, entering by nonrespiratory routes, can reach the lungs via systemic spread, and pneumonia may represent the final lethal event, as in overwhelming infections with herpesviruses or adenoviruses in immunocompromised neonates or AIDS patients.

Systemic viral infections such as measles generate a strong memory response and prolonged production of IgG antibodies, which protect against reinfections for life. In contrast, viruses that cause infection localized to the respiratory tract with little or no viremia, such as RSV or rhinoviruses, induce only a relatively transient mucosal IgA antibody response; hence, reinfections with the same or a somewhat different strain can recur repeatedly throughout life. In addition, numerous strains of viruses like influenza virus arising by antigenic drift may cause sequential episodes of the same disease in a single patient.

Whereas some viruses have a predilection for one particular part of the respiratory tract, most are capable of causing disease at any level, and the syndromes to be described below overlap somewhat (Fig. 36-1). Nevertheless, for ease of description we will designate six basic diseases of increasing severity as we descend the respiratory tract: rhinitis, pharyngitis, croup, bronchitis, bronchiolitis, and pneumonia (Table 36-1).

Rhinitis (Common Cold)

The classic common cold (coryza) is marked by copious watery nasal discharge and obstruction, sneezing, and perhaps a mild sore throat or cough, but little or no fever. All colds are viral. Rhinoviruses are the major cause, several serotypes being prevalent year-round and accounting for about half of all colds. Coronaviruses are responsible for about another 15%, mainly those occurring in the winter months. Certain enteroviruses, particularly coxsackieviruses A21 and A24 and echoviruses 11 and 20, cause febrile colds and

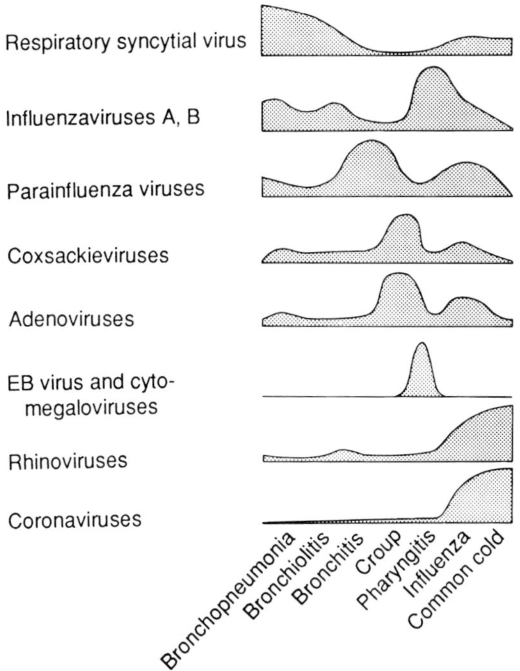

Fig. 36-1 Frequency with which particular viruses produce disease at various levels of the respiratory tract. (Courtesy Dr. D. A. J. Tyrrell.).

sore throats, especially in the summer. In children, respiratory syncytial virus (RSV), parainfluenza viruses, and the low-numbered adenoviruses are between them responsible for up to half of all upper respiratory tract infections (URTI).

Otitis media or sinusitis sometimes complicate URTI. Bacterial superinfection is generally involved, but viruses have also been recovered from the effusion. Respiratory infections with RSV, influenza, parainfluenza, adenovirus, or measles viruses predispose to otitis media. Indeed, repeated viral infections can precipitate recurrent middle ear infections, leading to progressive hearing loss.

Pharyngitis

Most pharyngitis is of viral etiology. Upper respiratory infections with any of the viruses just described can present as a sore throat, with or without cough, malaise, fever, and/or cervical lymphadenopathy. Influenza, parainfluenza, and rhinoviruses are common causes throughout life, but other agents are prominent in particular age groups: RSV and adenoviruses in young children, herpesviruses in adolescents and young adults. Adenoviruses, though not major pathogens overall, are estimated to be responsible for about 5% of all respiratory illnesses in young children. Pharyngoconjunctival fever is just one particular presentation, which was described in Chapter 19 together with the strange tendency of adenoviruses 4 and 7 to cause outbreaks of "acute respiratory disease" (ARD) in U.S. military camps. Primary infection with herpes simplex virus (HSV), if delayed until adolescence, presents as a pharyngitis and/or tonsillitis rather than as the gingivostomatitis seen principally in younger children; the characteristic vesicles, rupturing to form ulcers, can be confused only with herpangina, a common type of vesicular pharyngitis caused by coxsackie A viruses (see Chapters 20 and 23). Infectious mononucleosis (glandular fever) is usually seen in adolescents and young adults, and is marked by a very severe pharyngitis, often with a diphtheria-like membranous exudate, together with cervical lymphadenopathy and fever (Chapter 20). This syndrome is generally caused by Epstein-Barr virus (EBV), but occasionally by cytomegalovirus (CMV), especially if lacking the sore throat, swollen glands, and heterophil antibody.

Laryngotracheobronchitis (Croup)

Croup is one of the serious manifestations of parainfluenza and influenza virus infections. A young child presents with fever, a "barking" or "metallic" cough, inspiratory stridor, and respiratory distress, sometimes progressing to complete laryngeal obstruction and cyanosis. Parainfluenza viruses are responsible for about half of all cases, type 1 being commoner than type 2. Influenza viruses and RSV are important causes during winter epidemics.

Bronchitis

Influenza and parainfluenza viruses and RSV are the main viral causes of acute bronchitis. There is also evidence that chronic bronchitis, which is par-

Table 36-1
Respiratory Viral Diseases

Disease	Virus	
	Common	Less common
Rhinitis (common cold)	Rhinoviruses	RSV, parainfluenza, influenza
	Coronaviruses	Adenoviruses
		Coxsackie A21, A24; echo 11, 20
Pharyngitis	Parainfluenza 1–3	Rhinoviruses
	Influenza	Adenoviruses 1–7
	Herpes simplex	RSV
	Epstein-Barr virus	Cytomegalovirus
	Coxsackie A	
Laryngotracheobronchitis (croup)	Parainfluenza 1, 2	RSV
	Influenza	
Bronchitis	Parainfluenza 3	Parainfluenza 1, 2
	Influenza	
	RSV	
Bronchiolitis	RSV	Influenza A
	Parinfluenza 3	Parainfluenza 1, 2
Pneumonia	RSV	Parainfluenza 1, 2
	Parainfluenza 3	Adenoviruses 3, 7
	Influenza	Cytomegalovirus
		Measles
		Varicella

ticularly common in smokers, may be exacerbated by acute episodes of infection with influenza viruses, rhinoviruses, or coronaviruses.

Bronchiolitis

Respiratory syncytial virus is the most important respiratory pathogen during the first year or two of life, being responsible, during winter epidemics, for about half of all bronchiolitis in infants. Parainfluenza viruses (especially type 3) and influenza viruses are the other major causes of this syndrome. The disease can develop with remarkable speed. Breathing becomes rapid and labored, and is accompanied by a persistent cough, expiratory wheezing, cyanosis, a variable amount of atelectasis, and marked emphysema visible by X-ray. The infant may die overnight, and hence RSV is one of the causes of unexplained "cot deaths," otherwise known as the sudden infant death syndrome (SIDS).

Pneumonia

Whereas viruses are relatively uncommon causes of pneumonia in immunocompetent adults, they are very important in young children. RSV (Figs. 36-2A and 36-3A) and parainfluenza virus (mainly type 3) are between them responsible for 25% of all pneumonitis in infants in the first year of life. Influenza also causes a considerable number of deaths during epidemic years. Infections with adenoviruses 3 and 7 (Fig. 36-2B) are less common but can be severe, and long-term sequelae such as obliterative bronchiolitis or bronchiectasis may permanently impair lung function. Up to 20% of pneumonitis in infants has been ascribed to perinatal infection with cytomegalovirus (see Chapter 20). CMV may also cause potentially lethal pneumonia in immu-

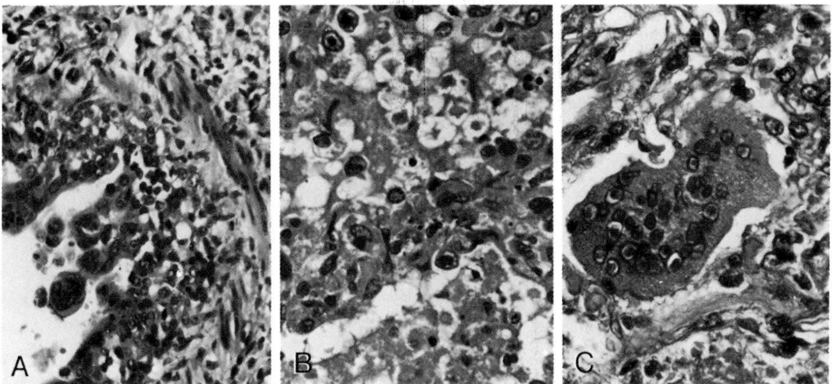

Fig. 36-2 Viral infections of the lung. Sections from fatal cases of pneumonitis in children, stained with hematoxylin and eosin. (A) Respiratory syncytial virus bronchiolitis (magnification: ×135). Note multinucleated giant cells being shed into a bronchiole which is surrounded by leukocytes. (B) Adenovirus, pneumonia (magnification ×126). Note characteristic intranuclear inclusions. (C) Measles giant-cell pneumonia (magnification: ×216). Note multinucleated giant cell with intranuclear inclusions. (Courtesy Drs. I. Jack and A. Williams).

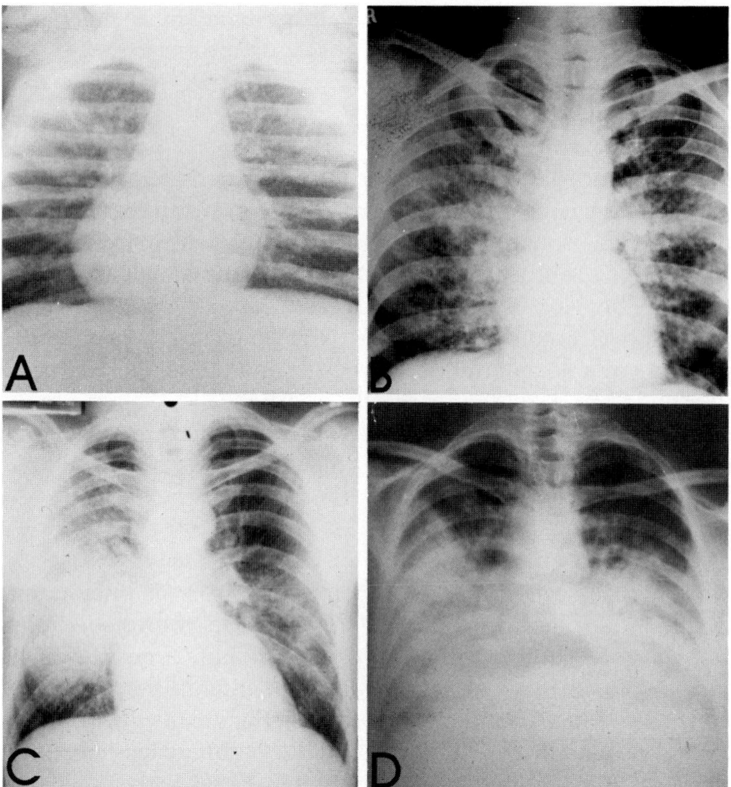

Fig. 36-3 Radiographs of viral pneumonitis. Note streaky, patchy, or nodular consolidation only. (A) Respiratory syncytial virus. (B) Varicella. (C, D) Influenza. (Courtesy Dr. J. Forbes.)

nocompromised patients, as may measles (Fig. 36-2C), varicella, and adenoviruses. Moreover, viral pneumonia not uncommonly develops in adults with varicella (Fig. 36-3B), and in military recruits involved in outbreaks of adenovirus 4 or 7, whereas measles is quite often complicated by bacterial pneumonia, especially in malnourished children in Africa and South America. In the elderly, particularly in those with underlying pulmonary or cardiac conditions, influenza is a major cause of death, either via influenza pneumonitis (Fig. 36-3C,D) or, more commonly, via secondary bacterial pneumonia attributable to *Staphylococcus aureus, Streptococcus pneumoniae,* or *Haemophilus influenzae.*

Viral pneumonitis often develops insidiously following URTI, and the clinical picture may be atypical. The patient is generally febrile, with a cough and a degree of dyspnea, and auscultation may reveal some wheezing or moist rales. Unlike typical bacterial lobar pneumonia with its uniform consolidation, or bronchopneumonia with its streaky consolidation, viral pneumonitis is usually confined to diffuse interstitial lesions. The radiologic findings are not striking; they often show little more than an increase in hilar shadows or, at most, scattered areas of consolidation (Fig. 36-3).

Viral Gastroenteritis

By no means do all viruses found in feces cause gastroenteritis. Some do, but others cause "silent" infections of the gastrointestinal tract, which is their portal of entry to the body (e.g., reoviruses and many enteroviruses and adenoviruses); some may then move on, usually via the bloodstream, to their target organ(s) in other parts of the body. For this reason, it has not been easy to pin down which viruses actually cause gastroenteritis, especially as enteritis is so very common and not always easy to distinguish from minor changes in bowel habits arising from time to time for dietary, psychologic, or other reasons.

Assiduous searches have revealed a fascinating range of viruses in feces, none of which are human parasites, including bacteriophages parasitizing enteric bacteria and plant or animal viruses from ingested food. There is much current interest in two small viruses with genomes consisting of two or three segments of double-stranded RNA, which because of similarities to the family *Birnaviridae* have been tentatively named picobirnaviruses and picotrirnaviruses, respectively. Miscellaneous other "small round viruses" and "parvovirus-like" agents have also been carefully described, as well as parvoviruses causing enteritis in cats and dogs, but a clear etiologic association has yet to be nailed down in humans. The same applies to the enteric coronaviruses, and perhaps enteric toroviruses, which have been frequently visualized by electron microscopy in feces from patients or even outbreaks of human gastroenteritis, especially in psychogeriatric patients, AIDS patients, and immunocompromised children, but not proved beyond a shadow of a doubt to cause the disease from which they were recovered. It is quite probable that some human enteric viruses are harmless passengers in most people most of the time but are capable rarely of causing diarrhea under certain circumstances in certain (especially immunocompromised) individuals. In Table 36-2 we list only those enteric viruses that unequivocally cause gastroenteritis in humans.

Gastroenteritis vies with upper respiratory infection for the mantle of commonest of all infectious diseases and is the greatest cause of death. It has been estimated that 5–10 million children die each year in the Third World from diarrheal diseases, rotavirus infections in malnourished infants being a major contributor. Rotaviruses infect young children, causing severe diarrhea which may last up to a week and lead to dehydration requiring fluid and electrolyte replacement. Most infections are sporadic, but nosocomial outbreaks occur frequently in hospital nurseries. Nearly all rotavirus infections are caused by group A serotypes and occur mainly in infants under the age of 2 years; group B rotaviruses have been asssociated with some very large waterborne outbreaks in China, affecting adults as well as children; group C rotaviruses cause only occasional zoonotic infections.

Enteric adenoviruses were first visualized in feces by electron microscopy and assigned to the correct family on the basis of their characteristic morphology; because of the copious numbers of virions excreted, they could be demonstrated by direct immunoassay to be distinct from other known members of that family, even though they themselves had never been cultured. Only later

Table 36-2
Viral Gastroenteritis[a]

Causative agent	Patient age groupings	Selected symptoms[b]		Incubation period	Duration of illness	Mode of transmission[c]
		Vomiting	Fever			
Rotavirus, group A	Infants and toddlers	Common	Common	1–3 days	5–7 days	Water, PTP, ?food, ?air, nosocomial, fecal–oral
Rotavirus, group B	Children and adults	Variable	Rare	56 hours (average)	3–7 days	Water, PTP, fecal–oral
Rotavirus, group C	Infants, children, and adults	Unknown	Unknown	24–48 hours	3–7 days	Fecal–oral
Adenovirus (enteric)	Young children	Common	Common	7–8 days	8–12 days	Nosocomial, fecal–oral
Calicivirus	Infants, young children, and adults	Common for infants; variable for adults	Occasional	1–3 days	1–3 days	Food, water, nosocomial, fecal–oral
Calicivirus (Norwalk virus)	Older children and adults	Common	Rare or mild	18–48 hours	12–48 hours	Food, water, PTP, ?air, fecal–oral
Astrovirus	Young children and elderly people	Occasional	Occasional	1–4 days	2–3 days; occasionally 1–4 days	Food, water, fecal–oral

[a] From Centers for Disease Control, Recommendations for collection of laboratory specimens associated with outbreaks of gastroenteritis, *Morb. Mortal. Wkly. Rep.* **39**, RR-14 (1990).
[b] Diarrhea is common and is usually loose, watery, and nonbloody when associated with gastroenteritis.
[c] PTP, Person-to-person; ?, not confirmed.

were suitable techniques developed for growing the "fastidious" or "enteric" adenoviruses, types 40 and 41, and later higher-numbered serotypes so far found only in AIDS patients. The sequence of events leading to the discovery of the other groups of viruses causing human enteritis has been essentially the same. These particular serotypes of human adenoviruses (but not the numerous other types that replicate in the respiratory and/or gastrointestinal tract but cause no disease in the gastrointestinal tract) have turned out to be common causes of gastroenteritis, especially in young children, for example, in outbreaks in day-care centers.

The "typical" enteric caliciviruses are also common, especially in young children. In contrast, the Norwalk group of viruses, now classified as members of the family *Caliciviridae,* tend to infect older children and adults, often in common-source outbreaks. The illness consists of an explosive episode of nausea, vomiting, diarrhea, and abdominal cramps, sometimes accompanied by headache, myalgia, and/or low-grade fever.

Astroviruses display many of the epidemiologic and clinical features of rotaviruses but are not as common and not as virulent. They appear to be endemic worldwide, with occasional epidemics, causing a relatively mild form of enteritis with watery diarrhea, mainly in young children; outbreaks have also occurred among immunosuppressed and institutionalized geriatric patients.

Viral Diseases of the Central Nervous System

Most meningitis and almost all encephalitis is of viral etiology (Table 36-3). Infections of the CNS arise, in the main, as a rare complication of a primary infection established elsewhere in the body which fortuitously spreads to the brain, usually via the bloodstream. Sometimes they occur following reactivation of a latent herpesvirus or papovavirus infection, particularly following immunosuppression. Overwhelming disseminated infections acquired perinatally may also involve the brain.

Certain viruses have a predilection for particular parts of the CNS, and the clinical signs of the resulting disease often reflect this. For example, most enteroviruses do not go beyond the meninges, but polioviruses invade the anterior horn of the spinal cord and the motor cortex of the cerebrum, whereas rabies singles out Ammon's horn, herpes simplex virus the temporal lobes, and so on. Some viruses lyse neurons directly, and there is abundant evidence of inflammation in the brain (Fig. 36-4A); others do their damage in more subtle ways, leading to demyelination of nerves (Fig. 36-4B), sometimes involving immunopathologic processes.

One must distinguish between neurovirulence, that is, the ability to cause neurologic disease, and neuroinvasiveness, that is, the ability to enter the nervous system. Mumps virus, for example, displays high neuroinvasiveness, in that evidence of very mild meningitis accompanied by changes in the cerebrospinal fluid (CSF) are detectable in about half of all infections, but low neurovirulence, in that it rarely causes much damage. In contrast, herpes simplex virus (HSV) displays low neuroinvasiveness, in that it rarely invades the CNS, but high neurovirulence, in that when it does it often causes devastating damage. Thus, *neurotropism,* the ability to infect neural cells, is the product of neuroinvasiveness and neurovirulence. Moreover, not all neurotropic viruses are neuronotropic, that is, able to infect neurons, as are rabies virus, polioviruses, togaviruses, flaviviruses, and bunyaviruses; some viruses, such as the polyomavirus JC, preferentially replicate in nonneuronal cells like oligodendrocytes, causing demyelination. Destruction of neurons has the most serious consequences as lost neurons are not replaced.

The blood–brain barrier, which tends to exclude viruses from the CNS, also limits access of lymphoid cells, antibodies, complement, etc.; only when inflammation disrupts the blood–brain barrier does the immune response come into play. Thus the barriers that inhibit virus invasion also deter virus clearance, accounting for the high frequency with which persistent virus infections involve the CNS.

The many and varied neurologic syndromes caused by viruses include meningitis, encephalitis, paralytic poliomyelitis, myelitis, polyneuritis, and several unusual demyelinating and degenerative syndromes.

Meningitis

Viral meningitis is much commoner than bacterial meningitis but is much less severe. Only meningeal cells and ependymal cells are involved, and recovery is almost always complete. The patient presents with headache, fever, and neck stiffness, with or without vomiting and/or photophobia. Lumbar puncture reveals a clear CSF, perhaps under slightly increased pressure, with near normal protein and glucose concentrations, and only a moderate pleocytosis; the white cell count may range from normal ($<10/mm^3$) to over $1000/mm^3$, but is usually $30–300/mm^3$, with lymphocytes predominanting after the first day or so. This is what is generally called "aseptic" meningitis.

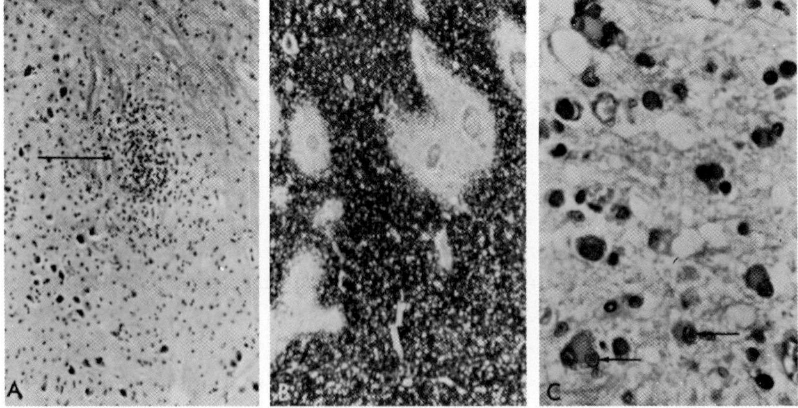

Fig. 36-4 Viral infections of the brain. Sections from fatal human cases of encephalitis. (A) Murray Valley (Australian) encephalitis (hematoxylin and eosin stain; magnification: ×55). Arrow indicates focal concentration of mononuclear leukocytes. (B) Measles, postinfectious encephalitis (luxol fast blue stain for myelin; magnification: ×20). Note areas of demyelination around vessels. (C) Measles, subacute sclerosing panencephalitis (hematoxylin and eosin stain; magnification: ×220). Arrows indicate intranuclear inclusions. (Courtesy Drs. I. Jack, R. McD. Anderson, and A. Williams.)

By far the most important etiologic agents are mumps virus (see Chapter 28) and numerous enteroviruses, including all the coxsackie B types, coxsackie A7 and A9, polioviruses (in countries where they still exist), and many echoviruses which were listed in Chapter 23. The herpesviruses, HSV, EBV, and CMV, are rare sporadic causes, whereas lymphocytic choriomeningitis virus can be acquired from laboratory or pet mice or hamsters.

Meningitis may be the only clinical evidence of infection with these viruses. For example, only half of all cases of mumps meningitis follow typical parotitis. Enteroviral meningitis often occurs during a summer/autumn epidemic in which others experience rashes, myositis, or other common manifestations of infection with the prevalent agent, but meningitis is often the sole presentation.

Paralysis

In countries from which polioviruses have not yet been effectively eliminated by vaccination, these viruses remain the major cause of both aseptic meningitis and paralytic poliomyelitis. Very rarely indeed, the oral poliovaccine itself can cause paralysis, mainly in immunocompromised individuals. Enterovirus 71 and coxsackievirus A7 are rare causes of a paralytic disease essentially indistinguishable from polio. The radiculomyelitis associated with enterovirus 70 infection is generally reversible.

Table 36-3
Viral Diseases of the Central Nervous System

Disease	Viruses[a]
Meningitis	**Enteroviruses**
	Mumps
	Lymphocytic choriomeningitis
	Herpes simplex; other herpesviruses rarely
Paralysis	**Polioviruses**
	Enteroviruses 70, 71; coxsackie A7
Encephalitis	**Herpes simplex**
	Mumps
	Arboviruses (togaviruses, flaviviruses, bunyaviruses; see Tables 25–2, 26-2, 33-2)
	Arenaviruses, rabies virus, enteroviruses, adenoviruses, other herpesviruses
Postinfectious encephalomyelitis	**Measles**, varicella, rubella, mumps, (vaccinia), others
Guillain-Barré syndrome	Cytomegalovirus, Epstein-Barr virus, HIV
Reye's syndrome	Influenza, varicella
Subacute sclerosing panencephalitis	Measles, rubella
Progressive multifocal leukoencephalopathy	Polyomavirus JC
AIDS encephalopathy (AIDS dementia complex)	**HIV**
Tropical spastic paraparesis	HTLV-I
Subacute spongiform encephalopathy	**Prions**

[a] The commonest causal agents are in bold type.

Encephalitis

Encephalitis is one of the most serious of all viral diseases. The illness often begins like meningitis with fever, headache, vomiting, and neck rigidity, but alteration in the state of consciousness indicates that the brain parenchyma itself is involved. Initially lethargic, the patient becomes confused then stuporose. Ataxia, seizures, and paralysis may develop before the victim lapses into a coma and dies. Survivors may often be left with a pathetic legacy of permanent sequelae, including mental retardation, epilepsy, paralysis, deafness, or blindness.

Encephalitogenic mosquito-borne or tick-borne togaviruses, flaviviruses, and bunyaviruses, endemic to particular regions of the world, cause epidemics of encephalitis from time to time when the appropriate combination of ecologic circumstances develops. The ecology of each of these arboviruses and features of the disease(s) they cause were described in detail in Chapters 25, 26, and 33 (see also Tables 25-2, 26-2, 33-2). Encephalitis is also an irregular feature in certain hemorrhagic fevers (see Table 36-5). Rabies causes a uniformly lethal type of encephalitis, described fully in Chapter 29.

In most of the temperate regions of the world, mumps is the commonest cause of encephalitis, but it is generally a relatively mild meningoencephalitis with only rare sequelae, mainly unilateral deafness. Herpes simplex virus is the most commonly identified cause of severe sporadic encephalitis. This is a very unpleasant disease indeed, infecting both neurons and glia to produce a focal encephalitis generally localized to the temporal lobes in immune adults, but diffuse necrotizing encephalitis in the newborn, with a 70% case–fatality rate (see Chapter 20). In neonates or immunocompromised patients HSV and the other herpesviruses, and occasionally enteroviruses or adenoviruses, are also capable of causing encephalitis, generally as part of a widely disseminated and often fatal infection. *Chronic meningoencephalitis* is a fatal condition seen in children with the B-cell deficiency, X-linked agammaglobulinemia, or severe combined immunodeficiency. Enteroviruses are the usual causal agent (see Chapter 23). The majority of these children also have a condition known as *juvenile dermatomyositis.*

Postinfectious Encephalomyelitis

Postinfectious encephalomyelitis is a severe demyelinating condition of the brain and spinal cord which occurs as an occasional complication following a few days after any of the common childhood exanthemata (measles, varicella, rubella) or mumps. Prior to the eradication of smallpox, it also occurred as an occasional complication of vaccination against that disease, using live vaccinia virus. The pathology of postinfectious encephalomyelitis resembles that of experimental allergic encephalomyelitis, giving rise to the hypothesis that this is an autoimmune disease in which virus infection provokes an immunologic attack on myelin. Certainly there is little virus demonstrable in the brain by the time postinfectious encephalomyelitis develops, and the major histologic finding is perivenous inflammation and demyelination.

Guillain-Barré Syndrome

Guillain-Barré syndrome is an acute inflammatory demyelinating polyradiculoneuropathy which follows exposure to any one of several viruses. Epstein-Barr virus (which has also been associated with transverse myelitis and Bell's palsy) is most commonly implicated, the Guillain-Barré syndrome appearing 1–4 weeks after infectious mononucleosis. Partial or total paralysis develops, usually in more than one limb. Complete recovery occurs within weeks in most cases, but 15% retain residual neurologic disability. The syndrome is also seen with cytomegalovirus and early in some HIV infections.

An outbreak of Guillain-Barré syndrome in the United States in 1976 was traced to the introduction of a formalin-inactivated vaccine against the so-called Swine strain of influenza. The vaccine was withdrawn, and the syndrome has not been associated with any subsequent flu vaccine. The whole episode remains something of a mystery. It does prove, however, that live virus is not a necessary ingredient in the genesis of Guillain-Barré syndrome. This argues strongly for an immunologic basis of the demyelination.

Reye's Syndrome

Reye's syndrome is a postinfectious encephalopathy with a 25% case–fatality rate which follows influenza or chickenpox in children. There is cerebral edema but no evidence of inflammation. Fatty infiltration of the liver is the other major feature. An epidemiologic association with the administration of aspirin during the original fever has been noted.

Chronic Demyelinating Diseases

Certain of the rarer demyelinating diseases are known to be due to viruses. Subacute sclerosing panencephalitis (SSPE) or inclusion body encephalitis is a rare late sequel to measles, whereas progressive rubella panencephalitis is an even rarer but similar demyelinating persistent infection. The pathogenesis of SSPE was discussed in Chapter 10. Progressive multifocal leukoencephalopathy (PML) is a different type of demyelination seen when AIDS or immunosuppression for renal transplantation or malignancy reactivates infection with the human polyomavirus JC, which targets oligodendrocytes (see Chapters 10 and 18). These associations have quickened interest in the possibility that more common demyelinating diseases of unknown etiology, notably multiple sclerosis, might also be caused by viruses. However, despite suggestive epidemiologic evidence and many false alarms, no virus has yet been incriminated.

AIDS Encephalopathy (AIDS Dementia Complex)

The human immunodeficiency virus has suddenly emerged as the commonest agent of viral infection of the CNS. Like animal lentiviruses, HIV is highly neuroinvasive from early in the prolonged preclinical phase. Occasional cases of acute meningitis and of Guillain-Barré syndrome can occur early in the course of infection. However, only after immunodeficiency becomes severe years later does the extent of the potential neurovirulence of HIV become manifest. The presentation can be protean, but over 50% of all patients develop progressive dementia with cerebral involvement, myelopathies, or sensory neuropathies (see Chapter 35).

Tropical Spastic Paraparesis

Infection with the human T-cell leukemia virus type 1 (HTLV-1) is usually subclinical. Rarely, however, after an incubation period of up to 40 years, a subacute disease of the thoracic spinal cord can develop involving progressive paralysis of the legs together with impotence and incontinence.

Subacute Spongiform Encephalopathy

The reader is referred to Chapter 10 for a detailed discussion of the role of putative subviral infectious agents known as prions in degenerative diseases of the brain classified as subacute spongiform encephalopathies, of which scrapie in sheep is the paradigm. Kuru was the first human model of these intriguing diseases to be unraveled, but the family has now been extended to include Creutzfeldt-Jakob disease (CJD), Gerstmann–Sträussler–Scheinker syndrome, and fatal familial insomnia, in the last two of which an inherited mutation in a particular gene can produce a disease clinically identical to that seen in kuru or in most cases of CJD where an infectious agent is responsible.

The search is now on for viruses or "subviral" agents as possible etiologic agents of much more common degenerative diseases of the CNS, such as amyotrophic lateral sclerosis, Parkinson's disease, Alzheimer's disease (presenile dementia), and even schizophrenia.

Viral Skin Rashes

Many viruses involve the skin in one way or another (Table 36-4). Some, such as papillomaviruses, poxviruses, and recurrent herpes simplex, produce relatively localized crops of lesions and few if any systemic symptoms. Others, such as those causing the childhood exanthemata, produce a generalized rash as part of a wider clinical syndrome that follows a systemic infection. These rashes vary greatly in their anatomic distribution and in the morphology of the individual lesions. They are classified for convenience into maculopapular, vesicular, nodular, and hemorrhagic rashes (Fig. 36-5).

Macules are flat, colored spots; papules are slightly raised from the surface of the skin but contain no expressible fluid. Virus is not shed from the lesions of maculopapular rashes. Many such rashes may in fact result from a hypersensitivity response to the virus growing in cells of the skin or capillary endothelium.

The differential diagnosis of maculopapular rashes is difficult, not only because many rashes are of toxic, allergic, or psychogenic origin, but also because they are a common feature of countless infectious diseases caused by bacteria, rickettsiae, fungi, protozoa, and metazoa as well as viruses! The rash

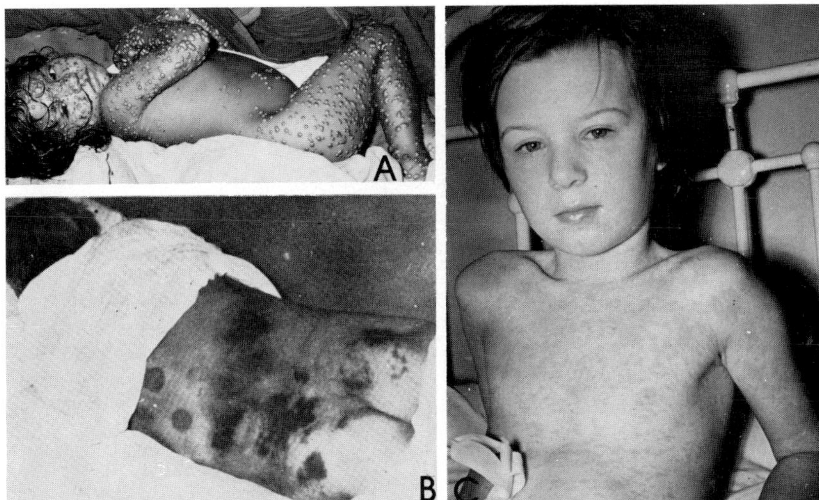

Fig. 36-5 Three distinct types of viral rashes. (A) Vesiculopustular rash of smallpox. (B) Hemorrhagic rash of hemorrhagic fever. (C) Maculopapular rash of measles. (A, Courtesy Prof. A. W. Downie; B, courtesy Dr. A. Shelekov; C, courtesy Dr. J. Forbes.)

itself is rarely pathognomonic; the whole clinical syndrome must be taken carefully into account.

The classic standards of reference against which other rashes are compared are the so-called morbilliform rash of measles and the rubelliform rash of rubella. The exanthem of measles (Fig. 36-5C) consists of flat reddish brown macules which coalesce to form rather large blotches; after the rash fades on day 5 or 6 the skin retains a brownish stain for a time then undergoes desquamation. In contrast, the exanthem of rubella consists of much smaller (pinpoint) pink macules which tend to remain discrete, giving the rash a fine or erythematous appearance; it usually disappears after 2–3 days.

Numerous unrelated viruses produce rashes almost indistinguishable from one or another of these two prototypes. Infections with literally dozens of different enteroviruses can present as a maculopapular rash, generally in children, often during late summer epidemics. These exanthems are usually ephemeral and nonpruritic. They are mainly rubelliform or morbilliform, but can be erythematous, petechial, urticarial, or vesicular in character. Space does not allow description of the syndromes associated with each of the 30-plus enteroviruses involved. Suffice it to note that the serotypes most frequently responsible for cutaneous eruptions are echoviruses 4, 9, and 16 and coxsackieviruses A9, A16, and B5.

Erythema infectiosum, or fifth disease, now known to be caused by a parvovirus, B19, is recognized for its unique rash. The child first develops flushed red cheeks, contrasting with pallor around the mouth, then a rubelliform eruption on the limbs which develops a lacelike appearance as it fades (see Fig. 17-4). Exanthem subitum, otherwise known as roseola infantum or sixth disease, is a universal exanthem of infants caused by human herpesvirus 6, although the classic rash is not always seen. About 10% of

Table 36-4
Viral Skin Rashes

Rash	Viruses
Maculopapular	Measles
	Rubella
	Parvovirus B19
	HHV-6 (human herpesvirus 6)
	Echoviruses 9, 16, many others
	Coxsackie A9, A16, B5, many others
	Epstein-Barr virus, cytomegalovirus
	Dengue, chikungunya, Ross River, other arboviruses
	Hepatitis B
Vesicular	Varicella–zoster
	Herpes simplex 1, 2
	Coxsackie A9, A16; enterovirus 71; others
Pustular	Monkeypox
	Cowpox
	Vaccinia
Nodular	Papillomaviruses
	Molluscum contagiosum
	Milker's nodes
	Orf
	Tanapox

cases of infectious mononucleosis, whether caused by EBV or CMV, have a maculopapular rash, usually on the trunk. Many arthropod-borne togaviruses and flaviviruses, including dengue, chikungunya, Sindbis, o'nyong-nyong, Mayaro, West Nile, and Ross River viruses, also produce a maculopapular or scarlatiniform rash lasting 2–3 days. Finally, mention should be made of the urticarial rash that forms part of the serum sickness syndrome seen fleetingly in the prodromal phase of 10–20% of cases of hepatitis B.

Vesicles are blisters, containing clear fluid from which virus can readily be isolated. Vesicular rashes do not present a great diagnostic problem, particularly now that smallpox (Fig. 36-5A) has disappeared. A generalized vesicular rash in a febrile child today is usually chickenpox (varicella). The lesions occur in crops, initially concentrated on the trunk, then spreading centrifugally. Each vesicle progresses to a pustule and a scab which then falls off. In herpes zoster the lesions are largely (but not necessarily exclusively) confined to a particular dermatome (Fig. 20-7), as is also the case with the recurrent form of herpes simplex (Figs. 20-5 and 20-6). However, in the case of disseminated herpes simplex or zoster, as seen in newborn infants or immunocompromised patients, the lesions may be widespread throughout the body. Something of a curiosity is the condition known as hand–foot-and-mouth disease caused by certain coxsackieviruses, in which vesicles or even bullae occur on the palms, soles, and buccal mucosa. Coxsackie A viruses also produce a similar type of vesicular enanthem on the mucous membrane of the throat and palate ("herpangina").

Poxviruses preferentially infect the skin, producing multiple pustular or nodular lesions, as in human monkeypox and molluscum contagiosum, respectively, or usually single lesions such as occur in the zoonotic infections caused by orf, milker's nodes, cowpox, and tanapox viruses (see Fig. 21-5).

Papillomavirus infections were described in Chapter 18. The papilloma, or wart, is a benign hyperplastic growth, usually multiple, occurring in crops on the skin or mucous membranes. Dermatologists classify them in various ways but generally recognize common warts, flat warts, plantar and palmar warts, epidermodysplasia verruciformis, and genital warts (see Fig. 18-3), all of which are clinically distinct and tend to be caused by different human papillomavirus types (see Table 18-2).

Viral Hemorrhagic Fevers

Although not an entirely homogeneous group of diseases, the hemorrhagic fevers (Table 36-5) share the common characteristic of widespread hemorrhage from the body's epithelial surfaces, including internal mucosae such as the gastrointestinal tract as well as the skin. The skin "rash" is often a mixture of pinpoint hemorrhages (petechiae) and massive bruising (ecchymoses), as depicted in Fig. 36-5B. The pathogenesis of these important diseases is generally not well understood. Thrombocytopenia and leukopenia are almost always present, but no general mechanism has been discovered to explain the hypovolemic shock without major blood loss which may lead to death within hours in dengue or Lassa fever, for example. Severe liver damage, extensive bleeding, and disseminated intravascular coagulation may be the key to the high mortality in the African hemorrhagic fevers, Crimean hemorrhagic fever, and the hemorrhagic form of Rift Valley fever. Encephalopathy and/or pneumonia can also be prominent in all the hemorrhagic fevers, whereas renal tubular necrosis and severe oliguria are distinctive features of Hantaan virus

Table 36-5
Viral Hemorrhagic Fevers[a]

Virus	Family	Distribution	Disease
Yellow fever	*Flaviviridae*	Africa, South and Central America	Yellow fever
Dengue 1–4	*Flaviviridae*	Widespread	Dengue shock syndrome
Lassa	*Arenaviridae*	Africa	Lassa fever
Marburg	*Filoviridae*	Africa	Hemorrhagic fever
Ebola	*Filoviridae*	Africa	Hemorrhagic fever
Crimean–Congo HF	*Bunyaviridae*	Africa, Eastern Europe	Crimean HF
Hantaan[b]	*Bunyaviridae*	Asia, Europe	Rodent-borne nephropathy
Rift Valley fever	*Bunyaviridae*	Africa, Middle East	Rift Valley fever
Omsk HF	*Flaviviridae*	Central Russia	Omsk HF
Kyasanur Forest	*Flaviviridae*	India	Kyasanur Forest disease
Junin	*Arenaviridae*	Argentina	Argentine HF
Machupo	*Arenaviridae*	Bolivia	Bolivian HF
Guanarito	*Arenaviridae*	Venezuela	Venezuelan HF

[a] HF, Hemorrhagic fever.
[b] And other members of the genus *Hantavirus*, such as Belgrade and Seoul viruses.

infection. Overall, the hemorrhagic fevers are protean in their presentation. Detailed descriptions of the clinical features and epidemiology of each of the dozen major hemorrhagic fevers (plus yellow fever, which could also be so regarded) were discussed in connection with flaviviruses (Chapter 26), filoviruses (Chapter 30), arenaviruses (Chapter 32), and bunyaviruses (Chapter 33).

The African filovirus hemorrhagic fevers have the highest case–fatality rates, but dengue hemorrhagic fever, Hantaan hemorrhagic nephrosonephritis, yellow fever, Rift Valley fever, and Lassa fever are the most prevalent on a world scale. The problem in Western countries is that the disease is likely to be completely outside the experience of the clinician who first sees it, and may also be a mild or atypical case with little to show other than an undifferentiated fever, possibly acquired abroad. The alternatives are limited by the traveler's recent itinerary, with Africa providing the most options. False alarms are frequent, especially in countries where expensive facilities for transporting, nursing, and diagnosing Class 4 pathogens have been established (Fig. 32-3), but "discretion is the better part of valor" in such circumstances. Barrier nursing and laboratory identification of the etiologic agent are essential.

Viral Genitourinary Infections

Two major viral sexually transmitted diseases (STD), genital herpes and genital warts, dramatically increased in frequency during the sexual revolution of the 1960s and 1970s. The painful itchy lesions of genital herpes (Fig. 20-6) and the accompanying local and systemic symptoms were described in Chapter 20. Dozens of recurrences, mainly attributable to HSV-2 but increasingly to HSV-1 also, may dominate the life of the hapless carrier. Genital warts, caused most commonly by the human papillomaviruses HPV-6 and HPV-11, can take the form of prolific excrescences on the external genitalia, perineum, vaginal introitus, penis, or anus (known as condyloma accuminatum (Fig. 18-3), or the form of a less conspicuous flat lesion on the cervix (condyloma planum); they are discussed in Chapter 18. Certain oncogenic HPV types, particularly types 16 and 18, produce cervical dysplasia which may progress over the course of many years to invasive cancer; the same HPV types are also etiologically associated with carcinomas of the male or female external genitalia and anus (see Chapter 11). Adenovirus type 37 is not uncommonly associated with cervicitis and urethritis. Molluscum contagiosum is also occasionally transmitted as an STD.

Several other very important human pathogens are shed in semen and in female genital secretions and are transmitted by sexual intercourse but cause no disease in the genital tract itself. Foremost among these, of course, are the human immunodeficiency viruses HIV-1 and HIV-2, but the list also includes the human T-cell lymphotropic viruses HTLV-1 and HTLV-2, hepatitis B and C viruses, and the herpesviruses, cytomegalovirus, and (probably) Epstein-Barr virus. Many more viruses are regularly conveyed between male homosexuals, depending largely on their particular sexual activities and number of partners. These include enteric viruses such as hepatitis A as well as those just listed.

Table 36-6
Viral Diseaes of the Genitourinary Tract[a]

Disease	Virus
Genital	
Genital herpes	Herpes simplex viruses (HSV-2 > HSV-1)
Genital warts	Human papillomaviruses 6, 11, and others
Genital carcinomas	Human papillomaviruses 16, 18, and others
Cervicitis	Adenovirus 37
Molluscum contagiosum	Molluscum contagiosum virus
Urinary	
Urethritis	Herpes simplex virus, adenovirus 37
Acute hemorrhagic cystitis	Adenovirus 11
Glomerulonephritis	Hepatitis B virus
Nephropathy	Cytomegalovirus
	Hantaan virus
Hemolytic–uremic syndrome	Enteroviruses?

[a] Many other important human pathogens causing major diseases not involving the genital or urinary tract clinically are nevertheless transmitted sexually. These include HIV-1 and -2, HTLV-1 and -2, hepatitis B and C viruses, cytomegalovirus, and probably Epstein-Barr virus.

Viruses rarely infect the urinary tract (Table 36-6). Urethritis can complicate infections with HSV. Acute hemorrhagic cystitis, an unusual disease of young boys, has been associated principally with adenoviruses 11 and (rarely) 21. Glomerulonephritis is sometimes observed as a manifestation of immune complex disease in chronic hepatitis B infections (see Chapter 22). It is safe to predict that future research may reveal that some cases of "idiopathic" glomerulonephritis are also caused by chronic persistent infections with other viruses yet to be identified (see Chapter 10). Cytomegalovirus persists asymptomatically in renal tubules (Fig. 20-8), from which cytomegalic cells as well as virus are shed into the urine. When primary infection or reactivation of CMV occurs during renal transplantation, rejection of the graft may be accelerated (see Chapter 20). The human polyomaviruses BK and JC (Chapter 18) also persist in the urinary tract and are reactivated by immunosuppression for renal transplantation, but they do not appear to play a role in rejection of the graft.

Clearly there is profound malfunction of the kidneys in hemorrhagic fever with renal syndrome, caused by the bunyavirus Hantaan virus. The clinical developments were described in Chapter 33, but the pathology is still unclear. Hemolytic–uremic syndrome is characterized by acute microangiopathic hemolytic anemia, intravascular coagulopathy, and impaired renal function; various enteroviruses have been isolated from family clusters of cases (see Chapter 23).

Viral Diseases of the Eye

It is not generally appreciated how frequently viruses can involve the eyes (Table 36-7). Conjunctivitis is a transient feature of a number of common childhood exanthemata such as measles (Fig. 36-5C), rubella, and certain

Table 36-7
Viral Infections of the Eye

Disease	Virus	Features
Conjunctivitis	Adenoviruses 3, 4, 7, others	Pharyngoconjunctival fever
	Sandfly fever	Dengue-like syndrome
	Dengue	Dengue-like syndrome
	Measles	Exanthem
	Rubella	Exanthem
	Marburg, Ebola	Hemorrhagic fever
Keratoconjunctivitis	Adenoviruses 8, 37, others	Epidemic
	Herpes simplex	Corneal ulceration
	Herpes zoster	Ophthalmic zoster
Acute hemorrhagic conjunctivitis	Enterovirus 70; coxsackie A24	Pandemics; ± radiculomyelitis
Chorioretinitis	Cytomegalovirus	Immunocompromised or congenital
	Rift Valley fever	
Cataracts, Glaucoma, Retinopathy, Microphthalmia	Rubella	Congenital rubella syndrome

enteroviral infections, and it is an important component of the dengue-like syndromes caused by many arboviruses, such as phlebotomus (sandfly) fever. Infections with adenoviruses, notably types 3, 4, and 7 in children, present as a bilateral follicular conjunctivitis or as pharyngoconjunctival fever.

Keratoconjunctivitis is potentially more dangerous, as it involves the cornea. Adenoviruses 8 and 37 are major causes of epidemic keratoconjunctivitis, which spreads readily by contact to adults, and usually involves only one eye, but may take up to a year to resolve. The main cause of sporadic keratoconjunctivitis, indeed the commonest infectious cause of blindness in the Western World, is herpes simplex virus (Fig. 20-5). Pathognomic "dendritic" or "geographic" ulcers develop on the cornea (Fig. 20-6), and if infection progresses to involve the stroma beneath, the immunologic reaction may lead to disciform keratitis, scarring, and loss of vision. Recurrent attacks are particularly damaging, as can be the application of corticosteroids. When herpes zoster involves the fifth cranial nerve, ophthalmic zoster (Fig. 20-7) can cause lasting damage to the eye.

Acute hemorrhagic conjunctivitis exploded on the world in 1969 and has since infected millions of people in a succession of pandemics. Subconjunctival hemorrhages, keratitis, and uveitis are quite common features; neurologic complications are rare. The etiologic agents are enterovirus 70 and coxsackievirus A24.

Retinitis, sometimes leading to permanent loss of central vision, was a feature of the 1977 epidemic of Rift Valley fever in the Nile Valley. Chorioretinitis is also a feature of cytomegalovirus infections in immunocompromised persons, such as recipients of organ grafts, and commonly causes blindness in AIDS patients (see Fig. 35-10C), as well as in congenitally infected babies with cytomegalic inclusion disease. Retinopathy, glaucoma, microphthalmia, and especially cataracts are the major eye abnormalities encountered in the congenital rubella syndrome (Fig. 36-6A); total or partial blindness may result. Finally, a rare accidental cause of zoonotic eye infection is autoinoculation with certain animal viruses, including Newcastle disease virus of chickens, seal influenza virus, or vaccinia virus.

Viral Arthritis

Arthritis, usually accompanied by fever and myositis, with or without a rash, is a common presentation of infections with many arboviruses of three families: the togaviruses, flaviviruses, and bunyaviruses (see Chapters 25, 26, and 33 and Table 36-8). The togaviruses chikungunya, o'nyong-nyong, and Ross River viruses, in particular, have caused huge epidemics of polyarthritis in Asia and Africa, Africa, and the Pacific islands, respectively. Arthritis is a somewhat less prominent feature of rubella but is common in adult females following either natural infection or rubella vaccine. Polyarthralgia is also an important feature of infection with the parvovirus B19, especially in women, and may smolder on for months. In all these diseases the polyarthritis tends to flit from one joint to another, involving principally the extremities such as the hands; only rarely does it persist for more than a few weeks. Much less frequently, ephemeral arthritis is seen in mumps, varicella, and coxsackievirus infection. The arthralgia sometimes observed in the prodromal stages of hepatitis B is immunologically mediated. Many people sense that rheumatoid arthritis may be of viral origin, but the quarry has proved to be elusive.

Viral Carditis

Coxsackie B viruses and certain other enteroviruses such as coxsackieviruses A4 and A16 and echoviruses 9 and 22 are now recognized to be the most important cause of carditis (see Table 36-9 and Chapter 23). The disease may present as myocarditis, pericarditis, or cardiomyopathy with a greatly dilated

Table 36-8
Viral Arthritis

Virus	Distribution	Features
Ross River	Australia, Pacific islands	Arboviral fevers and polyarthritis
Chikungunya	Africa, South Asia	
O'nyong-nyong	East Africa	
Sindbis	Africa, Asia, Europe	
Mayaro	South America	
Dengue	Tropics worldwide	
West Nile	Africa, Asia, Mediterranean	
Oropouche	Brazil	
Rubella	Worldwide	Especially in adult females
Parvovirus B19	Worldwide	Especially in adult females
Hepatitis B	Worldwide	Immunologically mediated

Table 36-9
Viral Carditis

Disease	Virus	Features
Myocarditis/pericarditis/cardio-myopathy	Coxsackie B and other enteroviruses	Recrudescences
Encephalomyocarditis syndrome	Coxsackie B, echovirus 11, others	Neonatal
Patent ductus arteriosus, pulmonary artery stenosis, septal defects	Rubella (congenital rubella syndrome)	Prenatal
Hydrops fetalis	Parvovirus B19	Prenatal
Endocardial fibroelastosis	Mumps	Prenatal

heart. Recrudescences quite often occur, leading to permanent myocardial damage, cardiomegaly, or congestive cardiac failure; this may have an autoimmune pathogenesis. The primary disease episode occurs at any age but especially in athletic adolescent or young adult males. Furthermore, coxsackie B viruses and echovirus 11 can infect the newborn prenatally, natally, or postnatally, resulting in the encephalomyocarditis syndrome (Fig. 23-5). The syndrome is characterized by fever, dyspnea, cyanosis, tachycardia, abnormal heart sounds, and electrocardiographic changes, and is often accompanied by meningoencephalitis; the case–fatality rate is high.

From time to time the heart may be infected in the course of systemic infections caused by many viruses, such as other enteroviruses, influenza viruses, cytomegalovirus, or Epstein-Barr virus. Moreover, congenital infection with rubella commonly damages the heart; the most common congenital abnormalities are patent ductus arteriosus, pulmonary artery stenosis, and septal defects. Prenatal mumps has been associated with endocardial fibroelastosis. Currently, there is some interest in a claim that herpes simplex virus and/or cytomegalovirus infection of arteries may contribute to the pathogenesis of atherosclerosis, but much more research is required before we can judge whether it has any validity.

Viral Hepatitis

Our knowledge of viral hepatitis has developed remarkably over the last two decades. Accordingly we have devoted a good deal of space elsewhere in this volume to the five major known agents, hepatitis A (Chapter 23), hepatitis B and D (Chapter 22), hepatitis C (Chapter 26), and hepatitis E (Chapter 24). Here, we simply produce a summary (Table 36-10) which brings together for easy comparison some of the main clinical and epidemiologic features of the five hepatitis viruses, that is, those whose main or only target appears to be the liver. It is remarkable that, although the acute diseases caused by these five viruses are clinically indistinguishable, the agents themselves are totally different, belonging in fact to five different families. The major generalizations that should be extracted from Table 36-10 are that (1) hepatitis A and E viruses are spread via the enteric route, whereas hepatitis B, C, and D viruses are transmitted parenterally, sexually, and (in one case at least) perinatally; and (2) only the latter subgroup, hepatitis B, C, and D viruses, establish

Table 36-10
Viral Hepatitis

	Hepatitis A	Hepatitis B	Hepatitis C	Hepatitis D	Hepatitis E
Virus family	*Picornaviridae*	*Hepadnaviridae*	*Flaviviridae*	Deltavirus	*Caliciviridae*
Transmission	Enteric	Parenteral, perinatal, sexual	Parenteral, sexual	Parenteral[a]	Enteric[b]
Acute disease	Mild or moderate	Moderate	Mild or moderate	Severe	Severe in pregnancy
Serodiagnosis[c]	IgM	HBsAg	IgM	IgM	IgM
Chronic carrier state	No	Yes (5–10%)	Yes (50%)	Yes (>50%)	No
Chronic hepatitis, cirrhosis	No	1–5%	20%	>50%	No
Liver cancer	No	Yes	Yes	No	No

[a] Coinfection with hepatitis B virus or superinfection of hepatitis B carrier.
[b] Especially waterborne.
[c] By enzyme immunoassay (or RIA) to identify specific antibody of the IgM class or in the case of hepatitis B virus, HBsAg. There are alternative diagnostic approaches (see Chapters 22, 23, 24, and 26).

persistent infections, and this enables them not only to cause chronic disease, including cirrhosis and cancer, but also to be passed on by the aforementioned routes over many years.

There is also a certain irony in the fact that, having waited so long for the identification of the causal agents of "infectious" hepatitis (A) and "serum" hepatitis (B), there seems to be an almost endless succession of non-A, non-B hepatitis viruses waiting in the wings to be discovered. Despite the recent identification of hepatitis C and E viruses, there is good evidence for the existence of at least one additional parenterally transmitted non-A, non-B virus, which was originally extracted from human factor VIII, has since been serially passaged in cynomolgus monkeys in which it causes hepatitis, yet is nonenveloped (as judged by chloroform resistance) and fails to cross-react serologically in either direction with any of the known hepatitis viruses.

It should be stressed, however, that hepatitis is an occasional feature of the clinical syndromes induced by several other viruses as well. This is hardly surprising, as so many of the infections that involve a viremic phase are characterized by amplification of virus in the reticuloendothelial system, including the liver. For example, all of the herpesviruses, especially HSV, EBV, and CMV, can affect the liver, and clinical hepatitis or elevated liver enzyme levels are occasionally seen with the adenoviruses, coxsackieviruses, and sometimes even the common childhood exanthemata, measles and rubella. Second, hepatitis may be prominent and severe in many of the hemorrhagic fevers, particularly in yellow fever (which is characterized by such a severe hepatitis that it takes its name from the jaundice it causes), but also in Marburg, Ebola, Lassa, Rift Valley fever, and Crimean–Congo hemorrhagic fever. Third, hepatitis is a major feature of most of the disseminated viral infections that overwhelm neonates (neonatal herpes simplex or varicella, cytomegalic inclusion disease, congenital rubella syndrome) or immunocompromised patients (herpes simplex, varicella, cytomegalovirus).

Viral Pancreatitis and Diabetes

Several viruses occasionally infect the pancreas in humans. Mumps, for example, can be complicated by severe pancreatitis, and coxsackie B viruses or various other enteroviruses have been incriminated also. Of greater research interest is the question of whether viruses may trigger juvenile diabetes mellitus of the insulin-dependent type (IDDM). Children born with the congenital rubella syndrome quite often develop IDDM before the age of 20. Mumps infections often affect the β cells of the pancreas. Mumps virus, reovirus, and coxsackie B viruses have all been demonstrated to induce diabetes in mice, but there is still no proof that these viruses induce IDDM in humans. The association of IDDM with particular human leukocyte antigen (HLA) types has encouraged a hypothesis that pancreatic infection with any of perhaps several viruses may trigger autoimmune destruction of β cells.

Chronic Fatigue Syndrome

In recent years there has been an upsurge of reports that the chronic illness characterized by extreme fatigue, known for many years as "neurasthenia," is associated with a recent viral infection. Coxsackie B viruses, EBV, CMV, HHV-6, and HTLV are among the many viruses to have been isolated from such patients, but no cause–effect relationship has been established. Immunologic abnormalities have also been recorded, such as mild IgA deficiencies and elevated levels of circulating immune complexes. A significant proportion of patients have histories of depression or susceptibility to mental illness prior to the development of "chronic fatigue syndrome." Some virologists and clinicians are skeptical about the existence of this disease as an entity. The present consensus is that infection with, or immunologic disorders due to, any of a variety of different viruses may contribute to at least some of these cases.

Congenital and Perinatal Viral Infections

Numerous viruses can cross the placenta; some of these infect the fetus, and some may precipitate a miscarriage. Parvovirus B19, which replicates only in cycling cells such as embryonic cells or bone marrow, appears to be responsible for the death of at least a proportion of the fetuses that are miscarried or stillborn with the syndrome known as hydrops fetalis; the gross generalized edema is usually ascribed to severe anemia and congestive cardiac failure. Much more important, however, are two viruses that do not normally kill the fetus but which do cause serious congenital abnormalities. Rubella during the first 3–4 months of pregnancy inflicts severe teratogenic effects; the congenital rubella syndrome (Fig. 36-6A) and its pathogenesis was described in con-

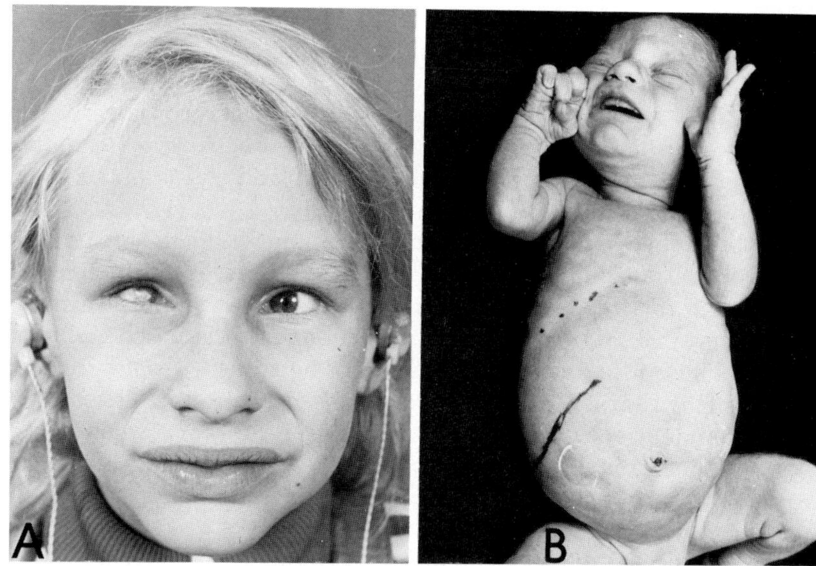

Fig. 36-6 Viral embryopathies. (A) Rubella syndrome. (B) Cytomegalic inclusion disease. (Courtesy Dr. K. Hayes.)

siderable detail in Chapter 25. Prenatal infection with cytomegalovirus induces cytomegalic inclusion disease, described in Chapter 20 (see also Fig. 36-6B). The congenital varicella syndrome (Chapter 20) is extremely rare.

In contrast to these congenital (prenatal) infections, several other viruses may infect the fetus during or shortly after birth (Table 36-11). Such perinatal (alias neonatal, natal, or intrapartum) infections may be acquired during passage of the baby through an infected birth canal (herpes simplex, cytomegalovirus) or by contamination with feces (coxsackie B, echovirus 11). Disseminated neonatal herpes, disseminated varicella–zoster, and myocarditis of the newborn (Fig. 23-5) are all overwhelming generalized infections with high case–fatality rates; these are usually primary maternal infections, and because

Table 36-11
Congenital and Perinatal Viral Infections

Time of infection	Virus	Disease
Prenatal	Rubella	Congenital rubella syndrome
(transplacental)	Cytomegalovirus	Cytomegalic inclusion disease
	Varicella	Congenital varicella syndrome
Intrapartum	Herpes simplex	Herpes neonatorum
	Coxsackie B	Myocarditis of newborn
	Varicella	Disseminated varicella–zoster
	Cytomegalovirus	Subclinical or pneumonia
Postnatal	Hepatitis B	Hepatitis B carrier state
	Hepatitis C	Hepatitis C carrier state
	HIV-1, HIV-2	AIDS
	HTLV-1	Subclinical or leukemia

they have occurred around the time of parturition, the baby is not protected by maternal antibody. The various congenital syndromes may be difficult to distinguish without recourse to virus isolation and IgM serology. Unequivocal laboratory confirmation is important, first because progressive damage can occur after birth and some viruses may be amenable to chemotherapy, and second because appropriate medical, social, and educational measures should be initiated as early as possible.

Still other viruses are commonly acquired within the first few weeks of life (postnatal infections) Newborn babies of hepatitis B virus carrier mothers almost always themselves become lifelong HBsAg carriers within the first few months of life; possible routes of infection were discussed in Chapter 22. The probability of perinatal transmission of hepatitis C from a carrier mother to her infant is much lower. However, the risk of transmission of the human immunodeficiency virus from an HIV-positive mother to her newborn baby is substantial. Again the route of postnatal (or natal) transmission is not known for sure but was discussed in Chapter 35. A similar situation obtains with HTLV-1 and HTLV-2.

At risk of stretching the definition of postnatal infection, we should for completeness remind the reader that the herpesviruses cytomegalovirus, Epstein-Barr virus, and HSV-1 tend to be acquired "vertically" from the mother via saliva (CMV, EBV, HSV-1) or milk (CMV) very early in life, particularly in the Third World, whereas others such as rotavirus and respiratory syncytial virus may be picked up by "horizontal" transmission, even by nosocomial spread in hospital nurseries. In general, such infections acquired from the mother shortly after birth are subclinical, having been acquired under the "umbrella" of maternal antibodies. However, there are two main circumstances under which such transmission is fraught with danger: (1) when the mother has experienced her primary infection so recently that the baby is not protected by antibody or (2) when the baby suffers from some form of congenital immunodeficiency or is significantly premature or sickly.

Viral Infections in Immunocompromised Patients

In Chapter 8 we discussed the role of the various arms of the immune response in controlling viral infections, and in Chapter 10 the reactivation of persistent infections by immunosuppression. Table 10-6 listed the viruses commonly reactivated in immunocompromised patients.

In the context of this final overview of viral disease it is sufficient merely to add that reactivation of latent herpesviruses, in particular, has become commonplace as a result of three advances in modern medicine: (1) organ transplantation, requiring as it does, profound immunosuppression to prevent rejection of the allograft, (2) chemotherapy of cancer using highly cytotoxic drugs and/or radiotherapy, (3) rescue of children with profound congenital immunodeficiency syndromes who would not previously have lived. To these three iatrogenic changes should be added two major naturally occurring immunosuppressive diseases, AIDS and cancer (particularly lymphomas). The dangers are enhanced by the fact that the massive blood transfusions (or hemodialysis), so often an integral part of the life-saving therapeutic regime, carry the risk of iatrogenic transmission of exogenous viruses such as cytomegalovirus or Epstein-Barr virus, which may overwhelm a patient already desperately ill; the once considerable danger of transmitting HIV, hepatitis B, and hepatitis C in this way has almost disappeared as a result of the introduction of universal screening of blood, blood products, and organ donors (not yet completely for hepatitis C). Severely burnt patients are also highly vulnerable to invasion by herpes simplex viruses.

The magnitude of the problem is dramatically illustrated by the observation that it is quite standard for more than one, and sometimes all six, of the herpesviruses (HSV-1 and HSV-2, VZV, CMV, EBV, and HHV-6) to be reactivated in bone-marrow transplant recipients; in some series, up to 25% of all such patients have died of CMV pneumonia alone! Similarly, AIDS patients characteristically suffer the consequences of successive reactivation, sometimes in a roughly predictable order as their $CD4^+$ T cell count drops, of any or all of the herpesviruses, plus polyomavirus, papillomaviruses, adenoviruses, and hepatitis B virus (see Chapter 35). The pathogenesis, clinical manifestations, and management of immunocompromised patients by antiviral chemotherapy, active or passive immunization, and appropriate virologic and immunologic screening were discussed for each of the herpesviruses individually in Chapter 20; reactivation of human polyomaviruses was addressed in Chapter 18 and that of adenoviruses in Chapter 19.

Further Reading

Anderson, L. J., Patriarca, P. A., Hierholzer, J. C., and Noble, G. R. (1983). Viral respiratory illnesses. *Med. Clin. North Am.* **67,** 1009.

Blacklow, N. R., and Greenberg, H. B. (1991). Viral gastroenteritis. *N. Engl. J. Med.* **325,** 252.

Bradley, D. W., ed. (1993). Hepatitis Viruses. *Semin. Virol.* **4**(5), 269.

Braude, A. I. (1993). "Microbiology: Basic Science and Medical Applications." Saunders, Philadelphia, Pennsylvania.

Centers for Disease Control. (1990). Viral agents of gastroenteritis. Public health importance and outbreak management. *Morb. Mortal. Wkly. Rep.* **39,** RR5.

Centers for Disease Control. (1990). Recommendations for collection of laboratory specimens associated with outbreaks of gastroenteritis. *Morb. Mortal. Wkly. Rep.* **39,** RR-14, 1.

Christensen, M. L. (1989). Human viral gastroenteritis. *Clin. Microbiol. Rev.* **2,** 51.

Cosgriff, T. M.,. ed. (1989). International Symposium on Hemostatic Impairment Associated with Hemorrhagic Fever Viruses. *Rev. Infect. Dis.* **11**(Suppl. 4), S669–S896.

Feigin, R. D., and Cherry, J. D., eds. (1992). "Textbook of Pediatric Infectious Diseases," 3rd Ed. Saunders, Philadelphia, Pennsylvania.

Fields, B. N., Knipe, D. M., Chanock, R. M., Hirsch, M. S., Melnick, J. L., Monath, T. M., and Roizman, B., eds. (1990). "Fields Virology," 2nd Ed. Raven, New York.

Gorbach, S. L., Bartlett, J. G., and Blacklow, N. R., eds. (1992). "Infectious Diseases." Saunders, Philadelphia, Pennsylvania.

Hoeprich, P. D., and Jordan, M. C. (1989). "Infectious Diseases," 4th Ed. Lippincott, Philadelphia, Pennsylvania.

Hollinger, F. B., Lemon, S. M., and Margolis, H. S., eds. (1991). "Viral Hepatitis and Liver Disease." Williams & Wilkins, Baltimore, Maryland.

Johnson, R. T. (1982). "Viral Infections of the Nervous System." Raven, New York.

Johnson, R. T. (1994). Nervous system viruses. *In* "Encyclopedia of Virology" (R. G. Webster and A. Granoff, eds.), Vol.2, p. 907. Academic Press, London.

Kass, E. H., Weller, T. H., Wolfe, S. M., and Tyrrell, D. A. J., eds. (1993). "Handbook of Infectious Diseases." Dekker, Burlington, Ontario.

Krugman, S., Katz, S. L., Gershon, A. A., and Wilfert, C. N., eds. (1992). "Infectious Diseases of Children," 9th Ed. Mosby, St. Louis, Missouri.

McCormick, J. B., and Fisher-Hoch, S. (1990). Viral hemorrhagic fevers. *In* "Tropical and Geographic Medicine" (K. S. Warren and A. A. F. Mahmoud, eds.), 2nd Ed., p. 700. McGraw-Hill, New York.

Mandell, G. L., Douglas, R. G., and Bennett, J. E., eds. (1990). "Principles and Practice of Infectious Diseases," 3rd Ed. Churchill-Livingstone, New York.

Pavan-Langston, D. (1993). Ocular viral diseases. *In* "Practical Diagnosis of Viral Infections" (G. J. Galasso, R. J. Whitley, and T. C. Merigan, eds.), p. 69. Raven, New York.

Peters, C. J., Johnson, E. D., and McKee, K. J. (1991). Filoviruses and management of viral hemorrhagic fevers. *In* "Textbook of Human Virology" (R. B. Belshe, ed.), 2nd Ed., p. 699. Mosby, St. Louis, Missouri.

Remington, J. S., and Klein, J. O., eds. (1990). "Infectious Diseases of the Fetus and Newborn Infant," 3rd Ed. Saunders, Philadelphia, Pennsylvania.

Saif, L. J., and Theil, K. W., eds. (1990). "Viral Diarrheas of Man and Animals." CRC Press, Boca Raton, Florida.

See, D. M., and Tilles, J. G. (1991). Viral myocarditis. *Rev. Infect. Dis.* **13,** 951.

Shulman, S. T., Phair, J. P., and Sommers, H. M., eds. (1992). "The Biologic and Clinical Basis of Infectious Diseases." Saunders, Philadelphia, Pennsylvania.

Tesh, R. B. (1982). Arthritides caused by mosquito-borne viruses. *Annu. Rev. Med.* **33,** 31.

ter Meulen, V., ed. (1991). Viral Cell Interactions in the Nervous System. *Semin. Neurosci.* **3**(2).

Valenti, W. M. (1991). Nosocomial viral infections. *In* "Textbook of Human Virology" (R. B. Belshe, ed.), 2nd Ed., p. 216. Mosby, St. Louis, Missouri.

Warren, K. S., and Mahmoud, A. A. F., eds. (1990). "Tropical and Geographic Medicine," 2nd Ed. McGraw-Hill, New York.

Webster, A., and Griffiths, P. (1990). Virus infections in immunocompromised patients. *In* "Topley and Wilson's Principles of Bacteriology, Virology and Immunity (L. H. Collier and M. C. Timbury, eds.), 8th Ed., Vol. 4, p. 151. Arnold, London.

White, D. O. (1994). Respiratory viruses. *In* "Encylopedia of Virology" (R. G. Webster and A. Granoff, eds.), Vol. 2, p. 1219. Academic Press, London.

World Health Organization. (1980). Viral respiratory diseases. *WHO Tech. Rep. Ser.* **642.**

Index

D

I

J

K

L

M

N

O

S

T

U

V

W

X

Y

Z